Nanostructures: Synthesis, Functional Properties and Applications

NATO Science Series

A Series presenting the results of scientific meetings supported under the NATO Science Programme.

The Series is published by IOS Press, Amsterdam, and Kluwer Academic Publishers in conjunction with the NATO Scientific Affairs Division

Sub-Series

I. Life and Behavioural Sciences	IOS Press
II. Mathematics, Physics and Chemistry	Kluwer Academic Publishers
III. Computer and Systems Science	IOS Press
IV. Earth and Environmental Sciences	Kluwer Academic Publishers
V. Science and Technology Policy	IOS Press

The NATO Science Series continues the series of books published formerly as the NATO ASI Series.

The NATO Science Programme offers support for collaboration in civil science between scientists of countries of the Euro-Atlantic Partnership Council. The types of scientific meeting generally supported are "Advanced Study Institutes" and "Advanced Research Workshops", although other types of meeting are supported from time to time. The NATO Science Series collects together the results of these meetings. The meetings are co-organized bij scientists from NATO countries and scientists from NATO's Partner countries – countries of the CIS and Central and Eastern Europe.

Advanced Study Institutes are high-level tutorial courses offering in-depth study of latest advances in a field.
Advanced Research Workshops are expert meetings aimed at critical assessment of a field, and identification of directions for future action.

As a consequence of the restructuring of the NATO Science Programme in 1999, the NATO Science Series has been re-organised and there are currently Five Sub-series as noted above. Please consult the following web sites for information on previous volumes published in the Series, as well as details of earlier Sub-series.

http://www.nato.int/science
http://www.wkap.nl
http://www.iospress.nl
http://www.wtv-books.de/nato-pco.htm

Series II: Mathematics, Physics and Chemistry – Vol. 128

Nanostructures: Synthesis, Functional Properties and Applications

edited by

Thomas Tsakalakos
Department of Ceramic & Materials Engineering,
Rutgers University, Piscataway, U.S.A.

Ilya A. Ovid'ko
Institute of Problems for Mechanical Engineering,
Russian Academy of Sciences, St. Petersburg, Russia

and

Asuri K. Vasudevan
Materials Division,
Office of Naval Research Arlington, Virginia, U.S.A.

Springer Science+Business Media, B.V.

Proceedings of the NATO Advanced Study Institute on
Synthesis, Functional Properties & Applications of Nanostructures
Crete, Greece
29 June–9 July 2001

A C.I.P. Catalogue record for this book is available from the Library of Congress.

ISBN 978-1-4020-1753-7   ISBN 978-94-007-1019-1 (eBook)
DOI 10.1007/978-94-007-1019-1

TABLE OF CONTENTS

Preface

The Advanced Study Institute on Synthesis, Functional Properties and Applications of Nanostructures, held at the Knossos Royal Village, Heraklion, Crete, Greece, July 26, 2002 – August 4, 2002, successfully reviewed the state-of-the-art of nanostructures and nanotechnology. It was concluded that Nanotechnology is widely agreed to be the research focus that will lead to the next generation of breakthroughs in science and engineering. There are three cornerstones to the expectation that Nanotechnology will yield revolutionary advances in understanding and application:

- Breakthroughs in properties that arise from materials fabricated from the nanoscale.
- Synergistic behavior that arise from the combination of disparate types of materials (soft vs. hard, organic vs. inorganic, chemical vs. biological vs. solid state) at the nanoscale.
- Exploitation of natural (e.g. chemical and biological) assembly mechanisms that can accomplish structural control at the nanoscale.

It is expected that this will lead to paradigms for assembling bio-inspired functional systems that accomplish desirable properties that are either unavailable or prohibitively expensive using top-down approaches.

While the general principles are clear, systematic and comprehensive studies of the effects of processing and the resultant evolution of structure on the overall mechanical properties, performance, and thermal stability, under complex loading conditions, have thus far not been performed in sufficient depth. This situation is further compounded by the fact that standardized procedures do not exist for the testing and analysis of size-dependent mechanical and functional properties at the nanoscale which are essential for the design, modeling, and life assessment of nanomaterials for biological and biomedical applications.

The NATO ASI on Synthesis, Functional Properties and Applications of Nanostructures emphasized the development of useful implementations and applications of nanotechnology. One key issue in nanotechnology is how to access, from the macroscopic world, the extremely high information density of nanostructured systems. The lecturers addressed this question by using bio-inspiration – techniques where they applied lessons learned from living systems to design new materials with localized feedback mechanisms. The meeting focused on multifunctional nanomaterials for use in biomedical and information technology applications. Lectures and discussions included embedded, or natural sensing, information processing and actuation of responses to the physical environment. The scientific program was structured with the following:

> SYNTHESIS AND SELF-ASSEMBLY OF NANOSTRUCTURES
> DEFORMATION OF NANOSTRUCTURES
> FUNCTIONAL NANOSTRUCTURES
> APPLICATIONS OF NANOSTRUCTURES
> NANOCLUSTERS AND NANOWIRES
> MAGNETIC CHARACTERIZATION OF NANOSTRUCTURES
> PROCESSING OF NANOSTRUCTURES
> OPTICAL PROPERTIES OF NANOSTRUCTURES

QUANTUM DOTS
CHARACTERIZATION OF NANOSTRUCTURES
NANOSTRUCTURED FILMS AND COATINGS
CARBON NANOTUBES
DEFORMATION AND MODELLING OF NANOSTRUCTURES
MECHANICAL BEHAVIOR OF NANOSTRUCTURES
FUNCTIONAL APPLICATIONS OF NANOSTRUCTURES

There was a general consensus, that the NATO ASI on Nanostructures was one of the most successful meetings in the field and served admirably, the scientific community and society at large. The lectures, invited talks, and discussions were outstanding both in breadth and depth.

We would like to acknowledge the significant contribution of the following organizers :

- Dr. Gan-Moog Chow, Professor, Materials Sci Dept, Univ of Singapore, Singapore.
- Dr. Lawrence Kabacoff, Office of Naval Research, USA.
- Prof. R.A. Andrievski, Russian Acad of Science, Moscow, 142432 Russia
- Prof. A. Misiuk, Institute of Electronic Technology, Warsaw, Poland.
- Dr. D. Niarchos, Local Organizer, Inst of Matls Sci NRC Demokritos, Greece

We gratefully acknowledge the major financial support of this ASI by: NATO Physical and Engineering Science and Technology (PST)

We also like to express our gratidute to the Co-Sponsors :

- Office of Naval Research (USA)
- Office of Naval Research International Field Office
- US Army Research Laboratory - European Research Office
- USAF, European Office of Aerospace Research and Development

We would like to thank Mr. Panos Prodromitis and Mrs Maria Prodromitis, the group of Select Holidays, and Knossos Royal Village administration and staff for their invaluable assistant with the travel and lodging for the participants. They showed the highest degree of professionalism and hospitality. We also thank Ms. Ioanna Tsakalakos, secretariat of the NATO ASI, for her outstanding work not only during the preparation but also during the conduction of the meeting. Finally we thank Ms Kathleen Paige for the outstanding editorial assistance with the proceedings.

Thomas Tsakalakos
ASI Director
Rutgers University

Ilya A. Ovid'ko
ASI co-Director
Russian Academy of Sciences

A.K. Vasudevan
ASI Chair, Publication Committee
Office of Naval Research, USA

"Synthesis, Functional Properties & Applications of Nanostructures"

List of NATO ASI Participants

Tarek Abdel-Fattah (USA)
Dept of Bio, Chemical and Environ Science, Christopher Newport University,1 University Place, Newport News, Virginia, USA 23606
Tel: (757) 594-7606, Fax: (757)594-7209, E-mail: fattah@cnu.edu

Pulickel Ajayan (USA)
MRC 112, Dept of Materials Science and Engineering, Rensselaer Polytechnic Institute, Troy, New York, USA 12180-3590
Tel: 518 276 2322, Fax: 518 276 8554, E-mail: ajayan@rpi.edu

R.A. Andrievski (Russia)
Institute of Chemical Physics Problems, Russian Academy of Sciences (RAS), Chernogolovka, Moscow Region, 142432, Russia
Tel/ Fax: (7-096)522-3577, E-mail: ara@icp.ac.ru

Demetrios Bakoyannakis (Greece)
Lab of General and Inorganic Chemical Tech, Aristotle University of Thessaloniki, Thessaloniki, 54006 Greece
Tel: +0310997766, Fax: +0310997759, E-mail: Elndbak@chem.auth.gr

Csaba Balazsi (Hungary)
Research Institute for Technical Physics and Materials Science, Ceramics and Refractory Metals Dept, Hungarian Academy of Sciences
1121 Budapest XII., H-1525 Budapest, P.O.Box 49, Konkoly-Thege u. 29-33. Hungary
Tel: +36-1-3922222 (3817) Fax: +36-1-3922226, E-mail: balazsi@mfa.kfki.hu
Web Site: http://www.mfa.kfki.hu

Marie Isabelle Baraton (France)
Faculty of Sciences, SPCTS - UMR 6638 CNRS,123 Avenue Albert Thomas, F-87060 Limoges (France)
Tel: + 33 555 45 7348, Fax: + 33 555 77 8100, E-mail: baraton@unilim.fr

Nathalie Le Bars (France)
CEA/DRT/DECS/SE2M Bat 451, CEN Saclay Gif sur Yvette 91191 France
Tel: 331 6908 2962, Fax: 3316908 9175. E-mail: nathalie.lebars@cea.fr

Chris Binns (United Kingdom)
Department of Physics & Astronomy, University of Leicester University of Leicester, LEI 7RH, United Kingdom
Tel: +44 116 2523585, Fax: +44 116 2522770, E-mail: cb12@le.ac.uk

Ennio Bonetti (Italy)
Dept of Physics, University of Bologna and INFM v.le Berti Plchat 6/2, Bologna 40127 IT
Tel: +39 051 2095107, Fax: +39 051 2095153, E-mail: bonetti@df.unibo.it

Roberto Buzzoni (Italy)
Catalytic Process Department, POLIMERI EUROPA SpA, "Istituto Guido Donegani"
Via G. FAUSER 4, I-28100, NOVARA (ITALY)
Tel.: +39-0321-447534, Fax: +39-0321-447506,
E-mail: roberto.buzzoni@polimerieuropa.com

Albano Cavaleiro (Portugal)
Departamento de Engenharia Mecanica, FCTUC - Universidade de Coimbra,Pinhal de Marrocos, 3030
Coimbra, Portugal
Tel: 351 239 790794 / 00, Fax: 351 239 790701, E-mail - albano.cavaleiro@dem.uc.pt

Moog Chow (Singapore)
Dept of Matls Science, National University of Singapore, Kent Ridge, Singapore 119260, Republic of
Singapore
Tel: +65 6 874 3325, Fax: +65 6 776 3604, E-mail: mascgm@nus.edu.sg

Joseph Christodoulides (USA)
Naval Research Laboratory, 4555 Overlook Ave, SW, Building 3, Room 106, Washington, DC 20375, USA
Tel: 2027674393, Fax: 2024048849, E-mail: joec@anvil.nrl.navy.mil

Zorica Crnjak Orel (Slovenia)
National Institute of Chemistry, Hajdrihova 19, Ljublijana, Slovenia
Tel: ++386 1 4760 236, Fax: ++386 1 4760300, E-mail: zorica.crnjak.orel@ki.si

Marian Deanko (Slovakia)
Institute of Physics, Slovak Academy of Sciences,84228 Bratislava, Slovakia
Tel: +421 2 5941 0570, Fax: ++421 2 5477 6085, E-mail: fyzidean@savba.sk
Web Site: www.savba.sk/rq

Jeffrey A. Eastman (USA)
Materials Science Division, Argonne National Laboratory, 9700 S. Cass Ave., Bldg. 212, Argonne, IL 60439,
USA
Tel: 630-252-5141, Fax: 630-252-4289, E-mail: jeastman@anl.gov
Web Site: www.msd.anl.gov/groups/im

Sergey Vasilievich Egorov (Russia)
Institute Applied Physics Russian Academy Sciences, Ulyanova 46, Nizhny, Novgorod, Golybera Str. 8-72,
603600, Russia
Tel: 8312 38-43-00, E-mail: egr@appl.sci-nnov.ru

Valery Fedosyuk (Belorus)
Institute of Solid State Physics and Semiconductors, Belorussian Academy of Sciences, P.Brovki 17, Minsk
220072, Belorus;
Tel: +375-17-2842791, Fax: +375-17-2840888, E-mail: fedosyuk@ifttp.bas-net.by

Janos Fendler (USA)
Distinguished CAMP, Clarkson University, CAMP - Box 5814, Potsdam, New York 13699-5814 USA
Tel: 315-268-7113, Fax:315-268-4416, E-mail: fendler@clarkson.edu

Emmanuel Giannelis (USA)
Department of Materials Science and Engineering , Cornell University,247 Bard Hall, Ithaca, NY 14853 USA
Fax: (607) 255-2365 ,E-mail: emmanuel@msc.cornell.edu

Prof. A. Glezer (Russia)
Kurdyumov Institute of Metal Physics & Functional Materials, Bardin State Science Center for Ferrous
Metallurgy, 2-nd Baumanskaya st. 9/23, Moscow 105005, Russia;
Tel/Fax: (7-095) 777-9350, E-mail: glezer@imph.msk.ru

Maria de Jesus Gomes (Portugal)
Dept. Fisica, Universidade do Minho. University of Minha, Portugal
Tel: +351-253-604327/0, Fax: +351 253-6778981, E-mail: mjesus@fisica.uminho.pt

Steve Goheen (USA)
Pacific Northwest National Lab. (P8-08) PO Box 999, Richland, WA 99352, USA
Tel: 509 376 3286, Fax: 509 376 2329, E-mail: Steve.goheen@pnl.gov

Prof. A.I.Gusev (Russia)
Institute of Solid State Chemistry, RAS, Ekaterinburg, Russia
Tel: 7 3432 747306, Fax: 7 3432 744495, E-mail: gusev@ihim.uran.ru

George Hadjipanayis (USA)
Dept. of Physics, University of Delaware, 223 Sharp Laboratory, Newark, DE 19716 USA
Tel: (302) 831 2736, Fax: (302) 831-1637, E-mail: hadji@udel.edu

Rolf Hempelmann (Germany)
Physikalische Chemie, Universität des Saarlandes, Im Stadtwald, Bau 9.2 D-66123 Saarbrücken, Germany
Tel: 0681 / 302-4750, Fax: 0681 / 302-4759, E-mail: hempelmann@mx.uni-saarland.de
Web Site: www.uni-saarland.de/fak8/hempelmann

Jeff De Hosson (Netherlands)
University of Groningen, Applied Physics, Nijenborgh 4, 9747 AG, The Netherlands
Tel: +31503634898, Fax: +31503634881, Email: hossonj@phys.rug.nl

Eliza Hutter (USA)
Clarkson University, Box 5814, Postsdam, NY 13699, USA
Tel: (315) 268-7149, Fax: (315) 268-4416, E-mail: huttere@clarkson.edu

Lawrence Kabacoff (USA)
Office of Naval Research, 800 N. Quincy Street, Arlington, VA 22217-5660, USA
Tel: 703 696 0283, Fax: 703 696 0934, E-mail: Kabacol@onr.navy.mil

A.I. Kharlamov (Ukraine)
Institute for Problems of Materials Science, 3 Krzjzanovski Str, 252680 Kiev, Ukraine
Tel: +38-044-444-02-56, E-mail: dep73@ipms.kiev.ua

George Kiriakidis (Greece)
Materials Group, IESL/FORTH, Greece
Tel:+30-81-391271/2, Fax:+30-81-391295, E-mail: kiriakid@iesl.forth.gr

Irina Kleps (Romania)
National Institute for Research and Development in Microtechnologies (IMT),
P.O. Box 38-160, Bucharest 72225, Romania
Tel.: 4021.4908412 ext. 33, Fax: 4021.4908238, E-mail: irinak@imt.ro

A. Kolesnikova (Russia)
Institute of Problems of Mechanical Engineering, Russian Academy of Sciences, Bolshoj 61, Vas.Ostrov,
St.Petersburg 199178, Russia
Fax: +7(812)321 4771, E-mail: koles@def.ipme.ru

Vladimir G. Konakov (Russia)
Institute of Silicate Chemistry of RAS Ul. Odoevskogo 24-2, St. Petersburg 199406, Russia
Tel: 7-812-351-08-07, Fax: 7-812-350-59-47, E-mail: konakov@isc1.nw.24,

J. Kovacik (Slovakia)
Institute of Materials and Machine Mechanics, Racianska 75, 831 02 Bratislava 3, Slovak Republic
Tel: +421 2 492 68 275, Fax: +421 2 442 53 301, E-mail: ummsjk@savba.sk

Lynn Kurihara (USA)
Naval Research LabCode 640, Washington D.C. 20375 USA
Tel: 2027672563, Fax: 2024048533, E-mail: Kurihara@anvil.nrl.navy.mil

Alexandros Lappas (Greece)
Inst of Elec Struct & Laser,

S. Litvinenko (Ukraine)
Kiev University, Volodimirska 64, Kiev, Ukraine
Tel: (38044) 2959206, Fax: (38044) 2656744, E-mail: litvin@uninet.kiev.ua

Sanjay Mathur (Germany)
Institute for New Materials, Saarland University, Saarbruecken, Germany
Tel: +49-681-9300-274 Fax: +49-681-9300-279, E-mail: smathur@inm-gmbh.de

Lhadi Merhari (France)
64 avenue de la Liberation, 87000 France
Tel: +33 683710 550, E-mail: Ceramec@wanadoo.fr

M. Miglierini (Slovakia)
Department of Nuclear Physics and Technology, Slovak University of Technology, Ilkovicova 3, 812 19 Bratislava, Slovakia;
Tel: +421 2 602 91 167 , Fax: +421 2 654 27 207, E-mail: bruno@elf.stuba.sk

Andrezej Misiuk (Poland)
Institute of Electron Technology, Al. Lotnikow 46, 02-668 Warsaw, Poland
Tel: +(4822) 5487792, Fax: +4822) 8470631, E-mail: misiuk@ite.waw.pl

Mamoun Muhammed (Sweden)
Dept of Mat'l Science and Engineering, Royal Institute of Technology, 100 44 Stockholm, Sweden
Tel: +468-7908158, Fax: +468-7909072, E-mail: mamoun@matchem.kth.se

Dimitri Niarchos (Greece)
NCSR Demokritos, 15310 Aghia Parasveki, Athens, Greece
Tel: 3010-6503385, Fax: 3010-6519430, E-mail: dniarchos@ims.demokritos.gr

Heinz-Georg Nothofer (Germany)
Sony International Europe, Materials Science Laboratories,.Heinrich-Hertz Str. 1, Stuttgart 70327 GER
Tel: +49 711 5858 178, Fax: +49 711 5858 484, E-mail: Nothofer@sony.de

Thomas Orlando (USA)
Georgia Institute of Tech, Boggs Chemistry Building.Atlanta, Georgia 30332-0001, USA
Tel: 404 894 4012, Fax: 404 894 7452, Email: thomas.orlando@chemistry.gatech.edu

Ilya Ovid'ko (Russia)
Institute of Problems for Mechanical Engineering, Russian Academy of Sciences, Bolshoj 61, Vas.Ostrov, St.Petersburg 195297, Russia
Fax: +7(812)3214771, E-mail: ovidko@def.ipme.ru

Chandra.S.Pande (USA)
Code 6325 Naval Research Laboratory, , Washington D.C. 20375, USA
Tel: 202-767-2744, Fax: 202-767-2623, E-mail: PANDE@ANVIL.NRL.NAVY.MIL

Tasos Papadopoulos (Greece)
National Technical University of Athens, Greece
Tel: 3010-8259114, E-mail: tpapad@mail.ntua.gr

Dimitri Petridis (Greece)
NCSR "Demokritos", Institute of Materials Science, 15210 Aghia Paraskevi, Attici, Athens, Greece
Tel: 3010-653345, Fax: 3010-6519430, E-mail: dpetrid@ims.demokritos.gr

Antoine Phelippeau (France)
Laboratoire MSS/Mat, Ecole Centrale Paris, Grande voie des vignes 92295, Chatenay Malabry FRANCE
Tel: 01-41-13-16-54, Fax: 01-41-13-14-30, Email: phelippeau@mssmat.ecp.fr

Erhan Piskin (Turkey)
Hacettepe University, Turkish Academy of Science,
Tel: 90 312 2977400, Fax: 90 312 2992124, E-mail: piskin@hacettepe.edu.tr

Katerina Polychronopoulou (Greece)
National Technical University of Athens, Greece, Karaiskaki 53-13231-Petroupolis-Athens
Tel: 30 10 5017105, E-mail: mm99028@mail.ntua.gr

Sylvie Pommier (France)
MSSMAT Ecole Centrale Paris Grande Voie des Vignes, Chatenay Malabry Cedex, 92295
Tel: 33 1 41 13 16 58, Fax: 33 1 41 13 14 30, E-mail: Sylvie@mssmat.ecp.fr

Virgil Provenzano (USA)
NIST, 100-Bureau Drive, Gaithersburg, MD 20999, USA
Tel: 001-301-975-6042, Fax: 001-301-975-4553, E-mail: Virgil12@nist.gov

L. Pustov (Russia)
Moscow Institute of Steel & Alloys, Leninsky prosp. 4, 119991, Moscow, RU
Tel/ Fax: 007-095-2304595, E-mail: pustov@mail.ru

Andrey Ragulya (Ukraine)
Frantsevich Institute for Problems in Materials Science, NAS of Ukraine, 3, Krzhizhanovski St., 03142 Kiev.
Ukraine
Tel. +38-044-444-1533, Fax:+38-044-444-2131, E-mail: ragulya@materials.kiev.ua

B.B.Rath (USA)
Naval Research Laboratory, Washington DC 20375-5341, USA
Tel: 202 767 2538, Fax: 202 404 120, E-mail: rath@utopia.nrl.navy.mil

Anna B. Reizis (Russia)
Institute of Problems of Mechanical Engineering, RAS, Vas. Ostrov, Bolshoj pr., 61. St. Petersburg 199175,
Russia
Tel: +7(812)3214764, Fax: +7(812)3214771, E-mail: reizis@def.ipme.ru

S.E. Romankov (Kazakhstan)
Institute of Physics and Technology, Ministry of Education and Science, 480082, Almaty 82, Kazakhstan
Tel/Fax: +7(3272) 545 224, E-mail: romankov@sci.kz romankovs@mail.ru

Alexei E. Romanov (Russia)
Ioffe Physico-Technical Institute, RAS, Polytekhnicheskaya 26, St.Petersburg 194021, Russia
Tel: 7 812 3065353, Fax: 7 812 247-10-17, E-mail: aer@mail.ioffe.ru

Maria Samaras (Switzerland)
Paul Scherrer Institute, CH-5232 Villigen-PSI Switzerland
Tel: 41 56 310 4184, Fax: 41 56 310 3131, E-mail: Maria.samaras@psi.ch

University of Basel, Dept. of Chemistry,
Klingelbergstrasse 80, CH-4056, Switzerland
Tel: +41-61-267 3801, Fax: +41-61-267 3855, E-mail: marc.sauer@unibas.ch

James Scott (United Kingdom)
Earth Sciences Dept., Cambridge University, Downing Street, Cambridge CB23EQ, United Kingdom
Tel: 44 1223 333461, Fax: 44 1223 333450, E-mail: Jsco99@esc.cam.ac.uk

Robert W. Shaw (United Kingdom)
Chemistry and Materials Science, European Research Office, Edison House, 223 Old Marylebone Road, London, NW1 5TH, UK
Tel: +44 (0) 20 7514 4909, Fax: +44 (0) 20 7724 1433,
E-mail: rshaw@usardsguk.army.mil

Olga Shikimaka (Moldova)
Academy of Sciences of Moldova, Institute of Applied Physics 5, Academy str. MD-2028, Chisinau, Moldova
Tel: (044) 73 81 09, E-mail: mech.prop@phys.asm.md

Athanasios Simopoulos (Greece)
NCSR "Demokritos", Institute of Materials Science, 15210 Aghia Paraskevi, Attici, Athens, Greece
Tel: 30-1-6544637, Fax: 30-1-6519430, E-mail: asimop@ims.demokritos.gr

B. Straumal (Russia)
Institute of Solid State Physics, Chernogolovka, Moscow District, 142432, Russia
Tel/ Fax: 007 095 23 82326, E-mail: straumal@issp.ac.ru

N. Sulitanu (Romania)
Department of Solid State Physics, Faculty of Physics, "Al.I.Cuza" University of Iasi, 11 Carol I Blvd., RO-6600 Iasi, Romania
Tel: 00 40 232 201173, Fax: 00 40 232 201150, E-mail: sulitanu@uaic.ro

Subra Suresh (USA)
Department of Materials Science & Engineering, MIT, 77 Massachusetts Avenue, Cambridge, MA 02139, USA
Tel: 617-253-3320, Fax: 617-253-0868, E-mail: ssuresh@mit.edu

Peter Svec (Slovakia)
Institute of Physics, Slovak Academy of Sciences, 84228 Bratislava, Slovakia
Tel: -4212 5941 0561, Fax: -4212 5477 6085m, E-mail: fysisvec@savba.sk
Web Site: www.savba.sk/rq

Helena Van Swygenhoven (Switzerland)
Paul Scherrer Institute, CH-5232 Villigen-PSI, Switzerland
Tel: 41 56 310 2931, Fax: 41 56 310 3131, E-mail: Helena.vs@psi.ch

E.Tabachnikova (Ukraine)
Institute for Low Temperature, Physics and Engineering, 47 Lenin Ave., Kharkov, 61103, Ukraine
Tel: 380 (572) 300 331, Fax: 380 (572) 322 370, E-mail: tabachnikova@ilt.kharkov.ua

Sigitas Tamulevicius (Lithuania)
Institute of Physical Electronics, Kaunas University of Technology, Savanoriu 271, LT-3009 Kaunas, Lithuania
Tel: 3707313432, 3707351128, Fax.: 3707314423, 3707456472;
E-mail: stamul@fmf.ktu.lt, Web Site: http://www.ktu.foi.lt

M. Taylor (United Kingdom)
Associate Director, Materials Science, ONR International Field Office, 223 Old Marylebone Road, London NW1 5TH, United Kingdom
Tel:44 (0)20 7514 4963, Fax:44 (0)20 7723 6359, E-mail: mtaylor@onrifo.navy.mil

Loucas Tsakalakos (USA)
General Electric - Global Research, 1 Research Circle, Bldg KW Room C1426, Niskayuna, New York 12309
Tel: 515 387 5715, Fax: 518 387 6030, E-mail: Tsakalakos@crd.ge.com

Thomas Tsakalakos (USA)
Dept of Ceramic & Materials Engineering, College of Engineering, Nanostructured Materials, Rutgers University, 607 Taylor Road, Piscataway NJ 08854-8065 USA
Tel: (732) 445 2888, Fax: (732) 445 3229, E-mail: tsakalak@rci.rutgers.edu
Web Site: http://web.rutgers.edu/nanostructures

V. Tsukruk (USA)
Materials Science & Eng. Dept., Iowa State University, 3155 Gilman Hall Ames, Iowa 50011 USA
515 294 6904, fax: 515 294 5444, E-mail: vladimir@iastate.edu

Victor Ustinov (Russia)
Ioffe Physico-Technical Institute, Russian Academy of Sciences, Polytekhnicheskaya 26, St.Petersburg 194021, Russia
Tel/ Fax: +7(812) 2473178, E-mail: vmust@beam.ioffe.rssi.ru

Ruslan Valiev (Russia)
Institute of Physics for Advanced Materials, Ufa State Aviation Technical University, Ufa 450000, Russia
Tel/Fax: 007-3472-233422, E-mail: RZValiev@mail.rb.ru

A.K. Vasudevan (USA)
Materials Division, ONR-332, 800 N. Quincy Str, Arlington, Va 22217
Tel: 703-696-8181, Fax: 703-696-0934, E-mail: Vasudea@ONR.NAVY.MIL

Daniel Vrbanic (Slovenia)
Faculty of Chemistry and Chemical Technology, , University of Ljubljana, Askerceva 5, Ljubljana, Slovenia
Tel. +386 1 477 3313, Fax.+386 1 251 9385, E-mail: daniel.vrbanic@uni-lj.si

R. Stanley Williams (USA)
HP Fellow and Director, Quantum Science Research. Hewlett-Packart Laboratories, 1501 Page Mill Rd., 1L 14, Palo Alto, CA 94304 USA
Tel: 650-857-6586, Fax: 650-813-3312, E-mail: stan_williams@hp.com
http://www.hpl.hp.com/research/qsr/index.html

Steve D Wolbach (USA)
Rutgers University, 1037 10th Ave. Folson, PA 19033, USA
Tel: 732 580 8650, E-mail: stevenwo@eden.rutgers.edu

Etienne Wortham (Greece)
Inst of Elec Struct & Laser, Greece

NANOSTRUCTURES AND NANOTECHNOLOGY: PERSPECTIVES AND NEW TRENDS

THOMAS TSAKALAKOS
Department of Ceramic and Materials Engineering
Rutgers University, 607 Taylor Road, Piscataway NJ 08854-8065

Abstract. Nanotechnology is widely agreed to be the research focus that will lead the next generation to breakthroughs in science and engineering. Nanotechnology will yield revolutionary advances in understanding and application at three cornerstones of expectation:

- Breakthroughs in properties that arise from materials fabricated from the nanoscale.
- Synergistic behaviour that arise from the combination of disparate types of materials (soft vs. hard, organic vs. inorganic, chemical vs. biological vs. solid state) at the nanoscale.
- Exploitation of natural (e.g. chemical and biological) assembly mechanisms that can accomplish structural control, at the nanoscale, are expected to lead to paradigms for assembling bio-inspired functional systems that accomplish desirable properties, either unavailable or prohibitively expensive, using top-down approaches.

1. Introduction

While general principles of nanotechnology are clear [1-39], systematic and comprehensive studies of the effects of processing and the resultant evolution of structure on the overall mechanical properties, performance and thermal stability, under complex loading conditions, have thus far not been performed in sufficient depth. This situation is further compounded by the fact that standardized procedures, essential for the design, modelling, and life assessment of nanomaterials for novel applications, do not exist for the testing and analyses of size-dependent mechanical and functional properties at the nanoscale. This review will emphasize the "on development" of useful implementations and applications of nanotechnology. One key issue in nanotechnology is how to access, from the macroscopic world, the extremely high information density of nanostructured systems. One way to address this question is by using bio-inspiration – techniques, where we apply lessons learned from living systems, to design new

T. Tsakalakos, et al. (eds) pgs 1 - 36
Nanostructures: Synthesis, Functional Properties and Applications;
© 2003 Kluwer Academic Publishers.

2

materials with localized feedback mechanisms. This approach leads naturally to the development of multifunctional nanomaterials for use in applications through basic research of bio-inspired nanostructural materials. Ultimately, this will include embedded or natural sensing, information processing, and actuation of responses to the physical environment. The potential to fit these candidate systems to a wide range of societal applications is not only intriguing but very challenging. Nanotechnology encompasses such similar diverse areas with different degrees of maturity. Specifically, we will evaluate the potential applications as components of advanced technological systems and as materials tailored for a great variety of special applications.

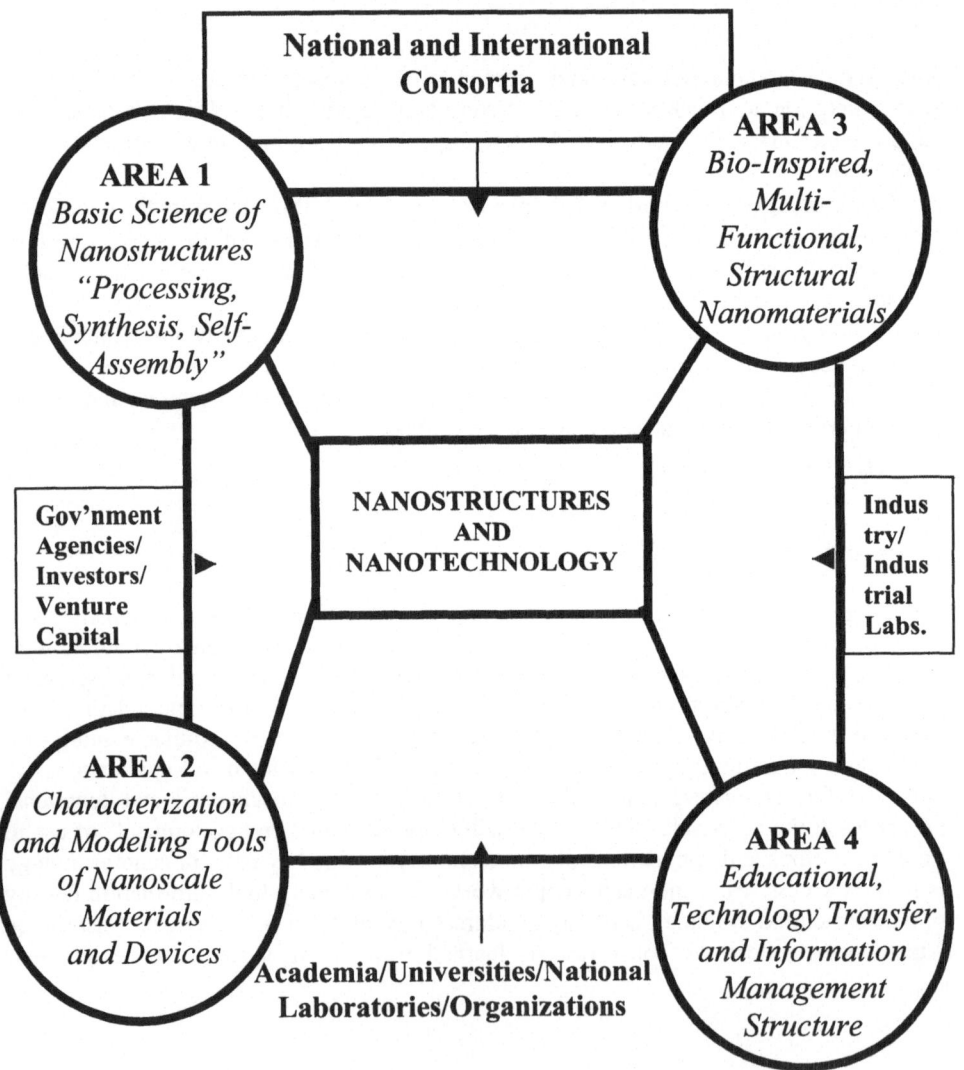

Figure 1. Schematic illustrating the overall scope of a Nanotechnology program.

There is a general consensus among scientists and engineers in the field, that technological advancements can be achieved by task descritization into several phases. One plausible scenario might, for example, involve the following Phase 1 which could focus on basic research necessary before creating and analyzing, or characterizing functional, biologically inspired, and structural, or any other type of nanomaterials. This stage could consist of constant synthesis and characterization followed by material redesign where, most likely, progress toward potential applications appears. Constant feedback could be carried out by teams of researchers with expertise in bio-inspired functional or structural materials synthesis, characterization, modelling and self - assembly. It is natural to expect breakthroughs in material synthesis to take place during this phase, however, achieving the desired functionality of nanostructured systems could be the major goal of these research teams. Phase 2, of research, could be to harness the basic building blocks of nanomaterials and to combine them into bio-inspired, functional, or structural nanostructured material systems which will create technologies for every conceivable societal application. This type of multi-phase, and often concurrent research & development approach, is very essential to the nanotechnogy in order to harvest the full benefits of the field. A synergistic approach of Nanotechnology R&D is shown in Figure 1.

Synthesis and processing methods include recent advances in preparation and consolidation of nanostructured powders by rapid condensation and quenching of ceramic melt to form metastable powders. One of the main areas for synthesis of nanocomposite materials is to develop ways to effectively incorporate the nanomaterials in polymer based nanocomposites.

Biomimetic, or bio-inspired materials, is a model by which we can examine nature's approach to environmental challenges in long-term environmental applications where long duration extremes in temperature, pressure, radiation exposure, and other physical properties require an unconventional approach. While the same materials used in nature may not apply to other areas, the methods by which these materials are assembled, the types of composites used in nature, and the application of these concepts, need careful consideration. New materials, such as carbon nanotubes, have been investigated with promising characteristics as structural materials of the future, however, many biomaterials are similar to carbon nanotubes, possessing strength, low density, and resistance to irradiation damage. Keratin, the major protein in hair, is one example withstanding well over 5-6.5 Gy without significant damage. Through the establishment of unique teams, several innovative and novel processing methods have been successfully developed to investigate such Bio-Inspired nanomaterials.

Recently, unique experiments have also been proposed for the in-situ and ex-situ characterization of the structure, internal stresses, and fracture and fatigue response, by recourse, to Advanced Photon Source, Scattered Intensity Tomographic Profiling, newly developed methods for testing and interpreting mechanical properties including: nanoindentation and microindentation of interlayers and coatings; microtesters and microfatigue testers. Moreover, bionanomaterials are typically evaluated for strength, stiffness, fracture toughness, and erosion resistance. The fundamental understanding derived from such studies are usually used to develop general constitutive models which will form the basis for multi-scale modelling of deformation and failure, spanning the atomistic to the continuum length scales and for formulating life prediction methodologies. Models of structural and functional performance have been also developed to provide basic insights about the ultimate potential of these new materials for a number of applications.

Another major effort has been recently dedicated to the goal of establishing the limits and promise of inorganic and bio-inspired functional nanomaterials in a device environment for magnetic, electronic, and optical applications. Manifestations include: optical materials with tuneable Bragg gratings, magnetic composites for absorbing aerospace materials and sophisticated self-assembled piezoelectric devices synthesized to monitor multi-function operation in electronic industries. Mechanically-linked actuation is another desirable property for nanomaterials. From the example of micro-electro-mechanical systems (MEMS), one can extrapolate to postulate nano- electro-mechanical systems (NEMS). Most of these efforts consist of a two-fold approach to the implementation of NEMS-type capabilities in nanomaterials: a "top-down" approach in which the idea is to push the state of the art in MF-MS to understand scaling limits in signal to noise and system integration and a "bottom-up"'" approach to investigate new materials with desirable nanomechanical properties, characterizing their performance and developing methods for integration.

Nanotechnology continues the trend, expected to remain in the foreseeable future, to develop novel materials and technologies in one of largest critical areas for national and global growth. This field is an essential core component of science and engineering activities to many national interests and goals. Because of its emerging strength, the structural and functional Nanomaterials areas could have a significant impact on related job growth rate and training worldwide. The Nanotechnology's R&D outcomes, as well as its educational and technology transfer resources, could have a significant impact on many nation long-term targets, needs and national economies.

A schematic representation of current and future trends in the various areas of science and technology is shown in Figure 2. It is clearly demonstrated that nanotechnology and nanomaterials, together with biomedical technologies, will continue the trend and growth in the next few decades (after L. Dutta and H. Hoffman, 2003[40]).

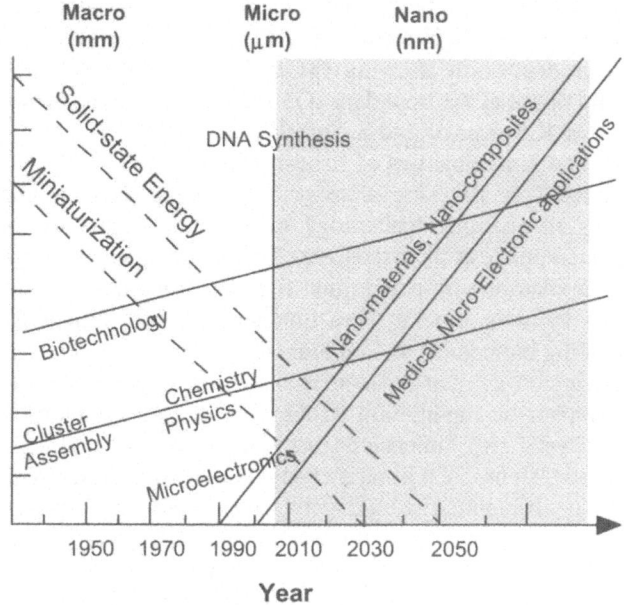

Figure 2. Evolution of science & technology and the future

2. Nanostructures and Nanotechnology

Nanostructure Science and Technology field encompasses a great number of diverse areas with different degrees of maturity. Nanomaterials are synthetic and metastable materials characterized by microstructures in which at least one of their dimensions is less than one hundred nanometers.

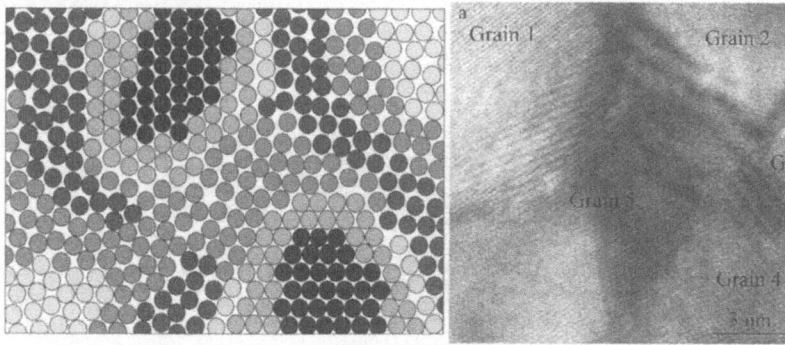

Figure 3. Scematic of nanocrustalline material showing the position of atoms with respect to gain boundaries.

Figure 4. High resolution STEM images of the nanocrystalline-Ni showing (a) multiple grain junctions that reveal no evidence of second phase at the boundary (G3 grain 3), and a clean atomically-faceted grain boundary.(after K.S. *Kumar and S. Suresh)*

6

A schematic representation of a bulk (consolidated) nanomaterial is shown in Figure 3, whereas, Figure 4 depicts High resolution STEM images of the nanocrystalline-Ni produced by electrodeposition showing (a) multiple grain junctions that reveal no evidence of second phase at the boundary (G3 grain 3) and a clean atomically-faceted grain boundary (after K.S. Kumar and S. Suresh et al.). [41]

The properties and functionalities of nanostructured materials depend upon the size and vary substantially as the size increases from nanometer to micrometer scale. Nanotechnology is an enabling technology by a great number of tools to observe, characterize, and manipulate at the nanoscale. Thereby, Nanotechnology is an emerging field that can manipulate the properties and functionalities through the formation and ensembly of nano building blocks for a number of novel structural and functional applications, including biomedical and biological.

The above definition of nanostructured materials implies the existence of four categories of characteristic modulation dimensions in nanostructures: one dimension, such as nanomultilayers; two dimensions, such as carbon nanotubes and nanowires or fibers; and three dimensions, such as nanocrystalline bulk materials (with a grain size of less than 100 nm). Nanoclusters, quantum dotes, and other similar structures are usually considered as zero modulation dimension nanostructures.

Another classification schema for nanomaterials, according to their chemical composition and the dimensionality (shape) of the crystallites (structural elements) forming the nanomaterials, has been proposed by Gleiter [1] as shown in Figure 5.

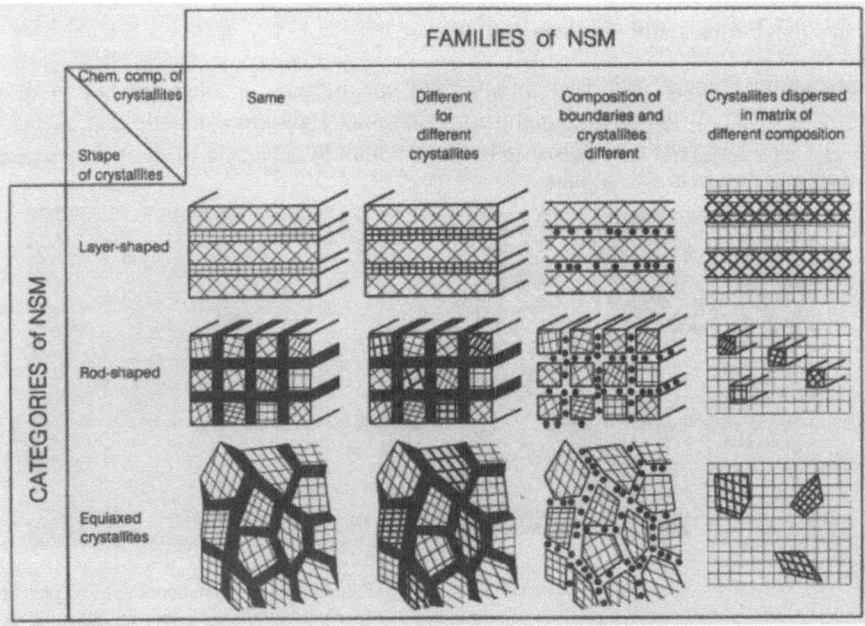

Figure 5. Classification of nanostructured Materials based on shape (dimensionality) and composition as defined by Gleiter.

In Figure 5, indicated in black, the boundary regions of the first and second family of nanomaterials emphasize the different atomic arrangements in the crystallites and in the boundaries. The chemical composition of this boundary and the crystallites is identical in the first family. In the second family, the same boundaries are the regions where two crystals of different chemical composition are joined together causing a steep concentration gradient.

As shown in Figure 5, the latter three categories can be further grouped into four families.

- In the most simple case (first family in the Fig.5), all grains and interfacial regions have the same chemical composition. Eg. Semicrystalline polymers (consisting of stacked lamellae separated by non-crystalline region), multilayers of thin film crystallites separated by an amorphous layer (a-Si:N:H/nc-Si), etc.

- As the second case, we classify materials with different chemical composition of grains, possibly quantum well structures are the best example of this family.

- In the third family, all materials, that have a different chemical composition of its forming matter (including different interfaces) eg, are included; ceramic of alumina with Ga in its interface.

- The fourth family includes all nanomaterials formed by nanometer sized grains (layers, rods or equiaxed crystallites) dispersed in a matrix of different chemical composition. Precipitation hardened alloys, or spinodally decomposed structures, typically belong to this family.

Nanotechnology is a broad and interdisciplinary area of research and development activity that has been growing explosively, worldwide, in the past few years. It has the potential for revolutionizing the ways in which materials and products are created and the range and nature of functionalities that can be accessed. Nanotechnology has been a significant commercial impact that will continue to manifest in the future.

Key to the success of nanomaterials, is the nanostructuring and nanoengineering of building blocks such as nanoclusters, nanoparticles, nanotubes, etc. Some of the unusual, and often novel properties of nanomaterials, are directly related to the extraordinary nature and properties of the nanoelements. For example, when building nanoclusters from close-packed structures, it is shown that a large number of atoms or molecules are at the surface of the clusters. Thus, it is expected that from these clusters the properties of both individual clusters and consolidated materials will be influenced by such arrangements. Figure 6, a typical close-packed cluster, shows the percentage of atoms at the surface for a nanocluster of size 5 nm and about half of the atoms/molecules are still at the surface. As shown schematically in Figure 1, these atoms will become grain boundary atoms after consolidation. A variety of properties (such as magnetic, optoelectronic and mechanical) will depend on the number of atoms near interfaces or grain boundaries. Advanced characterization methods, such as Moossbauer spectroscopy, EXAFS etc, which are sensitive to local molecular arrangements, can be used to provide quantitative information of these atomic arrangements and relate to both modelling and physical properties of nanomaterials.

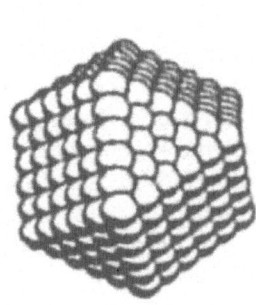

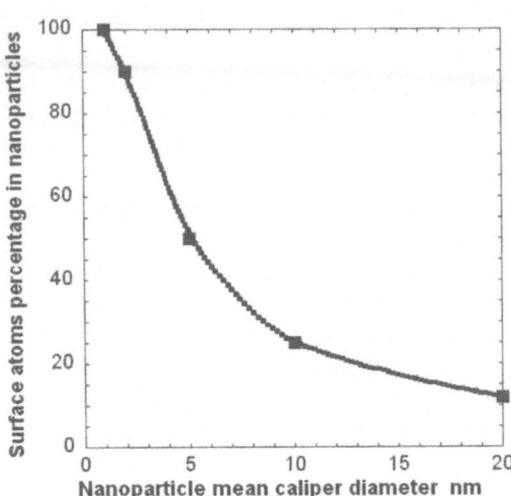

Figure 6. Close- packed nanocluster and percentage of surface atoms as a function of average particle size (mean calliper diameter). G. Jimbo,[42]

The particle (cluster) size is calculated from the mean calliper diameter (mean size averaged for all 3-D directions as defined by Stereological concepts). As seen even for the 20 nm particle size, there is a substantial number of atoms at the surface of the nanoparticle, approximately 12% (7680 surface atoms out of 64000 total number of atoms in the 20 nm nanocluster).

The first scientist to predict and appreciate the magnitude and breadth of the field of nanotechnology was the Physics Nobel Laureate Richard Feynman. In his latest lectures, he predicted that nanoengineering and nanofabrication will become tools to manufacture extraordinary objects as nature does, with cells, in biological systems. He also touched upon data storage and predicted that all information produced by mankind can be stored in a pamphlet, carried in a hand, and that it would be possible to place atoms, one by one, in extraordinary arrangements thus creating a new form of matter with novel properties.

3. Synthesis and Processing

During the past two decades, considerable research work has been directed towards understanding the synthesis, structural, and functional properties of nanomaterials where nanometer size scale components constitute the basic building blocks. Through appropriate processing methods, factors, such as the atomic structure, thickness, volume fraction, chemical composition, as well as the mean size of the building blocks in these nanostructured solids, can be optimized such that a wide range of desirable structural and functional properties can be realized in all materials systems [1]. For example, under certain conditions: the resistance to plastic flow in nanocrystalline materials can far exceed that of microcrystalline solids, and nanostructured surface layers can impart pronounced enhancements in hardness, wear-resistance and corrosion-resistance.

Consequently, there is an opportunity to engineer a variety of nanostructured materials, with enhanced damage tolerance and failure resistance, that may not be readily obtained in conventional microcrystalline aggregates.

A careful survey of available literature on nanostructured materials readily reveals that different synthesis, processing methods, characterization tools and testing methods have been employed to produce and study the mechanical and functional properties. However, in each material or material class, systematic and comprehensive studies of the effects of processing and underlying structure (including the effects of density, size, grain boundary structure and composition, defect type and population, and porosity on the overall mechanical and functional properties) have not been carefully investigated in sufficient depth. In addition, the tools employed, to date, to explore the local structure and properties of nanophase materials in the nanoscale size, have not been sufficiently quantitative to yield detailed insights into the mechanisms and novel properties. Developing materials to meet the societal challenges of the future will require new strategies. One such strategy is focused on meeting that demand by using biological design concepts to meet materials design challenges. Desired properties, in specific, materials, include: toughness, low density and durability. Depending on the application, some materials will need to withstand irradiation and/or very low pressures as well as dramatic temperature changes and other hostile environments. Biological systems are designed to perform efficiently and effectively in a changing environment. Organisms, with an inefficient design, are overcome by predators and become extinct. By examining the design characteristics of these systems, we can investigate new and effective approaches to materials design. The physical properties of biological materials often exceed those of current synthetic materials. Biomimetic, or bio-inspired materials, offer a model by which we can examine nature's approach to environmental challenges. While the same materials used in nature may not apply to space flight, the methods by which these materials are assembled, the types of composites used in nature and the application of these concepts need careful consideration.

10

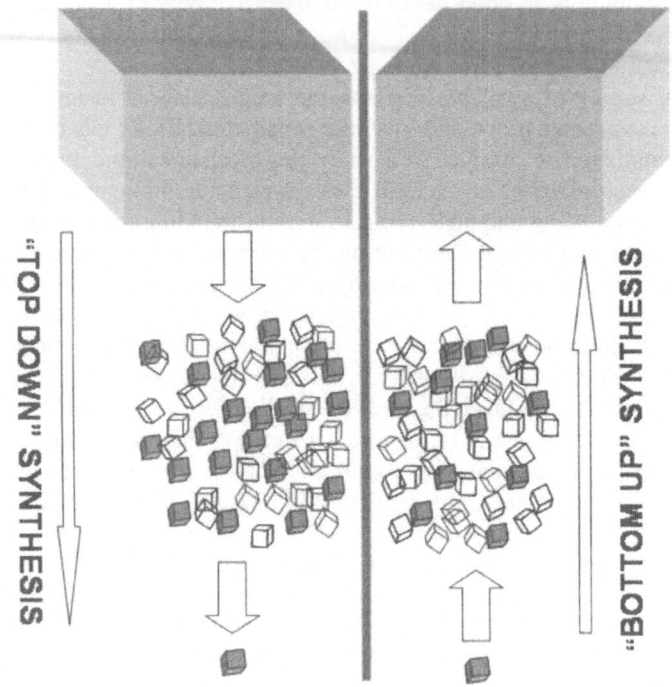

Figure 7. Schematic representation of top down or bottom up syntheses roots.

In general, two major strategies are used to produce nanostructured materials: the "bottom up" and "top down" method, as shown schematically in Figure 7. In the "bottom up" case, a structure is built up of nano building blocks. The "top down" method starts with large building blocks subsequently reduced in size to a finished nanoscale structure. 3-D nanostructures are usually a combination of top down and bottom up methods. Mechanical attrition (ball milling) and lithographic techniques, commonly used by the semiconductor industry, are top-down strategies. Bottom-up methods, where building blocks of nanoparticles or clusters are first prepared by an appropriate technique, include gas phase, liquid phase and solid state reactions. A summary of synthesis methods is shown in Table 1. These building blocks are then assembled or consolidated to produce nanostructured materials. An excellent review of the synthesis methods can be found in a chapter of this book, authored by Muhammed.

Table 1.   Summary of Synthesis Methods

NAME	ACRONYME	SUBPROCESSES	MATERIAL	EXAMPLES
Wet Chemical Synthesis	WC	Sol/gel process	all	Fe-Pt
Liquid Solid reactions:	LS		all	Ferrite Magnetite
Gas Phase synthesis	GP	Furnace/Arc	all	Oxides,alloys
Flame assisted ultrasonic spray pyrolysis	FAUSP		all	Fumed silica, titania
Gas Condensation Processing	GCP		all	Alumina, zinc oxide, titania
Chemical Vapour Condensation	CVC		Metal, oxides	Zirconia, alumina
Plasma CVD	PCVD		Nonoxides metals allloys	silicon
Microwave Plasma Processing	MPP		all	Zirconia, alumina
Laser ablation	LA		Oxides nonoxide	Barium tatanate
Vapor liquid growth	VL		Nooxides	SnO_2
Particle co-precipitation	PCP		Alloys, oxides,	Fe_3O_4
Mechanical attrition	MA	cryomilling	Metal, alloys oxides	Mg alloys
Flat-flame combustor process	FFCP		all	Alumina, titania
Melt-quenching process.	MQP		Ceramic, alloys	Alumina, zirconia
Pressure assisted sintering and polymerization	PAS *OR* PAP		all	Fullerenes, Diomonite, fullerite

In the following examples, recent advances in the preparation and consolidation of nanostructured powders, to produce a new class of nanocomposite ceramics (NCCs) for applications in lightweight vehicles, will be exploited. Nano-ceramic powders can be prepared by (1) rapid condensation of ceramic precursor species, from a supersaturated vapour and (2) rapid quenching of a homogeneous ceramic melt. The resulting metastable powders are consolidated by a novel high pressure sintering method that enables complete densification without causing significant grain growth. Al2O3-, ZrO2- and Carbon-base NCCs were produced by high pressure sintering and by superplastic forming developed by Kear and Mayo.

3.1 VAPOR CONDENSATION

This process involves controlled thermal decomposition of one or more metalorganic precursors in a low pressure flame followed by rapid condensation of the products of precursor decomposition in a cooling gas stream or on a chilled substrate [43,44]. In its optimal configuration (Figure 8), the process employs a flat-flame burner which ensures uniform thermal decomposition of the precursor feed over the entire surface of the burner. Hence, the resulting nanopowder product has a narrow particle size distribution. Gases, such as hydrogen, methane or acetylene are burned in oxygen to generate a steady state combustion flame. The flat flame, extending a few millimeters out of the burner, provides a uniform heat source with short residence time for efficient thermal decomposition and reaction of the precursor/carrier gas stream. The substantial heat release in the flame allows the burner to support a high precursor flow rate at pressures of 5-50 mbar which ensures that the nanoparticles are minimally aggregated. The process has been used to produce a variety of single-component ceramic nanopowders, including Al_2O_3, TiO_2, Y_2O_3 and ZrO_2, starting from readily available metalorganic precursors.

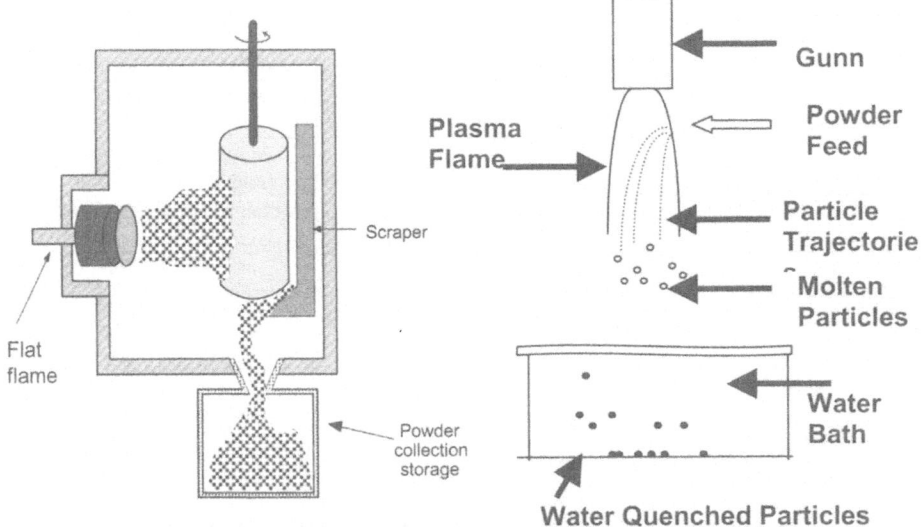

Figure 8. Schematic of the flat-flame combustor process metastable nano-ceramic powders.

Figure 9. Schematic of the melt-quenching process.

Recent research has demonstrated that this same method can be used to prepare multi-component oxide ceramic nanopowders. In a specific case, Al sec-butoxide and Zr tertiary-butoxide were decomposed simultaneously in the flat flame burner and the product species were rapidly condensed as nanoparticles on a chilled substrate. The product powder was composed of a metastable tetragonal ZrO_2 solid solution phase, despite the fact that the material contained 7 wt.% Al_2O_3, far in excess of the equilibrium solid solubility(Kear and Mayo)[45].

3.2 MELT QUENCHING

This process involves feeding an aggregated ceramic powder feed (typically 30 micron diameter) into a high enthalpy plasma jet so that the particles undergo melting prior to rapid quenching in cold water [3]. In practice, because of different particle trajectories in the plasma jet, incomplete melting of some of the feed particles often occurs (Figure 5). In such cases, we have found that *double* plasma spraying into cold water is necessary to ensure that *all* the particles undergo homogeneous melting prior to quenching. We have used this method to produce metastable powders of a variety of multicomponent oxide and non-oxide ceramics.

Under normal circumstances, the solubility of a ceramic within another ceramic is very limited. But, by rapid quenching, such as that achieved by spraying into water, the otherwise immiscible materials can be forced into solid solution. Examples of materials that can be processed in this way to form metastable phases are $Al_2O_3/13TiO_2$ and $ZrO_2/3Y_2O_3/20Al_2O_3$. This is an important result since optimum properties of a nanocomposite ceramic should correlate with a uniform distribution of its constituent, nanophases, which are easily produced by controlled heat treatment of the metastable ceramic starting material (Figure 10).

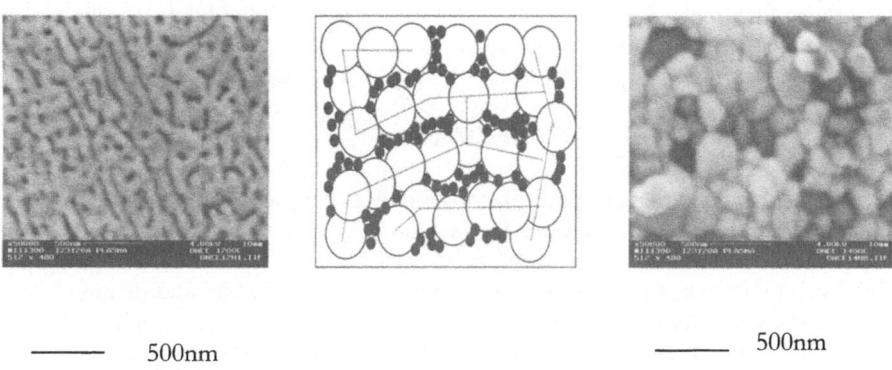

——— 500nm   ——— 500nm

Figure 10. FESEM images of nano-nano-ceramic YSZA composite formed by multiphasic TAC process. *Left*) decomposed at 1200°C/1 hr; *Right*) decomposed at 1400°C/8 hrs; *Center*) schematic drawing showing growth mechanism for chain morphology at 1200°C. The white circles are zirconia (≈70nm) while the black circles are the alumina grains (≈40nm).

4. Nanopowder Consolidation Methods

4.1 PRESSURE-ASSISTED SINTERING

This newly developed process involves high pressure (1-8GPa) compaction and sintering of a metastable ceramic nanopowder at low temperatures (often as low as 0.3-0.4 T_m) [45,46]. During compaction and sintering, nucleation of the stable phase occurs leading to increased density, enhanced sintering kinetics, and minimal grain growth. The great reduction in grain growth is achieved through control of nucleation events. Specifically, the temperature is kept low (minimizing diffusion) and the pressure is kept high (maximizing nucleation). This combination produces a sufficient number of nucleation events of the stable phase to produce a sintered product with a nanoscale grain size.

`A second achievement in this area has been the consolidation of a melt-quenched metastable powder (micro-sized particles) to produce an NCC structure directly. The net result of this pioneering work has been the ability to produce single phase bulk ceramics with densities < 99% and with grain sizes as small as 22 nm in TiO_2 and 40 nm in Al_2O_3, 3 MPa) typical of conventional hot pressing technology. Hence, there is no obstacle to producing large samples of NCCs for mechanical testing purposes and for structural applications. An interesting aspect of this new technology is that the resulting NCCs are inherently superplastic so that near-net shape fabrication of complex-shaped ceramic parts can be accomplished and, probably, at a competitive cost.

4.2 PRESSURE-ASSISTED POLYMERIZATION

A particularly interesting example of consolidation has been the identification of an optimal pressure-temperature regime for the consolidation of C_{60} fullerenes apparently, by a three-dimensional polymerization mechanism, Figure 11. The resulting material, called "diamonite", has properties intermediate between graphite and diamond, i.e, an exceptionally hard like diamond and a good electrical conductor like graphite [47,48].

Another intriguing finding has been that fullerenes make ideal infiltrants for porous ceramics or woven preforms which opens opportunities for creating a new generation of lightweight ceramic composites. For example, fullerenes or fullerene/nanotube mixtures densified at pressures as low as 0.1-0.3 GPa, can be used as a binder phase for a *scaleable* process to consolidate different grades of diamond grit or other superhard, low density materials, such as B_4C and TiB_2. In principle, large sheets of relatively inexpensive diamonite-bonded superhard phases can be produced by this new method with properties that can only be imagined at this time.

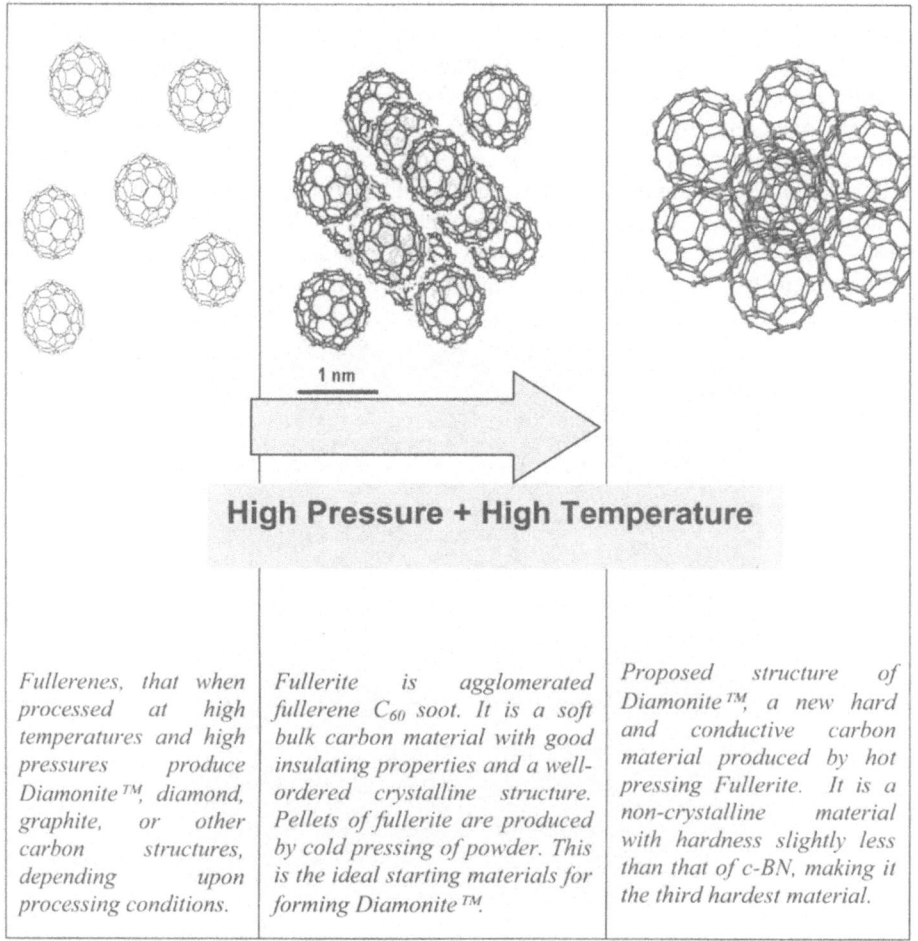

High Pressure + High Temperature

| Fullerenes, that when processed at high temperatures and high pressures produce Diamonite™, diamond, graphite, or other carbon structures, depending upon processing conditions. | Fullerite is agglomerated fullerene C_{60} soot. It is a soft bulk carbon material with good insulating properties and a well-ordered crystalline structure. Pellets of fullerite are produced by cold pressing of powder. This is the ideal starting materials for forming Diamonite™. | Proposed structure of Diamonite™, a new hard and conductive carbon material produced by hot pressing Fullerite. It is a non-crystalline material with hardness slightly less than that of c-BN, making it the third hardest material. |

Figure 11. Structure of Diamonite™ (Kear and Mayo)[47].

5. Synthesis Biomimetic and Bio-Inspired Materials.

A final example of novel synthesis roots, for future technological applications, is biomemetic or bioinspired materials. The physical properties of biological materials often exceed those of current synthetic materials. Biomimetic or bio-inspired materials offer a model by which we can examine nature's approach to environmental challenges. While the same materials used in nature may not apply to same type of applications, the methods by which these materials are assembled, the types of composites used in nature, and the application of these concepts need careful consideration.

Developing materials to meet technological challenges of the future will require new strategies. Nature has allowed biological organisms to adapt to complex environments. In several applications, the use of a biomimetic approach to design individual or

16

composite materials is adapted. Desired properties in structural materials include toughness, low density, and durability. If structural materials are layered, each layer may need very different properties. Biological systems are designed to perform efficiently and effectively in a changing environment. Organisms, with an inefficient design, are overcome by predators and become extinct. By examining the design characteristics of these systems, we can investigate new and effective approaches to materials design, for example, a bird's wing consists of multiple layers such as feathers, skin, muscle, and bone. Bones themselves are complex multi-component systems with a porous interior and non-porous exterior. This design optimizes strength while minimizing density. The tissue surrounding the bone consists of a complex array of blood vessels, muscle cells, and cartilage, are all designed for optimal performance

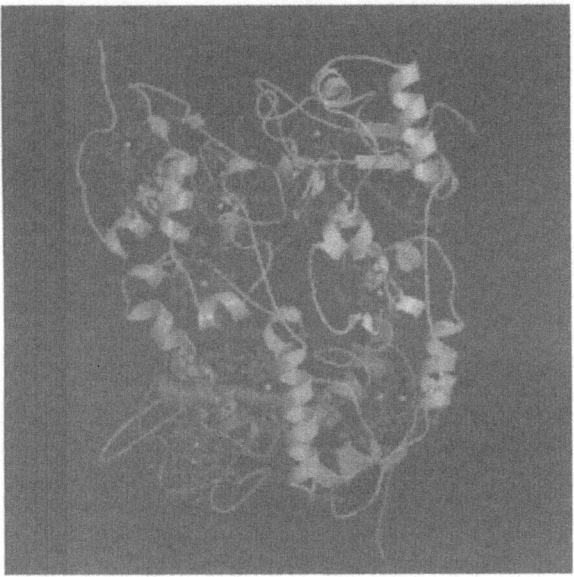

Figure 12: The three-dimensional structure of cytochrome c. obtained from the website: URL: http://www.rcsb.org/pdb/cgi/explore.cgi?pdbId=1AKK&page=20&pid=25921034898434&job=graphics (see S. Goheen chapter in this book)

Proteins have been known to unfold or denature on various rigid surfaces. In solution, when proteins are heated above a certain temperature or if they are introduced to a flexible surface, they can unfold very quickly. On a rigid support the kinetics appear to be much slower. The three-dimensional structure of cytochrome c is depicted in Figure 12.

Some new materials, such as carbon nanotubes, have been investigated with promising characteristics as structural materials of the future, however, these materials often require complex orientation properties to be most effective. Many biomaterials are similar to carbon nanotubes, possessing strength, low density, and resistance to irradiation damage. Keratin, the major protein in hair, is one example withstanding well over 5-6.5 Gy without significant damage. Yet, it is the failure of the follicle, not the hair itself, that causes hair to fail in irradiation therapy. An additional property, the darkness of the hair, can help protect it from (UV) radiation damage. Very small protein fibers, such as those in microtubules (although some cytoskeletal filaments are more resistant than others) appear to be less resistant to irradiation damage than larger fibers.. The resistance of other biopolymers to irradiation damage, such as cotton (cellulose) and carbon nanotubes, has also been studied. The relative resistance of these bio and nano materials, to withstand irradiation damage in comparison to synthetic and conventional materials, should be examined. Therefore, whether these biological, carbon-based or any bio-nanostructured hydrocarbons have a place in future applications, need further research and development.

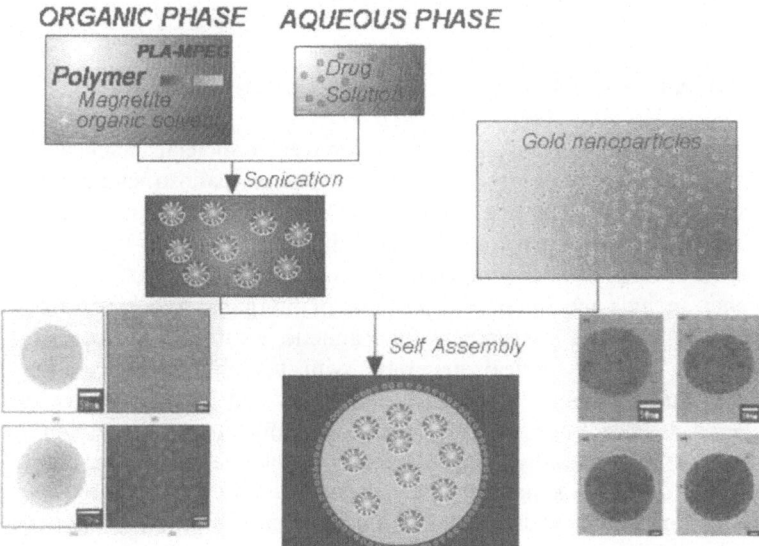

Figure 13. Schematic of synthesis root of magnetic nanosphers for drug delivery with TEM images of polymer-based nanocomposites loaded with drug, functionalized ferritic nanoparticles and various surface coverage Au nanoparticles for stability and contolled drug delivery(Kim, Muhammed and B. Bjelke)[50-52].

The synthesis of magnetic nanosphers for drug delivery (see Figure 13) involves:
- Ring Polymerization (ROP) of Poly(L-lactic acid) PLLA-mPEG methoxy-poly (ethylene glycol) amphiphilic diblock copo lymer biodegradable in the form of nanospheres.
- Water-in- Emulsion-Solvent Diffusion method.
- Entrapment of Fe_3O_4 superparamagnetic nanoparticles.
- Au nanoparticle coating on the nanospheres for better stabilization and drug release rate.
- Nanosphere size controlled by the PLLA concentration [50,51].

Biomimetics also apply to the design of solar panels. Solar panels should be able to survive heat stress, irradiation damage, and impact from space debris. One of the goals in the design of solar panels is to identify materials with good electrical conductivity at the appropriate low temperatures while maintaining physical integrity under anticipated mechanical stresses. Our skin protects us from excessive solar damage while promoting the conversion of provitamin D (ergosterol) to vitamin D2 (ergocalciferol) through the action of UV irradiation. In solar panels, light promotes the transfer of electrons across a barrier in a similar manner to the activation of vitamin D. Bio-photoactivation processes might be better approaches to solar power. Living organisms insulate the transport of electrons with a myelin sheath surrounding nerves. For electronic devices, we mimic a myelin sheath by insulating with air or polymeric materials. Yet, myelin is much lighter and more efficient insulator than most polymers. Near 0°K, a greater variety of materials may conduct electrons more efficiently than the traditional (dense) gold, copper, and other metals used in ambient conditions. An excellent review of this topic is given by S. Goheen by a chapter of this book.

6. Current and Future Applications of Nanostructures

Developing multifunctional nanomaterials for use in societal technical and industrial applications, through basic research of bio-inspired nanostructural materials, is of utmost relevance to the mission of many governments and nations, worldwide. These can include embedded or natural sensing information processing and actuation of responses to the physical environment. Functional applications, such as engineering components, include the limits and promise of inorganic and bio-inspired functional nanodevices in a device environment for magnetic, electronic and optical applications. Manifestations include: optical materials with tuneable Bragg gratings, magnetic composites for absorbing aerospace materials, and sophisticated self-assembled piezoelectric devices synthesized to monitor multi-function operation in industrial applications. Mechanically linked actuation, Nano- electro-mechanical systems (NEMS) with desirable nanomechanical properties and performance, and developing methods for integration in industrial productions of advanced components, has been proven to be extremely useful.

Nanostructured ceramics and ceramic barrier coatings, capable of imparting thermal, corrosion or wear resistance, are expected to provide considerable performance enhancement and/or cost savings in a variety of engine components including burner cans and rotating bearing sleeves. High strength ceramic and/or cermet components made of nanostructured coatings, show promise for use in advanced aircraft landing gear components as replacements for hard chrome-plated parts. Wear-resistant sleeves, in many rotating elements of pumps, air compressors and valves (if successfully made with nanostructured materials), could offer the possibility for marked improvement in performance.

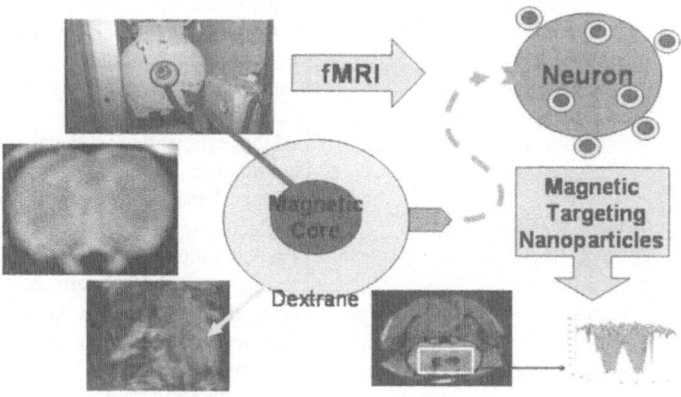

Figure 14. Magnetic Nanoparticles in Biomedical Applications [50-52]]

The design and fabrication of biochemically functionalized superparamagnetic iron oxide nanoparticles is of particular interest. Target-oriented drug release, by drugs encapsulated in polymeric nanocapsules, presently is the most prevalent research area in biomedicine (Figure 14). In Figure 14, a schematic demonstrates the usage of functional-MRI to image drug delivery through the nanocapsules on brain of epileptic rat. The lower right portion indicates quantitatively the mapping of the drug delivery system in the targeted area (D. Kim, M. Muhammed and B. Bjelke)[52].

Superior strength and rigidity derived from two interlocked, oriented, immiscible co-continuous polymer phases, offer new approaches to components such aircraft parts. Patented methods for IMPB's and other nanocomposites using ceramic nanoparticles and carbon nanotubes, have also shown to promote far superior resistance to contact damage and cracking compared to conventional components. These nanomaterials have demonstrated unprecedented reductions, in mass, combined with robust specific mechanical performance such strength, modulus and toughness. Such basic understanding is also anticipated to have a broader impact in the future development and application of nanostructured materials both in the commercial and military arena.

Some typical examples of current and future applications is shown in Table 2 (adapted from WTEC Panel on 'nanostructure science & technology- a worldwide study',1999, http://itri.loyola.edu/, pg. XX).

In order to demonstrate the impact, the degree of complexity as well as the synergistic approach to strategy development for the solution of technological and fundamental problems in nanotechnology, we will focus on the application of nanotechnology to magnetic data storage.

Table.2 Some examples of present and potential applications with significant technological impact.

Technology	Present Impact	Potential Impact
Dispersions and Coatings	Thermal barriers Optical (visible and UV) barriers Imaging enhancement Ink-jet materials Coated abrasive slurries Information-recording layers	Enhanced thermal barriers Multifunctional nanocoatings Fine particle structure Super absorbant materials (Ilford paper) Higher efficiency and lower contamination Higher density information storage
High Surface Area	Molecular sieves Drug delivery Tailored catalysts Absorption/desorption materials	Molecule-specific sensors Particle induced delivery Energy storage (fuel cells, batteries) Grätzel-type solar cells, Gas sensors
Consolidated Materials	Low-loss soft magnetic materials High hardness, tough WC/Co cutting tools Nanocomposite cements	Superplastic forming of ceramics Materials Ultrahigh-strength, tough structural materials Magnetic refrigerants Nanofilled polymer composites Ductile cements
Bio-medical aspects	Functionalised nanoparticles	Cell labelling by fluorescent nanoparticles Local heating by magnetic nanoparticles
Nanodevices	GMR read heads	Terabit memory and microprocessing Single molecule DNA sizing and sequencing Biomedical sensors Low noise, low threshold lasers Nanotubes for high brightness displays

7. Applications of Nanotechnology Both Current and Future

- **Magnetic Applications:**
    - magnetic data storage memories, magnetorestrictive materials, soft magnetic alloys such as Finemet, high-energy products high-temperature superconductors using nanoparticles
- **Optoelectronics:**
    - nanocrystalline silicon for optoelectronic chips and color displays, quantum dots with a voltage-controlled, tunable output color, electrically conducting nanoceramics
- **Optics:**
    - graded refractive index materials, plastic lenses,filters.

- **Energy storage**
  - o solar cells, batteries, nanocrystalline hydrogen storage materials(fullerenes),magnetic refrigerators
- **Gas sensors:**
  - o gas sensors, UV sensors,
- **Catalysis:**
  - o photocatalyst air and water purifiers, fuel cells, precursors for a new type of catalyst
- **Nanostructured coatings:**
  - o wear and corrosion protection materials, scratch-resistance top-coat using hybrid nanocomposite materials
- **Biomeedical appplications:**
  - o drug release and drug delivery , medical implants, immunomagnetic nanoparticles.

8. Nanotechnology for Magnetic Data Storage.

With the new demands of the information and media technologies, the data storage capacity for future applications is likely to approach tens or hundreds of terabytes. Obviously, the data rate transfer rates need to keep pace with the data storage consumption requirements. For instance, interactive 3-d video will require an areal density of 10 $TB.in^{-2}$ and data transfer rates of 100 Gb/sec. The success of the hard disk drives originates from a consistent enhancement in storage capacity and performance combined with significant reductions in price per Gigabyte.

The focus of the commercial market's data storage requirements is changing continuously. Today most of the revenue is generated from the 3–10 Gigabytes desktop hard drives. In the near future, the demand for the hard drives will grow in the mobile laptops and highend servers market. By year 2003, the revenue of the hard disk drive industries is expected to reach 100 billion US dollars (1998 Disk/Trend Report). These are key issues, challenges and limitations of the nanotechnology that affect the magnetic performance and the enhancement of storage capacity.

8.1. AREAL DENSITY GROWTH

Over two decades, the areal density of various data storage devices has grown substantially. In small personal computers, the need for storage is growing from hundreds of megabytes to gigabytes and in larger systems, from tens of gigabytes to terabytes. Correspondingly, the data rate has increased from hundreds of kilobytes per seconds to megabytes per seconds in PCs and from hundreds of megabytes per seconds to gigabytes per second in larger systems. To achieve high areal densities and data transfer rates at lower cost, commercial market utilizes magnetic hard disc, magnetic flexible tape and optical disc devices[57-72].

Although several competing technologies for data storage are available, magnetic hard disc drives continue to be the primary, high performance storage device of choice.

In the past five years, areal density of storage has grown at 60% annually (Figure 15). Growth is spurred by the introduction of the giant magnetoresistive (GMR) head technology and by proportionally reducing all dimensions. In the hard disc drive storage arena, technologies in all aspects of recording are continuously being stretched to the limit to sustain the 60% compound annual growth rate (CGR) (5). The size of the recording bit, necessary for higher linear bit and track densities, is achieved by scaling down the dimensions of the pole tip structure to nanometer dimensions. The bit aspect ratio reduces from 20 to 4 with an increase in areal density from 0.1 to 100 Gb.in-2. Typical lengths and widths of recording bits in 1, 10 and 100 Gb.in-2 areal densities, are 3.0 x 0.2 mm, 0.8 x 0.066 mm and 0.3 x 0.075 mm, respectively. Changes in various head, disc, and interface related parameters, to achieve increasing areal densities for hard disc drives, require extraordinary materials development.

The form factor (size) of the recording head shrinks with the areal density. Head design parameters, such as read gap, reader track width and stripe height, are also scaled down, accordingly, to the nanometer dimensions. The magnetic spacing decreases from 75 to 12 nm as areal density increases from 1 to 40 Gb.in^2. Thickness of the wear coatings on head and disc are reduced to zero at 20 Gb. in^{-2} areal density. Key technologies influencing high areal density are: the new GMR heads, positioning servo-mechanisms for high track densities, low noise magnetic media with high coercivity, and low magnetic spacing head-disc interface for high linear bit densities. Fabrication of GMR heads for high areal density would require the characterization, deposition and patterning of nanolayers with controlled thickness in subnanometer range.

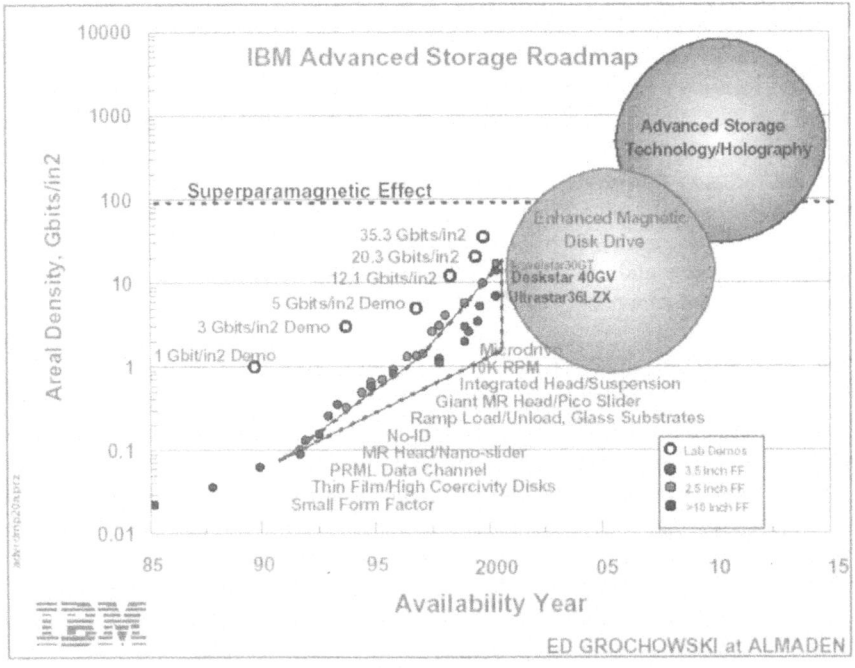

Figure 15. The areal density growth as a function of time by IBM.

8.2. GIANT MAGNETO-RESISTIVE (GMR) HEADS

Unknown until 1988, the main emerging technology for the recording head is Giant Magneto-Resistive (GMR) Technology. The GMR heads will enable areal densities to 15 Gb.in^{-2} and beyond. The primary advantage of the GMR head is greater sensitivity to the magnetic field generated by the recorded bit on the disk. The increased sensitivity makes it possible to detect smaller recorded bits and to read those at higher data rates. In the GMR head, resistance changes occur in the thin magnetic films separated by a thin conductor. One film has a variable magnetic orientation and is influenced by the disk's magnetic field. The second film has a fixed or pinned orientation.

The GMR effect is based on the exchange coupling between a multiplicity of thin films of a ferromagnetic material (Fe) separated by thin films of a nonmagnetic material (Cr). The structure of a GMR head is schematically shown in Fig. 16. The GMR structure consists of a multilayer sandwich of CoFe and Cu with a pinned layer (FeMn, NiMn, PtMN, PdPtMn, IrMn, or NiO) at the top. The GMR coefficient of a multilayer structure depends on the thickness of conducting Cu layer and the number of the bilayers (9). . Figure 17 depicts a sectional view Read-Rite's dual spin valve structure (after Menon and Gupta).[70]

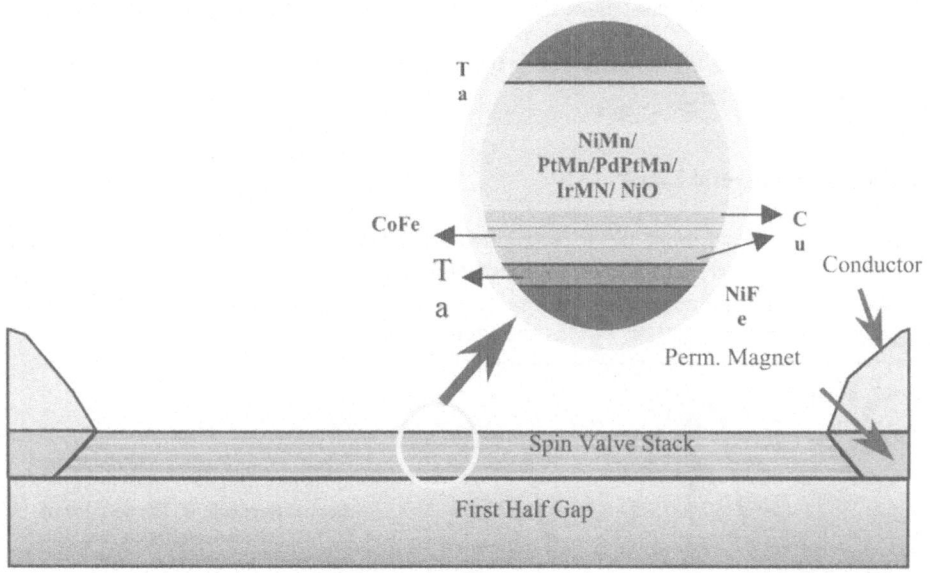

Figure 16. Cross section of the GMR structure

This behaviour demonstrates that the GMR coefficient is strongly influenced by the number of bilayers and the thickness of the conducting interlayer. GMR structure requires deposition and control of thin films contained in the multilayer nanostructure. Future challenges to GMR fabrication involve control of nanometer dimensions and

24

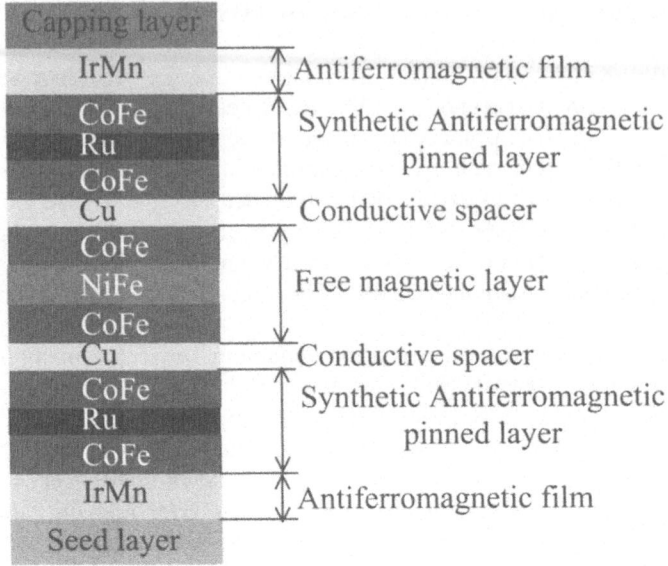

Capping layer

IrMn — Antiferromagnetic film

CoFe
Ru — Synthetic Antiferromagnetic
CoFe   pinned layer

Cu — Conductive spacer

CoFe
NiFe — Free magnetic layer
CoFe

Cu — Conductive spacer

CoFe
Ru — Synthetic Antiferromagnetic
CoFe   pinned layer

IrMn — Antiferromagnetic film

Seed layer

Fig. 17. Sectional view Read-Rite's dual spin valve structure (Data Storage, Sept. 1999).[70]

tolerances of ultrathin films, particularly the conducting spacer film which can be only 10–15 atomic layers thick.

9. Superparamagnetic Limit

Increase areal density must be accompanied by a particle size. However, when the particles become too small their magnetization becomes unstable due to thermal fluctuations. The time constant for these fluctuations is given by $\tau=\tau_0 e^{(K_u V/k_B T)}$ where $E_{anisotropy}= K_u V$ and $E_{thermal} = k_B T$ as shown in Figure 18.

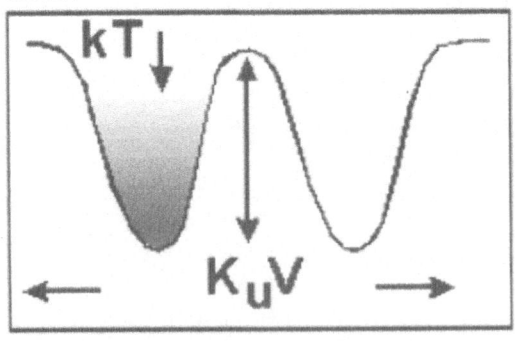

Figure 18. Schematic representation of the supermagnetic effect due to thermal fluctuations. The definion of the magnetic anisotropy constant, the volume of nanoparticles V and the energy barrier to reversal are also depicted.

$$\tau = \tau_0 e^{K_u V / kT}$$

There is some finite probability that, in a time t, the thermal disturbance will be sufficient to induce a reversal of the equilibrium direction causing the decay of the recorded dibits. If a number of these reversals occur during the time of a magnetization measurement, the magnetic bit (Nanoparticle) is said to be "superparamagnetic". Superparamagnetic behavior can be avoided if the energy barrier to reversal, E_B, is much greater that kT, where k is Boltzmann constant and T is the absolute temperature. Magnetic thermal stability limits are determined by the superparamagnetic effects of the media.

The key fundamental limits of the recording are media noise versus thermal stability of the recorded bit, smallest magnetic spacing achievable, and maximum switching speed of current head/media materials. Some of the quasi-fundamental limits include: scaling of magnetic behavior to submicron head geometries, signal detection at very low signal to noise ratio and precision of single stage actuator positioning over data tracks. The ultimate limit for room temperature magnetic storage is thermal demagnetization of the recorded bits. In a grain of the magnetic material, the direction of each magnetic moment oscillates in an energy well about its equilibrium direction. There is some finite probability that, in a time t, the thermal disturbance will be sufficient to induce a reversal of the equilibrium direction causing the decay of the recorded dibits. If a number of these reversals occur during the time of a magnetization measurement, the grain is said to be "superparamagnetic". Superparamagnetic behavior can be avoided if the energy barrier to reversal, E_B, is much greater that kT, where k is Boltzmann constant and T is the absolute temperature. Magnetic thermal stability limits are determined by the superparamagnetic effects of the media. This occurs when the volume V, of the basic switching unit, is not adequate to prevent thermal energy from demagnetizing the media. The limit is being approached because, in order to maintain constant media signal to noise ratio, as the bit size is reduced, it is necessary to make the media with even smaller and more isolated grains. Thermal decay and increase in the dynamic coercivity are correlated and inseparable. The instabilities can be improved by optimizing the particle size distribution and the anisotropy constants.

High anisotropy materials with anisotropy constant $K_u > 10^7$ ergs/cc, such as CoPt, FePt, $SmCo_5$, etc., have been developed. The grains are thermally stable even below 5 nm. Other ways to attack supermagnetism include: Exchange coupled media which effectively increases V for stability while maintaining S/N current 100Gbit/in2 media, perpendicular recording,Reduce demagnetizing influence of adjacent bit fields, minimizing transition parameter, potential increase in media thickness, patterned media, discrete features greatly reduced SNR., regular arrays of grains leads to accurate positioning of recording heads, others' techniques and a combination of above techniques.

Self-Ordered Magnetic Arrays of FePt (SOMA) (Figure 19)[73] have demonstrated that Heat Assisted Magnetic Recording (HAMR) could make it possible to write on FePt. With single particle/bit recording and 3nm stable particles, the potential areal density is 50Tbpsi. Today, 10 Gb.in^{-2} recording heads are in large-scale production. In order to achieve 40 Gb.in^{-2} areal density, it is necessary to have thin film media grain size of roughly 10 nm, magnetic spacing of 30 nm, read gap of 150 nm and track width of 450 nm. Evolution of grain size and coercivity of magnetic film, on the disk with areal density, is illustrated in Fig. 5. A system to manufacture 10 Gb.in^{-2} drives is

26

feasible with today's technology and will likely be in production by year 2000. 40 Gb.in^{-2} recording requires sub 5 nm thin film media grain size. Current media materials will not permit stability of the recorded dibit while maintaining sufficient signal to noise ratio. Micromagnetic and Monte Carlo analysis showed that transition, from stable to unstable dibits, occurs around 100 Gb.in^{-2} in the current head/media materials for longitudinal recording. Ultrathin carbon wear coating are applied on head and media to protect the magnetic films against corrosion, due to environmental exposure and against mechanical wear, due to head dragging over the disc surface, prior to take-off and after landing. Currently, a thin nanolayer (5–10 nm thick) of amorphous carbon (also called diamond-like-carbon (dlc)) is applied on the head and media surfaces. An overcoat of minimum thickness is needed to achieve high readback signal amplitude without sacrificing the friction and wear properties of the head-disk interface. Characterization of the head and disk surfaces, at nanometer dimensions, is crucial for optimizing the tribological performance (7). Mechanical properties of the carbon overcoats on the head and disk surfaces influence the friction and wear performance of the head-disk interface [70].

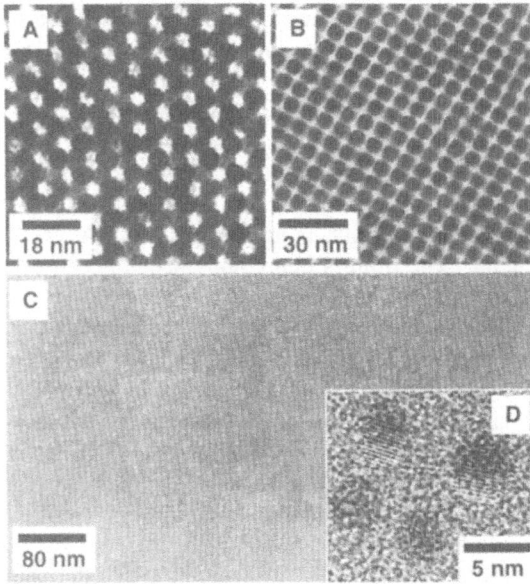

Figure 19. Ordering of 4 nm Fe/pt nanoparticles, IBM[73].

Requirements for Ultra-High Density Recording:
- Particle size below 10 nm (Thermal stability effects)
- High magnetocrystalline anisotropy, (K $\sim 10^7$ erg/cm^3)
- Coercivity Hc $\sim$ 4 -10 kOe
- Isolated particles (To reduce noise)
- Thermal Stability Relaxation time = τ =10-9EXP(K$_U$V / K$_B$T)
- T= 3 X 10^9 years for K$_U$V / K$_B$T = 60
- Magnetic annealing process to orient the nanoparticles to the c axis of the fct phase.

10. Multi-Scale Modelling

While controlled and systematic experiments provide valuable insights and a basic framework for developing fundamental knowledge of the principles in nanostructured materials, the development of a broad quantitative predictive capability, useful for practical design purposes, is not possible without parallel modelling efforts. Such efforts should inevitably seek to include the entire spectrum of length scales from the atomistic/molecular, to the continuum levels, through the dislocation and crystalline levels [2-5,37-39]. The microstructurally and physically based modelling efforts can especially be effective in providing a systematic and quantitative methodology for optimal micro-structural design. Usually, the major objectives in multi-scale modelling are to:

- Obtain quantitative understanding of the nanoscale processes in terms of length-scale in nanocrystalline materials and to establish a systematic approach for extracting useful information for continuum models,
- Establish a physically based multi-scale modelling methodology which builds upon the quantitative information, obtained through atomistic study analysis, and extends to scale-dependent continuum constitutive models. Implement the constitutive models in Finite Element (FE) codes.
- Obtain a quantitative understanding of the localized effects within the context of length-scale and microstructural issues which is crucial for improving nanomaterials in technological applications,

Atomistic and molecular dynamics simulations have been carried out using a variety of codes developed and potentials. Continuum as well as discrete deformation processes to study the meso-scale properties of nanostructured materials have been used very successfully. It is expected that these simulations, along with multi-scale modelling efforts as shown in Figure 20, will collectively enable the identification, visualization, and quantification of processes from the nanoscale to the macroscopic dimensions.

For metallic systems, well-characterized interatomic potentials, such as the embedded-atom-method, have been used to study the stability of nanomaterials.

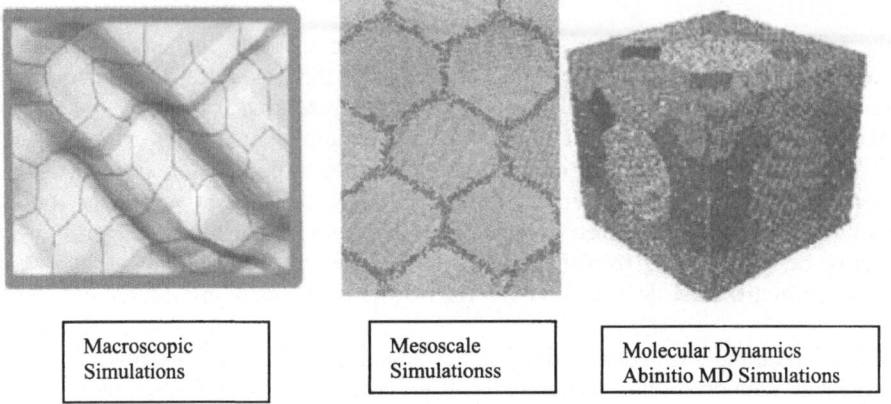

| Macroscopic Simulations | Mesoscale Simulationss | Molecular Dynamics Abinitio MD Simulations |

Figure 20. Various scale simulations.

11. Characterization Tools in Nanotechnology.

Technological advancements in nanoscience and engineering can be achieved through constant innovation and creation of novel nanostructures. However, analyzing or characterizing functional, biologically inspired, and structural or any other type of nanomaterials is an essential component of success. This stage could consist of constant synthesis and characterization followed by material redesign where, most likely, progress toward potential applications appears. In the following, a brief description of two or three methods very sensitive to local atomic or molecular environment will be discussed to demonstrate the depth and breadth of characterization methods in the nanofield. These include Mössbauer Techniques and Synchrotron X-ray Absorbtion Spectroscopy in Fe/Pt nanoscale systems which is critical in the nanotechnology of magnetic data storage.

11.1 MÖSSBAUER TECHNIQUES

Key to magnetic storage nanotechnology is the advaced characterization of magnetic nanoparticles by Mossbauer spectroscopy. An excellent review of this technique is provided in a chapter of this book, by Miglierini.

Mössbauer spectrometry, complementary to diffraction techniques, is often proven decisive in the area of nanotechnology because of its local behaviour that enables it to be an atomic scale sensitive tool while its time of measurement allows investigation of relaxation phenomena and dynamic effects. It is important to emphasise that ^{57}Fe is a valuable isotope that can be encountered in different kinds of materials including nanocrystalline alloys.

Recoilless nuclear gamma resonance - the Mössbauer effect [75] - takes place between the same type of nuclei located in a source and an absorber (sample). Radiation from the source is modulated by a Doppler effect which eliminates differences between energy levels of the source and the absorber. Mössbauer spectrum is a plot of intensity of gamma radiation accepted by the detector versus velocity. Differences in energy levels provide information on the character of nearest neighbourhoods of resonant atoms both from the point of view of structure and hyperfine interactions.

Conversion electron (CEMS) and transmission (TMS) Mössbauer spectroscopy have been used for Fe/Pt studies. In a recent study of Fe/Pt nanomultilayers, an atomic plane-by-plane magnetism study (see Figure 21) revealed the extraordinary depth of the method, (Simopoulos, Tsakalakos et al)[75]. The main advantage of TMS is its accessibility in any Mössbauer lab and the convenience for temperature variation experiments. It suffers from small absorption efficiency in comparison to CEMS. Another method often used is the enrichment of one to two monolayers in the Fe layer with Fe^{57}. This is an efficient technique and allows the probing of individual monolayers according to their distance from the interface. However, it has the drawback of the cost and the possibility of diffusion between ^{56}Fe and ^{57}Fe. As demonstrated in Figure 22, a systematic study of Fe/Pt system, with varying Fe thickness and good quality spectra, can be done with TMS in non-enriched samples and lead to the determination of hyperfine parameters at the monolayer level.

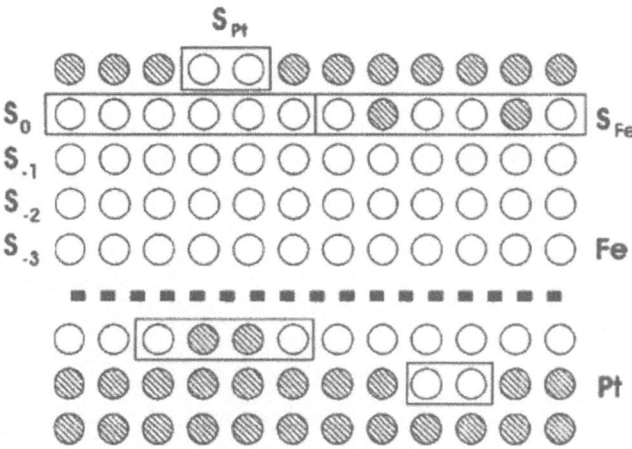

Figure. 21. Assumed structural model of the Fe/Pt structure used in the analysis of Mossbauer spectra. Open circles denote Fe atoms and shaded circles Pt atoms.

3 Å Fe/X Å Pt multilayers. These samples consist of 1–2 Fe monolayers and N Pt monolayers, N varying between 5 and 19. Figure 18 shows Mossbauer spectra for the 3/9 and 3/19 samples. Spectra for the 3/39 sample were not possible to take since the signal to noise ratio is too small for this sample. First, we notice that the 3/9 sample displays magnetic hyperfine spectra, already at RT, while the 3/19 sample is paramagnetic at this temperature displaying magnetic hyperfine structure below 240 K. This difference indicates the reduction of the interlayer exchange interaction due to the greater nonmagnetic Pt thickness. The magnetization of both samples is out of the plane as witnessed by the reduced intensity of the $\Delta m=0$ lines. The angle between the direction of the hyperfine field H_{hf} and the normal to the plane, as determined by the ratio $\Delta m_{\pm 1} / \Delta m_0$, is 39° for the 3/9 sample and 20° for the 3/19 sample. The latter result

is consistent with the linear temperature variation of the magnetization of the 3/19 sample indicating a two-dimensional behaviour for this sample. Thus, both superparamagnetic behaviour as well as perpendicular magnetic anisotropy in the Fe/Pt nanomultilayers have been observed and measured quantitatively by Mossbauer spectroscopy.

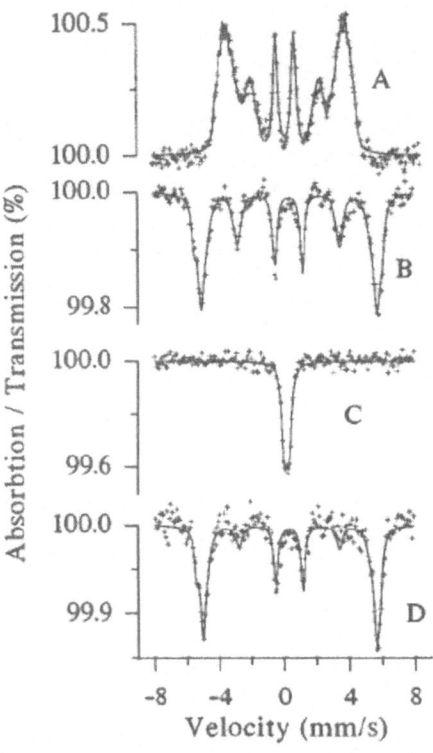

Figure 22. Mossbauer spectra of the ultrathin Fe/Pt samples. (a) CEMS of the 3/9 sample at RT. (b) Transmission spectrum of the 3/9 sample at LHe temperature. (c) RT spectrum of the 3/19 sample.(d) L_{He} spectrum of the 3/19 sample. Solid lines represent least square fits.

In the present case, each Fe monolayer faces, at least from one side, a Pt layer. Furthermore, some interdiffusion is expected giving rise to a distribution of the hyperfine fields as shown in Figures 21-26. From the spectra in Figure 23, in each monolayer, the Fe magnetism (Hyperfine fields) is measured. At nanoscale, the magnetic properties of each layer of Fe are plotted in Figure 25. Furthermore, a correlation between the magnetic properties of each layer from the interface, as a function of lattice d-spacing, reveals linear dependence (Fig. 26).

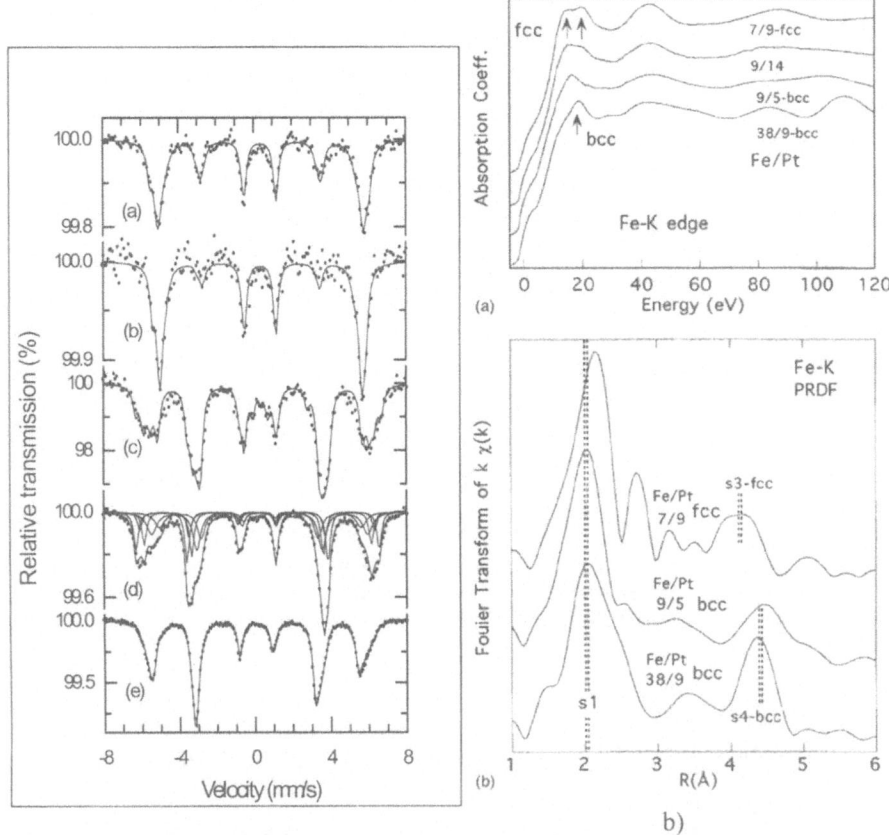

Figure 23. (a) Mössbauer spectra of Fe/Pt multilayers (x Fe thickness in A/ y Pt thickness in A) at 4.2 K. (a) 3/9 (b) 3/19 (c)7/9 (d) 9/9 and (e) 25/9. Individual components are shown for sample 9/9.

Figure 24. (a) The Fe-K spectra of 7/9, 9/14, 9/5, and 38/9 Fe/Pt samples. (b) The Fe-K pseudoradial-distribution-function (PRDF) results for 7/9, 9/5, and 38/9 Fe/Pt samples.

X-ray-absorption spectroscopy [75]. Fe-K-edge XAS measurements were carried out on beamline X-19A at the Brookhaven National Light Source. The beamline optics consists of a double crystal Si-111 and Si-220 monochromator along with vertical-parallelization and horizontal-focusing parabolic Rh mirrors. The data were collected in the fluorescence mode using a Canberra Si-PIPS detector. The XAS measurements revealed a systematic crossover in the Fe-layer structure from fcc, in the thin Fe-layer thickness limit, to bcc in the thick Fe-layer limit. [75]

The spectra of the 3/9 and 3/19 samples were analyzed with 3 magnetic components with B_{hf} values of 362(352), 339(333) and 314(310) kOe, and isomer shift (i.s) values of 0.39(0.36), 0.45(0.45) and 0.45(0.40) mm/s where the number in parenthesis refer to the the 3/19 samples. The i.s. values are larger than the bulk Fe value (0.10 mm/s at 4.2K) indicating the influence of the neighbour Pt atoms [75].

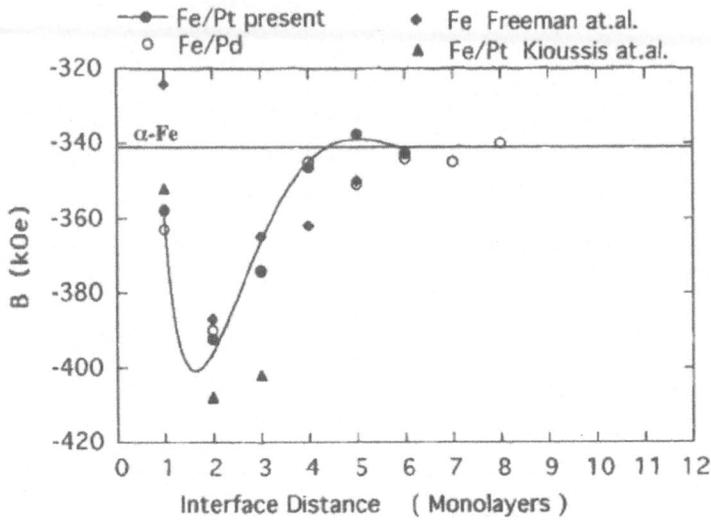

Figure 25. Variation of the hyperfine field B assigned to each Fe monolayer with the distance from the interface. Experimental data are for Fe/Pt of the present study ~full circles! and for Fe/Pd taken from Ref. 19. The solid line represents least square fit to the Fe/Pt data. Diamond and triangle symbols represent theoretical calculations. [75]

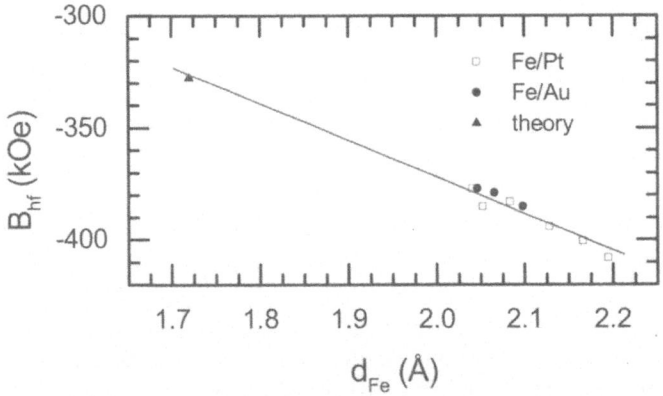

Figure 26. Variation of the hyperfine field B_{hf} of the 1st atomic layer under the interface with the average lattice spacing d_{Fe} of the Fe layer for the Fe/Pt samples. Fe/Au are also shown. A value from the calculation of ref. [75] is included.

Figures 21-26 demonstrate the determination of the hyperfine field associated with each Fe mono-layer in the Fe slab of the Fe/Pt nanoscale system by employing standard Fe [57] Mossbauer spectroscopy. The quality of the Fe/Pt nanomultilayers, studied in this work, was a main factor for this achievement. The results show that the hyperfine field values follow an oscillatory variation as we move from the interface to the center of the

Fe layer (see Figure 25). This variation seems to depend on two factors: the hybridization of the 3d and 5d electron wave functions and the variation of the interplanar spacing within the Fe layer (Figure 26). The importance and role of each of these two factors remain to be further clarified. However, it is very clear that the synergistic characterizations techniques, such as Mossbauer spectroscopy, X-ray absorption spectroscopy, EXAFS, AMF, STEM, etc., can provide extremely important information in Nanotechnology. In the present study, a systematic analysis of the Mossbauer spectra of samples with increasing Fe layer thickness, allowed the determination of the magnetic hyperfine field for each Fe monolayer within the Fe layer slab. Hyperfine fields larger than the bulk Fe value, appear in all samples with Fe layer thickness larger than 3 Å and display an oscillatory dependence on the distance of the corresponding Fe monolayer from the interface. These hyperfine field values scale linearly with the average interplanar distance of the Fe layer derived from the refinement of the XRD data for each sample. Fe atomic magnetic moments, determined from superconducting quantum interference device magnetometry and Rutherford backscattering spectroscopy measurements, are also larger than the bulk Fe value approaching it for large Fe layer thickness. The parameters determining the enhancement of magnetization in the Fe/Pt nanoscale system could be used to create new forms of nanomaterials for magnetic storage applications.

12. Conclusions

Nanotechnology is a broad and interdisciplinary area of research and development activity that has been growing explosively worldwide in the past few years. It has the potential for revolutionizing the ways in which materials and products are created, and the range and nature of functionalities that can be accessed. Nanotechnology is already having a significant commercial impact which will assuredly increase in the future. Key to the success of nanomaterials is the nanostructuring and nanoengineering of building blocks such as nanoclusters, nanoparticles, nanotubes etc.

A comparative assessment of the various nanotechnologies, from a perspective of performance, cost and maturity point of view, was discussed. The industry remains very vibrant primarily in data storage applications but the growth in technology areas, such information technology, is putting pressure on all aspects of nanotechnology to create nanomaterials and nanostructures with novel applications. The current outlook is that the nanotechnology will reach a few hundred billion dollars in revenues within the next few years.

Natural assembly mechanisms that can accomplish structural control at the nanoscale, are expected to lead to paradigms for assembling bio-inspired functional systems that accomplish desirable properties, either unavailable or prohibitively expensive, using top-down approaches.

34

13. References

1. H. Gleiter, *Acta Mat., The Millennium Special Issue,* 48 (2000)1-29.
2. A. J. Zarur and J. Y. Ying, *Nature,* 403 (2000) 65-67.
3. A. J. Zarur, H. H. Hwu, and J. Y. Ying, *Langmuir,* 16 [7] (2000) 3042-3049.
4. Chen and Wang, 2000.
5. M. D. Fokema, E. Chiu, and J. Y. Ying, *Langmuir,* 16 [7] (2000) 3154-3159.
6. C.-C.WangandJ. Y. Ying, *Chem. Mater.,* 11 [11] (1999) 3113-3120.
7. B. H. Kear, *Nanostr. Mater., 6* (1995) 227.
8. J. Karthikeyan, C. C. Berndt, J. Tikkanen, J. Y. Wang, A. H. King, and H. Herman, *Nanostr. Mater.,* 8 [1] (1997) 61-74.
9. U. Erb, *Nanostr. Mater.,* 6 (1995) 533.
10. A. H. Choi, E. S. Ahn, and J. Y. Ying, unpublished results, MIT.
11. H. Hahn and R.S. Averback, *J. Appl. Phys., 67* [2] (1990) 1113-1115.
12. A. M. Gallois, R. Mathur, .R. Lee, and J. Y. Yoo, *Mater. Res. Soc. Proc.,* 132, (1989) 49.
13. Y. J. Liu, H. J. Kim, Y. Egashira, H. Kimura, and H. Komiyama, *J. Am. Ceram. Soc., 79 [5]* (1996) 1335.
14. G. Soyez, J. A. Eastman, L. J. Thompson, R. J. DiMelfi, G.-R. Bai, P. M. Baldo, A. W. McCormick, A. A. Elmustafa, M. F. Tambwe, and D. S. Stone, *Appi. Phys. Lett.,* 77 [8] (2000) 1155-1157.
15. J. T. McCue and J. Y. Ying, unpublished results, MIT.
16. K.F. Jensen, in *Thin Film Processes II,* ed. J.L. Vossen and W. Kern, Academic Press, Boston, (1991) 283-368.
17. J. N. Musher and R.G.Gordon, *J. Electrochem. Soc.,* 143 [2] (1996) 736.
18. 3. A. Eastman, M. R. Fitzsimmons and L. 3. Thompson, *Phil. Mag. B, 66 [5].(1992)* 667-696.
19. P. G. Sanders, 3. A. Eastman, and 3. R. Weertman, *Acta Mat.* 45 [10] (1997) 4019-4025.
20. 0. E. Fougere, J. R. Weertman, and R.W. Siegel, *Nanostr. Mater.,* 5 [2] (1995) 127-134.
21. 3. A. Eastman and M. R. Fitzsimmons, *J. Appi. Phy.,* 77 [2] (1995) 522-527.
22. M. N. Rittner, 3. R. Weertman, and l.A. Eastman, *Acta Materialia,* 44, no.4, 127 1-1286 (1996).
23. 3. A. Eastman, M. A. Beno, 0. S. Knapp, and L. J. Thompson, *Nanostr. Mater., 6 (1995)* 543-546.
24. M. S. Choudry, M. Dollar, and J. A. Eastman, *Mater. Sci. Eng. A, 256,* (1998) 25-33.
25. R. Z. Valiev, A. V. Korznikov, and R. R. Mulyukov, *Mater. Sci. Eng. A,* A168 (1993) 141-148.
26. R. Z. Valiev, T. C. Lowe, and A. K. Mukherjee, *J. Metals,* April (2000) 37-40.
27. P. 0. Sanders, 0. E. Fougere, L. J. Thompson, J. A. Eastman, and J. R. Weertman, *Nanostr. Mater.* 8 [3], (1997) 243-252.
28. 3. A. Eastman, L. J. Thompson, and D. J. Marshall, *Nanostr. Mater.,* 2, (1993) 377-382.
29. K. Taketani, et al., *J. Mater. Sci.,* 29 (1994) 6513-65 17.
30. W. Liu and W.L. Johnson, *.1. Mater. Res.,* 11 [9] (1996) 2388-2392.
31. A. Inoue, A. Makino, and T. Masumoto, *Mater. Sci. Forum,* 179-18 1 (1995) 497-505.
32. C. Fan, A. Takeuchi, and A. Inoue, *Mater. Trans. JIM,* 40 [1] (1999) 42-51.
33. A. Inoue, et al., *Mater. Sci. Eng. A,* 217-218 (1996.) 401-406.
34. H. Gleiter, *Nanostruct. Mater., 6* (1995) 3-14.
35. 0. B. Stephenson, J. A. Eastman, 0. Auciello, A. Munkholm, C. Thompson, P. Fuoss, P. Fini, S. P. DenBaars, and J. S. Speck, *MRS Bulletin,* 24 [1] (1999) 21-25.
36. T. Tsakalakos and M. Croft, in Nanostructured Films and Coatings, 0. M Chow et.al. (eds), (2000) 223-230, Kluwer Academic Publishers, the Netherlands.
37. A. E. Giannakopoulos and S. Suresh, *Scripta Mat.,* 40(1999) 1191-1198.
38. T. A. Venkatesh, K. J. Van Vliet, A. E. Giannakopoulos, and S. Suresh, *Scripta Mat.,42* (2000) 833-839.

39. S. Suresh and A. E. Giannakopoulos, *Acta Mater.*, 46 (1998) *5755-5767.*

40. L. Dutta and H. Hoffman, to be published (2003).

41. K.S. Kumar, S. Suresh , M.F. Chisholm , J.A. Horton , P. Wang, Acta Materialia 51 (2003) 387–405

42. G. Jimbo, Proc. 2nd. World Congress on Particle Technology, (1990)19-22, September 1990, Kyoto, Japan

43. W. Chang, G. Skandan, H. Hahn, S.C. Danforth, and B.H. Kear, Nanostru. Mater. 4, 345, 1994.

44. N. Glumac, Y.-J. Chen, G. Skandan, and B.H. Kear, Mater. Lett. 34, 148, 1998.

45. J. Colaizzi, W.E. Mayo, B.H. Kear, S.-C. Liao, Metal Powder Industries Federation Outstanding Technical Paper Award-2000, Intn'l J. of Powder Metallurgy, 37, 45-54, 2001

46. S.-C. Liao, W.E. Mayo and K.D. Pae, Acta Mater. 45(10), 4027-4040, 1997.

47. S.-C. Liao, Y.-J. Chen, B.H. Kear and W.E. Mayo, Nanostruct. Mater. 10, 1063-1079, 1998.

48. O.A. Voronov, G.S. Tompa, B.H. Kear, W.E. Mayo, S.-C. Liao, R.K. Sadangi, K.J. Livi, R.O. Loutfy, "High Pressure-High Temperature Consolidation of Carbon Nanotubes for Structural and Other Applications," Report DMI-41023-Final, for US Army Aviation & Missile Command, 1998, 58 pages.

49. O.A. Voronov, G.S. Tompa, B.H. Kear, K.J. Livi, R.O. Loutfy, W.E. Mayo, "High Pressure Sintering of Fullerene Soot," Abstract No. 1457, The 1999 Joint Intern'l Meeting of Electrochemical Society, Honolulu, Hawaii, Oct. 17-22, 1999.

50. Kim D. K. (2001) Superparamagnetic Iron Oxide Nanoparticles for Biomedical Applications. ISBN 91-7283-126-X.*Royal Institute of Technology*, Sweden

51. Kim D. K., Mikhaylova M., Zhang Y. and Muhammed M. (2003) Protective coating of superparamagnetic iron oxide nanoparticles, *Chem. Mater.* **15** 1617-1627

52. Bjelke, Kim D. K., Mikhaylova M., Zhang Y. and Muhammed M. (2003)

53. Adapted from WTEC Panel on 'nanostructure science & technology- a worldwide study',1999, http://itri.loyola.edu/, pg. XX)

54. Barrett, R. C. and Quate, C. F. (1991) J. Appl. Phys. (1991) 70, pp. 2725- ?.

55. Binning, G., Despont, M., Drechsler, U., Haberle, W., Lutwyche, M., Vettiger, P., Mamin, H. J., Chui, B. W., and Kenny, T. W. (1999) Ultrahigh-Density Atomic Force Microscopy Data Storage with Erase Capability, Appl. Phys. Lett. 74, pp. 1329-1331.

56. Burr, G. W. (1999) Holographic Storage, presented in 10th Annual Symposium on Information Storage and Processing Systems, ISPS'99, Santa Clara.

57.Carley, L. R. (1999) Web site of Center for Highly Integrated Information Processing and Storage Systems (CHIPS), www.chips.ece.cmu.edu.

58. Chou, S. Y. (1997) Patterned Magnetic Nanostructures and Quantized Magnetic Disks, Proc. IEEE, 85, pp. 652-671.

59. Egelhoff, W. F., Chen, Jr. P. J., Powell C. J., Stiles, M. D., McMichael R. D., Lin, C-L., Sivertsen, J. M., Judy, J. H., Takano K., Berkowitz, A. E., Anthony, T. C., and Brug, J. A. (1996) Optimizing the Giant Magnetoresistance of Symmetric and Bottom Spin valves, J. Appl. Phys. 79.8, 5277-5281.

60. Fontana, R. E., MacDonald, Jr. S. A., Santini, H. A. A., and Tsang, C., (1999) Process Consideration for Critical Features in High Areal Density Thin Film magnetoresistive Heads: A Review, IEEE Trans. MAG. ??.

61. Gallagher, W. J., Parkin, S. S. P., Lu, Yu, Bian, X., Marley, A., Roche, K. P., Altman, R. A., Rishton, S. A., Jahnes, C., Shaw, T. M., Xiao, G., (1996) Microstructured Magnetic Tunnel Junctions, J. Appl. Phys. 81, 3741-3746.

62.Gibson, G., Kamins, T. I., Keshner, M. S., Naberhuis, S. L., Perlow, C. M., and Yang, C. C. (1996) Ultrahigh Density Storage Device, US Patent 5,557,596.

63.Hosaka S., Koyanagi, H., and Kikukawa, A, (1993), Jap. J. Appl. Phys. 32, L464

64.Howell, T., Ehrlich, R., and Lippman, M. (1999) TPI Growth is the key to Delaying Superparamagnetic's Arrival, Data Storage, 6.10, pp. 21-30.

65. Jin, S., Tiefel, T. H., McCormack, M., Fastnacht, R. A., Ramesh, R., and Chen, L. H., (1994) Thousandfold Change in Resistivity in magnetoresistive La-Ca-Mn-O Films, Science, 264, 413-415.

66. Kado, H. and Tohda, T. (1995) nanometer-Scale recording on Chalcogenide Films with an Atomic Force Microscope, Appl. Phys. Lett. 66, pp. 2961-2963.

67. Kryder, M. H., (1997) Outlook for Magneto-Optical Recording, in Handbook of Magneto-Optical Data Recording, (ed. By McDaniel And Victora), pp. 895-919, Noyes Publications, Westwood, NJ.

68. Mamin, H. J., Ried, R. P., terris, B. D., Rugar, D., (1999) High-Density Data Storage Based on the Atomic Force Microscope, Proceedings of IEEE, 87, 1014-1027.

69. Menon, A. K. and Gupta, B. K., (1999) Nanotechnology: A Data Storage Perspective, Datatech, 2nd ed. pp. 13-24.

70. Vettiger, P., Brugger, J., Despont, M., Drechsler, U., Durig, U., Haberle, W., Lutwyche, M., Rothuizen, H., Stutz, R., Widmer, W., and Binning, G. (1999) Ultrahigh Density, High Data Rate MEMS-based AFM Data Storage System, IEEE J. Microelectron Eng., in press.

71. Zech, R. G. (1992) Volume Hologram Optical Memories, Optics and Photonics news, Aug. issue, pp. 16-2

72. adapted from IBM's web site

73. D.M. Dabbs and I.A. Aksay, "Self-Assembled Ceramics Produced by Complex-Fluid Templation," Annu. Rev. Phys. Chem. 51 601-22 (2000)

74. A.Simopulos, M.Croft and T. Tsakalakos "Structure and enhanced magnetization in Fe/Pt multilayers" Phys. Rev. B., 54, 9931-9942,(1996)

ENGINEERING OF NANOSTRUCTURED MATERIALS

MAMOUN MUHAMMED
Materials Chemistry Division
Royal Institute of Technology (KTH), SE 100-44 Stockholm, Sweden

1. Introduction

The essence of Nanoscience and Nanotechnology is the ability to fabricate and engineer materials, structures and systems where the manipulation of the properties and functionalities is a result of the control of the material's building blocks whose dimension is in the nanometer regime. Nanostructured materials can be advantageously engineered by the controlled assembly of several suitable nano-objects as building blocks. While materials properties are determined by their atomic and molecular constituents and structure, their functionalities emerge when the microstructure of these early ensembles are in the nanometer regime. The properties and functionalities of these ensembles may be different as their size grows from nano-regime to the micron regime and bulk structures. Nanotechnology offers a unique possibility to manipulate properties through the fabrication of materials using the nano-objects as building blocks. Thereby, Nanotechnology is considered an enabling technology by which existing materials, virtually all man-made materials, can acquire novel properties and functionalities, is suitable for numerous novel applications varying from structural and functional to advanced biomedical *in-vivo* and *in-vitro* applications.

In the last decades, nanoparticles have attracted the attention of an increasing number of researchers from several disciplines of science. Nanoparticles have been the most commonly used building blocks while other entities with different morphology can also be used. However, several nano-objects can be used as suitable building blocks for the fabrication of nanostructured materials. For example, nano-sized wires, rods or tubes e.g., the variation of the object morphology, e.g., have a large variation of aspect ratios, or hollow structures, that may give rise to properties that are significantly different from the corresponding spherically shaped particles.

The term 'nanoparticle' is generally used in materials science to specify particles with diameter less than 100 nm. Clusters are considered as small nanoparticles with number of atoms up to few thousands and are of great interest for fundamental understanding of the evolution of particle properties as well as for potential applications. Clusters and nanoparticles have a tremendous increase of surface-to-volume ratio that results in dramatic changes to their properties. Besides the increase of the surface area, as a result of reducing particle size, the properties of the surface layer (one or two atomic layers) may be different from the inner layers as surface layers are less ordered and contain higher density of defects. Having a size between a molecular and bulk

T. Tsakalakos, et al. (eds) pgs 37 - 79
Nanostructures: Synthesis, Functional Properties and Applications;
© *2003 Kluwer Academic Publishers.*

solid-state structures causes nanoparticles to have hybrid properties. Some examples of these properties are a lower melting temperature, increased solid-solid phase transition pressure, lower effective Debye temperature, higher self-diffusion coefficient, and changed thermo-physical properties.

Nanostructured materials produced by the consolidation or the assembly of building blocks with a size range from ~1 (molecular scale) to ~50 nm, possess physical, chemical, and biological properties that are significantly enhanced or changes as compared to bulk materials [1]. The compaction of nano-powders is commonly used for the production of bulk three dimensional nanostructured materials with a domain size less than 100 nm. The control of the processing of nanoparticles, with given properties, such as, chemical (composition of the bulk, interaction between the particles, and surface charge) and structural (crystalline or amorphous structure, size, and morphology), is the main feature in engineering the properties of the nanostructured materials. Novel aspects for the fabrication of building blocks, as well as controlled formation of materials' ensemble, are the main challenges for the engineering of nanostructured materials, structures and systems. As a result, an increase in research for the development of supramolecular, biomolecular, and dendrimer chemistries for engineering substances of ångström and nanometer scale has emerged with such disciplines as nanoengineered materials, nanoelectronics, nanobioelectronics, and other nanobiomedical devices that require suitably sized and functional building blocks to construct their architecture and devices [2].

The size evolution of the physical properties from atoms to clusters, to nanoparticles, and finally to bulk single crystal or polycrystalline materials, may also be related to the variation of the surface-to-volume ratio. However, at such small sizes, quantum-mechanical properties of electrons play an important role; i.e. quantum-size effects. For this purpose, clusters have been extensively studied which requires a paradigm shift" in our understanding of the properties of nanosized particles and developing methods for the exploitation of the novel properties of such building blocks. Table 1 is a summary of the most common views that require different considerations for the understanding of properties of clusters and nanoparticles, for example, quantum mechanical effects predominate in particles with dimensions comparable to the wave length of electrons or phonons.

Table 1. New considerations for understanding Nano Science and Technology.

Current View	Think Different
Classical continuum physics	Quantum mechanics
Solid state properties	Binding properties
Volume dominating	Surfaces dominating
Homogeneous Materials	Composite-inhomogeneous
Statistical ensemble	Individual particles
Miniaturization	Self-organization/Assembly

2. Nanoparticles Engineering

During the past few years there has been a significant progress in the development of different methods for the fabrication of nanoparticles. Currently, nanoparticles of many common materials, such as; metals, ceramics (including oxides, e.g., silica, titania, alumina, iron oxide, zinc oxide, ceria, and zirconia, or non-oxide e.g., TiN), semiconductors and magnetic, are commercially available [3]. Other types of nanoparticles, with composite structures, e.g., WC-Co composites, are reported and commercially available. However, intensive studies are under way to develop methods to fabricate nanoparticles with more complex, yet exact, composition.

Modern technological achievements are concerned with the development of advanced, multifunctional, and greater"smart" materials for specific applications in highly integrated mechanical, optical, and electronic devices, sensors, and catalysts. Analogous achievements, in the field of biomedical applications, require novel interdisciplinary concepts emerging at the intersection of material science, chemistry, physics as well as molecular biotechnology. Such studies are closely associated with surface chemistry and physical properties of inorganic nanoparticles, and the topics of bioorganic and bioinorganic chemistry, as well as various aspects of molecular biology, recombinant DNA technology and protein expression, and immunology.

2.1 EVOLUTION OF NANOPARTICLES ENGINEERING

The inert-gas evaporation technique [4] has been the most widely used method for the preparation of clusters and nanoparticles. With respect to the surface of the cluster or the nanoparticles, several metallic nanoparticles have been prepared with high purity This method was especially suitable for the production of metallic nanoparticles although minor modifications have been developed for the passivation of the surfaces by e.g. controlled oxidation of few monolayers of the particles [2]. Other compositions, e.g., oxide and carbide nanoparticles, have been obtained by replacing the inert gas with oxygen and methane respectively. On the other hand, nanoparticles of several ceramic oxide and semiconductors materials have been prepared predominately by wet chemical methods. Chemical methods are simpler and easily scaled up for high throughput production [5]. Several other techniques, including sputtering, thermal plasma synthesis, mechanical alloying, etc., have been developed for the fabrication of different types of nanoparticles.

The *"First Generation Nanoparticles"*,defined as those generated at the early stage of the development of nanostructured materials, generally have a simple composition of one or two compounds with a uniform distribution of different constituents, e.g., metallic nanoparticles with one or two metals, ceramics of single or double oxide systems. Primarily, the interest has been focused on thepreparation of the nanoparticles with given compositions and small sizes as well as narrow particle size distribution. + On the other hand, methods, such as, sputtering and mechanical alloying, produce dispersion of nanoparticles in a given matrix either as thin films or bulk.

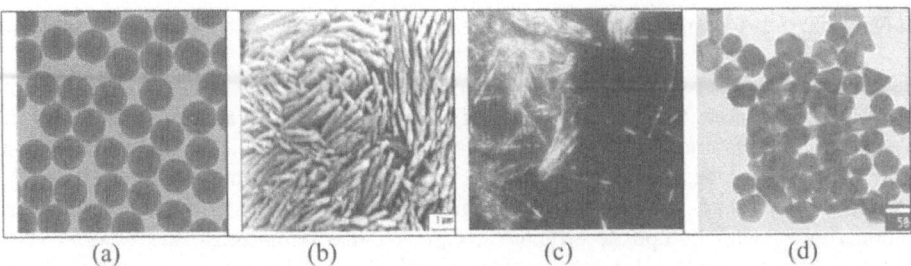

Figure 1. (a) TEM image of perfectly spherical polystyrene nanospheres, (b) SEM image of ZnO precursor, (c) Optical microscope image of Au nanowire with high aspect ratio and (d) TEM image of Au nanoparticles with different morphologies.

An important aspect for the preparation of firstgeneration nanoparticles has been the manipulation of the particle morphology [5]. Although tariation of the morphology of the nanoparticles may alter their properties, e.g., the magnetic properties of nanoparticles is dependant on shape anisotropy rendering them suitable for certain applications. For example, studies have been undertaken to manipulate the morphology of magnetic nanoparticles through the formation of particles under the influence of a magnetic field [6]. Other methods relied on the use of external additives to enhance the growth of particles across specific acrystalline phase, thus producing an asymmetric particle shape [7]. Wet chemical methods were developed using surfactant as a template for the synthesis of nanoparticles with varying aspect ratios, [8]. based on the use of certain surface active agents (surfactants) that form- in aqueous solutions-micelles with specific shape. The nanoparticles with micelles' shape are obtained as the nanoparticles are formed inside these micelles by a suitable precipitation method.

Figure 1 exhibits some examples of the "firstgeneration nanoparticles" where nanoparticles with different morphology have been prepared with different techniques. In most cases, the chemical composition of the particle is simple, i.e., one or two constituents.

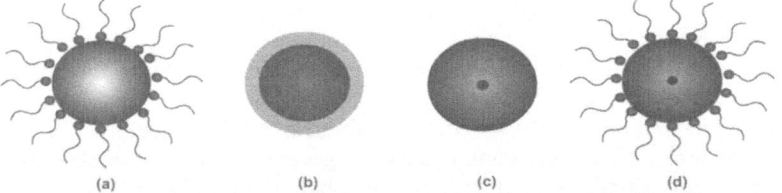

Figure 2. Evolution of second-generation nanoparticles. (a) Nanoparticles coated with surfactant for forming stable suspension, (b) Nanoparticle coated with thin metallic layer, (c) Small nanoparticle coated with porous ceramic layer and (d) dispersion of core-shell combination of (a) and (c) for stable suspension.

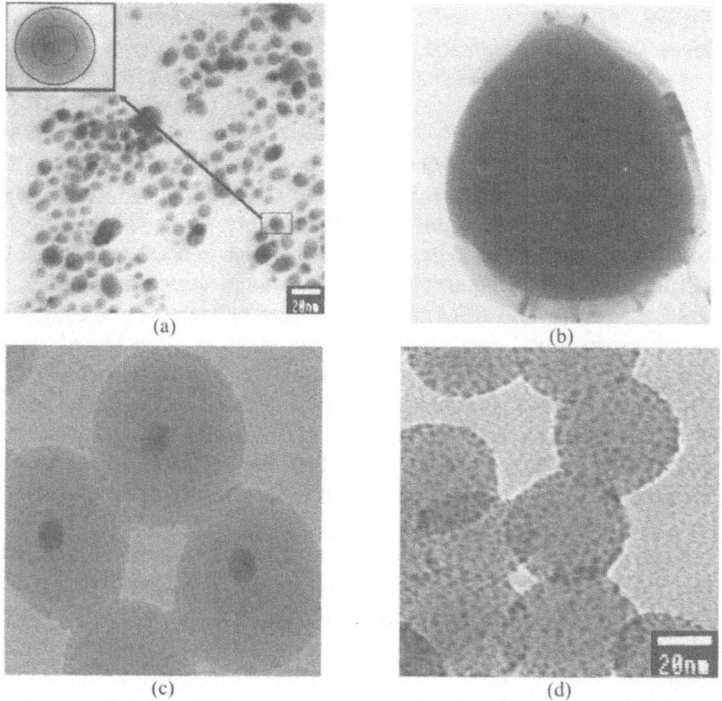

Figure 3. TEM images of second-generation nanoparticles. (a) Au coated magnetite nanoparticles prepared by sequential microemulsion processing, (b) Au coated SiO_2, (c) SiO_2 coated Au nanospheres prepared by sol-gel method, and (d) self-assembled Au nanoparticles on SiO_2 nanospheres.

As a result of an increasing degree of complexity and sophistication needed for the engineering of nanostructures for advanced applications, second generation nanoparticles have emerged. A key aspect, the need of multi functionality of these materials, results in several properties combined together to achieve a specific function. For example, a certain property could be achieved with the reduction of building blocks to the nanometer regime; e.g., thereby, superparamagnetism could be observed in magnetic nanoparticles when their size is below 10-12 nm. Proper suspension of these nanoparticles, in a suitable media, produce ferrofluids which have fluid properties of a liquid and the magnetic properties of a solid. Ferrofluids are useful as active components for enhancing the performance of several devices; e.g., mechanical (seals, bearings and dampers) or electromechanical (e.g., loudspeakers, stepper motors and sensors). However, combining the superparamagnetic properties of ferrofluids with a certain functionality on their surfaces enables advances in applications, e.g., in the biomedical area: magnetic targeted drug delivery.

The properties of functional nanostructured materials, made of the consolidation of nanoparticles, rely on the properties of the building blocks. Surface modification of nanoparticles, before consolidation, allows the design of the interfacial layer of nanostructured materials rendering different properties. For example, the introduction of a conducting layer on the surface of nanoparticles of non-conducting ceramic nanoparticles may result in a dramatic increase in the conductivity of the nanostructured materials. In addition, other important physical properties can be dramatically altered in a similar manner..

The surface layers of second-generation nanoparticles' (few or several monolayers) are distinctly different from the core material (composition or structure. Therefore, these are categorized as core-shell structures. The thickness of the surface layer may be a thin or thick layer depending on the functionality required. Figure 2 shows a schematic representation of the different types of first generation nanoparticles with surface modifications and nanoparticles with shell-core structure together with TEM images of gold, magnetite, and silica coated with heterogeneous composite (Figure 3).

In a broad prospective, these particles can also be considered as composite nanoparticles. However, the term nanocomposite generally refers to materials consisting of a dispersion of nanoparticles into a suitable matrix. The most common example of nanocomposites is the dispersion of inorganic nanoparticles into organic polymer matrix. It is interesting to note that the fundamental properties of the polymeric materials can be dramatically altered as a result of the dispersion of few percent of inorganic nanoparticles.

Advanced generation nanoparticles engineering and use in nanotechnology has recently emerged as the need for the fabrication of nanoparticles with higher degree of complexity. In addition to the the core-shell structure nanoparticles, with structure similar to the nanocomposites, have been fabricated (nanobeads). In these nanobeads, the single bead

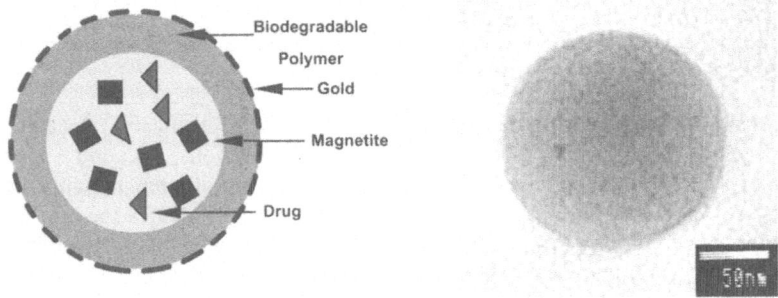

Figure 4. Schematic representation and TEM image of target-oriented drug release by drugs encapsulated in polymeric nanocapsules.

consists of a nanocomposite core with one or more smaller nanoparticles dispersed into the matrix. Several possible combinations of organic and inorganic particles can be dispersed in the matrix of the core structure. Each dispersed component can be selected to achieve a specific particle function or property. The surface layer can combine both physical (e.g. diffusion control) and chemical (e.g. allowing certain conjugation chemistries) functionality to the particles. Thereby, it is possible to "program" nano-beads with multiple functionalities suitable to perform certain tasks that can be triggered under specific conditions. For example, it is possible to fabricate nano-beads that can be magnetically moved or localized for a controlled drug release. The release of the drug can be controlled by the diffusion of the matrix of the bead or through the porosity of a suitable shell layer on the surface of the bead. Nanobeads can be programmed to be responsive to the environment e.g. small variation of temperature, or pH. The fabrication of these advanced generation nanoparticles requires the use of comprehensive and detailed procedures of its preparation. The design and fabrication of biochemically functionalized superparamagnetic iron oxide nanoparticles and near-infrared light absorbing nanoparticles is of particular interest. Target-oriented drug release by drugs encapsulated in polymeric nanocapsules presently is the most prevalent research area in biomedicine (Figure 4).

The recent development in nanoparticles engineering, and its application in Nanotechnology has placed a great emphasis on development of *"bottom-up"* strategies concerned with the self-assembly, as well as the directed assembly of (macro) molecular and colloidal building blocks, to construct larger, functional structures. On the other hand, perfectly fabricated nanoparticles can beconsidered as "artificial atoms' because they are obtainable as highly reproducible perfect nanocrystals which can be used as building blocks for the assembly of larger two- and three dimensional structures. A combination of *"top-down"* processes, produced by electron beam lithography and micro-contact printing (μCP), followed by directed assembly of different types of nanoparticles, allow the production of high 2-D and 3-D hierarchical structures [9].

The following section briefly describes the engineering of different types of nanoparticles as well as their assembly.

3. Fabrication of Nanoparticles

In the field of Nanostructured Materials, one of the most important challenges is the controlled synthesis and processing of materials at the nanometre scale. Numerous syntheses and processing techniques have been developed for the fabrication of nanomaterials.or the fabrication of nanoparticles with a , higher degree of complexityit is frequently necessary to tailor the synthesis method specifically to the material to be produced which may be. considered to be more restrictive than more conventional production methods.

Fabrication of nanostructured materials is considered to proceed through two main strategies the "bottom up" and "top down" approach. In bottom-up approach,, a structure is built up of small "Lego" units whereas n top-down, a larger unit is reduced in size to a finished structure. The fabrication of patterned 3-D nanostructure can be achieved through a combination of a top down and a bottom up approach. In the first stage, a lithographic technique is used to produce the pattern followed by the assembly of building blocks to form a hierarchical system.

The top-down approach comprises a few viable fabrication techniques that stem from experience and technology developed in other industries. Mechanical attrition (ball milling) has been widely utilized by the metallurgical industry for particle size reduction. Similarly, lithographic techniques, commonly used by the semiconductor industry, can be used in a top down approach to fabricate nanostructures.

Many fabrication methods use the bottom-up approach where building blocks of nanoparticles or clusters are first prepared by an appropriate technique, including gas phase, liquid phase and solid state reactions. Afterwards, the building blocks are assembled (e.g. into composites, coatings or layers) or consolidated (into bulk) under well-controlled conditions. During assembly, it is important to observe the coherent length scale of the building blocks in order to preserve the properties produced by nanostructuring.

For the fabrication of hierarchical 3-D structures on patterned surfaces, a combination of the two methods can be utilized. Lithographic techniques can be used for the patterning of the surfaces and suitable method of building block assembly is used to construct required structures. Attractive methods utilizing self-assembly and self-organization of the building blocksoffer an efficient way to fabricate complex nanostructures with a very high degree of precision. Directed assembly, using chemical functionalization of the surface has proven to be especially important for the fabrication of nanostructures using core-shell nanoparticles.

4. Gas Phase Synthesis

For a number of years, the formation of nanoparticles, through gas phase reaction, has been a frequently used technique for the fabrication of a wide range of nanoparticles. Several techniques have since been developed to make use of gas phase evaporation and condensation, or thermal decomposition. Detailed studies have been undertaken to determine the mechanisms for nanoparticles nucleation, and growth in the gas phase with varying compositions and under a variety of conditions of temperature and pressure. Inert gas evaporation techniques have been a most popular method for the synthesis of (mostly metal) nanoparticles because the surfaces of these particles can be obtained with a very high degree of purity.

Several variations of the gas phase synthesis route have been used, e.g., thermal decomposition (including PVD and CVD based methods), microwave plasma processing, combustion flame synthesis, etc. While gas phase synthesis methods are considered to be versatile techniques for the preparation of clusters and nanophase particles, they can be used for the fabrication relatively simple compositions. However, a proper combination of the starting materials (or precursors) and processing methods, it is possible to produce a wide variety of different nanoparticles. The following scheme demonstrates the possibility of producing different materials with various gas phase synthesis methods.

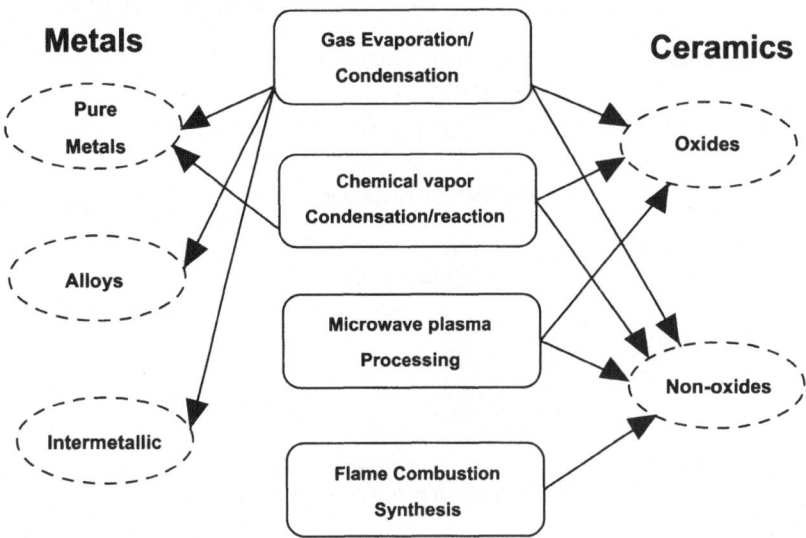

Figure 5. The suitability of different gas phase synthesis methods for production of various types of nanophase materials. The arrows indicate the types of nanoparticles that can be prepared by alternative techniques.

5. Chemical Synthesis of Nanoparticles

Advances in solution processing have demonstrated that different solid particulates can be prepared with precise control of the composition and size. By manipulating the different reactions in solution, it is possible to exert a high degree of control on the solid formation process on the molecular level (nucleation and growth) to produce particles well in the nanometre regime [5, 10-12]. Colloidal particles have been produced via solution processing for more than a century. Since the reactions, and hence the mixing of the different components, is performed on the molecular level, it is possible to achieve an accurate control of the composition and stoichiometry as well as uniformity of distribution of the different constituents within a single grain. Additionally, by controlling the reactions and their sequence, it is possible to produce nanoparticles with distinctly different core - shell structures. The layer-by-layer technique is made possible as different chemical processing steps can be employed sequentially to produce the required structure. Solution phase synthesis depends mostly on the chemistry and chemical reactions taking place in the reaction media leading to the formation of the solid particulates. In some techniques, powder is produced merely by the removal of the solvent, e.g., spray drying (where the solution is evaporated at very high rate) or freeze-drying, (where the solvent is frozen and evaporated from dissolved salts). In most cases, solution methods can advantageously utilize chemical reaction for the formation of the solid with required composition. A listing of most common solution chemical methods is given below.

Various solution phase methods for Nanoparticles synthesis

Precipitation
Homogenous precipitation
Co-precipitation
Hydrolysis
Oxidative hydrolysis
Reductive precipitation
Electrochemical reduction

Condensation
Sol-gel technique
Macro-molecular chemistry

Evaporation
Spray-drying
Spray-pyrolysis
Freeze-drying
Aerosol technique

Templates
Precipitation in microemulsion
Precipitation in presence of surfactants

Others
Sono-chemical reactions

5.1. CO-PRECIPITATION

Advanced solution chemical theories make it possible to obtain a solid particulate forms through precipitation of the required constituents form solution. Chemical co-precipitation can provide uniform nucleation, growth and aging of the particles in the solution. The size and morphology of the particles can be manipulated by controlling the different chemical reaction parameters, a technique generally inexpensive to perform thatrelies on simple coordination chemistry allowing the synthesis of the required solid "compound" with the desired composition and at high uniformity. On the other hand, the process of establishing and controlling the precipitation conditions are quite complex.

A rational approach has been developed to synthesize a pure phase of nanomaterials based on the quantitative analysis of different reaction equilibrium involved in the precipitation from aqueous solutions. Modern theories of coordination chemistry in solutions provide a powerful tool for the thermodynamic modeling of complex systems. In this way, it is possible to simulate different chemical reactions taking place in the system of concern and determine optimum operation conditions to obtain the solid powder with required phases. Although solution chemistry theories are well developed and are frequently utilized in other disciplines of science such as mineralogy, geochemistry, etc., is much less known in Materials Science. Availability of computer codes permits an easy simulation of a very complex precipitation process. Besides the prediction of the conditions for the formation of the solid nanoparticles, such simulation can also estimate the extent of the dissolution of the solid particles under different conditions. This is important e.g., where there is a need for the synthesis of nanoparticles with complex structures via subsequent processing in the layer-by-layer approach.

The advantage of this rational approach has been demonstrated for the fabrication of different systems of nanoparticles with various composition and structures by a novel method for the synthesis of nanoparticles of a metastable precursor $MgO-Al_2O_3$ using a controlled chemical co-precipitation route in aqueous solution based on thermodynamic modeling. The precursor could be consolidated and sintered to pure spinel $MgAl_2O_4$ ceramics [13]. A more complex system for high temperature superconductor oxides has been prepared by chemical co-precipitation in aqueous solutions and in microemulsion system. In this case, a rational approach, based on simulation of the precipitation process, has been proven invaluable in deducing and controlling the precipitation conditions required for obtaining nanoparticles with the exact composition. A High Temperature Superconductor (HTSC) is known to be point compounds where the superconductivity properties are very sensitive for any composition or phase changes. Wang et. al. [14] have reported the synthesis of nanoparticles of the 123- HTSC system $(YBa_2Cu_3O_{7-\delta})$ which was prepared both in aqueous solution as well as in microemulsion systems. In a different study, the synthesis of nanostructured $CoSb_3$ intermetallic system has also been reported based on the use of the process simulation [15, 16].

In these systems, the exact composition of the nanoparticles is a pre-requisite for obtaining the required phase needed for the functionality of the nanostructured materials. Figure 6 demonstrates the usefulness of the process simulation as it depicts the extent of the precipitation of the different constituents required for obtaining the functionality of the nanostructured materials. Figure 7 shows a comparison of the agreement of the simulation of the co-precipitation process with the experimental data determining the extent of the precipitation processes.

The fabrication of these particles could be carried out in either in aqueous solution or in microemulsion system. The chemical reaction conditions responsible for the co-precipitation reactions and the formation of the solid compounds are the same as indicated. The use of microemulsion system has the specific advantage of limiting the particle size since the reaction takes place inside the cavities of the microemulsion (this will be explained). In this system, the oxalate (either as sodium oxalate and oxalic acid) is used as the precipitation reagent. The oxalate coprecipitation of Y, Ba and Cu from nitrate solution was carried out at the optimum operating conditions, especially the pH range and concentrations of $Y/Ba/Cu/C_2O_4^{2-}$ estimated from thermodynamic modeling. Oxalate coprecipitation in stoichiometry of Y:Ba:Cu=1:2:3 is obtained in a pH range 3-4. Phase diagrams of several multicomponent systems: hydrocarbon-aerosol OT (surfactant)-water, in the absence and presence of nitric/oxalic acids and nitrates, has been systematically investigated. The thermodynamic modeling of the chemical system for the coprecipitation of magnetite phase from aqueous solutions containing suitable salts of Fe^{2+} and Fe^{3+} has been reported [17]. The results of the thermodynamic modeling, together with available kinetic information of relevant reactions, have been used for the selection of the optimum experimental conditions for the synthesis of magnetite from aqueous solutions. In this way, it is possible to establish, on a rational basis, the conditions (chemical, including the composition of the reaction media pH, redox potential, concentration of the different salts used, etc.) as well as the boundary conditions in which magnetite could be formed reproducibly as a pure phase.

In another system, the modeling of the precipitation process was used to establish the conditions of the formation of ZnO nanoparticles with different morphology. Experimental results have indicated that ZnO can be obtained as rod-shaped nanoparticles and that the formation of the metastable species $Zn_5(OH)_6(CO_3)_2$ (s) is responsible for the formation of rod-shaped nanoparticles. This was explained by the existence of NH_3 on the particle surface of this intermediate compound through - presumably - hydrogen bonding. As seen from Figure 8, the formation of this intermediate compound predominates at a specific pH range in the aqueous system (under specific solution concentration conditions).

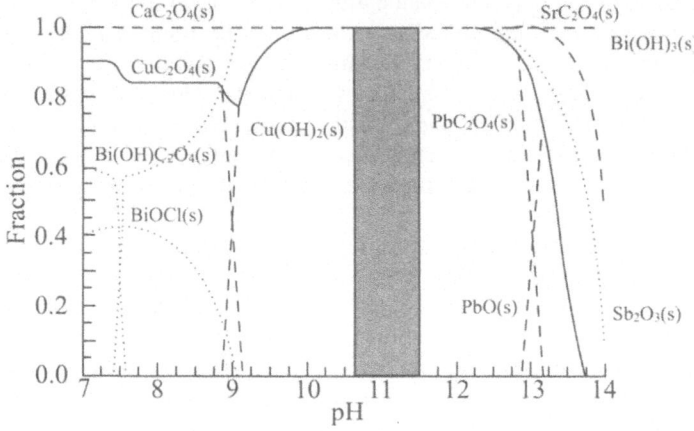

Figure 6. Computer modeling showing the fraction of precipitates of the different metals vs. pH, for a given initial concentrations to produce solid compound with the stoichiometric composition (Bi+Pb+Sb):Sr:Ca:Cu=2:2:2:3. Shaded area shows possible operational pH range.

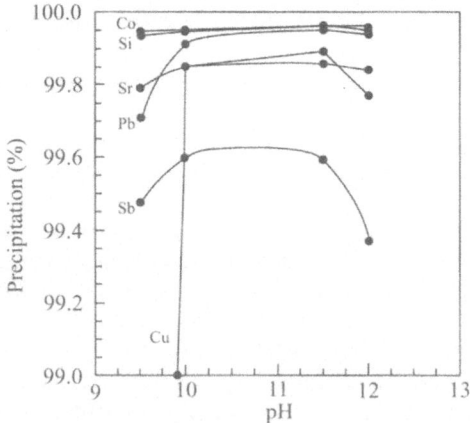

Figure 7. Experimental verification of the efficiency of the precipitation of the different metal compounds at various pH solutions compared to the results modeling.

The thermodynamic modeling for the fabrication of nanoparticles of magnetite is used to illustrate the simulation process. Magnetite is formed through the co-precipitation of Fe^{2+} and Fe^{3+} in a molar ration of 1:2. Under proper conditions the concurrent precipitation of FeO and Fe_2O_3 leads to the formation of pure magnetite phase (Fe_3O_4). The simulation of the co-precipitation requires the evaluation of the thermodynamics of the aquatic system: Fe^{2+}, Fe^{3+}, and OH^-. Additionally, the variation of the redox

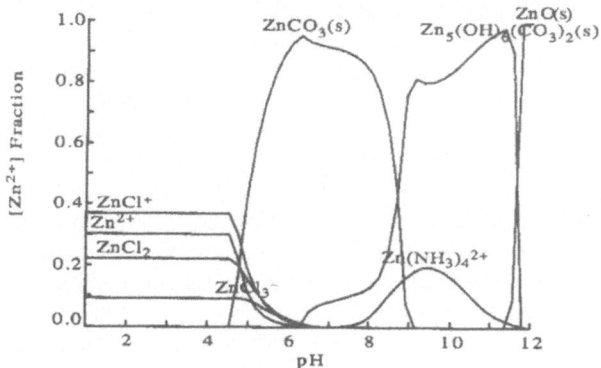

Figure 8. Distribution of different zinc species in the system: $ZnCl_2$- ammonium carbamate at various pH range.

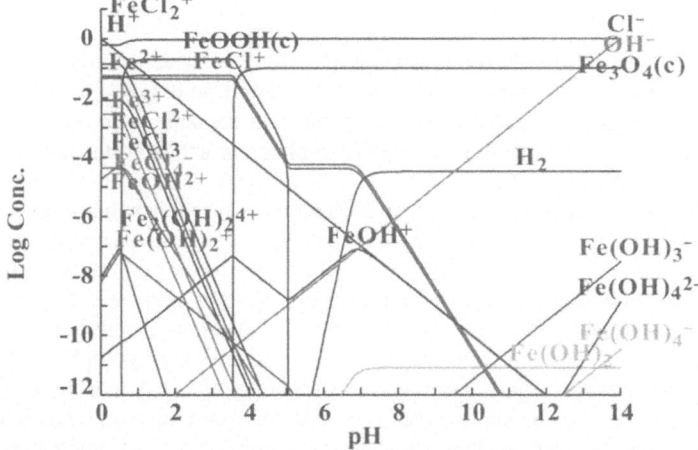

Figure 9. Thermodynamic calculations of the formation of different species in solution of $FeCl_2$ and $FeCl_3$.

potential in the system, since Fe^{2+} can be easily oxidized by atmospheric oxygen to form Fe^{3+} ,must be taken into consideration. Thermodynamic modeling of the chemical system for the coprecipitation of magnetite phase from aqueous solutions containing suitable salts of Fe^{2+} and Fe^{3+} has been reported [17]. The results of part of the modeling are given in Figure 9, where the formation of different species in this – relatively simple system- can be estimated.

Under closed system or/and surrounding conditions, thermodynamic modeling of the different phase formation can be calculated from the changes of the free energy, ΔG, of the reaction under given conditions. Iron exists in three oxidation states, therefore, redox reactions must be taken into consideration. The formation of the different iron species are correlated through different redox reactions.

$$Fe\ (s) \rightarrow \quad Fe^{2+} \quad \rightarrow \quad Fe^{3+} \tag{1}$$

Magnetite can be obtained by controlled oxidation of Fe^{2+} in solution according to the following equation:

$$3\ Fe^{2+} + O_2 + 2\ OH^- \rightarrow \quad Fe_3O_4 + 2\ H^+ \tag{2}$$

However, the kinetics for the oxidation of Fe^{2+} is slow and is difficult to control, therefore, it is more practical to control the formation of magnetite through the use of a mixture of Fe^{2+} and Fe^{3+} salts. In this case, the oxidation of Fe^{2+} should be avoided.

$$Fe^{2+} + 2\ Fe^{3+} + 2\ OH^- \rightarrow \quad Fe_3O_4 + 2\ H^+ \tag{3}$$

In the thermodynamic modeling of the precipitation of the magnetite, we define the system to be investigated by the following main components:

$$Fe^{2+}, Fe^{3+} - (NO_{3,}\ Cl) - H_2O \tag{4}$$

Both chloride and nitrate salts can be considered in order to evaluate the effect of their presence on the formation of magnetite. A plot of the concentration of the reaction products of all possible chemical reactions that can take place in the system is given in Figure 9. As shown, Fe_3O_4 can be formed over a wide pH range for given concentrations of Fe^{2+} and Fe^{3+}. The concentration of the starting solutions and the precipitation conditions can be evaluated from these calculations.

5.2. NANOEMULSIONS

Microemulsions (μE) are thermodynamically stable solutions containing nanosized droplets of an immiscible liquid dispersed in another liquid [18]. Emulsions can be formed by mixing two immiscible liquids. turbid and unstable with droplet size in the micrometer range. The stability of the microemulsion is achieved through the addition of a suitable surfactant where the drop size is reduced in the nanometer range. The stabilization of the microemulsion by the surfactant, as it reduces the interfacial tension and reinforcing the interfacial film, in turn prevents the coalescence of the droplets. Since the droplet size is less than 100 nm (with droplet size of 1 micron or larger the emulsion solution becomes turbid), Microemulsions are transparent and, therefore, should be properly referred to as "nano emulsions". Surfactants (surface active agents) are made of molecules that have one hydrophobic end and one hydrophilic end, thereby localized at the aqueous-organic interface. Long chain alcohols; e.g. butanol, octanol, etc, are the most common surfactants used. In micro- and nanoemulsion systems, several phases can be formed depending on the composition and concentration of the individual constituents. The emulsion systems can consist of either the aqueous or organic phases as the dispersed phase. Water-in-oil-system is the useful configuration for the synthesis of nanoparticles, therefore, for the use of nanoemulsion system, the phase diagram has to be considered. The major advantage of nanoemulsion, the nanosized cavity used as the maximum volume for the particle formation, imposes a physical limitation on the particle growth. Moreover, the shape of the nanoemulsion droplets can be controlled by the use of specific surfactants with e.g. cylindrical shape and imposes a shape-memory effect of the particles formed within the droplets to produce e.g. non-spherical or elongated particles with high aspect ratio.

The surfactants generally do not participate in the chemical reactions taking place in the aqueous phase and, consequently, the chemistries and the corresponding thermodynamic modeling are applicable to aqueous solutions and valid for nanoemulsion systems. Nanoemulsions are dynamic systems where the making and breaking of the droplet is continuously taking place in the solution. Therefore, if two nanoemulsion systems (water-in-oil) are brought together, spontaneous mixing of the aqueous phase occurs driven by the chemical potential. The reactions in nano-cavities can therefore be achieved by simply mixing two (or more) nanoemulsion systems simultaneously in a manner similar to that of ordinary solutions whereby the nanoparticles will be formed within the nano-reactors (Figure 10).

The synthesis of nanoparticles by micro and nanoemulsions has been reported by several groups [19-21]. The particle size of the powder obtained is smaller than that obtained in conventional precipitation process in solutions.

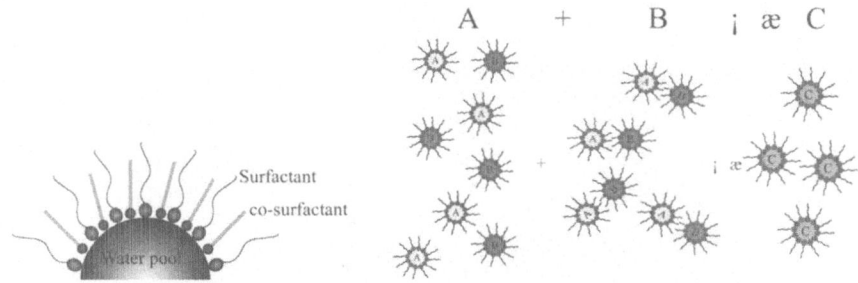

Figure 10. Illustration of the formation of nanoemulsion water in-oil droplets and the formation of nanoparticles inside the droplets.

Under suitable conditions, it is possible to keep the powder formed in nanoemulsion stably suspended in solution (mixture of aqueous and non-aqueous) which allows further processing of the particles. It is possible to form nanoparticles of a given composition prior to subsequently carry out another precipitation processing in the presence of the existing particles. In this case, the surface of the nanoparticles will act as nucleation sites that facilitate the heterogeneous precipitation rather than homogenous precipitation that normally required more energy. This results in the formation of a surface layer, or coating of the formed nanoparticles, which allows the formation of layer-by layer structure of nanoparticles as schematically illustrated in Figure 11 which is an example of two types of nanoparticles that have been prepared by this method.

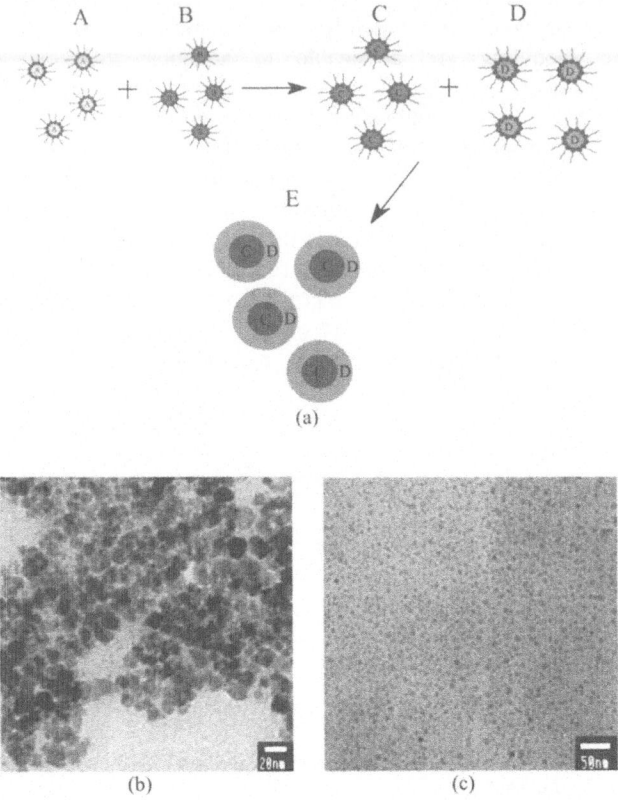

Figure 11. (a) Schematic of multi processing using nanoemulsion, (b) Au-coated magnetite and (c) NiO coated Co nanoparticles.

5.3. SYNTHESIS OF NANOPARTICLES IN FLOW INJECTION SYSTEM

The major problems with fabrication of the nanoparticles, using solution chemical method, are related to the uncontrolled particle growth and particle agglomeration. Agglomeration can be reduced by the use of suitable surfactants that caps the surface of the nanoparticles. On the other hand, during the course of the precipitation process, the particles are formed and normally stay in the reaction mixture during the processing which leads to unavoidable particle growth. Figure 12 is an example of the increase of the particle size during the precipitation process. Energetically, the nucleation process favors heterogeneous nucleation rather than homogeneous nucleation and therefore existing small particles grow larger as the precipitation process continues. Ostwald ripening is another important factor that contributes to the growth of nanoparticles kept in solution. In the precipitation process, the formation of the solid phase in solutions is governed by the solubility product. The solubility product does not vary significantly with particle size for large particles (with size over 1-10 micrometers). However, the

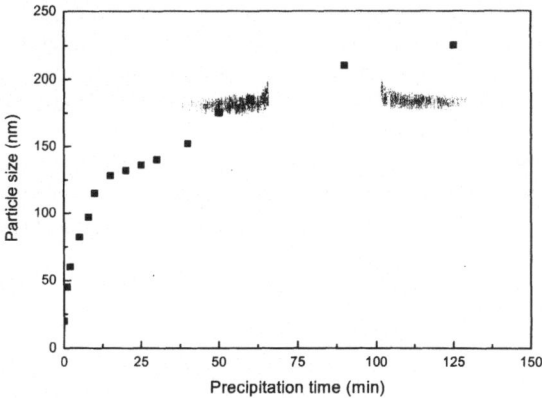

Figure 12. The dependence of the particle size on the reaction time.

solubility product of nanoparticles of a given solid phase composition is smaller than the larger particle of the same composition.

This leads to increased solubility of the nanoparticles causing a supersaturation in the solution and subsequent re-precipitation that takes place preferably on the surface of other particles.

In nanoemulsion systems, nanoparticles are confined into the aqueous droplet surrounded by the surfactants which minimize agglomeration and particle growth. However, in some cases it may be difficult to remove the strongly adsorbed surfactants molecules from the surfaces of the nanoparticles. The use of confined zone, in a flow system, is a novel way to minimize particle growth after the formation of the solid particles [22, 23].

Flow injection analysis is a well-developed technique in analytical chemistry for high through-put analysis [24]. In a non-turbulent laminar continuous flow of a fluid, at a suitable velocity, it is possible to inject a certain volume of another miscible fluid which can be confined to a zone of the moving flow with limited mixing, however, sufficient mixing takes place around the point of injection. To illustrate the flow injection principle, consider that pure water is continuously pumped through a narrow tube (few mm in diameter) and the absorbance of the water is recorded using a flow cell and a spectrophotometer. At a steady flow, a flat base line with no absorption will be recorded. If an aqueous solution containing a suitable dye is injected into the flow, during a short period of time, the dye will be diluted locally but confined into a zone that is carried along with the flow to the spectrophotometer. An absorption peak will be observed with a peak at the point of injection, after which the absorption will return to the base-line value. Repeating the injection will result in multiple absorption peaks as shown in the schematics in Figure 13.

The system can be simply modified to include two or more synchronized injections at the same point and at the same time (Figure 13b). Also, in this case, the multi-injections will be confined to the same zone and carried out through the stream. The synthesis of nanoparticles can thus be achieved by injecting different reacting solutions

54

thatare mixed and form nanoparticles confined to the same moving zone. After the reaction has been completed (zone life time > reaction time), the particles are recovered from the solution, e.g. by filtration. Also, in this case, the precipitation chemistry for precipitation in the solution is applicable for the reaction in the confined zone. The process can be further modified to allow an additional surface modification or coating steps can be included as an on-line operation.

Nanoparticles produced using flow injection technique show smaller particle size compared to those obtained by bulk solution processing but comparable to those obtained using nanoemulsion technique. The particles have a narrower particle size distribution.

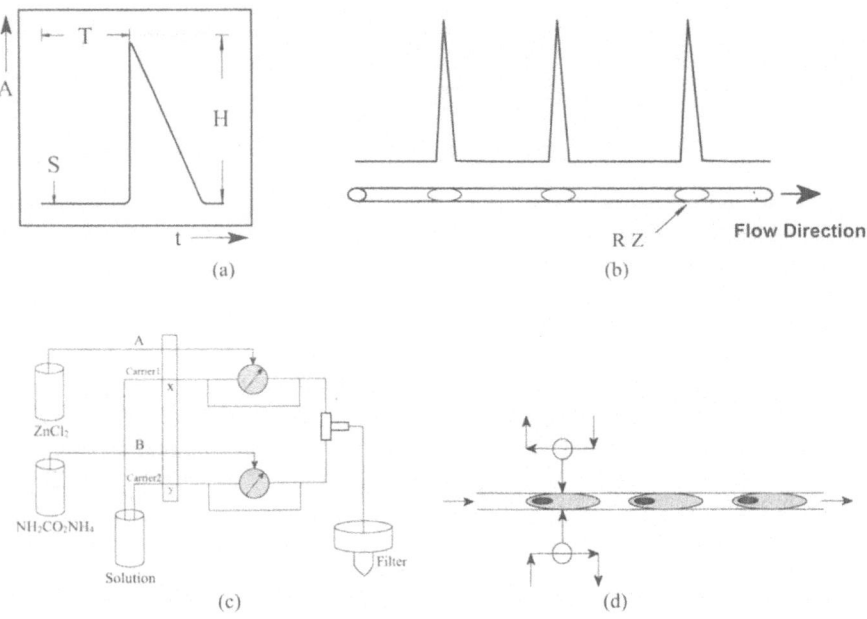

Figure 13. a) A typical FIA peak from recorder output (S, sample injection; H, peak height; T, residence time), b) a schematic representation of confined zone in the carrier stream (RZ, reaction zone), c) a schematic representation of the FIS system, and d) the synchronous merging of two zones in a FIS system.

5.4. SOL-GEL PROCESS

Sol-gel process is an important method for the synthesis of nanomaterials primarily for thin films and also for nanoparticles. The process has shown to be especially useful for the modifications of surfaces. During recent years, an improved understanding of sol-gels has been useful in developing several thin films suitable for a variety of commercial applications in optics, acoustics, electronics, insulation, and sensors. Sol-gel processing allows a high degree of control of the porosity (to tune refraction index) of sol-gel thin films, multilayer coatings help to improve reflectivity of optical devices such as, solar mirrors, and protective overlayer coating of integrated circuits to reduce unwanted electrical characteristics. Sol-gel processing can be useful for the manufacture of chemical sensors and gas-separation filters.

Aerogels (that are mainly possible to make by sol-gel processing) are foam-like transparent materials that can consist of 99% air, making them ideal heat insulators for many conventional and advanced applications, e.g. insulators for damping sounds.

Sol-gel process for the synthesis of nanoparticles incorporates several steps including hydrolysis, condensation, and polymerization. The precursor for the sol-gel process consists of metal-organic compounds. Generally, metal alkoxides are quite popular precursors for the sol-gel process since they have an organic ligand such as •OCH_3 or •OC_2H_5 binding to a metal or metalloid atom and can react easily with H_2O. For example, the *hydrolysis* of TEOS (tetraethyl orthosilicate, $Si(OC_2H_5)_4$) occurs because of the attachment of hydroxyl ion to metal ion is

$$Si(OR)_4 + H_2O \rightarrow HO\text{-}Si(OR)_3 + ROH \tag{5}$$

The complete hydrolysis reaction can be achieved when an adequate amount of H_2O and a suitable catalysis (acid or base) is added to the solution

$$Si(OR)_4 + 4H_2O \rightarrow Si(OH)_4 + 4ROH \tag{6}$$

After the hydrolysis reaction, *condensation* reaction takes place and the partially hydrolyzed molecules are crosslinked together.

$$(OR)_3Si\text{-}OH + HO\text{-}Si(OR)_3 \rightarrow (OR)_3Si\text{-}O\text{-}Si(OR)_3 + H_2O \tag{7}$$

By repeating the condensation reaction, the silicon molecule grows to a macromolecule by the *polymerization*. Figure 14 shows the TEM images of silica nanospheres with different particle size prepared by sol-gel process. An important application of sol-gel processing of silica is the possibility to prepare different zeolites. Silica sol was prepared by dissolution of tetraethyl orthosilicate, TEOS ($Si(OC_2H_5)_4$ in 2-propanol, $(CH_3)_2CHOH$ dried over activated molecular sieve Zeolite 4A and the addition of distilled water [25].

Sol-gel processing allows the synthesis of nanoparticles (or films) with a very complex composition. The well known precursor of alumina for sol-gel process, inorganic salt, such as $Al(NO)_3$ and metal alkoxide such as $Al(OC_4H_9)_3$, can be used for the preparation of alumina nanoparticles as well as for the preparation of several compositions of alumino-silicate ceramics with meso- and nano-pores.

The control of the size and composition of composite particles can be realized through suitable sol-gel preparation. SiO_2/CdS-nanoparticle composite films (SiO_2:CdS=85:15, 80:20, 75:25 and 70:30) were prepared by the sol-gel route [26]. 0.1N Hydrochloric acid was added as catalyst. A solution of $Cd(NO_3)_2 \cdot 4H_2O$ and thiourea (NH_2CSNH_2as sources for Cd and S, respectively, was also prepared separately in 2-propanol and distilled water. This solution was slowly added to the silica sol under vigorous stirring. The sols were used for spin coating on properly cleaned soda lime glass substrates. They reported a blue shift of the band gap energy with decreasing size of the nano-crystallites.

Sol-gel processing may be carried out using other salts than alkoxide which makes it suitable to produce nanoparticles of conventional ceramic materials. For example, ZnO nanoparticles, in the size range from 2 to 7 nm, could be prepared by addition of LiOH to an ethanolic zinc acetate solution [27].

56

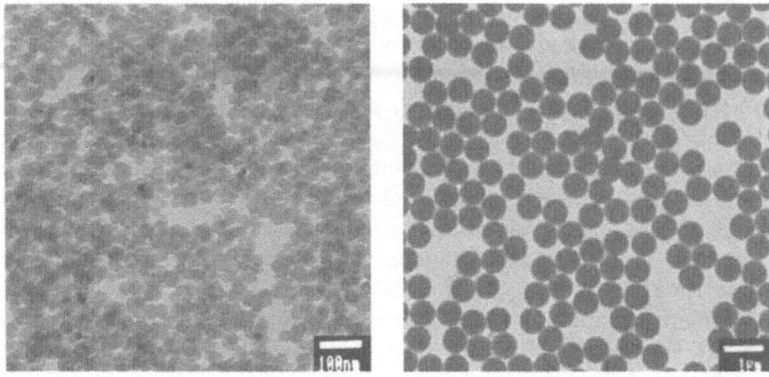

Figure 14. TEM images of silica spheres with a different particle size that is used for the self-assembly.

An important application of the sol-gel processing is in the preparation of mono-sized nanoparticles of silica. The size of the particles can be determined through a careful control of the hydrolysis and polymerization process leading to the formation of the spheres. In this way, it is possible to reproducibly prepare particles with a varying size e.g. 10 to 500 nm with a very narrow particle size distribution.

It is possible to prepare alkoxide for almost all known metals, however, their chemistry and processing may be complicated and the handling of these materials requires extreme care to avoid water or moisture from the air, which limit their practical applications.

5.5. MICROWAVE SYNTHESIS

Microwave synthesis a less commonly used technique for the synthesis of nanoparticles that has to be tailored to specific systems. In microwave synthesis, for example, heat is generated from inside of the sample in contrast with conventional heating methods. This internal heat allows a decrease in processing time and energy cost while allowing for new material synthesis to be possible. In conventional synthesis, heat is transferred to the reaction mixture from external heat sources. In the microwave synthesis technique, heat is transferred from the solution to the reaction mixture. Moreover, the components of the system are selected so that it produces heat as a result of interaction with the microwave radiation, due to its dielectric properties. This can make microwave heat treatment much more effective than conventional thermal treatment.

Nanocrystalline metal (Cu, Hg, Zn, Bi, Pb) sulfides, with different shapes and different particle sizes, were prepared in a formaldehyde solution of metal salt and thioacetamide (TAA) by microwave irradiation. In a typical procedure, an appropriate amount of metal salt was dissolved in formaldehyde. Then, an appropriate amount of TAA was added into the solution. The mixture solution was reacted in a microwave refluxing system for 20 min with suitable power output. Normally, the radiation is initiated in cycles to avoid excessive temperature increase and an explosive reaction to occur (typically few seconds power on and similar period power off). As a result of this reduction [28], metal particles are formed in the solution. In similar manner, ferric hydroxide (FeOOH) nanoparticles, with a size of 50 nm, could be synthesized from Fe^{3+} solution. Fe^{3+} solutions were placed in beaker and inserted into the microwave cavity (Figure 15 b). During the irradiation of microwave in the cavity, the sample was rotated with turntable allowing a homogeneous absorption of microwave. A special arrangement for the experiment was constructed. Microwave irradiation was absorbed by the molecules and the electromagnetic energy converted to thermal energy. In a chemical reaction process, microwave heating improves reaction kinetics, yields, and significantly reduces the overall processing time and temperature compared to conventional heating mode. The internal heating by microwave is due to the interaction between microwave and absorbing medium. Magnetic materials offer superior microwave coupling properties due to the effect of ferromagnetic resonance. Synthesis of aciculate ferric hydroxide nanoparticles,

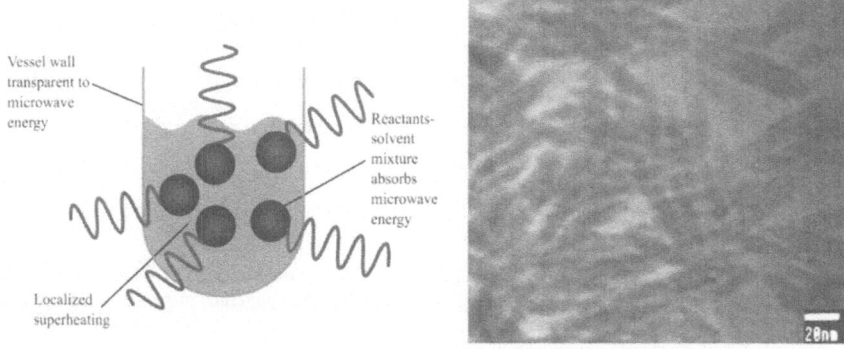

Figure 15. (a) The microwaves couple directly with molecules that are heating and creating rapid temperature rise and (b) TEM image of ferric hydroxide (FeOOH) nanoparticles prepared by microwave irradiation.

by microwave heating, can decrease the processing time and cost significantly in comparison with normal process.

Polar molecules can absorb microwave radiation causing an increase of the temperature of the reaction media. Non-polar molecules, on the other hand, cannot absorb microwave. The presence of polar water molecules causes a rapid and homogenous increase of the temperature of the reaction media in the first stage. This heat is transferred to Fe^{3+} ions, which is then hydrolyzed and converted to ferric hydroxide. Fe^{3+} ions do not absorb the microwave radiation but ferric hydroxide does absorb microwave radiation. Therefore, in the second stage on the reaction, the solution will be homogenously heated by the formed ferric hydroxide nucleation sites formed throughout the solution phase. As nanometer-sized ferric hydroxide seeds, magnetic medium could significantly accelerate the rate of microwave heating thus forming more nucleation sites rather than allowing particle growth. Microwave processing produces smaller particle size with a narrow particle size distribution while the processing time is significantly reduced [29].

Microwave processing can be adopted to initiate solid-state reaction, e.g., for the fabrication of the particles of MgB_2 superconductors. A microwave oven, which could be operated over the entire power range by alternating from maximum to zero power in

58

timed cycles (duty cycles), was used. A small tube furnace was specially designed from a high temperature insulating material (Kaowool, 1700°C). A silicon carbide susceptor, a material with a very high dielectric constant that is microwave absorptive and employed for hybrid heating, was placed inside the furnace to create a homogenous field effect for heating of the sample.. This This Metallic powders of Mg and B were mixed homogenously with a stoichiometric ratio of 1:2. The mixture was placed in a quartz tube, of diameter ca. 8 mm that was evacuated and sealed. Thereafter, this sealed-tube was placed in an alumina tube with a diameter of 2 cm and placed in the designed Kaowool (alumina board) holder

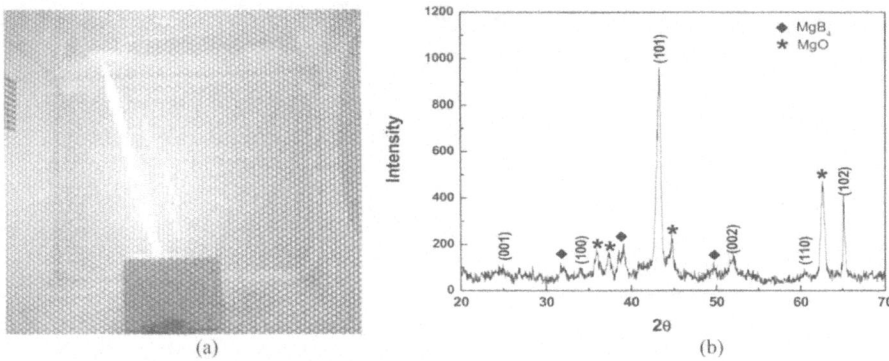

(a)   (b)

Figure 16. Microwave synthesis of MgB$_2$ superconductor in conventional microwave oven (a) Photo was taken during the reaction of Mg + B powders in qualtz tube and (b) XRD patterens of microwave synthesized MgB$_2$.

that can stand up to 1700 °C. A thermocouple was directly inserted into the vicinity of the reaction zone and the temperature was monitored and maintained at 850 °C. Figure 16 shows the microwave cavity during the synthesis of MgB$_2$ and XRD patterns of MgB$_2$ sample prepared by microwave heating. The produced particles were a larger size , however,the crystallite size was in the order of 100 nm [30].

6. Engineering of Hierarchical Structures Using Nanoparticles

In recent years, a great deal of attention has been drawn to the fabrication of composite micro- and nanoparticles that consists of either organic or inorganic cores coated with shells of different chemical composition. Colloidal particles represent very attractive building blocks to create ordered and complex materials. The use of nanoparticles as building blocks offers great possibility to design the materials interior on the nanometer scale. The nanoparticles can be engineered with a wide variety of composition and structuresI for the possibility to design the core of the material with several different constituents that offer multi-functional properties. These particles can then be processed to form different structures suitable for specific applications. Figure 17 presents an overall scheme representing different possible applications of nanoparticles.

The composition and homogeneity of the individual nanoparticle can be tailored to the desired requirement. In some cases, a highly uniform composition is required, while in other applications, composite materials, containing several phase uniformly distributed, may be the required target. For the production of bulk polycrystalline materials, the nanoparticles are consolidated by a proper processing technique to form fully dense materials. The surface of the nanoparticles may be treated or modified depending on the required characteristic of the grain boundaries of the consolidated materials. This allows a fine tuning of the properties of the material since the grain-boundaries constitute a significant volume of the consolidated materials. On the other hand, a major problem with the consolidation of nanostructured

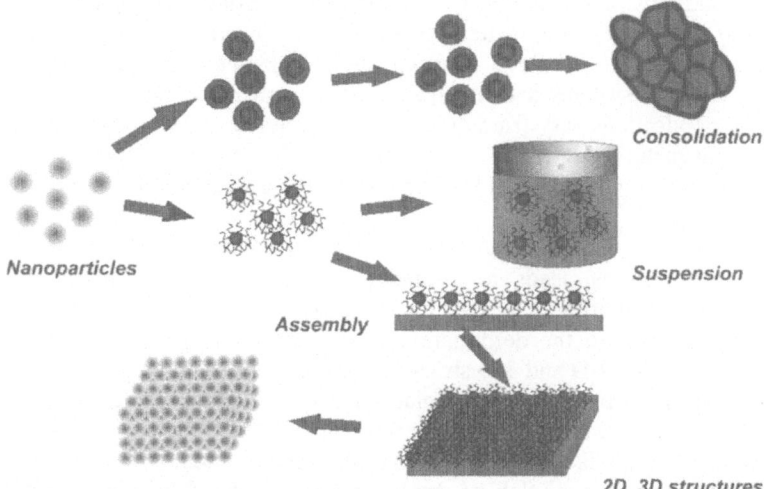

Figure 17. A general overview of the construction of 2D and 3D nanostructures with nanoparticles as building blocks.

materials is the rapid grain growth of nanoparticles during sintering. It is not the intention of this paper to discuss methods of nanoparticles consolidation, however, it is important to note that particle growth during sintering of nanoparticles can be greatly reduced by the addition of grain growth inhibitors (Ggi) at the grain boundaries. Therefore, a common practice is to add Ggi on the surface of the nanoparticles prior to their consolidation.

Another important application of nanoparticles is their use as suspensions. The stability of the suspension is important for the use of the nanoparticles. The dispersion of nanoparticles with high surface energy requires a good understanding of the different forces affecting particle agglomeration and aggregations [31, 32]. The modification of the surface properties can be achieved by several different methods. Characteristics, such as; solubility, resonant optical properties, electrical and mechanical aspects of nanoparticle-constituent materials, and even agglomeration, are influenced by chemical functionalization of nanoparticle surfaces [32-35]. For example, the surface layer (or coating) can alter the charge, functionality, and reactivity of the surface. This can enhance the stability and dispersibility of the colloidal core. Magnetic, optical, or catalytic functions can readily be imparted to the dispersed colloidal matter depending on the properties of the coating [36]. Encapsulating colloids, in a shell of different composition, may also protect the core from harsh chemical and physical changes.

Several stable suspensions of nanoparticles have found important applications, e.g. ferrofluids, paint, etc. Stable suspensions of slurry consisting of several types of nanoparticles are used for the fabrication of consolidated composite materials (organic/organic or inorganic/inorganic particulates). Nanoparticles are frequently used in, cosmetics, dyes, inkjet printing technology and recently under investigation,in the use of magnetic inks for printing [37]. Magnetic colloidal supports have become increasingly important because of their widespread use in fields such as biotechnology, i.e. in bioseparations or immunoassays. Nanoparticles offer high promise for several novel *in-situ* and *in-vivo* biomedical applications. Several in-vivo applications of nanoparticles are under intensive research and development including magnetically target drug delivery systems and monitoring of biological functions inside the living bodies. The applications vary from drug delivery to tissue engineering or as a contrast agents for magnetic resonance imaging [38]. Stable suspensions of magnetic nanoparticles are being considered for hypothermia as a treatment for tumors in living body [39].

The fabrication of bulk material through consolidation of nanoparticles as polycrystalline materials, leads to a random compaction and arrangement of the particles not in a given order. Compaction and sintering of nanoparticles to produce the dense materials, leads to the deformation of the particles. On the other hand, the engineering of ordered 2-D and 3-D structure requires the positioning of the particles in given order and pattern. Mechanical placement of the nanoparticles on surfaces or layers to form ordered structure is possible through the use of "nano-manipulators" such as AFM or STM. This is a very slow and costly process not considered practical for high throughput applications. Self-assembly, or directed assembly, of nanoparticles can be utilized for the arrangement of the nanoparticles in a required pattern. This can be done on a variety of surfaces including spheres, flat or patterned surfaces. Self-assembly has been demonstrated for the formation of monolayers of given molecules on surfaces. In this case, the molecules having binding groups that are capable to conjugate on the surface. A similar approach can be used to direct the assembly of the particles on the surfaces using coupling reagents (bi-functional molecules) that conjugate to the surface of the substrates and the surface of the particles and, as required,for assembly of ordered hierarchical structure nanoparticles with defined shape and size..

Nanoparticles can also be used as the core materials for colloidal template. The creation of core-shell particles is attracting a great deal of interest because they exhibit improved physical and chemical properties over their single-component counterparts, making them potentially useful in a broader range of applications. Particle characteristics can influence chemical properties, reactivity, selectivity, rate of dissolution, and bio-efficiency. These characteristics, primarily surface phenomena and tail the surface properties of particles, can be achieved by coating or encapsulating particles within a shell of preferred materials [40].

These core-shell structures can be employed in medicine, pharmaceutics, and materials science. They find diverse applications in the encapsulation of products (for the controlled release of drugs, cosmetics, inks and dyes), the protection of light sensitive components, catalysis, and coatings. There are several studies on the preparation of core-shell particles by layer-by-layer (LBL) assembly. The method is based on the alternative assembly of oppositely charged molecules onto the colloidal templates to fabricate composite core/shell structures. This technique allows the stepwise adsorption of various components as the layer growth is governed by electrostatic attraction allowing the formation of multilayer shells. Submicron composite particles with multilayer arrays of electrolytes, Fe_3O_4 and TiO_2, have been produced by this method [41, 42]. A general strategy for the formation of solid-core/metal nanoshell particles has been suggested by Oldenburg, et al., using a combination of self-assembly and colloid reduction chemistry [43].

Magnetic colloidal suspensions can be used as building blocks for the formation of ordered arrays/patterns on surfaces via manipulation using magnetic fields. We have developed an approach similar to that proposed by Oldenburg, et al., to prepare metal shells and demonstrate the applicability of self-assembly and colloidal template methods for the production of a novel type of metallic capsules with size in the sub-micron range containing a magnetic core and/or other constituents.

6.1. FABRICATION OF CORE-SHELL STRUCTURES NANOPARTICLE BY SELF–ASSEMBLY

In recent years, a great deal of attention has been drawn to the fabrication of composite micro- and nanoparticles that consist of either organic or inorganic cores having a surface layer as coating of different chemical composition. Such colloidal particles represent very attractive building blocks to create ordered and complex materials. Thus, it is possible to use nanoparticles e.g. with perfect spheres as a template upon which a suitable shell layer, with different composition, can be assembled. The creation of core-shell particles is attracting a great deal of interest because they exhibit improved physical and chemical properties over

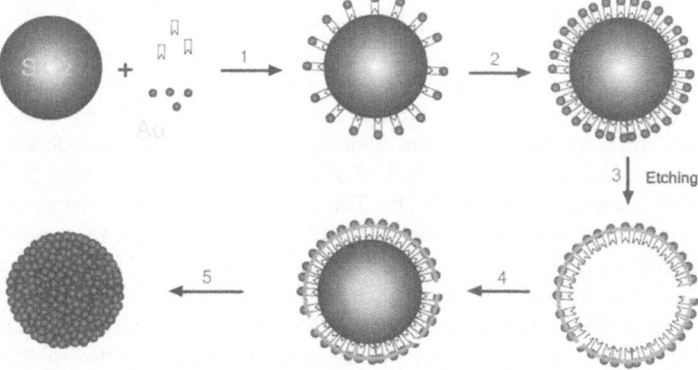

Figure 18. A schematic representation of the different steps used for the preparation of core-shell nanocomposites. (1) Activation of SiO_2 nanospheres with APTMS and attachment of colloidal gold nanoparticles. (2) In-situ formation of gold particles that are assembled as a surface layer. (3) Removal of the SiO_2 core with HF solution. (4) Precipitation of new nanoparticles inside the nanoshell, and (5) Further assembly of gold particles to produce dense non-porous shell layer encapsulating the core material.

their single-component counterparts making them potentially useful in a broader range of applications. Particle characteristics can influence chemical properties, reactivity, selectivity, rate of dissolution, and bio-efficiency. These structures combine characteristics of both core materials as well as that of the surface layer. Tailoring the surface properties of particles can be achieved by coating or encapsulating particles within a shell of preferred material which modifies the surface to fit specific environmental conditions and/or produce the desired surface properties.

The preparation of shell-core nanoparticles could be achieved using colloidal template approach and directed assembly of a surface shell layer using functional coupling reagents. In this method, we first prepare size-selected nanoparticles of silica, of which the size could be carefully controlled by controlling the polymerization process, leading to the formation of SiO_2. The surface of the silica particles is modified through a treatment with bi-functional coupling reagent containing two different binding groups. One of these two groups attaches to the silica surface and the other binding group has a high affinity for binding to e.g. gold particles. Gold particles are then formed in the solution in which the silica particles are suspended leading to the assembly of a layer of gold particles on the surface of the silica particles. The thickness and porosity of the gold layer could be controlled by the chemical process. The formation of this layer is continued until it attains a thickness with sufficient mechanical integrity and contains pores that allow water solutions to pass through the gold surface layer. Next, the silica core could be removed by the addition of hydrofluoric acid producing hollow gold spheres. These hollow spheres can be filled by suitable materials using suitable precipitation chemistries leading to the formation of the solid nanoparticles inside the hollow sphere. In the final stage, further deposition of gold on the surface is carried out in order to produce a more dense coating with no or very much reduced porosity. A conceptual presentation of the process for the fabrication of the core-shell nanoparticles is given in Figure 18.

Details of the experimental conditions used for these procedures are published elsewhere [29], however, a short summary is given here.

6.1.1. *Preparation of Silica nanoparticles Template*
Silica nanoparticles with an average diameter of about 300 nm were prepared using a variation of the method developed by Stöber et al. [44]. Starting materials were reagent-grade TEOS, ammonium hydroxide solution, and ethanol. An ethanol solution of TEOS was added to the ammonium hydroxide/ethanol solution to obtain a solution with the concentrations 0.36 mol NH_3/L, 0.27 mol TEOS/L and 4 mol H_2O/L, which were stirred at room temperature for 1h. The precipitated powder was centrifuged and dried at room temperature. The functional groups at the surface of these unmodified silica nanospheres were predominantly Si-OH or ethoxy ($Si-OCH_2CH_3$) groups. These groups were treated with trialkoxyorganosilane to modify the surface terminal functional groups [45]. Treatment with APTMS produces a surface terminated with amine groups. The functionalization was carried out by mixing the solution of silica nanospheres dispersed in ethanol with approximately five equivalents of organosilane (sufficient to provide five monolayer coatings of the silica nanoparticles). After the solution was left to react overnight, it was held at low boil for 1 h to promote covalent bonding of the organosilane to the surface of the silica nanospheres [46, 47]. Aliquots

of ethanol were added to maintain a constant volume during the boiling. The solution was then centrifuged and re-dispersed in ethanol at least five times to remove the excess silane.

6.1.2. Preparation of Gold Shell Layer and Hollow Spheres

The Au nanoparticles have been prepared by employing sodium citrate as a stabilizer. A solution of 3-5 nm diameter particles was prepared by adding 1 mL of 1 % aqueous $HAuCl_4.H_2O$ to 100 mL of H_2O while stirring vigorously. This was followed 1 min later by addition of 1 mL of 1 % aqueous trisodiumcitrate. After an additional minute, 1 mL of 0.075% $NaBH_4$ in 1 % trisodiumcitrate was added. The solution was stirred for 5 min and then stored at 4 °C until needed [48]. The concentration of SiO_2 nanospheres has been calculated as 10^{11} per mL. The total surface area of silica nanospheres have been calculated using this value. The concentration of Au particles has been calculated assuming the complete reduction of gold and an average diameter of 4 nm. The amount of Au solution required to coat the surface area of silica nanoparticles is then added to a surface activated silica solution in small portions at pH 5-7. Addition of Au nanoparticles to the solution of amine-coated silica nanospheres led to attachment of the Au nanoparticles to the silica nanospheres. Further coating of silica surface was achieved by addition of a required amount of $HAuCl_4$ solution and Na_2CO_3. In-situ reduction of Au was attained by using $NaBH_4$. The newly formed Au particles decorated the silica nanospheres and acted as nucleation sites, facilitating the growth of the Au shell around the nanospheres. The silica core in the composite structure was etched with a 5 % HF solution, thus forming hollow gold shells. The etched particles were filtered through a membrane having a pore diameter of 40 nm, depositing the hollow Au shells on the membrane. A treatment of Fe^{3+}-Fe^{2+} solution penetrated through the shell and deposited an iron core inside. Subsequent treatment with NaOH produced magnetite particles within the hollow spheres. Excess solution and the particles formed on the outer surface were washed away with water and

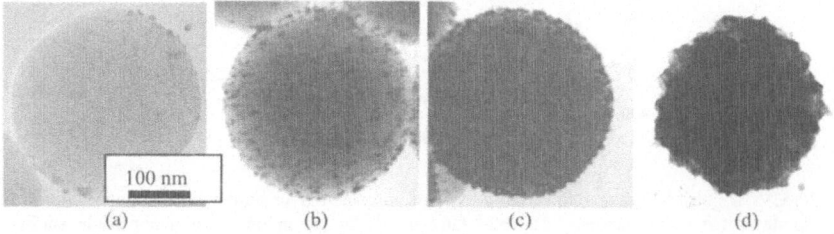

(a)   (b)   (c)   (d)

Figure 19. TEM images of nanoshell growth on silica nanospheres. a) Initial Au-colloid decorated silica nanoparticle. (b,c) Growth of Au layers on silica nanoparticle surface. (d) Completed growth of metallic nanoshell.

acetone through the porous membrane. The magnetite core-Au shell particles were then transferred to a quartz beaker and in-situ reduction of the Au with $NaBH_4$ was performed to block the pores in the hollow shells, trapping the magnetic nanoparticles.

The core-shell composite particles could be easily collected after the synthesis by using a magnet and decanting the solution. The gold shell formation around the silica particles is shown in Figure 19. The sequence of TEM images illustrates the progress in the metal nanoshell growth that occurs during the reduction.

Initially, the few colloid gold particles are assembled on the surface of the silica layer. As the reduction of gold proceeds more gold particles are assembled on the silica surfaces as seen in (b and c) increase in size as reduction continues (b, c). Successive assembly of gold particles on the surface of the silica is shown as the surface of the particles gets darker in the TEM images (d). The change in shell structure gives rise to changes in the optical signature of the formation of the gold layer on the surface of the silica layer as shown on Figure 20. Initially, the gold particles attached to the surface of SiO_2 have characteristic plasmon absorption at 520 nm. The absorption maximum shifts to higher wavelengths (red shift) as the growth progresses. The observed absorption differs slightly from the observation reported earlier by Oldenburg et. al., which states that as the nanoshell growth gets complete it gives rise to a nanoshell resonance at higher wavelengths [49]. The difference in observed absorption spectra comes from the porous nature of the gold shells around the silica prepared in this work, which does not obey the trend suggested by the Mie scattering theory. However, even porous shell growth shifted the absorption peak of gold from 520 nm to 590 nm. TEM images for the process of synthesis of magnetite within the metallic hollow spheres are given in Figure 21. The sequence shows the process; (a) the silica core was etched to produce the hollow Au shell, (b) magnetite precipitation within the

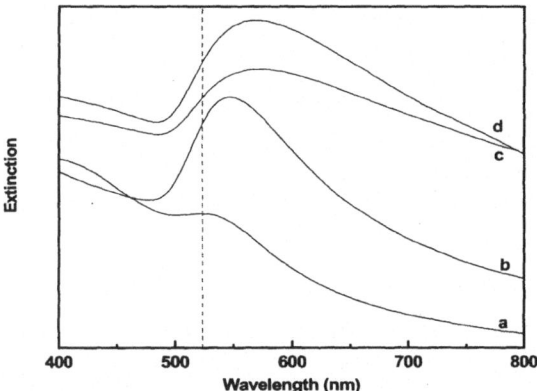

Figure 20. UV-VIS absorption spectra of the growth of gold shell layer on the silica nanoparticles a) Initial Au-colloid decorated silica nanoparticle. (b,c) Growth of Au layers on silica nanoparticle surface. (d) Completed growth of metallic nanoshell

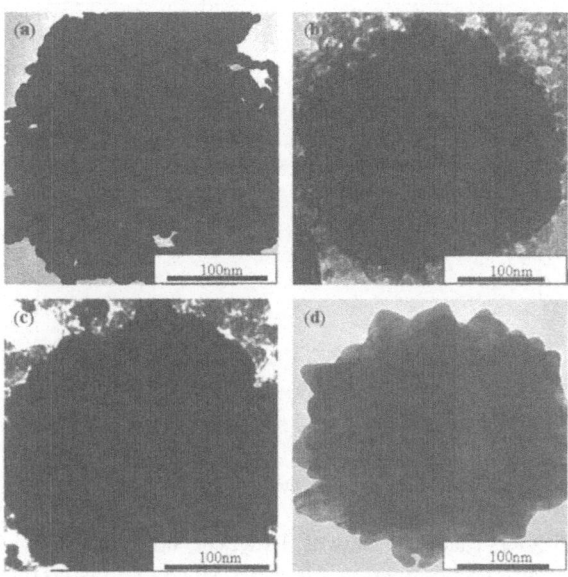

Figure 21. TEM images showing steps involved in the synthesis of magnetic core-Au shell particles. a) Hollow Au shells prepared by etching the silica core with HF, b) Fe$_3$O$_4$ particles precipitated inside the Au core, c) Complete coverage with Au to trap magnetite particles inside the core, d) The core/shell structure after final cleaning of excess reactants.

porous shell, and (c) complete entrapment of the magnetite by the Au shell formation (c). Surface roughness of the final particles is due to the fast in-situ reduction of Au on the magnetic core Au shell structures and can be controlled better by the adjustment of the HAuCl$_4$ to NaBH$_4$ ratio.

In order to demonstrate the dramatic change of the properties of the nanoparticles as a result of the encapsulation we investigated the oxidation of the magnetite phase. DSC analysis of the colloidal particles at different processing stages is given in Figure 22. The two lower thermograms (a and b) show the behavior of silica and the Au coated silica nanospheres. No significant change was observed for the bare silica particles and Au coated silica nanospheres, except for the endothermic peak resulting from the evaporation of water around 150 °C, observed in the thermogram of the colloidal silica. In order to investigate the effect of the Au shell around SPION, naked SPION and SPION core-Au shell nanocomposites were analyzed by the DSC separately (c and d). In the case of naked SPION, the first broad exothermic transition peak around 370 °C is attributed to the phase transition from magnetite to maghemite. The second much sharper exothermic transition peak represents the oxidation of maghemite to hematite. The existence of these phases after each transition was verified with separate experiments by performing phase analysis after

66

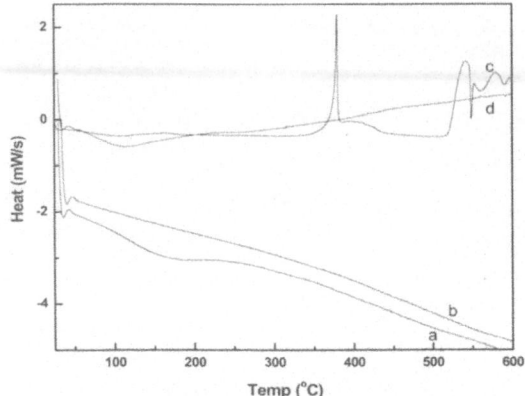

Figure 22. DSC analysis of bare and coated magnetite nanoparticles. (a) Bare silica, and (b) Au coated silica nanospheres. (c) Magnetite, and (d) Au coated magnetite.

the transition with XRD (data not shown). However, no transition was observed in the case of magnetite particles confined in Au shells, since there is no direct contact of the magnetite with the external atmosphere.

6.2. FABRICATION OF 2-D AND 3-D HIERARCHICAL STRUCTURES

The ability to engineer small structures or to scale down existing structures from the micrometer scale to the nanometer scale is a major subject in modern science and technology. There are many opportunities that might be realized by making new types of smaller structures, or by downsizing existing structures. It has been observed that many interesting new phenomena, such as quantum size effect [50, 51], occur at nanometer dimensions. Microfabrication, the generation of small structures, is essential for modern science and technology; supports information technology, and permeates society through its role in microelectronics and optoelectronics. The manufacture of such nanostructures often requires fabrication of a certain pattern from materials with a specific structure and properties. Different problems are encountered when processing these structures on the nanometer scale. The making of nano-patterns with dimensions in the order of 10-200 nm represents a technological challenge. Modern microelectronics and integrated circuitry have been developed based on the use of semiconductor materials and technology. Even in these materials, there is a need for modification of the existing technology or develop new technology to produce the required nanostructures from different materials. Nanoparticles can be used as building blocks for the fabrication of these nanostructures. This can be achieved by first making the nanopattern using suitable techniques on a given substrate and then using precise assembly of the nanoparticles on the pattern using directed assembly

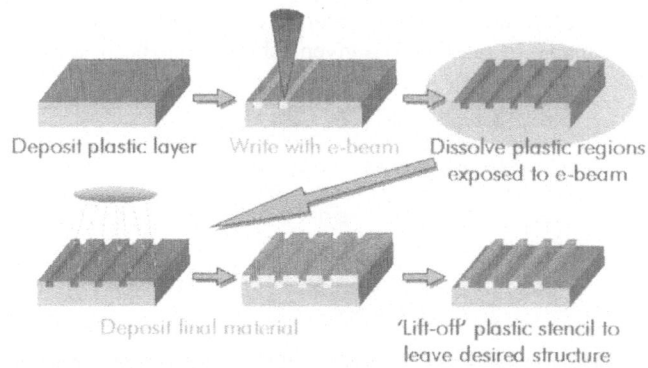

Figure 23. Schematic view of the procedure of e-beam lithography

methods. This approach is briefly discussed and demonstrated below.

The patterning required in microfabrication is usually carried out with photolithography. Nonetheless, photolithography has disadvantages, e.g. the sizes of the features it can produce are limited by optical diffraction and the high-energy radiation needed for small features requires complex facilities and technologies. Also, photolithography is expensive, it cannot be easily applied to nonplanar surfaces, it tolerates little variation in the materials that can be used; and it provides almost no control over the chemistry of patterned surfaces, especially when complex organic functional groups of the sorts needed in chemistry, biochemistry, and biology are involved. Non-photolithographic microfabrication methods are needed to complement photolithography. These techniques would ideally overcome the limits of photolithography, provide access to three-dimensional structures, and tolerate a wide range of materials and surface chemistries. Non-photolithographic microfabrication methods must be inexpensive, experimentally convenient, and accessible to molecular scientists. One such method, called soft lithography, employs a patterned elastomer as the mask, stamp, or mold. Soft lithography is able to provide routes to form high-quality patterns and structures with lateral dimensions of about 30 nm to 500 nm in systems presenting problems in topology, materials, or molecular-level definition that cannot (or at least not easily) be solved by photolithography [52].

Two-dimensional arrays of high refractive index structures can be fabricated using a combination of e-beam lithography (Figure 23) for pattern definition and electrochemical deposition for structure formation. The potential of this method was demonstrated for CdSe where nanopillars, mushrooms, walls, and crosses, are prepared. Such arrays have potential in optical device applications such as photonic crystals and waveguides [53].

Self-assembly or directed assembly of particles on flat surfaces can be achieved using an approach similar to those adopted for the preparation of core-shell structured nanoparticles. For simplicity we first consider the assembly of organic molecules on a given substrate.

68

The formation of a self-assembled monolayer (SAM) of molecules containing functional group capable of binding to a given substrate has been demonstrated [53]. It has been

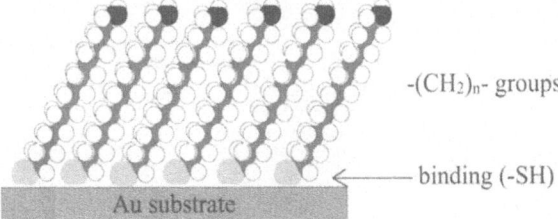

-(CH$_2$)$_n$- groups

binding (-SH)

Au substrate

Figure 24. Representation of a highly ordered monolayer of alkanethiolate formed on Au surface after [53].

shown that a highly ordered monolayer of alkanethiolate, CH$_3$-(CH$_2$)$_n$-SH, a hydrocarbon where the –SH is the functional group), can be formed on a Au surface through the binding of the sulfur atom to the gold surface. The thickness of the layer is determined by the number of methylene groups (n) in the alkyl chain. The surface properties of the monolayer can easily be modified by changing the binding group, e.g. amine group. The alkyl chains (CH$_2$)$_n$ extend from the surface in a nearly trans configuration. Figure 24 demonstrates a schematic representation of the SAM formation.

In order to assemble nanoparticle on substrate surfaces, coupling reagents are used. Coupling reagents, normally similar to the example above, consist of hydrocarbon chain with two functional groups each located on one end of the molecules. The functional groups are selected to have conjugation chemistries suitable to the surfaces considered. There are several types of functional groups that can be used to form a suitable bonding with different surfaces; S and N binding groups are suitable for several metallic surfaces while silanes are suitable to silica surface, etc. Examples of the chemical composition of some of these reagents are given

H$_2$N. ⌇⌇⌇⌇ NH$_2$

(a) diamino dodecane

HS. ⌇⌇⌇⌇ SH

(b) dodecandithiol

```
        OCH3                          OCH3
         |                             |
H3CO—Si—OCH3                  H3CO—Si—OCH3
         |                             |
        CH2                           CH2
         |                             |
        CH2                           CH2
         |                             |
        CH2                           CH2
         |                             |
        NH2                           SH
```

(c) 3-aminopropyltrimethoxysilane (APTMS)   (d) 3-mercaptopropyltrimethoxysilane (MPTMS)

6.2.1 *Nanolithography and Microcontact Print*

The combination of the nano-pattern writing and directed assembly of nanoparticles on the pattern is a powerful tool for fabricating nanostructures on large area for high throughput and low cost. The approach is based on the method of nanoimprint lithography first presented by Chou et al. [54]. This technique, known as soft lithography, offers great potential for the fabrication of 2-D and 3-D patterned structures using a variety of patterns and materials.

Soft lithographic techniques currently used include: microcontact printing (μCP), replica molding, microtransfer molding, micromolding in capillaries, and solvent-assisted micromolding. Soft lithography generates micropatterns of self-assembled monolayers (SAMs) by contact printing where self-assembly is defined as the spontaneous organization of molecules into stable, well-defined structures.

First, the pattern is made using a suitable e-beam lithography technique, a normally cumbersome and time consuming operation that requires extensive time to write relatively small areas. Once the pattern is made, it can be used as a master to transfer the pattern to a mould which can be used for transferring the pattern to the substrate surface by contact (microcontact, μ-contact). The pattern is transferred to the substrate through printing using "ink-like" fluid which is done by wetting the pattern of the mould with the liquid containing the coupling reagent. When the mould is applied to the substrate surface the coupling reagent will be transferred to the surface of the substrate as one end of the molecule will conjugate to the surface of the substrate. This results in the formation of the pattern on the substrate in a form of a highly ordered single monolayer of the coupling agent in a way similar to that given in Figure 24. The pattern is used to assemble through the binding of the surface of the nanoparticles to the second functional group of the coupling reagent. A schematic presentation of the process is given in Figure 25.

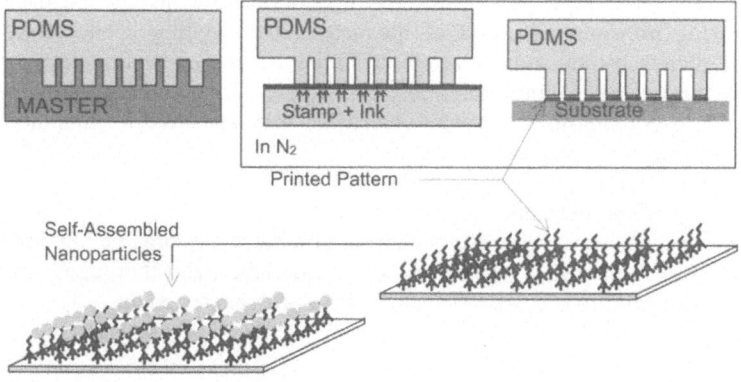

Figure 25. A schematic illustration of the procedures for μCP using a coupling agent (sulfur - based) on a silicon wafer: Printing on a planar surface with a planar stamp (I: printing of the pattern on PDMS stamp. II: wetting the pattern with ink (reagent dissolved in ethanol) III: printing (deposition of ink in the required pattern). IV: assembly of nanoparticles on the coupling reagent.

μCP is a flexible and efficient method to form patterned SAMs containing regions terminated by different chemical functionalities with submicron lateral dimensions. The printed SAMs may be used as resists to protect the underlying substrate from etching or deposition. A precise contact between the stamp and the surface of the substrate is very important for accurate transfer of the pattern without distortion. Once the stamp is fabricated, multiple copies of the pattern can be produced using straightforward experimental techniques. μCP, an additive process wherebymaterials waste is minimized and it also has the potential to be used for patterning large areas.

The experimental details for the fabrication of self-assembled nanoparticles of gold on silicon wafer is given elsewhere [55, 56], however, a brief description of few experimental details are given here.

Chemical needed: 3-Aminopropyltrimethoxysilane (APTMS), ethanol, $HAuCl_4$, tetrahydroxy phosphonium chloride (THPC), $FeCl_2 \cdot 4H_2O$, $FeCl_3 \cdot 6H_2O$, and NaOH. PDMS stamps were fabricated by mixing an elastomer with a curing agent at a ratio of 10:1. This mixture was poured on the masters which was prepared by lithographic technique. Dots and photolithographically defined parallel lines, 900 nm in width spaced at 200-600 nm whichwere already printed on the masters. After curing at 60 °C for 3 h, the stamps were peeled away from the master. The stamps were inked using a contact inking technique [57] in which the patterned stamp is not exposed to liquid ink; rather it is only exposed to an ink-impregnated PDMS block that mediates the transfer of silane, i.e.APTMS, from solution to the elastomeric pattern. In practical terms, the inker pad is equilibrated for long times (~12 h) with a solution of silane (10 mM) in ethanol. This pad is withdrawn from solution, dried with a stream of N_2, and contacted by the patterned face of the stamp. The elasticity and surface characteristics of the inker and the patterned stamp promote good conformal contact between the two elastomers. A similar texture of both PDMS surfaces in contact also favors the formation of a continuous interface through which the homogeneous transfer of silane is possible. Patterns on the stamps are not subjected to capillary forces (wetting and dewetting of the pattern by the ink) nor to N_2 streams required during the inking and drying steps. Contact inking prevents distortion of the pattern by swelling. The stamp is then transferred to the substrate and contacted for 5 min. Shorter contacting times resulted in incomplete transfer of ink molecules to the substrate surface. All printing work was performed in a N_2 atmosphere, since silanes are very sensitive to humidity and can easily polymerize in the presence of water.

6.2.2 Assembly of nanoparticles:
Silicon wafer substrates were carefully cleaned to remove any organic impurities on the surface and to homogeneously grow a silicon oxide layer, and thoroughly rinsed with de-ionized water prior to use. APTMS molecules were transferred to

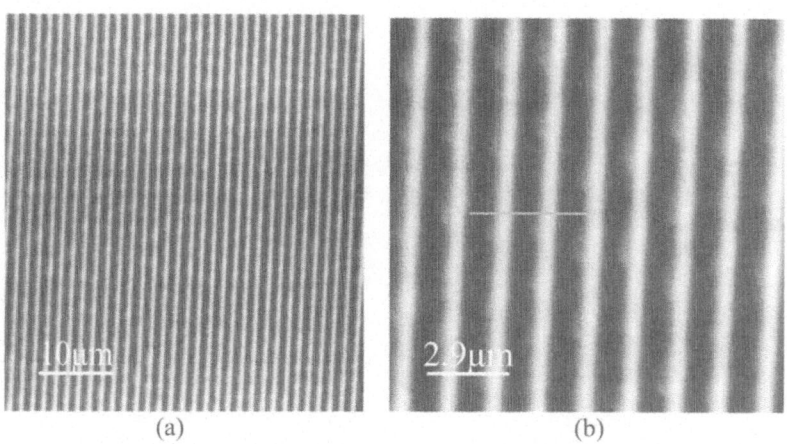

Figure 26 (a) AFM image of PDMS stamp. (b) a magnified region of (a).

the substrate surface using the μCP technique with prepared PDMS stamps. Two different particles were used in this study: gold nanoparticles and magnetite nanoparticles coated by a thin gold layer (core-shell structured nanoparticles). Subsequent functionalization of the substrate surface with Au or Au-coated magnetite colloids was easily achieved by dipping the substrate in the colloid solution (pH = 5-7) for 6 h. The functionalization successfully yielded a metallization of the APTMS printed surfaces through covalent bonding between the Au coated magnetiteparticles (prepared by *μE* method) and –NH$_2$ groups on the substrate. Substrates were then sonicated to remove particles attracted to the inactive surface, i.e. the part that is not silanized.

The size of the nanoparticles was determined from the TEM and found to be 10 nm. However, from the imaging of nanoparticles by AFM, the mean particle diameter was calculated to be 9 nm, using an image analysis program based on lognormal distribution. However, there may be due to artifacts in the AFM measurements resulting from the use of pyramidal type with micron size at the base although the tip is in the nanometer size. The prepared stamps were also investigated before use by AFM analysis. Several line patterned PDMS stamps, with lines of 900 nm in width and the line was separated by 200-600 nm, have been used for the printing work. APTMS is very diffusive and did not give any patterned SAMs with narrowly separated line stamps. 600 nm separations gave the best patterning with APTMS. Figure 26 shows AFM images of the stamps prepared using a master of regular parallel lines, with a line width of 900 nm separated by 600 nm gap. The regularity of the stamp surface was confirmed by performing a cross sectional analysis.

The pattern was successfully transferred to the substrate surface via μCP, and a highly ordered SAM of APTMS was formed at the contact regions. A contact time of 5 min, without external pressure, was found to be enough to form patterns with sufficient coverage.

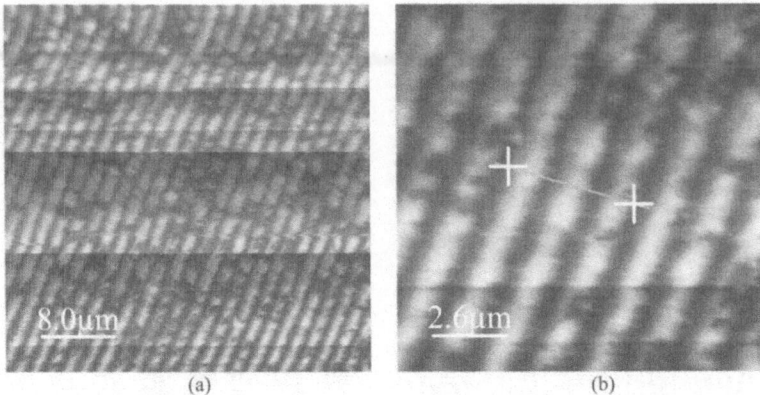

Figure 27. (a) AFM images of the substrate after printing the ink (APTMS) molecules, using a stamp having parallel lines, and assembly of Au coated magnetite colloid. (b) A magnified region of (a). The bright regions are those functionalized by APTMS and subsequently Au coated magnetite particles were deposited; the dark regions are bare Si/SiO_2 substrate.

Shorter printing times resulted in incomplete ink transfer to the substrate surface. Selective assembly of the nanoparticles was achieved through specific interaction between the Au surface and the terminating $-NH_2$ groups on the substrate. Figure 27 shows AFM images of APTMS printed and that with particles assembled on the surface with Au coated magnetite using the line patterned stamp. The image reveals the regularity in the transferred pattern and subsequent metal-coated nanoparticles deposition on the SAM surfaces. A cross sectional analysis of the magnified region showed wider lines than the original stamp, due to the diffusion of silane molecules on the surface of the substrate. (Figure 28). Closer examination of the lines showed the existence of some defects along the lines. The defects were attributed to imperfections on the stamp surface, which were transferred to the surface during activation of the surface.

A comparison of the cross section of the depth profile of the line of the stamp and line of the transferred after the assembly of the nanoparticles confirm that the pattern of the stamp was successfully transferred to the substrate. It can also be noticed that the width of the lines of the transferred pattern is slightly larger of the original stamp with ca. 20% increase.

This can be a result of in accurate contact printing, diffusion of the ink on the substrate or artifact due to the limitation of the AFM technique. The AFM tip used was pyramidal shaped with few tens of nanometer width of the tip and few microns at the base. From the AFM imaging we could not accurately calculate the height of the formed line by the self-assembly of the particles. On the other hand the gaps between the lines are maintained also in the printed pattern.

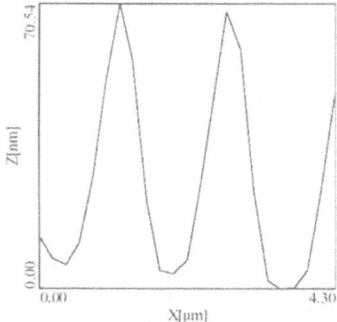 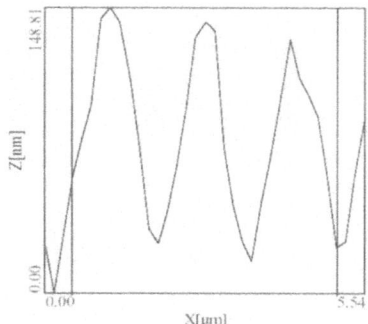

Figure 28. A cross-section analysis of the depth profile of the AFM image of (a) the surface of the stamp of the area marked with the bar in Figure 26b and (b) the surface of the transferred pattern after the assembly of the nanoparticles in the area marked with the bar in Figure 27b.

6.2.3 Fabrication of Multi-layered Patterns

Complex 3-D patterned structure can be fabricated using similar principles. Multi-layer structure can be made using micro-contact printing and self-assembly of nanoparticles through the use of suitable coupling agents. After the formation of the first layers the procedures can be repeated to assemble subsequent layers. The method is schematically illustrated in Figure 29.

The process is in principal very similar to that of that described earlier. First the stamp is prepared as described earlier and wetted with ink for transferring the pattern. For the transfer of the pattern MPTMS molecules is transferred to the substrate surface using the micro-contact printing. The first layer of the nanoparticles is then assembled on the ink sites. Also here the chemistry of the solutions should be made to enhance the conjugation of the nanoparticles on the surface of the substrate. However, the chemistry should be compatible with the further preparation procedures. Depending on the nanoparticles to be assembled as the subsequent layer, suitable coupling reagent should be selected. . Several di- functional molecule can be used three of which are given in Figure 30. One end of the coupling reagent should conjugate to the surface of the assembled particles and the second functional group at the other end of the molecule should couple to the next layer of nanoparticles to be assembled. The conjugation chemistry applied here should be compatible with the system. There is no need to align the particles in special direction, since the molecules align themselves according to their structure. The selection of different structure of the alkyl chain; e.g. length and branching, affect the distance between the assembled layers and the degree of coverage of the surface by these molecules. The process is then repeated to produce a subsequent assembled layer of nanoparticles.

For the assembly of multi-layers on pattern that does not cover the entire surface, it is important to mask the surface of the substrate to prevent the assembly of the coupling

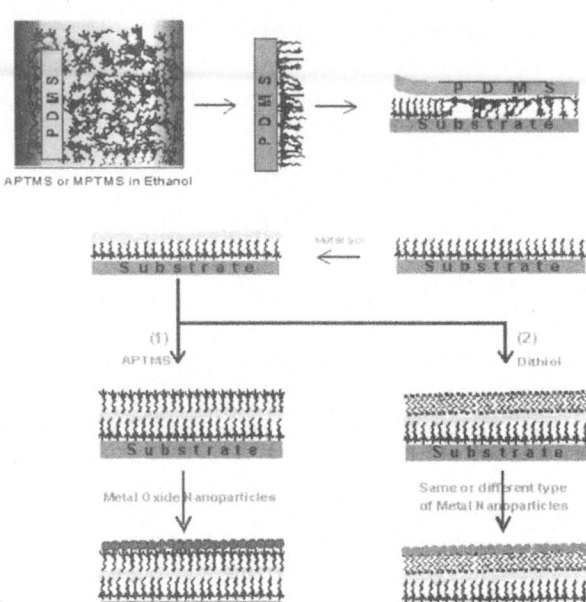

APTMS or MPTMS in Ethanol

Figure 29. A scheme of the functionalization of surfaces with contact inking for multilayer structure formation.

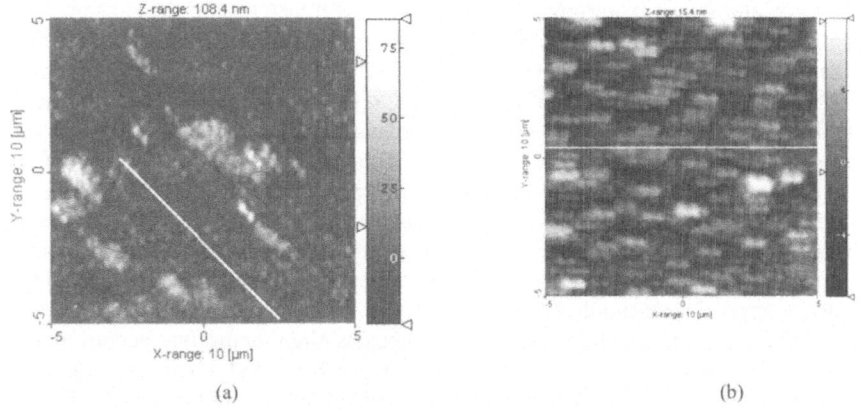

(a)   (b)

Figure 30. AFM images of the Au/Fe$_3$O$_4$/Au film formed by micro-contact printing technique. (A) first Au layer assembled on the substrate. (B) Image of the upper most layer of Fe$_3$O$_4$.

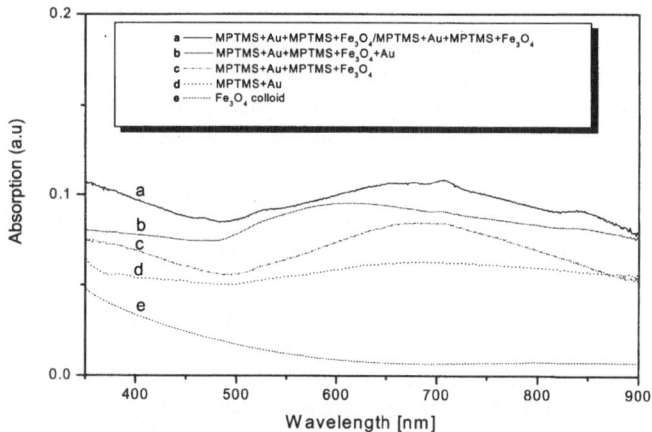

Figure 31. UV-Vis absorption spectra for film consisting of Au/Fe₃O₄ multilayers with different thickness.

reagent. The blocking molecule will have one functional group that allows the binding to the substrate, while the other end of the molecule does not have any binding functionality; e.g. silane group on one side and methyl group on the other end of the molecule. The coupling reagents used for the assembly of the second and subsequent layers of nanoparticles will not bind to the blocking molecule and consequently no assembly of particles takes place at these sites.

Figure 31 shows an AFM image of a multilayer assembly of a composite thin film of Au/Fe₃O₄/Au on silica substrate [55]. The substrate was first assembles was gold nanoparticles, followed by the assembly of magnetite nanoparticles. The film consisted of totally four layers of the different nanoparticles. From the cross sectional measurement it was found that the size of the Au particles is about 200 nm and the size of the Fe₃O₄ particles is about 250 nm. This indicates that the assembled layer is of agglomerate since the primary nanoparticles are about 10 nm in size. This agglomeration may have been caused by the coupling agent. The optical properties of the assembled film have shown to vary with the thickness of the film. From Figure 31, it is evident that new absorption satellites emerged as the thickness of the film is increased.

7. Conclusions

Nanoparticles are important building blocks for the fabrication of materials, structures and systems. The properties may be different from their bulk counterparts. The use of nanoparticles as small "Lego" units to build up larger nanostructures with high complexity offerstremendous potential for constructing nano-systems with novel functionalities. During the last two decades, interest and the significance of these building blocks have been the driving force of dramatic evolution of nanoparticles research and technology.

Methods for the fabrications of first generation nanoparticles with small size and relatively simple composition have dominated the research efforts at early stage of the development of the nano-era. Several novel methods for the fabrication of different types of nanoparticles have been developed and perfected for the production of several types of nanoparticles including metallic, ceramic and composite materials. Second generation nanoparticles have emerged as a result of increasing demand for multi-functionalities and more advance applications. Next generation nanoparticles emerged as the need for the fabrication of nanoparticles with high degree of complexity of structures that can be programmed for specific functions.

Several methods have been developed to fabricate nanoparticles. Gas phase methods have been developed in a variety of methods and can be used for the production of several types of nanoparticles. These methods show a good control of the particle surfaces and can produce particles with relatively narrow particle size distribution. Chemical synthesis methods have also been developed and successfully used for the fabrication of several different types of nanoparticles. They have the distinct advantage of being capable of producing nanoparticles with complex and heterogeneous structures. Besides, they are easily scalable which render these methods suitable for large-scale production. We have briefly reviewed the co-precipitation method for the fabrication of nanoparticles and demonstrated its usefulness especially for obtaining complex structures through layer-by-layer processing. The use of a co-precipitation approach in a confined zone, such as, micro-emulsions and in fluid flow system, have been presented. A rational approach for the development of solution chemical processing methods could be done through the use of thermodynamic modeling of the co-precipitation process. These methods offer extended control on the size, morphology and the possibility of sequential processing to build core-shell structured nanoparticles.

The consolidation and assembly of nanoparticles in higher hierarchical systems is an important application for nanoparticles. Besides, the use of nanoparticles in suspensions has shown to have several applications. The development of methods for the assembly of nanoparticles is essential for the construction of nanostructures and systems.

Nanoparticles with different core and shell structure could be fabricated by the self-assembly using colloidal template approach. Example for the preparation of core-shell nanoparticles was carried out by the self-assembly of gold nanoparticles on silica nanospheres. By the dissolution of the silica core hollow metallic spheres were obtained, which could be filled by another materials as required. This approach allows the fabrication of core-shell nanoparticles with numerous combinations of core and shell materials.

The assembly of 2-D and 3-D structure through the assembly of nanoparticles offers the possibility of engineering organized hierarchical structures with a high degree of accuracy. This is possible through the combination of micro-contact printing of given pattern of flat substrate surfaces. Micro-contact printing approach permits a high throughput pattern production at low costs. Specific chemical molecules acting as coupling reagents, are used as the printing ink. The coupling agents bind to the substrate surface at one end while the other end is amendable for conjugation with surfaces of the nanoparticles. In this way, it is possible to have an accurate positioning of the nanoparticles at the required pattern. By proper selection of the coupling molecules, it is possible to allow assembly of sequential layers to for large-scale structures.

The future holds very high promise for novel and sophisticated applications of nanotechnology, however, the success of these applications depends on our ability to fabricate building blocks with a very high degree of control over their composition, properties and the way they are arranged. Therefore, there is a need for the further development of sophisticated methods for the engineering of the building blocks and the methods of their assembly. Programmable nanoparticles will certainly be an area of future research and development, thereby, chemical methods will certainly be valuable tools for this purpose. Moreover, methods for arranging the building blocks in given patterns are of equal importance. The self-assembly or directed assembly has the highest odds to achieve these goals. Current achievements in nanomaterials and nanotechnology are representative of only the tip of the iceberg.

8. Acknowledgement

The work reported has been undertaken by several members of a group that deserves proper acknowledgement. The author also acknowledges financial support of Swedish funding agencies (VR (TFR), SSF and Vinnova/Nutek), and several EU programs.

9. References

1. (a) Proceeding of the 4th International Conference on Nanostructured Materials (Nano98), *Nanostruct. Mater.* **11** (1999), (b) Proceeding of the 5th International Conference on Nanostructured Materials (Nano2000), *Scripta Mater.* **44 (8/9)**, and (c) Schoonman J. (2000) Nanostructured materials in solid state ionics, *Solid State Ionics* **135**, 5-19
2. Edelstein A. S. and Cammarata R. C. (1998) Nanomaterials: Synthesis, Properties and Applications, *Institute of Physics Publishing*, London
3. Schmidt H. (2001) Nanoparticles by chemical synthesis, processing to materials and innovative applications, *Appl. Organometal. Chem.* **15** 331-343
4. Granqvist C. G. and Buhrman R. A. (1976) Ultrafine metal particles, *J. Appl. Phys.* **47** 2200-2219
5. Wang L. and Muhammed M. (1999) The Synthesis of Zinc Oxide Nanoparticles with Controlled Morphology, *J. Mater. Chem.* **9** 2871-2878
6. Rahman I. Z., Razeeb K. M., Rahman M. A. and Kamruzzaman Md. (2003) Fabrication and characterization of nickel nanowires deposited on metal substrate, *J. Magn. Magn. Mater.* **262** 166-169
7. Mohamed M. B., Ismail K. Z., Link S., and El-Sayed M. A. (1998) Thermal Reshaping of Gold Nanorods in Micelles, *J. Phys. Chem. B.* **102** 9370-9374
8. Nikoobakht B., Wang Z. L., and El-Sayed M. A. (2000) Self-Assembly of Gold Nanorods, *J. Phys. Chem. B.* **104** 8635-8640
9. M. S. Toprak, D. K. Kim, M. Mikhaylova and M. Muhammed (2002) Nanopatterning 2D metallic surfaces by soft lithography, *Nanopatterning; From Ultralarge-Scale Integration to Biotechnology, Mat. Res. Soc. Proc.* **705** Y.7.22.1-Y7.22.6
10. Kim D. K., Zhang Y., Voit W., Rao K.V. and Muhammed M. (2001) Synthesis and Characterization of Surfactant Coated Superparamagnetic Monodispersed Iron Oxide Nanoparticles *J. Magn. Magn. Mater.* **25** 30-36
11. Toprak M. S., Zhang Y. and Muhammed M. (2003) Chemical Alloying and Characterization of Nanocrystalline Bismuth Telluride, *Mater. Lett.* **4460** 1-7
12. Adamopoulos O., Zhang Y., Croft M., Zakharchenko I., Tsakalakos T. and Muhammed M. (2001) The Characterisation and Reactivity of Nanostructured Ce-Cu-O Composites for Environmental Catalysis, *Synthesis, Functional Properties and Applications of Nanostructures, Mat. Res. Soc. Proc.* **676**

78

13. Wang M. S. and Muhammed M. (1997) Synthesis and characterization of $MgAl_2O_4$ spinel ceramic precursor, *Synthesis and properties of mechanically alloyed and nanocrystalline materials, pts 1 and 2 - ismanam-96, materials science forum,* **235** 241-247

14. Wang L. N., Zhang Y. and Muhammed M. (1995) Synthesis of Nanophase Oxalate Precursors of YBaCuO superconductor by Coprecipitation in Microemulsions, *J. Mater. Chem.* **5** 309-314

15. Bertini L., Toprak M., Williams S., Platzek D. and Mrotzek A., Zhang Y., Gatti C., Müller E. and Muhammed M., (2003) Michael Rowe Nanostructured $Co_{1-x}Ni_xSb_3$ skutterudites: Synthesis, thermoelectric properties, and theoretical modeling, *J. Appl. Phys.* **93** 438-447

16. Stiewe C., Toprak M., Platzek D., Muller E. and Muhammed M. (2002) Improvement of Thermoelectric Properties of Nano-grained CoSb₃ Skutterudites Doped with Ni(n) or Fe(p), *Proceeding of the VII European Workshop on Thermoelectrics of the European Thermoelectric Society Pamplona*

17. Kim D. K., Mikhaylova M., Zhang Y. and Muhammed M. (2003) Protective coating of superparamagnetic iron oxide nanoparticles, *Chem. Mater.* **15** 1617-1627

18. Ward A. J. I. and Friberg S. (1989) Preparing Narrow Size Distribution Particles from Amphiphilic Association Structure, *MRS Bul.* 41-48

19. Yadav O. P., Palmqvist A., Cruise N. and Holmberg K. (2003) Synthesis of platinum nanoparticles in microemulsions and their catalytic activity for the oxidation of carbon monoxide, *Colloid Surface A* **221** 131-134

20. Berkovich Y., Aserin A., Wachtel E. and Garti N. (2002) Preparation of Amorphous Aluminum Oxide-Hydroxide Nanoparticles in Amphiphilic Silicone-Based Copolymer Microemulsions *J. Colloid Interface Sci.* **245** 58-67

21. Fang J., Stokes K. L., Joan A. Wiemann, Zhou W. L., Dai J., Chen F. and O'Connor C. J. (2001) Microemulsion-processed bismuth nanoparticles, *Mat. Sci. Eng. B-Solid* **83** 254-257

22. Wang L. and Muhammed M. (1999) The Synthesis of Zinc Oxide Nanoparticles with Controlled Morphology, *J. Mater. Chem.* **9** 2871-2878

23. Jongen N., Donnet M., Bowen P., Lemaitre J., Hofmann H., Schenk R., Hofmann C., Aoun-Habbache M., Guillemet-Fritsch S., Sarrias J., Rousset A, Viviani M., Buscaglia M. T., Buscaglia V., Nanni P., Testino A. and Herguijuela J.R. (2003) Development of a continuous segmented flow tubular reactor and the "scale-out" concept - In search of perfect powders, *Chem. Eng. Technol.* **26** 303-305

24. Hansen E. H. and Ruzicka J. (1979) Principles of flow injection analysis as demonstrated by 3 lab exercises, *J. Chem. Educ.* **56** 677-680

25. Murt M., Amokrane A., Bastide J. P. and Montanaro L. (1992) Synthesis of zeolites from thermally activated kaolinite-some observations on nucleation and growth, *Clay Miner.* **27** 119-130

26. Bhattacharjee B., Bera S. K., Ganguli D., Chaudhuri S. and Pal A. K. (2003) Studies on CdS nanoparticles dispersed in silica matrix prepared by sol-gel technique, *Eur. Phys. J. B* **31** 3-9

27. Meulenkamp E. A. (1998) Synthesis and growth of ZnO nanoparticles, *J. Phys. Chem. B* **102** 5566-5572

28. Liao X. H., Zhu J. J. and Chen H. Y. (2001) Microwave synthesis of nanocrystalline metal sulfides in formaldehyde solution, *Mat. Sci. Eng. B-Solid,* **85** 85-89

29. Kim D. K. (2001) Superparamagnetic Iron Oxide Nanoparticles for Biomedical Applications. ISBN 91-7283-126-X.*Royal Institute of Technology*, Sweden

30. Köseoglu Y., Aktas B., Yildiz F., Kim D. K., Toprak M. and Muhammed M. (2003) ESR study on high T_c superconductor MgB_2, *Physica C.* **390** 197-203

31. Christmas K.G., Gower L. B., Khan S. R. and El-Shall H. (2002) Aggregation and dispersion characteristics of calcium oxalate monohydrate: Effect of urinary species, *J. Colloid Interface Sci.* **256** 168-174

32. Dzwinel W. and Yuen D. A. (2002) Mesoscopic dispersion of colloidal agglomerate in a complex fluid modelled by a hybrid fluid-particle model, *J. Colloid Interface Sci.* **247** 463-480

33. Hu Z. S., Oskam G., Penn R.L., Pesika N. and Searson P. C. (2003) The influence of anion on the coarsening kinetics of ZnO nanoparticles, *J. Phys. Chem. B* **107** 3124-3130

34. Portales H., Saviot L., Duval E., Gaudry M., Cottancin E., Pellarin M., Lerme J. and Broyer M. (2002) Resonant Raman scattering by quadrupolar vibrations of Ni-Ag core-shell nanoparticles, *Phys. Rev. B*, **65** 165422

35. Park J. H. and Jana S. C. (2003) The relationship between nano- and micro-structures and mechanical properties in PMMA-epoxy-nanoclay composites, *Polymer*, **44** 2091-2100

36. Viau G., Brayner R., Poul L., Chakroune N., Lacaze E., Fievet-Vincent F. and Fievet F. (2003) Ruthenium nanoparticles: Size, shape, and self-assemblies, *Chem. Mater.* **15** 486-494

37. Nikles D.E., Huh J. Y. and Woo T. (2001) Effect of silane coupling agents on rheological properties of solventless magnetic ink, *Abstracts of papers of the American Chemical Society* **222**: 232-PMSE Part 2

38. Matsumura M., Yamada K., Fujimaki M., Sugihara H. and Nakagami H. (1999) Proton relaxation caused by magnetic resonance imaging contrast agent, oral magnetic particles, *Chem. Pharm. Bull.* **47** 727-731

39. Le B., Shinkai M., Kitade T., Honda H., Yoshida J., Wakabayashi T. and Kobayashi T. (2001) Preparation of tumor-specific magnetoliposomes and their application for hyperthermia, *J. Chem. Eng. Jpn.* **34** 66-72

40. Ming M., Chen Y. and Katz A. (2002) Synthesis and characterization of gold-silica nanoparticles incorporating a mercaptosilane core-shell interface, *Langmuir* **18** 8566-8572

41. Zhang Y. J., Yang S.G., Guan Y., Cao W.X. and Xu J. Fabrication of stable hollow capsules by covalent layer-by-layer self-assembly, *Macromolecules*, **36** 4238-4240

42. Zhang H. L., Evans S. D. and Henderson Jr. (2003) Spectroscopic ellipsometric evaluation of gold nanoparticle thin films fabricated using layer-by-layer self-assembly, *Adv. Mater.* **15** 531-534

43. Westcott S. L., Oldenburg S. J., Lee T. R. and Halas N. J. (1998) Formation and adsorption of clusters of gold nanoparticles onto functionalized silica nanoparticle surfaces, *Langmuir* **14** 5396-5401

44. Stöber W., Fink A. and Bohn E. (1968) Controlled Growth of Monodisperse Silica Spheres in the Micron Size Range, *J. Colloid Interface Sci.* **26** 62-69

45. Badley R. D., Ford W.T., McEnroe F. J. and Assink R. A. (1990) Surface modification of colloidal silica, *Langmuir* **6** 792-801

46. Blaaderen A. V. and Vrij A. (1993) Synthesis and characterization of monodispersed colloidal organo-silica spheres, *J. Colloid Interface Sci.* **156** 1-18

47. Waddell T. G., Leyden D. E. and DeBello M. T. (1981) The nature of organosilane to silica-surface bonding, *J. Am. Chem. Soc.* **103** 5303-5307

48. Grabar K. C., Allison K. J., Baker B. E., Bright R. M., Brown K. R., Freeman R. G., Fox A. P., Keating C. D., Musick M. D. and Natan M. J. (1996) Two-dimensional arrays of colloidal gold particles: A flexible approach to macroscopic metal surfaces, *Langmuir* **12** 2353

49. Oldenburg S. J., Averitt R. D., Westscott S. L. and Halas N. J. (1998) Nanoengineering of optical resonances, *Chem. Phys. Lett.* **288** 243-247

50. Weller H. (1993) Colloidal semiconductor Q-particles – Chemistry in the transition region between solid-state and molecules, *Angew. Chem. Int. Edit.* **32** 41-53

51. Claeson T. (1992) Single electron tunneling, Single Electronicslikharev KK, *Sci. Am.* **266** 80-85

52. Xia Y. N. and Whitesides G. M. (1998) Soft lithography, *Angew. Chem. Int. Edit.* **37** 551-575

53. Su Y. W., Wu C. S., Chen C. C. and Chen C. D. (2003) Fabrication of two-dimensional arrays of CdSe pillars using e-beam lithography and electrochemical deposition, *Adv. Mater.* **15** 49

54. Chou S.Y., Kraus P. R., and Restorm P. J. (1996) Imprint lithography with 25-nanometer resolution, *Science* **272** 85-87

55. Toprak M. S. (2003) Engineering nanostructures and Thermoelectric Nanomaterials, Doctoral Thesis, *Royal Institute of Technology*, Sweden

56. Toprak M. S, Kim D. K., Mikhaylova M. and Muhammed M. (2003) Multilayer structures by self-assembly, Nanomaterials for Structural Applications, *Mat. Res. Soc. Proc.* **740**

57. Libioulle L., Bietsch A., Schmid H., Michel B. and Delamarche E. (1999) Contact-inking stamps for microcontact printing of Alkanethiols on gold, *Langmuir* **15** 300-304

SIZE AND SHAPE OF EPITAXIAL NANOSTRUCTURES

Influence of Kinetics and Thermodynamics

R. STANLEY WILLIAMS
Hewlett-Packard Lab, 1501 Page Mill Road, Palo Alto, CA USA
GILBERTO MEDEIROS-RIBEIRO
Laboratório Nacional de Luz Síncrotron, Campinas Brazil

1. Introduction

One of the primary goals in nanomaterials science is to be able to design and grow structures with specific sizes, shapes and spatial orientations, since many active device applications that are anticipated for nanomaterials require a significant degree of control over their properties. An example of such an active device can actually be found in nature. Magnetotactic bacteria have developed the capability to grow and asssemble nanocrystals of magnetite inside themselves, which they then use as navigational aids for swimming along magnetic field lines in order to optimize their search for nutrients [1]. Each species of magnetotactic bacterium has its own specific crystal habit and size, which are very uniform within a particular species. Human beings have not yet developed the ability to grow and assemble nanostructures as well as magnetotactic bacteria, and so this example raises some significant questions. How do nanocrystals form in nature? Can humans design and grow nanocrystals for technological applications? How can we control the shape, size and the uniformity of nanocrystals? How stable are the different morphologies? How important are kinetic and thermodynamic effects in determining the final nanocrystal product?

This chapter will not supply definite answers to these questions. It will outline the present understanding of the factors that control the growth and evolution of ensembles of nanocrystals that are grown epitaxially on a single-crystal substrate and provide a guide to the literature or those interested in detail. Although island formation that results when one material is deposited on top of another has been observed and described many times in the past, interest in the nature of epitaxial nanocrystals was piqued considerably in 1990 by two reports that appeared describing islands of Ge that form on Si(001). In this system, both the overlayer and the substrate have the same diamond crystal structure, but the Ge has ~4.2% larger lattice constant than the Si. The first three or so monolarers of Ge that are deposited on Si(001) form a coherently strained overlayer, where the Ge is forced to adopt the lattice constants of the substrate

T. Tsakalakos, et al. (eds) pgs 81 - 93
Nanostructures: Synthesis, Functional Properties and Applications;
© *2003 Kluwer Academic Publishers.*

in the plane of the film. However, the deposition of additional Ge results in the growth of islands on top of the strained overlayer. Eaglesham and Cerullo [2] observed coherently strained (defect free) dome-shaped nanocrystals using Transmission Electron Microscopy (TEM). Mo and Lagally [3] observed 'hut'-shaped nanocrystals using Scanning Tunneling Microscopy (STM) for similar deposited Ge amounts and substrate temperatures. Later, many other researchers demonstrated that both the hut and dome shaped islands could coexist on a surface [4], and that in fact the huts were smaller in volume and therefore were the precursor to the domes [5] (see Fig. 1). Further studies showed that the elongated huts of [3] were metastable, with the more stable shape being a square-based pyramid [6] bounded by {105} facets.

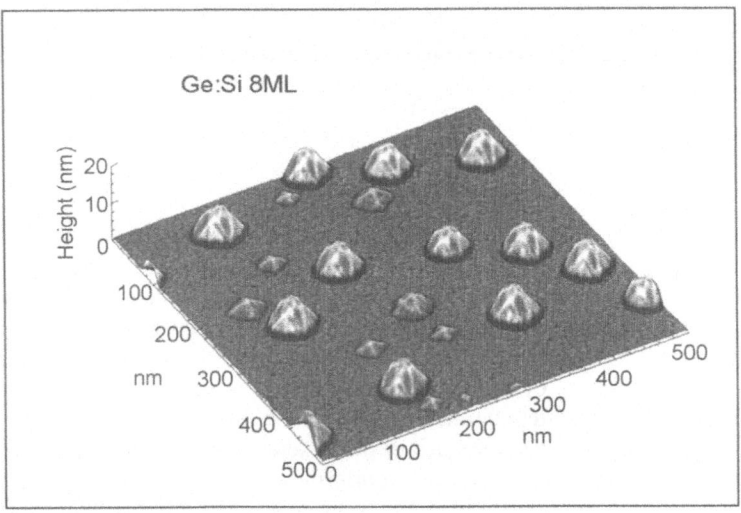

Figure 1. An *in situ* STM topograph of a Si(001) surface,onto which 8 equivalent monolayers of Ge were deposited by physical vapor deposition at a substrate temperature of 600°C [5].

Although many other systems display similar behavior, this chapter will utilize only examples of strained epitaxial nanocrystals on Si(001) to illustrate the concepts that will be discussed. To set the stage, brief introductions to the kinetics of size control, Ostwald ripening, and the thermodynamics of strained nanocrystals will be presented. The types of nanocrystal size distribution that result from each of these limiting behaviors will be described and compared to experimental results in order to obtain insight into the Ge on Si(001) system. The influence of a surfactant on the size and shape of Ge nanocrystals on Si(001) will be briefly described, as well as a strategy for utilizing broken symmetry to control the nanocrystal shape. Finally, strategies for controlling size and shape of epitaxial nanocrystals via both kinetic and thermodyanamic means will be proposed.

2. Kinetic Control of Size Distributions

In 1951, Reiss published a landmark paper [7] in which he showed that with the proper control of solution kinetics, it was possible to grow a "uniform colloidal dispersion," e.g. a colloid in which all of the particles were nearly the same size. This process is commonly used now to grow uniform micron-sized polymer spheres for use as calibration standards and in experiments on colloidal crystals. There were two key insights in this paper. The first was that the process of diffusion created a gradient of monomers in a spherical region around a growing particle in solution, and the resulting growth rate for a single isolated particle would have the general form

$$\partial r/\partial t = \alpha/r^n ,$$ (1)

where r is the radius of the growing particle, α is a constant that is related to the background concentration of monomer and the diffusion constant, and n depends on the specific growth mechanism of the particle. The key is that in general, larger particles will grow slower than smaller particles as long as the background concentration of monomers is larger than the equilibrium concentration (kinetic control), and thus the smaller particles will catch up in size to those that nucleated earlier. The second insight is that by limiting the time during which particles can nucleate, the focusing property of the growth law Eq. 1 will cause the *relative* size distribution of the particles to become narrower with time. In order for Eq. 1 to remain valid, the concentration of the growing particles should be low enough so that they do not interact with each other and the monomer concentration should be higher than the equilibrium value. When the desired particle size is reached, the system can be quenched to prevent further growth or broadening of the distribution.

Ngo and Williams [8] applied this idea to the case of three-dimensional particles growing on a two-dimensional surface. In order to limit the nucleation time, they considered the case of heterogeneous nucleation, where the islands grow on defects or other special sites on a surface, which gives rise to the nucleation rate law in Eq. 2,

$$\partial N/\partial t = k(N_\infty - N[t]) = kN_\infty \exp[-kt],$$ (2)

which is known as the declining nucleation model, where $N[t]$ is the surface density of particles as a function of time, N_∞ is the area density of sites on which islands can grow, and k is the particle nucleation rate, which depends on the monomer surface density and diffusion constant. In this case, the nucleation rate decreases exponentially with time, which provides a temporal limit for the nucleation. The time variable can be eliminated between Eqs. 1 and 2 to yield the size distribution for the particles. For the specific case of n = 3 in Eq. 1, which is commonly observed experimentally, the *volume* distribution of particles on a surface becomes

$$\partial \tilde{N}\partial V = bN_\infty \exp[b(V-V_{max})],$$ (3)

where V is the volume of the particles, V_{max} is the volume of the largest particle, and b is a geometrical factor times k/α. Thus, the volume distribution function increases exponentially with V up to the value of the largest particles, which were those that nucleated first. This simple distribution does not account for fluctuations in local growth rates that result from statistical variations in arrival flux of material on the surface, etc., so that a real distribution should be somewhat broadened with respect to Eq. 3. However, one issue that should be common to nearly any kinetic size distribution is the long tail to smaller sizes, e.g. the negative skewness of the disribution. Although the relative standard deviation of this distribution decreases with growth time, it converges rather slowly [8], since the absolute width of the volume distribution is constant, Fig. 2 (the absolute *radial* distribution does get narrower, as expected from Eq. 1!). Thus, this procedure for growing islands on an array of predefined nucleation sites may produce very uniform micron-scale particles, but it will not likely yield a very narrow distribution of nanoparticles.

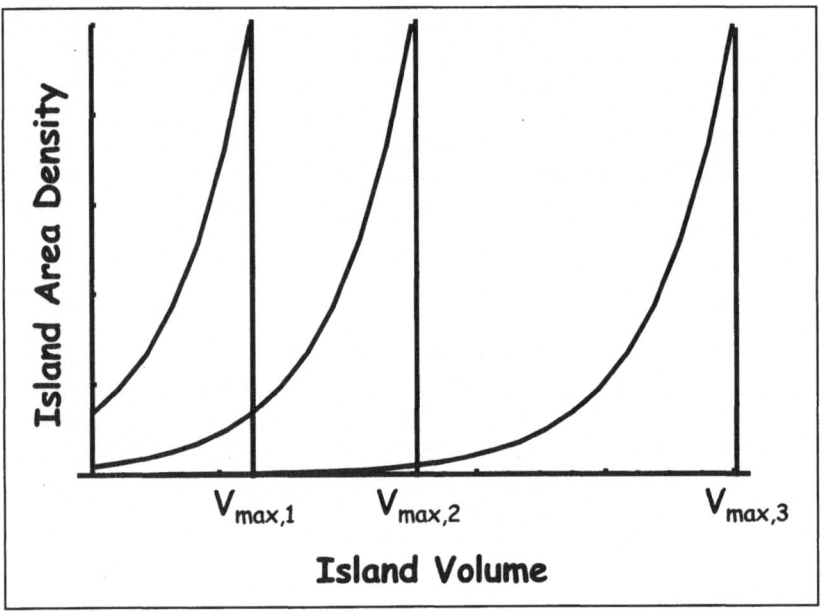

Figure 2. The island volume distribution at various times for growth controlled by kinetics, where the island nucleation is determined by Eq. 2 (declining nucleation model) and the growth rate is determined by Eq. 1 with n = 3.

Gorer and Penner [9] improved on this process with a non-uniform temporal supply of atoms to a set of growing islands. First, they created nucleation sites on a graphite substrate with an STM tip. Then, they deposited metal onto the surface electrochemically. By pulsing the supply of atoms to the surface, they were able to grow a narrow distribution of nano-island sizes. This strategy works because by pulsing the supply of atoms, the density of free atoms on the surface is kept below that required for nucleation of new islands. However, the supply of atoms is kept above the equilibrium atomic density for the ensemble of growing islands. Independently, Alivisatos et al. [10] discovered that by pulsing reactants into a flask, they were able to synthesize a narrow range of nano-particles in solution.

3. Ostwald Ripening

Ostwald ripening is a familiar process in which an ensemble of particles will coarsen with time, e.g. larger particles in the distribution will grow at the expense of smaller particles. An excellent review of Ostwald ripening by Zinke-Allmang et al. [11] already exists, so the coverage here is brief. A significant mathematical formulation of Ostwald ripening was due to Lifshitz and Slyozov [12]. Their realization was that in an ensemble of aging particles, the background concentration of atoms or monomers will eventually fall to the equilibrium concentration of the *ensemble*. At this time, the particles compete with each other for atoms from the surroundings. Particles that have a higher effective vapor pressure or activity, e.g. a smaller radius, will lose atoms to particles with a larger radius. At any given time, there will be particles with a critical radius r_C that neither grow nor shrink. This idea is embodied in Eq. 4 by incorporating r_C into Eq. 1 to obtain the more general form,

$$\partial r / \partial t = -\beta / r^m (1/r - 1/r_C), \tag{4}$$

where m depends on the dimensionality of the system and the specific mechanism for particle growth and r_C is a function of time. Under various symmetrizing assumptions, such as mean field theory, this equation can be solved to yield the temporal dependence of the size distribution of the ripening ensemble of particles. Chakraverty [13] extended these ideas to the specific case of islands on a surface, and the functional form for the volume distribution is as follows:

$$f(\mathbf{V}, t) = (3/(3 + \mathbf{V}^{1/3})^{7/3} (-3/2(\mathbf{V}^{1/3} - 3/2))^{11/3} \exp[\mathbf{V}^{1/3} / (\mathbf{V}^{1/3} - 3/2)] / (1 + t/\tau)^{4/3}, \tag{5a}$$

$$\mathbf{V} = V/V_C(t), \text{ and} \tag{5b}$$

$$V_C(t) = V_0(1 + t/\tau), \tag{5c}$$

where V_0 and τ are the characteristic island volume and time constant, respectively, for the system of interest. A distinguishing characteristic of the Chakraverty distribution is the fact that in a manner similar to a strictly kinetic growth process, the skewness of the size distribution is negative. Both the mean island volume and the width of the volume distribution increase linearly with time, as illustrated in Fig. 3.

Kamins and Williams [14] showed that the same type of behavior occurs for polygonal islands, given the usual types of assumptions about surface atom diffusion, attachment at straight edges and de-attachment at corners. They also presented a simple numerical model that could easily be modified to examine the influence of kinetic barriers on ripening.

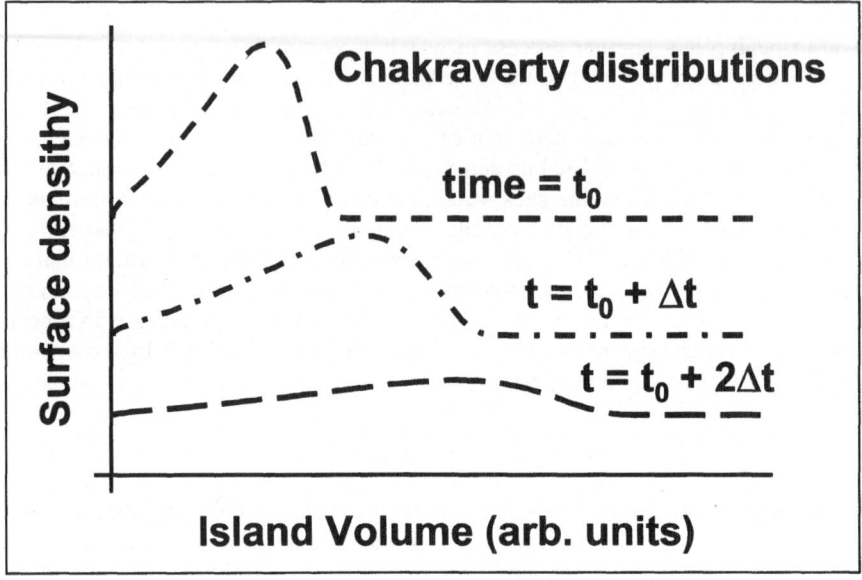

Figure 3. An illustration of the time evolution of a Chakraverty volume distribution of 3D ripening particles on a surface.

4. Thermodynamic Stabilization

After asking the interesting question "What is the solid-state analogy of a micelle?" Whetten and Gelbart [15] published a very general statistical mechanical treatment of the stabilizing influence of surfactants on the size and shape of nanocrystals growing in solution. The ideas presented in this paper were soon put to the test by Leff et al. [16], who showed that very narrow size distributions of Au nanoparticles could be grown in solution in the presence of a surfactant, and furthermore that the characteristic sizes of the Au nanoparticles depended on the ratio of surfactant to Au salt, as expected from the equilibrium theory. Simultaneously, Shchukin et al. [17] asked similar questions with regard to the growth of strained epitaxial nanocrystals on a single crystal surface. Surprisingly, they showed that in a narrow range of structural conditions, the surfaces of the strained nanocrystals could be stabilizing rather than destabilizing, as is usually assumed in nucleation theory and for Ostwald ripening. A simple (and only mildly incorrect) way to view this is that the surface free energy of certain stabilizing facets on strained epitaxial nanocrystals is lower than on the surface of the substrate or the strained layer on the substrate, which is used as the reference energy of the system, and is thus negative. This makes a close analogy with the case of surfactants on nanocrystals in solution. The nanocrystal edges contribute a positive or destabilizing term to the free energy, which becomes substantial and dominates at very small sizes. However, in the case of epitaxial nanocrystals, Shchukin realized that there is an additional complication; the strain fields in the substrate around each of the nanocrystals repel each other, and thus there is another positive energy contribution to the free energy of the ensemble of islands that depends on the amount of material present. Since the nanocrystals interact with each other, this is a non-ideal thermodynamic system.

Daruka and Barabasi [18] realized that the interplay of the negative stabilizing energy of facets and the positive repulsive energy of the nanocrystals would yield a fairly complex behavior, and they showed that depending on the lattice mismatch and the amount of material deposited, the Shchukin theory could display a wide range of nanocrystal growth modes. Williams et al. [19] simplified the Shchukin theory somewhat, but then added an explicit entropy term to account for finite temperature effects to construct a thermodynamic "shape diagram" to model the Ge on Si(001) system. This simplification allowed an understanding of the nanocrystal shape transition and the coexistence of different morphologies that withstand long annealing treatments [6]. The basis of the model is indicated schematically in Figure 4.

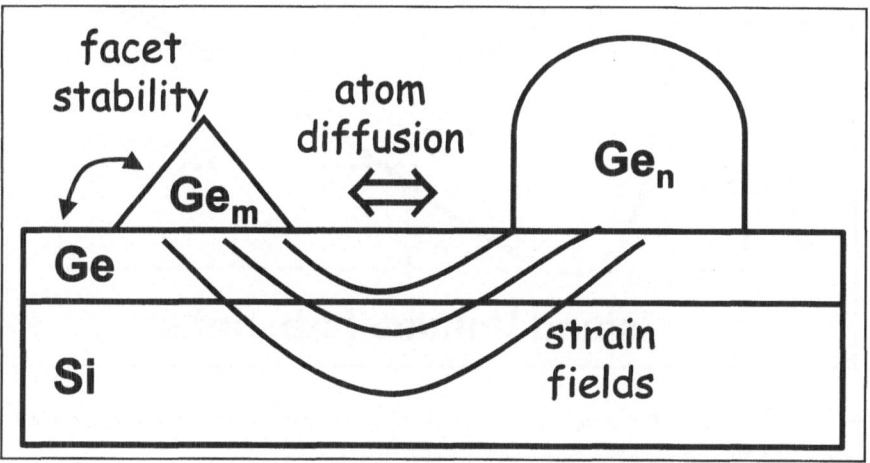

Figure 4. A schematic representation of the thermodymanic model for strained epitaxial nanocrystals growing on a single crystal substrate. The islands communicate with each other both via adatom diffusion along the surface and also via the strain fields that propagate through the substrate. The interaction of the stabilizing effect of nanocrystal facets, the mutual repulsion of islands, and the fact that more than one shape of nanocrystal may be stabilized yields a rich and complex system.

In this model, the energy of the strained islands as a function of their volume V is parameterized in the following fashion:

$$E(V) = AV + BV^{2/3} + CV^{1/3}, \text{ or} \tag{6a}$$
$$E(V)/V = A + BV^{-1/3} + CV^{-2/3}, \tag{6b}$$

where A is a parameter that characterizes the energy of atoms in the 'bulk' of the nanocrystal, B is the surface energy, and C is the edge energy of the nanocrystals. One of Shchukin's significant realizations was that the B parameter can be a negative number under certain restricted conditions (see, for instance Raiteri *et al.* [20]). When looking at the free energy of an *ensemble* of nanocrystals, one must examine the free

88

energy per atom or per volume of the nanocrystals. This normalized free energy of Eq. 6b is the appropriate thermodynamic function for the system, and if the parameter B is negative, the resulting free energy for a particular shape of nanocrystal will have a minimum value (see Fig. 5). Thus, there will be a most stable nanocrystal size, and if an ensemble of nanocrystals have grown at a finite temperature, there will be a width to the size distribution that depends on the curvature of the free energy near the minimum and the growth temperature, essentially a Boltzmann broadening of the nanocrystal sizes [5], as illustrated by the dotted line in Fig. 5.

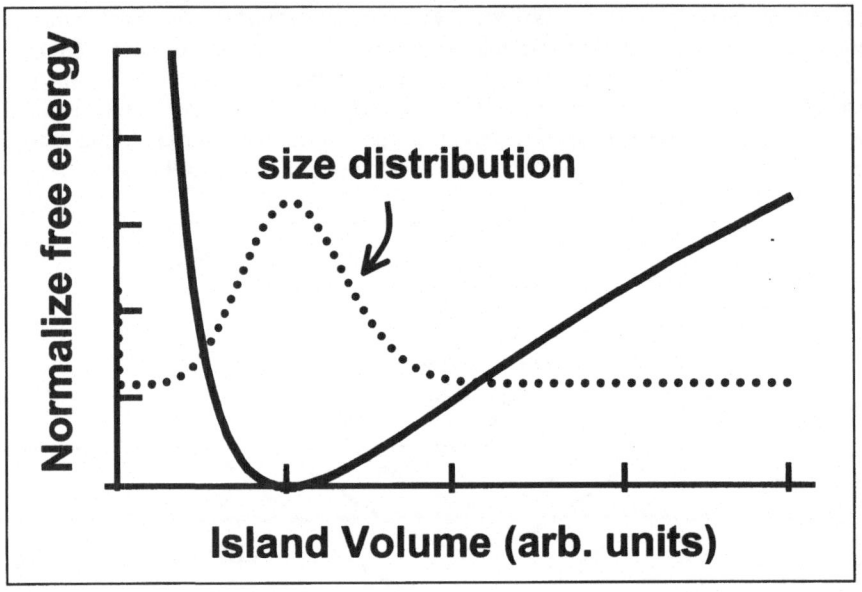

Figure 5. The normalized free energy for an ensemble of strained epitaxial nanocrystals (Eq. 6a) and the resulting volume distribution for growth at a finite temperature.

The purpose of a thermodynamic model is to be able to compute as many properties of a system using as few parameters as possible. This does not mean prediction – rather the intent is data compression. The idea is to perform a sufficient number of experiments to be able to reliably determine a set of parameters, and then to use those parameters to compute the properties of the system for an arbitrary set of conditions and obtain those properties with an accuracy equivalent to what would have been achieved if they had been measured. A sufficiently good thermodynamic model allows one to design a system that is optimized given the physical constraints of the system. In Ref. 21, a thermodynamic model in which the parameters B and C in Eq. 6 are considered to be linear functions of the total amount of Ge deposited on a surface has been presented and discussed at length.

5. Experimental Results for Ge nano-Islands on Si(001)

There are a very large number of experimental investigations of the Ge on Si(001) system, and it is still an extremely popular system to study so the number of publications is increasing rapidly. Williams et al. have provided a brief review of the system up to the year 2000 [21]. Most early authors argued that the growth of strained Ge nanocrystals was a kinetic process, with the relatively narrow size distributions that were observed being the result of kinetic barriers. However, the behavior of this system is very complicated, and attention must be paid to the actual growth conditions to be able to decide if growth is thermodynamic or kinetic and what kinetic effects are at play – ripening, alloying or slow diffusion of Ge. The possibilities are shown schematically in Figure 6.

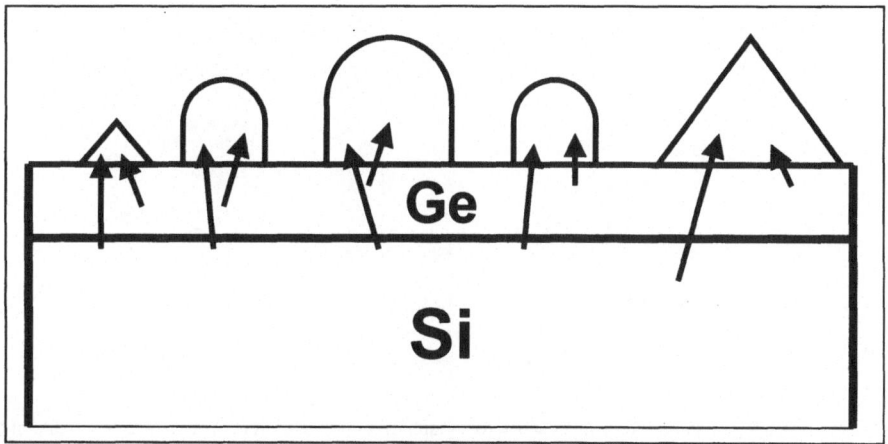

Figure 6. A schematic representation of various kinetic processes that influence or limit the growth of epitaxial nanocrystals. Two that were not considered in early analyses were slow diffusion of Ge at low temperatures, causing the wetting layer to overgrow at the expense of islands, and fast diffusion of Si at high temperatures, causing the nanocrystals to alloy. This can even lead to a reverse shape transition from domes with a low Si content to pyramids with a high Si content.

Most of the early analyses that led researchers to conclude that Ge on Si(001) was a kinetically controlled system were qualitative. This interpretation found its strongest representation in the work of Ross et al. [22], who performed *in situ* transmission electron microscopy studies of Ge island growth and compared their results to a model of anomalous ripening, e.g. two different ripening distributions that cross over each other as a function of increasing size such that the initial shape is pyramids and then switches over to domes. However, one significant problem with the experiments and analysis performed by those who concluded that Ge on Si(001) was a kinetic system was the fact that they utilized growth temperatures of 650°C or higher. Several groups have now shown that at these temperatures, there is significant diffusion of Si into the Ge islands that results in alloy formation [23,24]. This in turn causes the lattice mismatch to change, i.e. to become smaller, as growth proceeds and thus shifts the characteristic size of the nanocrystals to larger volumes. In fact, with sufficient annealing, a new shape transition is seen in which domes transform back to pyramids as the fraction of Si in the alloy becomes large enough. Thus, there certainly were kinetic effects at work, but they were not the ones usually identified as being responsible for nanocrystal evolution.

However, when careful attention is paid to the growth conditions to limit Si diffusion and alloying, the detailed volume distributions of the Ge nanocrystals as functions of amount of Ge deposited and growth temperature can be fit extremely well to the simple thermodymanic model [21], as shown in Figure 7.

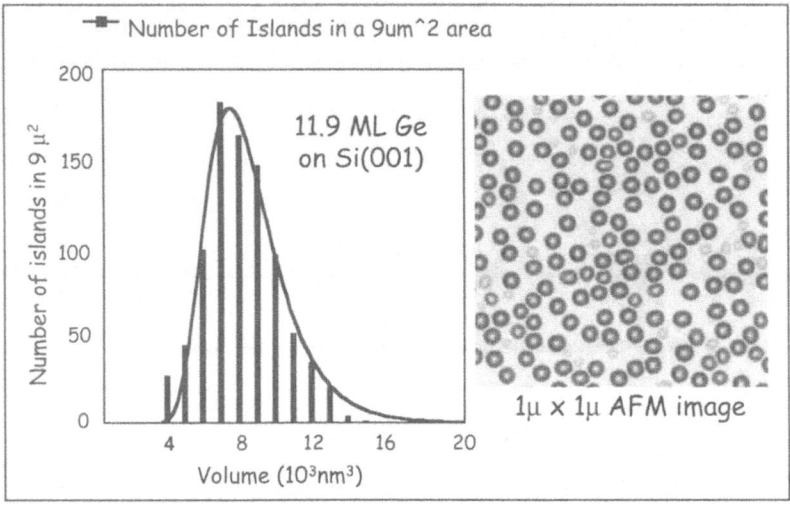

Figure 7. Right, an atomic force microscope topograph of a one square micron area of an ensemble of Ge islands on Si(001) grown by chemical vapor deposition at 600°C. Left, a comparison of the dome volume distribution measured from AFM topographs with a fit to the thermodynamic model Ref. 21. This agreement between a measured distribution and the thermodynamic model is typical. Note particularly the positive skewness of the distribution.

6. Influence of Surfactants

Recalling that control of size and shape of nanocrystals in solution is obtained with the use of surfactants [15], one can inquire into the effect of adding a surfactant to a system of epitaxial islands. The first such study by Kopel et al. [25] utilized As as a dopant/surfactant on a growing surface of Ge on Si(001). In this case, island formation was suppressed completely, which at that time was considered a major victory, since the major emphasis then was growing uniform films of arbitrary thickness. Recently, Kamins et al. [26] have looked at the influence of P as the dopant/surfactant on thin Ge layers on Si (001). In this case, they observed that an entirely new set of island shapes and sizes resulted from the growth. There were still different characteristic shapes, but they were different from the case of neat Ge growth and they were also smaller. In this case, it appears that the P stabilizes a different set of nanocrystal facets, in particular (111)-type facets, to produce a new set of nanocrystal shapes with smaller characteristic volumes.

7. Symmetry Breaking

Another consideration for epitaxial nanocrystals is the symmetry of the crystal lattice of the nanocrystal compared to that of the substrate on which they are grown. Nearly all of the early systems that were studied were similar to Ge on Si in that both overlayer and substrate had the same crystal structure, and therefore the same symmetry. This leads to nanocrystals with a high degree of symmetry, such as the four-fold symmetric pyramids and domes of Ge on Si(001). However, it should be possible to choose overlayer materials with a different symmetry from the substrate to stabilize different nanocrystal shapes. A significant illustration of this principle was supplied by Chen et al., [27, 28] who deposited submonolayer amounts of rare earth metals such as Er on a Si(001) surface and heated to form reaction products. The most stable silicides of many rare earth metals have a hexagonal lattice structure in which one axis matches that of Si along a [110] axis to within 2% and the perpendicular axis has a lattice mismatch of greater than 4%. In such cases, the silicide nanocrystals that grow as a result of the reaction between the rare earth and the substrate are limited to nanometer dimensions along the axis of large lattice mismatch but can grow to many microns without dislocations in the direction of small mismatch. If the growth conditions are carefully controlled, the resulting nanowires are perfectly straight, because they are oriented along crystal axes of the substrate, and they contain only point defects, e.g. surface vacancies, or edge kinks. Long time or high temperature anneals are required to form misfit dislocations in these nanowires [28].

8. Kinetic and Thermodynamic Control of Epitaxial Nanocrystals

In this chapter, we have seen that there are both kinetic and thermodynamic means to influence the size distributions of strained epitaxial nanocrystals. The details of the materials systems and the constraints they impose limit the actual size distributions that can be obtained in practice. However, we have seen how to guide or adjust the distributions that are achieved.

The details of how material is deposited can be used to kinetically steer a distribution. The first requirement for obtaining a narrow or a regularly spaced nanocrystal distribution is to create the nuclei for the nanocrystals by some means, for example by lithographically defining a set of nuclei, such as demonstrated by Kamins et al. [29]. Subsequent growth of nanocrystals should be performed by depositing material onto the surface to maintain the background concentration of diffusing atoms higher than the equilibrium value for the ensemble to prevent Ostwald ripening but lower than the density required to homogeneously nucleate new islands, if these two concentrations do in fact have the appropriate relation to each other. Experience has shown that it is best to pulse the depositing species to focus the kinetic distributions and achieve the narrowest distributions [9].

Thermodynamic control of nanocrystal growth is illustrated schematically in the diagram of Figure 8.

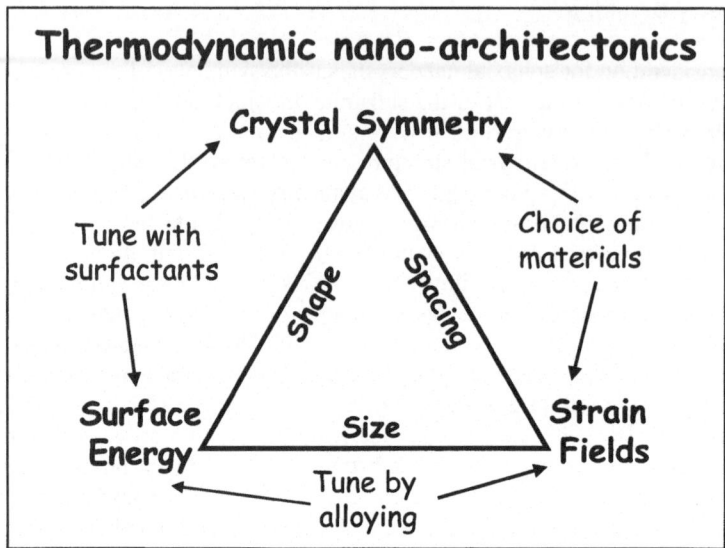

Figure 8. A schematic diagram for designing an ensemble of epitaxial nanocrystals within a thermodynamically stable system. One desires to control the size, shape and the spatial orientation of the nanocrystals. However, thermodynamics does not provide independent control over these properties. Crystal symmetry, strain fields and surface energies are the relevant materials properties that must be controlled to create a particular distribution. However, these are controlled by the choice of materials (and this is usually a significant constraint), the ability to adjust lattice constants by alloying if possible, and the use of surfactants.

The materials involved must have the appropriate set of properties to cause a stabilization of surface facets, which then leads to a local minimum in the normalized free energy for the nanocrystals. In practice, this may be a fairly rare occurrence, since the materials properties have to be fine-tuned. On the other hand, it may be that if a stabilization can occur, many systems will find the appropriate set of crystal facets to achieve nanocrystal stabilization. Once such a system is identified, the nanocrystal sizes and shapes are limited by the minima in the free energy curves, and the widths of the distributions are limited by the curvature of the free energies near their minima and the growth temperatures. However, there is the possibility to adjust the surface energies of different facets by the use of surfactants, which can stabilize different shapes and make the widths of the distributions narrower. Finally, one does have the choice of symmetry by choosing the appropriate substrate surface and overlayer material along with the lattice mismatches in order to stabilize particular shapes.

9. References

1. Frankel, R.B., Blakemore, R.P., and Wolf, R.S. (1979) Magnetite in Freshwater Magnetotactic Bacteria, *Science* **203**, 1355-56.
2. Eaglesham, D.J. and Cerullo, M. (1990) Dislocation-Free Stranski-Krastanow Growth of Ge on Si(100), *Phys. Rev. Lett.* **64**, 1943-46.
3. Mo, Y.W., Savage, D.E., Swartzentruber, B.S., and Lagally, M.G. (1990) Kinetic Pathway in Stranski-Krastanov Growth of Ge on Si(001), *Phys. Rev. Lett.* **65**, 1020-23.

4. Tomotori M., Watanabe, K., Kobayashi, M., and Nishikawa, O. (1994) STM Study of the Ge Growth Mode on Si(001) Substrates, *Appl. Surf. Sci.* **76/77**, 322-28.

5. Medeiros-Ribeiro, G., Bratkovski, A.M., Kamins, T.I., Ohlberg, D.A.A., and Williams, R.S. (1998) Shape transition of germanium nanocrystals on a silicon (001) surface from pyramids to domes, *Science* **279**, 353-55.

6. Medeiros-Ribeiro, G., Kamins, T.I., Ohlberg, D.A.A., and Williams, R.S. (1998) Annealing of Ge nanocrystals on Si(001) at 550°C: Metastability of huts and the stability of pyramids and domes, *Phys. Rev. B* **58**, 3533-36.

7. Reiss, H. (1951) The Growth of Uniform Colloidal Dispersions, *J. Chem. Phys.* **19**, 482-87.

8. Ngo, T.T. and Williams, R.S. (1995) Kinetic routes to the growth of monodisperse islands, *Appl. Phys. Lett.* **66**, 1906-08.

9. Gorer S. and Penner R.M. (1999) "Multipulse" electrochemical/chemical synthesis of CdS/S core/shell nanocrystals exhibiting ultranarrow photoluminescence emission lines, *J. Phys. Chem B* **103**, 5750-53.

10. Peng, X.G., Wickham, J., and Alivisatos, A.P. (1998) Kinetics of II-VI and III-V colloidal semiconductor nanocrystal growth: "Focusing" of size distributions, *JACS* **120**, 5343-44.

11. Zinke-Allmang, M., Feldman, L.C., and Grabow, M.H. (1992) Clustering on surfaces, *Surf. Sci. Repts.* **16**, 377-463.

12. Lifshitz, I.M. and Slyozov, V.V. (1961) The kinetics of precipitation from supersaturated solid solutions, *J. Phys. Chem. Solids* **19**, 35-50.

13. Chakraverty, B.K. (1967) Grain Size Distribution in Thin Films – 1. Conservative Systems, *J. Phys. Chem Solids* **28**, 2401-12.

14. Kamins, T.I. and Williams, R.S. (1998) A model for size evolution of pyramidal Ge islands on Si(001) during annealing, *Surf. Sci.* **405**, L580-86.

15. Whetten, R.L. and Gelbart, W.M. (1994) Nanocrystal Microemulsions – Surfactant-Stabilized Size and Shape, *J. Phys. Chem.* **98**, 3544-49.

16. Leff, D.V., Ohara, P.C., Heath, J.R., Gelbart, W.M. (1995) Thermodynamic Control of Gold Nanocrystal Size - Experiment and Theory, *J. Phys. Chem.* **99**, 7036-41.

17. Shchukin, V.A., Ledentsov, N.N., Kop'ev, P.S., and Bimberg, D. (1995) Spontaneous Ordering of Arrays of Coherent Strained Islands, *Phys. Rev. Lett.* **75**, 2968-71.

18. Daruka, I. and Barabasi, A.L. (1997) Dislocation-free island formation in heteroepitaxial growth: A study at equilibrium, *Phys. Rev. Lett.* **79**, 3708-11.

19. Williams, R.S., Medeiros-Ribeiro, G., Kamins, T.I., and Ohlberg, D.A.A. (1998) Equilibrium Shape Diagram for Strained Ge Nanocrystals on Si(001), *J. Phys. Chem B* **102**, 9605-09.

20. Raiteri P., Migas, D.B., Miglio, L., *et al.* (2002) Critical role of the surface reconstruction in the thermodynamic stability of {105} Ge pyramids on Si(001), *Phys. Rev. Lett.* **88**, 256103.

21. Williams, R.S., Medeiros-Ribeiro, G., Kamins, T.I., and Ohlberg, D.A.A. (2000) Thermodynamics of the Size and Shape of Nanocrystals: Epitaxial Ge on Si(001), *Annu. Rev. Phys. Chem.* **51**, 527-51.

22. Ross, F.M., Tersoff, J., and Tromp, R.M. (1998) Coarsening of self-assembled Ge quantum dots on Si(001), *Phys. Rev. Lett.* **80**, 984-87.

23. Kamins, T.I., Medeiros-Ribeiro, G., Ohlberg, D.A.A., and Williams, R.S. (1998) Dome-to-pyramid transition induced by alloying of Ge islands on Si(001), *Appl. Phys. A – Mater.* **67**, 727-30.

24. Chaparro, S.A., Zhang, Y., Drucker, J., Chandrasekhar, D. and Smith, D.J. (2000) Evolution of Ge/Si(100) islands: Island size and temperature dependence, *J. Appl. Phys.* **87**, 2245-54.

25. Copel, M., Reuter, M.C., Kaxiras, E. and Tromp, R.M. (1989) Surfactants in Epitaxial Growth, *Phys. Rev. Lett.* **63**, 632-35.

26. Kamins, T.I., Ohlberg, D.A.A., Williams, R.S. (2001) Effect of phosphorus on Ge/Si(001) island formation, *Appl. Phys. Lett.* **78**, 2220-22.

27. Chen, Y., Ohlberg, D.A.A., Medeiros-Ribeiro, G., Chang, Y.A., and Williams, R.S. (2000) Self-assembled growth of epitaxial erbium disilicide nanowires on Silicon (001), *Appl. Phys. Lett.* **76**, 4004-06.

28. Chen, Y., Ohlberg, D.A.A., Medeiros-Ribeiro, G., Chang, Y.A., and Williams, R.S. (2002) Growth and evolution of epitaxial erbium disilicide nanowires on Si (001), *Appl. Phys. A.* **75**, 353-61.

29. Kamins, T.I., Ohlberg, D.A.A., Williams, R.S., Zhang, W., and Chou, S.Y. (1999) Positioning of self-assembled, single crystal, germanium islands by silicon nanoimprinting, *Appl. Phys. Lett.* **74**, 1773-75.

MOLECULARLY ASSEMBLED INTERFACES FOR NANOMACHINES

VLADIMIR V. TSUKRUK
Materials Science and Engineering Department, Iowa State University, Ames, IA 50011, USA

Abstract. State-of-the-art development of the field of ultrathin organic and polymeric molecular coatings is briefly reviewed from the prospective of long-term applications for microelectromechanical systems, microfluidic devices, and microanalytical instrumentation. A possible evolution of the field of protective/lubricative coatings towards intelligent (responsive, sensing, self-repairing and self-reinforcing) molecular interfacial assemblies acting as an adaptive nanointerface between core elements of nanodevices and molecular environment is discussed.

1. Introduction

The field of organic and polymeric coatings is traditionally focused on the fabrication of relatively thick and robust films for surface protection of large-scale parts and devices. Classical examples are automotive paints and lubrication layers [1,2]. In these kinds of applications, surface coatings act as passive (although, very complicated) and protective layers that, among many other functions, can inhibit electrochemical reactions, prevent direct physical contact of metal surfaces, distribute mechanical stresses, and dissipate mechanical/thermal/light energy.

Modern developments in the field of microscale devices and their prospective applications ignite new activities in the coating fabrication and study that go far beyond their traditional role [3-5]. Current developments in this field include such topics as the fabrication of nanometer thick lubricant protection of high-density magnetic discs and microelectronic packaging of electronic circuits [3]. New organic and polymeric coatings shield electromagnetic field in microelectronic packaging, selectively reflect light in antireflective coatings, change hydrophobic/hydrophilic surface balance in antifogging layers, and reduce friction/adhesion on micromotor surfaces [3-8]. These recent developments explore new ways of how active, "built-in", molecular functions of molecular materials can be exploited for the functional molecular coating fabrication. For example, selective adsorption or scattering of a specific wavelength by

T. Tsakalakos, et al. (eds) pgs. 95 - 109
Nanostructures: Synthesis, Functional Properties and Applications;
© 2003 Kluwer Academic Publishers.

molecules/nanoparticles can be used as an active element of antireflective layers [9]. Polar/non-polar balance and molecular flexibility of long chain molecules with reactive terminal groups can be used in the fabrication of robust grafted lubrication layers [8]. Current and future developments in device minituarization and the creation of molecular-based machines and mechanisms will require very different kinds of protective/interactive interfaces.

In this paper, we focus on one particular development, as we see it, namely, molecularly assembled interfaces for nanotribological applications. These interfacial nanoassemblies can be designed to serve as a sophisticated buffer that controls molecular-scale mechanical and energetic interactions between nanoscale-sized mating and interacting objects.

Current developments that are directly related to the topic discussed here include molecular protective coatings for microelectromechanical systems (MEMS), where micrometer-sized parts, fabricated from solids like silicon, mechanically interact with each other in the course of their sliding, vibration, rotation, and stop-go motion [3-5]. Another example is microfluidic systems (MFS), where microfluid is a subject of constrained flow within microchannels that create significant shearing stresses at solid-liquid interfaces [10]. Finally, it is worth to note that we only focus on organic and polymer-based interfacial films, as applied to nanodevices and put aside other possible developments, such as hard coatings or new microfabrication routines, which can be instrumental for some nanoscale applications.

An efficient nanofluidic device provides flow control and precision mixing on a nanoscale. An efficient micro/nanoelectromechanical system (MEMS/NEMS) is capable of precise, nanoscale motion of nanogram parts over millions of cycles. These devices rely on very localized surface molecular processes at solid-solid and solid-liquid interfaces with predictable molecular behavior under shearing and compression stresses, selective chemical interactions, and conformational reorganizations under external stimuli [9-13].

Current developments in the MEMS field include the design of new sophisticated micromotors, turbines, comb-like drivers, and pumps with increasingly smaller critical parts. For micrometer-sized parts, many elementary processes such as localized solid-solid surface interactions of movable parts or fluid-solid interactions occur at a scale well below 100 nm. Indeed, for a beam of 1 μm diameter interacting with a flat surface, the contact area for intermediate loads usually does not exceed 50 nm. Further decreasing of the dimensions of microdevice parts to less than 100 nm (200 nm dimension is a modern standard lithographic resolution) will result in significant shrinkage of interacting surface contact areas to less than several nanometers. System behavior in such a situation depends upon localized surface interactions, which involves only a few atomic groups, molecules, or unit cells.

Modern microdevices fabricated from a traditional electronic material (silicon) display constrained motion of microparts due to high surface energy Because of the brittleness of an oxide surface layer with nanograin surface morphology, they also have a short lifespan [11]. These problems will be even more critical for devices with nanoscale contact areas due to much higher local stress and significant topographical contribution. The high surface energy of oxide layers constitutes a problem for future "wet" applications. MEMS/NEMS operation in fluids (such as blood or chemical fluids in microreactors) is highly restrained and virtually impossible. The presence of chemically active surrounding fluids can lead to either undesired chemical alternation of surface chemical composition or build-up surface layer as a result of uncontrolled adsorption. Both of these processes can prevent microdevices from functioning under wet conditions (e.g., microimplants in a human body). Biocompatibility is another issue that requires special molecular "treatment" of solids surfaces to mediate interaction with biomolecules.

From a material prospective, there are two main directions in the development of micro- and nanodevices. First, silicon-based microdevices will continue to get smaller and their surfaces will be modified to make them reliable and "compatible" with varying environmental conditions. Secondly, new materials will be gradually introduced to replace silicon for many "wet" applications. In any case, one cannot expect that surface properties of materials important for nanoapplications will be a simple derivative of their bulk properties. On the contrary, special efforts in tuning specific properties via surface modification routines will be required. Currently, self-assembling monolayers (SAMs) are used for silicon surface modification [5, 6]. This permits the wide-ranging control of surface properties, from completely hydrophobic to completely hydrophilic with greatly reduced friction forces. The chemical attachment (grafting) of the polymers with functional groups in the form of polymer brushes is explored for other applications. These developments substantiate the idea that surface modification can be an effective means of controlling the molecular surface properties.

2. Current Developments

High surface energy of such surfaces is due to the high concentration of silanol (Si-OH) groups on a silicon oxide surface that results in extremely high hydrophilicity (low contact angle, close to $0°$ for clean surfaces). This, in turn, leads to complete surface wetting under normal air conditions and very strong capillary forces being developed between micrometer-sized parts. On the other hand, high brittleness of silicon and silicon oxides contributes to the fast deterioration of micromachined silicon surfaces being the subject of high local shear stresses. The microcracks propagate via intergrain grooves with high local friction, as can be seen from lateral/friction force microscopy images (Figure 1). Several ways to improve microtribological properties via the chemical modification of such devices were proposed and tested to date.

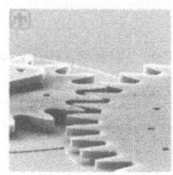 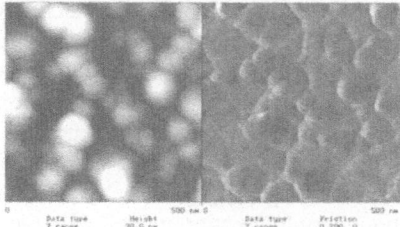

Figure 1. Left: SEM image of MEMS device (500x500 μm, microgears, Sandia National Lab, www.sandia.gov). Right: scanning probe microcopy images (500 x 500 nm) of the micromachined silicon surface of a microelectromechanical device that demonstrate nanograiny surface topography on topographical images (left) and uneven distribution of shearing forces on the lateral force image (right).

98

First approach includes the chemical treatment of the silicon oxide surface with hydrofluoric acid, which increases hydrophobicity of the treated surface by reducing the number of silanol surface groups and reducing oxide layer thickness [11]. After this treatment, the contact angle increases to 60-70°, this makes the surfaces partially wet and significantly reduces capillary interaction. Another approach explores self-assembling monolayers (SAMs) to modify surface properties [5,6,7,12]. Alkylsilane and thiol-based organic molecules with non-polar terminal groups can be used to fabricate SAMs on silicon oxide and gold surfaces respectively, and make these surfaces completely hydrophobic. Various versions of composite molecular layers with interesting nanotribological properties can be fabricated by polymer grafting to functionalized SAMs (see several examples in Figure 2) [12].

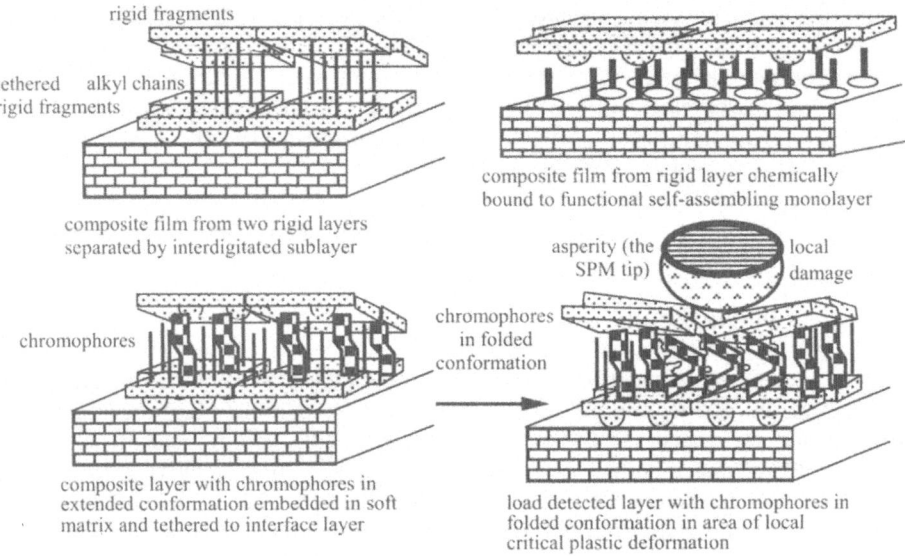

Figure 2. Several examples of nanocomposite molecular layers for surface modification.

Dense monomolecular layer chemically attached to the appropriate solid surface with functional terminal groups can serve as reactive interface for such build-ups (Figure 2). As the result of this fabrication, surface properties can be controlled in a wide range from completely hydrophobic (contact angle of 100-110°) with low capillary forces to partially/completely hydrophilic with high chemical reactivity [8, 12]. Interfacial shear strength of such modified surfaces decreases significantly due to the presence of thin molecular layer with low shear strength. As the result, the greatly reduced friction coefficient is usually observed. This approach, in fact, resembles classical boundary lubrication model, with the only difference being the firm chemical attachment of the organic monolayer to solid surface [2].

Finally, the chemical attachment (grafting) of the polymer layers onto the functionalized silicon surface has been explored in order to fabricate wear-resistant and super-elastic molecular coatings [13]. The latest development displayed that the tethered nanocomposite molecular coating can improve surface stability and enhance the shearing surface response (Figure 3).

Controlled binding to the surface with the presence of rigid polymer fragments or grafted elastic phase reinforced by the glassy nanodomains can significantly enhance wear resistance of such nanocomposite coatings in comparison with one-component low-molecular mass organic SAM layer [13]. These approaches have been introduced very recently and initially showed promising results, enhancing wear resistance and reducing surface friction [6-8, 12, 13].

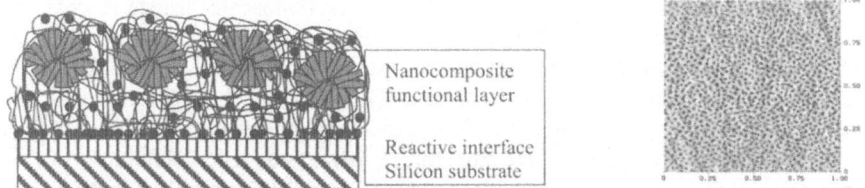

Figure 3. Nanocomposite/reinforced elastomeric layer chemically grafted to the silicon substrate via reactive interfacial SAM and bearing surface functional groups for further modification [13]: sketch on molecular structure (left) and (right) SPM image of 2D network of reinforcing nanodomains (<10 nm across).

Recently, we proposed another molecular design of well-defined nanoscale surfaces that include a highly elastic, reinforced rubber interlayer chemically grafted in-between a solid substrate and a hard top layer [14-16]. We suggested that such a combination will provide an effective mechanism for energy dissipation, facilitated by reversible elastic deformations of the chemically grafted/reinforced rubber matrix, enhanced by capping with a hard layer, which prevents the penetration of solid asperities through the compliant layer (Figure 4).

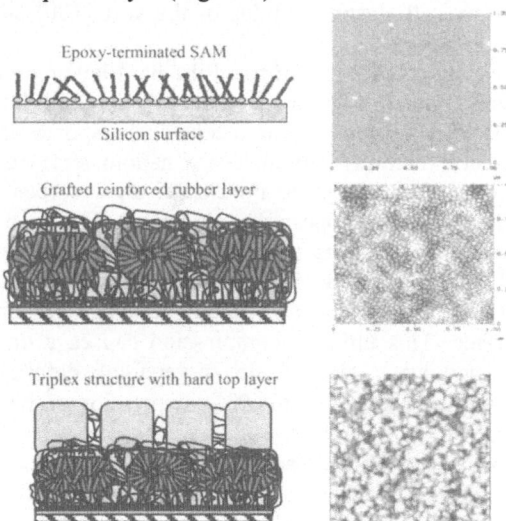

Figure 4. Left: assembling triplex nanocoating from top to bottom; right: corresponding AFM images of surface morphology. From Ref. 14.

Such "triplex", multilayered surface structures, with total thickness not exceeding several tens of a nanometer, can be fabricated via a combination of a directed multi-step self-assembly with UV-irradiation, as is described in detail elsewhere [16]. These triplex nanostructures demonstrated superior wear stability. Multilayered nanoscale structures such as these triplex coatings may be the logical next step in enhancing the reliability of microdevices via surface dissipation, relaxation, and healing.

Figure 4 shows the step-by-step assembly of the triplex surface nanolayers with a total thickness of 20-30 nm beginning with a bare silicon surface functionalized with epoxy-terminated SAM. Cartoons illustrate the microstructure of each layer, and scanning probe microscopy (SPM) images display the nanoscale surface morphology. This figure shows the molecularly smooth surface of epoxy-terminated alkylsilane SAM (0.7 nm thick) and the dense network of glassy nanodomains interlocked within the rubber layer (functionalized tri-block copolymer, about 9 nm thick). The fine, grainy surface microstructure of the rigid top photopolymerized nanolayer (methacrylate-based polymer, 5-20 nm thick) possesses a combination of hard nanodomains (<50 nm across) embedded in a less densely cross-linked matrix due to internal spatial heterogeneity of these polymer networks. The hard top layer was tethered to the compliant interlayer that, in turn, was grafted to the epoxy-terminated SAM. Such mutual interlayer tethering prevented the delamination of dissimilar molecular layers during large surface mechanical deformations. On the other hand, the preservation of the individual microstructures of the different layers with distinctive mechanical properties.

The micromechanical testing of each individual layer showed that the top hard layer possessed an elastic modulus of 2 GPa (typical for a hard plastic), while the rubber interlayer was highly compliant with an elastic modulus of 10 MPa (typical for a reinforced rubber). The elastic modulus of the epoxysilane SAM was estimated to be about 1 GPa. The overall elastic modulus of the stiff [100] silicon surface was 190 GPa.

The elastic response, indeed, confirmed the hard-layer/compliant layer/hard-layer mechanism at work, which was to be expected for well-defined triplex structure (Figure 5) [14,15]. When testing the nanomechanical response of this trilayer structure, very small surface deformations (within several nanometers) were initially carried out. Because these small deformations are almost entirely absorbed by the top hard layer, the composite elastic modulus appeared to be near 2 GPa. Upon further deformation, the rubber interlayer accepted the majority of the load and the "composite" elastic modulus dropped manifold (Figure 5). Finally, both compressed layers were squeezed and pushed against the less compliant bottom hard layer consisting of the SAM grafted to the silicon surface. This ultimate compression caused a dramatic increase of the apparent elastic modulus. Total elastic (reversible) deformation of the trilayer molecular structure was observed up to 60% compression of the initial thickness of the trilayer.

Wear resistance of the triplex coatings was tested for the contact of a steel ball and local pressures/velocities comparable with that of conventional MEMS operating conditions [16]. The variation of the apparent coefficient of friction as a function of the number of reciprocal sliding cycles (up to 20,000) was monitored. In this method of wear testing, a sharp increase of the friction forces indicates detrimental surface failure.

Also shown in this experiment for the purpose of comparison is the data for alkylsilane SAM, which is a common molecular lubricant for polysilicon surfaces of MEMS (Figure 5). At the low load of 0.3 N which corresponded to a local pressure of about 660MPa, both trilayer structure and SAM, showed excellent wear stability with the "apparent" friction coefficient being smaller for the trilayer surface structures (within 0.02-0.08 for both surfaces, which was several times lower than for bare silicon). Under severe loading conditions, the wear resistance mechanism was controlled by the ability of the surface to self-heal and restore itself, rather then by direct elastic resistance of the surface. All reference surfaces failed almost immediately (Figure 5). Alkylsilane SAM showed a higher wear resistance but failed after 900 cycles. Finally, the trilayer surface structure showed much higher wear stability, and was worn down only after 3,000-3,500 cycles due to the intensive thermo-oxidation occurring in the contact area.

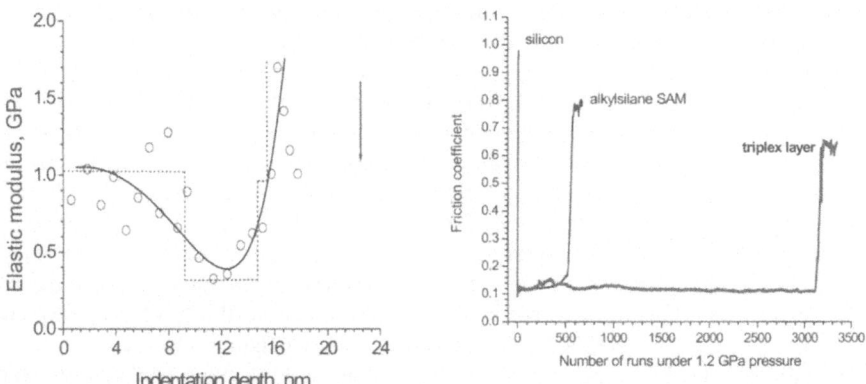

Figure 5. Depth profile of the elastic modulus of triplex nanocoating (left, arrows indicates the total thickness of the coating) and the wearing test of triplex nanocoating, silicon surface, and SAM coating under high normal load (right).

These results demonstrate that focused molecular design of the multilayer surface structures with built-in additional mechanisms for mechanical energy dissipation and modulated elastic resistance provides wear stability superior to that of conventional SAMs, which is the current choice for microdevices. This new materials technology can have a great impact on increasing the possibilities, and improving the operation in such microsystems. Chemical grafting through "wet chemistry" allows uniform coating of complicated 3D nanoscale topography with robust triplex films capable of huge localized deformations followed by restoring initial surface morphology (healing). We suggest that further optimization of molecular design focusing on the oxidation stability could result in significant progress towards the fabrication of nearly wearless surface nanoscale structures applicable for nanodevices with complex surface topography. However, further development is required to improve molecular coating performance for micro- and especially nano-scale devices and explore sophisticated molecular designs.

3. Trends

Here we will speculate on the question of what kind of surface coatings and why should they be developed for future nanomachines. We will overview current understanding of some critical issues in the nanotechnology development and what role will be devoted to the molecular coatings. We will briefly discuss fundamental understanding and possible impact of these developments on nanotechnology. We need to mention that the current discussion is too brief to be comprehensive and is based on the author's own experience in the field of organized molecular coatings as well as on a critical review of literature data available.

3.1. FUNDAMENTAL PROBLEMS AND APPROACHES

We are not going to discuss what kind of nanomachinery will be developed within next 10 years, or predict a new revolution in this field. Instead, for further discussion, we define two possible types of these devices, which can be created by using two very different approaches. The first type of nanoscale machinery can be developed by a gradual downsizing of current lithographical microfabrication technique [3]. The current trends and possible development of new light sources (like X-rays) will, probably, result in routine nanofabrication of nanodevices with spatial dimensions less than 20 nm. This is much smaller than current 100-1000 nm scale and will represent a critical step to a nanoscale operation. On this scale, the contact area of interaction between mating parts as well as sizes of surface asperities will be in the range from 1 to 5 nm. This truly atomic scale includes only several dozen interacting atoms and molecular groups. This number marks the limit of the applicability of the continuum mechanics, as we know, and the transition to molecular mechanics laws [5].

Obviously, the basic principles of the design of molecular coatings for such nanoscale parts should be reconsidered. Instead of relaying on laterally homogeneous organic layers with the thickness on a sub-microscopic scale, one should design molecular coatings with cluster/nanodomain lateral dimensions of several tens of molecules and the layer thickness not exceeding several nanometers. These molecular assemblies should still possess interfacial properties required for stable interfacial interactions and keep their integrity under substantial shear and normal stresses.

On the other hand, for nano-sized devices (as well as nanotubes, nanoparticles, etc.), the local radius of curvature of various areas of solid surfaces becomes so small that a planar molecular packing of symmetrical molecules in organic layer is not going to satisfy the conditions for dense and organized molecular surface ordering (Figure 6). Indeed, for molecular coatings of 2-5 nm thick on a surface feature with the radius of curvature of 5 - 10 nm, the mismatch between the surface area per molecule on solid substrate and layer surface can easily reach 100% (Figure 6).

Obviously, none of the current symmetrical molecular designs of protective/lubricant layers can satisfy these severe constrains. Other molecular structures should be thought of to match these extreme curved surfaces. One of the possible candidates is the recently introduced class of tree-like, dendrimer molecules, which can grow from one center and follow local surface curvature on a nanometer scale (Figure 6) [17]. These molecules and macromolecules can be considered be promising candidates for supramolecular assemblies of complex shape and high mobility.

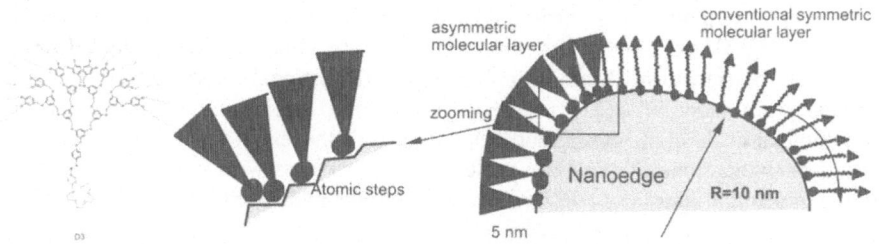

Figure 6. An example of the "nanoedge" (right) coated with conventional symmetric molecular layer (right side) and asymmetric molecular layer (left side) and "zooming" out crowded molecular assembly attached to "curved" surface with atomic steps (left). Far left is the chemical structure of a dendritic molecule [18].

Another aspect of the molecular coating design is brought in by atomistic, stepped structures of nano-beams/edges/channels at this spatial scale with inhomogeneous chemical/atomic structure across the atomic planes (Figure 4). No longer can solid asperities and parts be considered as having smooth homogeneous surface. This feature should play a crucial role in the mechanism of the molecular coating formation and greatly enhance their surface stability. The molecular behavior of such "stepped" assemblies is absolutely unknown. In this aspect, the direct molecular dynamics simulation of the molecular interactions on a spatial scale comparable to actual atomistic situation can bring invaluable insight [19].

Different stress distribution arises for molecular coatings tethered to nano-sized surfaces due to the type of the deformational contact of molecular assemblies during interfacial interactions (Figure 6). Unlike current molecular coatings, where the sliding behavior is determined by the shearing deformation of predominantly vertically oriented molecules, for highly curved coatings, splay deformational mode should play a dominant role in their nanomechanical behavior. Thus, the whole pattern of interfacial interactions of nanoscale surfaces will be changed and new, unknown, interfacial nanomechanical behavior can emerge.

A second scenario of nanomachine fabrication relies on supramolecular synthesis of complex functional molecules capable of specific functions [20]. Supramolecular interactions allow them to be organized or self-organized in the ordered array of molecular mechanisms capable of doing a specific function [8]. First examples of building such molecular mechanisms with scanning probe microscopy (SPM) assistance are demonstrated [21]. The role of supramolecular chemistry in the design of complex, stable, and functional molecules is crucial. However, even for these molecular mechanisms, some sort of interface with environment (such as atomistic lattice of supporting substrate) should be thought of. The ways to deal with such problems as the configurational/conformational behavior of molecular mechanisms constrained to specific atomistic interfaces should be explored. However, atomistic incompatibility of organic molecular mechanisms and inorganic solid substrates is obvious.

104

The ultimate solution for this dilemma could be complete abandoning solid-state platform for nanomachinery. In this scenario, "play field" should be transferred to mobile "soft matter" environment analogous to human cell structures. In the framework of this approach, molecular mechanisms with specific functions should be kept separately within/on, e.g., artificial membrane-like layers, which provide support and energy/signal exchange with environment. In this development, molecular coatings can be transformed into complicated multifunctional interfaces analogous to molecular membranes of cell structures. Apparently, the biomimetic approach will play a dominant role in the design and development of these systems. The fundamental question of interactions of synthetic molecular mechanisms with the environment via adaptive molecular interfaces will be a major task to be addressed.

3.2. TECHNOLOGICAL DEVELOPMENTS: FABRICATION, MEASUREMENT, AND MANIPULATION

From the technological point of view, the major question in the field of the development of molecular assembled coatings for nanomachinery is how these molecular coatings will be fabricated and how their interfacial behavior will be controlled. Current technologies for molecular coating fabrication and studies include Langmuir-Blodgett monolayer deposition and dip- and spin-coatings [5]. Although very versatile for model studies, these techniques are not useful for future nanoscale applications. The most probable candidates for the nanotechnological development is

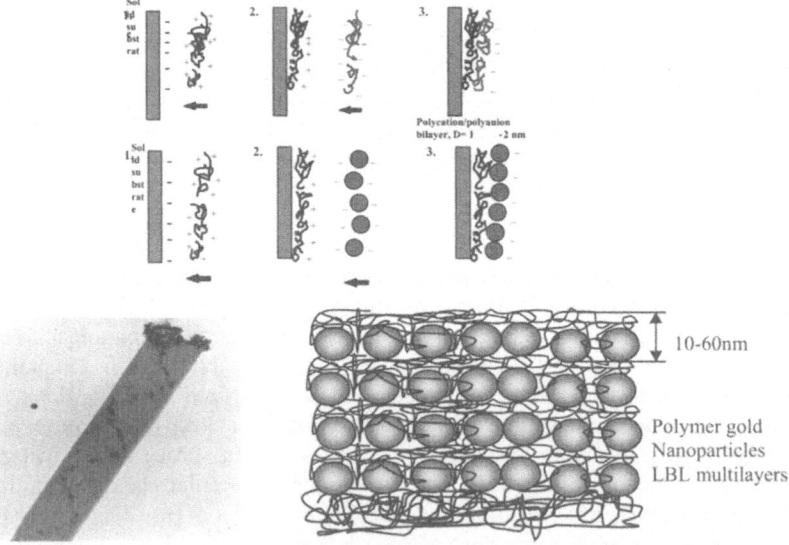

Figure 7. Principles of layer-by-layer deposition as applied to nanoparticles-polymer films (top, courtesy of Dr. Yu. Lvov) and resulting multilayered coatings (bottom, right). Transmission electron microscopic image of the lipid tubule (600 nm in diameter) covered with electrostatically assembled polymer-nanoparticles multilayers and silicon oxide nanoparticles in a patterned manner (see Yu. Lvov et al. [25] for detail discussion) (bottom, left).

molecular organization through self-assembly that is known in different versions such as chemical self-assembly for SAM fabrication or electrostatic self-assembly for layer-by-layer deposition (Figure 7) [22, 23].

We can speculate that specific conditions for the local adsorption of the specially designed molecules would require for further refining of these technologies. Exploiting additional weak interactions to organize molecular assemblies or external field assistance in aligning molecular ordering can be explored. Very complicated supramolecular structures can, in fact, be built. Two examples include 3D ordered nanodomain networks from associated dendrimers [24] or tubular superstructures from lipid complexes with helical surface patterns, spontaneously formed from self-assembled multilayer polyelectrolyte-nanoparticules complexes (Figure 7) [25]. Inorganic nanoparticle-polymer multilayered films can be formed by layer-by-layer deposition (Figure 7).

Such complex molecular assemblies as presented in the figure above, may serve several purposes simultaneously: reduce adhesion and friction, bring hydrophobicity to the silicon surface, provide selective molecular permeability through protein pore, and allow controlling mechanisms for the variation of surface properties and layer permeability through photoisomerization reaction of photosensitive molecules. In addition, asymmetrical shape of dendrimer molecules should assist in adaptation of this assembly to the highly curved nanostructures (Figures 6). These and other initial tests demonstrate that by exploiting the combination of self-assembling principles and sophisticated organic chemistry, one can nanofabricate very complicated superstructures and not just planar uniform layers or microscale-patterned planar surfaces. In the end, the nature relies only on multiple/competitive weak interactions to build, keep functioning, and repair complex and mobile nano- and microscopic structures of high complexity.

An additional challenge in monitoring molecular assembly behavior and measuring their properties down on a nanometer scale should be addressed for a successful implementation of this strategy. The natural candidate for such measuring and controlling tools is SPM technique introduced at the end of 80^s [27]. A combination of the latest SPM modes with other tools such as optical tweezers and near-field microscopy can constitute an effective nanomanipulation set. Compression and tensile measurements of molecular elasticity of individual macromolecules and their nanostructures on solid surfaces have been recently tested [28].

However, the level of sophistication should be brought to a much higher level in comparison with today's development. Much higher force sensitivity, multi-array probing capability, and truly molecular resolution should be achieved. Recent developments show the use of functionalized nanotubes as atomistic probes and demonstrate perspectives of this technique (Figure 8) [29].

Properties of single molecules and their clusters on solid substrates should be measured with high precision to understand how properties of individual molecules and their molecular aggregates at interfaces are different from that measured in dilute solutions and bulk state (Figure 9) [28].

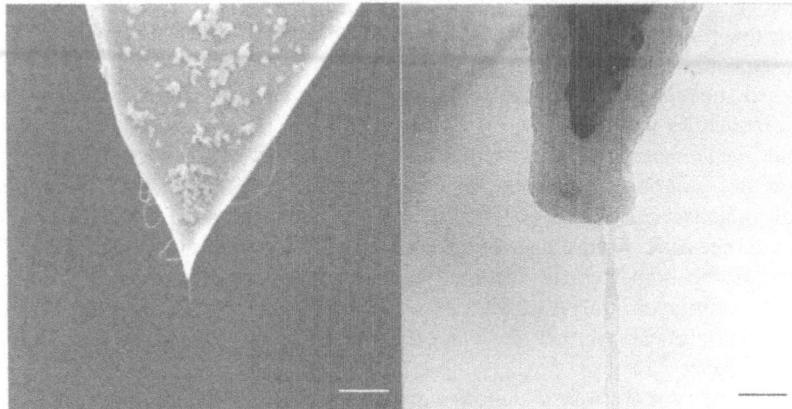

Figure 8. Nanotubes grown on the end of the SPM tip as a truly nanoscale tool for molecular and atomic manipulations. Left: SEM image of the tip, a scale bar is 500 nm; right: TEM image of the tip end, a scale bar is 20 nm. Courtesy of J. H. Hafner, C. Lieber et al. [29].

Finally, the effective mechanism of self-repair in these molecular assemblies must be incorporated. For interacting areas of several tens of molecules across, even one molecular defect can completely disturb normal functioning. The mechanism of finding and replacing such molecular defects should be thought of. Wide operational limits of molecularly assembled interfaces should be assured for stable and reliable functioning under fluctuating environmental conditions. Various bio-inspired schemes of molecular replication and multiple duplications can be explored.

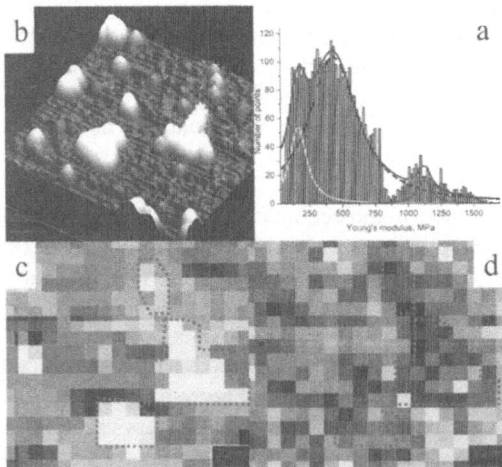

Figure 9. Micromapping of surface aggregates of dendritic molecules: a) overall statistics of elastic moduli collected from 1x1µm surface area showing three different levels of the surface stiffness corresponding to substrate, cluster edges, and molecules inside clusters; b) 3D image of selected surface are with three different aggregates marked by dotted lines; c) topographical map of the same surface area with three different aggregates marked by dotted lines; d) surface distribution of local stiffness with three different aggregates marked by dotted lines. (Ref. [28[).

4. Applications and Impact

Here, we will speculate on the impact that resulted from the development of molecular assembled coatings for nanotribological applications. We should emphasize that does not matter how great of a progress in nanomachinery fabrication will be achieved; this nanotechnology cannot work in vacuum (except, probably, some space-related projects and computer simulations). In real-life situation, application of these technologies must be accompanied by the development of sophisticated nanoscale interfaces to accommodate massive arrays of nanodevices, protect them from environment, and provide adequate exchange of energy between these devices and environment. Therefore, we see several major activities in the field of molecularly assembled coatings for task-oriented developments.

First, new molecular protective coatings will be developed for nanodevices and molecular mechanisms, which will integrate great sophistication and versatility of naturally occurred supramolecular assemblies. Such molecular coatings will have highly mobile and adoptive molecular organization that can change their molecular ordering at a complex nanointerface in response on external stimuli and adopt their properties to provide the highest survivability of the nanodevice protected. Such protective molecular assemblies will provide required structural stiffness and integrity in case of "soft matter"-based molecular mechanisms and compensate for constrained atomistic configurations of nanodevices. These assemblies will serve as an adaptive buffer between internal molecular motions and mobile environment providing reduced interfacial stresses and appropriate conformational mobility.

Second, in addition to their mostly protective functions, new generation of molecularly assembled coatings will serve as an active interface between internal organization of nanodevices and external stimuli/disturbance. These interfaces should be able to support control delivery of energy in/out nanodevices in the form of excessive heat dissipation and thermal equilibration with environment. Controlled permeability is required for functioning of chemical "nanofactory" and selective removal of "molecular waste". Timely transfer of external signal to nanomachinery and molecular feedback can be provided by this interface.

Third, the long-term development of such molecular interfacial assemblies should result in, and cannot be successful without, much deeper understanding of fundamental principles of local molecular/supramolecular organization at nanostructured surfaces/interfaces, role of weak multidirectional interactions in self-organization of nanoscale assemblies, and their adaptive/responsive behavior. New molecular and atomistic designs of organic molecules and macromolecules should be searched to make them compatible with nanomachinery. As we mentioned before, highly asymmetric molecular structures with carefully tailored intra/inter molecular/interfacial interactions should be designed for assembling at nanointerfaces.

Finally, we think that on the path towards such nanoassembled interfaces, we are going to get closer to self-consistent nanodevices resembling the efficient cell-like structures. And, inevitably, we should learn, adopt, and implement basic principles of self-organization of most sophisticated nanomachines, namely, a human cell. Only integration of these principles with the logistic of deterministic artificial nanodevices may lead to successful exploration of the nanoworld. And, in this world, the molecularly assembled interfaces will be considered as an integral part of nanodevices and not a post-added function.

5. Acknowledgement

The author's work in the field of surface molecular engineering and assemblies is supported by The US National Science Foundation, CMS-0099868 and DMR-0074241 Grants, AFOSR F496200210205 Contract, and NASA, NAG 1-102098 Grant. The author thanks members of his research group, who performed studies mentioned in this review: M. Lemieux, K. Larson, D. Julthingpiput, Dr. A. Sidorenko, M. Ornatska, H. Shulha, and X. Zhai.

We are grateful for collaboration with many colleagues who provided us with materials and microdevices to work with. MEMSs were provided by Dr. M. Dugger (Sandia National Lab), dendritic compounds were provided by Prof. D. McGrath (University of Arizona) and Perstorp Polyols. Part of Figure 7 was courtesy provided by Prof. Yu. Lvov (Louisiana Technical University) and Figure 8 from Ref. 25 was courtesy provided by C. Lieber and J. H. Hafner (Harvard University).

6. References

1. L. Lin, G. S. Blackman, R. R. Matheson, in: *Microstructure and Microtribology of Polymer Surfaces*, Eds. V. V. Tsukruk, K. Wahl, ACS Symposium Series, v. 741, 2000, p. 428.
2. *Fundamentals of Friction*, Eds. E. Singer, H. Pollack, Kluwer Acad. Press, 1992.
3. Muller, R. S., in: *Micro/Nanotribology and Its Applications*, B. Bhushan, Ed., Kluwer Press, 1997, p. 579. *Tribology Issues and Opportunities in MEMS*, Ed. B. Bhushan, Kluwer Academic Press, 1998.
4. M. T. Dugger, D. C. Senft, G. C. Nelson, in: *Microstructure and Microtribology of Polymer Surfaces*, Eds. V. V. Tsukruk, K. Wahl, ACS Symposium Series, v. 741, 2000, p. 428.
5. *Micro/Nanotribology and Its Applications*, B. Bhushan, Ed., Kluwer Press, 1997. Tsukruk, V. V. *Adv. Materials, 13*, 95, 2001.
6. K. Komvopoulos, *Wear, 200*, 305, 1996
7. V. V. Tsukruk, T. Nguyen, M. Lemieux, J. Hazel, W. H. Weber, V. V. Shevchenko, N. Klimenko, E. Sheludko, in: *Tribology Issues and Opportunities in MEMS*, Ed. B. Bhushan, Kluwer Academic Press, 1998, p. 608.
8. V. V. Tsukruk, *Progress in Polymer Science, 22*, 247, 1997
9. *Organic Thin Films*, Ed. C. W. Frank, ACS Symposium Series, v. 695, 1998.
10. G. Blankenstein, U. D. Larsen, J. Branebjerg, *SPIE Proceedings, 2998*, 2982, 1997.
11. M. R. Houston, R. T. Howe, R. Maboudian, *J. Appl. Phys., 81*, 3474, 1997.
12. V. N. Bliznyuk, M. P. Everson, V. V. Tsukruk, *J. Tribology, 120*, 489, 1998; V. V. Tsukruk, V. N. Bliznyuk, J. Hazel, D. Visser, M. P. Everson *Langmuir, 12*, 4840, 1996. I. Luzinov, D. Julthongpiput, H. Malz, J. Pionteck, V. V. Tsukruk, *Macromolecules, 33*, 1043, 2000.
13. V. V. Tsukruk, A. Sidorenko, H. Yang, *Polymer, 43*, 1695, 2002.
14. V. V. Tsukruk, H.-S. Ahn, A. Sidorenko, D. Kim *Appl. Phys. Lett., 80*, 4825, 2002. A. Sidorenko, Hyo-Sok Ahn, Doo-In Kim, H. Yang, V. V. Tsukruk *Wear 252*, 946, 2002.
15. D. A. Tomalia, *Adv. Materials, 6*, 529, 1994. J. M. Frechet, *Science, 263*, 1711, 1994. J. F. Jansen, E. M. de Brabander-van den Berg, E. W. Meijer, *Science, 266*, 1226, 1994. V. V. Tsukruk, *Advanced Matl., 10*, 253, 1998
16. M. Hashemzadeh, D. V. McGrath, *Polym. Prepr. 39(2)* 338, 1998.
17. Drexler, K.E., *Nanosystems: molecular machinery, manufacturing, and computation*, Wiley&Sons, 1992.

18. T. Cagin, J. Che, Y. Qi, Y. Zhou, E. Demiralp, G. Gao, W. A. Goddard III, *Journal of Nanoparticle Research, 1*, 51, 1999; T. Cagin, A. Jaramillo-Botero, G. Gao, W. A. Goddard, III, *Nanotechnology, 9 (3),* 143, 1998; J. Gao, W. D. Luedtke, U. Landman, *J. Chem. Phys., 106,* 4309, 1997.

19. D.M. Eigler, E.K. Schweizer, *Nature, 344,* 524, 1990; M. T. Cuberes, J. K. Gimzewski, R. R. Schlittler *Applied Physics Letters, 69,* 3016, 1996.

20. C. D. Bain, G. M. Whitesides, *J. Am. Chem. Soc., 111,* 7164, 1989. Y. Xia, G. M. Whitesides, *Angew. Chem., 37,* 550, 1998.

21. G. Decher, Yu. Lvov, J. Schmitt, *Thin Solid Films, 244,* 772, 1994. G. Decher, *Science, 277,* 1232, 1997.

22. V. Percec, C.-H. Ahn, T. U. Bera, G. Ungar, D. J. Yeardley, *Chem. Eur. J., 5,* 1070, 1999.

23. Yu. Lvov, R. Price, A. Singh, J. Selinger, M. Spector, J. Schnur, *Langmuir,* 2000, in press

24. L. Song, M. R. Hobaugh, C. Shustak, S. Cheley, H. Hayley, J. E. Gouaux, *Science, 274,* 1859, 1996.

25. G. Binnig, C.F. Quate, Ch. Gerber, *Phys. Rev. Lett. 12,* 930, 1986.

26. V. V. Tsukruk, H. Shulha, X. Zhai, *Appl. Phys. Lett., submitted*

27. J. H. Hafner, C. Li Cheung, C. M. Lieber, *J. Amer. Chem. Soc., 121,* 9750, 1999.

SHAPE FABRICATION OF COTTON-DERIVED INORGANIC HOLLOW RIBBONS

A. B. BOURLINOS, N. BOUKOS AND D. PETRIDIS
Institute of Materials Science, NCSR "Demokritos",
Ag. Paraskevi Attikis, Athens 15310, Greece

ABSTRACT. Cotton ribbons serve as unique templates for the shape fabrication of a wide range of woven inorganic hollow ribbons, like ceramic, semiconducting, magnetic and others. In principle, this novel shaping process is very simple and of low cost due to the nature of the cotton template. An interesting highlight of the present work concerns the morphogenesis of magnetic hollow ribbons of γ-Fe_2O_3 and its effect on its magnetic properties.

1. Introduction

During the past decade it has been well recognised that the template approach in designing a large variety of inorganic matters with controllable architecture and physico-chemical properties, is a powerful tool and, definitely, the central idea in many molecular engineering processes. According to this simple and elegant approach, well-shaped organic arrays play the role of the template upon which the desirable inorganic phase is properly deposited through sol-gel, precipitation and self-assembling techniques. Following, the organic template is removed by a chemical or thermal treatment to leave behind the inorganic phase as a replica of the template. In this way, numerous inorganic solids of high technological importance (ceramic, magnetic, semiconducting, metallic etc) have been fabricated in certain engineering forms, like porous inorganic oxides with a long range periodicity [1-3], ultralight foams [4], hollow inorganic fibres with unique structural features [5-7], shaped hollow capsules [8-10], ribbon-like materials [7,11] and others. In particular, very few metal oxide ceramic hollow fibers for photovoltaic, catalytic and other applications have been fabricated by a templating technique that uses cellulose and its derivatives as the controlling structural mold [12-14]. This work describes in a more general way a simple and readily realizable process for the morphogenesis of cotton-derived hollow inorganic ribbons. More specifically, the surface patterning of cellulose-based cotton ribbons with suitable inorganic precursors through hydrolysis, sol-gel, co-precipitation, ion-exchange and chemical modification reactions, followed by calcination in air of the as-made cotton derivatives, leads to a wide range of ultralight hollow inorganic materials, like ceramic (CeO_2, MgO, SiO_2), semiconducting (α-Fe_2O_3, SnO_2, TiO_2), magnetic (γ-Fe_2O_3, Co_3O_4)

T. Tsakalakos, et al. (eds) pgs 111 – 116
Nanostructures: Synthesis, Functional Properties and Applications;
© *2003 Kluwer Academic Publishers.*

and others ($NdFeO_3$), that inherit the morphology, dimensions and macroscopic appearance of the parent cotton template. This particular method is simple, general and of low cost since cotton is a natural raw material in plentiful occurrence. The composition of the inorganic shells in the hollow ribbons was identified by the XRD technique, while, their morphology and shell construction by SEM and TEM measurements. In addition, the effect of the ribbon-like morphology on the properties of the hollow materials, for instance the magnetic, was also examined.

2. Experimental

The α-Fe_2O_3, CeO_2 and Co_3O_4 hollow ribbons with crystalline shells were prepared as following: a piece of cotton (0.5g) was immersed in 40mL of isopropanol containing 0.3-0.35g of the corresponding salt precursor [$Fe(NO_3)_3 \cdot 9H_2O$, $(NH_4)_2Ce(NO_3)_6$ and $Co(NO_3)_2 \cdot 6H_2O$]. The mixture was boiled for 20min. During this time hydrolyzed particles from the decomposition of the corresponding salts were adhered on the cotton surfaces either through hydrogen bonding interactions between the hydroxyl groups of the cellulosic template and those of the hydrolyzed particles, or through chelation of the metal ions in the deposited particles by the hydroxyl groups of cellulose near their conducting surfaces. The dyed cotton was removed from the bath solution, well wrung and dried at 70°C and finally calcined in air at 600°C for 1h via steps of 1°C min^{-1}.

The perovskite $NdFeO_3$ hollow ribbons were prepared similarly, except of using an aqueous solution of equimolar amounts of the $K_3Fe(CN)_6$ and $Nd(NO_3)_3 \cdot 6H_2O$ salts as the dyeing bath [the precursor particles produced are $NdFe(CN)_6 \cdot xH_2O$], and a calcination temperature of 700°C [15]. For the sol-gel derived hollow ribbons (SiO_2, TiO_2), a piece of cotton (0.5g) was immersed in 40mL of isopropanol containing 0.5mL of either $(CH_3O)_4Si$ plus 3-4 drops of conc. HNO_3 or $(i\text{-}PrO)_4Ti$ and the mixture was boiled for 20min. During this time, amorphous silica or titania sol particles adhered on the cotton surfaces through hydrogen bonding interactions and eventually transformed to amorphous SiO_2 or anatase after calcination at 600°C. For the γ-Fe_2O_3 hollow ribbons, 0.5g of a cotton piece was immersed in 40mL of a hot aqueous solution containing trivalent and divalent iron in a molar ratio 2:1 (300mg $FeCl_3 \cdot 6H_2O$, 150mg $FeSO_4 \cdot 7H_2O$). Following, 0.5mL of conc. NH_3 was added to the hot solution and instantly a black solid (Fe_3O_4) was precipitated over the cotton surfaces [16]. The black derivative, after isolation and drying, was calcined in air at 450°C for 1h via steps of 0.5°C min^{-1}. Finally, the MgO and SnO_2 analogues were prepared through aqueous ion-exchange reactions between the slightly acidic protons of the cellulosic template and Mg^{2+} ions, and, partial capping of the surface exposed hydroxyl groups of cellulose with $(CH_3)_2SnCl_2$ in hot water [the organotin compound dissociates in water to give $(CH_3)_2Sn^{2+}$ ions capable of accepting oxygen-containing ligands in their coordination sphere]. Both derivatives, after calcination at 600°C, afforded the corresponding hollow ribbons with crystalline shells.

3. Results and Discussion

Cotton ribbons constitute excellent template frameworks that incorporate and organize on their surfaces metal-containing precursors that upon calcinations in air at 600°C afford hollow ribbons made up from a plethora of different inorganic materials. The hollow inorganic ribbons are free of the cellulose template as evidenced from IR spectra (cotton burns off above 400°C) and exhibit only the characteristic for the inorganic phase absorption bands, as demonstrated for example in Figure 1 for iron(III) loaded cotton ribbons. In addition, the derived inorganic ribbons are ultralight and inherit the woven texture of the parent cotton template, Figure 2. The SEM images of the starting cotton material and of the cotton-derived inorganic samples are presented in Figure 3. As can be seen in Figure 3a, the cotton material consists of twisted ribbons having a width of approximately 5-10μm and a thickness of 2-5μm. The compact cotton ribbons are folded in several places and their surface is relatively smooth. The morphologies of the α-Fe_2O_3, SiO_2, TiO_2 and SnO_2 samples are depicted in Figures 3b-d. Apparently, the inorganic materials inherit the morphology of the cotton template, i.e. they are twisted ribbons, folded in several places. Their size, within the experimental error, lies in the same range with that of the cotton ribbons, while, their surface roughness is more intense than that of the cellulose-based ribbons as a result of a random packing of the inorganic constituents within the shell. The inset photo in Figure 3b, exemplifying a typical cross-section of a ribbon, reveals the hollow structure of the ribbons.

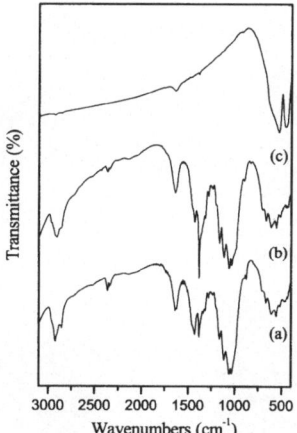

Figure 1. IR spectra of cotton ribbons (a), cotton ribbons coated with hydrolyzed $Fe(NO_3)_3 \cdot 9H_2O$ (b), and of the latter after calcination (c) [the sharp band at 1380cm^{-1} in spectrum b comes from NO_3^- ions present in the deposited particles, while, the absorption bands at 540cm^{-1} and 460cm^{-1} in spectrum c are typical of an iron(III) oxide network].

A similar picture also emerges for the γ-Fe_2O_3 hollow ribbons, where additionally, we observe a higher surface roughness than that in the inorganic ribbons derived from the hydrolysis, sol-gel, ion-exchange and chemical modification routes, Figure 4, upper photo. This can be ascribed to the rapid precipitation and, therefore, more random packing of the magnetite particles on the cotton surfaces during the coating procedure. The TEM study of the ribbon walls reveals that nearly spherical nanograins with sizes ranging between 10-20nm are densely packed in the shell as shows the lower photo in Figure 4. The XRD pattern of the sample (inset photo) shows that the major phase of the polycrystalline ribbon walls is γ-Fe_2O_3. A very small quantity of α-Fe_2O_3 is also detected by the presence of a weak reflection at $2\theta=33°$, as a result of heat treatment in the composition of iron oxide particles [17]. In the present case, we have noticed that smaller heating rates lead to less hematite formation. The effect of the ribbon-like morphology on the magnetic properties of the hollow γ-Fe_2O_3 ribbons could be of particular interest taking into consideration the significance of magnetic iron oxides in many technological, environmental and industrial applications [18]. To this aim, we recorded the magnetization versus applied field curves at room temperature for the γ-Fe_2O_3 hollow ribbons [19] and powdered γ-Fe_2O_3 as a blank sample [20]. The γ-Fe_2O_3 hollow ribbons and blank sample exhibit identical XRD patterns and mean particle sizes. The hysteresis loops are characteristic of superparamagetic particles exhibiting zero values of coercivity and remanence magnetization above their blocking temperature, Figure 5. But most surprisingly, the γ-Fe_2O_3 hollow ribbons show a considerably lower saturation magnetization (M_s=13 emu g^{-1}) than the powdered sample (M_s=70 emu g^{-1}). These preliminary results, although not well understood, clearly demonstrate a strong influence of the γ-Fe_2O_3 morphology on its magnetic properties.

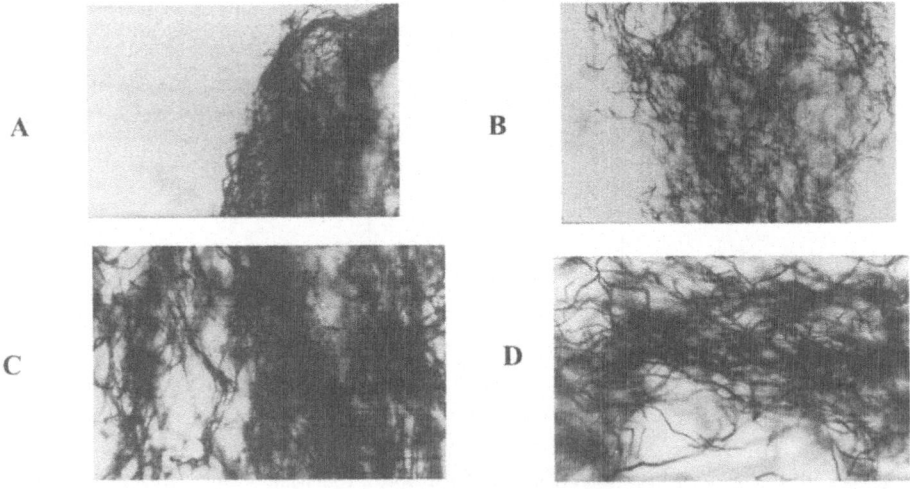

Figure 2. Optical microscopy pictures for CeO_2 (A), Co_3O_4 (B), $NdFeO_3$ (C) and MgO (D) hollow ribbons

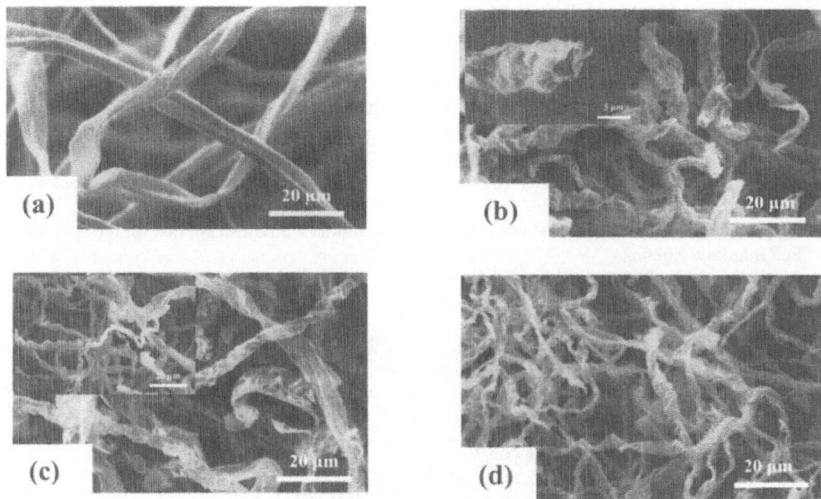

Figure 3. SEM images for cotton (a), and α-Fe₂O₃ (b), TiO₂ and SiO₂ as an inset (c), and SnO₂ (d) hollow ribbons.

4 Summary

Cotton ribbons were successfully employed as templates for the shape fabrication of high importance ultralight inorganic hollow materials with the morphology and woven texture of the parent template and possessing interesting shape-induced properties. The method is simple and of low cost, while the cellulosic composition of the cotton material enables its facile surface modification with diverse materials through dyeing-like processes. Noticeable, hair fibers are also promising for similar morphogenesis processes [21].

5. References

1. Beck, J. S.; Vartuli, J. C.; Roth, W. J.; Leonowicz, M. E.; Kresge, C. T.; Schmitt, K. D.; Chu, C. T-W.; Olson, D. H.; Sheppard, E. W.; McCullen, S. B.; Higgins, J. B.; Schlenker, J. L. *J. Am. Chem. Soc.* 1992, *114*, 10834.
2. Davis, S. A.; Burkett, S. L.; Mendelson, N. H.; Mann, S. *Nature* 1997, *385*, 420.
3. Göltner, C. G.; Berton, B.; Kramer, E.; Antonietti, M. *Chem. Commun.* 1998, 2287.
4. Chandrappa, G. T.; Steunou, N.; Livage, J. *Nature* 2002, *416*, 702.
5. Nakamura, H.; Matsui, Y. *J. Am. Chem. Soc.* 1995, *117*, 2651.
6. Jung, J. H.; Ono, Y.; Shinkai, S. *Chem. Eur. J.* 2000, *6*, 4552.
7. Jung, J. H.; Kobayashi, H.; Van Bommel, K. J. C.; Shinkai, S.; Shimizu, T. *Chem. Mater.* 2002, *14*, 1445.
8. Bergbreiter, D. E. *Angew. Chem. Int. Ed.* 1999, *38*, 2870.
9. Caruso, F. *Chem. Eur. J.* 2000, *6*, 413.
10. Caruso, F. *Adv. Mater.* 2001, *13*, 11.

116

11. Livage, J.; Bouhedja, L.; Bonhomme, C. *J. Sol-Gel Sci. Technol.* 1998, *13*, 65.
12. Imai, H.; Iwaya, Y.; Shimizu, K.; Hirashima, H. *Chem. Lett.* 2000, 906.
13. Caruso, R. A.; Schattka, J. H. *Adv. Mater.* 2000, *12*, 1921.
14. Shigapov, A. N.; Graham, G. W.; McCabe, R. W.; Plummer Jr, H. K. *Appl. Catal. A: Gen.* 2001, *210*, 287.
15. Gallagher, P. K. *Mat. Res. Bull.* 1968, *3*, 225.
16. Mayes, E. L.; Vollrath, F.; Mann, S. *Adv. Mater.* 1998, *10*, 801.
17. Caruso, F.; Spasova, M.; Susha, A.; Giersig, M.; Caruso, R. A. *Chem. Mater.* 2001, *13*, 109.
18. Zboril, R.; Mashlan, M.; Petridis, D. *Chem. Mater.* 2002, *14*, 969.
19. The composition of the magnetic hollow ribbons is purely that of an iron(III) oxide phase as evidenced from iron elemental analysis.
20. Powdered γ-Fe_2O_3 was prepared under identical conditions with those applied for the synthesis of the γ-Fe_2O_3 hollow ribbons.
21. Bourlinos, A. B.; Petridis, D. unpublished results.

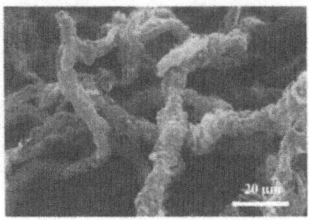

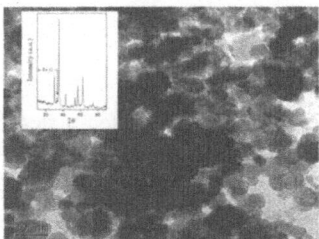

Figure 4. Upper photo, a SEM image of γ-Fe_2O_3 hollow ribbons. Lower photo, a TEM image of the magnetic shell and, as an inset, the corresponding XRD pattern.

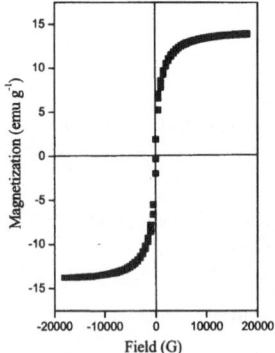
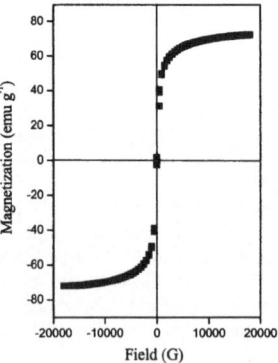

Figure 5. Magnetization versus applied field curves for the γ-Fe_2O_3 hollow ribbons (left) and powdered γ-Fe_2O_3 (right) at room temperature.

NATURE AND NANOTECHNOLOGY

S.C. GOHEEN, K. A. GAITHER, and A.R. RAYBURN
Pacific Northwest National Laboratory
902 Battelle Boulevard
Richland, WA 99352

Abstract. Scientists and engineers are rapidly developing techniques to produce nanostructures for various applications: nano-devices include electronic components, catalysts, and mechanical systems such as levers and motors. Biomedical applications include nanobots intended to function in various body fluids, pumps and drug delivery products, and implantable biosensors. Nature has been developing processes and products at the nano scale throughout the evolutionary process. There are several biochemical processes that involve energy-producing reactors, synthesis, and various mechanical processes. The relationships between the recent accomplishments of scientists and engineers in nanotechnology and natural processes are the subject of this paper. Numerous examples are provided in which the structure of a biological system at the nano-scale relates to some elaborate biological function. The structure and function for biological systems and biochemicals can in some cases be correlated to the structure and function of man-made nanomaterials. In this paper, we focus on proteins, but similar insight can be attained from other biomolecules and a wide variety of biological processes.

1. Structure and Function

Structure in nature is typically correlated to a function. For example, the Earth is round because it is the most compact form. A tree is tall so it can compete for sunlight. Red blood cells are biconcave discs so they can deform easily as they pass through a capillary, and transport oxygen to tissues. Humans have eyes near the top of their heads to enhance vision over long distances.

The implication of providing an explanation for structure is that biological systems were designed for optimal performance. Often there is no immediate or clear explanation for the structure of a biological system, so we search for one. Typically, an explanation is found and our curiosity is satisfied. An example could be skin color. We are all aware of differences in skin color, yet the biological "explanation" for the differences are not widely known. We claim that people of different skin color are distributed close to or away from the equator due to their susceptibility to carcinoma induced by exposure to the sun. And, humans are mobile so their location is often chosen rather than designed by nature. Another example is the shape of the leaves of a

T. Tsakalakos, et al. (eds.) pgs. 117 - 137
Nanostructures: Synthesis, Functional Properties and Applications;
© *2003 Kluwer Academic Publishers.*

plant. Why don't all plants have the same leaf shape, or, why is there so much diversity in plant species at any specific location? Why don't the most efficient plants take over some geographical location and eliminate all other plants? Does the structure of the plant leaf protect it in some way? We tend to ask these questions because most biological systems have a clear relationship with their surroundings so that it is often, yet not always obvious how structure and function are related.

If we observe the largest known structures in nature we can observe non-intuitive order. One example is the order of the galaxy (see Figure 1). The galaxies of the universe have both radial and mirror symmetry, as do nearly all stars and planets. To a first order approximation, nearly all macroscopic living organisms have either radial or mirror symmetry, or both. Reasons for this massive organization are not essential to this discussion. However, it is interesting that many natural objects, regardless of their size, have this symmetry. This together with our knowledge that entropy is always increasing, can cause us to ponder various explanations.

Figure 1. The largest structures we are aware of have radial and mirror symmetry. Shown is galaxy NGC 4414 photographed with the help of the Hubble Telescope from NASA. Similar symmetry can be observed from star clusters, solar systems, stars, and planets. Reproduced with permission from the Space Telescope Science Institute's website at http://oposite.stsci.edu/pubinfo/Pictures.html.

In smaller biological systems, we also see symmetry (see Figure 2). Many bacteria, protozoa, viruses, and even individual biochemicals have symmetry not unlike that of our galaxy. In fact, atoms themselves are symmetrical. One wonders whether atomic symmetry is responsible for biological, planetary, and galactic symmetry. But atomic symmetry cannot alone explain why proteins, for example, are so beautifully organized. This is especially puzzling because the amino acid sequence of a protein is typically asymmetrical.

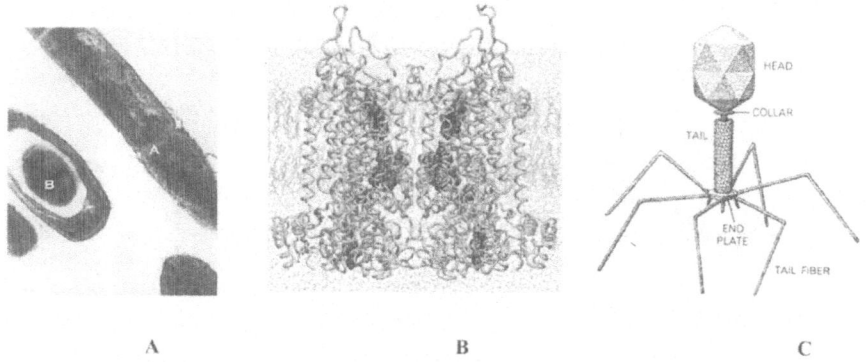

A   B   C

Figure 2. Symmetry of bacteria, viruses, proteins. (a) Transmission electron micrographic image of *Bacillus anthracis* from an anthrax culture, A: cell division, B: spores (from http://phil.cdc.gov/phil/default.asp). (b) Cytochrome bc1 complex (from http://www.ks.uiuc.edu/Research/smd_imd/bc1/) (c) Bacteriophage (from http://www.foresight.org/)

Mammals' size, shape, and strength correspond to their ability to compete and maintain their place in the ecosystem. Bones are stronger and more dense than muscleso they can help us to move and support our bodies. Our more susceptible organs are located near the center of our bodies so they will be protected from various insults. Individual tissues are also organized to enhance and streamline their function. It is easier to understand the relationship between structure and function for living systems because they have evolved over the millennia, perfected their various processes, and are functioning organisms.

The structure and symmetry of natural materials therefore extends from this almost incomprehensibly large dimension, to the atomic level, and certainly to the nanometer level. Biological symmetry can be examined over dimensions covering several orders of magnitude. One example is tendons. On the meter to millimeter scale, tendons are organized in bundles of collagen fibers. On the nanometer scale, collagen, the proteins that comprise the major structural component of tendons, are repeating units of the amino acids glycine-X-Y, where X is frequently proline, and Y is often hydroxyproline (see Fig. 3). These sequences form microfibrils, which provide strength along the length of the fibrils as well as the fibril bundles.

120

Tendon Hierarchical Structure
Adapted from Kastelic et al 1978 and Heinegard 1994

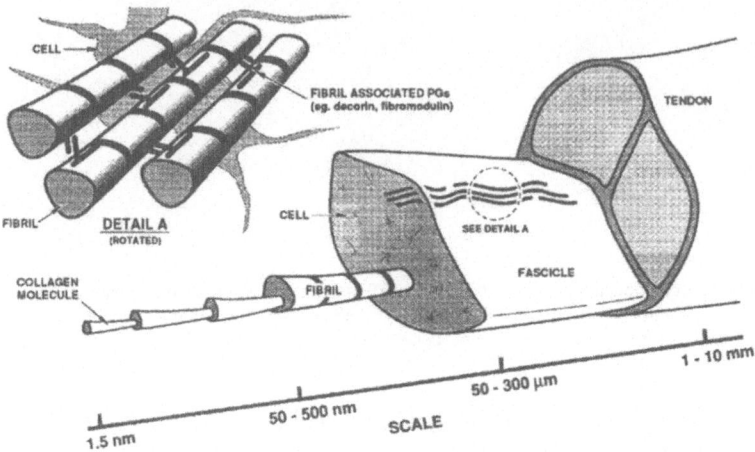

Figure 3. The structure of tendons. Note the ordered structure from the nanometer level to the mm level. Reproduced with permission from Derwin, KA. "A Quantitative Investigation of Structure-Function Relationships in a Tendon Fascicle Model" [dissertation]. Ann Arbor, MI: The University of Michigan; 1998.

Tendons need to be able to expand and contract in their plastic region thousands or millions of times without damage. Most synthetic materials cannot withstand this type of cyclic deformation. It is the complex structure of tendons that provide this ability to cycle into the plastic deformation region without damage. At the smallest level of organization, the (tropocollagen) protein molecule is in the shape of a right-handed triple helix consisting of two identical polypeptide chains (alpha-1), and one slightly different polypeptide chain (alpha-2). Each of the polypeptide chains themselves take on the structure of a left-handed helix with a pitch of 9.5 A°. The right-handed triple chain has a pitch of 104 A°. Each polypeptide chain (alpha-1, and –2) consists of about 1000 amino acids. Tropocollagen molecules are arranged in bundles of microfibrils and are discontinuous. That is, there are spaces between each molecule (called a hole zone) in the longitudinal direction. These spaces are very regular and can be seen in the electron microscope. They give the appearance of bands about 668 A° apart. The polypeptides are held together in the bundles by hydrogen bonds between the NH group of glycine in one chain and the $C = O$ group of proline (or other amino acid) in an adjacent polypeptide chain. The physical properties of this structure can be adjusted by adjusting the cross sectional area of the matrix. Between hydrogen bonds, helices, and overlapping, the overall structure is strong yet ductile. However, duplicating a structure with these properties cannot be easily carried out without resorting to cleverly designed nanostructures.

2. Applications

Rope approximates the structure of collagen. Rope often consists of natural fibers and typically contains bundles of long strands that are wound around other bundles to enhance strength. While rope is not as flexible as tendons, it serves its purpose well. At the nano-level, natural rope fibers consist of long strands of cellulose and lignin, both polymeric biomolecules. Those structures are nicely approximated when synthetic polymer chains such as nylon or polypropylene are used in place of natural fibers.

There are numerous other examples of biological structures that are so intricate and complex at many levels that it would be impossible to describe them all. Many of these systems remain undiscovered, even with modern medicine and our intricate knowledge of the Human genome.

Engineers, through nanomaterials technology, are not yet able to duplicate the most complex processes or fully understand biological processes relevant to the survival and performance of nano-scale devices. Early attempts may be in the repair of tissues. Although nanotechnology may not be able to fully replicate the collagen structure at this time, there have been attempts to develop nanomaterials that can help in the healing of injured tendons. Part of tendon structure also includes a sheath around the tendon to isolate the moving from the non-moving parts. During healing, the healed tendon sometimes adheres to the sheath, restricting motion. The strategy being developed involves smart bandages that can help isolate the tendon from its sheath. The bandage was developed at the University of Glasgow in Scotland by Chris Wilkinson and Adam Curtis. It consists of 10 micron grooves. Macrophages adhere to the collagen as well as the grooves in the bandage during healing. After the healing is nearly complete, the bandage dissolves and the collagen remains separated from the sheath [1].

Another smart bandage nanomaterials technology involves the healing of chronic wounds. Patients with poor circulation are sometimes plagued with open wounds that heal slowly, or not at all. These wounds contain fluid that is very high in proteolytic enzymes such as elastase. Elastase helps digest proteins such as collagen. The role of this protein in a non-healing wound is to help digest dead tissue. But too much of this enzyme will also cause living tissue to be digested. Elastase and other proteases can help digest other peptides such as growth factors that would otherwise accelerate healing. A strategy to help solve the problem is to use bandages that will selectively remove elastase or other proteases from the chronic non-healing wound [2].

One method to selectively remove elastase from the wound is to develop an antibody to elastase, bind it to a bandage, and allow the antibody to interact with the antigen (elastase) thereby removing it. However, antibodies are relatively unstable when placed near a sorbent and their ability to bind antigens can be enhanced if the antibodies are packed more closely together, holding them upright [3]. An alternative to using antibodies is to identify a small molecule that interferes with the activity of the enzyme, bind it covalently to the bandage material, and hope that the small molecule will selectively remove the enzyme from solution.

We examined the ability of the peptide inhibitor, valine-proline-valine-O-CH$_3$ (val-pro-val, the methyl ester of the tri-peptide) to sequester elastase when bound at either the amino, or carboxyl end. The orientation of the peptide made a large difference on its ability to sequester elastase. Elastase is rich in arginine, an amino acid containing two (positively charged) amino groups. The elastase is attracted to the carboxyl terminal group of the peptide, and then recognizes the val-pro-val sequence at its active site [4]. The combined attraction and recognition appear to sequester elastase irreversibly from solution, and may prove to be an outstanding wound dressing for those needing this specialized treatment.

In the elastase example, both the peptide val-pro-val and the elastase enzyme can be thought of as nanomaterials. One is native (elastase) and one is synthesized (peptide). The distribution of val-pro-val sequences on the bandage support is relatively unimportant, as long as the spacing between peptides does not cause significant interference between the elastase molecules. This is because the elastase is in solution in the wound fluid, and little structure is involved with the exception of the protein conformation itself. If the protein had no particular structure, there would be no recognition site and the peptide would not be able to bind elastase as strongly. However, if the peptides could be arranged in such a way as to maximize the packing of elastase molecules using an additional type of nanotechnology on the bandage material, the capacity of the dressing could be maximized. And, as we will see later, the folded or unfolded state of both nanoparticles can influence binding efficiency. Therefore, it is important to understand the structures and relationships between the nanostructures of the peptide and elastase well enough to design a coupling arrangement so this product will function most efficiently.

The ideal packing density of the peptide recognition sequence (val-pro-val) probably does not include a close-packed array. Elastase is much larger than the tri-peptide sequence of the recognition peptide. Therefore, steric hindrance will alone mean that a loose-packed array of peptides should bind elastase more efficiently than a tight-packed array [3].

3. General Discussion of Proteins

Proteins are similar to one another in that they all contain amino acids covalently linked by a peptide bond, but there are far more differences between them than there are similarities (see Figures 4 and 5). Proteins are diverse not only in size and shape, but also in amino acid sequence, presence or absence of covalently linked carbohydrates (glycosylation), phosphates (phosphorylation), or other moieties. Some are soluble in blood fluids or the cytosol (liquid phase) of cells. Some are bound to any of a number of membranes (nuclear, mitochondrial, cytoplasm, etc.). Recently, there have been numerous detailed structures published for a large variety of proteins (available through various websites such as The Enzyme Structures Database). The 3-dimensional structure of many aqueous soluble proteins have been well characterized. The structure of most membrane-bound proteins have yet to be determined, partially because when these proteins are purified their structure is compromised. But, nevertheless, the structure of several membrane-bound proteins have been determined (see for example, Figure 2B).

In materials science, we measure various physical parameters and correlate these with function. Examples are grain size and elasticity, or phase of the solid and conductivity. Some physical properties of proteins that can be measured include amino acid sequence, structure, and isoelectric point. The total number of charged groups can also be determined from the amino acid sequence. A relatively newly researched parameter is a protein's "melting point", or transition to a molten globule. A protein in the molten globule state is one that has an intact exterior and a fluid interior (Fink, 1999). The exterior shell is intact because it is hydrophilic and held in place with the assistance of surrounding water molecules. When proteins reach a certain temperature, they become more fluid and, some reach the molten glubule state. This transition has been studied for several proteins. But, not all proteins are known to undergo this transition. However, they can and often do undergo other structural modifications with temperature.

In this chapter, we will spend some time examining a newly identified physical property of proteins, the kinetics of surface-mediated unfolding. This parameter appears to be important in understanding the dynamic function of proteins. Examples will be given, but there are many functions of proteins that require flexibility or rigidity. These can be estimated by examining this new unfolding kinetics parameter.

Nearly all proteins exhibit remarkable symmetry. It is not yet clear why proteins commonly exhibit the radial and/or mirror symmetry discussed earlier, particularly since their amino acid sequences are different throughout. Certainly, structure and function are related, but it is not obvious whether the function of all proteins requires mirror, and/or radial symmetry. Could it be that proteins are more efficient when they are symmetrical, being twice as likely to collide and interact in the proper orientation to perform their intended function?

4. Surface-Mediated Unfolding of Proteins

Proteins have been known to unfold, or denature, on various rigid surfaces, but the kinetics have only recently been measured. In solution, when proteins are heated above a certain temperature, or if they are introduced to a flexible surface, they can unfold very quickly [5]. On a rigid support the kinetics appear to be much slower [6].

The unfolding of the protein lysozyme on a hydrophobic surface was reported about 15 years ago when the enzyme was exposed to a reversed-phase high-performance liquid chromatography (HPLC) support for varying periods of time [6]. In those experiments, lysozyme disappeared slowly as the enzyme was kept on the sorbent material. This process took several minutes before all lysozyme was lost. These studies were carried out on a fully saturated alkane surface. Or, an excessively hydrophobic material. Under these conditions, proteins are expected to unfold over much shorter periods of time as on surfactants in solution [5].

We have demonstrated losses of protein to sorbents (Goheen and Hilsenbeck 1998; Herbold, Miller et al. 1999) previously believed to resist protein unfolding [7]. From these results, we have been able to examine kinetics of protein unfolding on several different surface chemistries [8, 9]. This unfolding rate depends on the molecular weight of the protein and the surface chemistry.

Hydrophilic supports such as those designed for gel permeation chromatography, are not anticipated to unfold proteins when used under physiological conditions (37°C,

124

.015 M NaCl, pH 7.4) [10]. This well known property suggests that for water-soluble proteins, the most hydrophilic surfaces (typically those with hydroxyl or other polar groups) resist protein adsorption and unfolding, probably due to the thick hydration layer protecting the surface [11, 12].

The kinetics of the unfolding of proteins shown in Figure 6 has been studied for only a few water-soluble species. From Figure 6, if the percent of unfolded protein can be represented as **P**, leaving the following relation:

$$\mathbf{P} = \mathbf{m}(\log MW) \qquad (1)$$

Where **m** depends on surface chemistry, temperature, density of surface functional groups, pH, and other variables. But for a single, non-changing surface and otherwise constant conditions, **m** can be considered a constant. The surface conditions under which equation 1) was generated include adsorption to a rigid polymer matrix. If the surface consisted of a polymer with more flexibility, there may be some contribution of the surface fluidity along with the protein flexibility to assist in the unfolding process. Because, at the nanometer scale motion of polymer side chains should be relatively greater than larger objects at the macro scale, as a function of temperature and rigidity of the polymer matrix.

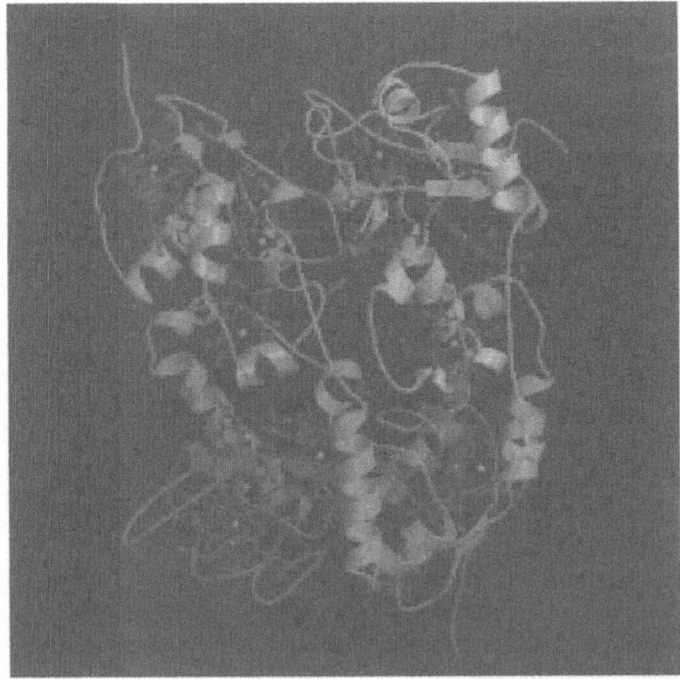

Figure 4: The three-dimensional structure of cytochrome c. obtained from the website: URL: http://www.rcsb.org/pdb/cgi/explore.cgi?pdbId=1AKK&page=20&pid=25921034898434&job=graphics.

It may be inappropriate to compare just a few proteins and their unfolding kinetics, especially when all these proteins are similar in that three out of four of them are from blood plasma. Cytochrome c (Figure 4), a protein of the cytosol in cellular fluid, also follows this trend (Figure 6). Yet, we have not yet compared proteins from other sources such as plants and bacteria to those from mammals. But, it is reasonable that much larger proteins will typically have a greater variety of exposed functional groups and will be more flexible than small proteins. So, we believe that all water-soluble proteins will experience this trend of larger proteins unfolding more rapidly.

Some of our experiments were designed to closely resemble the results one might expect when blood passes over a solid. But adsorption to more flexible substrates, such as other proteins or micelles, should more closely resemble adsorption behavior with the behavior in a cell with another protein or a membrane, or a receptor. Yet, the interactions with a solid support should resemble the interactions with a rigid implant. The interactions between blood proteins and solids should be slower than the adsorption processes between two nano-sized particles because of two factors. One of these is more rapid diffusion. The other is steric hindrance of the solid support.

The three most abundant proteins in blood are albumin, immunoglobulin G (IgG), and fibrinogen. For weakly cationic surfaces such as those derivitized with diethylaminoethyl- (DEAE), only about half of the protein IgG adsorbed in our laboratory [8]. Albumin desorbed from the same surface at conditions which were nearly physiological, and fibrinogen bound strongly, desorbing when the NaCl concentration was raised to ca. 0.2 M.

The dependence of unfolding kinetics on protein molecular weight is consistent with the observation that fibrinogen coats foreign materials. Because, although fibrinogen is 10X less concentrated in blood than albumin (3.94 vs 0.33g/100g plasma) [13], it unfolds much more rapidly and probably displaces any bound albumin [14]. Albumin (69,000 MW) and IgG (156,000 MW) are smaller and both have slower unfolding kinetics. This can be seen in the plot of Figure 6. In Figure 6, two different types of surface chemistries are compared. One consists of weak ion exchange (IEC), and the other is a weakly hydrophobic substrate (HIC). The slopes of these two lines indicate that surface chemistry influences protein unfolding kinetics. It is also noteworthy that the logarithmic relationship between molecular weight and unfolding kinetics appears to be preserved for two very different surface chemistries.

It is worth noting that a direct comparison between the unfolding kinetics in solution, on a fluid surface, and a solid support with identical surface chemistries has not yet been reported. This is an area for additional work to determine not only how various materials and surface chemistries cause proteins to unfold, but also how proteins themselves will behave under a variety of near-physiological conditions.

Of all the major blood proteins, fibrinogen unfolded more rapidly on all examined surfaces. This may be because it has the most amino acid side chains exposed to the environment, with both negative and positive charges as well as hydrophobic patches distributed throughout its hydrated surface. This makes us believe that the protein is so large and flexible that it can easily deform and adhere to almost any substance. Smaller proteins, such as cytochrome c, can also deform, but much more slowly [15].

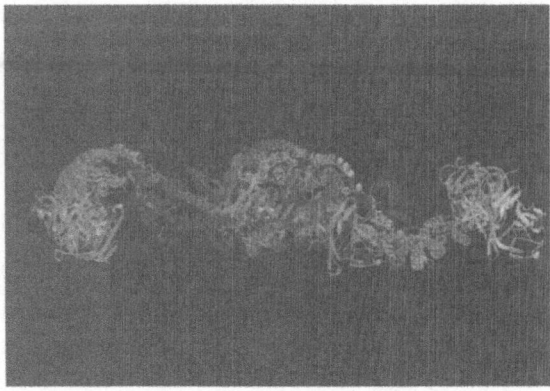

Figure 5. The detailed structure of fibrinogen is not yet fully known. However, this figure shows the 3-dimensional structure of the majority of this protein with less than optimal detail. Notice the symmetry and characteristic dumbbell shape. This figure was taken from the Internet at URL http:www.rcsb.org/pdb/cgi/explore.cgi?pdbId=1DEQ;page=0;pid=58861033687804;job=graphics&opt=show&size=500.

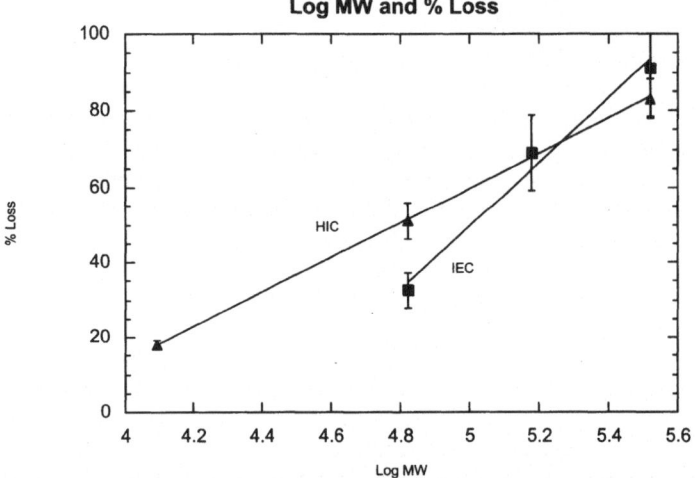

Figure 6. This shows the loss of protein with the log of the molecular weight for cytochrome c, albumin, IgG, and fibrinogen. The losses were measured at a holding time of 9 minutes. This figure is reproduced with permission from the authors [16].

5. Blood Coagulation

Blood contains not only proteins but also red and white cells, platelets, lipoproteins, various small organic molecules (such as hormones), and electrolytes. When foreign objects such as glass are introduced, not only do proteins adsorb, and a rejection process begins [17], but the rejection of implants can also be caused by an immune response (infection). These processes are initiated by the adsorption of proteins.

A material introduced into the body, adjacent to blood, can collide with all the available plasma proteins, blood cells, platelets, etc. at a frequency related to their size, surface area, and concentration. The components that remain attached to the material are determined by both the thermodynamics and kinetics of the process. From a thermodynamic perspective, either entropy or enthalpy terms predominate in the adsorption process. Enthalpy predominates in ionic interactions. Entropy prevails in hydrophobic intereactions. If the overall free energy is lowered by adsorption, the blood component will adsorb. There are often several components that will adsorb due to the complexity of blood for almost any surface chemical characteristics.

Although several components can adsorb at once, more typically, surfaces will select specific macromolecules out of blood with both high affinity and high concentration. And, in many cases, the first adsorbed species will be displaced by others with higher affinity for the surface. This displacement process, for proteins in blood, is known as the Vroman Effect [18-20]. In the Vroman Effect adsorbed proteins with a lower interfacial energy will displace other adsorbed proteins already on a surface. The Vroman Effect does not describe how this occurs, only that the end result is the displacement of adsorbed proteins.

The unfolding of fibrinogen on a surface (Chan and Brash, 1981) starts the blood clotting and rejection process (Lindon, et al., 1986). Fibrinogen becomes the structural component of clots when it is polymerized to form fibrin. It adsorbs and unfolds quickly to many different types of solids [16, 21-37]. This specialized design helps fibrinogen initiate surface reactions. If this protein coats the foreign material, it forms a solid anchor for the subsequent polymerization reaction, or clot formation (Linden, et al., 1986).

Fibrinogen is a dumbbell shaped soluble blood protein (Fig. 5) and has both radial and mirror symmetry. Fibrinogen's flexibility in solution may help it participate more efficiently in the blood coagulation process. Its flexible nature can help it conform to any wound configuration.

X-ray crystallographic structures of fibrinogen have not yet been obtained but segments have been crystallized and examined in detail (see Figure 5). Fibrinogen is "fluidic" due to its role in the blood, which seems to be partially to find foreign surfaces, adsorb, unfold, and initiate the blood clotting cascade [38]. This is an ideal protein to initiate such a cascade reaction because it is also involved in polymerizing with itself to form fibrin, the main structural component of the blood clot. The blood clot is a random matrix of fibrin strands that traps platelets and white cells to accelerate the healing process.

128

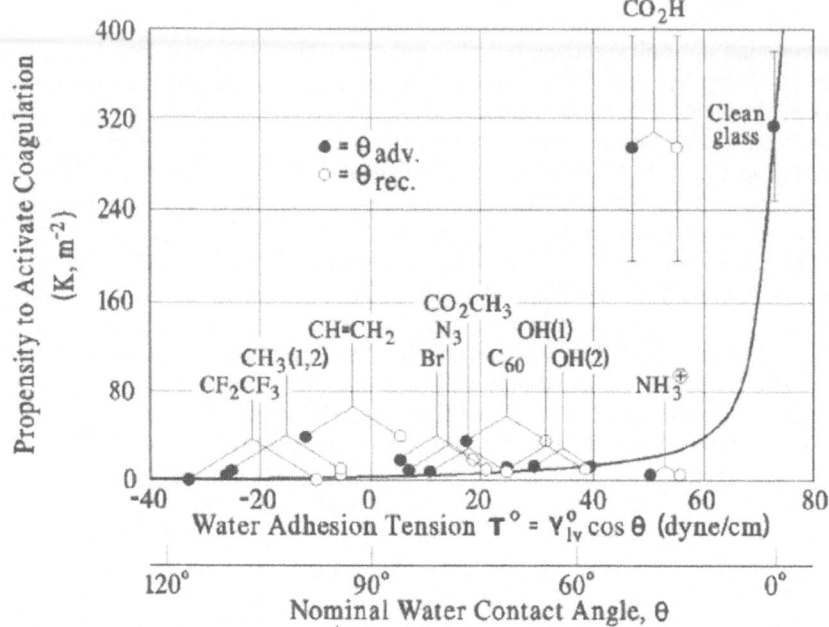

Figure 7. Exponential-like trend in the propensity to contact activate the plasma coagulation cascade (as measured through K in m^{-2}) with saline wettability T0 of a system f glass-disc procoagulants bearing close-packed, self-assembled saline monolayers terminated in different functional groups (filled circles = advancing contact angles; open circles = receding contact angles). Annotations identify terminating functional group, and the parenthesized numbers refer to different trials with the same surface. K was obtained as described in Volger, et al (1995). This figure was obtained with permission from the author, Dr. Erwin A. Volger who is located at the Materials Research Institute and Life Science Consortium, University Park, PA.

It can be imagined that surfaces could be designed with functional groups that resist the adsorption of soluble proteins. But it is not clear that these surfaces will also inhibit blood clot formation. Blood coagulation on a solid implant is believed to be initiated by protein adsorption [17]. When blood clot tests were performed on surfaces of various well-defined chemistries, the surface chemistry that caused the slowest rate of coagulation was hydrophobic (Figure 7). That causing the fastest coagulation was a surface terminated by a hydroxyl group [17]. This is the opposite of what was expected. Those surfaces that promote protein adsorption most aggressively appear to inhibit, not accelerate blood coagulation. All the surface chemistries tested (in Figure 7) were constructed using a self-assembled monolayer (SAMs). These are believed to be the cleanest, purest surface chemistries that can be produced [39-43]. SAMs are produced by starting with a clean (e.g. gold) surface, attaching a thiol group, and then adding an alkyl chain with some carefully selected terminal chemical group (see Fig. 8).

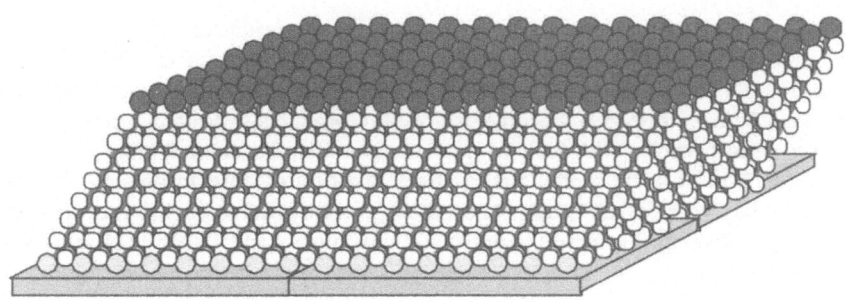

Figure 8. This depicts a self-assembled monolayer (SAM) on a gold surface. Light gray circles at the base of each acyl group depict a thiol linkage. White circles depict methyl groups in an alkyl chains, and dark gray circles at the top of each acyl group depict hydroxyl groups. This figure was obtained with permission from Dr. Barbara Tarasevich of the Pacific Northwest National Laboratory.

In Figure 8, a terminal OH group is shown. The alkyl chains on these sandwich structures are believed to smooth out any imperfections that may be present on the gold surface (although the lower surface does not need to be gold). And although these surfaces may not be perfect [43], they are cleaner than most other synthetic surfaces. When protein adsorption experiments have been performed on these, there has been no evidence that the acyl chains of the SAMs are disrupted by the sorbed material. One plausible explanation of Volger's data is that the hydrophilic surface binds the most hydrophilic region(s) of the various proteins in blood (such as fibrinogen), which more rapidly activates the clotting cascade than when the more hydrophobic regions are bound. This example suggests that the rate of protein adsorption may not be a rate-limiting step, or may have little to do with the overall rate of clot formation. The production of a clot requires a series of complex reactions [44] of which protein adsorption is a small but essential part.

To correlate this behavior with the behavior of nano-materials means we should understand the physical properties of nano-materials and how their physical properties correlate with their function when placed in a dynamic biological environment.

Since proteins are complex components of dynamic systems, it may be difficult to develop synthetic nano-materials that can avoid protein adhesion, fouling, and failure. Some significant efforts have been made to design coatings that will resist the adsorption of proteins. But, as we have seen with blood coagulation, this alone may not be sufficient. And, these have typically shown limited success. But the most successful coating strategies to resist protein adsorption include hydrophilic chemistries. These surfaces prefer to bind water more strongly than proteins and therefore squeeze out protein molecules. Such a coating strategy may help enhance the life of nano-biomaterials, but will probably be insufficient to allow them to have unlimited life.

Other than fouling, there are other reasons why proteins are important components in nano-materials. For example, there are new nano-technologies emerging in which proteins perform a major role. One example is the recent development of in vitro methods to transport large molecules on a flat surface using the proteins that are responsible for intracellular transport [45]. It is not yet clear how these will be used commercially, but being able to transport small objects at the nanoscale could have considerable value in future mechanical or biomedical applications.

Any synthetic nano-devices placed in the body or similar environment will likely encounter numerous biomolecules (lipids, carbohydrates, polysaccharides, etc). It is necessary to better understand these molecules, their interaction with solids (nano-materials) before they can become successful products. Proteins are beautifully designed nano-particles with their own natural symmetry, structure, and function. The natural function of some proteins is to attack and destroy foreign objects. Implanted nano-materials will be successful when these processes have been better understood and we have developed schemes to outsmart these ancient and elaborate processes.

6. Electron Transport

Cytochrome c (Fig. 4) and fibrinogen (Fig. 5) are both water-soluble proteins, but their functions, their size, and their shapes are dramatically different. Cytochrome c is a small protein, with a molecular weight around 12,400. Fibrinogen is much larger with a MW of about 340,000. Cytochrome c contains a porphyrin ring with an iron atom at its center. The iron fluctuates between oxidation states as it transfers an electron with cytochrome oxidase, a much larger membrane-bound protein.

At least two structural transitions have been reported for cytochrome c as a function of temperature. One near 53 °C, and a second at about 82 °C (Myer, 1968). Surface-mediated unfolding on an anionic substrate starts at about 40 °C [15], a lower temperature than the first structural transition in solution. In solution, the first transition (at 53°C) is the composite of the unfolding of the polypeptide chain and the uncoupling of the heme-helix interaction (Myer, 1968). If the transition in solution at 53°C is the same as the unfolding transition on an anionic support at ca. 40°C, then it would be reasonable to predict that the cytochrome molecule absorbs at the site where the polypeptide chain unfolds or the heme group is attached to the alpha helix. Correlations such as this help us gain insight in how the structure of a nanoparticle, cytochrome c, might be correlated with its function; transferring an electron to and from the porphyrin ring.

In contrast to fibrinogen, cytochrome c (Figure 4) is much smaller, has a very well defined structure, resists unfolding in solution, and is easily crystallized. Cytochrome c exchanges an electron with cytochrome oxidase, a membrane-bound protein. This interaction is a key process in energy production in the mitochondria. Cytochrome c only deforms on negatively charged surfaces, and only to a slight extent. This protein does not unfold unless it is near or above body temperature, and the reduced and oxidized forms interact differently [15], as one would expect if unfolding was an integral part of the electron exchange process.

Our data allows us to propose a model for the transfer of an electron with cytochrome oxidase in midochondria. That is, once the electron is transferred, cytochrome c appears to flex enough to separate the two molecules so the porphoryn rings are further apart and are no longer attached to one another. Then the cytochrome c goes back into solution to go through this cycle again.

7. Other Biomolecules

Biomolecules are not traditionally considered as nanomaterials. But they span the size range of nanostructures and can play a role in the naomaterials development. Nanotechnology is a broader discipline of which nanomaterials is a subset. In the world of materials science, products are generated from materials deigned by the materials scientist. The products of nanotechnology are no exception. In nanotechnology, conceptual models have been suggested for biological applications. Some of these applications require biocompatibility. Some require a thorough knowledge of complex biological processes. Almost all nano-sized objects will require careful evaluation and consultation by the materials community to determine the feasibility of producing these objects. This is because many of the components in nano-sized objects can contain only a small number of atoms. Such nano-machines cannot be machines as we know them today. They may, in fact, look much more like proteins, fragments of DNA, or other biomolecules. For these, we may turn to nature to find answers. And either by invoking existing biochemical processes, or through genetic engineering, this is where native nanotechnology (biotechnology), and materials science will merge.

Proteins, lipids, carbohydrates, DNA, RNA, etc. are all both biochemicals and nano-particles. Their size ranges from ca. 1.0 – 100 nm, but they are rarely thought of as structural elements to a materials scientist. Yet, many of these biomolecules perform structural functions in biotechnology. Nanostructures to those in the materials science field are more commonly nanometer-sized grains, or particles, thin films, or nano-tubes. These structural components are typically rigid, often enhancing the strength or durability of a material. Biomolecules can be thought of similarly, except biological structures are often much more fluid. Additionally, these natural nano-particles are very carefully designed so that they not only can act as structural components, but also typically have other complex chemical functions. Trans-membrane proteins are one example. They span the biological membrane, providing fluidity in the lateral dimension of a membrane (Fig 9) (structural proteins), while other proteins actively transport specific components across the membrane (performing their biochemical function).

In the red blood cell and many other membranes, the protein spectrin spans the interior of the cell to aid in transport throughout the cell and maintain the biconcave disc shape (Fig 9). These microtubules (actually in the nm size range) are attached to the cell membrane at specific membrane-protein sites [46].

132

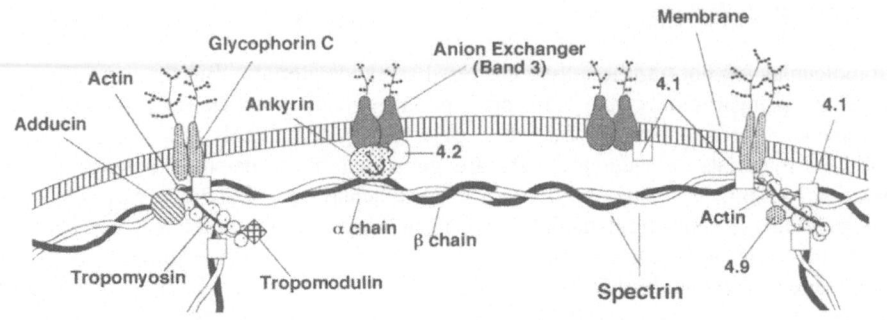

Figure 9. Membrane-cytoskeleton attachments in the erythrocyte. The diagram shows the major interactions among spectrin, ankyrin, adducing, band 3, protein 4.1, protein 4.2, dematin, glycophorin C, acin, and tropomyosin. Band numbers have been assigned due to their order of appearance in an electrophoresis gel. Note how the cytoskeleton protein (spectrin) is bound to some but not all of the trans-membrane proteins (glycophorin and the anion exchanger) through other linkages (ankyrin, band 4.1, 4.9, actin, and adducin). This cytoskeleton helps keep the erythrocyte shaped as a biconcave disk so it can transfer oxygen more quickly to peripheral tissues. This shape and these linkages also give the red cell adequate flexibility to squeeze through capillaries in the circulatory system. Reproduced with permission from Dr. Elizabeth J. Luna, Professor of Cell Biology at the University of Massachusetts Medical School [46].

These are just a few examples of how biomolecules form nanostructures in nature. Surfactants are commonly used in nano-materials science to help form micellular structures. These can aid us in the development of highly organized silicates and related porous silicates as well. Some proteins can be obtained in very high purity and with various shapes. These could also serve as a "soft" second-phase nanoparticle for materials which form at low temperatures such as polymers. Their presence could be used to create random or ordered imperfections to modify physical properties. Strands of collagen (protein) or cellulose (complex carbohydrate) should also be considered as potential additions to polymer matrices for structural integrity.

Biomolecules are often miraculously symmetrical. It is unlikely that we will ever be able to construct nano-materials as perfectly and as effortlessly as they are produced in nature, but understanding their structural role in biological systems can help us learn some of nature's secrets. These lessons can then be applied to the design of synthetic systems. Biological nano-particles can also themselves be used in synthetic systems. We already use some of these such as for paper, sponge, natural rubber, etc. But many of these applications were discovered by chance rather than design. As nanomaterials evolve, so will the incorporation of biomolecules in nanostructures and the adoption of biological design concepts in synthetic nanomaterials.

Nanomaterials, nano-sized objects, and nanostructures placed in close proximity with large proteins will likely collide and interact just as any surfaces. For very small, nano-sized objects (such as small grains or grain boundaries), it may only take a few proteins to cover the entire nanostructure. And for thin films, fouling, which depends on the resistance to adsorption and unfolding, may be one of the most important design considerations.

8. Conclusion

Can we benefit from understanding the structure of biological systems? Will our design of synthetic systems benefit from understanding the evolutionary development of biochemicals? We can only answer these questions once we fully understand the biological and biochemical processes, and why they are so efficient. Can we improve on the designs of nature, or can we only at best approximate these elaborate processes? Furthermore, can the natural processes be harnessed and used themselves, possibly enhanced through genetic engineering?

We as engineers can, and likely will design nano-materials that approach the eloquent characteristics of native proteins. To meet this goal, we will find ourselves consulting biotechnology, possibly synthesizing nometer-size products using genetically engineered bacteria. This takes advantage of nature's manufacturing process, but allows us to create a template (based on natural systems) for generating the desired product. Compromising the eloquence of natural macromolecules and learning only from the concepts is another approach. For example: When would an aqueous catalyst need to be rigid, and when fluid? Could we use what is known about the relationship between fluidity and MW to design more accessible active groups for such catalysts? How can recognition sites be optimized in sensors? How can we take advantage of the temperature/unfolding properties to keep the sensor array rigid and active? How should we design contact lenses so they do not attract even the most fluid protein or glycoprotein? How can fluidity of the surface moieties and hydration be utilized to keep foreign material from adsorbing? How can bandages be designed so they selectively remove unwanted components and contribute needed macromolecules during the healing process?

Nature has had millions, possibly billions of years to perfect various biological processes. Put another way, biological processes have been successful for millions of years because they are self-perpetuating, consume very little energy, and operate from the nano-scale to the grand scale, structured in layers complimenting each other's functions. The more thorough our understanding of biology at the nano scale, the more elaborate the structure appears to be.

It is easy to marvel at the eloquence of nature at all levels. The nano-scale is one dimension of nature that is not yet been fully explored. But as scientific methods improve, so do our means of viewing this tiny and intricate natural world. In it are the keys to life sustaining processes. The assembly of DNA, the folding of proteins, assembly and function of membranes, transport of molecules through a cell, are all nano-scale processes not yet fully understood. Perhaps as we explore nano-materials and nanotehnology we will teach ourselves how these processes work. And, as we investigate these fields, we will discover our limitations in using nano-materials and implementing nanotechnology in natural systems. Regardless of what we discover in the nanometer range, understanding the nanometer processes of nature will help us develop a more effective nanotechnology.

9. Acknowledgements:

The authors thank those who helped generate much of the data and subsequent publications that led to the development of this article. Those included Dr. Vince Edwards, Jaqueline Hilsenbeck, David Lu, and Craig Herbold.

10. References

1. Menezes, A.J., V.J. Kapoor, V.K. Goel, B.D. Cameron, and J.-Y. Lu, *Within a nanometer of your life.* Mechanical Engineering, 2001. **August**: p. 54-58.
2. Edwards, J.V. and S.C. Goheen, *Design, synthesis and affinity properties of biologically active peptide and protein conjugates of cotton cellulose.* Recent Research develoopments in Bioconjugational Chemistry, 2002. **1**: p. 113-121.
3. Prisyazhnoy, V.S., M. Fusek, and Y.B. Alakhov, *Synthesis of high capacity immunoaffinity sorbents with oriented immobilized immunoglobulins or their $F_{ab'}$ fragments for isolation of proteins.* J. of Chromatography, 1988. **424**: p. 243-253.
4. Edwards, J.V., S.L. Batiste, B.M. Gibbins, and S.C. Goheen, *Synthesis and activity of NH2- and COOH-terminal elastase recognition sequences on cotton.* Journal of Peptide Research, 1999. **54**: p. 536-543.
5. Roder, H., G. Elove, and S.W. Englander, *Structural characterization of folding intermediates in cytochrome c by H-exchange labeling and proton NMR.* Nature, 1988. **335**: p. 700-704.
6. Lu, X.M., K. Benedek, and B.L. Karger, *Conformational effects in the high-performance liquid chromatography of proteins: Further studies of the reversed-phase chromatographic behavior of ribonuclease a.* Journal of Chromatography, 1986. **359**: p. 19-29.
7. Regnier, F.E., *High-Performance Liquid Chromatography of Biopolymers.* Science, 1983. **222**: p. 245-252.
8. Goheen, S.C. and J.L. Hilsenbeck, *High-performance ion-exchange chromatography and adsorption of plasma proteins.* J. Chromatography A, 1998. **816**(1): p. 89-96.
9. Goheen, S.C. and B.M. Gibbins, *Protein losses in ion exchange and hydrophobic interaction HPLC.* Journal of Chromatography A, 2000. **890**: p. 73-80.
10. Goheen, S.C., *The Influence of pH and Acetonitrile on the High Performance Size Exclusion Profile of Proteins.* Journal of chromatography, 1988. **11**: p. 1221-1228.
11. McPherson, T., A. Kidane, I. Szleifer, and K. Park, *Prevention of protein adsorption by tethered poly(ethylene oxide) layers: Experiments and single-chain mean-field analysis.* Langmuir, 1998. **14**: p. 176-186.
12. McPherson, T.B., S.J. Lee, and K. Park, *Analysis of the Prevention of Protein Adsorption by Steric Repulsion Theory*, in *Proteins at Interfaces II: Fundamentals and Applications*, J.L. Brash, Editor. 1995, American Chemical Society: Washington, D. C. p. 395-404.

13. Tiselius, A., *A new apparatus for electrophoretic analysis of colloidal mixtures.* Trans. Faraday Soc., 1937. **33**: p. 524-531.

14. Malmsten, M., D. Muller, and B. Lassen, *Sequential adsorption of human serum albumin(HSA) Immunoglobulin G and Fibrinogen at HMDSO Plasma polymer surfaces.* Journal of Colloid and Interface Science, 1997: p. 88-95.

15. Herbold, C.W., J.H. Miller, and S.C. Goheen, *Cytochrome c unfolding on an anionic surface.* Journal of Chromatography A, 1999. **863**: p. 137-146.

16. Goheen, S.C., B.M. Gibbins, J.L. Hilsenbeck, C.W. Herbold, and J.V. Edwards. *Adsorption and surface-mediated unfolding of proteins.* in *National ACS Meeting.* 1999. New Orleans, Louisiana.

17. Volger, E.A., J.C. Graper, G.R. Harper, H.W. Sugg, L.M. Lander, and W.J. Brittain, *Contact activation of the plasma coagulation cascade. I. Coagulant surface chemistry and energy.* Journal of Biomedical Materials Research, 1995. **29**: p. 1005-1016.

18. Vroman, L., *What factors determine thrombogenicity".* Bull. N.Y. Acad. Med., 1972. **48**: p. 302-310.

19. Horbett, T.A., *Mass-action effects on competitive adsorption of fibrinogen from hemoglobin-solutions and from plasma.* Thromb. Haemostas, 1984. **51**(2): p. 174-181.

20. Slack, S.M. and T.A. Horbett, *The Vroman Effect,* in *Proteins at Interfaces II.* 1995. p. 112-128.

21. Brash, J.L., C.F. Scott, P.T. Hove, P. Wojciechowski, and R.W. Colman, *Mechanism of transient adsorption of fibrinogen from plasma to solid surfaces: Role of the contact and fibrinolytic systems.* Blood, 1988. **71**: p. 932-939.

22. Ryu, G.H., J. Kim, Z. Ruggeri, S.H. Han, J.H. Kim, and B.G. Min, *Effect of Shear Stress on Fibrinogen Adsorption and Its Conformational Change.* ASAIO Journal, 1995. **41**: p. M384-M388.

23. Shiba, E., J.N. Lindon, L. Kushner, M. Kloczewiak, J. Hawiger, G. Matsueda, B. Kudryk, and E.W. Salzman, *Conformational changes in fibrinogen adsorbed on polymer surfaces detected by polyclonal and monoclonal antibodies,* in *Fibrinogen 3: Biochemistry, Biological Functions, Gene Regulation and Expression: Proceedings of the International Fibriongen Workshop, Mulwaukee, Wisconsin, 13-15 June 1988.*, J.P. DiOrio, Editor. 1988, Excerpta Medica: New York. p. 239-244.

24. Chinn, J.A., J. Richard E. Phillips, K.R. Lew, and T.A. Horbett, *Tenacious Binding of fibrinogen and albumin to pyrolite carbon and biomer.* Journal of Colloid and Interface Science, 1996. **184**: p. 11-19.

25. Grunkemeier, J.M. and T.A. Horbett, *Fibrinogen adsorption to receptor-like biomaterials made by pre-adsorbing peptides to polystyrene substrates.* Journal of Molecular Recognition, 1996. **9**: p. 247-257.

26. Tsai, W.-B., J.M. Grunkemeier, and T.A. Horbett, *Human plasma fibrinogen adsorption and platelet adhesion to polystyrene.* Journal of Biomedical Materials Research, 1999. **44**: p. 130-139.

136

27. Horbett, T.A. and K.R. Lew, *Residence time effects on monoclonal antibody binding to adsorbed fibrinogen.* Journal of Biomaterials Science Polymer Edition, 1994. **6**(1): p. 15-33.

28. Grunze, M., P. Harder, and R. Dahint. *Adhesion of proteins on self-assembled organic model surfaces.* in *The 20th Annual Meeting of the Adhesion Society.* 1997. Hilhar Head Island, South Carolina.

29. Tang, L., *Mechanisms of fibrinogen domains: biomaterial interactions.* Journal of Biomaterials Science: Polymer Edition, 1998. **9**(12): p. 1257-1266.

30. Whitlock, P.W., S.J. Clarson, and G.S. Retzinger, *Fibrinogen adsorbs from aqueous media to microscopic droplets of poly(dimethylsiloxane) and remains coagulable.* Journal of Biomedical Materials Research, 1999. **45**: p. 55-61.

31. Balasubramanian, V., N.K. Grusin, R.W. Bucher, V.T. Turitto, and S.M. Slack, *Residence-time dependent changes in fibrinogen adsorbed to polymeric biomaterials.* Journal of Biomedical Materials Research, 1999. **44**(3): p. 253-260.

32. Bohnert, J.L. and T.A. Horbett, *Changes in adsorbed fibrinogen and albumin interactions with polymers indicated by decreases in detergent elutability.* Journal of Colloid and Interface Science, 1986. **111**(2): p. 363-377.

33. Fabrizium-Homan, D.J. and S.L. Cooper, *Competitive adsorption of vitronectin with albumin, fibrinogen, and fibronectin on polymeric biomaterials.* Journal of Biomedical Materials Research, 1991. **25**: p. 953-971.

34. Chan, B. and J.L. Brash, *Conformational change in fibrinogen desorbed from glass surface.* Journal of Colloid and Interface Science, 1981. **84**: p. 263-265.

35. Zembala, M., J.C. Voegel, and P. Schaaf, *Elution process of adsorbed fibrinogen by SDS: competition between removal and anchoring.* Langmuir, 1998. **14**: p. 2167-2173.

36. Ortega-Vinuesa, J.L., P. Tengvall, and I. Lundstrom, *Aggregation of HSA, IgG and fibrinogen on methlyated silicon surfaces.* Journal of Colloid and Interface Science, 1998: p. 228-239.

37. Lin, J.-C. and S.L. Cooper, *In vitro fibrinogen adsorption from various dilutions of human blood plasma on grow discharge modified polythylene.* Journal of Colloid and Interface Science, 1996: p. 315-325.

38. Furie, B. and B.C. Furie, *The molecular basis of blood coagulation.* Cell, 1988. **53**: p. 505-518.

39. Seigel, R.R., P. Harder, R. Dahint, M. Grunze, F. Josse, M. Mrksich, and G. Whitesides, *On-Line Detection of Nonspecific Protein Adsorption at Artificial Surfaces.* Analytical Chemistry, 1997. **69**(16): p. 3321-3328.

40. Chapman, R.G., E. Ostuni, S. Takayama, R.E. Holmlin, L. Yan, and G.M. Whitesides, *Surveying for surfaces that resist the adsorption of proteins.* Journal of the American Chemical Society, 2000. **122**(34): p. 8303-8304.

41. Prime, K.L. and G.M. Whitesides, *Adsorption of proteins onto surfaces containing end-attached oligo(ethylene oxide): a model system using self-assembled monolayers.* J. Am. Chem. Soc., 1993. **115**: p. 10714-10721.

42. Wirth, M.J., R.W.P. Fairbank, and H.O. Fatunmbi, *Mixed self-assembled monolayers in chemical separations.* Science, 1997. **275**: p. 44-47.

43.   Wood, L.L., S.S. Cheng, P.l. Edmiston, and S.S. Saavedra, *Molecular orientation distributions in protein films. II. Site-directed immobilization of yeast cytochrome c on thiol-capped, self-assembled monolayers.* Journal of the American Chemical Society, 1997. **119**: p. 571-576.

44.   Ratnoff, O.D. and H. Saito, *Coagulation factors and the role of surface in their activation,* in *Annals of the New York Academy of Sciences,* E.F. Leonard, Editor. 1977, New York Academy of Sciences: New York. p. 283.

45.   Hess, H. and V. Vogel, *Molecular shuttles based on motor proteins: active transport in synthetic environments.* Reviews in mMolecular Biotechnology, 2001. **82**: p. 67-85.

46.   Luna, E.J. and A.L. Hitt, *Cytoskeleton-Plasma Membrane interactions.* Science, 1992. **258**: p. 955-964.

GENESIS OF NANOSIZED PARTICLES, GRAINS AND INTERFACES IN THE RATE-CONTROLLED PROCESSES OF SYNTHESIS AND SINTERING OF CERAMICS.

ANDREY V. RAGULYA and VALERY V. SKOROKHOD
Frantsevich Institute for Problems in Materials Science NAS of Ukraine
3, Krzhizhanovski St., 03142 Kiev. Ukraine. tel. +38-044-444-1533;
Fax:+38-044-444-2131, E-mail: ragulya@materials .kiev.ua

1. Introduction

Genesis of nanostructure is the key feature inherent in preparation of nanocrystalline materials using powder methods. Structure of particles originates from synthesis conditions. Density of defects such as twins, domains, stacking faults, core-and-shells as well as particle aggregates and agglomerates, strongly depends on process parameters. These features of particles define behavior of powder during consolidation and affect on the structure of monolithic sample. Such heredity has been revealed in the past in coarse particles and micronsize-grained bulk materials [1]. Decreasing of particle or grain sizes to the nanoscale leads a material to increasing of its structural sensitivity. Today we enable to make powders of the same particle size distribution and shape. But different processes of synthesis commonly give different defect structure, and therefore, demonstrate different sintering behavior and properties of sintered bodies. The most important achievement of today's techniques of synthesis and consolidation of nanoparticles is flexible control over properties [2]. To illustrate this from experiments, the present review is compiled.

New ideology and experimental achievements of rate-controlled thermally activated processes (RCP) of synthesis and sintering is considered useful in manufacturing of nanosized powders and nanograined materials. The rate-controlled processes strongly differ from the conventional synthesis and sintering due to feedback established between transformation value (completeleness of chemical reaction or shrinkage) and instantaneous temperature. During RCP, the transformation value is the dependent parameter, whereas temperature is the independent parameter, contrary to conventional processes. Sintering, chemical synthesis and phase transformations proceed through the competition between densification and grain growth or new phase nucleation and nuclei growth. This competition becomes an important fundamental prerequisite for flexible particle and grain size control within the nanoscale range, where useful "size effect" takes place. Some features of size effect in ceramics are considered bellow.

T. Tsakalakos, et al. (eds.) pgs. 139 - 154
Nanostructures: Synthesis, Functional Properties and Applications;
© *2003 Kluwer Academic Publishers.*

2. Peculiarities of Size Effect in Ceramics

Size effect is one of the most attractive fundamental problems in nanocrystalline materials as soon as it directly answers the question whether these extremely fine-grained materials are worth to be obtained. Variations in properties with scale and dimensionality occur in all solids, because the relative contribution from surface and volume energies are both size-dependent and scale differently for particles, fibers, films and polycrystals. Let us consider several examples for ceramic materials.

The miniaturization of today's multilayer ceramic capacitors exploits the "size effect" in barium titanate (Fig. 1) and demands decreasing of dielectric layer thickness to $0{,}5 - 1$ μm. Structure of such a thin layer must correspond to several requirements: the layer must be polycrystalline, its grains must be 3-D constrained and grain boundaries must be well conjugated. All these requirements originate from physics of ferroelectricity and internal stress influence the dielectric properties of barium titanate ceramics. Thinning of dielectric layer becomes possible making smaller grains when starting from nanocrystalline barium titanate powder. Today's technological schemes still cannot accept nanoparticles of 10-20 nm in manufacturing of thin dielectric layers because of some problems, dealing particularly with sintering [3,4].

Another example of material demonstrating ferroelastic behavior is pure or partially stabilized zirconia. Tetragonal zirconia expires double stabilization effect from dopants like yttria or rare earth oxides and from surface or grain boundary tension. Tightly curved grain boundaries exert a compressive stress on the grains, thereby favoring phase with smaller specific volumes than the equilibrium phase. With nanocrystalline grain sizes pure zirconia (normally monoclinic) becomes cubic [5]. Critical grain size for phase transformation depends on clamping conditions on the grain boundaries as it was shown above for barium titanate.

Structural ceramics demonstrates direct and reverse Hall-Petch behavior (Fig. 2) near extreme point as it was found for a critical size of dislocation loop in [6].

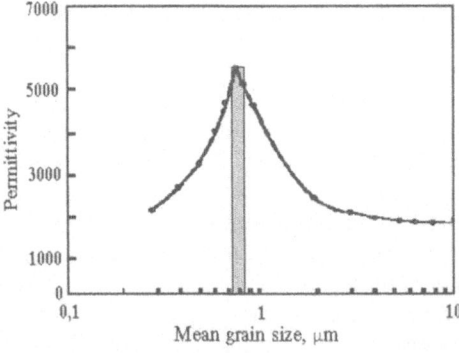

Figure 1. Arlt's grain size dependence of dielectric constant in barium titanate [4].

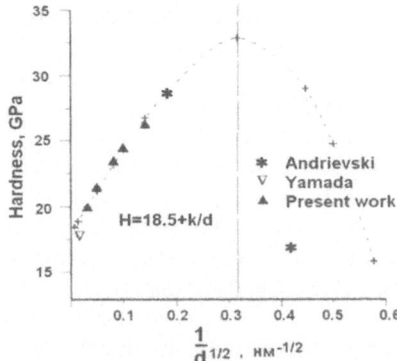

Figure 2. Hardness versus grain size for nano-TiN [6].

3. Synthesis

Chemical synthesis and phase transformations, including competition between new phase nucleation and nuclei growth, are often non-isothermal processes and therefore can be implemented under rate-controlled mode. In particular, the rate-controlled thermal decomposition of unstable precursor was assumed to result in the control of both particle morphology and size distribution of powder [3,7]. In this process, the transformation degree is the dependent parameter, whereas temperature is the independent one, contrary to conventional synthesis methods. Particularly, the oxalate-process is one of the best methods for BaTiO$_3$ powder preparation, which can be carried out under rate-controlled conditions for both pure and doped barium titanate. The same oxalate process was found available for yttria-stabilized zirconia. The behavior of the BaTiO$_3$ particle size as a function of the heating rate was investigated and the optimal temperature-time path has been calculated and experimentally checked to obtain an optimum powder with particles of 20-25 nm in diameter. More than two times difference in particle sizes for linear and rate-controlled modes was achieved in 3 mol.% yttria doped zirconia [7]. The smallest final particle size was around 6 nm. Important conclusion arises from the data presented here (Figs. 3,4): non-isothermal method of powder synthesis by decomposition of unstable precursors allows flexible control of mean particle size and dispersion of distribution. The advantage of the rate-controlled regime over linear heating rate regime is also shown in Figs. 3,4. The changes of product morphology and phase composition versus heating rate were studied in details stage-by-stage to define the mechanism of decomposition responsible for difference in particle size distribution of the final nano powder. It was found that the intermediate substances like polymerized complex carbonates of titanyl and zirconyl form dense or foam-like products under slow and fast heating rate regimes, respectively.

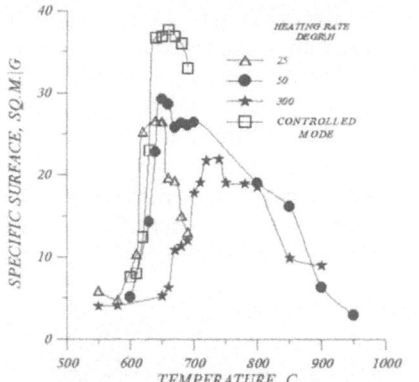

Figure 3. Temperature dependence of the specific surface area of barium titanate nanopowders [3].

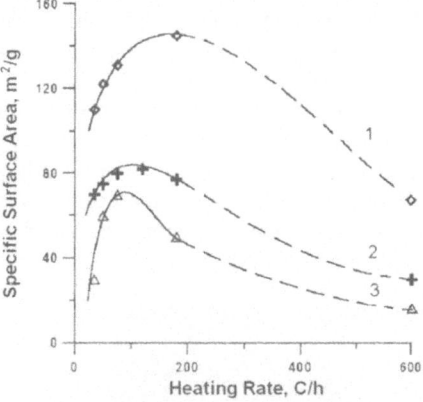

Figure 4. Temperature dependence of the specific surface area of yttria stabilized zirconia nanopowders [7].

4. Sintering

Four consolidation techniques were selected to make near fully dense nanostructured ceramics based on barium titanate, zirconia and titanium nitride: nonisothermal sintering with constant heating rate, rate-controlled sintering, high pressure sintering and quasi-hydrostatic compacting with following sintering. These methods provide different conditions for grain boundary formation in ceramics: diffusion controlled neck growth on sintering and deformation under high stresses on pressing.

Sintering with constant heating rate (CHRS). Precision dilatometry is the most powerful technique for sintering study of ultrafine and nanocrystalline powders, as it allows detailed study of densification kinetics on heating with constant and variable rate. Preliminary CHRS experiments with nano-barium titanate of two powders (BT_40, particle size of 25 nm and BT_125 - 80 nm) was carried out in two stages: rapid heating at 3000 °C/h up to temperature of 700 °C (temperature of shrinkage initiation) and subsequent slow constant heating rate stage of 100 ÷ 3000 °C/h to 1300 °C without isothermal hold. Preliminary sintering of YSZ nanopowder (of 10-12 nm) in two-step heating rate regimes without isothermal hold was carried out. Rapid heating of 20 °C/min up to 865 °C was changed by slow heating rate in the range from 0.5 to 10 °C/min. Final temperature of heating was determined as 1150°C. Sintering of nanosized TiN (15 nm) in vacuum was implemented in the temperature range upon 1500 °C with linear heating rates in the range from 0.1 to 3.0 °C/s without isothermal hold.

It is well studied that the nanocrystalline particles form agglomerates. Some quantity of agglomerates appears during synthesis and markedly increases during preliminary annealing of a powder. Agglomerates after annealing consist of strongly bonded particles owing to surface diffusion-controlled neck growth (Fig. 5). As-prepared particles have many degrees of freedom and enable to rotate with respect to each other and adjust lattices, forming "ideal" grain boundaries (Fig.6). Densification of the powder BT_40 is one stage process. Two densification stages were found for agglomerated powder labeled BT_125. Two-stage densification is a sequence of agglomeration due to rapid rate densification inside agglomerates on the first stage, and the subsequent sintering between agglomerates. This differential sintering is undesirable phenomenon, which usually results in nonuniform microstructure with a wide or multimodal distribution of pores and grains. This leads to the wrong conclusion that conventional pressureless sintering is inappropriate for densification of nanograined materials at all.

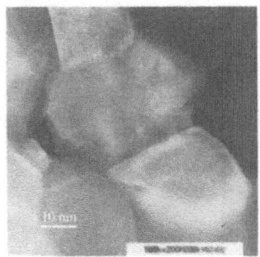 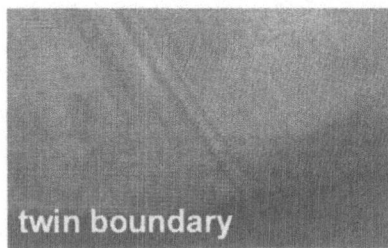

Figure 5. Agglomerate of coarse BT_125

Figure 6. Twin grain boundary in agglomerate of BT_125 as a result of annealing.

The grain growth during constant heating rate sintering is reversibly proportional to the heating rate (Fig. 7). Rapid sintering of a powder BT_40 with a heating rate of 3000 °C/h resulted in ceramics of 96 % dense and mean grain size of 0,5 μm (Fig. 8, grain growth factor was equal to 12), but slow heating rate of 100 °C/h gave 99 % density and 1,8 μm (factor is equal to 45). The intensive grain growth corresponds to the fractional density above 90 %, i.e. to the moment when open porosity disappears, the total number of pores in triple junctions rapidly decreases and the rest of pores unable to prevent grain boundaries from migration any more (Fig. 7). Such ceramics is unavailable for multilayer ceramic capacitors because of quite large residual porosity and size of grains. Rapid rate sintering gives smaller grain size, but porosity distribution through the volume is not homogeneous and pore size distribution is polymodal owing to temperature gradient and strong differential densification. Mayo in [8] showed and explained the dedensification effect in zirconia as a result of temperature gradient and inhomogeneous density. The best sintering regime with constant heating rate of 5 °C/min gave near fully dense zirconia (Fig. 9, 10).

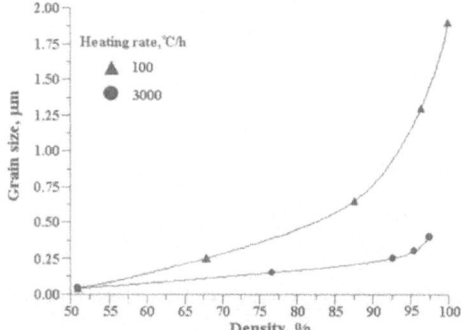

Figure 7. Grain size versus fractional density of BT_40 nanopowder of barium titanate at different heating rates

Figure 8. SEM image of the BT_40 sintered at heating rate of 3000 °C/h to 1300 °C.

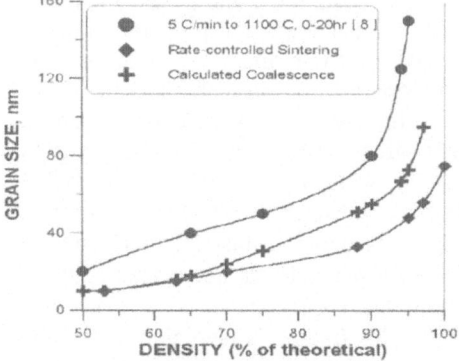

Figure 9. Grain size versus fractional density of 3YSZ at different sintering modes.

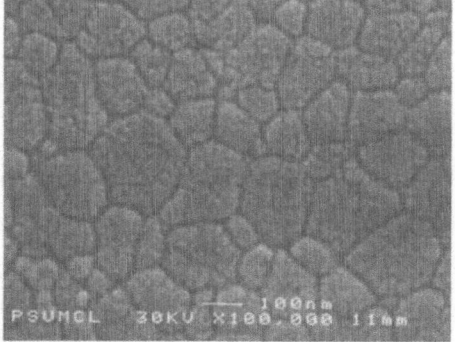

Figure 10. SEM image of the 3YSZ sintered at heating rate of 300 °C/h to 1100 °C.

The effects of differential sintering and dedensification can be overcome using the rate-controlled sintering, which allows minimization of diffusion controlled coalescence and equalizing densification rates inside and between agglomerates on intermediate stage of sintering. The shrinkage rate-controlled sintering allows us more flexibly affect the evolution of microstructure, to keep open porosity at total porosity of 7% and less.

Rate-Controlled Sintering (RCS). The best temperature time paths for sintering was calculated for nanosized powders of TiN, BT_40 and 3YSZ to determine the minimum final temperature of sintering suitable for achievement the maximum fractional density. This condition indirectly corresponds to the minimum grain growth. The calculation procedure can be carried out on the basis of the kinetic field of responses (Fig. 11), which is the geometrical image of the densification kinetics experimental data collected under constant heating rate conditions.

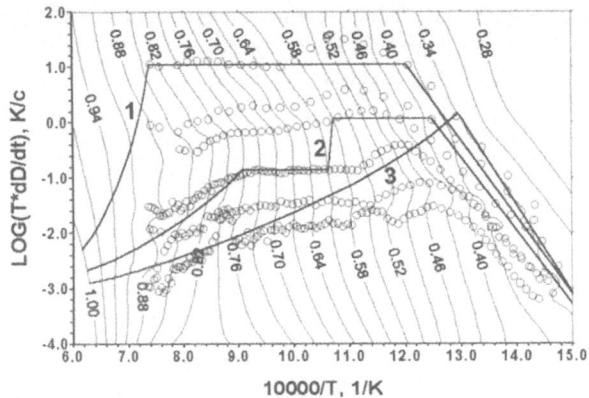

Figure 11. Kinetic field of responses (nano-Ni) and several examples of RCS protocols: constant shrinkage rate (1), classic (2), pick-shape (3); circles - experiment.

Obviously, the best RCS schedule can be defined as that, which results in full density and smallest grain size, if the nanograined ceramics is the target. Instead of grain size, one enables to optimize sintering by another important property, which is variable through the process. In this case, the schedule must be modified. The calculated non-linear temperature-time paths for sintering are presented in figs. 12, 14. The best one for barium titanate corresponds to the constant densification rate of 0,5 %/min on the initial and intermediate stages of sintering (fig. 12). Experimental verification of the calculated regime resulted in near fully dense (99.9%) BaTiO$_3$ ceramics with mean grain size of 0.35 μm and quite narrow grain size distribution (0,15 ÷ 0,6 μm). These grains are two times smaller compared to that after the constant heating rate sintering. The final temperature of RCS schedule did not exceed 1285 ± 5 °C, that is 15 ÷ 20 °C less than in constant heating rate regime and 40 ÷ 50 °C less than in conventional sintering technology of BaTiO$_3$. This finding is favorable for pinning the grain boundaries and indirectly shows uniform evolution of pore and grain microstructure on sintering (fig. 13).

Rate-controlled sintering of nanosized Y-TZP powders with various yttria content from 1.0 up to 3 mol% gave near fully dense ceramics ~99.9% with mean grain size of 75-80 nm (Fig. 13).

New opportunity of refinement the microstructure in the course of RCS was revealed for nano-TiN powder. Calculation of the kinetics field of response and thorough stage-by-stage study of porous structure showed that the sintering temperature can be markedly decreased on the final stage of densification as it is demonstrated in fig. 15. Obviously, making optimal the rapid rate sintering with shrinkage constant rate up to fractional density of 90%, it becomes possible to decrease sintering temperature from 1400 to 1250 °C and subsequently use an isothermal hold for 15-30 min to reach near fully dense ceramics (98%) and to retain grain size of 40-50 nm, i.e. two-three times smaller compared with that after CHRS regimes.

The detailed study of grain boundary structure as a function of consolidation parameters has been implemented. The term "structure of grain boundaries" denotes orientation of grain boundaries, orientation of grains, imperfections of grain boundary, and existence of the second (amorphous) phase along the plane of the boundary. The grain boundary misorientation in nanocrystalline materials cannot be thoroughly determined and practically impossible from the local diffraction patterns because of lack of Kikuchi lines, as the crystal is "insufficiently thick". Three couples of Kikuchi lines are to be defined for correct orientation of a boundary with respect to incident beam. Moreover, the fragments of transparent edge can be found as seldom as finer is the grain, and if overlapping of grains and moiré fringes is absent. Orientation of adjacent grains was found from bright field image of two-side fringe in both grains, which appears on tilting the specimen under incident beam. EELS technique was used for local chemical analysis. As the second phase has not been found in the sample, this

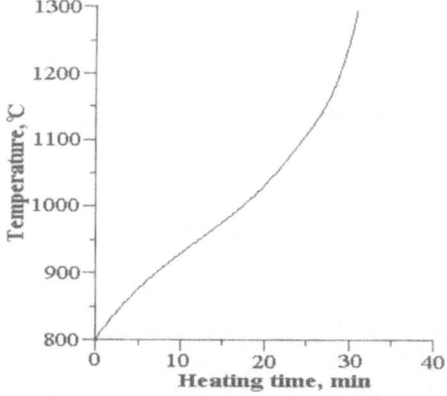

Figure 12. The temperature –time path for RCS of nanocrystalline barium titanate BT_40: constant sintering rate of 0,5%/min.

Figure 13. SEM image of sintered barium titanate after RCS of BT_40 powder.

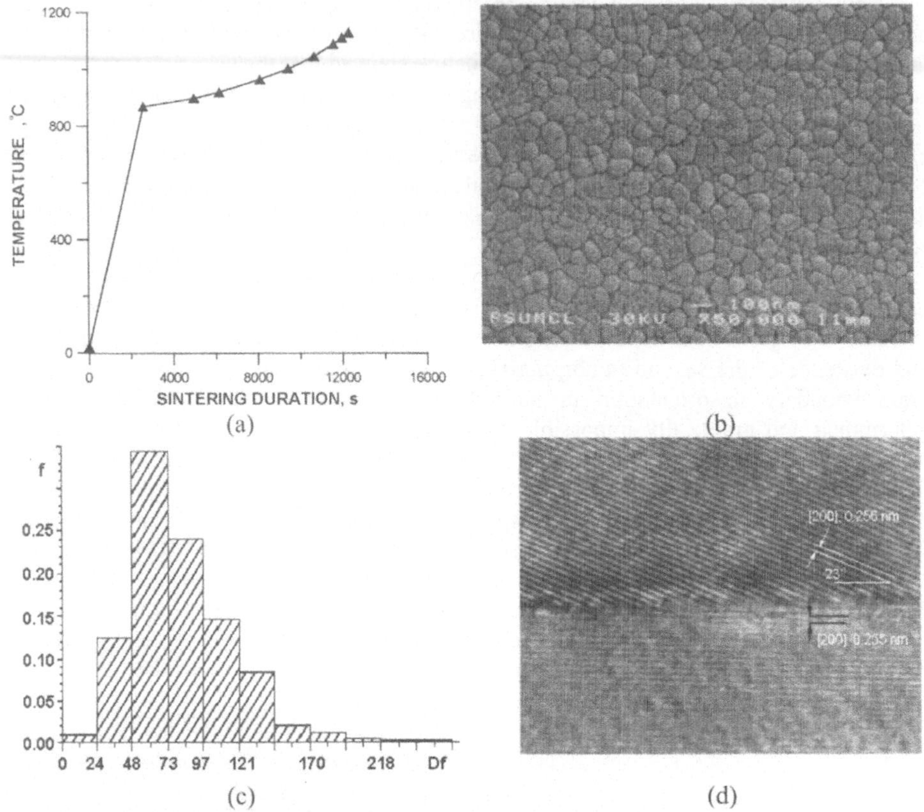

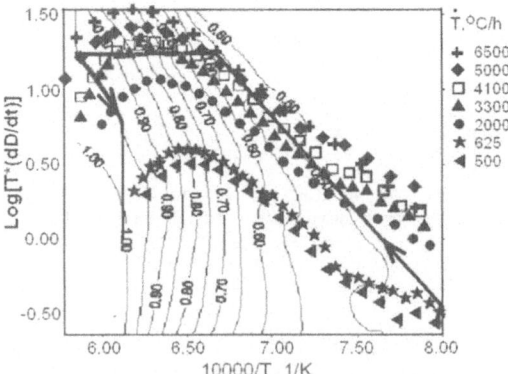

Figure 14. Regime of sintering and microstructure of 3 mol% stabilized ZrO_2:
(a) rate-controlled sintering temperature-time path for 3YSZ;
(b) thermally etched dense (99.9%) ceramics :
(c) grain size distribution, SEM image of equilibrium shaped grains
(d) HRTEM image of "pure" grain boundary.

Figure 15. Kinetic field of responses for sintering of nanosized TiN ceramics.

fact was considered the best evidence of high purity of the powder, ceramics and sintering atmosphere.

Grain boundaries in ceramics, sintered under rate-controlled regime of densification, are clean, perfect and large angle (Figs. 16 a,b). The grain boundary lattice is formed due conjugation of the lattices between the adjacent grains. Presently studied grain boundaries were found quite perfect (Σ = 5, 11, 17) with constant angle of misorientation along the facet. These properties correspond to equilibrium grain boundaries. The grain boundaries are fairly narrow (0.3 ÷ 0.5 nm) and do not contain amorphous phase. Such a boundary does not create charge barrier and corresponding depolarization field, which decreases the dielectric constant of nano ferroelectrics.

Thus, the rate-controlled sintered ceramics of barium titanate possesses high density, narrow grain size distribution and mean grain size of 0,35 μm and grain boundaries of high quality. All these parameters correspond to the requirements of advanced ceramic multilayer capacitors. However, further development of miniaturized capacitors, it becomes necessary to improve present technology of sintering, which would permit dense barium titanate ceramics with grain size of 80 ÷ 100 nm.

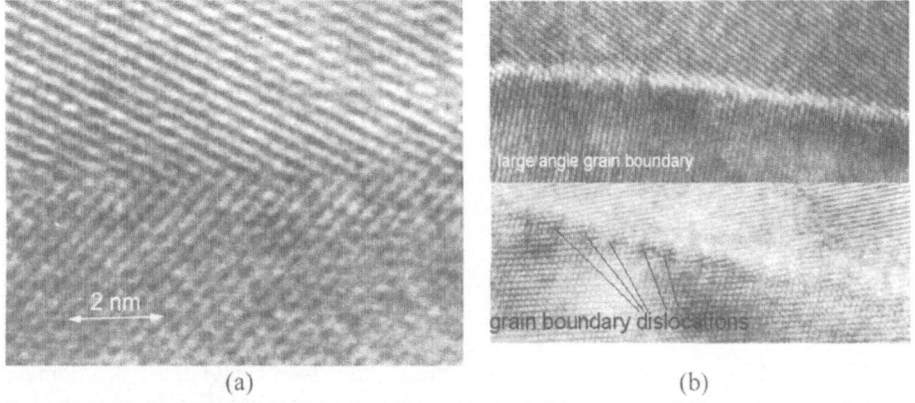

(a)   (b)

Figure 16. Examples of grain boundaries formed in ceramics during rate-controlled sintering of (a) barium titanate, (b) titanium nitride.

High pressure sintering. High pressure sintering (HPS) is one of the widely used technologies for consolidation of nanosized powders. The main feature of HPS is the consolidation of particles at moderate temperatures for several minutes, i.e. under conditions, which maintain the grain growth factor around 2-3.

The powder BT_40 was compacted under high pressure in two stages. The first stage was carried out at room temperature in steel die at low pressure of 500 MPa, and the second one was finished in high pressure cell at temperature 600 ÷ 1000 °C under quasihydrostatic compressing of 5 GPa. Density of samples sintered at room temperature is approximately proportional to the logarithm of applied pressure in the range from 500 MPa to 5 GPa and changes from 48 to 78.6 %. At 1000 °C and 5 GPa the BT_40 achieved 95 % of theoretical density. Increase of isothermal hold duration under pressure from 0.5 to 10 min resulted in negligible density increase up to 96,2 %. The residual porosity after high pressure sintering is closed and located in triple junctions. Incomplete densification of nanosized powders under high pressure is very good known fact earlier described elsewhere for barium titanate [9, 10], titanium nitride [11] etc. The final mean grain size around 50 nm was revealed after HPS, i.e. the grain growth factor was decreased to 1.5 times during this consolidation process (Fig. 17a).

Study of grain boundaries in HP-sintered ceramics has revealed two types of grain boundaries: coherent twinned and large angle multifaceted boundaries of variable misorientation. The former boundaries are equilibrium ones and the latter boundaries are non-equilibrium, incoherent, with low dense atomic structure, variable misorientation angle at constant angle of grain misorientation. Consequently, the multifaceted and rather fragmented grain boundaries have large free volume, imperfect boundary lattice (Fig. 17b). The width of such a boundary is approximately 1 ÷ 2 nm. Similar findings are known for other oxides: MgO, ZnO etc. [12]. The major part of grain boundaries have atomic structure with $\Sigma = 3$ и $\Sigma = 11$. Small curvature radius of nanograins is a reason for crystallographic orientation to change without preferable coincidence of cites.

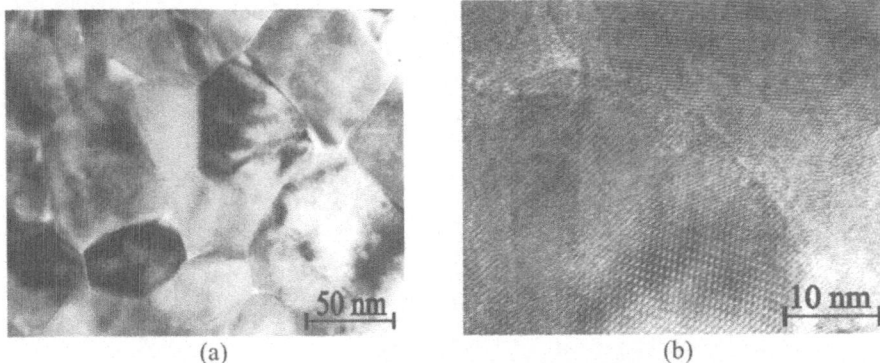

(a)   (b)

Figure 17. Microstructure of samples, obtained under HPS at 5 GPa and 1000 °C.

Barium titanate demonstrates low stress for twin migration or domain rearrangement of 0.01 MN/m^2. Therefore, twin boundaries appear (Fig. 18 b) with high probability. One of the most probable reasons for twin boundary formation is an adjustment of the primary nuclei contacting by the facets of the same crystallographic orientation and forming agglomerate on a stage of synthesis. The statistics of grain misorientation in nanostructured ceramics is defined by twin migration to the surface of pore or grain boundary during grain growth, and by grain adjustment with shift or rotation. Despite of high fractional density of specimens (95 ÷ 98 %), the grain boundaries are not completed and one can see rather mechanical contact than boundary lattice (fig. 18 a). The indirect evidence of imperfectness of such boundaries is their low strength: failure of sample occurs at low stresses or under incident electron beam in the column of HREM. Small pores of 2 nm in size locate in triple junctions between grains (fig. 18 a), proving lack of plastic deformation in the contact zone under high pressures. All the mentioned structural details of imperfect boundaries strongly influence the depolarization field increase, and therefore, dielectric constant decrease.

Nonequilibrium configuration of grain boundaries is able to become a reason of low mechanical strength of sintered material. Both pores and boundaries with large free volume are places of elastic energy relaxation. It would be fair to suggest the stress conditions quite different in defect structure compared to equilibrium porousless sintered ceramics with perfect grain boundaries. As soon as internal elastic stresses relax in part by breaking the intergrain contacts, the dielectric constant, the mechanical strength were found lower in HP-sintered samples, compared to sintered ones.

Though the technology of high pressure sintering allows keeping nanoscale grain structure, the obtained ceramics contains defects: residual porosity located in triple junctions, deviation from stoichiometry in oxygen sub-lattice, imperfect grain boundaries.

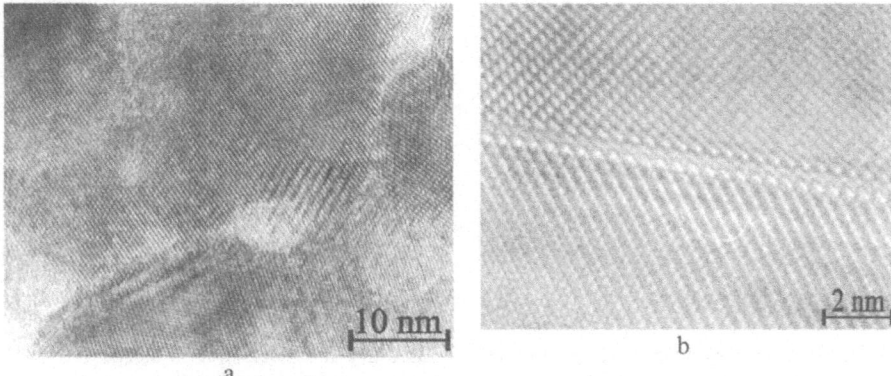

Figure 18. Grain boundaries in high-pressure sintered ceramics. a – fragmented grain boundaries with small pores in triple junctions; b – twin grain boundary.

All these defects markedly worsen the properties of ceramics. In particular, the dielectric properties of nano-ferroelectics will become substantially lower due to strong influence of depolarization field. The transformation toughening effect in partially stabilized zirconia ceramics will be weakened due to elastic energy relaxation. The combined method of consolidation was proposed to marry the advantages inherent in RCS and HPS in one process.

Combination of cold high-pressure compacting and RCS. Experiments were carried out with barium titanate nanopowder labeled BT_40. Cold consolidation under pseudo-hydrostatic pressure of 5 GPa at room temperature resulted in fractional density of 78 ÷ 80 %. Next step of ramp-and-hold sintering at heating rate of 1000 °C/h was carried out in dilatometer.

Densification of samples on heating proceeded in one stage. It is interesting to mention that the samples started shrinking at 950 °C, i.e. 200 °C higher than regularly observed for samples compacted at 500 MPa. Densification was completed at 1140 °C, that is 150 °C lower, than the final temperature of RCS regime.

As a result of such a combined technology, the ceramics with residual porosity of 1,4 %, and narrow grain size distribution in the range from 80 to 100 nm was obtained. The grain growth factor was of 2 ÷ 2,5 (Fig. 19 a).

The high resolution electron microscopy showed that the grain boundaries are clean and perfect similar to boundaries formed in RCS regime (fig.19 b). Such a microstructure of ceramics based on barium titanate has to satisfy the requirements to dielectric ceramics in multilayer capacitors.

Thus, the proposed combined method for consolidation of nanosized powders allowed obtaining near fully dense nanograined ceramics with perfect grain boundaries. Grain growth factor was around 5 or 6 that is two times less if rate-controlled sintering of nanosized particles has been used instead of ramp-and-hold one.

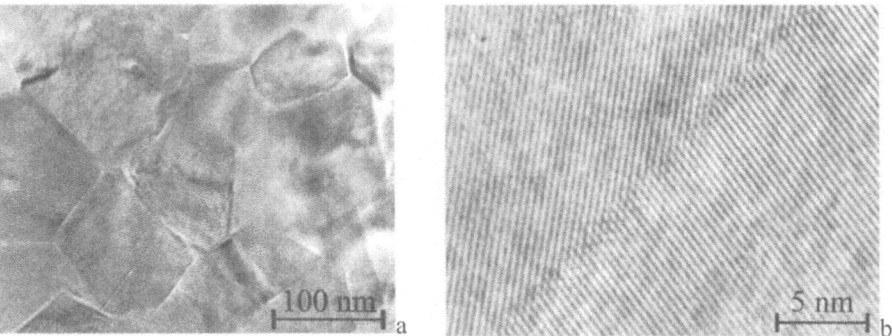

Figure 19. Microstructure of ceramics (a) and grain boundaries (b), obtained by combined technique.

The balance between surface and grain boundary mass transfer is suggested to be responsible for grain growth diminishing in the course of RCS. At the same time, this balance leads to perfect, well-conjugated grain boundaries in sintered ceramics. On the contrary, the deformation of particles prevails over diffusion-controlled accommodation at high pressure sintering and results in fragmented grain boundaries with large free volume. Obviously, the grain boundary structure dependent properties prove different.

5. Properties

Fig. 20 represents dielectric constant as a function of grain size at 70 °C for barium titanate ceramics, obtained by different consolidation. Dielectric constant extremely depends on grain size. Maximum K of 5000 was fixed for barium titanate with grain size of 0,7 μm. In submicron ceramics with grain size less than 1 μm, the dielectric constant strongly depends on consolidation technology. The barium titanate ceramics after the rate controlled sintering and combined compacting method, demonstrated the dielectric constant at least two times higher than high pressure sintered ceramics with the same grain size and fractional density. Presented extreme function of dielectric constant versus grain size is caused by competition of two processes: grain size decrease results in degradation in domain structure (from mosaic to single-domain microstructure), increase of internal elastic stresses and consequently, increase of dielectric constant, in opposite

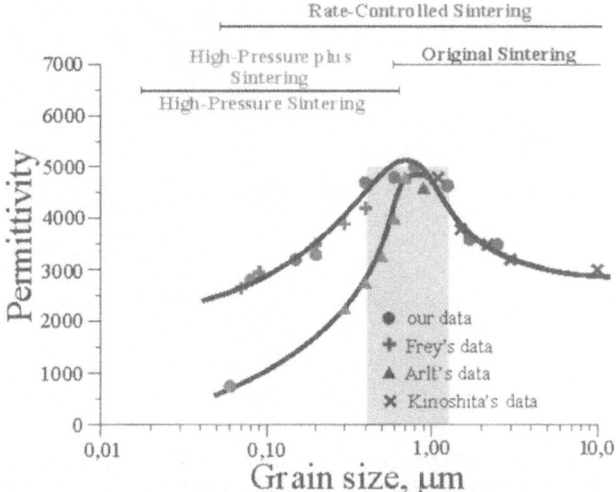

Figure 20. Dielectric constant of the dense barium titanate ceramics obtained by different consolidation methods versus grain size at 70 °C. Experimental points marked with one color, demonstrate that the ceramics was obtained by one method.

to dielectric constant decrease in consequence of growing volume of grain boundaries, which represents poor ferroelectrics. Reducing volume of grain boundaries, for instance, making them thinner, we enable to shift this maximum of dielectric constant to the area of smaller grains. It was revealed, that those processes, which provide diffusion-controlled mechanisms of grain boundary formation and flattening, are the most prospective for such a purpose. Sintering (RCS) and combined process of HP and RCS allowed the best diffusion-controlled accommodation of grains and formation of grain boundaries with minimal free volume. The prospective result of this research is "plateau" maximum of dielectric constant in the range of grain sizes of 0.3-1.1 μm instead of "peak" described by Arlt and Hennings in [13]. This result is more natural if the mentioned above competition acts. Presence of the "plateau" is much more important then a "peak" for the technology of multilayer capacitors.

Grain size distributions in the sintered zirconia stabilized by 1.0, 1.4, 1.7, 2.0 and 3 mol% Y_2O_3 are very similar. The microstructure, however, is absolutely different in the specimens where yttria content varies from 2 to 3 mol% and from 1 to 2 mol%. No martensitic transformations were observed in the first range of dopant concentration, tetragonal phase remains stable and the microstructure looks like in Fig. 14. Contrary, in the second range, the monoclinic phase becomes increasingly stable with respect to tetragonal. One can observe complex structure where larger grains are transformed into monoclinic phase (twins in the Fig. 21) and smaller grains are still tetragonal. The overstabilization coming from both grain size and dopant concentration in the first case does not work when yttria content reduced.

152

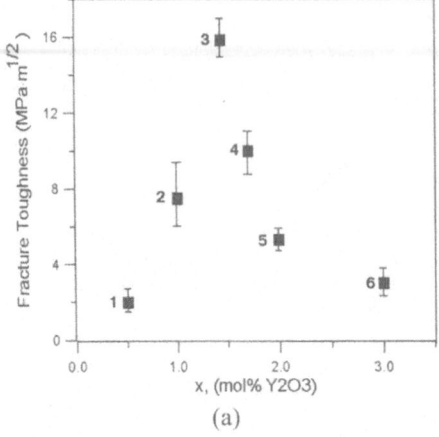

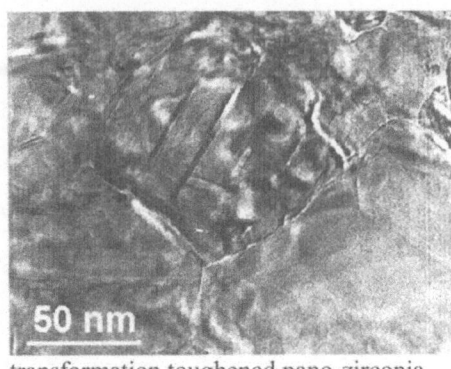

50 nm

transformation toughened nano-zirconia

(a)   (b)

Figure 21. a. Fracture toughness for nano-Y-TZP   *Figure 21 b.* Yttria stabilized ZrO_2 (1.4 mol% Y_2O_3)
vs. yttria content: 2,3,4,5,6 presently obtained data   TEM images of transformed and nontransformed
1,2,5,6 data from [6].   grains in the dense (99.8) ceramics

Grain size affects on fracture toughness of X mol% yttria stabilized zirconia ceramics
as it is shown in the Fig. 21 [14]. Rate-controlled sintered specimens of 3 mol% Y_2O_3
demonstrated the lowest fracture toughness. The same sintering process being applied
to nano-zirconia with reduced yttria content resulted in the highest fracture toughness
(Fig. 21; mean value of 15.5 MPa·m$^{1/2}$). This behavior has been predicted M.J. Mayo in
[14]. The mechanism of pinching off of pores due to tetragonal-into-monoclinic phase
transformation was discussed elsewhere. Note that the transformed volume should be
optimal for the best properties and can be thoroughly controlled by dopant
concentration and grain size distribution. This distribution is a function of sintering
technique. Another sintering regime dependent variable is a quality of grain boundaries,
which strongly affects on the relaxation of elastic energy during transformation. Thus,
making wide the size distribution of grains with perfect grain boundaries, we enable to
flexibly control the transformed volume and achieve highest fracture toughness of
ceramics.

Care about cleanliness of grain boundaries in TiN nanoceramics can be illustrated with
the aid of Fig. 2 and Table 1. Preliminary treatment of TiN powders in nitrogen and
hydrogen lead to different residual content of oxygen and presence of oxide phases on
boundaries. The highest hardness of the sintered nano-TiN has been achieved in sintered
ceramics with smallest grains and lowest content of oxygen.

TABLE 1. Mechanical properties of TiN nanocrystalline ceramics

Process and characteristics	TiN	TiN
Powders were preliminary treated in	nitrogen	hydrogen
Relative density	0,98	0,985
Lattice parametr, nm	4,2411	4,2362
Oxygen content, wt. %	2,95	2,34
Temperature of sintering, °C	1400	1250
Pore size, nm	100-120	~30
Grain size, nm	~150	50 - 70
Microhardness (P = 0.5 H), GPa	20 ±2,2	23 ±0,5
Nanohardness (P=120 mN), GPa	22,4±0,26	26,8±2,47
Fracture toughness, MPa m$^{1/2}$	2,4±0,1	4,2±0,4

6. Conclusions.

The conception of rate-controlled thermally activated processes has been presented. The rate-controlled thermal decomposition of unstable precursors results in nanoscale powders of barium titanate (25 nm) and yttria stabilized zirconia (10 nm).

Bulk ceramics of pure barium titanate, titanium nitride and yttria stabilized tetragonal zirconia have been obtained by rate-controlled sintering and high pressure sintering at different temperatures of nanocrystalline powders. Rate-controlled sintering of nanosized powders resulted in full densification (99.9% of theoretical density) and uniform grain structure, which have been studied under HRSEM and HRTEM. Atomic structure of grain boundaries in nanograined ceramics directly depends on sintering conditions. Nanocrystalline ceramics was characterized by fragmented grain boundaries after high pressure sintering and perfect boundaries as a result of diffusion controlled pressureless sintering. Combined RCS and HPS technique can be also applied to get nanograined ceramics with perfect grain boundaries.

The rate-controlled sintered ceramics with grain size of 300 nm, 75 nm and 50 nm for BaTiO$_3$, 3YSZ and TiN, respectively, demonstrated: quite high dielectric constant in barium titanate, around 5000±100 at room temperature, highest fracture toughness of 15.5±1.2 MPa·m$^{1/2}$ in zirconia stabilized by 1.4 mol% yttria, high hardness of 26±1.8

154

GPa. Presented results can be explained from the standpoint of so called "size effect" and high quality of grain boundaries.

7. References

1. Skorokhod, V.V., Solonin, Yu.M., Uvarova, I.V. (1990) *Chemical, diffusion and rheological processes in technology of powder materials*. Naukova Dumka, Kiev.
2. Skorokhod, V.V., Uvarova, I.V., Ragulya, A.V. (2001) *Physico-chemical kinetics in nanostructured systems*. Academperiodica, Kiev.
3. Ragulya, A.V. (1998) Rate-Controlled Synthesis and Sintering of Nanocrystalline Barium Titanate Powder *Nanostructured Mater.* **10**, 349-356.
4. Arlt, G., Hennings, D., de With, G. (1985) Dielectric properties of fine grained barium titanate ceramics *J. Appl. Phys.* **58**, 1619-1625.
5. Mayo, M.J. (1996) Processing of nanocrystalline ceramics from ultrafine particles. *Intern. Mater. Reviews* **41**, 85-115.
6. Zgalat-Lozynskyy O.B. (2002) *Regularity of structure and properties formation for nanocomposites based on high-melting titanium, aluminium and silicon nitrides under rate-controlled sintering* Ph.D. Thesis, IPMS, Kiev, Manuscript.
7. Vasylkiv, O., Sakka, Y. (2000) Nonisothermal Synthesis of Yttria-Stabilized Zirconia Nanopowder through Oxalate Processing: 1, Characteristics of Y-Zr Oxalate Synthesis and its Decomposition *J. Amer. Ceram. Soc.* **83** 2196-2202.
8. M.J. Mayo, (1998) *Nanocrystalline Ceramics for Structural Applications: Processing and Properties* Nanostructured Materials: Science and Technology, NATO-ASI series, Kluwer Academic Publishers, Dordrecht, 361-383.
9. Bykov, A.I., Polotai, A.V., Ragulya, A.V., Skorokhod V.V. (2000) Synthesis and sintering of nanocrystalline barium titanate powder under nonisothermal conditions: V. Non-isothermal sintering of barium titanate powder of different dispersion // *Powder Metallurgy and Metal Ceramics* **7/8** 88-98.
10. Frey, M.H. (1996) Grain Size Effect on Structure and Properties for Chemically Prepared Barium Titanate, Ph. D. Thesis, Urbana, Illinois.
11. Andrievski, R.A. (2000) *New superhard materials based on nanostructured high-melting compounds: Achievements and perspectives* // in NATO ASI Seria "Functional Gradient Materials and Surface Layers Prepared by Fine Particles Technology", Kiev, Ukraine 18-28 June
12. Ishida, Y., Ichinose, H., Kizuka, T., Suenaga, K. (1995) High-Resolution Electron Microscopy of Interfaces in Nanocrystalline Materials // *Nanostructured Materials* **6** 115-124.
13. Arlt, G., Hennings, D. and De With, G. (1985) Dielectric properties of fine-grained barium titanate ceramics *J. Appl. Phys.* **58** 1619-1625.
14. Cottom, B.A., Mayo, M.J. (1996) Fracture Toughness of Nanocrystalline ZrO_2 - 3 mol% Y_2O_3 determined by Vickers Indentation. *Scripta Mater.* **34** 809-814.

IMPACT OF GRAIN BOUNDARIES ON STRUCTURAL AND MECHANICAL PROPERTIES

H. VAN SWYGENHOVEN, P.M. DERLET, A. HASNAOUI, M.
SAMARAS
Paul Scherrer Institute, CH-5232 Villigen-PSI, Switzerland

1. Introduction

For some polycrystalline metals with grain sizes in the nano regime, experiments have suggested a deviation away from the Hall-Petch relation relating yield stress to average grain size [1]. The debate continues whether or not such deviations are a result of intrinsically different material properties of nanocrystalline (nc) systems, or due simply to inherent difficulties in the preparation of fully dense nc-samples and in their microstructural characterization. Nevertheless, it suggests that the traditional work hardening mechanism of pile-up of dislocations originating from Frank-Read sources may no longer be valid at the nanometer scale. In-situ deformation testing in the transmission electron microscope (TEM), performed on Cu and Ni_3Al nc samples, reveals a limited dislocation activity in grains below 50nm [2,3]. However, due to the presence of large internal stresses which make grain boundaries (GB) in TEM images difficult to observe, and also possible artifacts induced by thin-film geometry such as dislocations emitted from the surface [4], in-situ tensile tests did not until now, bring convincing evidence for abundant dislocation activity. Mechanical testing also revealed the issue of the "GB state" by means of a property dependence on thermal history and internal strains. It is shown that a substantial strengthening can be obtained by a short heat treatment. The cause of the strengthening is possibly associated with a reduction in internal strains and/or dislocation content produced by the annealing [5]. The effect of strengthening has been measured both on nc materials obtained by grain refinement techniques and those obtained by consolidation of clusters.

These observations strongly suggest that there is still a lack in understanding of the relationship between the structure of GB and triple junctions (TJ) and the overall mechanical properties. To complement ongoing experimental investigations large-scale molecular dynamics are used to study the structural and mechanical properties of nc metals. Despite the limitations in time and length scales, which impose high strain rates and short deformation times, as is discussed in [6], the MD technique is provides an invaluable detailed picture of the atomic scale processes during plastic deformation.

The present paper gives an overview of the role played by the grain boundary in the structural and mechanical properties of fully 3D nanocrystalline fcc metals observed in MD simulations. Results are shown for uni-axial tensile deformation, nano-indentation and irradiation simulations.

T. Tsakalakos, et al. (eds.) pgs. 155 - 167
Nanostructures: Synthesis, Functional Properties and Applications;
© *2003 Kluwer Academic Publishers.*

2. Simulation Technique

The samples used are constructed by beginning with an empty simulation cell with fully 3D periodic boundary conditions, and choosing randomly a number of positions. The number of positions is determined from the simulation cell size and the desired characteristic grain size. From each position and corresponding oreintation, an fcc lattice is constructed geometrically. At a point where atoms from one grain center are closer to the center of another grain, construction is halted. Eventually construction will cease throughout the entire sample, resulting in a three dimensional granular structure according to the Voronoi construction. At this stage atom pairs, each atom originating from a different crystallite, are inspected and where there is a nearest neighbor distance of less than 2 Angstrom, one atom is removed. Molecular statics is then performed to relax any local high potential energy configuration that may exist, followed by NPT MD at room temperature to further relax and equilibrate the structure. It has been previously shown that the samples resulting from this type of synthesis, have grain boundaries that are fundamentally not different from their coarse grained counterparts [7]. For all types of missorientation, a large degree of structural coherence is observed and misfit accommodation occurs in quiet regular patterns.

All molecular dynamics are performed within the Parrinello-Rahman approach with periodic boundary conditions and fixed orthorhombic angles. Unless otherwise states, the simulations are carried out at room temperature. We used the second moment (tight binding) potential of Cleri and Rosato for "model" fcc Ni [8].

3. Definitions of Grain Boundaries

In much of our previous work, the atomic visualization of grain and GB structures has been facilitated greatly by a medium range order analysis of all atoms within the sample. Such a local analysis ascribes a crystallinity class to each atom, and was first proposed by Honeycutt and Andersen [9]. This is performed by selecting the common neighbours of a pair of atoms separated by no more than a second nearest neighbour distance, and introducing a classification scheme for the corresponding topological structure in terms of bond pathways. Since each crystallinity symmetry has a unique topological signature, when all second nearest neighbour bond permutations are enumerated, a local symmetry label can be assigned to each atom. Using this classification scheme, we have defined four classes of atoms via a color-code system: gray=fcc, red=hcp, green=other twelve, and blue non-twelve coordinated atoms. Fig. 1 shows a black and white view of a nc sample, containing 100 10nm size grains, where grey atoms represent fcc and black atoms represent non-fcc atoms. Using this scheme, the grain and grain boundary regions can clearly be identified. In addition to this, a short range order analysis involving only first nearest neighbour pairs is employed yielding the coordination of each atom.

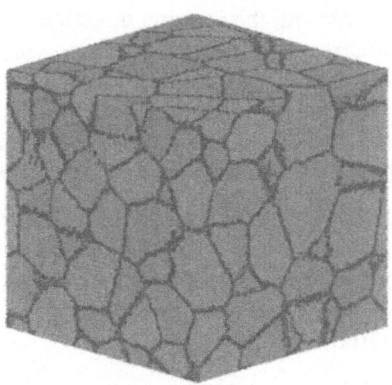

Figure 1. General view of a modelled nanocrystalline sample containing 100 grains with a mean grain diameter of 10nm.

This local atomic classification scheme allows the GB network and structure to be easily identified. A significant advantage of such a local crystallinity analysis is that single (111) HCP planes represent twin planes, and two neighbouring parallel (111) HCP planes represent an intrinsic stacking fault. The visualization of the twin planes has allowed for the easy identification of GBs containing structural units of a $\Sigma = 3$ symmetric boundary. In the case of stacking fault defects, this approach has given evidence for partial dislocation activity [6].

We now give further detail to our definition of the GB, by introducing an additional classification that indicates whether or not an atom is positionally disordered [10]. By this, we mean an atom is regarded as positionally disordered when its location cannot be attached to a lattice site of the nearby fcc grains. At first sight this may be considered a non-trivial task since in principle both the orientation and position of the neighbouring fcc lattices need to be determined. However when the local crystallinity is already known for the surrounding atoms, the task becomes straight forward. If a GB atom has at least one nearest neighbour atom that is fcc, then that GB atom must be at a lattice site of the fcc grain to which the neighbouring fcc atom belongs to. If all nearest neighbour atoms are non-fcc, then that GB atom will be positionally disordered with respect to the nearby fcc lattices. Such a definition is approximate but generally appropriate for the fully dense and locally relaxed GB interfaces considered in the present work.

Fig. 2a is a GB with the viewing orientation parallel to the normal of the GB plane. This GB arises from grains 1 and 14 in the 12nm sample, and has been studied before in the context of sliding under uni-axial tensile loading conditions [7] (see also figure 1 in [11]). This GB may be considered as a general GB. In fig. 2a, the light grey circles represent positionally ordered GB atoms, and the black circles, positionally disordered GB atoms. By viewing the grey atoms, an approximate dichromatic pattern is evidenced arising from the lattice positions of the two different fcc grain orientations. Such an observation is already an indication of a high level of structural order within the GB. The relatively small number of the black positionally disordered atoms confirms this. Furthermore their linear spatial arrangement suggests the presence of clearly defined regions of positional disorder. We note that an increased density of positionally disordered atoms around the edges of the GB, the TJ region, is evident.

158

Fig. 2b. is a view of the same GB with the viewing direction now in the plane of the GB. Again, the light grey circles represent positionally ordered GB atoms and the black circles, positionally disordered GB atoms. The atomic section constituting this figure is several nanometres thick along the viewing direction, and by displaying fcc atoms as small light grey dots, (111) planes in both grains can be easily visualised. In this figure we also display the full crystallographic directions of each grain, where the two orthogonal axes, [1̄10] and [11̄2̄] in the (111) planes, are projected onto the figure plane. By following the (111) planes of both grains a GB dislocation network can be easily identified, or equivalently, regions of coherence separated by regions of misfit can be seen. From this perspective we see that for this GB, the positionally disordered atoms are located in regions of misfit - the GB dislocation cores.

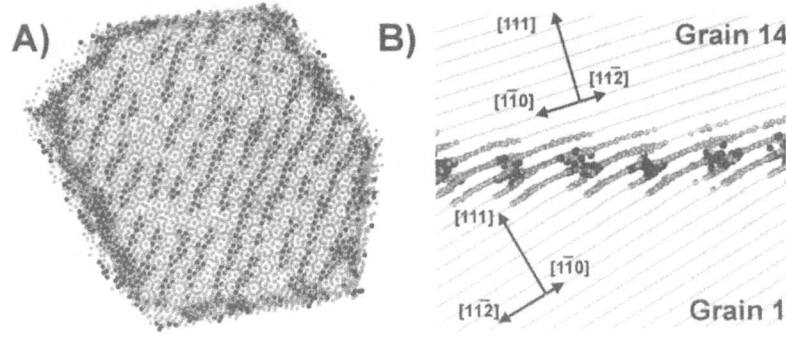

Figure 2 a) Non fcc atoms of a grain boundary 1-14. Grey atoms represent grain boundary atoms that are at lattice positions of one of the neighbouring fcc grains, whereas black atoms indicate those that are positionally disordered. b) View of grain boundary 1-14 with viewing direction within the grain boundary plane. Atoms are identified as in a), where in addition, (111) planes are visualised by displaying the fcc atoms of both grains as light grey dots.

4.   Plastic Deformation Processes Under Uni-axial Tensile Loading

In all samples with mean grain sizes up to 20nm, grain boundary sliding is observed as being the main contribution to the observed plasticity. Careful analysis of the GB structure during sliding under constant tensile load shows that sliding includes a significant amount of discrete atomic activity, either through uncorrelated shuffling of individual atoms or, in some cases, through shuffling involving several atoms acting with a degree of correlation [11]. In all cases, the excess free volume present in the disordered regions plays an important role. In addition to the shuffling, we have observed hopping sequences involving several GB atoms. This type of atomic activity may be regarded as stress assisted free volume migration. Together with the uncorrelated atomic shuffling they constitute the rate controlling process responsible for the GB sliding (GBS).

The atomic shuffling is shown in Fig.3 where a section of the grain boundary between grain number 1 and grain number 14 of the 12 nm sample is shown. The plane of the paper represents a (011) plane for grain 1 and a (-112) plane for grain 14. The atoms are shown at their positions prior to loading and the thickness of the section (perpendicular to the sheet of paper) is 4 atomic layers. In grain 1 the unit cell is highlighted in transparent grey, in grain 14 the (111) planes are indicated and also the crystallographic directions [110] and [1-11]. To visualize the relative displacement of atoms in the two grains resulting from approximately 2% plastic deformation, the displacement vectors have been calculated with respect to the local center-of-mass coordinate system of the atoms in grain 1. As can be observed GB sliding is approximately parallel to the GB plane and with a significant sliding component in the (111) plane. Within the GB some single events of much larger displacements across the grain boundary are observed, which can be identified as shuffling of atoms across the interface.

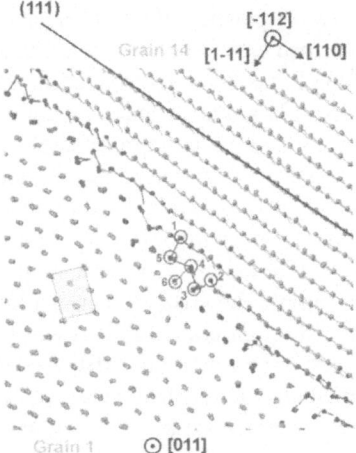

Figure 3. A section of the GB between grains 1 and 14 in the 12nm sample. Displacement vectors are shown indicating the change in position between two levels of strain during plastic deformation. Atomic shuffling between the grains can be observed.

In order to resolve the atomic activity in time, numbers are attributed to those columnar regions (with respect to the viewing direction) taking part in the atomic shuffling. When a load is applied, the atoms occupying regions 1 and 2 (constituting lattice positions in grain 14) slide away increasing the excess free volume at the connecting region between grains 14 and 1. When the atoms in grain 14 have slid by an amount equal to about 50% of the indicated sliding vector, two atoms sitting in region 3 of an (011) plane of grain 1, move in the direction of region 2, transferring free volume into grain 1. The new positions taken by these atoms are again lattice positions of an (011) plane in grain 1. Then a shuffle takes place to refill one of the vacant sites in region 3 by an atom sitting in region 4. Region 4 constitutes lattice positions common to both grain 1 and grain 14. Only a very short time later, a similar process takes place: an atom sitting in region 5 shuffles to region 1. This position is filled up with another atom in region 4, which then in turn is filled up by a shuffle of an atom in region 6. This atom sits closer to the GB plane since the GB plane is slightly inclined relative to the perpendicular of the viewing plane.

At larger grain sizes, dislocation activity is observed. In fully three dimensional GB networks, which have been modelled now up to 20nm grain sizes, only partial dislocations have been observed. Only in quasi 2D-samples with columnar diameters greater than 20 nm full dislocations have been observed [12]. A detailed study of the difference between a 2D columnar and a fully 3D network has been given in [13]. MD simulations have shown that a GB dislocation emits a partial lattice dislocation meanwhile changing the grain boundary structure and its dislocation distribution [14]. This mechanism is the reverse of what is often observed during absorption of a lattice dislocation, where the impinging dislocation is fully or partially absorbed in the GB, creating local changes in the structure and GB dislocation network .

Fig.4 shows the dislocation activity occurring in grain 13 at the interface with grain 12. Fig.4a shows a section of the GB 12-13, including part of the triple junction (TJ) involving grain 1 just at the onset of plastic deformation. Such a configuration will be referred to as the elastically deformed case. The view is along a [1-10] direction of grain 13, where for this grain the unit cell has been highlighted in transparent grey. The grain boundary plane is close to a (1,-1,13) plane of grain 13 and the tilt angle between the observed (111) planes in grain 13 and 12 is approximately 24° and a twist angle of approximately 18° is found. The GB structure has to accommodate the above-mentioned misfit through a GB dislocation (GBD) network. Two of these GBDs are indicated in fig.3a by grey circles. Others are not shown, since they can only be observed in other viewing orientations. For detailed information we refer to [14].

During deformation of this sample the GBD, which is closest to the TJ, dissociates into a Shockley partial, travelling through grain 13. This can be seen in fig.4b, showing the same GB after 2.3% plastic strain, in which this GBD has annihilated. The two (111) planes of darker HCP atoms indicate the intrinsic stacking fault left behind when the partial is travelling through grain 13. Since the Burgers vector of the lattice dislocation is not the same as the Burgers vector of the GBD, the emission of the partial occurs along with local changes in GB orientation, GB structure and GBD distribution.

The atomic mechanism behind the dislocation emission is the following: during deformation sliding is observed and free volume migrates from the nearby triple junction, and diffuses under the applied stress towards the GBD observed in fig.3a, or, equivalently, two atoms from the core region of the GBD migrate to nearby positions within the GB. Local shuffling around the GBD allows the creation of the necessary Burgers vector for the partial dislocation. The nucleation and propagation induces changes in the resultant GBD distribution and additional structural relaxation is observed in the GB and nearby TJ.

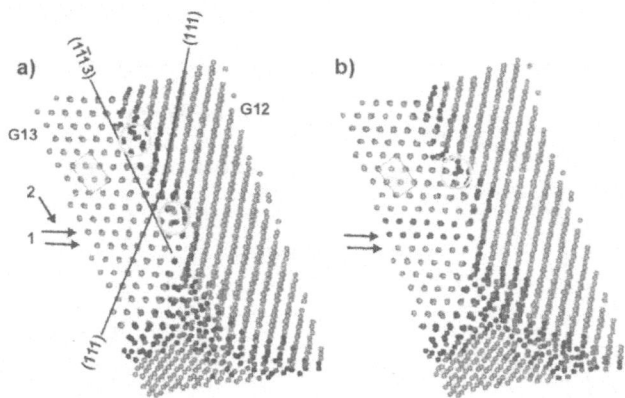

Figure 4. View of a grain boundary in the 12nm sample for a) the configuration at elastic loading and b) the configuration at a plastic strain of 2.3%. Grain boundary dislocations (of type A) that accommodate the misfit between the grains are highlighted by yellow circles.

It is shown in [14] that for increased grain diameters, an increase in partial dislocation activity is seen, but no full dislocations are observed probably due to subsequent structural relaxation after the emission of the partial. For the model Ni potential used, the room temperature stacking fault energy is 280 mJ/m^2 [15]. Such a value, together with a lack of full dislocation activity, seems to suggest that the stacking fault energy plays only a minor role in the issue concerning whether or not a full dislocation is seen. At smaller grain sizes, the same GBDs are observed. They undergo significantly more atomic scale activity resulting in climb of the GBD without the generation of partials, suggesting that GBD are more delocalised and therefore motion of GBDs by atomic shuffling is facilitated and association of GBDs into a partial dislocation is hindered.

t has also been observed that with increased structural order within the GB region, or equivalently the extent to which the GB is in equilibrium, there is a corresponding decrease in the level of plasticity under uniaxial tensile conditions and thus an effective increase in the strength of the nc material [16]. Fig.5 shows the deformation curve for a 12nm sample prepared according to the Voronoi construction (called as-prepared), and a sample that has been annealed at 800K prior to deformation in order to relax more the GBs and TJ's. Both are deformed at 300K. The annealed sample exhibits significantly less deformation compared to the as-prepared sample. To confirm that the decrease in strain does indeed arise from a reduction in plastic deformation activity at the GB, we unloaded the as-prepared and annealed samples to determine the residual strain. The difference in residual strain is 0.43%, which is only slightly, less than the 0.56% difference in strain under load conditions. Inspection of this figure also reveals that the strain rate for the as-prepared sample is 2.2×10^7 /s, whereas for the annealed sample it is 9.4×10^6 /s, which is a reduction of more than half. These strain rates where calculated using the last 30ps of deformation data before unloading and should be regarded as applicable only to the time scale of fig. 5. The deformation simulations were continued

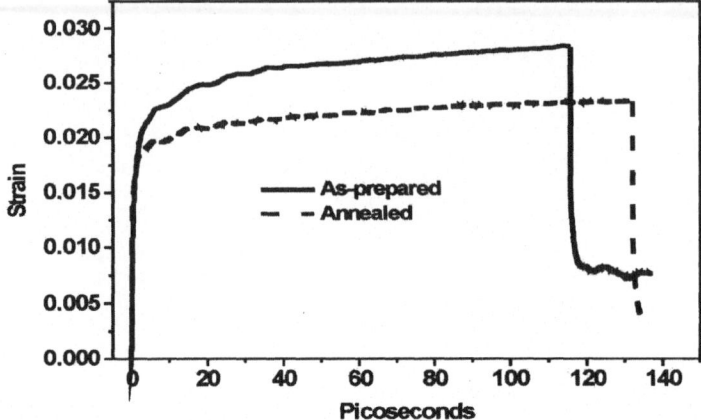

Figure 5. The deformation curves for the as-prepared and the annealed in which, all have been unloaded to determine the residual strain.

up to 0.3ns, by which time the strain rates became closer in value to approximately 3.8×10^6 for both the as-prepared and annealed samples.

5. Grain Boundaries and Nanoindentation

Nanoindentation offers one of the few experimental techniques to probe in *situ* high strain-rate plasticity at the atomic level [17]. Indeed discrete bursts displayed in experimental load-indentation depth curves have been attributed to the nucleation of individual dislocation events [18]. Atomistic simulation therefore provides a natural bridge to complement the experimental studies where, for example, it has been demonstrated that the structures resulting from onset plasticity are qualitatively similar to that seen in experiment [19]. These simulations have all been performed on perfect single crystal structures. In experiments however, thin layers are often nc and the resulting onset plasticity may be profoundly affected by the high density of GBs known to interact with lattice dislocations.

e performed atomistic simulation of nanoindentation into model nc-Au for two samples with average grain diameter of 5nm and 12nm, and compare the results with a simulation for a single crystal structure [20]. The simulations provided three major results: 1) there is a reduction in Young's modulus at the 5nm diameter grain size 2) the surrounding GB structure acts as a sink for the dislocation loops nucleated under the indenter 3) for indenter-substrate surface contact areas less than the grain size, most of the plastic deformation activity is confined within the grain directly below the indenter, whereas for larger contact areas intergranular motion is observed as a result of grain boundary sliding (GBS).

Nanoindentation simulations were performed on three samples: a single crystal model FCC Au(111) sample (1.1 million atoms) and 5nm and 12nm average grain size nc-Au samples with the grain below the indenter having a (111) surface orientation. The dimensions of all samples are 26 nm^3. The spherical indenter was modelled as a strongly repulsive potential $V(r) = A\Theta(R-r)(R-r)^3$ where R is the radius of the indenter, r the distance of the indenter center to the atom, A is a force constant and $\Theta(R-r)$ is the standard step function. Both molecular dynamics and conjugant gradient techniques are employed to investigate the structural relaxation as the indenter is lowered.

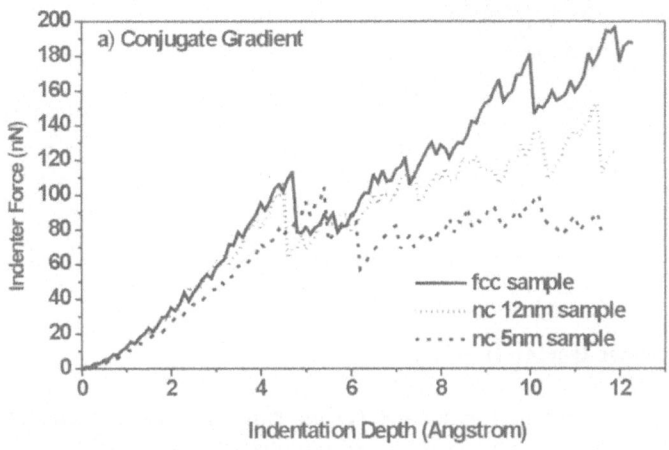

Figure 6: Force vs. indentation depth plots for bulk fcc Ni, and 5nm and 12nm average grain size nc-Ni.

Fig. 6 displays the force vs. indentation depth curves as a function of grain size obtained from conjugate gradient atomistic simulations. The curves all exhibit a distinctive yield point. The elastic modulus calculated by applying the isotropic Hertz model to the appropriate elastic region of each sample, resulted in (111) Young's modulus of 134GPa for the single crystal, 129GPa for the 12nm sample and 104GPa for the 5nm sample. Such a reduction in elastic modulus at the very small grain sizes has been recently reported in nc Ni-P samples with constant P content. Fig. 7 shows the CG atomic structure under the indenter for the 12nm grain size sample at an indentation depth of 12Å. The dark plane of atoms indicates the upper indentation surface. At this indentation depth the dislocation loops are attracted to neighboring GBs identified by the dark atoms below the surface (fcc atoms are not shown). Thus the GBs act as dislocation sinks, accommodating the associated slip across the grain by structural changes within the GBs.

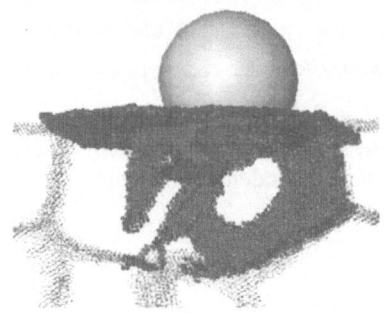

Figure 7: Zone beneath the indenter in CG for the 12nm average grain sample at a displacement of 11.9Å.

6. Grain Boundaries During Irradiation

It is well known that grain boundaries (GBs) can strongly influence the damage produced by irradiation. In nanocrystalline (nc) materials a considerable amount of atoms belong to or are affected by the GBs, therefore they are the ideal microstructure to investigate the effect of GBs. The presence of so many interfaces, has led to the belief that nc materials exhibit different responses under irradiation environments compared to the pollycrystalline counterparts. This is experimentally supported by a significantly smaller defect production in irradiatied nc materials than in coarser-grained samples [21].

Molecular dynamics (MD) computer simulations provide unprecedented information on the mechanisms and defect states of irradiated metals in the collisional time frame of a cascade. A primary knock on atom (PKA) is introduced and imparts its energy to the surrounding lattice, producing a molten and highly disordered region that subsequently cools. Replacement collision sequences (RCSs) move self interstitial atoms (SIAs) away from the cascade core region to a point where they cannot recombine with existing vacancies. Introduction of sinks such as GBs and triple junctions (TJ) [22] affect the damage state of the irradiated material by attracting SIAs. During the thermal spike phase, a larger number of SIAs are seen to form and move, via RCSs, to GB regions. Some SIAs remain in the grain as the cascade core region cools. These SIAs move to GBs via a sequence of 1D steps in 3D, towards nearby GBs. Although only a small number of SIA clusters remain in the grain, the 1D/3D motion has been recognised as typical of surviving SIA clusters. In single crystal simulations, clusters with more than 3-SIAs undertake 1D motion only. In our nc Ni work, larger SIA clusters are found to experience an initial reorientation before moving 1D to a nearby GB, however in our MD simulations 6-SIA clusters where observed to move also 1D/3D, emphasising the dominant role of GBs as sinks during cascade simulations [23]. The result of this effective removal of SIAs from the grain leads to a vacancy rich grain region.

The following results describe the SIA movement in a particular 5keV cascade simulation, the chosen example being representative of all other cascade simulations in the presence of GBs [23]. The development of the cascade at particular times during the simulation is presented in figures 8a to 8d, each figure showing a 'snapshot' at a particular time. Only non-fcc atoms are shown for ease of visualisation. In all of the figures, atomic displacement vectors between different times are superimposed as pink lines onto the figure in order to further clarify atomic movement.

Figure 8a presents the melt phase and movement of atoms at 0.3ps after the introduction of the PKA. On the lower right hand side an RCS (labelled as RCS1) to a nearby GB is evident from the atomic displacement vectors and the non-fcc atoms extending along its path. On the lower left hand side, a second RCS appears as a similar elongated non-fcc region extending out 2.5nm from the periphery of the cascade region (labelled as RCS2). Figure 8b represents a configuration at 0.7ps after the introduction of the PKA. The corresponding atomic displacement vectors are taken from this configuration and that before the cascade. The two SIAs corresponding to the two RCSs identified in figure 8a have now been labelled as SIA1 and SIA2. SIA1 had already arrived at the GB in figure 8a at 0.3ps, and by 0.7ps was observed to move further within the GB. The region to which SIA1 is attracted in the GB contains a GB dislocation where there is known free volume. At this snapshot, it is evident that not all SIAs are attracted by the GBs via RCSa. As the cascade cools, some SIAs are left within the grain. SIA2 which has moved out towards the GB is no longer part of the cascade core (SIA2), and the atoms in the region previously connecting it to the cascade core resolidifying and returning to their fcc positions, leaving SIA2 as a mono-crowdion SIA within the grain.

Figure 8c represents a configuration at 5.5ps after the introduction of the PKA with the corresponding atomic displacement vectors derived from this configuration and that before the cascade. In this figure the core has resolidified to reveal small vacancy defects as evidenced by the surrounding black non-12 coordinated atoms within the grain. The mono-SIA crowdion (SIA2) has changed direction and moved along another crowdion axis to a nearby GB within 3.1ps of the PKA. A 6-SIA crowdion cluster containing all remaining SIAs is also identified. Figure 8d represents a configuration at 11.9ps after the introduction of the PKA with the corresponding atomic displacement vectors derived from this configuration and that before the cascade. Here the 6-SIA crowdion cluster has moved in the direction of the GB, and by 13.1ps has arrived at the GB via 1D/3D motion along its crowdion axis - the cluster changing its direction close to the GB between 11.9ps and 13.1ps (not shown) to move along another close packed plane before being absorbed by the GB. After 13.1ps, the remaining defects in the grain are small vacancy clusters, since all SIAs have moved to the lower GB indicating that this GB may have a greater "sink strength" than the other surrounding GBs.

166

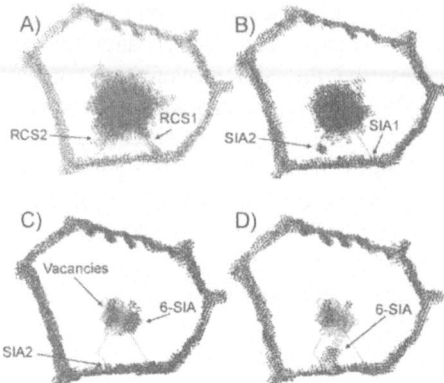

Figure 8: Development of a 5keV cascade in an nc grain. a) represents the atomic configuration 0.3ps, b) 0.7ps, c) 5.5ps and d) 11.9ps after the introduction of the PKA. Only non-fcc atoms are shown.

7. Conclusion

Uniaxial tensile loading of nc-Ni samples reveals the presence of two fundamental deformation mechanisms: 1) GB sliding triggered by atomic shuffling and to some extent stress assisted migration and 2) the emission of partial dislocations from GB that contain GBDs, resulting in a redistribution of GBD and additional structural relaxation. It is also observed that increased structural order within the GB region results in a decrease in the level of plasticity under uniaxial tensile conditions and thus an effective increase of strength of the nc material, suggesting a beneficial effect of structural disorder from the perspective of grain boundary sliding. Nanoindentation simulations reveal that grain boundaries act as efficient sinks for dislocation nucleation below the indenter and that intergranular sliding activity increases with increasing indenter contact area to grain size ratio. Further, we establish that at the 5nm grain diameter regime there is a decrease in the Young's modulus. Simulation of cascade production of the primary damage state are performed in fcc nano-crystalline Ni of average grain diameters of 5nm and 12nm demonstrate that the GB structure at the nano-scale regime strongly affects the primary damage state, with the GB acting as an interstitial sink and thus allowing for, after re-solidification, a vacancy dominated defect structure which in some cases can be completely removed by the surrounding GB/TJ regions. This is quite different from the single crystal result where the primary damage state must arise from a vacancy/interstitial defect distribution.

8. Acknowledgment

Work is supported by the Swiss NSF grants (2000-056835.99), TOPNANO21 (CTI 5443.2), and BBW (Grant No. 98.0098)

9. References

1. Weertman, J.R. (2002) Mechanical behaviour of nanocrystalline metals, *Nanostructured Materials: Processing, Properties, and Potential Applications*, William Andrew Publishing, Norwich.
2. Youngdahl, C.J., Hugo, R.X., Kung, H. and Weertman, J.R. (2002), TEM observation of nanocrystalline copper during deformation, *Structure and Mechanical Properties of Nanophase Materials- Theory and Computer Simulation vs. Experiment*, MRS Symposium Series Vol. **634**, B1.2.
3. McFadden, S.X., Sergueeva, A.V., Kruml, T., Martin, J-L. and Mukherjee, A.K. (2000), Superplasticity in nanocrystalline Ni3Al and Ti Alloys, *Structure and Mechanical Properties of Nanophase Materials- Theory and Computer Simulation vs. Experiment*, MRS Symposium Series Vol. **634**, B1.3.
4. Derlet, P.M. and Van Swygenhoven, H. (2001) The role played by two parallel free surfaces in the deformation mechanism of nanocrystalline metals: a molecular dynamics simulation, *Phil. Mag. A.* **82**, 1-15.
5. Volpp, T., Goring, E., Kuschke, W.M. and Arzt, E. (1997) Grain size determination and limits to Hall-Petch behaviour in nanocrystalline NiAl powders, *Nanostruct. Mater.* **8**, 855-865.
6. Van Swygenhoven, H. (2002) Polycrystalline materials: grain boundaries and dislocations, *Science* **296**, April 4, 66-67.
7. Van Swygenhoven, H., Farkas, D. and Caro, A. (2000) Grain-boundary structures in polycrystalline metals at the nanoscale, *Phys. Rev. B* **62**, 831-838.
8. Cleri F. and Rosato, V. (1993) Tight bindng potentials for transition metals and alloys, *Phys. Rev. B* **48**, 22-33
9. Honeycutt D. J. and Andersen H. C. (1987) Molecular dynamics study of melting and freezing of small Lennard-Jones Clusters, *J. Phys. Chem.* **91**, 4950-4963
10. Derlet, P. M. and Van Swygenhoven, H. (2002) Atomic Positional Disorder in Fcc Metal Nanocrystalline Grain Boundaries, *Phys. Rev. B.*, **67**, 014202-8.
11. Van Swygenhoven, H. and Derlet, P.M. (2001) Grain-boundary sliding in nanocrystalline fcc metals, *Phys. Rev. B* **64**, 224105-9.
12. Yamakov. V., Wolf, D., Salazar, M., Phillpot, S.R.and Gleiter, H. (2001) Length-scale effects in the nucleation of extended dislocations in nanocrystalline al by MD simulations, *Acta Mater.* **49**, 2713-2722.
13. Derlet, P.M. and Van Swygenhoven, H. (2002) Length scale effects in the simulation of deformation properties of nanocrystalline metals, *Scripta Mater.* **47**, 719-724.
14. Van Swygenhoven, H,. Derlet, P.M. and Hasnaoui, A. (2002) Atomic mechanism for dislocation emission from nanosized grain boundaries, *Phys. Rev. B* **66**, 024101-8.
15. Hasnaoui, A., Van Swygenhoven, H. and Derlet, P.M. (2002) Cooperative processes during plastic deformation in nanocrystalline fcc metals - a molecular dynamics simulation, *Phys. Rev. B* **66** 184112-8.
16. Hasnaoui, A., Van Swygenhoven, H. and Derlet, P.M. (2002) On non-equilibrium grain boundaries and their effect on thermal and mechanical behaviour: a molecular dynamics computer simulation, *Acta Mater.* **50**, 3927-3939.
17. Gerberich, W. W., Nelson, J. C., Lilleoddem, E. T., Anderson, P., and Wryobek, J. T. (1996) Indentation induced dislocation nucleation: the initial yield point, *Acta. Mater.* **44** 3585-3598.
18. Gouldstone, A., Koh, H.-J., Zeng, K.-Y., Giannakopoulos, and Suresh, S. (2000) Discrete and continuous deformation during nanoindentation of thin films, *Acta. Mater.* **48** 2277-2295.
19. Zimmermann, J. A., Kelchner, C. L., Klein, P. A., Hamilton, J. C., and Foiles, S. (2001) Surface step effects on nonoindentation, *Phys. Rev. Lett.* **87**, 165507-4.
20. Feichtinger, D., Derlet, P. M., and Van Swygenhoven, H. (2003) Atomistic simulations of spherical indentations in nanocrystalline Gold, *Phys. Rev. B.*, **67**, 024113-4.
21. Rose, M., Balogh, A. G., and Hahn, H. (1997) Instability of irradiation induced defects in nanostructured materials, *Nuc. Inst. Meth. Phys. Res. B* **127/128** 119-122.
22. Samaras, M., Derlet, P. M., Van Swygenhoven, H., and Victoria, M. (2002) On non-equilibrium grain boundaries and their effect on thermal and mechanical behaviour: a molecular dynamics computer simulation, *Phys. Rev. Lett.* **88**, 125505-4.
23. Samaras, M., Derlet, P. M., Van Swygenhoven, H., and Victoria, M. (2003) SIA Activity during irradiation of nanocrystalline Ni, *J. of Nucl. Mater.*, Submitted.

CURRENT STATUS OF GRAIN GROWTH MODELLING WITH SPECIAL REFERENCE TO NANOCRYSTALLINE MATERIALS

CHANDRA S. PANDE and ROBERT A. MASUMURA
Code 6325, Naval Research Laboratory
Washington D.C. 20375 USA

ABSTRACT. Current situations in the modeling of grain growth are critically reviewed. The models of Hillert, Lucke *et al.,* and Mullins are reviewed briefly and shown to be inadequate in describing one or more features of grain growth phenomenon. It is well known that models of Hillert and Lucke and co-workers do not give the correct description of the grain size distribution. Mullins' model, although it gives a correct distribution, still fails because it cannot even, in principle, explain the well known fact that the scaled distribution, obtained after the initial growth, is independent of the initial distribution. An approach based on the stochastic methods leads to a more realistic model. The application of these results, to the growth of nanocrystalline materials, is considered. Here it is shown, that the mobility of the triple junctions may play a significant role.

1. Introduction

Grain growth is a well known phenomenon of the evolution of microstructure in a deformed polycrystal, after recrystallization, resulting in the increase in average grain size by the motion of grain boundaries due to annealing at a certain temperature and time. The driving force for this growth is the reduction in grain boundary area, and hence, the total grain boundary energy of the system. Grain growth is a cooperative process involving simultaneously many individual grains of various sizes and shapes [1,2]. Much of our understanding of grain growth comes from the work of Cyril Smith [3-5]. In particular, the importance of grain boundary area/length reduction with time, motion by curvature, and the role of dihedral angles are now considered central to this understanding.

A detailed understanding of the phenomenon leading to an ability to control grain growth in materials is of technological importance. Many properties of materials, such as, yield stress in metals and critical currents in superconductors, depend on grain size. In addition, the phenomenon of grain growth exhibits features of great theoretical interest. Therefore, its theoretical understanding [1] is expected to provide further insight into future applications in nanosystems as well.

T. Tsakalakos et al. (eds.) pgs. 169 - 178
Nanostructures: Synthesis, Functional Properties and Applications,
© *2003 Kluwer Academic Publishers.*

170

The parameters used in modeling grain growth in two dimensions is shown in figure 1 where $2R$ is the grain diameter, or size, and n is the number of sides. This figure is from a computer simulation of grain evolution with time which is available at http://mstd.nrl.navy.mil/6320/6325/vertex.html

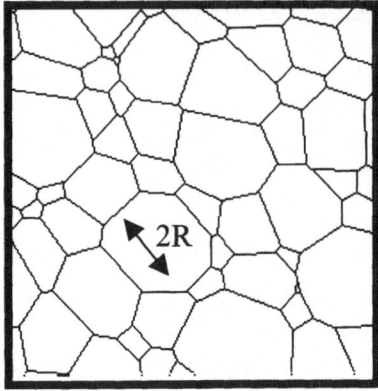

Figure 1. Simulated Grain Evolution of *n*-sided grains and of grain size *2R*.

2. Main Features of Grain Growth

These features, described in great detail by Kurtz and Carpay [2] are briefly:

Time exponent: During growth, the average grain size increases as $\sim (\text{time})^{1/2}$.

Independence from initial distribution: The final grain size distribution is more or less independent of the initial distribution.

Scaling: During growth, after sufficient time, the distribution of grain sizes remains self-similar if all sizes are scaled to average grain size.

Size Distribution: After sufficient time in grain growth, the distribution approaches approximately a log-normal form.

Any model of grain growth must be able to explain these features.

Due to the complexity of the grain structure of the individual grains, and the participation of a large number of grain boundaries in the process, developing a realistic model of grain growth is a formidable challenge and many conceptual and mathematical simplifications are made. Further simplification is possible by first considering grain growth in two dimensions only. This is possible because of the existence of the von Neumann law whose applicability to grain growth was proved by Mullins [22].

3. Von Neumann-Mullins Law

If $A_n(t)$ is the area of the grain, with n sides at any instant t, and M is a constant, then under the assumptions mentioned above, and further discussed later, the following relation is deduced [6, 22]:

$$\frac{dA_n(t)}{dt} = M(n - 6) \qquad (1)$$

This relation has played an important role in the development of the theory of grain growth.

In obtaining von Neumann-Mullins law, Mullins used the following three basic assumptions of the grain growth process:

(a) All the grain boundaries during grain growth move with a velocity proportional to its curvature.

(b) The triple junctions of the grain boundaries have infinite mobility during any change in number of sides of the grain, *etc.*

(c) The boundaries meet at grain boundaries at an angle of 120 degrees.

It should be realized that all the three assumptions are built in the von Neumann-Mullins law of grain growth and so the use of this law is equivalent to accepting these assumptions. If the last two conditions are relaxed, it can be easily shown that von Neumann-Mullins law will not hold exactly for each grain but it may hold in a statistical sense.

4. Models of Grain Growth

A brief review of the various attempts to understand the salient features of grain growth may be pertinent here. Almost all grain growth models are based on an idealized univariate system where the grain size is characterized by its radius, R, and all the grain boundaries are assumed to have the same energy. The grain size distribution is then expressed in terms of R only. With this starting point, a mean field model was developed in [7, 8] and subsequently more sophisticated stochastic models were constructed [9-15] for deriving the distribution function alluded to above and its time evolution.

4.1 MEAN FIELD MODEL

Hillert [7] has presented a theory of grain growth based on these ideas, which considers a single grain growing in an average environment. A mean field model such as the one due to Hillert or Mullins is characterized by the following:

(a) Assumes that a given grain grows in an average or 'mean' field of all other grains.

(b) Has limited predictive capability and is confined to long time behavior.

(c) Not suitable for transient growth, for example, as it occurs in welding.

(d) Fails to predict independence of final distribution from initial one.

172

(e) Assumes infinite triple junction mobility.
Thus the mean field does not give an adequate description.

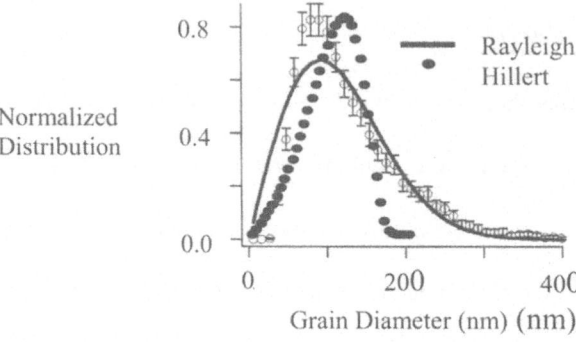

Figure 2. Comparison of Experiment with the Rayleigh and Hillert Distributions. Experimental Data from Carpenter *et al.* [18]

Figure 2 gives a comparison of grain sizes as predicted from Hillert model with the recent experimental results of Carpenter *et al.* [18]. A comparison is also made with the theoretical Rayleigh distribution which is some time used to describe grain size distribution. It is seen that Hillert's distribution fails badly. Distribution obtained by Lucke *et al.* [21] is identical with tht due to Hillert. Even the Rayleigh distribution fit is not adequate. However, the data fits best with a lognormal distribution. This distribution is discussed elegantly by Kurtz and Carpay [2] and is given by

$$F(R,\mu,\sigma) = \left[\frac{1}{\left(2\pi\sigma^2\right)^{1/2}}\right]\frac{1}{R}\exp\left\{\frac{-\left[ln(R-\mu)^2\right]}{2\sigma^2}\right\}.$$ (2)

where $\mu = ln(R_{median})$ and is a constant and σ is the standard deviation of $ln(R)$ for the distribution.

4.2 STOCHASTIC MODEL

It is thus clear the mean field models dealing with average or mean rate of growth of a given grain is not sufficient to provide a good description of grain growth. Since an individual grain does not grow in an average or constant environment. Adjacent grains share common boundaries resulting in an ensemble that is topologically connected. As Smith [5] has pointed out, normal grain growth results from the interaction between the topological requirements of space filling and geometrical needs of surface tension equilibrium. Any realistic theory of grain growth therefore must take both these features of grain growth into account. We contend that such a model can be developed by treating grain growth as a stochastic process.

Mathematically, a stochastic process [16] in its simplest form involves consideration of a function of two variables one of which is time t and involves, in general, both a deterministic and a random term. The other parameter for the present analysis is taken as grain radius R. In the development of this approach the deterministic equation for a rate of change of a given grain is made stochastic by addition of a random ('noise') term, *i.e.*, it is assumed that the macroscopic rate equation is only approximate, in reality individual grains depart from this behavior in a unpredictable way. In the stochastic model, it is assumed that during grain growth, an individual grain does not grow in an effective average environment because adjacent grains share common boundaries. Since growth is a dynamical cooperative process, a given grain is expected to grow in a stochastic way. The time evolution of such a random process is parameterized by R and time, t. The time rate of change, in addition to the average growth $A(R, t)$, is governed by fluctuations $B(R, t)$ around it and is given by the Langevin equation

$$\frac{dR}{dt} = A(R,t) + \sqrt{B(R,t)}\, T(t). \tag{3}$$

Here $T(t)$ represents the random process associated with the sharing of boundaries, assumed to be a Gaussian process [16]. The specific forms of these terms and their origins are, in our opinion, the central issues of the theory of grain growth. The nature and magnitude of the fluctuation may be estimated as in [17]. The next step is to define a grain size distribution $F(R,t)$ such that $F(R,t)dR$ gives the <u>number</u> of grains with radii between R and $R+dR$ at any time t. This distribution depends on time as does the total number of grains, $N(t)$, defined by

$$N(t) = \int_0^\infty F(R,t)\, dR. \tag{4}$$

Thus the distribution is normalized to a time-dependent function, $N(t)$. The Fokker-Planck equation obeyed by $F(R,t)$ is obtained by using the Langevin equation,

$$\frac{\partial F(R,t)}{\partial t} = -\frac{\partial}{\partial R}\{A(R,t)F(R,t)\} + \frac{1}{2}\frac{\partial^2}{\partial R^2}\{B(R,t)F(R,t)\}, \tag{5}$$

a stochastic equation. The physics of grain growth will be essentially contained in this equation, once the deterministic term $A(R,t)$ and fluctuation term $B(R,t)$ have been properly chosen.

It should be noted that unlike many other stochastic processes the fluctuation or 'noise' in this system is not superimposed externally. Rather it is an 'internal noise'. This internal noise is superimposed by the topological requirements and local environment and is inherent in the way the state of the system evolves, and cannot be separated from the kinetics of the system. One therefore needs to specify the nature of the internal noise more precisely or make some plausible assumption about their nature. We have so far taken the later alternative although it may be possible to justify these assumptions on some other grounds. This aspect of the problem is currently under investigation.

The physical picture embodied in the Langevin type, equation (3), is this: the random term in this equation gives unpredictable and sudden jumps creating a tendency for the grain size (or R) to spread out over an ever broadening range of values, while the first term (deterministic term) on the right hand side of equation (3) is akin to damping term in the classical Langevin equation which here tends to bring the grain size (or R) back to zero. The ultimate distribution of R (*i.e.*, grain sizes) is the outcome of these two opposing tendencies. In addition, all the appropriate boundary conditions such as constancy of specimen size must be satisfied.

A solution for grain size distributions *etc.*, using this formulation is discussed by Pande and his coworkers [11-13]. Thus, a more complete description of the phenomena has been developed in terms of a stochastic theory. This is based on the fact that a given grain grows or shrinks not in an average or 'mean' environment but in a more dynamic fashion, in an environment, varying from grain to grain. This is described mathematically by a Langevin equation and a corresponding Fokker-Planck (henceforth termed FP) continuity equation for the grain size distribution. This equation contains a drift term due to curvature effects corresponding to the term appearing in the 'mean field' theory mentioned above. In addition, this equation has a 'diffusion' contribution which takes into account the statistical evolution of the grain due to the surroundings. By assuming certain form for the drift and diffusion terms, the following results were deduced by Pande and Rajagopal [15]: scaling, independence of the long time solution from the initial distribution, kinetics and the grain size distribution.

5. Effect of Initial Distribution

Lifshitz and Slyozov (LS) [19] have shown that in case of Ostwald ripening irrespective of initial distribution, the final distribution after sufficiently long time approaches the same distribution, *i.e.*, the long time distribution is independent of the initial distribution. Experimentally the same condition prevails in grain growth. Grain growth under usual circumstances is known to approach a quasi-stationary distribution of grain sizes after a transient period. (For a review see [1].) An accurate description of the spatial and temporal evolution of a polycrystal from an initial stage, through the transient period and finally to the quasi-stationary state is still only poorly understood. The quasi-stationary state in a wide variety of materials exhibits a scaling property such that the grain size distribution has an invariant form when expressed in terms of the grain size scaled by its mean value. This is the so called normal grain growth when only the scale varies with some power of time, grain size distribution remaining self similar. Similar scaling regimes are found to occur also in phase coarsening (Ostwald Ripening) and in bubble growth. (For a detailed discussion of self similarity and its relation to the kinetics of the process, see [2].) What is even more remarkable is that this scaling regime is reached independent of the choice of the initial state.

Novikov [20] gives an interesting example of the independence of the steady state distribution on the initial distribution. In his computer simulation he took the initial grain size distributions to be different from log normal. The grain size distribution after a certain period of growth took one and the same shape irrespective of the parameters of the initial function. (The steady state distribution were however not exactly lognormal.) It was also found that if the steady state distribution was taken as initial, it remained practically the same distribution at a later time.

LS arguments, cannot be used in our analysis. Nonetheless, we can also show that for some restrictions on the term $F(R,t)$ and on the boundary conditions, which are easily satisfied in grain growth, any two solutions of our equation are identical, *i.e.*, the long time solution is independent of the initial distribution, every solution of equation (5) finally decays to the solution for $t \to \infty$.

The proof of this important result for stochastic equations is given by Pande and Rajagopal [15] and will not be repeated. But a graphical illustration of the argument is provided below. In figure 3, a function $H(t)$ is given which is afunction the two different solutions, $F_1(R,t)$ and $F_2(R,t)$ of the grain growth equation given as

$$H(t) = \int_0^\infty F_1(R,t) \; \ell n \left\{ \frac{F_1(R,t)}{F_2(R,t)} \right\} dR \,. \tag{6}$$

Pande and Rajagopal show that for the mean field case as $t \to \infty$, $H(t)$ does not vary with time, whereas in the stochastic case, $H(t)$ is always positive and decays to zero. Hence, for the stochastic model, as $t \to \infty$, as $H(t) \to 0$ implies that $F_1(R,t) = F_2(R,t)$. This remarkable result in their analysis means that independence from initial distribution is inherent in stochastic treatment only and is not present in any mean field treatment. By a similar argument, the scaling property can also be proven. These are strong arguments against the validity of mean field models of any type.

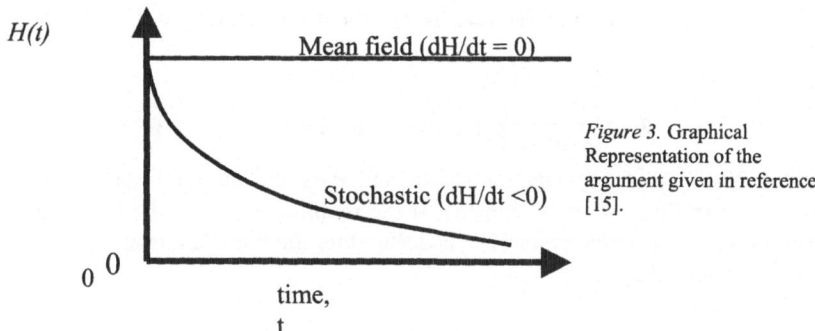

Figure 3. Graphical Representation of the argument given in reference [15].

The four main features of grain growth were discussed in Section 2. Table I lists a comparison among the several mean field and stochastic models. Only the stochastic formulation is able to describe all four of the required attributes for grain growth.

Table I. Comparison between the Mean Field and Stochastic Approaches.

Grain Growth Attributes	Mean Field			Stochastic
	Hillert [7]	Lucke *et al.* [21]	Mullins [8]	
Time Exponent	Yes	Yes	Yes	Yes
Initial Distribution Independence	No	No	No	Yes
Scaling	No	No	No	Yes
Correct Distribution	No	No	Yes	Yes

6. Grain Growth of Nanocrystalline Materials

In case of nanocrystalline materials two additional factors need to be taken into account:
  (a) Grain rotation (increases grain growth).
  (b) Finite triple junction mobility (retards grain growth).
These two mechanisms will operate in addition to curvature driven grain growth. Grain rotation has been considered in detail by Li [23] theoretically and by Haslam *et al.* [24] by simulation. We will not consider this any further except to state that in nanocrystalline materials this mode is quite possible especially if the grain growth is retarded by triple junction junction finite mobility. We will consider this in detail below.

6.1 TRIPLE JUNCTION MOBILITY AS FUNCTION OF THE NUMBER OF SIDES

Gottstein and Shvindlerman [25] have shown that a parameter, λ, defined as

$$\lambda = \frac{\text{(grainsize)(triple junction mobility)}}{\text{grain boundary mobility}} \qquad (7)$$

which involves the ratio of the mobilites of the triple junction to the grain boundary and can be expressed as

$$\lambda = \frac{\ell n[\sin(\theta)]}{2\cos(\theta) - 1} \text{ for } n > 6 \text{ and } \lambda = \frac{2\theta}{2\cos(\theta) - 1} \text{ for } n < 6, \qquad (8)$$

where 2θ is the triple junction angle and n is the number sides. This relation is illustrated in figure 4. It is seen that if the mobility is finite, the triple junction angle will always be different from 120° and therefore the von Neumann-Mullins law is only approximately obeyed.

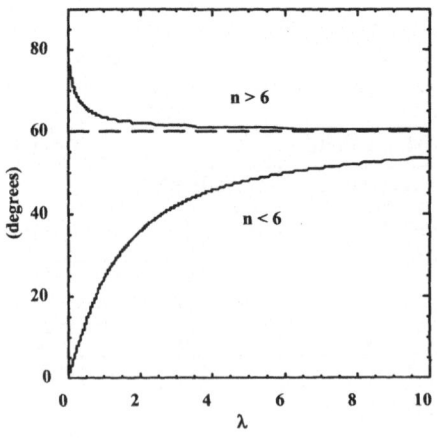

Figure 4. Triple Junction Angle dependence on λ.

6.2 MODIFIED VON NEUMANN-MULLINS LAW

When the triple junction mobility is infinite, λ is infinite and the triple junction angle, 2θ, is equal to $120°$. However, when λ is finite, the triple junction can deviate from $120°$. Thus, the von Neumann-Mullins law requires modification. Using the above results, we plot, dA_n/dt as a function of λ for various values of n in figure 5. It seen that for $1.43 \leq \lambda \leq 10.73$, dA_n/dt is positive for all n; thus, *all* the grains will tend to grow, which is not possible, since the total area of the polycrystal (for a two dimensional system) is conserved. Grain growth is only possible if some grains decrease in size and disappear. Thus for sufficiently small λ, grain growth will be retarded. Since λ involves the grain size, the smaller grain sizes will have relatively smaller λ. Thus in nanocrystalline materials grain growth could be retarded. This conclusion is strictly true for two dimensional systems only. However since three dimensional systems usually behave, at least qualitatively, in a similar fashion, it may be valid for these cases as well. Experimental evidence for grain retardation has been provided by Okuda *et al.* [26] for thin films

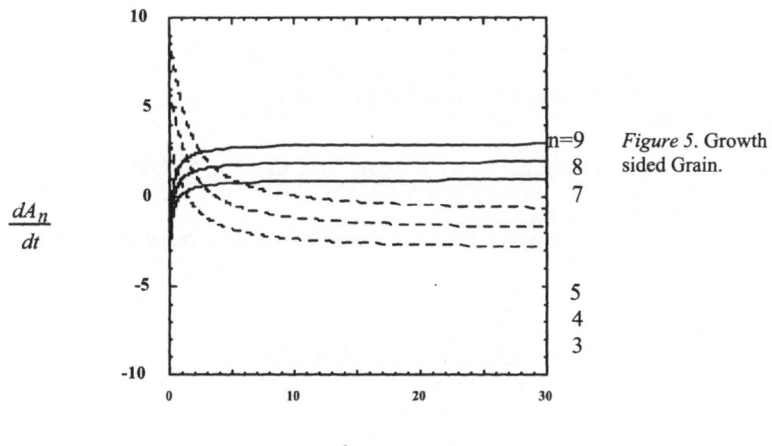

Figure 5. Growth Rate for n-sided Grain.

178

7. Conclusions

From this analysis we tried to show that a stochastic description of grain growth is essential for a more comprehensive understanding than provided by a limited mean field approach. The modified von Neumann-Mullins formulation to account for a finite triple junction moblity can in principle result in grain growth retardation in nanocrytalline materials.

8. References

[1] Atkinson, H. V., *Acta metall.,* 1988, **36**, 469.
[2] Kurtz, S. K. and Carpay, F. M. A., *J. Appl. Phys.,* 1980, **51**, 5725.
[3] Smith, C. S., *Transactions AIME*, 1948, **175**, 15.
[4] Smith, C. S., *Transactions AIME*, 1949, **194**, 755.
[5] Smith, C. S., *Acta Metall.,* 1953, **1**, 295.
[6] von Neumann, J. in *Metal Interfaces*, (Cleveland, OH: American Society for Metals), 1952, p. 108.
[7] Hillert, M. *Acta Metall*, 1965, **13**, 227.
[8] Mullins, W. W., *Acta Metall.,* 1998, **46**, 6219.
[9] Pande, C. S., *Acta Metall.,* 1987, **35**, 2671.
[10] Chen, I. W., *Acta Metall.,* 1987, **35**, 1723.
[11] Pande, C. S., and Dantsker, E., *Acta Metall.,* 1990, **38**, 945.
[12] Pande, C. S., and Dantsker, E., *Acta Metall.,* 1991, **39**, 1359.
[13] Pande, C. S., and Dantsker, E., *Acta Metall.,* 1994, **42**, 2899.
[14] Pande, C. S., Masumura, R. A., and Marsh, S. P., *Phil. Mag. A*, 2001, **81**, 2009.
[15] Pande, C. S., and Rajagopal, A. K., *Acta Metall.,* 2001, **49**, 1805.
[16] Gardiner, C. W., *Stochastic Methods,* 1985, Springer-Verlag. Berlin, Heidelberg, New York.
[17] Zhao, X., *Scripta Metall.,* 1995, **33**, 1081.
[18] Carpenter, D. T., Codner, J. R., Barmak, K., and Rickman, J. M., *Mat. Lett.,* 1999, **41**, 296.
[19] Lifshitz, I. M., and Slyozov, V.V., *J. Phys. Chem. Solids.* 1961, **19**, 35.
[20] Novikov, V. Y., *Acta Met.,* 1978, **26**, 1739.
[21] Lücke, K., Abbruzzese, G., and Heckelmann, I., *Proceedings of Recrystallization 90,* (TMS, Warrendale, PA), Ed: T. Chandra, 1990, p. 51.
[22] Mullins, W. W., *J. Appl. Phys.,* 1956, **27**, 900.
[23] Li, J. C. M., *Appl. Phys.,* 1962, **33**, 2958.
[24] Haslam, A. J., Phillpot, S. R. and Wolf, D. *Mat. Sci.&Engr A,* 2001, **318**, 293.
[25] Gottstein, G. and Shvindlerman, L. S., *Scripta mater.,* 1998, **38**, 1541.
[26] Okuda, S., Kobiyama, M., Inami, T. and Takamura, S., *Scripta mater.,* 2001, **44**, 2009.

DISCUSSION ABOUT SCALING EFFECTS IN METALS WITH A CONTINUUM MECHANICS APPROACH

Elasticity and Plasticity

S. POMMIER, A. PHELIPPEAU, C. PRIOUL
Ecole Centrale Paris, Grande Voie des Vignes
92295 Châtenay-Malabr, France

Abstract: The mechanical behavior of metals is well known to display scaling effects (e.g. Hall-Petch law). The scale that is usually referred to is the grain size, however, a new scale, larger than the grain size, is expected to also play a major role on the deformation mechanisms of a polycrystal.

Actually, in elasticity, a load percolation network forms through the polycrystal. This network has its own scale which can be much larger than the grain size. The load percolation effect, through the polycrystal, is inherited from the variability of the mechanical response of individual grains with respect to a given load direction. This variability and, consequently, the intensity of the percolation network, are all the higher that the elastic anisotropy of the grains is strong. As a result, before the yield point, the stress distribution within the polycrystal is heterogeneous (which is well-known) and self-organized at the scale of this load percolation network (which is a new result). After the yield point, if the material displays plastic softening, this self-organization controls the development of plasticity and lead to early plastic strain localization.

Now the propensity for a metal, to display hardening or softening after the yield point, was found by numerous authors to be size dependent. For instance, metallic materials exhibit a general trend for softening during nanoindentation tests and bulk nanocrystalline materials, for example, nanosteel for tire cords, often display a large softening effect at the onset of plasticity. That trend for softening is explained by the theory of geometrically necessary dislocations and predicted by atomic scale simulations for a few grains. However, the analysis of the collective behavior of a large number of grains is not yet possible with such approaches. A continuum approach would therefore be useful in order to identify the effects of the elastic percolation network on plastic flow but, classical constitutive models for continuum mechanics are unable to predict any grain size effect. The missing ingredients that should be included, in order to close the gap between continuum mechanics and the dislocation theory, are reviewed and discussed briefly. Strain gradient plasticity models are expected to be suitable to solve this type of problems.

T. Tsakalakos et al. (eds.) pgs. 179 - 204
Nanostructures: Synthesis, Functional Properties and Applications,
© *2003 Kluwer Academic Publishers.*

1. Introduction

The development of bulk ultrafine grained materials and of testing facilities at very small scales put into light the problem of the scale dependency of mechanical properties. "By mechanical properties" is meant elastic properties (§2) and plastic properties (§3). The elastic properties of a crystal are usually said not to be scale dependent. Conventional finite element calculations can be applied down to grain or volume sizes as small as a few hundreds of nanometers. On the contrary, plastic properties, if inherited from the flow of dislocations, are scale dependent since the grain boundary is a bound for dislocations. For very small scale, other flow mechanisms, (hich are not discussed) such as grain boundary sliding, may become predominant. Therefore, the grain size is usually considered to be the main internal scale of a metallic material and, scale dependency, (e.g. Hall Petch equation) is studied with respect to the grain size as observed on micrographs.

However, elasticity usually implies long-range effects and, thus, interesting to evaluate the "elastic grain size" of a material. In other words, the mechanical perturbation introduced in a mean stress field by a grain is not constrained inside that grain. The internal scale associated to the elastic long-range interactions, between grains, should be larger than the grain size itself. This problem is discussed in the first part of section 2 (§2.1) followed by possible implications for plastic flow are examined (§2.2). As a matter of fact, when a crystal is plastically deformed, the stress is still associated to the elastic part of strain through the elastic constants of the crystal and, therefore, long-range elastic interactions, between grains, should also appear during plastic flow.

2. Elasticity: Elastic Load Percolation Network

2.1. SELF-ORGANIZATION OF THE MECHANICAL FIELDS THROUGH THE POLYCRYSTAL IN ELASTICITY

2.1.1. *Experiments*

A few experiments have been conducted using the photostress technique in order to characterize the spatial distribution of stresses inside a polycrystal in elasticity. When a photoelastically coated sample is subjected to loads, the resulting stresses cause strains to exist over its surface. Because the photoelastic coating is bonded to the surface of the sample, the strains in the sample are transmitted to the coating. The stresses in the coating produce proportional optical effects, which appear as isochromatic fringes when viewed with a reflection polariscope.

The experiments were conducted on a duplex TA6V titanium alloy, for which sets of α nodules were shown to display the same crystallographic orientation over areas, so-called "macrozones" [1, 2], whose diameter is close to one millimetre. The conventional yield stress of this alloy is 850 MPa. In figure 1 are displayed pictures taken during a tensile test at various stress levels. At 356 MPa the sample is fully elastic, which was confirmed using micro-strain gauge measurements, however it is observed that inclined bands are crossing the specimen. Moreover, in the points indicated as black dots in figure 1, the principal strain directions were determined by photostress analysis. The local principal strain direction in the inclined bands is aligned with the load axis, showing that the observed inclined bands are not shear bands. At 800 MPa the bands are obvious and located at the same place of those at 356 MPa. If the specimen is loaded up to the yield stress (i.e. 850 MPa), the contrast is increased and particularly evident close to the surfaces. Though plastic strain has occurred at 850 MPa, the shape of the patterns observed on the surface is not significantly modified as compared with that observed at 356 MPa.

It can be concluded from these experiments that the scale associated with the strain heterogeneity in the TA6V titanium alloy is larger than the grain size, approaching 10 grains.

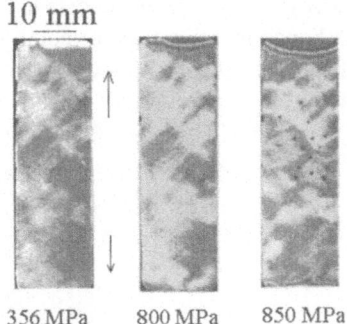

356 MPa   800 MPa   850 MPa

Figure 1. Observations of the strain at the surface during a tensile test using the photostress technique on a TA6V titanium alloy.

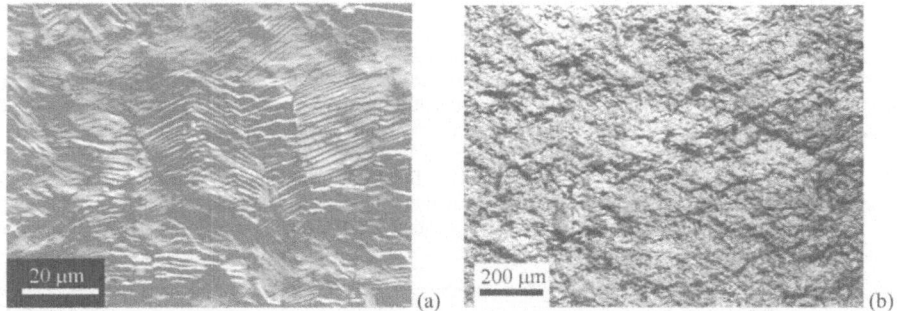

Figure 2. OFHC polycrystalline copper, cyclically creep tested, with σ_{min} =0 and with σ_{max} increasing by 2.5 MPa every 500 cycles up to failure. The observations are performed out of the striction zone. (a) slip lines at the surface of the sample revealing a grain size close to 20 μm. (b) patterns generated by cyclic creep at the surface of the sample.

The following experiments, conducted on polycrystalline copper, also reveal a large scale associated with the deformation process. Thin sheets of pure polycrystalline copper have been subjected at room temperature to a monotonic tensile test and to a cyclic creep test. In this test, the sample is tested with σ_{min} =0 and with σ_{max} increasing slowly by 2.5 MPa, every 500 cycles, up to failure and the grain size is close to 20 μm (figure 2 (a)). After a monotonic tensile test, the surface of the sample is rough due to plastic strain, however, it is not possible to distinguish any regular pattern on the surface with a scale larger than the grain size. On the contrary, after a cyclic creep test, fine inclined lines forming a regular pattern are observed at the surface of the sample (figure 2 (b)) revealing a scale for the heterogeneity of strain larger than one millimetre.

2.1.2. *Finite element analyses*

Finite elements calculations were performed in order to understand the above-mentioned effects observed in the experiments and to discuss their importance for the deformation process of polycrystalline materials.

A polycrystalline thin sheet was modelled by 3D FEM analysis and the grains were modelled as 3D regular hexagons, as shown in figure 3 (a). The number of "grains" in the model is 228 (figure 3 (b)). The elastic constants of the studied crystals are displayed in table 1. In each hexagon, the crystal orientation is set to be constant. The local orientations (1,2,3) of the hexagons are randomly selected for each hexagon in order to create an isotropic texture.

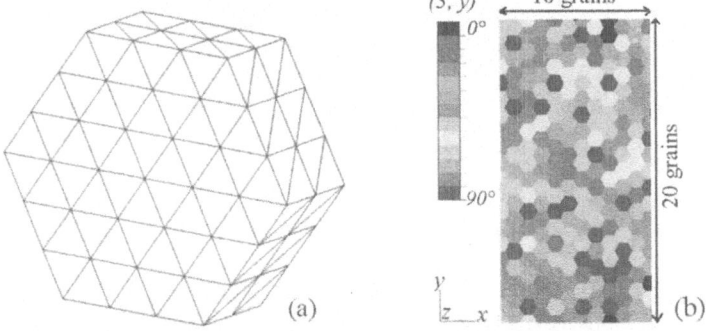

Figure 3. (a) Mesh of an individual grain by linear tetrahedrons, (b) finite element model of a thin sheet. The intensity map corresponds to the angle between the direction 3, of the local coordinate system attached to the crystal in each grain (1,2,3), with the direction y of the coordinate system of the model (x,y,z)

TABLE 1. Elastic constants (GPa) of the studied crystals. The elastic behaviour of the hexagonal crystals (Zr, Ti and Zn) is modelled by an isotropic transverse elasticity (5 independent elastic constants) and that of the cubic crystals (Al, Fe and Cu) is modelled by a cubic elasticity (3 independent elastic constants). $2C_{66}=C_{11}-C_{12}$.

Crystal	C11	C12	C13	C33	C66	C44
Zirconium	144.	72.8	65.3	165.	35.6	32.1
Titanium	162.	92.	69.	180.7	35.2	46.7
Zinc	165.	31.	50.	62.	67.	39.6
Aluminium	107.	60.8	60.8	107.	28.3	28.3
Iron	231.4	134.6	134.6	231.4	116.4	116.4
Copper	168.4	121.4	121.4	168.4	75.5	75.5

In figure 4, displayed are iso-contours of the maximum principal stress component in the model, in the case of copper. The same set of crystalline orientations is used in uniaxial extension, shear, and biaxial extension. The local stress is very heterogeneous which was expected from the high elastic anisotropy of the copper crystal (Table 1). The most interesting result is that regular patterns are found in the calculations. Though the distribution of the local orientations was defined to be random, the local stress distribution is definitely not random. In uniaxial extension it was checked that the direction of the maximum principal stress is close to the vertical axis everywhere in the sample, however, vertical links appear in the model that sustain higher stresses. With the same model, if the mean principal stress directions are rotated, the directions of the links follow the principal stress directions (e.g. see figure 4 (b)).

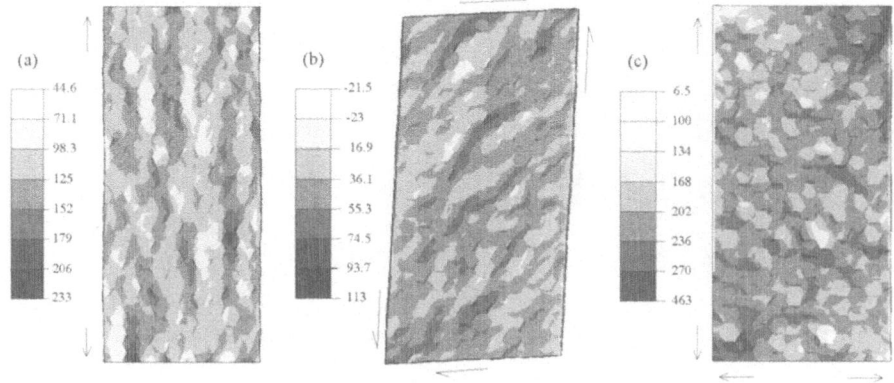

Figure 4. Intensity maps (MPa) of the maximum principal stress component in the case of copper. (a) uniaxial extension $\varepsilon_{yy} = 0.1$ %, (b) shear strain $\gamma_{xy} = 0.1$ %, (c) biaxial extension $\varepsilon_{xx} = \varepsilon_{yy} = 0.1$ %. The displacements are magnified by a factor 100.

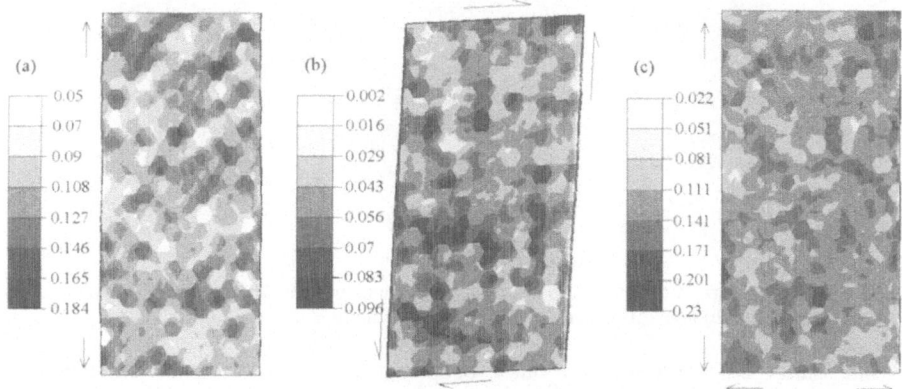

Figure 5. Intensity maps of the maximum principal strain component (in %) in the case of copper. (a) uniaxial extension $\varepsilon_{yy} = 0.1$ %, (b) shear strain $\gamma_{xy} = 0.1$ %, (c) biaxial extension $\varepsilon_{xx} = \varepsilon_{yy} = 0.1$ %. The displacements are magnified by a factor 100.

In figure 5, displayed are iso-contours of the maximum principal strain component in the model, in the case of copper. The same set of crystalline orientations is used in uniaxial extension, shear, and biaxial extension. As in figure 4, though the crystalline orientations of grains were defined to be random, the spatial distribution of strain in the model is not random, however, the patterns observed for the strain are different from those observed for the stress. There is roughly a rotation of 45° between the links of the maximum principal stress pattern and that of the maximum principal strain pattern. In uniaxial extension, for example, the intensity of maximum principal strain is higher within links inclined of about 45° with the *y* axis though the maximum principal strain direction is roughly aligned with the *y* axis. This result is consistent with the experimental observations on TA6V (figure 1).

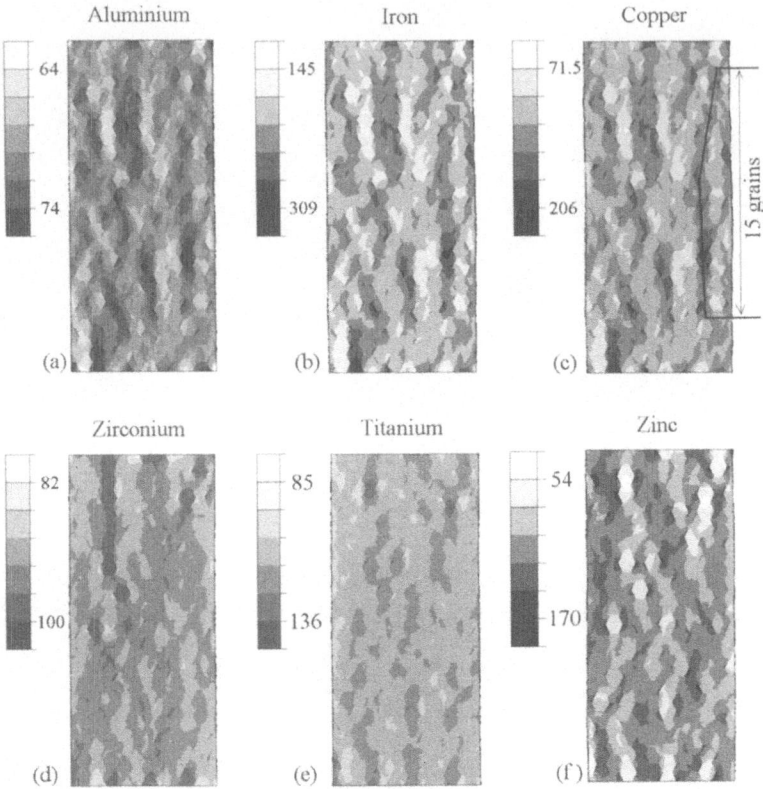

Figure 6. Intensity maps of the maximum principal stress component (in MPa) in uniaxial extension ε_{yy} = 0.1 %, (a) aluminium, (b) iron, (c) copper, (d) zirconium, (e) titanium, (f) zinc.

The same calculations have also been performed in uniaxial extension for aluminium, iron and copper, as well as zirconium, titanium and zinc. If the same set of orientations is employed, the spatial distribution of the maximum principal stress, in the three cubic crystals (aluminium, iron and copper), is similar (figure 6. a,b,c). The only difference is in the intensity of the pattern which is all the higher than the elastic anisotropy of the crystal is high. For example, while the relative difference between the maximum principal stress in overstressed, and understressed, links are lower than 15% in aluminium, it exceeds 70% in copper. The distribution of the maximum principal stress in zirconium, titanium, and zinc, is also self-organized (figure 6, d,e,f). As for the cubic crystals, the distributions are similar in the three cases but the intensity of the patterns differ. It is maximum in the case of zinc that displays the largest elastic anisotropy.

It is worth noticing, that the length of overstressed links is approaching 15 grains for example in copper or zinc. However, there is no proof that it would not be larger if the number of grains in the model was increased.

2.1.3. *Mechanism*

From experimental results and finite element calculations it was shown that the heterogeneous stress distribution, within a polycrystal in elasticity, is not random. The stress distribution is arranged as links that are roughly aligned with the principal stress directions and appear to result from a percolation process (e.g. figure 6 (f)).

As a matter of fact, the stress is heterogeneous within the polycrystal because it is constituted of grains with various crystalline orientations. In this way, though a polycrystalline metal is continuous, it is analogous to a granular media. Now, the self-organization of the mechanical field, through a granular media, is a well-known problem usually referred as the "arching effect" [3,4]. This "arching" effect is at the origin of numerous particularities in the deformation process of granular material (figure 7) among which can be distinguished three domains: the flow (arching, instabilities), the mixing (miscibility, size separation) and the vibration of particles (convection, segregation).

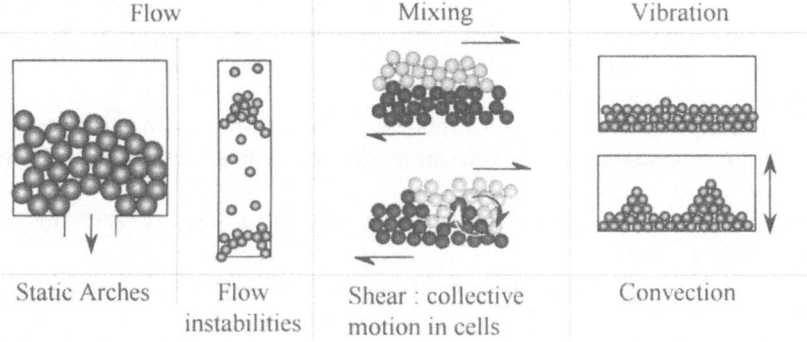

Figure 7. Illustration of some famous consequences of the so-called "arching effect" on the deformation of granular materials.

Early experimental evidences of that "arching effect" were provided in 1968 by Dantu [5] in a 2D granular media constituted of photoelastic cylinders. It was shown that the load is transferred through grains in mutual contact. As a result, a load percolation network, coincident with a percolating network of grains in mutual contact, is formed through the granular media. The links of that network follow the isostatic lines in the equivalent elastic continuum material [5]. The load percolation network is carrying a force larger than the mean one [6, 7]. The intensity of that network is all the higher that the disorder in the granular material is high. When the disorder is the highest, the links of the network can be as large as the specimen size. At the onset of the deformation process, arches are broken, which gives rise to macro-instabilities [3,4].

More generally, a percolation effect is usually observed when, on the one hand, the material is heterogeneous and when, on the other hand, the capability of the material to transfer a percolating quantity (load, heat, sound, electrical current...) varies significantly from one point to another [3].

186

Polycrystals are constituted of grains with various crystalline orientations, and therefore can be considered as granular materials, but they differ essentially from real granular materials (solid grains+fluid) since all the constituents of a polycrystal are able to transfer all the components of a stress field, while in a real granular material the fluid phase, for example, can hardly transfer the shear component of the mean stress field. However, according to the crystalline orientation of grains toward a given direction, their apparent compression modulus and shear modulus may vary very significantly if the elastic anisotropy of the crystal is high. Thus, though all the constituents of a polycrystal are able to transfer all the components of a stress field, their efficiency to do it varies very significantly and very suddenly at grain boundary. Therefore, a percolation effect should also appear through a polycrystal [8,9].

The main parameter controlling the scale of that load percolation network can be identified as follows: let consider a stress concentrator embedded within a polycrystal (figure 8) subjected to a uniaxial mean stress field. The isostatic lines turn in the vicinity of the stress concentrator, leading to a stress concentration. This rotation corresponds to a load transfer by shear. Now, if the elastic shear modulus of the material is high in the neighbourhood of the stress concentrator (figure 8 (a)), the homogenization length (l_H) is short. On the contrary, if the shear modulus is low (figure 8 (b)) the homogenization length (l_H) is large. Then if stress concentrators (e.g. grains with a high rigidity) are dispersed throughout the polycrystal, their elastic perturbation of the mean stress field might percolate (figure 8 (c)), provided that the homogenization length is large enough (l_H).

Consequently, the typical scale of the elastic percolation network is related to the probability to find grains with a low shear modulus around grains with a high rigidity, with respect to the principal stress directions [8,9] (figure 8). If the texture of the material is random, this probability can be calculated directly from the elastic constants of the crystal.

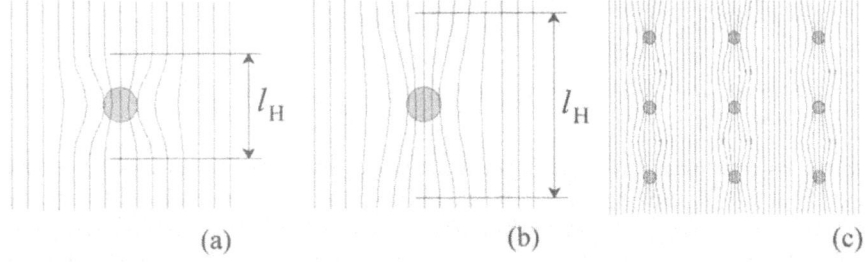

Figure 8. Illustration of the mechanism controlling the scale of the load percolation network within a polycrystal [8,9].

The main conclusion of this first section is that the elastic anisotropy of the metallic crystals lead to the formation of a load percolation network through a polycrystal, analogous to that observed in a granular material. This network possesses an intrinsic scale, larger than the grain size, which is related to the probability of a low shear modulus with respect to a given direction.

Thus, static arches develop themselves inside a polycrystal within its elastic domain. This is important for the nucleation of fatigue cracks or for the initiation of brittle failure, for example, since these mechanisms occur usually within that elastic domain.

It is also interesting to examine the possible effects of the formation of such arches during plastic flow, since it is the regime in which real granular material exhibit the most particular behaviours.

2.2. CONSEQUENCES FOR THE ONSET OF PLASTICITY

2.2.1. *Threshold for Plastic Yielding*

In order to quantify the importance of that elastic load percolation network for the onset of plasticity, statistic calculations have first been performed. As a matter of fact, two conditions have to be fulfilled in order to exceed the yield point inside a grain. On the one hand, the grain should be heavily loaded and on the other hand its crystalline orientation should be favourable for slip. It is interesting, to check if one condition is of prime importance as compared with the other.

TABLE 2. Maximum principal stress at the centre of the grain, with a crystalline orientation as follows: $(\varphi_1, \psi, \varphi_2) = (0,0,0)$ and located at the centre of the model. The results have been calculated from 70 random configurations of its neighbours and uniaxial mean extension $\langle \varepsilon_{yy} \rangle = 0.1\%$. Comparison with the distribution of the Schmid factor (SF) [9].

	Al	Fe	Cu	Zr	Ti	Zn
Minimum value (MPa)	64.0	137.9	67.9	92.6	105.8	79.2
Maximum value (MPa)	72.6	210.7	128.3	104.6	122.5	126.9
Mean Value (MPa)	68.1	179.9	101.8	97.6	114.1	103.6
Distribution's width (%) *	± 7	± 24	± 35	± 8.5	± 9.25	± 35
—	—	—	—	—	—	—
Symmetry	FCC	BCC	FCC	HCP	HCP	HCP
Mean value of SF	0.462	0.462	0.462	-	-	-
SF distribution's width				-	-	-
Upper bound (%)	+ 7.6	+ 7.6	+ 7.6			
Lower bound (%)**	- 23.5	- 23.5	- 23.5			

* for a Gaussian distribution 99.865 % of the results are lower than the mean value of the distribution plus three times the standard deviation of the distribution

** only 4% of the results are over the upper bound and under the lower bound.

The first condition was examined as follows: a single hexagon located at the centre of the FE model (figure 3 (b)) was set to have a fixed crystalline orientation, while the crystal orientations of the other hexagons in the model were randomly selected before each calculation. Then, for each one, the maximum principal stress component was determined at the centre of the grain for which the crystalline orientation is fixed. The statistic distribution of this value was calculated for seventy random configurations of its neighbours. This distribution is found to be a Gaussian. Therefore the distribution width is calculated as three times the standard deviation divided by the mean value of the distribution. The results of these computations are gathered in Table 2. The maximum principal stress in aluminium is found to vary of +/- 7 % around the mean value. This variability is rather low as compared with copper and zinc for which the variation is of +/- 35 %. The variability of the maximum principal stress found for iron, is also very high, i.e. +/- 25 %.

This variability should be compared with that inherited from the second condition, i.e. the effect of the crystalline orientation of the grain itself. As a matter of fact, two conditions have to be fulfilled in order to exceed the yield point inside a grain. On the one hand, the grain should be heavily loaded and on the other hand its crystalline orientation should be favourable for slip. For this purpose, the statistic distribution of the Schmid factor (SF) was calculated according to the crystalline orientation of the grain, for a random texture. For each orientation, the maximum Schmid factor was calculated over the 12 slip systems of the FCC and of the BCC systems. The statistic distributions of the maximum Schmid factor are very similar for the FCC and BCC systems. These distributions are not centred on their mean value. Low values of the maximum Schmid factor are not excluded, but their probability is very small. The mean value of the Schmid factor is found to be high (<SF>=0.462). Therefore, in grains for which the Schmid factor is at its maximum value (SF=0.5), the Schmid factor is only 8% higher than the mean Schmid factor in the polycrystal.

In comparison, in copper, for a fixed crystalline orientation of the grain and according to the configuration of its neighbours, the maximum principal stress can be higher by 35 % than the mean value. Therefore, in iron and copper, the yield point in each grain should be controlled by the configuration of its neighbours rather than by its own crystalline orientation. In aluminium on the contrary, the two effects appear to be of the same magnitude (Table 2).

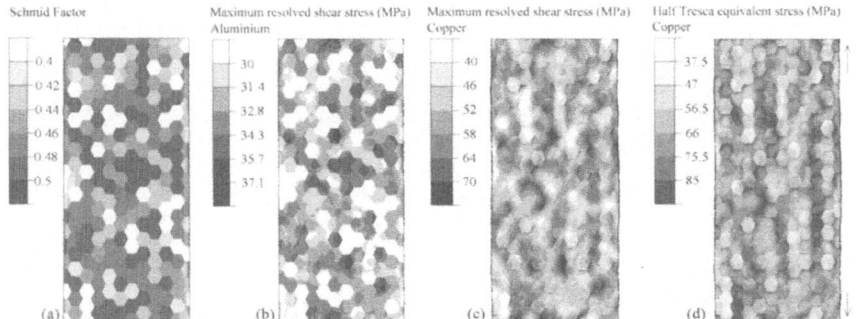

Figure 9. (a) Schmid factor intensity maps for aluminium and copper. (b,c) Maximum resolved shear stress intensity maps (MPa), over the twelve (111)<110> slip systems of the FCC crystal (b) in aluminium, (c) in copper; (d) Half Tresca equivalent stress in copper. The model is subjected to a uniaxial extension ε_{yy}=0.1 %.

In order to illustrate this discussion the following calculations have been performed. In each point of the model (figure 9), the maximum resolved shear stress over the twelve slip systems of the FCC crystal was calculated as follows, using the stress tensor obtained from the FEM:

$$\tau_{max} = \frac{1}{\sqrt{6}} \underset{s=1\,to\,12}{Max} \left[\langle \bar{1}10 \rangle \sigma^{FE} \begin{pmatrix} 1 \\ 1 \\ 1 \end{pmatrix}, \langle 10\bar{1} \rangle \sigma^{FE} \begin{pmatrix} 1 \\ 1 \\ 1 \end{pmatrix}, \; etc... \right] \tag{1}$$

The model is subjected to a uniaxial extension and, if the material is set to have an isotropic elastic behaviour, τ_{max} is proportional to the Schmid factor. The yield point in each grain is only determined by the crystalline orientation of the grain itself. When the elastic anisotropy of the crystal increases, a load percolation network appears within the polycrystal and the value of τ_{max} in each grain is sensitive to the location of the grain toward that network. In order to visualize which effect is dominant on the spatial distribution of τ_{max} in the polycrystal, the maps for τ_{max} are plotted out for the same set of crystalline orientations and compared with the intensity maps of the Schmid factor and with the Tresca equivalent stress which measures the maximum possible shear stress (figure 9). That map is plotted for copper only since the spatial distribution of the Tresca equivalent stress is very similar in aluminium, except that the intensity of the network is much lower.

It is obvious from figure 9 that, in aluminium, τ_{max} is mostly dominated by the crystalline orientation of the grain itself, while in copper, τ_{max} is mostly determined by the location of the grain with respect to the load percolation network.

It can be concluded from these results, that the importance for the onset of plasticity of the formation of such a load percolation network through the polycrystal depends on the elastic anisotropy of the material.

When the elastic anisotropy is high the load percolation network should control the onset of plasticity. This is the case for copper, iron and zinc for example. For such materials the spatial distribution of the first yielding points should therefore be self-organized. On the contrary, when the elastic anisotropy is low, the spatial distribution of the first yielding points should be more or less random, at least if the spatial distribution of crystalline orientation is random.

2.2.2. Early Plastic Strain Localization

It was explained above that if the elastic anisotropy of grains is high (e.g. copper, iron and zinc), the spatial distribution within the polycrystal of the first yielding grains, should be self-organized (figure 9), as a consequence of the development of an elastic percolation network.

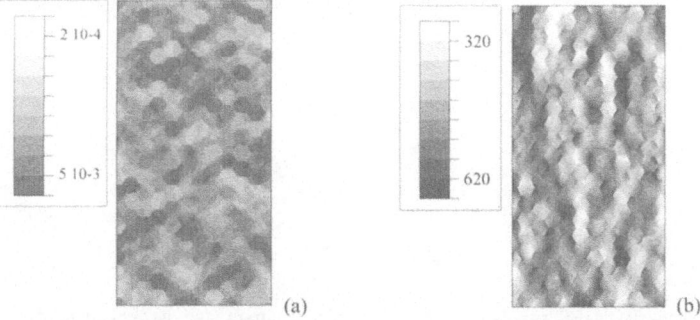

(a)   (b)

Figure 10. Elastic FE simulation. A uniaxial mean strain $\langle\varepsilon\rangle = 0.4\%$ is applied to the model. Elastic constants of copper. (a) distribution of the maximum principal strain component, (b) of the maximum principal stress component (MPa).

After that yield point, two cases have to be distinguished: on the one hand the case of plastic hardening and on the other hand the case of plastic softening. As a matter of fact, whatever the details of the mechanical behaviour of grains, if plastic softening occurs, plastic strain localizes around the first yielding points. In such a case, the existence of a self-organized stress field within the polycrystal, before the onset of plasticity, should be of primary importance. This effect is illustrated in figure 10, 11

190

and 12. A set of crystalline orientations was selected and the distribution of strain and stress was calculated at first in elasticity for this given set of orientations (figure 10). The elastic constants are those of copper.

Then, with the same set of orientations elasto-plastic, calculations have been performed either with plastic hardening (figure 11) or with plastic softening (figure 12). The constitutive behaviour of the grains was said to obey the Von Mises criterion with a yield stress varying in agreement with equation (2).

$$R = R_o + Q\left(1 - \exp^{-b.p}\right) \ with \ p = \int_{\tau=0}^{\tau=t} \sqrt{\frac{2}{3} d\varepsilon_p : d\varepsilon_p} \qquad (2)$$

The saturation stress $(Ro+Q)$ is equal to 200 MPa in both cases, while the initial yield stress (Ro) is equal to 100 MPa in the case of plastic hardening $(Q = 100$ MPa$)$ and to 300 MPa in the case of plastic softening $(Q = -100$ MPa$)$. The rate of strain hardening or softening b, with the cumulated plastic strain p, is set to be equal to 100.

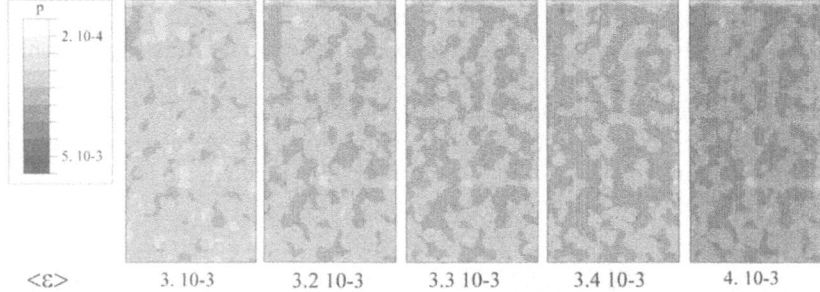

Figure 11. Elasto-plastic FE simulations. A uniaxial mean strain $<\varepsilon>$ is applied to the model. Elastic constants of copper. Plastic hardening, Von Mises Criterion. Spatial distribution of the cumulated plastic strain in the model.

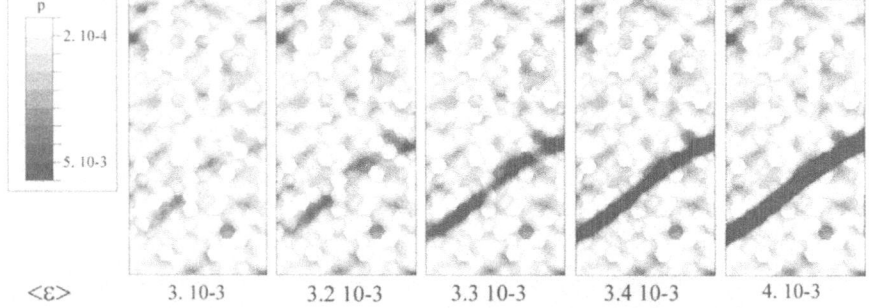

Figure 12. Elasto-plastic FE simulations. A uniaxial mean strain $<\varepsilon>$ is applied to the model. Elastic constants of copper. Plastic softening, Von Mises Criterion. Spatial distribution of the cumulated plastic strain in the model.

When grains display plastic hardening (figure 11), the spatial distribution of plastic strain within the polycrystal is not homogeneous, therefore, there is a significant variability of the plastic strain from grain to grain. However, the polycrystal can still be considered as a homogeneous material since that heterogeneity does not modify the overall deformation of the polycrystal. As a matter of fact, the polycrystal deforms globally as a homogeneous material with a local variability around the mean value.

On the contrary, when grains display plastic softening (figure 12), the strain heterogeneity is not reduced to a variability around the mean strain level calculated with considering the polycrystal as a homogenized continuum material. Here, the heterogeneity modifies the overall deformation of the polycrystal. As a matter of fact, in this case, early plastic strain localization occurs as a result from the self-organization of the mechanical field already present in the elastic domain (figure 10). The most interesting result is that the localization band (figure 12) forms by coalescence of areas that have been plastically deformed at the very beginning of the deformation process, inside a band of the elastic percolation network (figure 10 a).

2.3. SUMMARY

The previous paragraphs can be summarized as follows into five main points:

1 an elastic percolation network, analogous to that observed in granular media is formed through a polycrystal within its elastic domain.

2 this network possesses an intrinsic scale, larger than the grain size, which is related to the statistic distribution of the shear modulus with respect to the load direction.

3 the intensity of this elastic percolation network and its importance for the onset of plasticity is all the higher that the elastic anisotropy of the crystal is high.

4 in such a case, the spatial distribution of the first yielding grains in the polycrystal is determined by the elastic percolation network, and as a result, is self-organized.

5 finally, if grains display plastic softening, early plastic strain localization occurs by coalescence of initial yielding areas within the links of the elastic percolation network.

The main prospect of this first part is to study the collective behaviour of grains within a polycrystal, for which crystallites display plastic softening. In such a case, some characteristics of the deformation of granular materials might also be observed, for instance macro-instabilities during the deformation process and "convection" movements.

3. Plasticity: Modeling Size Effect

In order to examine the collective behavior of grains during the plastic deformation of a metallic polycrystal, at least a few cells of the elastic percolation network that was described above should be considered. A model should therefore include no less than a hundred grains. This leads to prioritize a continuum description of the crystallites rather than a description at the atomic scale or a semi-discrete description by the dislocations dynamics. Moreover, the most interesting case is the case of plastic softening, for which the elastic percolation network should dominate the deformation process. It is thus important to have a good understanding of the mechanisms leading to plastic softening at the onset of plasticity of metallic crystallites, and a good description of the phenomenon in the constitutive behavior.

Now, the propensity for a metal to display plastic hardening or softening after the yield point was shown by numerous authors to be scale dependent [10-13]. A general trend for softening is observed under one micrometer, in nanohardness experiments, micro-bending tests [14] and micro-torsion tests [15].

In metallic materials, a scale reduction often produces two major effects, on the one hand a drastic increase of the yield strength, which is better known as the Hall Petch effect, and on the other hand a tendency for plastic softening after the yield point.

In this section, these effects are illustrated by the example of nanosteel wires in the first place, then the physical origin of such size dependency of the mechanical behavior is briefly discussed and finally continuum models available in the literature are recalled.

3.1. SIZE DEPENDENCY IN THE MECHANICAL BEHAVIOR OF NANO-STEELS FOR TIRE CORDS

Tire cords wires are obtained by deep drawing of eutectoid steels. The drawing strain, (i.e. $\ln(S_o/S)$ where S_o is the initial section and S the final section), is commonly up to 3.5. During this process, the internal scales of the material are drastically reduced, which are: the diameter of the wire, the grain size and size of the ferrite lamellae. At the beginning of the process, the steel is pearlitic with a lamellae thickness approaching 150 nm (figure 13 a) and a grain size of 20 µm. After a cold drawing strain of 3.5, the thickness of the ferrite lamellae is less than 20 nm (figure 13 b), the wire diameter is of 200 µm, and the grain size is assumed to be less than one micrometer.

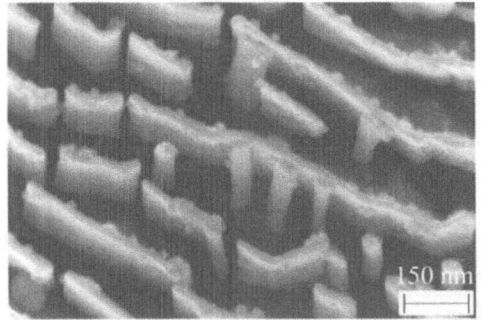 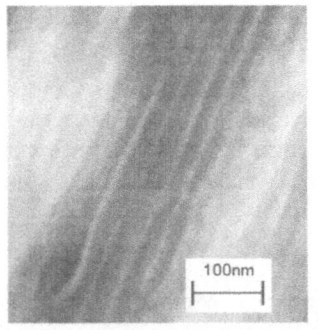

Figure 13. 0.8 % C steel. (a) pearlitic structure in the initial microstructure (etching Nital 3%) (b) ferrite lamellae after a cold drawing strain of 3.5 (etching 0.5% HF)

The evolution of the grain size during cold drawing is not easy to determine, since in such heavily refined eutectoid steels, grain boundaries are not easy to distinguish from lamellae. Therefore, a 0.1 % C steel with an initial grain size of 20 µm was cold drawn in order to characterize the evolution of the grain morphology (figure 14 a and b). In the initial microstructure, grains are equiaxed. After cold drawing, the morphology of grains was characterized in the drawing direction (figure 14 a) and in the transverse directions (figure 14 b). It appears clearly that the evolution of the grain morphology is not homothetic of that of the wire. Grains are elongated in the drawing direction but they are not equiaxed in a transverse section. In a cross section (figure 14 b), their shape is complex, elongated and curled around the wire axis. This effect results from the inadequacy of the axisymmetric deformation imposed by the drawing process and of the cubic symmetry of the ferritic phase [16]. The main consequence of this phenomenon is that the shortest dimension of grains diminishes during the drawing process with a higher rate than the wire diameter. Above $\varepsilon=1.5\text{-}2$ the texture is stabilized.

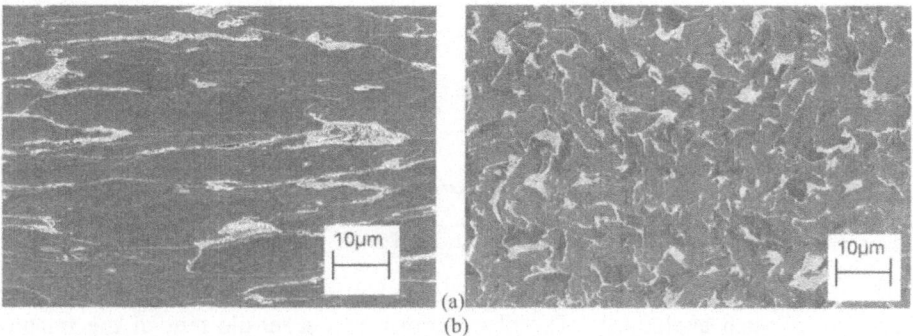

Figure 14. Grain morphology after a cold drawing strain of 0.8 of a 0.1% carbon steel with an initially equiaxed microstructure (a) longitudinal section, (b) cross section (pictures by J. D. Capdeville)

That reduction of the internal scales of the material is associated with a drastic increase of the strength of the wires. The tensile strength of 0.8%C steel wires, for example, is multiplied by a factor 2.5 by cold drawing up to 3.4 (figure 15 a). Simultaneously the plastic elongation (Ap%) is reduced by nearly a factor 4. Before a cold drawing strain of 1, the evolution of Ap% is a little chaotic. In that domain, at least two phenomena are superimposed. On the one hand there is a reduction of the material scales in the transverse directions and on the other hand there is also a strong evolution of the crystallographic texture toward a fiber texture.

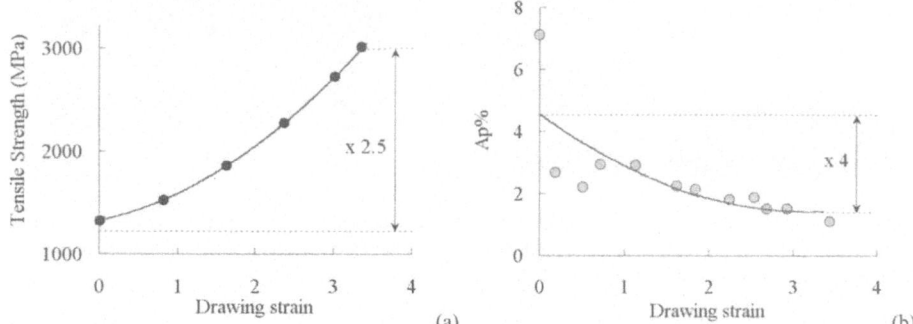

Figure 15. Consolidation of a 0.8% C steel by cold drawing. (a) tensile strength of wires (MPa), (b) elongation of wires (%).

The evolution of the mechanical strength with cold drawing can also be characterized by the mean of hardness test. A surprising result was obtained with a set of 0.8% C steel wires at various drawing levels: while the tensile strength of the wire is increased by a factor 2.6 by cold drawing, the micro-hardness is only increased by a factor 1.6.

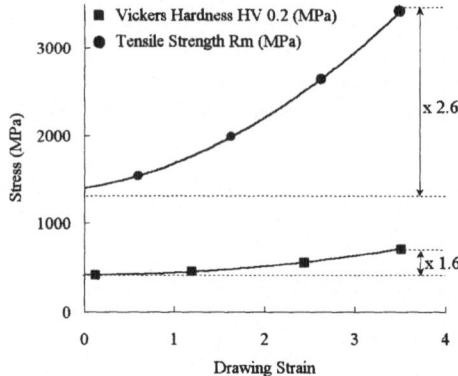

Figure 16. Consolidation of a 0.8% C steel by cold drawing. Tensile strength and micro-hardness (MPa).

Experiments have been conducted in order to better understand the mechanisms at the origin of such evolutions. The observations after a tensile test, of the fracture surfaces of a 0.7%C steel wire cold drawn at various levels between $\varepsilon=0$ and $\varepsilon=3.5$, revealed that the failure is typically ductile. Though the elongation Ap% is very small, the striction is always very pronounced (figure 17 a).

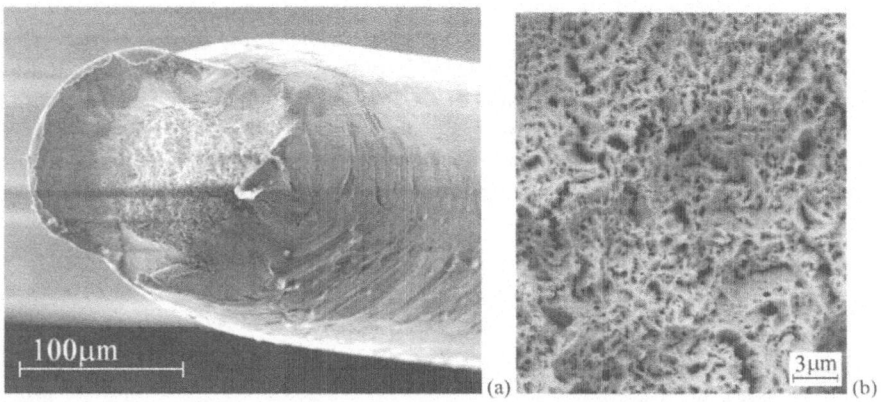

Figure 17. Fracture surface of a 0.7% C steel wire, cold drawn at $\varepsilon=3.5$, after a tensile test at room temperature. (a) striction area, $\ln(\phi o^2/\phi^2)=0.55$. (b) detail of the fracture surface, typical ductile failure with curly and elongated voids on the fracture surface.

The reduction of the wire's ductility with cold drawing (figure 15 b) does not correspond to an evolution of the failure mechanism from a ductile failure toward a brittle failure.

On the fracture surface, voids are curled and elongated (figure 17 b, figure 18 a) like grains in a cross section (figure 14 b). In order to better understand the formation of voids, the external surface of wires was mechanically polished, and the development of slip bands was studied. Below a drawing strain of about 1.5, slip bands can either be constrained inside ferrite lamellae or cross a whole colony of lamellae of the pearlitic structure. Above $\varepsilon=1.5$, the lamellar structure is no more a barrier for dislocations. For instance, for a drawing strain of 3.5 (figure 18 b), the interlamellar spacing is less than 30 nm (figure 13 b) but slip bands are continuous over domains that can be as large as 2 micrometers. Moreover, it was confirmed that voids appear at the intersection between slip bands and grain boundaries (figure 18 b).

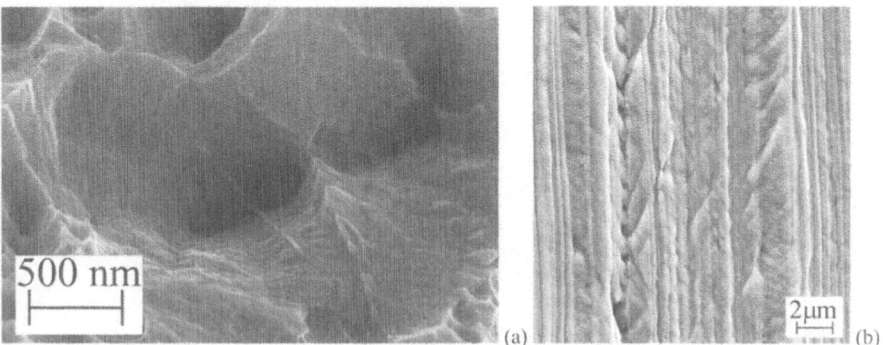

(a)　(b)

Figure 18. Tensile test at room temperature of a 0.7% C steel wire, cold drawn at $\varepsilon=3.5$. (a) detail of a void on the fracture surface. (b) slip bands on the cylindrical surface of the wire, polished before testing.

A last set of experiments provides the explanation of the apparent contradiction in figure 16. Nanohardness tests were performed for three drawing levels, $\varepsilon=0$, $\varepsilon=2$ and $\varepsilon=4$. Five experiments have been conducted in each case, though only one is plotted in figure 19. The results are very reproducible.

In the micro-hardness regime, on the right side of the graph, the hardness is not depth dependent. In that regime, a factor 1.7 is observed between the hardness in the initial state and after a cold drawing strain of 4. This is in reasonable agreement with the results displayed in figure 16.

Below that micro-hardness regime, the hardness becomes depth dependent [12,13]. A very pronounced softening effect is observed when the indentation depth increases. This depth dependency appears only for $\varepsilon=4$. At its peak value, the hardness is 2.5 higher after a cold drawing strain of 4 than in the initial microstructure. This factor 2.5 is precisely what is found for the increase of the tensile strength due to cold drawing (figure 16).

The apparent contradiction in figure 16, results from the fact that a striction can appear during a tensile test but not during a hardness test. During the indentation, the indenter imposes the displacements at the surface, and a large softening effect can be measured. During a tensile test, a softening effect can hardly be observed since it causes plastic strain localization and the modification of the shape of the sample.

Consequently, in figure 16, the evolution of the tensile strength that is reported corresponds to the peak strength of the material, before plastic softening had occurred. On the contrary, the evolution of the micro-hardness measures the strength of the material after plastic softening had occurred.

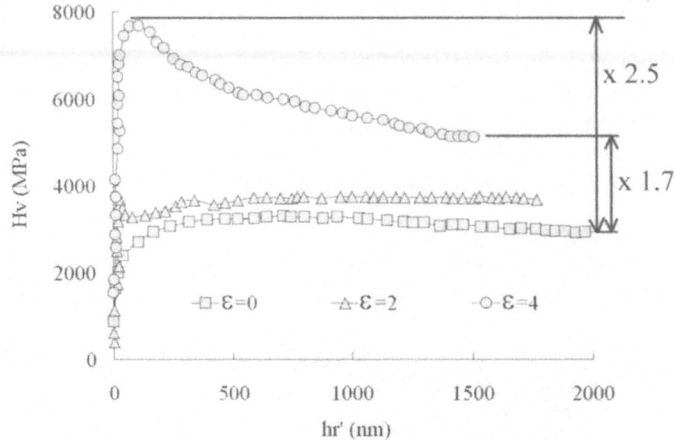

Figure 19. Nanohardness tests conducted at room temperature for a 0.7% C steel wire, cold drawn at ε=0, ε=2 and ε=4. The hardness is plotted in MPa versus the indentation depth.

These results obtained in nanosteel wires can be summarized as follows:

1 deep drawing causes a drastic increase of the tensile strength of wires, a moderate increase of the micro-hardness, and a pronounced reduction of their elongation.

2 the failure is typically ductile, voids forms at grain boundary as a result of intergranular plastic strain incompatibilities. A pronounced striction is always found.

3 the drop of ductility corresponds to a premature striction, not to a more brittle character of failure. That premature striction results from the appearance of a softening effect at the onset of plasticity for large drawing strains.

4 the moderate increase of the micro-hardness with cold drawing corresponds to the increase of the saturation stress of the material, after plastic softening had occurred.

5 the drastic increase of the tensile strength with cold drawing corresponds to the increase of the peak strength of the material, at the onset of plasticity, before plastic softening had occurred. Then, plastic strain localization occurs and the sample fails before the saturation stress (after plastic softening at the onset of plasticity had occurred) could be reached.

6 in nanohardness experiments, the peak hardness corresponds to indentation depths smaller than 150 nm. This confirms that the scale controlling the extraordinary mechanical properties of these nanosteels is the nanoscale.

3.2. SIZE DEPENDENCY OF PLASTIC PROPERTIES.

A large number of micro-indentation studies [10-13], and more recently micro-bending [14] and micro-torsion experiments [15] have demonstrated that the plastic properties of crystalline metals are size dependent. The phenomenon whereby the indentation hardness decreases with increasing the indentation depth is called the indentation size effect and was reported very early (see e.g. [12] for a review). Various artifacts were suspected to be responsible for this effect, such as the existence of thin oxides films on the surface, an evolution of the friction between the indenter and the specimen with the depth, a higher importance of elastic recovery at small indentation depth or, at constant indentation rate, the variation of the strain rate with the indentation depth. These effects are likely to contribute to the indentation size effect. However, Ashby showed in 1970, that the nucleation of geometrically necessary dislocations is sufficient to produce such a size effect [17]. He examined the problem of bending [17] and the indentation [18]. More recently, simulations at the mesoscale [19,20], or at the atomic scale [21-23] have successfully reproduced the indentation size effect. Micro-bending [14] and micro-torsion [15] experiments have finally demonstrated that this effect is an intrinsic size effect and not only related to the indentation experimental procedure.

We will try to give an insight into this problem through a simple example, e.g. the plastic deformation of a cylinder with a unique slip system:

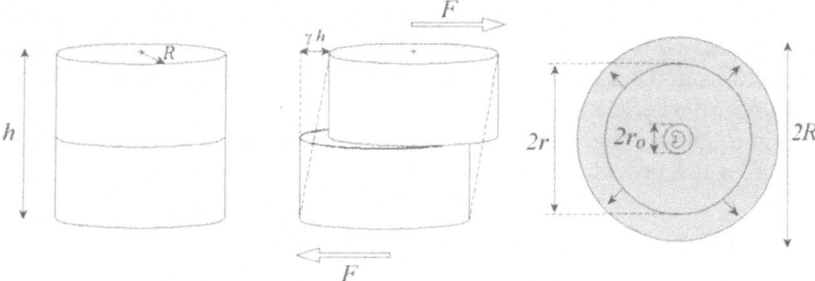

Figure 20. Illustration of the plastic deformation of a cylinder with a unique slip system.

Let say that the strain level γ is fixed, we aim at determining the shear stress τ associated with this strain level, as a function of the dimension of the cylinder. The external power as applied to this system is as follows:

$$P_{ext} = F.\dot{u} = \left(\pi R^2 \tau\right).\left(\dot{\gamma}h\right) \tag{3}$$

Now, the physical mechanism at the origin of plastic shear is the nucleation and flow of geometrically necessary dislocations. At a given strain level γ corresponds a given number of geometrically necessary dislocation. In this case γ is inherited from the complete flow of n dislocations through the cylinder and to the partial flow of one dislocation with a radius $r < R$ (figure 20):

$$n = \left[\gamma h/b\right] \quad and \quad \pi r^2 \Big/ \pi R^2 = \gamma h/b - \left[\gamma h/b\right] \tag{4}$$

Now, the internal power is the sum, on the one hand of the heat dissipation Q due to the friction (τ_c) of the flowing dislocation on the slip plane and on the other hand of the rate of variation of the free energy $d\Psi/dt$ stored by dislocations. If interactions between dislocations are neglected (for this rough example), that free energy Ψ is the product of the free energy per unit length of dislocation $\Psi_\perp$ by the total dislocation length in the sample, then:

$$Q = \left(2\pi r.b.\tau_c\right)\frac{dr}{dt} \quad and \quad \frac{d\Psi}{dt} = 2\pi\frac{dr}{dt}\Psi_\perp \tag{5}$$

The flow rate of the dislocation, dr/dt is calculated with equation (4) and introduced into (5). This provides the expression of the internal power:

$$P_{int} = Q + \frac{d\Psi}{dt} = 2r\frac{dr}{dt}\pi b\tau_c + 2\pi\frac{dr}{dt}\Psi_\perp = \pi R^2 h.\dot{\gamma}.\tau_c + \pi R^2 h.\dot{\gamma}.\frac{\Psi_\perp}{br} \tag{6}$$

The thermodynamic balance finally provides the expression for the shear stress:

$$P_{int} = P_{ext} \Rightarrow \tau = \tau_c + \frac{\Psi_\perp}{br} \tag{7}$$

At this stage, it becomes clear that if the elastic energy stored in dislocations is taken into account, a length scale parameter should appear in the yield stress. As a matter of fact, the lowest is the value of r, the highest should be the contribution to the yield stress of the free energy stored in dislocations. Moreover, if the shear strain direction is reversed, stored dislocations will be able to flow back, restoring their elastic energy. This will result in a reduction of the threshold for yielding in the reverse direction. This effect is well known as a kinematic hardening effect. Consequently, kinematic hardening is also an intrinsically scale dependent phenomenon and the back stress derives from the amount of geometrically necessary dislocations (GNDs) stored in the volume of material considered.

Now, in equation (7), r can be replaced by its expression as a function of the strain level (eq. 4), which provides the evolution of the stress versus strain:

$$\tau = \tau_c + \Psi_\perp\left(b.R\sqrt{\frac{\gamma h}{b}} - \left[\frac{\gamma h}{b}\right]\right)^{-1} \tag{8}$$

This evolution is plotted in figure 21 for R=500nm, R=1μm and R=10μm, h=R, b=2.47A° and $\Psi_\perp$= 11.10^{-9} J/m. These last two values correspond to iron. During the calculations, data corresponding to values of r lower than $10b$ have been suppressed. For these calculations τ_c was assumed to obey:

$$\tau_c = 50 + 5\mu b.\rho^{1/2} \tag{9}$$

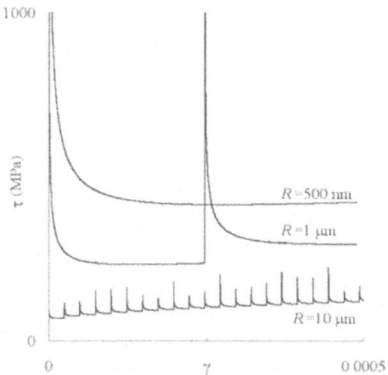

Figure 21. Illustration, stress-strain curve at different scales, R=500 nm, R=1 μm and R=10 μm.

This illustration shows that a length scale dependency appears in the yield strength if geometrically necessary dislocations (GNDs) stored in the crystal are considered. In the case of iron, the scale dependent term exceeds 200 MPa for $\gamma=10^{-4}$, as soon as R is below one micrometer.

3.3. MECHANICAL MODELING WITH A CONTINUUM APPROACH.

It has now become clear that the yield strength of a given volume of crystal depends on the one hand on the flow of geometrically necessary dislocations (GNDs), through a friction term τ_c, and on the other hand on the amount of GNDs stored in that volume (back stress). Moreover, it should be considered that there might be a flux of GNDs from one side of the volume of material to the other.

With a continuum approach, dislocations can be described as a dislocation density. GNDs density tensors have been introduced originally and about at the same time by Kondo [24] and Nye [25]. Then a large number of researchers have examined this problem and recently Cermelli [26] provided a new expression for the density of GNDs for large deformations. This density G is introduced as a tensor, since a dislocation is determined by two vectors, e.g. the direction s of its burgers vector and its line direction l. G measures the total burgers vector per unit area [26]:

$$G = \sum_{\alpha} \rho_{\odot}^{\alpha} s^{\alpha} \otimes s^{\alpha} + \rho_{\perp}^{\alpha} l^{\alpha} \otimes s^{\alpha} \tag{10}$$

This tensor can be expressed as the sum of the contributions of screw and edge dislocations on the α slip systems of the crystal.

Then dislocations and strain have to be associated (see §3.2 equation 4). It is clear that the strain rate is associated with the flux of GNDs. The density itself of GNDs stored inside the volume of material is to be calculated from the plastic strain gradient on each slip system, as it is illustrated in figure 22. Let consider the emission of a dislocation from a source at the center of the crystal (figure 22 a), as long as the dislocation is inside the volume there is a plastic strain gradient (figure 22 b). Then when the dislocation is out of the volume, plastic strain is homogeneous (figure 22 c).

The GND's density tensor G can therefore be expressed [26] from the plastic strain gradient $\nabla\gamma_\alpha$ on each slip system α, which consist in a slip plane m_α and a slip direction s_α:

$$G = \sum_\alpha (\nabla\gamma_\alpha \times m_\alpha) \otimes s_\alpha \qquad (11)$$

Therefore, models including a strain gradient should be able to account for the scale effects associated with the elastic energy stored in GNDs, that is believed to be at the origin of the scale dependency of the plastic properties of metals.

Three main classes of continuum approaches of plasticity include a strain gradient. The first approach [27,28] is based on Cosserat media. In Cosserat media, internal loads are modeled by stresses and distributed torques, which counterparts are strains and strain gradients. The second and the third approaches are not based on Cosserat media. The second approach, developed par Fleck and Hutchinson [29], introduces the strain gradient in the constitutive equations. Since the strain is a second order tensor the strain gradient is a higher order tensor. The third approach, developed originally by Aifantis [30], introduces the gradient of plastic strain on each slip system (which is a vector) in the equations of the constitutive behaviour.

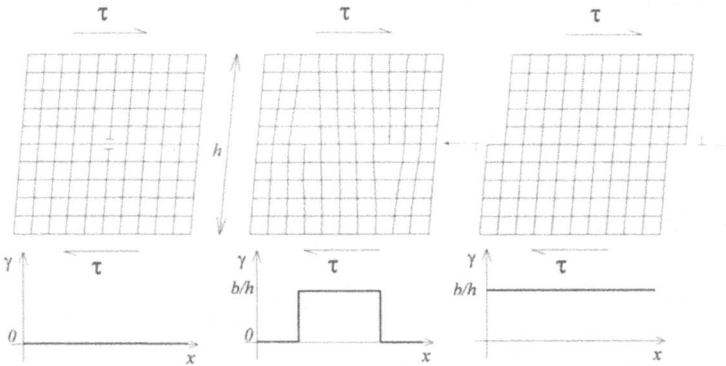

Figure 22. Illustration, (a) nucleation of a dislocation look on a source s, (b) flow of that dislocation inside the volume, with a plastic strain gradient and an elastic distorsion associated with the dislocation inside the volume, (c) flow of the dislocation out of the volume, the plastic strain becomes constant and the elastic distorsion associated with dislocations is released.

The model developed by Gurtin [31,32] belongs to the same category. However, this last model considers the fact that dislocations coming from outside can flow inside the volume of material. Therefore a supplementary equation is introduced, which is a balance equation for the density of GNDs. Basically it says that the variations in the density of GNDs consist in nucleation or annihilation of dislocation on the one hand and in flux of dislocations through the frontiers on the other hand.

This balance equation is very important. As a matter of fact, in a polycrystal, dislocations stored on one side of the grain during loading, are able to flow back at unloading. This effect is responsible for the existence of a back stress. Usually this back stress is simply introduced as an internal variable in the constitutive equations. In this case, the back stress results from the balance equation.

Central to the model developed by Gurtin [32] is the idea that should be accounted the working associated with each kinematical process, namely:

a - overall elastic lattice distortion: E_e (figure 22 a)

b - elastic distortion around GNDs: $\nabla\gamma_\alpha$ (figure 22 b)

c - slip of GNDs: γ_α

A system of forces is introduced as a counterpart of the system of kinematical process:

a - lattice stress T (figure 22 a), which projects into τ_α on a slip system α.

b - distributed Peach and Köhler force ξ_α (figure 22 b)

c - friction stress π_α

The internal power is thus written as follow, the two first terms are classical, the third one accounts for GNDs:

$$P_{int} = \int_V T\dot{E}_e dv + \int_V \sum_\alpha \pi_\alpha \cdot \dot{\gamma}_\alpha + \xi_\alpha \cdot \nabla\dot{\gamma}_\alpha \qquad (12)$$

Then, three principles are invoked: the principle of virtual power, the second principle of thermodynamics, and the GNDs balance equation. Without introducing any constitutive behavior, these considerations leads to a yield criterion dependent on the flux of peach and Kohler force $(div\xi_\alpha)$, which corresponds to a flux of GNDs:

$$\tau_\alpha = \pi_\alpha - div\xi_\alpha \qquad (13)$$

In equation 13, π_α plays the role of the effective stress, while $(div\xi_\alpha)$ plays the role of the back stress. Here, the back stress (kinematic hardening), is not added in the constitutive equation, as it is usually done, but results directly from the balance equations, in particular from the balance equation of GNDs.

3.4. SUMMARY

This section can be summarized into four main points:

1 Plastic properties of metallic materials are scale dependent. A scale reduction has two main consequences, on the one hand an increase of the flow stress, on the other hand an evolution toward strain softening. Evidences have been given in the case of nanosteel for tire cords.

2 This scale dependency is related to the activity of geometrically necessary dislocations (GNDs). If the working associated with the lattice distortion by GNDs is taken into account, the high flow stresses observed in nanocrystalline materials can be explained. The back stress, inherited from stored dislocations, is expected to be also highly scale dependent.

3 With a continuum approach, GNDs can be characterized by the GNDs density tensor G. This tensor can be calculated from the plastic strain gradient on each slip system.

4 Strain gradient plasticity models appear therefore to be suitable to deal with small scales problems, in particular those of Aifantis [30] and Gurtin [31,32]. It is worth to mention that in the model of Gurtin, the back stress is not introduced

in the constitutive behaviour, but results directly from the GNDs balance equation.

4. Conclusions and Prospects

In this paper, some scaling effects associated with elasticity and plasticity were discussed. In elasticity, an elastic percolation network, analogous to that observed in granular media forms through a polycrystal. This network possesses an intrinsic scale, larger than the grain size, which is related to the elastic anisotropy of the crystal. It was shown that, if after the yield point grains display strain softening, the self-organisation of the mechanical field inherited from the elastic percolation network, leads to shear banding at the onset of macro-plasticity.

Now, the propensity for a material to display strain softening is scale dependent. Strain softening appears during nanoindentation test and in nanocrystalline bulk materials. Thus, it is believed that the deformation of bulk nanocrystalline metals, with a high crystalline elastic anisotropy, should be influenced by the existence of that elastic percolation network. In particular, shear banding at the onset of macro-plasticity is expected.

When plasticity occurs by a dislocation-based mechanism, the free energy stored by dislocations (e.g. elastic distorsion of the lattice around dislocations) introduces a scale dependency. As a matter of fact, the power associated with the flow of a dislocation on a slip plane consist in a dissipation by friction, that is proportional to the area and in a variation of the free energy stored by dislocations, that is proportional to the defect length. Thus the "power per unit area", and consequently the "yield stress", consists in a friction term, which is not scale dependent and a defect free energy term that is scale dependent with a negative power. This scale dependency becomes non-negligible when the scale is lower than one micron. Including the free energy associated with geometrically necessary dislocations into a finite element model should therefore reproduce efficiently that scale dependency, with lower computation times than atomic scale or meso-scale simulations. The plastic strain gradient on each slip system is a measure of the density of geometrically necessary dislocations, and thus, strain gradient plasticity models, such as those developed by Aifantis [30] or Gurtin [31], can be used to include scale dependency in a finite element modelling.

Our main prospect is to examine, using strain-gradient plasticity, the collective behaviour of grains within a bulk nanocrystalline material with a high elastic anisotropy (iron, copper, zinc) and to see whether some characteristics of granular media would not be observed. This track seems interesting since, for instance in 1998, Carsley et al [33] have studied bulk nanocrystalline iron alloy and found that from 77 to 470 K the unique deformation mechanism was shear banding. The elongation of grains indicated that plasticity occurred by a dislocation-based mechanism. Malow and Koch [34] obtained similar results. It has also been reported by Segal [35] for example, that during processing the flow was continuous at moderate strain but becomes discontinuous at large strain (i.e. for ultra fine grains). The mesoscale organisation in shear bands is not yet understood.

5. Acknowledgement

The authors would like to thank E. Depraetere and P. Lescœurs of the Michelin Company providing materials and many fruitful discussions.

6. References

1. Le Biavant, K., Pommier, S. and Prioul, C. (1999) Ghost structure effect on fatigue crack initiation and growth in a Ti-6Al-4V alloy, in Goryin, I.V. and Ushkov, S.S. (eds), *Titane 99: Science and technology*, Saint Petersburg, Russia, pp 481-487.
2. Le Biavant, K., Pommier, S. and Prioul, C. (2002) Local texture and fatigue crack initiation in a Ti-6Al-4V Titanium alloy, *Fat. Fract. Engng. Mater. Struct*, **25**, 527-545.
3. Guyon, E. and Troadec, J.P. (1994) *Du sac de billes au tas de sable*, eds. Odile Jacob, Paris.
4. Savage, S.B, (1997), Problems in the static and dynamics of granular materials, in Behringer, R.P. and Jenkins, J.T. (eds.), *Powder and grains*, Rotterdam, pp. 185-194.
5. Dantu, P. (1968). Etude statistique des forces intergranulaires dans un milieu pulvérulent, *Géotechnique*, **18**, 50-55
6. Radjai, F., Wolf, D.E., Roux, S., Jean, M. and Moreau, J.J. (1997) Force networks in dense granular media, in Behringer, R.P. and Jenkins, J.T. (eds.), *Powder and grains*, Rotterdam, pp. 211-214.
7. Roux, J.N. (1997) in Behringer, R.P. and Jenkins, J.T. (eds.), *Powder and grains*, Rotterdam, pp. 215-218.
8. Pommier, S. (2002) "Arching" effect in elastic polycrystals, *Fat. Fract. Engng. Mater. Struct*, **25**, 331-348.
9. Pommier, S. (2002) Variability in fatigue lives: an effect of the elastic anisotropy of grains ?, in ESIS STP (eds), *6th International Conference on Biaxial/Multiaxial Fatigue and Fracture*, Elsevier, accepted for publication.
10 Samuels, L.E. (1986) Microindentation in Metals, in P. J. Blau and B. R. Lawn (eds) *Microindentation Techniques in Materials Science and Engineering, ASTM STP 889*, Philadelphia, pp.5-25
11. Swadener, J.G., Misra, A., Hoagland, R.G. and Nastasi, M. (2002) a mechanistic description of combined hardening and size effect, *Scripta Mat*, **47**, 343-348
12. Elmustapha, A.A. and Stone, D.S. (2002) Indentation size effect in polycrystalline FCC metals, Acta Materiala, **50**, 3641-3650
13 Nix, W.D., Gao, H. (1998) Indentation size effect in crystalline materials: A law for strain gradient plasticity, *J. Mech. and Physic of Solids*, **46**, 411-425
14 Stölken, J.S. and Evans, A.G. (1998) a micro-bend test method for measuring the plasticity length scale, *Acta Materiala*, **46**, 5109-5115
15 Fleck, N.A., Muller, G.M., Ashby, M.F. and Hutchinson, J.W. (1994) Strain gradient plasticity, theory and experiments, *Acta Metall. Mater*, **42**, 475-487.
16 Sevillano, J.G., Matey Munoz, L. and Flaquer Fuster, J. (1998) Ciels de Van Gogh et propriétés mécaniques, *Journal de Physique IV*, France, **8**, 155-165.
17 Ashby M.F. (1970) The deformation of plastically non-homogeneous alloys, *Phil. Mag.*, **21**, 399-424
18 Poole W.J., Ashby M.F. and Fleck N.A (1996) Micro-hardness of annealed and work-hardened copper polycrystals, *Scripta Materialia*, **34**, 4, 559-564.
19 Cleveringa, H.M.M., Van der Geissen, E. and Needlemann, A. (1998) Discrete dislocation simulations and size dependent hardening in single slip. *Journal de Physique IV France, 8*, 83-92.
20 Fivel M. C., Robertson C. F., Canova G.R. and Boulanger L. (1998) Three dimensionnal modelling of indent-induced plastic zone at mesoscale, *Acta Mater*, **46, 17**, 6183-6193.
21 Zimmerman J.A, Gonzalez M.A.; De La Figuera J.; Hamilton J.C.; Rojo J.M.; De La Fuente O. Rodriguez and Pai Woei Wu (2002) Dislocation emission around nanoindentations on a (001) fcc metal surface studied by scanning tunneling microscopy and atomistic simulations, *Physical Review Letters*, **88, 3**, 36101/1-36101/4.
22 Kelchner C.L., Plimpton S.J. and Hamilton J.C. (1998) Dislocation nucleation and defect structure during surface indentation, *Physical Review Letters*, **58**, 11085-11088
23 Tadmor E.B., Miller R., Phillips R. and Ortiz, M. (1999) Nanoindentation and incipient plasticity, *J Mater Res*, **14, 6**, 2233-2250

24 Kondo, K. (1952) On the geometrical and physical foundations of the theory of yielding. in: 2nd Japan Nat. Congress Appl. Mech.; 2, pp. 41-47.

25 Nye, J.F. (1953) some geometrical relations in dislocated solids, Acta Metal A, 153-162.

26 Cermelli, P. and Gurtin, M.E. (2001) On the characterization of the geometrically necessary dislocations in finite plasticity, JMPS, 49, 1539-1568.

27 Forest, S., Cailletaud, G. and Sievert, R. (1997) A Cosserat Theory for elastoviscoplastic single crystal at finite deformation, Arch. Mech., 49, 705-736.

28 Forest, S., Barbe, F. and Cailletaud, G. (2000) Cosserat modeling of size effects in the mechanical behaviour of polycrystals and multi-phase materials, International J. of Solids and Struct., 37, 7105-7126.

29 Fleck, N.A. and Hutchinson, J.W. (1997) Strain Gradient Plasticity, Advances in Applied Mechanics, 33; 295-361

30 Zbib, H., Aifantis, E.C. (1992) On the gradient dependent theory of plasticity and shear banding, Acta Mechanica, 92, 209-225.

31 Gurtin, M.E. (2002) A gradient theory of single-crystal viscoplasticity that accounts for geometrically necessary dislocations, JMPS, 50, 5-32.

32 Gurtin, M.E. (2000) On the plasticity of single crystals: free energy, microforces, plastic strain gradients, JMPS, 48, 989-1036.

33 Carsley J.E., Fisher A., Milligan, W.W. and Aifantis, E.C. (1998) Mechanical behavior of a bulk nanostructured iron alloy, Metal. and Mater. Trans A, 29 A, 9, 2261-2271.

34 Malow, T.R. and Koch, C.C. (1998) Mechanical properties, ductility and grain size of nanocrystalline iron produced by mechanical attrition, Metal. and Mater. Trans A, 29 A, 9, 2261-2271.

35 Segal V. M. (2002) Severe plastic deformation: simple shear versus pure shear, Materials Science and Engineering, A338, 331-344.

DEFECTS AND DEFORMATION MECHANISMS IN NANOCRYSTALLINE MATERIALS

I.A. OVID'KO
Institute of Problems of Mechanical Engineering, Russian Academy of Sciences, Bolshoj 61, Vasil. Ostrov, St. Petersburg 199178, Russia

Abstract. We provide a brief overview of theoretical models of defects and plastic flow processes in nanocrystalline materials with focuses placed on their specific deformation mechanisms being commonly not effective in conventional coarse-grained polycrystals. In particular, we will consider triple junction diffusional creep and rotational deformation mode in nanocrystalline materials. Also, the notion of nano-dislocations is introduced and used in a description of plastic flow localization in nanocrystalline materials. Finally, we discuss deformation mechanisms releasing misfit stresses in nanocrystalline films and coatings.

1. Introduction

Nanocrystalline materials exhibit the unique mechanical behavior due to nano-scale and interface effects; see e.g. [1-3]. High-density ensembles of grain boundaries (GBs) in nanocrystalline materials serve as effective obstacles for lattice dislocation slip which dominates in coarse-grained polycrystals. In addition, the image forces acting in nanometer-sized grains cause the difficulty in forming lattice dislocations in such nano-grains [4,5]. In these circumstances, the grain refinement hampers the lattice dislocation slip. At the same time, GBs provide the effective action of deformation mechanisms being different from the lattice dislocation slip. As a corollary, the grain refinement leads to the competition between the lattice dislocation slip and deformation mechanisms associated with the active role of GBs. It is believed that nanocrystalline matter represents the arena for the discussed competition which causes unique mechanical properties of nanocrystalline materials [6-15]. In this context, identification of effective deformation mechanisms and their contributions to plastic flow is very important for understanding the fundamentals of the mechanical behavior of nanocrystalline materials as well development of technologies based on plastic forming of nanostructures.

T. Tsakalakos et al. (eds.) pgs. 205 - 215
Nanostructures: Synthesis, Functional Properties and Applications;
© *2003 Kluwer Academic Publishers.*

The deformation mechanisms in nanocrystalline materials, in many cases, cannot be unambiguously identified with the help of contemporary experimental methods because of high precision demands on experiments at the nano-scale. Therefore, theoretical modeling of defects and plastic deformation processes is a very important constituent of both fundamental and applied research of nanostructured materials. This paper briefly reviews theoretical models of deformation mechanisms and defects being carriers of plastic flow in nanocrystalline materials. For brevity, we will concentrate our consideration on final results of theoretical models while their mathematical details will be omitted.

2. Deformation Mechanisms in Nanocrystalline Materials

Let us consider plastic deformation mechanisms in nanocrystalline materials. The specific deformation behavior of nanocrystalline materials, in particular, is exhibited in the so-called abnormal Hall-Petch effect which manifests itself as the softening of a material with reducing the grain size d [16,17]. The most theoretical models relate the abnormal Hall-Petch dependence to the competition between deformation mechanisms in nanocrystalline materials, e.g., [6-15]. In the framework of this approach, deformation mechanisms associated with the active role of GBs dominate over conventional lattice dislocation motion in nanocrystalline materials, in which case, the grain refinement leads to a softening of a material. In doing so, generally speaking, standard and new deformation mechanisms associated with the active role of GBs can contribute to plastic flow in nanocrystalline materials. The standard deformation mechanisms are the GB sliding and Coble creep related to enhanced diffusional mass transfer along GB planes [6-13]. This idea is originated, in fact, from the classical theory of creep and superplasticity of microcrystalline materials where these mechanisms play the essential role; see, e.g., [18,19]. We think that, together with these two standard deformation mechanisms associated with the active role of GBs, non-conventional (new) deformation mechanisms are capable of effectively contributing to plastic flow in nanocrystalline bulk materials and coatings. In particular, such specific deformation mechanisms are creep associated with enhanced diffusion along triple junctions of GBs [14] and rotational deformation mode [20,21].

First, let us consider the triple junction diffusional creep. In general, in recent years, triple junctions of GBs have been recognized as defects with the structure and properties being essentially different from those of grain boundaries that they adjoin [22]. For instance, from experimental data and theoretical models, it follows that triple junctions play the role as enhanced diffusion tubes [23,24], nuclei of the enhanced segregation of the second phase [25,26], strengthening elements and sources of lattice dislocations [8,27-29] during plastic deformation, and drag centers of grain boundary migration during re-crystallization processes [30]. Also, as it has been shown in experiments [23], creep associated with enhanced diffusion along triple junctions contributes to plastic flow of coarse-grained polycrystalline aluminum. Actually, since triple junction diffusion coefficient D_{tj} highly exceeds grain boundary diffusion coefficient D_{gb} [23,24], enhanced diffusional mass transfer along triple junction tubes, occurs under mechanical stresses and contributes to plastic forming of a mechanically loaded material. The contribution in question is characterized by a rather specific dependence of plastic strain rate on grain size d. More precisely, grain size exponent is -4, in this case [23], in contrast to GB diffusional creep (Coble creep) and bulk diffusional creep characterized by grain size exponents -3 and -2, respectively:

$$\dot{\varepsilon}_{tj} \propto D_{tj} \cdot d^{-4} \tag{1}$$

$$\dot{\varepsilon}_{gb} \propto D_{gb} \cdot d^{-3} \tag{2}$$

$$\dot{\varepsilon}_{bulk} \propto D_{bulk} \cdot d^{-2} \tag{3}$$

Here $\dot{\varepsilon}_{tj}$, $\dot{\varepsilon}_{gb}$ and $\dot{\varepsilon}_{bulk}$ are plastic strain rates that correspond to triple junction diffusional creep, GB diffusional creep and bulk diffusional creep, respectively. D_{bulk} denotes the bulk self-diffusion coefficient.

Fedorov, et al [14]. suggested a theoretical model describing the yield stress dependence on grain size in fine-grained materials based upon competition between conventional dislocation slip, GB diffusional creep, and triple junction diffusional creep. It has been shown, that the contribution of triple junction diffusional creep increases with reduction of grain size causes a negative slope of the Hall-Petch dependence in the range of small grains (see curve 1 in Fig. 1). The results of model [14] are compared with experimental data [31-35] from copper and shown to be in rather good agreement (Fig. 1).

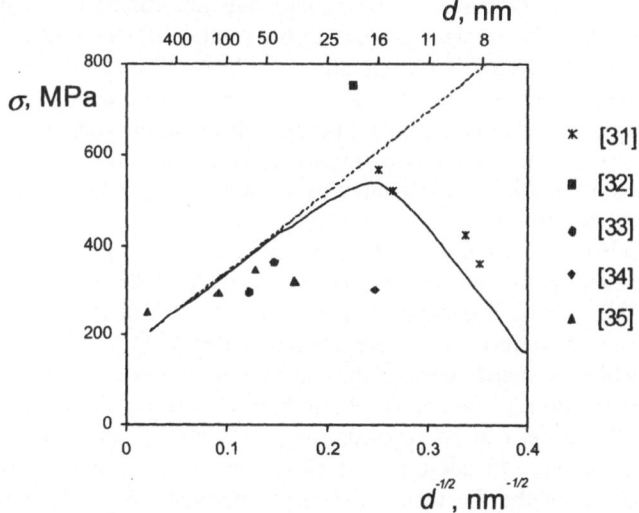

Figure 1. Dependences of yield stress σ on grain size d and d^{-1/2}. Dotted and solid lines represent respectively classical Hall-Petch dependence and theoretical [14] dependence taking into account competition between lattice dislocation slip, grain boundary diffusional creep and triple junction diffusional creep. Also, experimental data [31-35] from copper are presented.

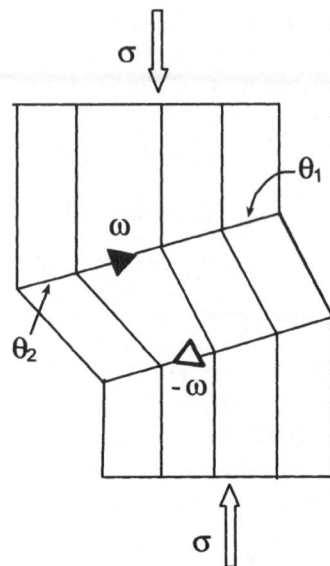

Figure 2. Rotational deformation occurs through movement of a dipole of grain boundary disclinations. Grain boundary fragments separated by disclination of strength ω are characterized by misorientation parameters θ_2 and θ_1, where $\omega = \theta_2 - \theta_1$.

Now let us turn to a discussion of a rotational deformation mechanism in nanocrystalline materials. The primary carriers of the rotational plastic deformation in solids are believed to be dipoles of GB disclinations [36]. A GB disclination represents a line defect that separates two GB fragments with different misorientation parameters (Fig 2). That is, GB misorientation exhibits a jump at disclination line. A disclination dipole consists of two GB disclinations of opposite signs. Motion of a disclination dipole causes plastic deformation accompanied by crystal lattice rotation (Fig. 2). It is called as the rotational deformation mode.

Conventional electron microscopy experiments are indicative of the essential role of the rotational plastic flow in coarse-grained materials under high-strain deformation (for a review, see [37,38]). It is also supported by experimental observations of grain rotations in superplastically deformed microcrystalline materials [19,39]. The rotational deformation is capable of being very intensive in nanocrystalline materials [20]. Actually, the volume fraction of GBs is extremely high in nanocrystalline materials, in which case, GB disclinations can be formed, roughly speaking, in every point of a mechanically loaded sample. In addition, the elastic energy of a disclination dipole rapidly diverges with rising the distance between disclinations [36]. Therefore, such dipoles are energetically permitted, mostly for GB disclinations, that are close to each other. It is the namely case of nanocrystalline materials where the interspacings between neighbouring GBs are extremely small. Finally, nanocrystalline materials contain high-density ensembles of triple junctions where the crossover from the conventional GB sliding to the rotational deformation effectively occurs. Murayama, et al's atomic-level observation of diclination dipoles in deformed nanocrystalline materials [40], supports theoretical statement [20] and provides experimental evidence for the action of rotational deformation mechanism in nanostructures.

3. Nano-dislocations and Localization of Plastic Flow in Nanocrystalline Materials

Now let us discuss models of the experimentally observed [41-45] localization of plastic flow in nanocrystalline materials. One of model explanations of plastic flow localization in shear bands in nanocrystalline materials is based on the concept of cellular dislocations, which originally has been suggested and exploited in the theory of grain growth and superplasticity of conventional microcrystalline materials [46,47]. Cellular dislocations are topological defects in a regular array of cells or grains. Figure 3 illustrates, schematically, a cellular dislocation with pair of 5- and 7-sided grains representing its core. Movement of such cellular dislocations results in a local plastic flow associated with grain sliding along the shear surface. This grain boundary sliding assumes correlated grain movement and grain shape change with the cellular dislocation core. Similar to conventional crystal lattice dislocations, movement of cellular dislocations results in a relative displacement of material blocks at the shear surface. Theoretical representations on cellular dislocations are indirectly supported by experimental observation of inhomogeneous plastic shear along shear bands in microcrystalline materials [46,47]. However, in the framework of the approach discussed, the driving force for plastic flow localization is not identified. In general, cellular dislocations may be distributed homogeneously within a deformed sample causing homogeneous plastic flow.

Another model has been suggested, in the works [6,7]. describing plastic flow localization in nanocrystalline materials as the phenomenon associated with GB migration. The model suggests that GB migration occurs during plastic deformation and results in the formation of a zone where all GB planes are tentatively parallel each other (Fig. 4). The GB sliding is enhanced in this zone which, therefore, develops into a shear band where plastic flow is localized. However, model [6,7] does not identify a mechanism for the specific GB migration which makes GB planes to be parallel each other.

In this context, we suggest a model explanation [48] of the plastic flow localization which, in fact, combines and modifies representations of models [6,7,46,47] discussed. We treat that development of a shear band in a nanomaterial occurs through generation and to motion of partial cellular dislocations with local GB migration and sliding being localized in their cores (Fig. 5). The dislocation core is characterized by length scale being in the order of grain size. In these circumstances, the discussed defect – a partial cellular dislocation – in a nano-grained material can be naturally denoted as a *nano-dislocation*. Movement of a nano-dislocation is followed by formation of a stacking fault in a regular array of grains. The stacking fault is the namely shear band region where GB planes are becoming tentatively parallel each other during plastic deformation. This facilitates further GB sliding, in the stacking fault region, causing plastic flow localization. Thus, the nanodislocation movement in local regions of a mechanically, loaded nanocrystalline material, gives rise to the softening of these regions which, therefore, serve as nuclei of experimentally observed shear bands.

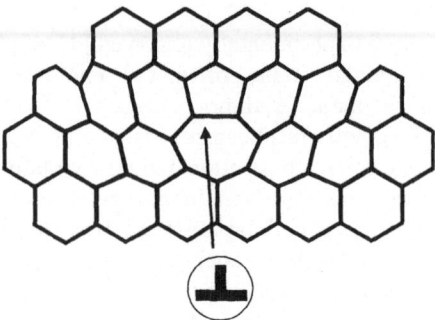

Figure 3. Cellular dislocation in a model array of hexagonal nanograins (cells).

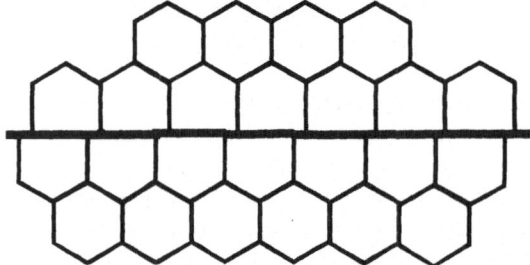

Figure 4. Formation of shear band via local migration of grain boundaries in model array of hexagon nanograins. Grain boundary planes in the shear band region are parallel each other, facilitating enhancement of grain boundary sliding.

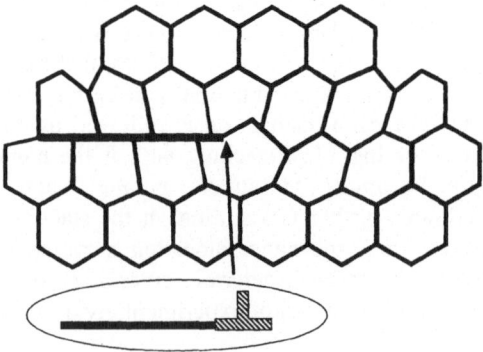

Figure 5. Nanodislocation – a partial cellular dislocation – and stacking fault region (solid segment) formed behind it. Grain boundary planes are parallel each other in the stacking fault region, causing enhancement of grain boundary sliding.

4. Defects and Plastic Deformation Releasing Residual Stresses in Nanostructured Films and Coatings

Let us discuss the role of defects and plastic deformation in the internal stress relaxation in nanostructured films and coatings. In general, an interphase film/substrate boundary serves as a source of internal stresses occurring due to lattice parameter mismatch, thermal expansion mismatch, elastic modulus mismatch and plastic flow mismatch between the adjacent phases. The internal stresses strongly influence the structure and the functional properties of nanocrystalline films, in which case, it is very important to identify mechanisms for stress relaxation in such films.

Any interphase boundary between a single crystalline substrate and a nanocrystalline film is featured by the existence of many boundary fragments bordered by junctions of the interphase boundary and grain boundaries of the nanocrystalline film (Fig. 6). In these circumstances, one of effective micromechanisms for stress relaxation, inherent to nanocrystalline films, is the formation of a partly incoherent interphase boundary, that is, interphase boundary consisting of both coherent and incoherent fragments [49,50]. In doing so, coherent fragments induce mismatch stresses in the film while incoherent fragments, characterized by low adhesion, do not induce such stresses. The fraction of incoherent fragments, of the interphase boundary, increases with rising the film thickness in order to provide the effective relaxation of stresses in growing film.

In addition, due to the existense of high-density ensembles of grain boundaries, relaxation of stresses in nanocrystalline films effectively occurs via the formation of grain boundary dislocations as misfit defects [50-53]., These grain boundary dislocations induce stress fields that compensate for, in part, internal mismatch stresses and are located at grain boundaries (Fig. 7). ,The formation of grain boundary dislocations as misfit defects does not induce any extra violations of coherency of interphase boundaries and, therefore, does not lead to degradation of their functional properties used in applications.

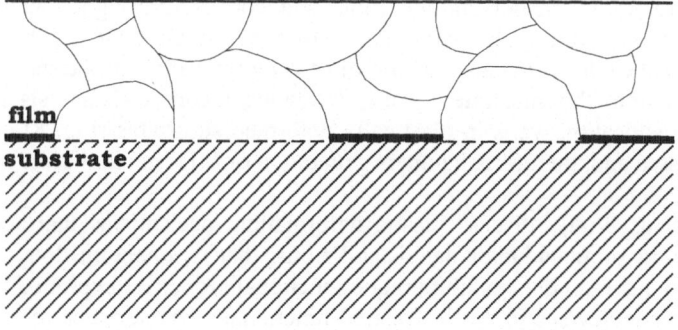

Figure 6: Partly incoherent interphase boundary between nanocrystalline film and single crystalline substrate. It consists of coherent (solid) and incoherent (dashed) fragment.

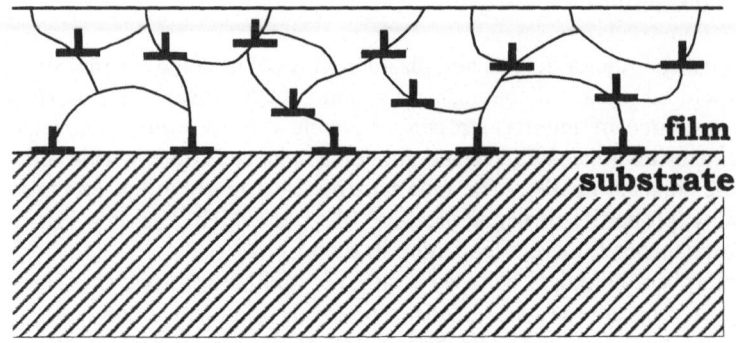

Figure 7: Grain boundary dislocations as misfit defects in nanocrystalline film.

Thus, internal stresses in nanocrystalline films and coatings effectively relax via the formation of grain boundary dislocations as misfit defects and the formation of incoherent fragments of interphase boundaries and/or grain boundaries. These micromechanisms are inherent to namely nanocrystalline films and coatings, and provide essentially more effective relaxation of residual stresses in such films and coatings, compared to their coarse-grained counterparts where the formation of misfit dislocations at interphase boundaries is the dominant relaxation mechanism; see, e.g., [54-58]. This explains the experimentally documented capability for producing very thick nanocrystalline coatings, in contrast to conventional coarse-grained coatings, where stress buildup essentially limits the coating thickness, see discussion in paper [59].

In general, there are technological methods allowing to provide effective relaxation of internal stresses in nanocrystalline films and coatings through human-controlled modification of substrates. In particular, plastic deformation of a substrate creates defects in the substrate which are capable of causing relaxation of internal stresses in a film deposited onto the substrate. Thus, following theoretical analysis [60,61], the formation of dislocation walls in plastically deformed substrates (Fig. 8) compensates for misfit stresses and, therefore, hampers generation of MDs in nano-scale multilayered films (Fig. 8).

Finally, let us briefly discuss plastic deformation mechanisms that accommodate misfit stresses in semiconductor nanoisland films, that is, quantum dots. In general, self-assembled quantum dots exhibit unique functional properties exploited in electronic and optoelectronic devices [62-70]. Desired functional characteristics of quantum dots crucially depend on their structure and geometry. In particular, the formation of misfit dislocations releasing stresses in quantum dots leads to dramatic degradation of their functional properties [63]. In this context, knowledge of critical parameters of quantum dots, at which the formation of MDs is energetically favourable, is of utmost importance for applications of such dots.

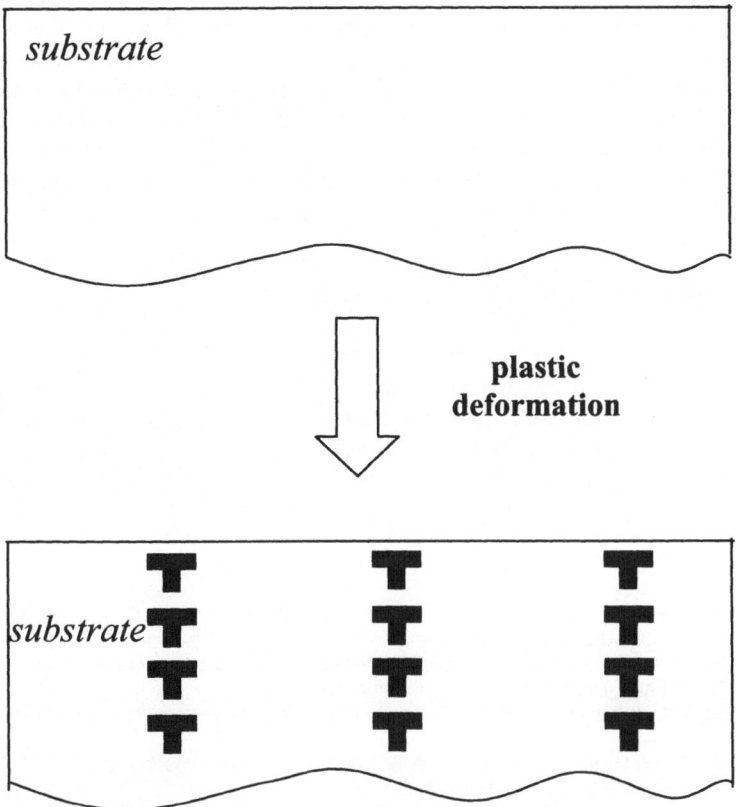

Figure 8: Plastic deformation of a substrate creates dislocation walls whose stress fields are capable of hampering the formation of misfit dislocations in single- and multilayered films deposited onto the substrate.

The standard deformation mechanism leading to a partial relaxation of misfit stresses in quantum dots is treated to be the generation of perfect misfit dislocations [62,71-73]. Recently, a new deformation mechanism in quantum dots has been suggested. It is the generation of partial and split misfit dislocations [74]. According to theoretical analysis [74], the generation of partial misfit dislocations effectively competes with that of conventional perfect dislocations in germanium pyramid-like quantum dots on silicon substrate. In doing so, different partial dislocation structures are energetically preferred in different regions of the interface. Single partial dislocation is generated near lateral free surface of a nanoisland. Then, during its motion towards the nanoisland base center, the second partial dislocation is generated.

5. Concluding Remarks

Thus, owing to interface and nano-scale effects in nanostructured materials, the set of deformation mechanisms in these materials is richer than that in conventional coarse-grained polycrystals. In particular, triple junction diffusional creep and rotational deformation are capable of essentially contributing to plastic flow of nanocrystalline bulk materials and coatings. Localization of plastic flow, in shear bands in nanocrystalline materials, is effectively described as a process associated with generation and evolution of nano-dislocations being partial cellular dislocations. The formation of partial and split misfit dislocations causes plastic deformation effectively releasing misfit stresses in semiconductor quantum dots. These deformation mechanisms should definitely be taken into account in further experimental and theoretical research of plastic flow processes in nanostructured materials.

6. Acknowledgments

This work was supported, in part, by the Office of US Naval Research (grant N00014-01-1-1020), the Office of US Naval Research, International Field Office, Europe (grant N00014-02-1-4045), the Russian Fund of Basic Research (grant 01-02-16853), St.Petersburg Scientific Center of Russian Academy of Sciences, and the "Integration" Program (grant B0026).

7. References

1. Nalwa, H.S. (ed.) (1999) *Handbook of Nanostructured Materials and Nanotechnology*, Vol.1-5, Academic Press, San Diego.
2. G.-M. Chow, I.A. Ovid'ko, and T. Tsakalakos (eds.) (2000) *Nanostructured Films and Coatings*, NATO Science Ser., Kluwer, Dordrecht.
3. M.C. Roco, R.S. Williams, and P. Alivisatos (eds) (2000) *Nanotechnology Research Directions*, Kluwer, Dordrecht.
4. Gryaznov, V.G., Polonsky, I.A., Romanov, A.E., Trusov, L.I. (1991) *Phys.Rev. B* **41**, 42.
5. Romanov, A.E. (1995) *Nanostruct.Mater.* **6**, 125.
6. Hahn, H., Mondal, P. and Padmanabhan, K.A., (1997) *Nanostruct.Mater.* **9**, 603.
7. Hahn, H., and Padmanabhan, K.A. (1997) *Philos.Mag. B* **76**, 559.
8. Kostantinidis, D.A. and Aifantis, E.C., (1998) *Nanostruct.Mater.* **10**, 1111.
9. Masumura, R.A., Hazzledine, P.M. and Pande, C.S. (1998) *Acta Mater.* **46**, 4527.
10. Swygenhoven, H. van, Spavzer, M., Caro, A., and Farkas, D. (1999) *Phys.Rev. B* **60**, 22.
11. Swygenhoven, H. van, Spavzer, M., and Caro, A., (1999) *Acta. Mater.* **473**, 117.
12. Kim, H.S., Estrin, Y. and Bush, M.B. (2000) *Acta Mater.* **48**, 493.
13. Yamakov, V., Wolf, D., Phillpot, S.R., and Gleiter, H., (2002) *Acta Mater.* **50**, 61.
14. Fedorov, A.A., Gutkin, M.Yu. and Ovid'ko, I.A. (2002) *Scr.Mater.* **47**,51.
15. Gutkin, M.Yu., Ovid'ko, I.A. and Pande, C.S., (2001) *Rev.Adv.Mater.Sci.* **2**, 80.
16. Siegel, R.W., and Fougere, G.E. (1995) *Nanostruct. Mater.* **6**, 205.
17. Padmanabhan, K.A. (2001) *Mater.Sci.Eng.* **A304-306**, 200.
18. Pilling, J. and Ridle, N., (1989) *Superplasticity in Crystalline Solids*, The Institute of Metals, London.
19. Zelin, M.G. and Mukherjee, A.K., (1996) *Mater.Sci.Eng.* A **208**, 210.
20. Ovid'ko, I.A. (2002) *Science* **295**, 2386.
21. Gutkin, M.Yu. Kolesnikova, A.L., Ovid'ko, I.A. and Skiba, N.V., (2002) *J.Metast.Nanocryst.Mater.* **12**, 47; (2002) *Philos.Mag.Lett.*, in press.
22. King, A.H., (1999) *Interf. Sci.* **7**, 251.
23. Rabukhin, V.B., (1986) *Poverkhnost* **7**, 126 (in Russian).
24. Bokstein, B., Ivanov, V., Oreshina, O., Peteline, A., and Peteline, S. (2001) *Mater.Sci.Eng.* A**302**, 151.
25. Yin, K.M., King, A.H., Chen, F.-R., Kai, J.J., and Chang, L. (1997) *Microsc.Microanal.* **3**, 417.
26. Gutkin, M.Yu., and Ovid'ko, I.A. (1994) *Philos.Mag.* A **70**, 561.
27. Kaibyshev, O.A. (1999) *Mater.Sci.Forum* **304-306**, 21.
28. Owusu-Boahen, K. and King, A.H., (2001) Acta Mater. **49**, 237.
29. Fedorov, A.A., Gutkin, M.Yu., and Ovid'ko, I.A., *Acta Mater.*, in press.
30. Gottstein, G., King, A.H., and Shvindlerman, L.S. (2000) *Acta Mater.* **48**, 397.

31. Chokshi, A. H., Rosen, A., Karch J., and Gleiter, H. (1989) *Scr. Metal.* **23**, 1679.
32. Youngdahl, C. J., Sanders, P.G., Eastman, J. A.,. Weertman, J.R. (1997) *Scr. Mater.* **37**, 809.
33. Suryanarayana, R., Frey, C. A., Sastry, S.M.L., Waller, B.E., Bates, S.E., Buhro, W.E. (1996) *J. Mater.Res.* **11**, 439.
34. Sanders, P. G., Eastman, J. A., Weertman, J. R. (1997) *Acta Mater.* **45**, 4019.
35. Sanders, P. G., Eastman, J. A., Weertman, J. R. (1996) in: C. Suryanarayana, J. Singh, F. H. Froes (eds.), *Processing and properties of NC materials* , TMS, Warrendale, 397.
36. Romanov, A.E., and Vladimirov, V.I. (1992) in: F.R.N. Nabarro (ed.), *Dislocations in Solids*, vol.9, North-Holland Publ., Amsterdam,191.
37. Klimanek, P., Klemm, V., Romanov, A.E., and Seefeldt, M. (2001) *Adv.Eng.Mater.* **3**, 877-884.
38. Seefeldt, M. (2001) Rev. *Adv.Mater.Sci.* **2**, 44.
39. Valiev, R.Z., and Langdon, T.G. (1993) *Acta Metall.* **41**, 949.
40. Murayama, M., Howe, J.M., Hidaka, H., and Takaki, S. (2002) *Science* **295**, 2433.
41. Niemann, G.W., Weertman, J.R. and Siegel, R.W. (1991) *J.Mater.Res.* **6**, 1012.
42. Witney, A.B., Sanders, P.G., Weertman, J.R. and Eastman, J.A. (1995) *Scr. Metall.Mater.* **33**, 2025.
43. Carsley, J.E., Milligan, W.W., Hackney, S.A. and Aifantis, E.C. (1995) *Metall. Mater. Trans.* **26A**, 2479.
44. Andrievskii, R.A., Kalinnikov, G.V., Kobelev, N.P., Soifer, Ya.M., and Shtansky, D. (1997) *Phys.Sol.State* **39**, 1661.
45. R.A.Andrievskii (1998), in G.-M. Chow and N.I. Noskova (eds.), *Nanostructured Materials: Science and Technology*, Kluwer Academic Publ., Dordrecht.
46. Zelin, M.G., and Mukherjee, A.K. (1995) *Acta Metall.Mater.* **43**, 2359.
47. Zelin, M., Guillard, S., and Mukherjee, A. (2001) *Mater.Sci.Eng.* A**309-310**, 514.
48. Ovid'ko, I.A., submitted.
49. Ovid'ko, I.A., (2000) *Nanostructured Films and Coatings*, NATO Science Ser., G.-M. Chow, I.A. Ovid'ko, and T. Tsakalakos (eds.), Kluwer, Dordrecht, 231.
50. Ovid'ko, I.A. (2000) *Rev. Adv. Mater. Sci.* **1**, 61.
51. Ovid'ko, I.A. (1999) *J. Phys.: Condens. Matter.* **11**, 6521.
52. Ovid'ko, I.A. (2001) *J. Phys.: Condens. Matter.,* **13**, L97.
53. Ovid'ko, I.A., and Sheinerman, A.G. (2001) *J. Nanosci. Nanotechnol.* **1**, 215.
54. Fitzgerald, E.A. (1991) *Mater. Sci. Rep.* **7**, 87.
55. Merve, J.H. van der (1991) *Crit. Rev. Sol. State and Mater. Sci.* **17**, 187.
56. Jain, S.C., Willis, J.R., and Bullough, R. (1990) *Adv. Phys.* **39**, 127.
57. Jain, S.C., Harker, A.H., and Cowley, R.A. (1997) *Philos. Mag. A* **75**, 1461.
58. Hosson, J. Th. M. de, and Kooi, B.J. (2001) in H.S. Nalwa (ed.), *Handbook of Surfaces and Interfaces of Materials*, vol. 1, Academic Press, NewYork, pp.1.
59. Kabacoff, L.T. (2000), in G.-M. Chow, I.A. Ovid'ko, and T. Tsakalakos (eds.) *Nanostructured Films and Coatings,* NATO Science Ser., Kluwer, Dordrecht, 373.
60. Ovid'ko, I.A., and Sheinerman, A.G. (2001) *J. Phys.: Condens. Matter* **13**, 7937.
61. Ovid'ko, I.A., and Sheinerman, A.G. (2002) *Phys.Sol.State* **44**, 298.
62. Shchukin, V.A. and Bimberg, D. (1999) *Rev. Mod. Phys.* **71**, 1125.
63. Ledentsov, N.N., Ustinov, V.M., Shchukin, V.A., Kop'ev, P.S., Alferov, Zh.I., and Bimberg, D. (1998) *Semiconductors* **32**, 343.
64. Ustinov, V. (2000), in G.-M. Chow, I.A. Ovid'ko, and T. Tsakalakos (eds.), *Nanostructured Films and Coatings,* NATO Science Ser., Kluwer, Dordrecht, p. 41.
65. Sutter, P. and Lagally, M.G. (2000) *Phys. Rev. Lett.* **84**, 4637.
66. Bourett, A. (1999) *Surf.Sci.* **432**, 32.
67. Jesson, D.E., Kaestner, M., and Voigtlaender, B. (2000) *Phys. Rev. Lett.* **84**, 330.
68. Kamins, T.I., Karr, E.C., Williams, R.S., Rosner, S.J. (1997) *J. Appl. Phys.* **81**, 211.
69. Chaparro, S.A., Drucker, J., Zhang, Y., Chandrasekhar, D., McCartney, M.R., Smith, D.J. (1999) *Phys. Rev. Lett.* **83**, 1199.
70. Chaparro, S.A., Zhang, Y., Drucker, J., and Smith, D.J. (2000) *J. Appl.Phys.* **87**, 2245.
71. Pehlke, E., Moll, N., Kley, A., and Scheffler, M. (1997) *Appl. Phys. A* **65**, 525.
72. Johnson, H.T., and Freund, L.B. (1997) *J. Appl. Phys.* **81**, 6081.
73. Kukta, R.V., and Freund, L.B. (1997) *J. Mech. Phys. Solids* **45**, 1835.
74. Ovid'ko, I.A. (2002) *Phys. Rev. Lett.* **88**, 046103.

SCALE DEPENDENT ANELASTICITY AND MECHANICAL BEHAVIOR: THE CASE OF NANOCRYSTALLINE METALS

E. BONETTI, L. PASQUINI, L. SAVINI
Department of Physics, University of Bologna and INFM v. Berti-Pichat 6/2 40127 Bologna Italy

Abstract A lot of experimental evidence exists now, demonstrating that under suitable circumstances the materials size reduction below critical values leads to physical properties which are significantly modified with respect to those commonly exhibited by conventional materials. Most of the dimensionality effects derive from structural constraints to which a specific physical mechanism is subjected. Critical conditions the scale length of a specific property. Typical examples of structural dimensions are offered by the grain size in bulk materials or the thickness in thin films. The mechanical behavior in the whole range encompassing elastic to plastic regime via anelasticity does not constitute exception to the general rule. Anelastic relaxation processes have been recently observed which are properly described phenomenologically taking into account dimensionality. In nanocrystalline materials, the crystallite size the significant volume fraction of interfaces and the modified dislocation dynamics constitute key parameters which must be carefully controlled to tailor the mechanical behavior. Specific items of scale dependent anelasticity are connected with the possibility offered to combine anelastic relaxation processes depending on atomic diffusion with the grain size. Anelasticity in thin metal films offers also the possibility to explore cases where dislocation dynamics is confined and consequently scale dependent. Interfaces mechanical relaxation is also expected to be modified entering the nanoregime. This item is linked to other relevant features of the mechanical behavior including enhanced microplasticity and superplasticity. This review will report and discuss some recent experimental results obtained by mechanical spectroscopy investigations on nanocrystalline metals and alloys prepared by a variety of synthesis procedures. The results are critically compared to those early obtained on similar conventional materials aiming at understanding some specific aspects of confinement effects on mechanical behavior.

T. Tsakalakos, et al. (eds.) pgs 217 -237
Nanostructures: Synthesis, Functional Properties, and Applications;
© *2003 Kluwer Academic Publishers.*

1. Introduction

All branches relating to the investigation of materials: materials physics and chemistry, materials science, materials technology, differentiate mainly with respect to the spatial range in which they apply, going from interatomic distances up to macroscopic ones. The unifying aspect of these branches is based on the concepts of length scales and their interaction. With particular reference to properties depending on the microstructure it is possible to identify several dimensional parameters strongly influencing the physical properties of materials. From this point of view it appears that the modification of a set of properties achieved through an accurate microstructural control, allows the synthesis of materials with physical-chemical properties new or modified with respect to those exhibited by conventional materials.

An important aspect to be taken into account when studying the influence of dimensionality on the material properties in general and specifically on the mechanical properties is represented by the possibility, offered by some new technologies, to assembly materials on a miniaturized scale, or to control the microstructure features in bulk materials with extreme accuracy. At present the controlled manipulation of the microstructure at an atomic level, constituting the main goal of nanoengineering and nanotechnology represents one of the emerging interdisciplinary scientific branches. In the big family of nanomaterials, where at least one typical dimension is reduced down to the nanometer range, the modified or new properties can be determined by confinement effects (reduced linear dimension), high surface area, and dimensionality effects (dimensionality < 3, as in thin films, nanotubes, quantum dots). In the present context we will concentrate mainly on confinement effects in the class of bulk nanomaterials, i.e., bulk materials with a microstructure modulated on a nanometer scale, and in thin films.

Most of size effects on the materials behavior derive from *dimensional* or *microstructural* constraints to which a specific materials property is subjected. In the framework of condensed matter physics it is possible to find a number of characteristic *intrinsic* length scales L_c in the nanometer range, associated with specific phenomena and properties. These characteristic length scales can match a *confinement* length L, corresponding to a particular dimension of the material. In this sense, *dimensional* constraint arise when the diameter L of an isolated nanoparticle becomes similar to L_c (Fig. 1). On the other hand, in materials science, dealing with the spatial range in between the atomic and macroscopic ones, materials etherogeneity in the nanorange can be important sources of *microstructural* constraints (Fig. 1). The strong interplay between these two lengths (confinement vs. intrinsic) determine the new phenomena and related properties of nanomaterials. From this side, a *mesoscopic condition* [1] expressed as:

$$L_c \approx L \qquad (1)$$

defines unambiguously a nanomaterial. Mesoscopic properties are therefore associated with the structural modification of a given material in such a way as to induce strong overlapping of a characteristic length with a size parameter.

As a well known example linked to the thermodynamics of nanomaterials, one can consider the role of thermal fluctuations in confined systems. The reduction of the melting temperature of small metal particles demonstrates that thermal behavior can be strongly altered as a consequence of size confinement [2]. The mechanical properties in small dimensions do not constitute exception to the general rule [3]. Significant deviations from bulk scaling laws can be observed when some microstructural confinement length (e.g., grain size) or dimensional confinement length (e.g., thin film thickness) enters the length scales of structural defects or defects interaction and dynamics, responsible for mechanical behavior (anelasticity, plasticity, etc.). Deviations of the mechanical properties from bulk scaling behavior have been observed at relatively large length scale typically in the sub-micrometer range. Examples are provided by thin films: in Ag it has been experimentally proved that, below a thickness of 400 nm, the average distance between pinning points along dislocations scales approximately linearly with thickness [4]. This result provides the input for a phenomenological description of the generally reported strength increase in fcc metallic films with respect to their bulk counterparts [5].

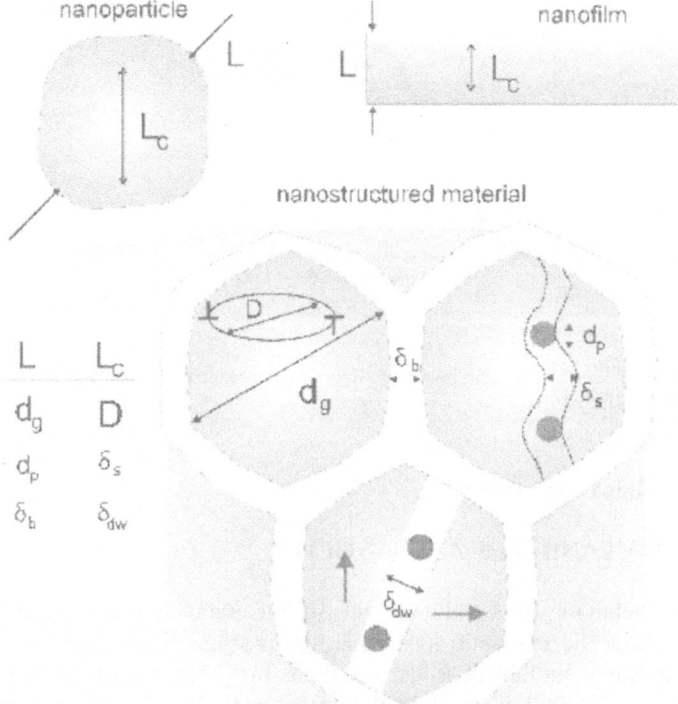

Figure 1. Dimensional constraints: diameter of a nanoparticle, thin film thickness. Microstructural constraints: grain size d_g, diameter of precipitates d_p, and grain boundary thickness δ_b. Characteristic lengths: dislocation loop diameter D, separation between dislocation partials δ_s, and width of a magnetic domain wall δ_{dw}.

Another example of confinement and interface effects can be found in the lattice vibration spectrum of metals with an ultrafine grain size. Fig. 2 compares the phonon density of states (DOS) measured in bulk iron and in nanocrystalline iron prepared by inert-gas condensation with a mean grain size of 6.6 nm. The DOS has been experimentally determined by nuclear inelastic absorption of synchrotron radiation [6].

Two features of the nano-DOS are remarkable: (1) an increased population of low-energy modes, i.e., a global softening of the spectrum, which has been ascribed to vibrational modes at the interfaces [6], and (2) a broadening of the DOS peak at about 36 meV. This second feature can be described quite well by a damped oscillator model, and phonon decay distances consistent and correlated with the grain size are derived. This result hints at phonon confinement in nanocrystallites: the role of the microstructural confinement length L in this case is played by the grain size d_g, whereas the intrinsic length L_c would correspond to the phonon mean free path in the infinite crystal.

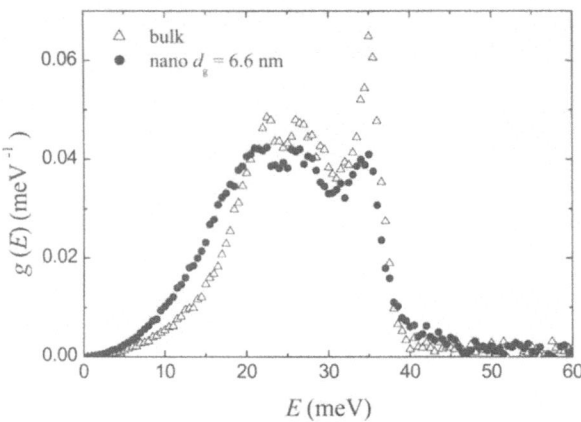

Figure 2. Phonon DOS of bulk and nanocrystalline iron, measured by nuclear inelastic scattering of synchrotron radiation (from [6]).

2. Anelastic Relaxation Behavior

2.1 GENERAL MEANING OF ANELASTICITY

The anelastic behavior of a solid is one of the many irreversible thermodynamic processes -such as the magnetic hysteresis, the passage of electric current through a resistor, the irregular motion of a viscous fluid- in which energy (in this case, elastic energy) is converted into heat. It is not by chance that the quantity perhaps most representative of this behavior is called *internal friction*: it measures the rate at which the elastic vibration energy is dissipated due to relaxation processes occurring in the material. The definition of anelastic behavior follows directly from that of elastic behavior, relieving the requirement of instantaneous relation between stress and strain. The essential features of anelasticity are therefore: (a) to each level of applied stress corresponds a *unique* equilibrium strain value, and vice versa; (b) the stress-strain relation is linear; (c) the equilibrium is reached after the passage of a certain time. To be noted the existence of a unique equilibrium state in response to the external field: the anelastic solid is then a thermodynamic system whose state is fully described by the value of the stress only (at constant temperature, magnetic field, etc.), which is not true in cases where plastic or viscous deformation occurs. Moreover, it is also evident that this postulate implies complete recoverability of the anelastic deformation upon removal of the applied stress [7].

From the microscopic viewpoint, the existence of a time-dependent response means that stress and strain are coupled through internal variables, e.g. an order parameter, the electronic density, the magnetisation, etc., that relax towards the equilibrium by kinetic processes, such as diffusion. For this reason, the term *anelastic relaxation* is used to indicate this class of phenomena. To put these assertions into a mathematical basis, let us consider, for the sake of simplicity, the case where this coupling is mediated by only one of such variables, namely ξ. To each stress σ corresponds to a unique equilibrium value $\xi = \bar{\xi}$ that, by virtue of linearity, may be written as:

$$\bar{\xi} = \kappa\sigma \qquad (2)$$

where κ is a constant and, without loss of generality, we have chosen $\bar{\xi} = 0$ for $\sigma = 0$. The strain is now a function of both σ and ξ, in the form:

$$\varepsilon(\sigma,\xi) = \varepsilon_e(\sigma) + \varepsilon_a(\sigma,\xi) = J_U\sigma + \chi\xi \qquad (3)$$

where $\varepsilon_e(\sigma) = J_U\sigma$ is the elastic strain and J_U is the *unrelaxed compliance*; $\varepsilon_a(\sigma,\xi) = \chi\xi$ is the anelastic time-dependent strain, χ being a constant. The equilibrium strain is a function of the stress only:

$$\bar{\varepsilon}(\sigma) = \varepsilon_e(\sigma) + \bar{\varepsilon}_a(\sigma) = J_U\sigma + \chi\kappa\sigma \equiv (J_U + \delta J)\sigma \qquad (4)$$

where we have introduced the *relaxation of the compliance* $\delta J \equiv \chi\kappa$; the ratio between equilibrium strain and stress is called *relaxed compliance* $J_R \equiv J_U + \delta J$. Up to now, we did not make assumptions on the form of the relaxation kinetics. The simplest and most important type of anelastic relaxation is obtained with the restriction of first-order kinetics for the variable ξ:

$$\dot{\xi} = -(\xi - \bar{\xi})/\tau_\sigma \qquad (5)$$

where τ_σ is a characteristic *relaxation time at constant stress*. From Eqs. (2)-(3) we have $\xi - \bar{\xi} = (\varepsilon - J_U\sigma)/\chi - \kappa\sigma$; inserting this in the derivative of (3), and taking into account (5), we get after some algebra:

$$\varepsilon + \tau_\sigma\dot{\varepsilon} = J_R\sigma + J_U\tau_\sigma\dot{\sigma} \qquad (6)$$

Note that in this first-order differential equation we have eliminated the reference to the internal variable ξ, and to the microscopic coupling constants κ and χ; stress and strain are related through the macroscopic quantities J_R and J_U, and the relaxation time. A solid which obeys the simple Eq. (6), independently of the specific relaxation mechanism, is called *Standard Anelastic Solid* (SAS).

In complete analogy, one can assume the strain fixed and look at the time-dependent stress. The analogous of Eq. (4) for the equilibrium stress is:

$$\bar{\sigma}(\varepsilon) = (M_U - \delta M)\varepsilon \equiv M_R\varepsilon \qquad (7)$$

where M_U is the *unrelaxed modulus*, which gives the instantaneous (elastic) stress; δM is known as *relaxation of the modulus* and M_R as *relaxed modulus*. The relaxation kinetics will be now rate-controlled by a *relaxation time at constant strain* τ_ε; the differential stress-strain equation obtained in this case is:

$$\sigma + \tau_\varepsilon \dot{\sigma} = M_R \varepsilon + M_U \tau_\varepsilon \dot{\varepsilon} \tag{8}$$

Let us now look at the relations between the quantities we have defined up to know. First, due to the uniqueness and linearity of the equilibrium stress-strain relationship, it is clear that $\bar{\varepsilon}(\sigma)/\sigma = \varepsilon/\bar{\sigma}(\varepsilon)$, and therefore:

$$J_R = 1/M_R \tag{9}$$

On the other hand, at very short time scales the material behaves as an ideal elastic solid, so that:

$$J_U = 1/M_U \tag{10}$$

A very important quantity in the description of anelastic behavior is the *relaxation strength* Δ, defined as the ratio between the equilibrium anelastic strain and the elastic strain; from Eq. (4) it is immediately obtained that:

$$\Delta \equiv \frac{\bar{\varepsilon}_a}{\varepsilon_e} = \frac{\delta J}{J_U} = \frac{\delta M}{M_R} \tag{11}$$

where the last equality follows directly by substitution of Eqs. (9)-(10).

2.2 STATIC AND DYNAMIC RESPONSE FUNCTIONS

Typical experiments aimed at the investigation of anelastic relaxation processes can be subdivided into quasi-static and dynamic, depending on the type of external stress (or strain) imposed to the specimen.

In a quasi-static measurement, the stress (or strain) undergoes a step-like change. Afterwards, the strain (or stress) relaxes and it approaches some equilibrium value. In creep experiments, a constant stress σ_0 is applied to the sample at time $t=0$ and is maintained for $t>0$, while the variation of the strain $\varepsilon(t)$ as a function of time is monitored and recorded.

Assuming a relaxation process that depends on a single relaxation time (simply τ in the following), the creep compliance is given by:

$$J(t) \equiv \varepsilon(t)/\sigma_0 \equiv J_U + J_{an}(t) = J_U + \delta J[1 - \exp(-t/\tau)] \tag{12}$$

where J_U and $J_{an}(t)$ represent the unrelaxed elastic and the time-dependent anelastic response, respectively. For $t \gg \tau$, $J_{an}(t)$ approaches the equilibrium value δJ. Upon unloading after a load time t_{load}, the sample undergoes a sudden strain decrease and eventually approaches zero strain as described by the creep recovery compliance:

$$J_{rec}(t') = \delta J[1 - \exp(-t_{load}/\tau)]\exp(-t'/\tau) \tag{13}$$

where $t' \equiv t - t_{load}$. The anelastic behavior exhibited by real materials most often involves a distribution of relaxation times $X(\ln \tau)$, with $\int_{-\infty}^{+\infty} X(\ln \tau)d(\ln \tau) = \delta J$. Eqs. (12)-(13) then transform to:

$$J(t) = J_U + \int_{-\infty}^{+\infty} X(\ln \tau)[1 - \exp(-t/\tau)]d(\ln \tau) \tag{14}$$

$$J_{rec}(t') = \int_{-\infty}^{+\infty} X(\ln \tau)[1 - \exp(-t_{load}/\tau)]\exp(-t'/\tau)d(\ln \tau) \tag{15}$$

In creep experiments, the loading compliance generally comprises a plastic component. For this reason, it is preferable to determine $X(\ln \tau)$ from the recovery compliance. In the hypothesis of a smooth distribution of relaxation times, $X(\ln \tau)$ can be approximated by [7]:

$$X(\tau) \approx -(dJ_{rec}/d \ln t')/(1 - \exp(-t_{load}/\tau))_{t'=\tau} \tag{16}$$

The purely anelastic compliance can thus be reconstructed by combining Eqs. (14) and (16), and subtracted from the total experimental creep to obtain the plastic response.
In most cases, the relaxation process is thermally activated:

$$\tau = \tau_0 \exp(H/k_BT) \tag{17}$$

If the distribution of relaxation times arises mainly from a distribution of pre-exponential factors τ_0 rather than of energy barriers, then the effective activation energy of the anelastic creep can be determined from the time shift between J_{an} curves measured at different temperatures:

$$H = k_B\left(\partial \ln t/\partial T^{-1}\right)_{J_{an}=\text{const}} \tag{18}$$

where t is the creep time necessary to attain a fixed value of the anelastic response J_{an}.
In a dynamic measurement, the applied stress is oscillating with time, i.e., $\sigma = \sigma_0 \exp(i\omega t)$, where ω is the angular velocity. Due to the relaxation process, a phase shift between stress and strain occurs. This shift is usually represented by a complex mechanical susceptibility, or *compliance*, $J^* \equiv J_1 - iJ_2 \equiv \varepsilon/\sigma$. J_1 and J_2, known as the *storage* and *loss compliance*, give the in-phase and out-of-phase strain, respectively. The ratio J_2/J_1 represents the elastic energy dissipation of the material, and is commonly known as *internal friction, specific damping factor*, or simply Q^{-1}. For a single-time process, Q^{-1} assumes the form of a Debye peak:

$$Q^{-1} \equiv J_2/J_1 = \Delta\omega\tau/(1 + \omega^2\tau^2) \tag{19}$$

where the *relaxation strength* Δ is equal to $\delta J/J_U$. The temperature-dependence of the internal friction is contained in the Arrhenius relation for the relaxation time. In particular, the peak temperature, T_p, corresponds to the condition $\omega\tau=1$ which, by virtue of Eq. (17) becomes:

$$\ln \omega + \ln \tau_0 + H/k_BT_p = 0 \tag{20}$$

Following Eq. (20), the inverse peak temperature is usually plotted as a function of the logarithm of the frequency to determine the relaxation parameters H and τ_0.
In addition to specific relaxation peaks, the internal friction in polycrystalline metals generally exhibits a so-called *background damping*, which increases monotonically with increasing temperature and can be described by the equation:

$$Q^{-1} = (A/\omega^n)\exp(-nH_{bg}/k_BT) \tag{21}$$

where A is a constant and $n \leq 1$ a correction factor which arises from the presence of a distribution of relaxation times [8]. The true activation energy H_{bg} must be evaluated from the temperature shift between Q^{-1} curves measured at different frequencies:

$$H_{bg} = -k_B \left(\partial \ln \omega / \partial T^{-1} \right)_{Q^{-1}} \tag{22}$$

The correction factor n is then calculated by comparison with the slope of the background:

$$[\partial \ln(Q^{-1}) / \partial T^{-1}]_\omega = -nH_{bg} / k_B \tag{23}$$

3. Point Defects Relaxation

3.1 GENERAL THEORY OF POINT DEFECTS RELAXATION

As a general example of anelastic relaxation, we shall now describe the mechanical behavior due to point defects whose symmetry class is lower than that of the host crystal. A vacancy in a cubic crystal clearly doesn't lower the symmetry of its environment. Therefore, since all vacant sites are equivalent (i.e., have the same free energy) the application of a stress doesn't cause the vacancy to migrate. Let us consider instead an interstitial defect in a bcc crystal. The interstitial atoms in Fig. 3, for instance, have an octahedral environment and therefore possess a tetragonal symmetry. Atoms labeled with indexes i=1, 2, 3 in Fig. 3 correspond to different orientations of the octahedron axis. Each interstitial site produces a lattice distortion with tetragonal symmetry and axis parallel to one of the cube edges. In complete analogy with the case of dielectric relaxation, these defects can be viewed as *elastic dipoles*, which possess a permanent strain and are reoriented by an external field. In the same way in which permanent electric dipoles rotate trying to align their axis with the applied electric field, the interstitials jump to those sites whose associated strain tensor is aligned with the applied stress. This site re-population process requires thermally-activated jumps over free energy barriers, and therefore gives rise to a time-dependent anelastic strain. The internal friction peak associated with this atomic mechanism was first observed in α-iron containing small amounts of carbon or nitrogen and is known as Snoek relaxation [9]. This type of relaxation is now recognized to be a general phenomenon that originates from the distorted character of interstitial solute atoms in the bcc structure. We may remark immediately that in fcc and hcp metals the symmetry class of the interstitial sites is the same as the crystal symmetry, with the result that isolated solute interstitials in these metals do not constitute orientationally-distinguishable defects and do not produce a relaxation of the Snoek type. However, anelastic relaxations in fcc and hcp crystals may originate from other type of defects, such as divacancies, solute pairs, etc. The necessary conditions for the occurrence of a relaxation process are: (i) a symmetry class of the defect lower than the crystal symmetry, and (ii) a lattice distortion associated with the defect. Turning to the bcc structure, two distinct families of sites (octahedral and tetrahedral) are possible candidates for interstitial occupancy. However, both possess tetragonal symmetry and without loss of generality we may confine our attention to sites of just one type [7], as in Fig. 3.

For a dilute solution of non-interacting interstitials of mole-fraction C_0, the equilibrium concentrations $\overline{C}_i$ in each orientation i are given by Boltzmann statistics as:

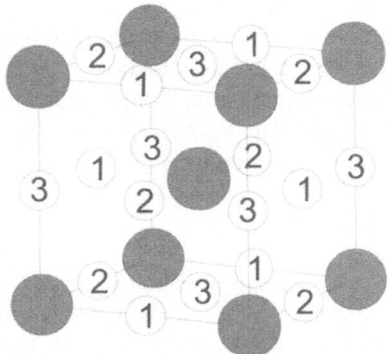

Figure 3. Octahedral interstitial sites in a bcc cell. The labels 1,2, and 3 correspond to different orientations of the octahedron axis.

$$\frac{\overline{C_i}}{C_0} = \frac{\exp(-g_i/k_B T)}{\sum_i \exp(-g_i/k_B T)} \tag{24}$$

where g_i are the free energy levels, or site-occupation energies. In the stress-free crystal, the free energy levels $g_i(0)$ of all orientations are equivalent, and the equilibrium concentrations $\overline{C_i}$ are equal to $C_0/3$. The application of a stress leads to a redistribution of interstitials which is mediated by atomic jumps and to the development of an anelastic strain whose magnitude depends on the lattice distortion associated with each interstitial atom.

The kinetics of the Snoek relaxation, i.e. the relation between the atomic jump rate w and the relaxation time for dipole reorientation τ_R are relatively straightforward. If the three orientations in Fig. 3 are populated with non-equilibrium concentrations C_1, C_2 and C_3, the time derivative of C_1 can be written:

$$\dot{C_1} = -4wC_1 + 2w(C_2 + C_3) \tag{25}$$

where the factor 4 represents the four nearest-neighbours sites available for a jump from a site 1, and the factor 2 the two sites with orientation 2 or 3 from which an atom can jump to site 1. Upon using the conservation condition $C_0 = C_1 + C_2 + C_3$ and the equilibrium condition $\overline{C_1} = \overline{C_2} = \overline{C_3} = C_0/3$, Eq. (25) becomes:

$$\dot{C_1} = -6w(C_1 - \overline{C_1}) \tag{26}$$

By analogy with Eq. 4, it emerges that:

$$1/\tau_R = 6w \tag{27}$$

It is now worthwhile to exploit the link between the relaxation time and the coefficient D for diffusion of interstitial atoms. If there are z different directions in which the jump w can occur (so that the total jump rate is zw), and the mean jump distance is s, then from the theory of random-walk diffusion one has:

$$D = zws^2/6 \tag{28}$$

for octahedral sites, $z=4$ and $s=a/2$ (with a lattice parameter). Combination of Eqs. (27)-(28) gives:

$$\tau_R = a^2 / 36D \qquad (29)$$

The activation energy of the Snoek relaxation is the activation energy for diffusion of the interstitial species, and the precise determination of H and τ_0 using both dynamic and static measurement has represented one of the principal applications of the Snoek relaxation in bulk materials [7,10]. The Snoek relaxation has the potential for both the identification and the semi-quantitative analysis of interstitial solutes in bcc metals.

3.2 THE GORSKY RELAXATION

In contrast to the point defects relaxation examples referred to in section 3.1, all relating to reorientation of elastic dipoles (short range diffusion), the Gorsky relaxation is intrinsically connected to long range diffusion, arising when a stress gradient, produced for example by bending a thin film sample, induces a spatial gradient in the chemical potential of the point defects (Fig. 4).

This, in turn, determines a diffusion of defects proceeding until the chemical potential gradient of the diffusing species across the film thickness is eliminated by virtue of the established concentration gradient. The relaxation time for the Gorsky relaxation is given by [11] :

$$\tau_G = d^2 / \pi^2 D \qquad (30)$$

where d is the macroscopic diffusion distance, playing in this case the role of confinement length and D is the diffusion coefficient of the defect species. If, as usual, one deals with transverse atomic diffusion of a bent beam, the Gorsky effect must be investigated by quasistatic anelastic relaxation tests [11]. The reason is that the relaxation time τ_G is much larger than the typical relaxation time for point defects reorientation relaxation (see Eq. 29).

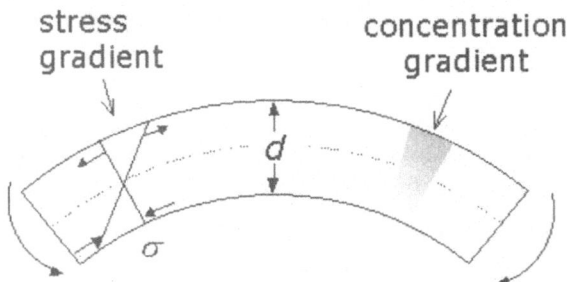

Figure 4. Representation of a bent beam of thickness d in presence of solute interstitials. The dotted line represents the neutral plane, i.e., the plane of zero stress. The stress σ attains the maximum value (either in tension or compression) at the sample surface. This stress gradient leads to a gradient in the equilibrium solute concentration which is achieved via long range diffusion.

3.3 INTERCRYSTALLINE GORSKY RELAXATION IN ULTRAFINE-GRAINED MATERIALS

The above discussion of the Gorsky effect applies to elastically homogeneous materials. On the other hand, diffusional relaxation mechanisms can set up between elastic inhomogeneities, if these exist in the sample. The macroscopic diffusion distance of Eq. (30) is then no longer the sample thickness, but rather the typical separation between inhomogeneities. If these are due to misfit stresses between elastically anisotropic grains in a polycrystal, then the similarity $d \approx d_g$, with d_g the grain size, should apply. The relaxation process in this situation can be named *intercrystalline Gorsky effect*.

With reference to the relaxation times for Gorsky (Eq. 30) and reorientation relaxation (Eq. 29), in a polycrystal it should be possible to observe and discriminate on the same kinetic scale the two processes, if the confinement length d –i.e., the grain size d_g - is selected properly. In fact, taking into account that both processes are thermally activated and obey in the simplest case the Arrhenius Eq. (17), it follows that for the same type of diffusing species, the respective τ_0 values should scale according to the ratio d^2/a^2, whereas the activation energy H may well remain the same.

A clear experimental confirmation of an intercrystalline Gorsky effect in a nanocrystalline ZrCuNiAl alloy, due to hydrogen diffusion, was recently reported [12]. The alloy contained $CuZr_2$ grains with size $d_g \approx 30$ nm. As shown in Fig. 5, two distinct damping peaks were detected. From Eq. (20), the activation energy was determined to be 0.5 eV for both peaks. The low temperature peak is the usual elastic dipole reorientation (Snoek-like) relaxation. The high temperature peak is a novel damping phenomenon peculiar of the ultrafine-grained nanostructure of the material. The τ_0 ratio for the two peaks was indeed find to agree well with the expected a/d_g ratio, with the lattice constant of $CuZr_2$.

The combination of a highly diffusing species: hydrogen, with a small grain size, thus allowed the simultaneous observation in a dynamic measurement of the intercrystalline Gorsky and closely related elastic dipole reorientation processes for hydrogen.

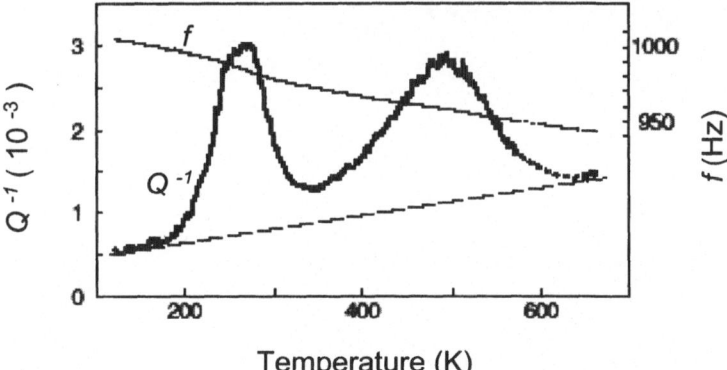

Figure 5. Damping Q^{-1} and resonance frequency f ($=\omega/2\pi$) of a nanocrystallized and hydrogenated ZrCuNiAl alloy. The dashed line indicates the linear background (from [12]).

4. Dislocation Damping

4.1 DISLOCATION RELAXATION

Anelasticity and anelastic relaxation processes measurements, can yield information about the mechanism which control the mobility of dislocations and hence dislocation mediated plasticity.

With reference to section 2.1, Eqs. (4)-(5) may be rewritten in order to connect anelasticity to dislocations motion as follows:

$$\varepsilon = J_U \sigma + \Lambda b u \tag{31}$$

$$B\dot{u} = -(Ku - \sigma b) \tag{32}$$

Eqs. (31)-(32) correspond to Eqs. (4)-(5) respectively with: $\xi = u$, $\chi = \Lambda b$, and $\tau = B/K$. In the above equations Λ is the dislocation density, b the Burgers vector, u the mean distance traveled by dislocations; σb gives the force per unit length acting on the dislocation and K is a kind of spring constant. Viscous motion is assumed, so that the dislocation velocity is proportional to the net force $\sigma b - Ku$ through a damping or dragging coefficient B. Now using the SAS differential stress-strain Eq. (6), one can obtain the expressions for dislocation damping :

$$Q^{-1} = \Delta \, \omega\tau/ (1+ \omega^2\tau^2) \qquad \text{with: } \Delta = \Lambda b^2/KJ_U \tag{33}$$

$$(J_1(\omega)-J_U)/J_U = \Delta/(1 + \omega^2\tau^2) \quad \text{with: } \tau = B/K \tag{34}$$

Generally the restoring force is due to the dislocation line tension γ, which can be written as:

$$\gamma \approx \frac{b^2}{2J_U} \tag{35}$$

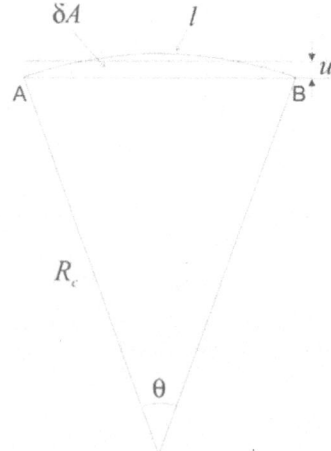

Figure 6. Bowing of a dislocation segment pinned at the points AB (length *l*) under an applied stress. δA is the area swept by the dislocation and R_c the radius of curvature; *u* represents the distance traveled by the dislocation, defined such that $\delta A \approx lu$ in the limit of small θ.

When the dislocation segment is pinned at the points A and B and acted by an applied stress σ, one obtains the following expression for the radius of curvature R_c :

$$R_c = \gamma / \sigma b \qquad (36)$$

and of course at the equilibrium it must be:

$$Ku = \sigma b \qquad (37)$$

With reference to Fig. 6, for small applied stresses the radius of curvature is large and one can write for the area δA, between the arc of length $l = R_c \theta$ and the segment AB the following expression:

$$\delta A \approx R_c^2 \theta^3 / 12 \approx lu \qquad (38)$$

Combination of Eqs. (35)-(38) yields:

$$K = 12\gamma / l^2 = 6b^2 / l^2 J_U \qquad (39)$$

and, substituting in equations (33) and (34), Δ and τ become:

$$\Delta = \Lambda l^2 / 6 \text{ and } \tau = Bl^2 / 12\gamma \qquad (40)$$

These general expressions may apply to a number of situations. Of interest for low dimensional systems are the cases where the dislocation loop length ,which play the role of characteristic length L_c , is reduced by constraints, such as pinning points of different structural nature. In the simple case of point defects, if n_p is the number of pinning points (defects) arriving at a dislocation segment at time t, the string model gives the following expression for K:

$$K = 12 \gamma (n_0 + n_p(t))^2 / \Lambda^2 \qquad (41)$$

Where $l = \Lambda / n_0$ is the dislocation loop length before the addition of n_p point defects.

In the hypothesis that in the equations for damping $\omega\tau \ll 1$, the following well known expressions are obtained:

$$Q^{-1} \propto \Delta\tau \propto l^4 \propto K^{-2} \text{ and } (J_1(\omega)-J_U)/J_U \propto \Delta \propto l^2 \qquad (42)$$

These expressions indicate a strong sensitivity of the anelastic response functions to the presence of pinning points.

4.2 CONFINEMENT EFFECTS IN THIN FILMS

In the case of nanocrystalline metals significant modifications of the dislocation anelasticity are predicted on account of confinement effects on the average link length distribution, due to crystallite size reduction. The mesoscopic condition in this case becomes: $l = d_g$ or $l = \delta_b$ for intragrain or intergrain (extrinsic) dislocations, respectively. In thin films, confinement due to the reduction of film thickness may occur. Recent anelasticity measurements on Cu films [13], and transmission electron microscopy observations [14], have been explained on the hypothesis of dislocation relaxation mechanisms conditioned by the film thickness. The dislocation model for the anelastic relaxation peak observed [13] is schematically shown in Fig. 7.

Dislocation segments pinned between the film-surface and film-rigid substrate boundaries, are generated by the large stresses arising in the film as a consequence of temperature increase. The damping peak observable in Fig. 8 at around 600K is attributed to the short range dynamics of these dislocation segments [13].

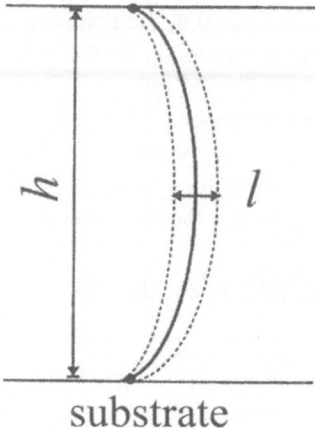

substate

Figure 7. Vibrating string dislocation model for the damping in thin films of thickness h. The dislocation segment (solid line, length l) is created by plastic deformation and oscillates between dashed lines when a harmonic stress is applied during dynamic anelastic measurements.

The decrease of the relaxation strength Δ with decreasing film thickness evident in Fig. 8 can be explained by a simple string model with the additional confinement condition $l \approx h$. In fact, using Eq. (40), one obtains [13]:

$$\Delta = \Lambda h^2 / 6 \qquad (43)$$

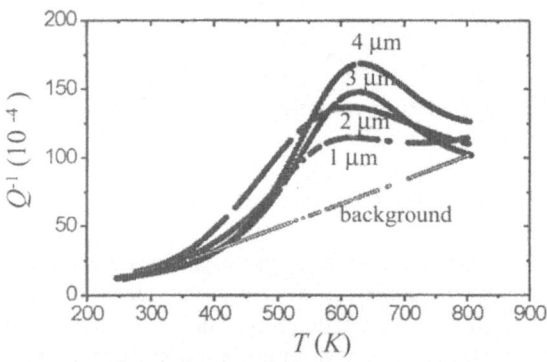

Figure 8. Internal friction versus temperature in copper films of different thickness in the 1-4 μm range (from [13]).

5. The Grain Boundary Relaxation

5.1 THE ZENER MODEL

The possibility that anelastic relaxation behavior might be associated with grain boundaries (GB) or other internal interfaces in metals was first predicted by Zener, based on experimental evidences that viscous sliding could occur between two adjacent crystals. In Zener's picture the shear stress σ initially acting across a boundary is gradually relaxed through time-dependent viscous slip while the grain corners sustain more and more of the total shearing force, as depicted in Fig. 9. The slip offset is elastically accommodated by the build up of a back stress at the grain corners that leads to a reversal of the slip on unloading. When recovery is complete, the distortions at the grain corners are eliminated and the specimen returns to its original conditions. This behavior indicates that an idealized sample containing a collection of identical boundaries would behave like a standard anelastic solid. In practice, however, a polycrystalline sample contains a wide variety of boundary types so that the corresponding relaxation peaks are significantly broadened with respect to a single-time Debye peak.

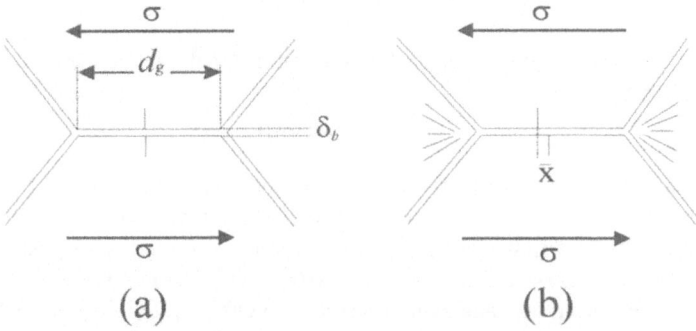

Figure 9. Schematic picture of grain boundary sliding in polycrystals: (a) initially, the shear stress σ is uniformly distributed along the grain boundary of linear dimension d_g and thickness δ_b ; (b) the shear stress along the grain boundaries is relieved by slip while stress accumulation occurs at the grain corners.

The relaxation strength associated with grain boundary relaxation processes is generally quite high in comparison for example with those typical for point defects relaxation. For a fine-grained polycrystalline aggregate, Zener calculated the maximum relaxation strength in tension by comparing the elastic strain energy under uniform stress with that under conditions of zero stress across the boundary but for the same average stress in the sample [15]. Under the simplifying assumption that each grain can be represented as an elastically isotropic sphere, he found that the relaxation strength depends only on the Poisson's ratio χ of the material:

$$E_R / E_U = \frac{1}{2}(7 + 5\chi)/(7 + \chi - 5\chi^2) \qquad (44)$$

where E_R and E_U are the relaxed and unrelaxed Young modulus, respectively. For the typical value $\chi=1/3$, one gets $E_R / E_U \approx 0.64$ and hence $\Delta \approx 0.56$. The relaxation strength experimentally observed for the grain boundary peak in many materials is often lower. It is indeed important to note that Zener's calculations is based on the fact that the sliding can occur entirely across the boundary, therefore relieving the stress completely. If the sliding were stopped by obstacles spaced at distances much less than the grain boundary length, then the relaxation strength would be strongly reduced. A simple continuum model can be used to calculate the relaxation time τ_σ. We assume that the development of the slip offset x across the boundary can be described by a first-order kinetics as in Eq. (5), i.e.:

$$\dot{x} = -(x - \bar{x}) / \tau_\sigma \qquad (45)$$

Since the anelastic (i.e. time-dependent) strain x/d_g along the boundary must completely relieve the initial elastic strain $\varepsilon_e = \sigma/\mu$, (where μ is the shear modulus) the equilibrium displacement takes the following value:

$$\bar{x} = \sigma d_g / \mu \qquad (46)$$

The slip velocity v_0 at time $t=0$, when the local shear stress along the boundary is equal to the applied stress σ, can be expressed in terms of the boundary viscosity η:

$$v_0 = \sigma \delta_b / \eta \qquad (47)$$

While from Eq. (45), one has directly:

$$v_0 = \bar{x} / \tau_\sigma \qquad (48)$$

Combination of Eqs. (46)-(48) leads to the expression for the relaxation time in terms of the microscopic parameters:

$$\tau_\sigma = \eta d_g / \delta_b \mu \qquad (49)$$

It should be noted that Eq. (49) predicts that the relaxation time should vary linearly with the grain size and exhibit a temperature dependence governed by the grain boundary viscosity. The first example of relaxation associated with grain boundaries was given by Kê in polycrystalline aluminium wires [16]. Using a torsion pendulum at 1 Hz frequency, he found an anelastic relaxation peak located around 300 °C. The relaxation strength, taking into account the peak broadening due to the dispersion of grain boundary sizes and types, was in fair agreement with Zener's calculations. Moreover, the peak shifted to high temperatures with increasing grain size, in agreement with Eq. (49), and was not present in single crystal specimens. Since that paper, several works have been carried out on GB sliding by means of internal friction techniques. A critical sum-up of the most important results obtained over the last 50 years has been recently proposed in a review paper by Kê himself [17].

5.2 GRAIN BOUNDARY RELAXATION IN MESOSCOPIC SYSTEMS

With reference to the pioneering Kê work on pure Al [16], and until the more recent works on thin films [18,19], one of the most intriguing issues was connected with the activation energy of the internal friction peaks observed in all these systems and attributed to GB sliding. In polycrystalline Al, Kê measurements yielded an activation energy close to the one for lattice self diffusion (1.4-1.5 eV), whereas in thin films [18,19] values closer to those for GB diffusion (0.5-0.6 eV) have been obtained. The different structure of the interfaces resulting from strongly reduced grain size was postulated as a possible reason for the difference.

In real polycrystals, the intrinsic viscosity of the interfaces influences the sliding rate only over small distances, before the building up of internal stresses opposes further sliding. As indicated by Ray and Ashby [20], this distance may be expressed by $x \approx d_g \sigma / \mu$, (see Eq. (46)). Values in the range 10^{-2}- 10 nm are consistent with grain sizes from

10^{-3} to 1 mm. Different structural features (ranging from atomic to macroscopic scale) along the interfaces, may be responsible for diffusion-accommodated sliding. The time dependence of GB sliding is governed by the diffusive atomic motion, necessary to accommodate all these structural incompatibilities, which constitute a deviation from ideal interfacial planarity. Therefore in thin films the anelastic relaxation mechanisms are associated with grain boundary sliding while grain boundary migration is dominating in bulk polycrystals [16]. This effect may be further retarded by impurity dragging. Assuming a pure sliding mechanism, in thin films one should expect an activation energy corresponding to the one for GB diffusion, as actually observed for Al [18,19]. Additionally, a linear scaling of the relaxation time pre-exponential factor τ_0 with the grain size d_g, as predicted by Eq. (49), should occur. This has indeed been verified on thin Al films [19] as reported in Figure 10.

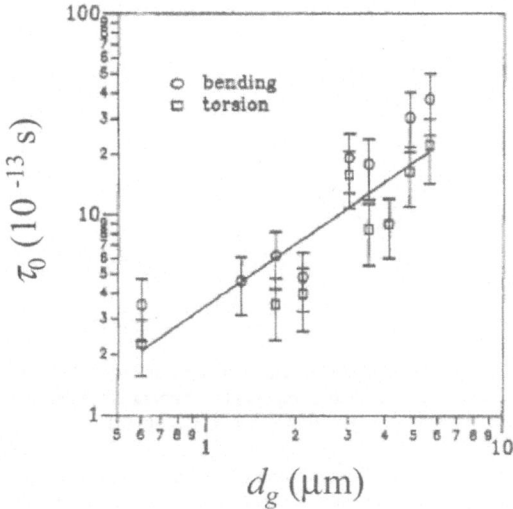

Figure 10. Pre-exponential factors τ_0 for the grain boundary peak in pure Al thin films of different grain sizes. Results obtained both in torsion and bending vibrations are shown. The line represents a linear relation $\tau_0 \propto d_g$ (from [19]).

5.3 DAMPING AND ANELASTIC CREEP LINKED TO INTERFACES IN NANOCRYSTALLINE METALS

Dynamic anelasticity measurements on several n-metals have revealed, with respect to coarse-grained counterparts, an excess temperature-dependent background damping. This characteristic behavior has been reported for Pd [21], Ni[22], Fe[23] and Cu[24]. The underlying relaxation processes have been investigated by performing measurements at different frequencies spanning the 0.01-2000 Hz range on n-Ni and n-Fe [22,25]. Both materials were prepared by ball milling and had an average grain size of 14 nm (n-Fe) and 20 nm (n-Ni) [25]. In the case of n-Ni (Fig. 11), an internal friction peak was also detected superposed to the background damping. Both components (peak and background) exhibited a thermally activated character, as shown by the temperature

234

shift between curves taken at different frequencies (Fig. 11).

The background activation energy H_{bg} and the correction factor n have been determined according to Eqs. (22)-(23), whereas the activation energy H_P and pre-exponential factor τ_{0P} for the peak have been derived by the Arrhenius relation, Eq. (20). The results are summarized in Table I and compared to literature diffusion data. The values of H_{bg} and H_P are within those reported for grain boundary diffusion, H_{GB} (Table I), well below the value of H_L, activation energy for lattice self diffusion.

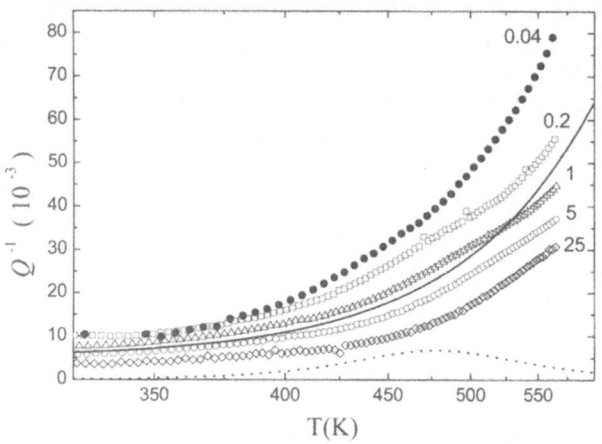

Figure 11. Internal friction Q^{-1} in n-Ni after 30 minutes annealing at T=573 K to ensure thermal stability. The measurement frequency is indicated. The two components: temperature-dependent background (solid line) and Q^{-1} peak (dotted line) are shown for the measure at 0.2 Hz (from [25]).

TABLE I. Activation energy H_{bg} and correction factor n of the background damping, and activation energy H_P and time factor τ_{0P} of the internal friction peak in nanocrystalline Fe and Ni samples. Literature values of the activation energies for lattice (H_L) and grain boundary (H_{GB}) diffusion in coarse-grained Fe and Ni are also reported.

Sample	H_{bg} (eV)	n	H_P (eV)	τ_{0P} (s)	H_L (eV)	H_{GB} (eV)
n-Fe	1.4 – 1.8	≈ 0.2			2.5 [26]	1.4-1.9 [27]
n-Ni	1.1 – 1.4	≈ 0.2	1.2±0.1	$10^{-13\pm1}$	3.0 [28]	1.15-1.35 [29]

In this respect it is also worth to examine the quasi-static behavior. Fig. 12-a shows a creep test carried out on n-Ni, exhibiting both irreversible (viscoplastic) and reversible (anelastic) deformation on loading. The separation of these two components, pictured in the figure, has been carried out by analyzing the recovery compliance through Eqs. (14) and (16). By applying Eq. (18) to the temperature dependence of the sole anelastic deformation $J_{an}(t)$ (reported in Fig. 12), an effective activation energy in agreement with H_{bg} is obtained.

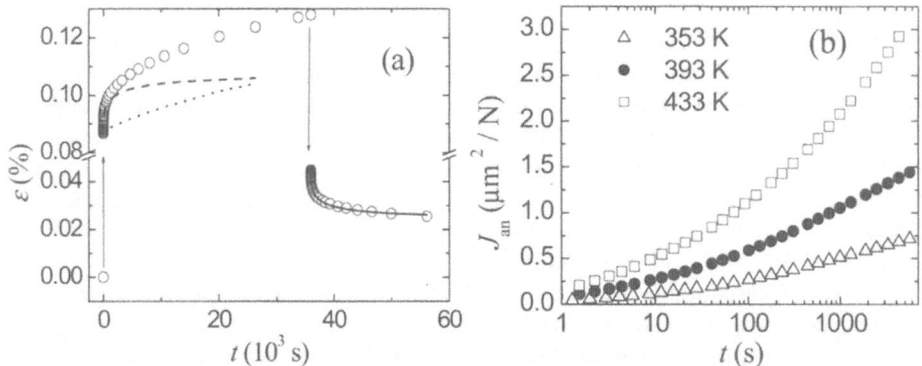

Figure 12. (a) Strain ε in n-Ni during creep and creep recovery (initial part) at T=433 K, σ_0=78 MPa. Continuous line: modeling of the recovery based on Eqs. (15)-(16). Dashed line: anelastic strain reconstructed from creep recovery, Eqs. (14),(16). Dotted line: plastic strain. (b) Temperature dependence of the anelastic creep compliance in n-Ni, at a stress σ_0=112 MPa.

As concerns the damping peak in n-Ni, its temperature and relaxation strength are different from those of internal friction peaks detected in pure polycrystalline Ni above room temperature [30]. The attribution of this peak to a specific mechanisms is not straightforward. However, the value of τ_{0P} indicates that the relaxation process is rate-limited by single atom jumps, while H_P suggests that these jumps may take place at the grain interfaces.

The factor n, close to 0.2, recalls the high temperature behavior of pure metals and indicates that the background is indeed anelastic, as opposed to pure viscoelastic, in character [8]. These values of n much less than 1 are typical for a wide distribution of relaxation times arising from a distributed configuration of crystalline defects. In the case of polycrystalline metals, a spread of dislocation link lengths within grains is present, and the activation energies H_{bg} correspond to dislocation cross-slip or climb. Such values, close to those for lattice self-diffusion, are significantly higher than reported in Table I for the background damping of n-metals. Furthermore, in n-metals, experimental evidence suggests that dislocations are rarely present and scarcely mobile [31].

The common anelastic (recoverable) character and the similar activation energies indicate that the background damping and creep originate with the same microscopic mechanism. The observation that $H_{bg} \approx H_{GB}$, and recent studies showing that H_{GB} is similar in nanocrystalline and coarse-grained polycrystalline metals [32], support the hypothesis that the relaxation process is mediated by atomic jumps at the grain interfaces. In summary, present experimental evidence suggests that the enhanced anelastic response of n-metals, demonstrated both by excess dynamic damping and increased creep deformation, originates with sliding mechanisms at the grain interfaces.

236

6. Conclusions

From the foregoing presentation it follows that the anelastic behavior in materials may display significant modifications due to confinement effects when a dimensional or microstructural feature of the material approaches a critical size. These deviations, starting from the elastic regime where phonon confinement has been experimentally observed in different nanostructured materials, extend to the time-dependent elasticity, involving both specific anelastic relaxation processes or more generally wide spectrum processes. Confinement on dislocation dynamics, such as dislocation anelastic relaxation in thin films, and the intercrystalline Gorsky effect determined by atomic diffusion in nanostructured polycrystalline metals, are examples of Debye-like anelastic relaxation due to a reduced dimension: respectively the thickness of the film and the grain size of a polycrystal. The enhanced background damping and noticeable anelastic creep observed respectively in dynamic and quasi-static experiments on bulk nanostructured metals, provides a further proof of the relevance of anelasticity in small scale materials, connected with the increasing importance of interface dynamics. These last experiments are intimately correlated to specific aspects of the mechanical behaviour in the plastic regime, such as the reported superplastic behavior of nanoscale systems.

Finally, it is worth to mention that other aspects of anelasticity, not treated in this chapter, are expected to be affected by confinement and dimensionality in mesoscopic systems. Among these, thermoelastic behavior and magnetoelastic effects, are particularly relevant in a wide class of technological applications such as sensors, actuators and microelectronic devices.

7. References

1. H. Dosch, Appl. Surf. Sci. **182**, 192 (2001)
2. P. Buffat, J.P. Borel, Phys. Rev. A **13**, 2287 (1976)
3. S.P. Baker, R.P. Vinci, T. Arias, MRS Bull. **27**, 26 (2002)
4. M.J. Kobinsky, G. Dehm, C.V. Thompson, E. Arzt, Acta Mater. **49**, 3597 (2001)
5. W.D. Nix, Metall. Trans. **20A**, 2217 (1989)
6. L. Pasquini A. Barla, A.I. Chumakov, O. Leupold, R. Rüffer, A. Deriu, E. Bonetti, Phys. Rev. B **66**, 073410 (2002)
7. A.S. Nowick, B.S. Berry, *Anelastic Relaxation in Crystalline Solids*, Academic Press, New York, (1972)
8. G. Schoeck, E. Bisogni and J. Shyne, *Acta Metall.*, **12**, 1466 (1964).
9. J.L. Snoek, Physica **6**, 591 (1939)
10. R.W. Powers, M.V. Doyle, J. Appl. Phys. **30**, 514 (1959)
11. B. S. Berry, W.C. Pritchet, Mechanical relaxation behavior of hydrogenated metallic glasses, in G. Bombakidis and R.C. Bowman (eds.), *Hydrogen in Disordered and Amorphous Solids*, Plenum Publishing Corporation (1986) pp. 215-236
12. H. -R. Sinning, Phys. Rev. Lett. **85**, 3201 (2000)
13. M. Weller, Mat. Sci. Forum **366-368**, 549 (2001)
14. G. Dehm, E. Arzt, Appl. Phys. Lett. **77**, 1126 (2000)
15. C. Zener, Phys. Rev. **60**, 906 (1941)
16. T.S. Kê, Phys. Rev. **71**, 533 (1947)
17. T. S. Kê, J. Mater. Sci. Technol. **14**, 481 (1998)
18. B. S. Berry, W.C. Pritchet, J. de Physique **42**, C5-1111 (1981)
19. H.G. Bohn, M. Prieler, C.M. Su, H. Trinkaus, W. Schilling, J. Phys. Chem. Solids **55**, 1157 (1994)
20. R. Raj, M.F. Ashby, Metall. Trans. **2**, 113 (1971)

21. J. Diehl, M. Weller, H.-E. Schaefer, Philos. Mag. A**63**, 527 (1991)
22. E. Bonetti, E.G. Campari, L. Pasquini, E. Sampaolesi, J. Appl. Phys. **84**, 4192 (1998)
23. E. Bonetti, L. Del Bianco, L. Pasquini, E. Sampaolesi, Nanostruct. Mater. **10**, 741 (1998)
24. W.N. Weins, J.D. Makinson, R.J. De Angelis, S.C. Axtell, Nanostruct. Mater. **9**, 509 (1997)
25. E. Bonetti, E.G. Campari, L. Del Bianco, L. Pasquini, E. Sampaolesi, Nanostruct. Mater. **11**, 709 (1999)
26. J. Kučera, B. Million, J. Ružickova, V. Foldyna and A. Jakobova, *Acta Metall.*, **22**, 135 (1974).
27. H. Gleiter, B. Chalmers, *High Angle Grain Boundaries*, Pergamon Press (1972).
28. M.B. Bronfin, G.S. Bulatov and I.A. Drugova, *Fiz. Met. i Metalloved.*, **40**, 363 (1975).
29. N.W. De Reca, C.A. Pampillo, *Scri. Metall.*, **9**, 1355 (1975).
30. A. Rivière, J. Woirgard, Scripta Metall. **17**, 269 (1983)
31. R.W. Siegel, G.E. Fougere, Nanostruct. Mater. **6**, 205 (1995)
32. B.S. Bokstein, H.D. Brose, L.I. Trusov, T.P. Khvostantseva, Nanostruct. Mater. **6**, 873 (1995)

BULK NANOSTRUCTURED SPD MATERIALS WITH ADVANCED PROPERTIES

R.Z. VALIEV, I.V. ALEXANDROV
Institute of Physics of Advanced Materials, Ufa State Aviation Technical Universit K. Marx st. 12, 450000 Ufa, Russia.

Abstract. When severe plastic deformation (SPD), i.e., intense plastic straining under high imposed pressure, is applied to crystalline solids, the processed bulk materials can possess nanostructures and exhibit novel properties. This paper presents several results from recent investigations of SPD materials focussing on two main objectives: modelling and experimental works on SPD techniques, aimed at producing homogeneous nanostructures in bulk large-size billets, and to process hard-to-deform and low-ductility materials; establishing the influence of SPD (strain amount, temperature, applied pressure, etc) and microstructure (types of grain boundaries, defect structures) parameters on the enhancement of properties in as-processed materials.

Keywords: nanostructures, severe plastic deformation, strength and ductility.

1. Introduction

It is well established that plastic deformation can have essential effects on the microstructure and material properties. For example, microstructure refinement and the forming of cells, sub-grains, and fragments take place during heavy rolling, or drawing, and it can lead to improvement of some properties [1-3].

However, materials produced by these methods usually have a decreased ductility due to the resulting microstructures having mostly low-angle dislocation boundaries., At the same time, it has been recently shown that the formation of ultrafine-grained (UFG) nanostructures is possible using severe plastic deformation (SPD), i.e., intense plastic straining under high imposed pressure of several GPa [4-5]. Unique physical and mechanical properties, e.g., high strength and ductility [4, 6], low temperature and/or high strain rate superplasticity [4, 7-11] and others, can correspond to such UFG nanostructures.

However, processing of UFG nanostructures, by SPD techniques, is a non-trivial problem. It requires special experimental and theoretical investigations of plastic flow mechanics, thorough characterization of the processed UFG nanostructures, and establishing the processing guidelines.

T. Tsakalakos, et al. (eds.) pgs. 239 - 249
Nanostructures: Synthesis, Functional Properties and Applications;
© *2003 Kluwer Academic Publishers.*

This paper presents the results of recent studies of SPD processed nanostructured materials, performed at this laboratory in cooperation with our collaborators, and focuses on the following main objectives:
- modelling and experimental works on developing SPD methods, namely: high pressure torsion and equal-channel angular pressing, aimed to producing homogeneous nanostructures in bulk large-sized ingots;
- development of SPD processing for hard-to-deform metals and alloys, e.g. Ti, W and their alloys;
- determination of critical SPD processing parameters and physical mechanisms resulting in the formation of nanostructures.

2. Investigations and Development of SPD Processing

High pressure torsion (HPT) (*figure 1a*) and equal-channel angular (ECA) pressing (*figure 1b*) refer to the techniques which were used in the pioneer works devoted to UFG structures formation in metals and alloys [9, 10] as a result of large deformations with true strain of 10 and more, without damage to the samples. Recently, these methods were further developed.

2.1. HIGH PRESSURE TORSION

Recent investigations have demonstrated that there is a possibility to produce homogeneous nanostructures with a grain size of about 100 nm and less having high-angle grain boundaries by means of HPT [5, 6, 9]. They have allowed to consider this method as the new technique for bulk nanostructured materials processing.

Samples processed under high pressure torsion deformation are disk-shaped (*figure 1a*). In this process, the sample is being put between anvils and compressed under the applied pressure (P) of several GPa. The lower anvil turns and friction forces result in shear straining of the sample. As a result, the deforming sample does not break notwithstanding the high strains of deformation [9].

Samples processed by HPT are typically simple disk-shaped with a diameter from 10 to 20 mm and a thickness of 0.2÷0.5 mm. Essential structure refinement is observed after half- or one complete (360°) turn deformation [12, 13]. However, to create a homogeneous nanostructure, deformation by several turns is necessary (*figure 2*). The important role of applied pressure, in the formation of the homogeneous nanostructured state during the HPT process, is shown in the recent work on Ni [14].

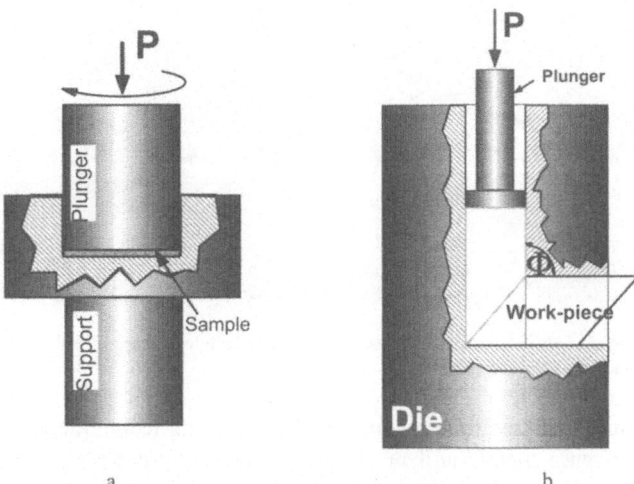

Figure 1. 1. Severe plastic deformation technique principles: a – high pressure torsion, b – ECA pressing.

HPT has been successfully applied for microstructure refinement in metals, alloys and not long ago in composites and in semiconductors [15, 16]. The important advantage of this technique is the possibility to adjust cumulative strain, applied pressure and deformation speed. This possibility makes HPT a very convenient technique for investigation of the influence of different parameters on structure and properties evolution during SPD. However, we have to notice that samples processed by HPT are small-sized. The topical problem is the enlargement of their size. Another important problem is an increase in the HPT sample microstructure homogeneity. It is connected with the different cumulative strains in the center and on sample periphery.

2.2. ECA PRESSING

In the early 1990s, Valiev, with his co-workers [9, 10], developed and applied for the first time ECA pressing [17] as an SPD technique for processing of microstructures with submicro- and nanometer grain sizes. In these early experiments, original ingots,s with square or round cross sections, were cut from rods of length from 70 up to 100 mm. The diameter of their cross section or their diagonal did not exceed 20 mm.

During ECA pressing implementation, the ingot is being pressed in a special die through two channels with equal cross section intercrossing usually at an angle of 90° (*figure 1b*). In the case of hard-to-deform materials, the deformation is realized at elevated temperatures, or with increased channel crossing angles, if it is necessary. In that case, there are some particular requirements with respect to heat resistance and to the die durability. Each pass imparts a supplementary strain, approximately 1, for the most often applied channel crossing angle of 90°.

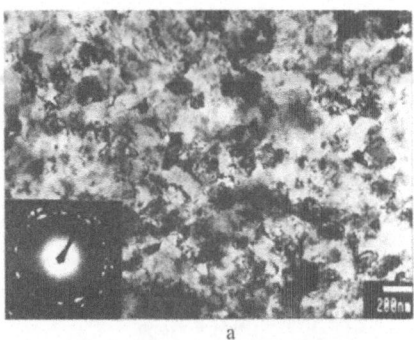 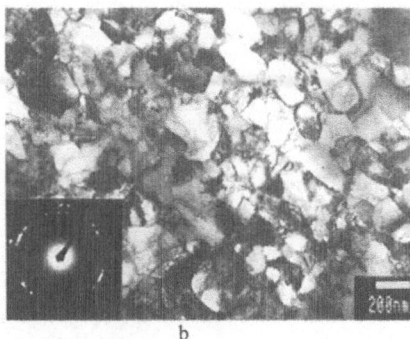

a
b

Figure 2. TEM images of UFG Cu nanostructures formed at room temperature by HPT (P = 6 GPa, 5 turns) (a) and by ECA pressing (12 passes) (b).

A strong microstructure refinement, by ECA pressing technique, can quite easily be obtained both in pure metals and in alloys. However, producing homogeneous UFG nanostructures having high-angle grain boundaries by ECA pressing technique is a special problem. It is well known that the number of passes, and the selected ECA pressing route, are very important parameters of the processing.

The influence of the number of passes (cumulative strain) and ECA pressing routes on the microstructure evolution for a number of metals and alloys, namely Al-Mg alloys [8] and pure Ti [18], has been investigated in detail. It was shown that a homogeneous microstructure in alloys can be observed just after 4-6 passes following route B_C. During this process, the billet is turned between consecutive passes in one direction around its axis by an angle of 90°. The analysis of the shearing characteristics, for different processing routes, indicates that route B_C leads to reconstruction of the shape of an initially cubic element in the unpressed sample after 4n (n being an integer) passes through the die and it lead to a homogeneous equiaxed structure formation [8].

According to similar investigations of the influence of ECA pressing routes on Ti microstructure it is preferable to follow route B_C in order to form an equiaxed grain structure and obtain a higher quality of ingot shape and surface, although the mechanical properties in this case may not be the best ones [18].

The experimental and theoretical modelling of the mechanics of ECA pressing, e.g. in the stress-deformed state, contact stresses in the die is one of the important tasks for developing SPD processing. Such investigations of the influence of the friction coefficient between the deforming billet and the die walls, determination of the contact stress in the die walls have shown [19, 20] that the shear plastic strain during ECA pressing could be essentially non-uniform along the deformed sample. At the same time a significant sensitivity of plastic deformation uniformity to friction conditions between ingot and die was found (*figure 3*) [20]. The approaches of enhancement of ECA pressing uniformity by optimization of friction conditions based on the obtained results of experimental and field-emission microscopy (FEM) were worked through. On this basis a die was fabricated and bulk billets with uniform ultrafine-grains of hard-to-deform W and Ti were obtained [20, 21]. Here a maximum size of the ingots equal to 60 mm in diameter and 200 mm in length (*figure 4*) was reached.

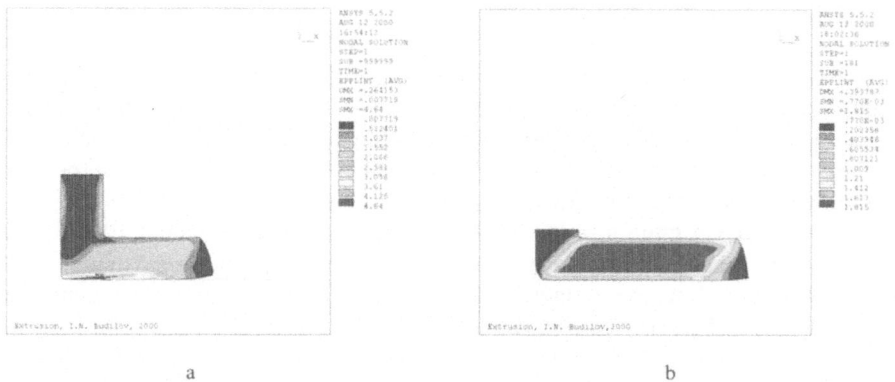

a   b

Figure 3. The distribution of plastic deformations intensity in the case of different values of the friction coefficient k: k=0.2 (a), k=0 (b) [20].

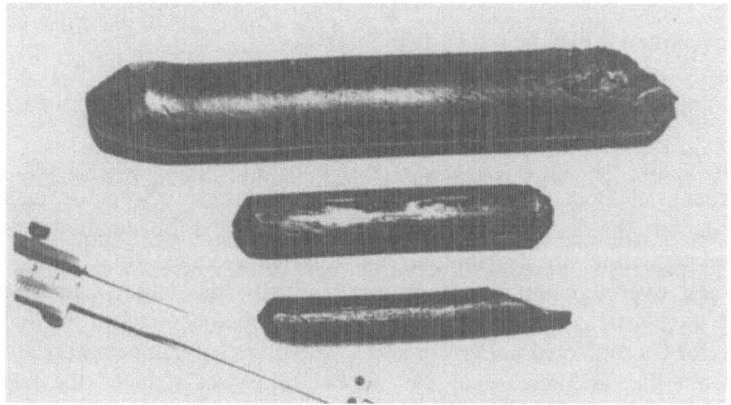

Figure 4. Bulk Ti billets, processed by ECA pressing.

Another way which can improve the homogeneity of UFG nanostructures is using back pressure during ECA pressing. This was studied for processing nanostructured Cu billets, where ECA pressing was performed with different back pressures [22].

In spite of a considerable recent progress in SPD techniques development for processing homogeneous UFG nanostructures in bulk samples of different metals and alloys, problems of further grain refinement down to the nano range, sample size enlargement, especially in low ductile materials, remain topical. Designing new SPD techniques in order to increase technological effectiveness is also a topical issue. The SPD applications to polymeric materials are of great interest as well. Polymers are widely used as functional, structural and packaging materials and their improvement by SPD techniques appears to be a promising task for ongoing investigations.

3. Formation of Ultrafine Grains

Careful microstructure characterization of SPD materials is a non-standard task. It is connected with very high defect density, considerable internal stresses and distortions of crystal lattice and also with limited applicability to such materials of common techniques for structural investigations. As a result, the analysis of SPD-processed nanostructures needs special modifications of these techniques.

As a rule the mean grain size in SPD-processed pure metals is about 100-200 nm [4]. However, in many alloys or metal samples, processed by SPD consolidation of powders after ball milling, the grain size can reach 15-20 nm [23]. The mean grain size processed by HTP is typically smaller than after ECA pressing.

Earlier investigations have allowed to determine several important structural features of UFG nanostructures, processed by SPD techniques [4, 5]. The appearance of non-equilibrium grain boundaries with extrinsic grain-boundary dislocations (EGBD) of high density is one of these features. These EGBD assemblies create long-range elastic stress fields resulting in considerable static and dynamic atomic shifts in the crystal lattice. These shifts are especially large near grain boundaries. The dislocation density in grains interior may be not large in this case.

Recent structural investigations have confirmed this fact. The most interesting results were obtained using new approaches, recently developed and applied in the X-ray structural analysis of SPD nanostructured Cu [24-27].

For example, the use of the modified Williamson-Hall plot and of the modified Warren-Averbach method, considering anisotropic contrast effects of dislocations, allowed to calculate the dislocations density, the median and the standard deviation of the log-normal crystallite size distribution, two different averages of the grain size (over the volume and over the number of grains), and also the lattice parameter value, depending on the elastic constants of the crystal and of the character of the dislocations in the crystal, for Cu subjected to ECA pressing with different numbers of passes.

It turned out that an increase in the number of passes reduced the mean grain (crystallite) size and also that its distribution became more narrow (*figure 5*) [24]. Besides, it was determined that the average dislocation character became more edge than screw during the increase of straining. This probably originated from the easier screw dislocation annihilation. On the other hand, the increase of edge dislocations role is in accordance with other experimental data indicating a misorientation angle increase and high-angle misorientations formation between neighbouring areas in the crystal, as the strain increased [28].

The change in the dislocation density dependence, with the reduction in the grain size, is an interesting phenomenon. In the case of a grain size larger than 100 nm, the dislocation density increased but for smaller grain sizes it began to decrease [26, 27]. Probably the latter effect is connected with a dynamic recovery of the microstructure during SPD processing.

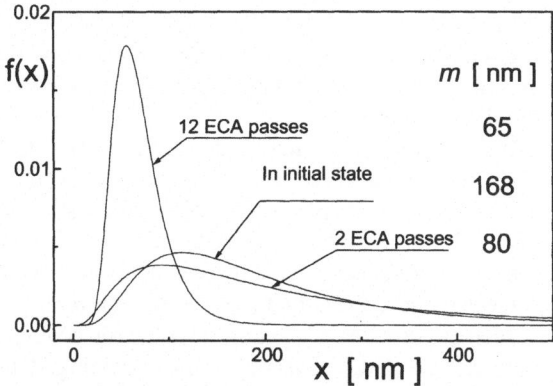

Figure 5. Cu. The size distribution functions of subgrains in the extruded initial state and the state deformed by 2 ECA passes, and of grains in the state deformed by 12 ECA passes [24].

Low-temperature X-ray investigations allowed to estimate the atomic shifts and the Debye temperature in grains interior and in grain-boundary areas [29]. It turned out that the atomic shifts are very large and the Debye temperature is strongly decreased in the grain boundary area. In particular, the averaged atomic shifts in the grain-boundary areas of nanostructured Ni is 0.18 Å, which is equal to 7.2% of the shortest interatomic distance in Ni at a temperature of 295 K. The Debye temperature in the Ni grain-boundary areas equals 127 $\pm$1 K; it means that it is considerably lower (almost by 200 K) than the value for coarse-grained Ni. A similar Debye temperature decrease was found during Mössbauer investigations of SPD nanostructured Fe [30].

Extremely large atomic shifts near grain boundaries indicate an enhanced atom energy there. It may explain the high grain boundary diffusivity which was revealed in SPD nanostructured materials [5].

In spite of an evident progress in the structural characterization of SPD nanostructured materials, some problems still should be scrutinised by investigators. One may refer to the issues concerning a deeper understanding of grain refinement mechanisms, attaining a minimum grain size, the analysis of atomic defects in nanostructures of different SPD materials and others.

Recent results in the physics of large plasticity appear to be very important for understanding the microstructure refinement mechanisms. As it is now well established, there are five stages of plastic deformation, differing by the work hardening parameter $\Theta = d\tau/d\gamma$ where τ is shear stress and γ is shear strain [27, 31-33]. A shear strain γ approximately equal to 1-2 corresponds to the first three work hardening stages. A work hardening decrease takes place in stage III, it remains invariable or increases a little in stage IV and it begins to decrease and can be equal to zero in stage V. Numerous experimental investigations allowed to conclude that the transition from stage III to stage IV is associated with a progressive decrease of the dislocation walls thickness and an increase of misorientation angles between the neighbouring cells. Electrical resistivity measurements results and X-ray data indicate a decrease in the dislocation density due to their annihilation [33] which takes place in stage V. As it was mentioned

246

above, a similar effect was revealed in Cu during SPD. In this case the strain increase resulted in a monotonous decrease in the grain size and in a dislocation density increase. The latter changed into a decrease in the dislocation density after reaching some threshold value of the grain size [26, 27]. These observations are well interpreted in the framework of the mentioned theories about the dislocation density modifications during the transition from stage IV to stage V.

Precise X-ray analysis results indicated that the transition to stage IV makes considerable changes in the dislocation configuration character [34, 35] from a polarized dipolar wall (PDW) structure to a polarized tilt (or twist) wall (PTW) structure (*figure 6*). This approach allows to explain experimentally the observed misorientation increase between the neighbouring cells with increasing strain. As it was mentioned above, this effect also takes place in SPD materials. It is important to take into account deformation-induced vacancies creation [32] caused by the dislocations interaction for a reliable quantitative explanation of the observed work hardening. This issue should be investigated in future for SPD nanostructured materials as well.

Recent investigations [8, 18] also showed that the final size of subgrains during heavy plastic straining is smaller if several deformation modes are applied and this may be connected with an involvement of new slip systems in the deformational process. This in principle may be useful for developing new SPD techniques using combined deformation modes [36].

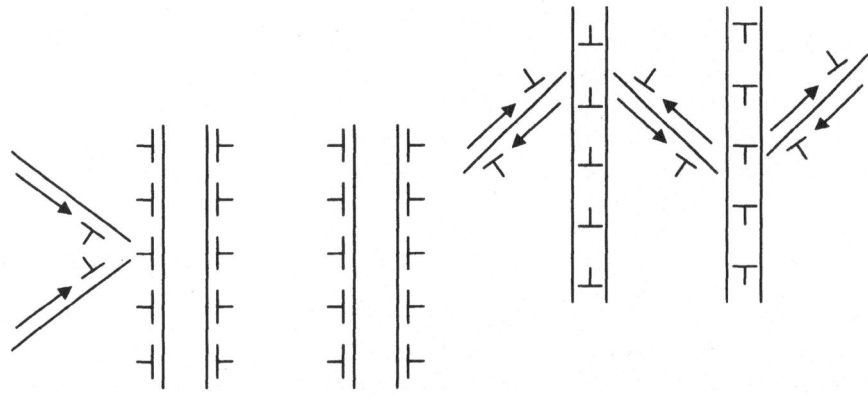

Figure 6. Sketch of a polarized dipole wall ("PDW", left), and a polarized tilt wall ("PTW", right) [35].

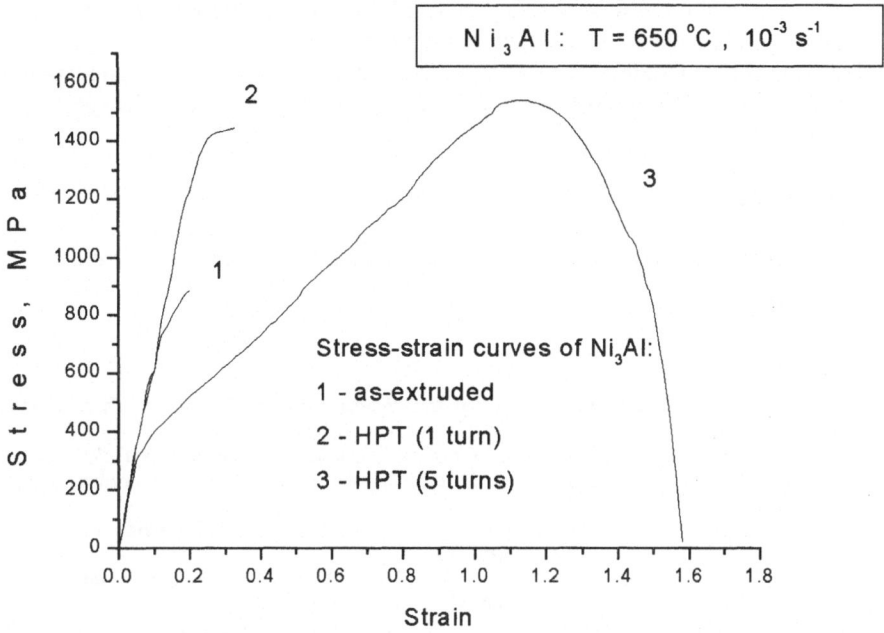

Figure 7. Tensile engineering stress-strain curves of Ni$_3$Al tested at 650°C. The strain rate - 10^{-3} s^{-1}. The processing conditions for each curve are listed in the figure.

The microstructure evolution during the SPD process from the cell substructures to UFG nanostructures can result in fundamental changes of their properties. For example, the intermetallic compound Ni$_3$Al during tension at 650C° (*figure 7*) [7] shows a typical low ductility (curve 1). After 1 complete turn of HPT its strength increases but its ductility remains low (curve 2), while only a cell substructure forms in the microstructure. However, after 5 turns of HPT when an UFG nanostructure with a grain size of about 70 nm is formed, a radical change in mechanical behaviour is observed in Ni$_3$Al. The material demonstrates very high strength and ductility. This phenomenon, recently observed in several SPD materials, which was termed the paradox of severe plastic deformation [7], can result in principally new possibilities of material properties enhancement.

4. Conclusions

The results of recent investigations have demonstrated that SPD techniques can be successfully used for fabrication of homogeneous microstructures with ultrafine grains in various bulk metallic materials. SPD technique development allowed to enlarge the dimensions of severely deformed billets and to increase the number of materials treated by SPD. In particular, a successful application of ECA pressing for low-ductility and hard-to-deform materials has been shown. The development of microstructure

248

characterization methods allows to study in detail the microstructure evolution character during SPD and to reveal the nanostructure formation mechanisms.

It has been shown that the microstructure evolution during SPD from dislocation cell substructures to ultrafine-grains can result in radical material properties changes.

5. Acknowledgements

This work was supported in part by the Russian Foundation for Basic Research.

6. References

1. Schmid, E. and Boas, W. (1968) *Plasticity of Crystals, with Special Reference to Metals*, Chapman and Hall, London, UK.
2. Bridgman, P.W. (1952) *Studies in Large Plastic Flow and Fracture*, McGraw-Hill, New York, USA .
3. Brandes, E.A. and Brook, G.B. (1992) *Smithells Metals Reference Book*, Butterworth-Heinemann Ltd., Oxford, UK.
4. Valiev, R.Z., Islamgaliev, R.K., and Alexandrov, I.V. (2000) Bulk nanostructured materials from severe plastic deformation, *Prog. Mat. Sci.* **45 (2)**, 103-189.
5. Lowe, T.C. and Valiev, R.Z. (eds.) (2000) *Proceedings of the NATO ARW on Investigations and Applications of Severe Plastic Deformation*, 80, Kluwer Publishers, Dordrecht.
6. Valiev, R.Z., Alexandrov, I.V., Zhu, Y.T., and Lowe, T.C. (2002) Paradox of strength and ductility in metals processed by severe plastic deformation, *JMR* **17 (1)**, 5-8.
7. McFadden, S.X., Mishra, R.S., Valiev, R.Z., Zhilyaev, A.P., and Mukherjee, A.K. (1999) Low-temperature superplasticity in nanostructured nickel and metal alloys, *Nature* **398**, 684-686.
8. Langdon, T.G., Furukawa, M., Nemoto, M., and Horita, Z. (2000) Using equal-channel angular pressing for refining grain size, *JOM* **52 (4)**, 30-33.
9. Valiev, R.Z., Korznikov, A.V., and Mulyukov, R.R. (1993) Structure and properties of ultrafine grained materials by severe plastic deformation, *Mater. Sci. Eng.* **A186**, 141-148.
10. Valiev, R.Z., Krasilnikov, N.A., and Tsenev, N.K. (1991) Plastic deformation of alloys with submicron-grained structure, *Mater. Sci. Eng.* **A137**, 35-40.
11. Kim, H.S., Hong, S.I., Dubravina, A.A., and Alexandrov, I.V., submitted to *Mater. Sci. Eng. A*.
12. Valiev, R.Z., Ivanisenko, Yu.V., Rauch, E.F., and Baudelet, B. (1996) Structure and deformation behaviour of armco iron subjected to severe plastic deformation, *Acta Mater.* **44 (12)**, 4705-4712.
13. Tarakanova, A.A. and Alexandrov, I.V. (1999) *Abstracts of NATO ASI "Multiscale Phenomena in Plasticity: from Experiments to Phenomenology, Modelling & Materials Engineering"*.
14. Zhilyaev, A.P., Lee, S. , Nurislamova, G.V., Valiev, R.Z., and Langdon, T.G. (2001) Microhardness and microstructural evolution in pure nickel during high-pressure torsion, *Scripta Mater.* **44**, 2753-2758.
15. Alexandrov, I.V., Zhu, Y.T., Lowe, T.C., Islamgaliev, R.K., and Valiev, R.Z. (1998) Microstructures and properties of nanocomposites obtained through SPTS consolidation of powders, *Metall. Trans.* **A29**, 2253-2260.
16. Islamgaliev, R.K., Kuzel, R., Mikov, S.N., Igo, A.V., Burianek, J., Chmelik, F., and Valiev, R.Z. (1999) Structure of silicon processed by severe plastic deformation, *Mat. Sci. Eng.* **A266**, 205-210.
17. Segal, V.M., Reznikov, V.I., Drobyshevskiy, A.E., and Kopylov, V.I. (1981) Equal angular extrusion, *Russian Metals* **1**, 99-105.
18. Stolyarov, V.V., Zhu, Y.T., Alexandrov, I.V., Lowe, T.C., and Valiev, R.Z. (2001) Influence of ECAP routes on the microstructure and properties of pure Ti, *Mat. Sci. Eng.* **A299 (1-2)**, 59-67.
19. Alexandrov, I.V. and Valiev, R.Z. (2001) Developing of SPD processing and enhanced properties in bulk nanostructured metals, *Scripta Mater.* **44**, 1605-1608.
20. Zhernakov, V.S., Budilov, I.N., Raab, G.I., Alexandrov, I.V., and Valiev, R.Z. (2001) A numerical modelling and investigations of flow stress and grain refinement during equal-channel angular pressing, *Scripta Mater.* **44**, 1765-1769.

[21]. Alexandrov, I.V., Raab, G.I., Valiev, R.Z., Shestakova, L.O., and Dowding, R.J. (2000) Microstructure refinement in tungsten by severe plastic deformation, in *Proc. of 2000 Intern. Conf. on Tungsten, Hard Metals and Refractory Alloys*, **5**, pp. 27-33.

[22]. Raab, G.I., Krasilnikov, N.A., Alexandrov, I.V., and Valiev, R.Z. (2000) Structure and properties of copper after ECA-pressing in conditions of elevated pressures, *The Physics and Technique of High Pressures* **10 (4)**, 73-77.

[23]. Shen, H., Li, Z., Guenther, B., Korznikov, A.V., and Valiev, R.Z. (1995) Influence of powder consolidation methods on the structural and thermal properties of a nanophase Cu-50wt%Ag alloy, *Nanostr. Mater.* **6**, 385-388.

[24]. Ungár, T., Alexandrov, I., and Hanák, P. (2000) Grain and subgrain size-distribution and dislocation densities in severely deformed copper determined by a new procedure of X-ray line profile analysis, in T.C. Lowe, R.Z. Valiev (eds.), *Proceedings of the NATO ARW on Investigations and Applications of Severe Plastic Deformation*, Kluwer Publishers, Dordrecht, **80**, pp.133-138.

[25]. Ungár, T. (2000) Size distribution of grains or subgrains, dislocation density and dislocation character by using the dislocation model of strain anisotropy in X-ray line profile analysis, in T.C. Lowe and R.Z. Valiev (eds.), *Proceedings of the NATO ARW on Investigations and Applications of Severe Plastic Deformation*, Kluwer Publishers, **80**, pp.93-102.

[26]. Alexandrov, I.V. (2000) X-ray studies and computer simulation of nanostructured SPD metals, in T.C. Lowe and R.Z. Valiev (eds.), *Proceedings of the NATO ARW on Investigations and Applications of Severe Plastic Deformation*, Kluwer Publishers, **80**, pp.103-108.

[27]. Ungar, T., Alexandrov, I., and Zehetbauer, M. (2000) Ultrafine-grained microstructures evolving during severe plastic deformation, *JOM* **52 (4)**, 34-36.

[28]. Hansen, N. and Juul Jensen, D. (1999) Development of Microstructure in FCC Metals during Cold Work, *Phil. Trans. R. Soc. London* **A357**, 1447.

[29]. Zhang, K., Alexandrov, I.V., Valiev, R.Z., and Lu, K. (1998) The thermal behavior of atoms in ultrafine-grained Ni processed by severe plastic deformation, *J. Appl. Phys.* **84 (4)**, 1924-1927.

[30]. Valiev, R.Z., Mulyukov, R.R., Ovchinnikov, V.V., and Shabashov, V.A. (1991) Moessbauer investigations of submicron grain-size Fe, *Scr. Metall. Mater.* **25**, 2717.

[31]. Zehetbauer, M.J. (2000) Strengthening processes of metals by severe plastic deformation, in T.C. Lowe and R.Z. Valiev (eds.), *Proceedings of the NATO ARW on Investigations and Applications of Severe Plastic Deformation*, Kluwer Publishers, **80**, 81-91.

[32]. Zehetbauer, M., (1993) *Acta Metall. Mater.* **41**, 589.

[33]. Zehetbauer, M. and Seumer, V. (1993) *Acta Metall. Mater.* **41**, 577.

[34]. Müller, M., Zehetbauer, M., Borbély A., and Ungár, T. (1996) *Scr. Mater.* **35 (12)**, 1461.

[35]. Ungár, T. and Zehetbauer, M. (1996) *Scripta Mater.* **35 (12)**, 1467.

[36]. Zehetbauer, M. and Les, P. (1998) *Kovové Materiály* **36 (3)**, 153.

HEAVILY DRAWN EUTECTOID STEEL: A NANOSTRUCTURED MATERIAL

A. PHELIPPEAU, S. POMMIER, C. PRIOUL, M. CLAVEL
Laboratoire de Mécanique des Sols, Structures et Matériaux,
École Centrale Paris, Grande voie des vignes,
92295 Châtenay-Malabry, Cedex, France

1. Introduction

Heavily drawn eutectoid steel wires are mainly used by the cable industry for applications such as tyre reinforcement, suspension cables or springs. Cables are made by a torsion assembly of wires, which have very high tensile strengths up to 4000 MPa and a level of ductility that has to be sufficient to undergo the torsion assembly process. Needless to say that theses mechanical properties are exceptional, as shows the comparison in figure 1 with those of carbon fibres, well known for their structural reinforcement properties.

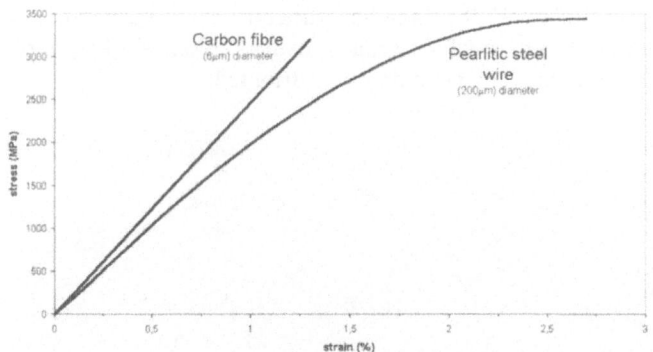

Figure 1. Tensile curves of a carbon fibre and an heavily drawn pearlitic steel wire

The mechanical properties of this material have early been related to the fine lamellar microstructure, which characteristic length is of a few tens of nanometers, developed during the drawing process [1,2].

T. Tsakalakos, et al. (eds.); pgs 251- 270
Nanostructures: Synthesis, Functional Properties and Applications;
© *2003 Kluwer Academic Publishers.*

The purpose of this article is to present an extensive literature review about this subject, completed by some of our own results. We will try to show that this material, although produced by a classic process for a lot of decades, belongs to the group of nanostructured materials.

In a first part we will describe the evolution of the lamellar microstructure during the drawing process and put it in relation with the mechanical properties in a second part. In these two parts, the identification and the contribution of the various internal scales of this material are discussed. The last part will deal with the ageing mechanisms that occur after the drawing process.

2. The Pearlite Microstructure and Its Evolution During the Drawing Process

In this part, we will in the first place describe the initial microstructure, then the main characteristics of the drawing process and finally the evolution of the microstructure with cold drawing.

2.1 THE PEARLITIC MICROSTRUCTURE

Pearlite results from the decomposition below 727 °C of the austenite phase with the eutectoid composition (0.77 C wt%) into alternate lamellae of ferrite (Fe α, bcc, 88 vol.%) and cementite (Fe_3C, orthorhombic, 12 vol.%). In the industrial context, the carbon content of the steels ranges between 0.7 and 0.9 wt.% so that a little fraction of pro-eutectoid ferrite or cementite can occur.

As shown in figure 2, the lamellar structure nucleates at the austenite grain boundaries leading to the growth of groups, named colonies, of parallel lamellae. The lamellae orientation in the bulk is randomly distributed.

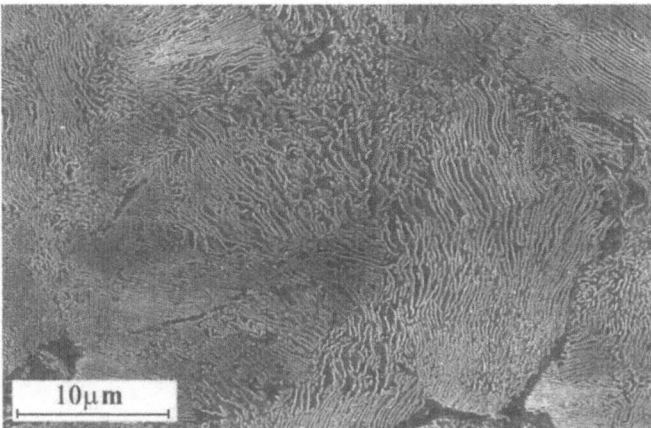

Figure 2. SEM image of the initial pearlitic microstructure showing alternate lamellae of cementite (in white) and ferrite (etched with Nital, in black).

Depending on the cooling rate, the interlamellar spacing ranges from 100 nm (fine pearlite) up to 500 nm (coarse pearlite). In the application studied, the initial spacing is about 150 nm.

2.2 THE DRAWING PROCESS

The drawing process consists in the reduction of the wire diameter by passing through of conical dies in series. The reduction of cross section area can be as high as 98 %. The true strain ε is defined by :

$$\varepsilon = 2 \cdot \ln\left(\frac{d_0}{d}\right) \tag{1}$$

where d_0 and d are respectively the initial and the final diameter.

In the industrial context, there are three main steps of the process to consider. First, pearlitic wires are drawn in dry conditions, i.e. no lubricant, to reduce the diameter from about 5 mm to a final one that ranges between 1 and 2 mm. Then a heat treatment, called patenting, is applied to recover the pearlitic initial microstructure and a small layer of brass is deposited on the surface. This layer is used as a lubricant in the second step of drawing made in a bath of water at room temperature. This wet drawing step reduces the wire diameter down to 200 µm. The strain rate $\dot{\varepsilon}$ can reach values up to 10^4. As a consequence, the plastic work inside a die leads to a temperature rise that was evaluated, by FE analyses, as about 100°C. The temperature rise during the complete process is difficult to determine, since the thermal exchanges with the bath are unknown. However, the final temperature is assumed to exceed 400°C.

2.3 EVOLUTION OF THE DRAWN MICROSTRUCTURE

2.3.1 *Evolution of the morphology*
As shown in figure 3, the drawing process leads to two principal morphologic evolutions that depend on the level of drawing strain: the orientation of the pearlite lamellae along the drawing axis and the diminution of the interlamellar spacing.

254

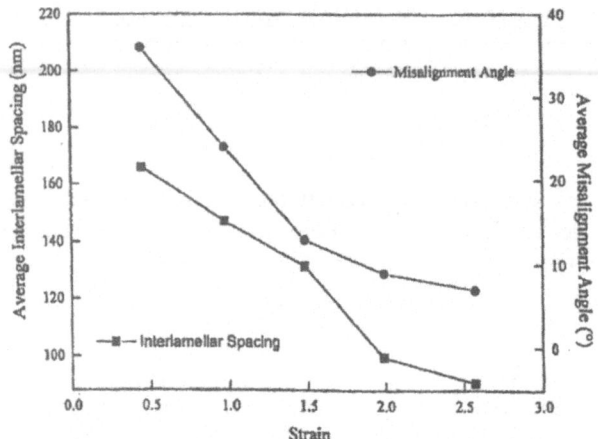

Figure 3. Evolution of the average interlamellar spacing and misalignment angle with the drawing strain [3].

First detailed observations of drawn pearlitic wires showed that lamellae tend to incline toward the drawing axis for deformation larger than $\varepsilon = 0.7$ [1]. This alignment gets more and more pronounced for drawing strains higher than 1.5 [4], the average misalignment angle of lamellae with the wire axis being less than 10° for $\varepsilon > 2$ [3]. Langford [5] argued that this alignment involves a plastic deformation of cementite, and showed that a size effect in plastic deformability of thinner cementite lamellae exists.

The interlamellar spacing is reduced from an initial value higher than 100 nm to values in the order of 10 or 20 nm for $\varepsilon \geq 3.5$, as shown in figure 4.

Figure 4. TEM micrograph of the as drawn pearlitic structure ($\varepsilon = 3.5$) in a longitudinal section [6].

The observation of the microstructure in the transverse direction before (Fig. 5 (a)) and after the drawing (Fig. 5 b) shows a curling of the lamellae. This type of morphology is due to the development of a <110> ferrite fiber texture along the drawing direction that leads to plane strain of the lamellae compensated by a bending of the colonies to maintain the global axially symmetric deformation of the wire [7].

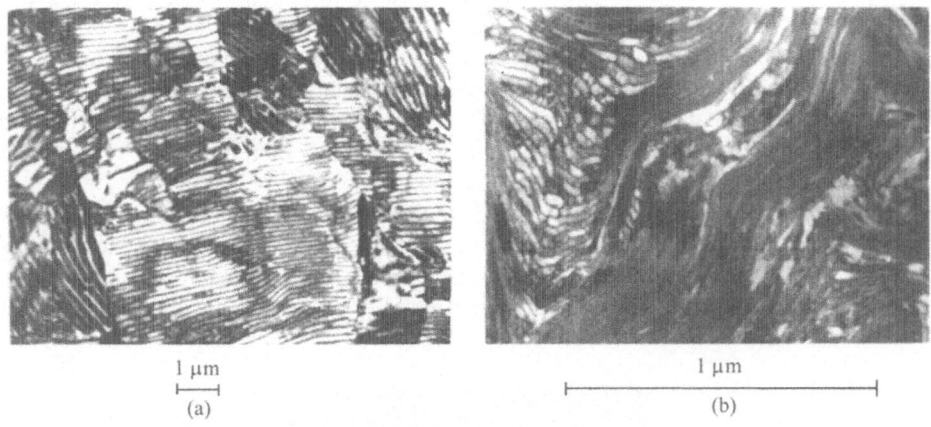

1 μm
(a)

1 μm
(b)

Figure 5. TEM micrographs in transverse section of a pearlitic structure non drawn (a) and drawn to ε = 4.22 (b) [5]

2.3.2 *Crystallographic Textures*

The deep drawing of bcc metals is known to induce a <110> fibrous texture along the wire axis [7-8]. In the case of pearlitic steel wires, the <110> direction of ferrite is aligned along the drawing direction and a preferred crystallographic orientation along the radial direction **r** is heterogeneously developed [9,10]:

- Near the surface area (0.8 R<r<R; R radius of the wire), the ferrite texture reported is a smooth <110> fiber texture in the case of wet drawing and a {112}<110> circular texture for dry drawing.
- In the intermediate zone (0.5 R<r<0.8 R), a {110}<110> circular texture occurs, related to the shear stress components in the die.
- In the central zone, the texture is mainly a fibrous <110> one due to the uniaxial stress state.

The relative extension of these zones depends on drawing parameters like the die angle or the lubrication conditions.

2.3.3 *The Cementite Dissolution*

Decomposition of several tens of percent of cementite due to cold drawing have early been studied by means of TEM observations, thermomagnetic analyses and Mössbauer spectroscopy in the 70's and a first review of this phenomena was proposed by Gridnev *et al.* [11]. A dissolution of 50% of cementite by cold drawing to ε = 1.96 has also been measured [12]. This phenomenon has also been reported for other deformation process like torsion under quasi-hydrostatic pressure [13]. According to these works, cementite dissolution would be attributed to the fact that the interaction energy between a carbon atom and a dislocation exceeds the binding energy between carbon and iron atoms in cementite. Dislocations created by the drawing, mainly located at the ferrite/cementite interface, would drag and trap a large amount of the cementite carbon atoms.

Languillaume *et al.* [14] observed the cementite dissolution on a pearlitic wire drawn to ε = 3.5 with different experimental techniques. The evolution of neutron diffraction spectrum with the temperature of annealing indicates qualitatively that a weak proportion of cementite is present in the as-drawn state. HRTEM micrographs show that cementite is fragmented into small nanosized crystallites. This point was confirmed by dark field TEM (see figure 6) and HRTEM observations by Hong *et al.* [15] on a ε = 4.22 drawing strain state. An electron energy loss spectroscopy (EELS) analyses indicates that the difference of carbon concentration between ferrite and cementite is between 7 and 11 at.%, that is to say much less than the 25 at.% theoretical value in cementite [14].

Arguing that the mechanism proposed by Gridnev *et al.* [11] can not explain the high level of dissolution, Languillaume et al. [14] proposed a model based on a thermodynamical destabilization. The refinement and the fragmentation of cementite lamellae would increase substantially the interfacial energy per unit volume of cementite, leading to the dissolution.

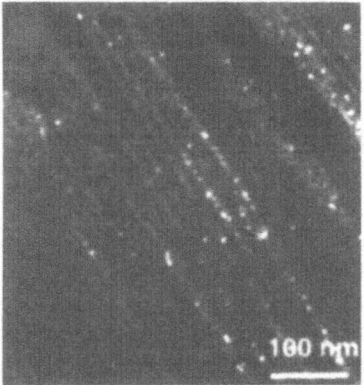

Figure 6. Dark field TEM micrograph of a drawn eutectoid wire (ε = 4.22) in a longitudinal section [15].

Developments during the last decade of the field ion microscopy (FIM) such as the Atom Probe (APFIM) or the Tomographic Atom Probe (TAP) allow a quantitative chemical analysis of the cementite dissolution at the nanometer scale [6,15-17]. As shown in figure 7, the dissolution leads to non stoichiometric cementite and, as a consequence, a non uniform carbon concentration in ferrite ranging between 0.2 and 3 at.%, i.e. the ferrite is oversaturated.

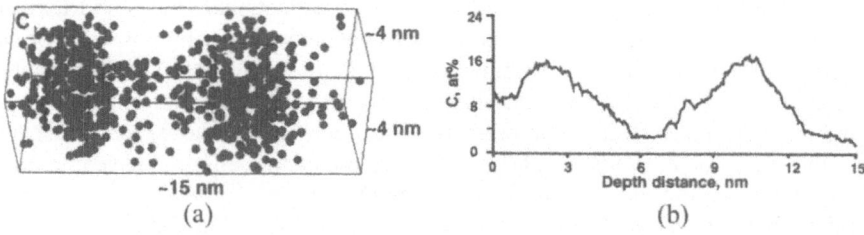

Figure 7. (a) 3-DAP map of carbon in a pearlitic wire (ε = 4.22). (b) The corresponding carbon concentration profile across the ferrite/cementite interface. [15]

The level of dissolution seems to depend on the initial interlamellar spacing: cementite dissolution is all the higher that the initial interlamellar spacing is small [15, 18]. There is a distribution of interlamellar spacing over the colonies in the initial state and as this distribution tends to widens with the drawing increase [5]. As a result, the cementite dissolution is not only non uniform inside the lamellae but also at the scale of colonies.

Hono *et al.* [19] showed, in the case of a 0.97 wt.% carbon steel, that the increase of drawing strain can induce a complete cementite dissolution: a $\varepsilon = 3.6$ drawn wire still presents a variation of carbon concentration related to the "ex"-cementite lamellae (Figure 8 a) whereas in a $\varepsilon = 5.1$ drawing state the carbon concentration is nearly constant (Figure 8 b). Nevertheless, the chemical position of carbon atoms (interstitial, substitution atom, segregated...) remains unknown. The variation of the $(110)_\alpha$ lattice parameter measured by X-Ray diffraction between the non deformed state and the $\varepsilon = 5.1$ state is less than 2‰. Referring to the empirical equation proposed by Fasika *et al.* [20] which gives the relationship between the ferrite lattice parameter and the concentration of interstitial carbon atoms, they estimated the carbon content in ferrite to 0.7 at.% which is, on the one hand, much more than the 0.2 at.% maximum carbon solubility in ferrite and, on the other hand, much less than the 4 at.% dissolved. Among the explanations proposed by Ref. [19] we can quote the formation of a martensitic lattice and the segregation of carbon atoms at dislocations in ferrite at ferrite subgrains. No experiments are yet available to support these hypothesis.

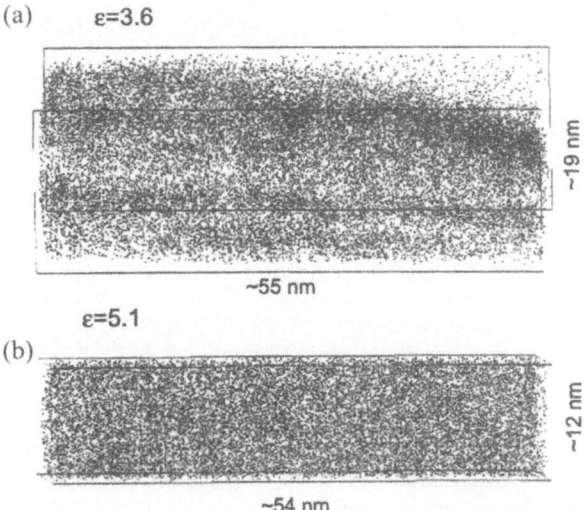

Figure 8. 3-DAP carbon maps obtained from the pearlitic steel wires with (a) $\varepsilon = 3.6$ and (b) $\varepsilon = 5.1$ [19]

This phenomenon of phase dissolution and possible oversaturated solution is also observed in other nanocomposite alloys like heavily drawn Cu/Nb wires [21]. To explain what seems to be a specificity of nanograins materials, Gleiter *et al.* [22] invoked the enhancement of the solute solubility due to the space charge in the vicinity of the interphase boundaries, which volume fraction is higher in nanograins materials.

3. Deformation and Mechanical Properties

As mentioned in the introduction, heavily drawn pearlitic steel wires exhibit a very high tensile strength up to 4000 MPa. In this part, we will focus on the relations between the microstructure evolutions detailed in the previous part and the mechanical properties. At least three contributions to this high mechanical properties can be *a priori* taken into account: the strain hardening, the effect of residual stresses and solid solution hardening. Here, we will more focus on the first two contributions, the last one being treated in the next part concerning the ageing mechanisms.

3.1 THE STRAIN HARDENING MODELLING

The main characteristic of drawn pearlitic wires is their hardening behaviour that increases exponentially with the drawing strain [1, 2]. As shown in figure 9, the flow stress determined by the tensile strengths measured at different intermediate drawing strains, follows a exp($\varepsilon/4$) evolution.

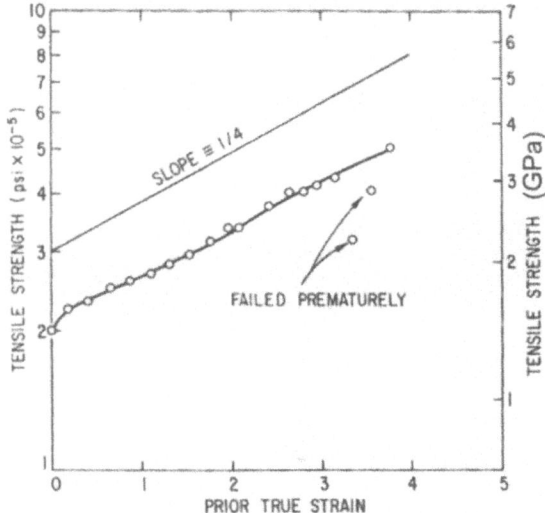

Figure 9. Exponential increase of pearlitic steel wires tensile strengths as a function of drawing strain [2].

As the microstructure looks like ferrite and cementite lamellae aligned along the wire axis, drawn pearlite has often been described as a uniaxial composite material with a flow stress σ given by the rule of mixture:

$$\sigma = f_a \cdot \sigma_a + f_\vartheta \cdot \sigma_\vartheta \tag{2}$$

where f_α and f_β are respectively the volume fraction of ferrite and cementite; σ_α and σ_β the respective flow stress. Fine lamellar cementite is supposed to behave like whiskers with its flow stress σ_β in the order of the theoretical stress $\mu_\theta/10$, where μ_θ is the cementite shear modulus. This value is generally evaluated to be in the order of $\mu_\theta = 80$

GPa [2, 23]. The flow stress of a pure iron wire drawn to ε = 4 is about 1000 MPa [24]. The use of these numerical values in combination with (2) leads to an underestimation of the experimental flow stress. Indeed, this too simple description does not take into account that the plasticity of ferrite is scale dependent.

Nishida *et al.* [23] invoked a work hardening due to the increase of stored dislocations substructures. Nevertheless, this hypothesis seems highly impossible for spacing of a few nanometers [25]. Other modelling involves a "Hall-Petch" like mechanism with pile-up of ferrite dislocations against the cementite interface [5]. This hypothesis was already refute by Embury *et al.* [1] because the size of the lamellae is to small to allow such pile-up mechanisms.

Some authors [25, 27] proposed an alternative strengthening mechanism assuming that the flow stress needed to propagate a ferrite dislocation between two non penetrable cementite lamellae, is the stress that can bow a dislocation to a semi-circular critical shape, as shown in figure 10 (*a, b,e* and *f*). This stress can be explained as:

$$\tau = \frac{\mu_f b}{2\pi d_f} \ln\left(\frac{d_f}{b}\right) \tag{3}$$

where b is the Burgers vector, d_f the ferrite width and μ_f the ferrite shear modulus.

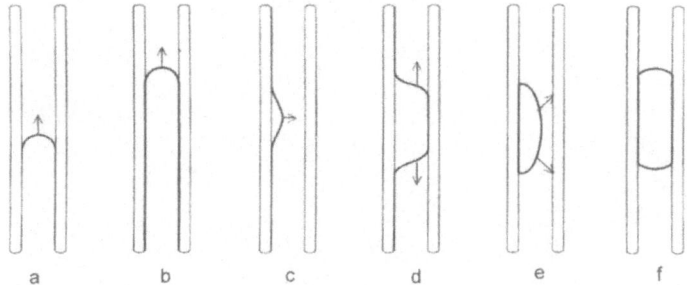

Figure 10. 3 different modes of nucleation and propagation of dislocations [28]: the hairpin-shaped dislocation from a conventional source (a and b) and the bulging sources of type-1 (c and d) and type-2 (e and f).

As mentioned by Janecek *et al.* [28], the stress in expression (3) tends to reach the theoretical stress when interlamellar spacing is equal to a few nanometer whereas the critical shear stress is often measured in nanofilms at lower values. They proposed the existence of two types of bulging sources along the ferrite/cementite interfaces: Type-1 corresponds to the nucleation of a half-loop from a dislocation at the interface (Fig. 10 b) that results in two z-shaped dislocations or superkinks (Fig. 10 d) that could propagate even in narrow spacing without increasing the dislocation length. Type-2 bulging source corresponds to the nucleation of a loop at the interface (Fig. 10 e and f) that propagates by bowing of its mobile segments like in figure 10 (b).

The main characteristic of these models is to consider cementite lamellae as non penetrable obstacle and to elude the plastic deformation of cementite even though recent works about cementite dissolution (§ 2.3.3) show that cementite has no more a continuous monocrystalline lamella shape and can be far from stoichiometry. The relevance of the relation between mechanical properties and nanoscale based on the cementite barriers is not so clear. This is enforced by the results of Hono et al. [19] partly shown in part §2.3.3 showing that cementite is partially dissolved for ε = 3.6 (Fig. 8 a) and totally dissolved for ε =5.1 (Fig. 8 b) whereas the tensile strength is still increasing respectively from 3840 to 5710 MPa. So, the role of cementite in the strengthening mechanism needs to be examined.

This fact is enforced by our own observations when the external surface of wires is mechanically polished and then observed with a SEM after tensile testing. The development of slip bands can be observed with a typical length that ranges between 500 nm and 2 μm, as shown in figure 11 for a ε = 3.5 drawing strain. These observations are valid both in the homogeneous zone and in the striction zone although they are more clearly evidenced enhanced in the latter. As, the length of the slip bands is between 50 and 100 times higher than the interlamellar spacing, it can be concluded that cementite doesn't act as a non penetrable barrier to the dislocation slip.

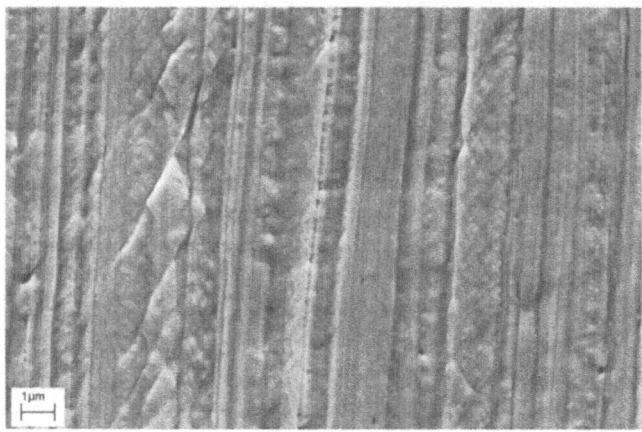

Figure 11. Slip bands at the micrometer scale on the cylindrical surface of a drawn wire (ε = 3.5), polished before tensile testing (vertical axis).

3.2 RESIDUAL STRESSES

The strain field heterogeneity inherited from the drawing process and from the biphasic nature of the microstructure leads to the existence of internal stresses. Before reporting the results available in the literature, we shall first give a brief definition of the different types of residual stresses appearing in heterogeneous materials.

3.2.1 *Definitions*
Residual stresses result from the heterogeneity of the strain field: three types of internal stresses are usually defined in the case of a single-phase material, depending on the characteristic scale of the heterogeneities [29]:

- The first order residual stresses, noted σ^I, homogeneous over many grains and self-equilibrate over an entire cross section of the body. They are mainly due to plastic or thermal strain heterogeneities related to forming process.

- The second order residual stresses, noted σ^{II}, homogeneous in a grain and due to heterogeneous deformation between adjacent grains or different elastoplastic behaviours in a multi-phase material.
- The third order stresses, noted σ^{III}, homogeneous on a few lattice distances and related to crystalline defects (dislocations, interstitial atoms, precipitates, coherency stresses between phases...).

In the case of small grain materials ($<10\mu m$), second and third order stresses tends to be non discernable and are called microstress, in contrast to the first order stresses called macrostress. Taking into account the two-phase nature of the material, residual stresses in one grain of a phase, e.g. α, is decomposed in a first order term σ_α^I and a microstress term σ_α^m. Finally, a first order stress σ^I is defined as a volume average of the σ_α^I and σ_β^I components in a volume containing a characteristic number of grains of each phase. This gives access to a continuous first order internal stress as in a single phase material, useful for mechanical calculations of structures. As a consequence, we can also define the pseudo-macrostress as $\sigma_\alpha^{MII} = \sigma_\alpha^I - \sigma^I$ in the α phase and also σ_β^{MII}, replacing the α subscript by the β subscript.

3.2.2 Results

The analysis of the few results available in the literature needs to pay attention to the techniques used and the type of stress measured. These results are difficult to compare between each other because they differ by the methods used, the type of stress and components measured and by the level of drawing deformation of the samples. We present, here, the results by type of stress: ferrite first order stress σ_α^I, pseudo-macrostress σ^{MII} and microstress σ^m.

Measurements of ferrite first order stress σ_α^I are made by X-Ray diffraction. The stress are determined locally in a shallow volume under the wire surface because of the small depth penetration of X-Ray in the material. Zolotorevsky et al. [30] measured residuals stresses on a 0.7 C wt% steel drawn to $\varepsilon = 3.1$. They find the following macrostresses: $\sigma_{\alpha,ax}^I = 511$ MPa, $\sigma_{\alpha,t}^I = 69$ MPa and $\sigma_{\alpha,rad}^I = -113$ MPa, the subscripts ax, t and rad referring respectively to the axial, the orthoradial (tangent) and the radial wire directions. These results differ from those of Van Acker et al. [30], especially concerning the axial components. They find a compressive stress state of the $\sigma_{\alpha,ax}^I$ component, -114 and –156 MPa for $\varepsilon = 1.96$ and 2.59 respectively, and the $\sigma_{\alpha,t}^I$ component, -56 and –48 MPa for the respective drawing strains.

Neutron diffraction allows to access to the mean value across the wire section of the σ_{ax}^{MII} stress of both ferrite and cementite phases because of the low attenuation of neutrons in steel. Van Acker et al. [30] measurements, focused on the as-drawn $\varepsilon = 1.96$ and $\varepsilon = 2.52$ states and $\varepsilon = 2.52$ heat treated to 450°C for 10h, are summarized in Table 1:

TABLE 1. Summary of pseudo-macroscopic stresses given in MPa [31]. The error is given between brackets.

Diffraction plane	$\varepsilon = 1.96$		$\varepsilon = 2.59$		$\varepsilon = 1.96$ heat treated	
α Fe (110)	-152	(30)	-122	(33)	50	(9)
α Fe (200)	-127	(36)	-114	(32)	9	(4)
Fe_3C (301)	2270	(310)	1730	(170)	-28	(14)
Fe_3C (131)	2050	(110)	1770	(50)	0	(18)

Tensile stresses about 2000 MPa are reported in cementite whereas compressive stresses are measures in ferrite. These values are slightly decreasing with the increasing drawing strain. Measurements on a $\varepsilon = 1.96$ wire that has been chemically etched to different diameters would indicate that there is no significant variation of the σ_{ax}^{MII} stress component over the cross section, i.e. it would be quite homogeneous. We can notice that the level of pseudo-macrostresses is reduced to low values by the heat treatment that induces a spheroidised pearlitic microstructure. Lukáš et al. [32] studied by neutron diffraction the evolution of the mean ferrite lattice parameter during in situ tensile tests on wires initially not drawn, i.e. patented, or drawn to $\varepsilon = 1.4$. Figure 12 (a) shows that the mean ferrite lattice parameter of a patented steel decreases after tensile tests to failure. The associated pseudo-macrostress σ_α^{MII} is in the order of -400 MPa and could probably be related to the difference between elastoplastic behaviour of the ferrite and cementite phases. On the contrary, the ferrite lattice parameter doesn't evolve after tensile test of an as-drawn wire. This doesn't mean that there is no more residual pseudo-macrostresses but that they don't evolve during the tensile test. They proposed to explain these differences between the patented and the as-drawn state by a change of deformation mechanisms. They invoked grain boundary sliding processes typical of superplastic or nanocrystalline materials.

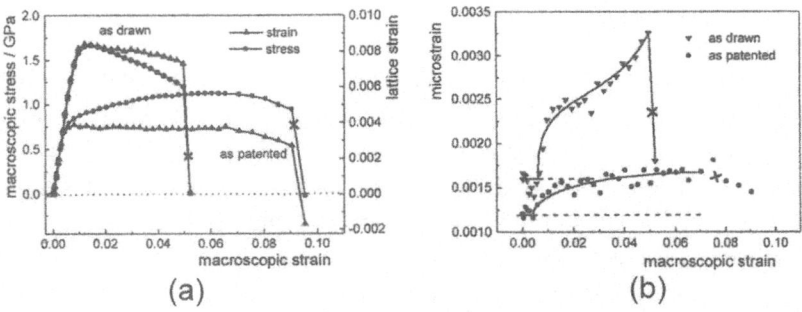

Figure 12. Neutron diffraction measurement of the evolution during a tensile tests of the mean ferrite lattice parameter (a) and the microstrains $<\varepsilon^2>^{1/2}$ on patented and drawn wires ($\varepsilon = 1.4$).

The analysis of the diffraction peaks broadening due to the microstresses σ_α^m allows to estimate the root-mean-square microstrains. Lukáš et al. [32] have measured it during their in situ tensile tests (Figure 12 b). It is shown that these strains saturates to 0.15% after tensile test of the patented wire whereas, in the as-drawn case, they start at this 0.15% value and reach a value higher than 0.3% during straining before being relaxed down to 0.15% by the failure. These values are in the same order than those given by Languillaume *et al.* [14] which show , in the case of $\varepsilon = 3.5$ drawn wire, that the related root-mean-square microstrains $<\varepsilon^2>^{1/2}$ are about 0.25 % [14]. This leads to a rough estimation of the mean microstress value $<\sigma_\alpha^{m\,2}>^{1/2}$ in the order of 520 MPa (taking a Young's modulus of 210 GPa for ferrite). Zolotorevsky *et al.* [30] evaluated the axial ferrite microstress $\sigma_{\alpha,ax}^m$ to 300 MPa ($\varepsilon = 3.1$).

This review of the available data about residual stresses of drawn pearlitic wires shows that they are just few and also difficult to compare because of the measurements methods used, the different types of residual stresses and also variety of drawing. Therefore, it is still difficult to draw reliable relationships between the residual stresses and the high mechanical strength.

4. Post Drawing Ageing

The study of the ageing properties of this material is a major concern in the industrial context: an evolution of the mechanical properties of heavily drawn pearlitic steel wires is observed at room temperature or when they are subjected to heat treatments such as the rubber vulcanization process.

Room temperature ageing can sometimes reduce so much the ductility that it can prevent the further torsion assembly process of cables. The use of heat treatments in the industrial context is known to enhance the mechanical strength or the ductility depending on the required properties. Figure 13 shows the evolutions of mechanical properties such as yield and tensile strength (a) or total elongation (b) as a function of temperature, after Languillaume *et al.* [33].

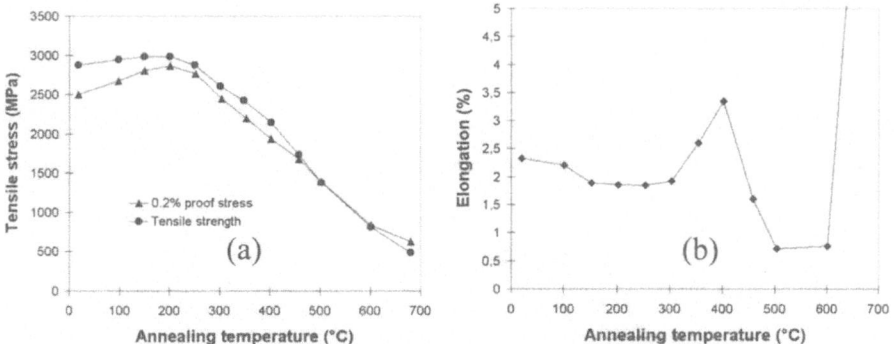

Figure 13. Room temperature evolution of the strength. (a) and the elongation (b) for drawn pearlitic wires ($\varepsilon = 3.5$) annealed at various temperature for 1h [33].

From these investigations we can distinguish 4 temperature ranges:

- The first one between room temperature and 250°C characterized by an increase of both the yield and tensile strength from a few hundreds of MPa and a small decrease of ductility,
- The second one between 200°C and 400°C corresponding to an important softening; the tensile strength decrease from about 1000 MPa while the elongation rises from less than 2% up to 3.5%,
- The third one from 400°C to 600°C in which both strength and ductility decrease,
- The fourth one above 600°C where the mechanical properties turns back to those of a non work hardened material.

As temperatures of industrial processes are mainly in the first temperature range, we will focus on this temperature range. The other temperature ranges will shortly be described as the related microstructure evolutions are quite well identified.

4.1 AGEING FROM ROOM TEMPERATURE TO 250°C

This temperature range has been studied for decades and, as a consequence, some early studies don't take into account the cementite dissolution induced by cold drawing. Nevertheless, it should be reminded that cementite dissolution leads to a highly out-of-equilibrium microstructure. As mentioned previously in section 2.3.3 the carbon concentration in ferrite is highly heterogeneous and can reach values up to a few atomic percentage which doesn't match standard ferrite. Large microstrains in ferrite have also been reported (§ 3.2.2). As a consequence, a return to equilibrium or at least an evolution to a more stable microstructure is expected. Our DSC investigations, shown in figure 14, proves that energy relaxation occurs during heating from room temperature to 600°C. It can be seen that there is an energy relaxation peak between 100 and 250-300°C, which correspond to an energy of the order of 4.5 J/g. This result confirms previous work by Watté et $al.$ [34]. It is then important to understand how the nanostructure and carbon supersaturation do evolve during further heat treatments in this temperature range.

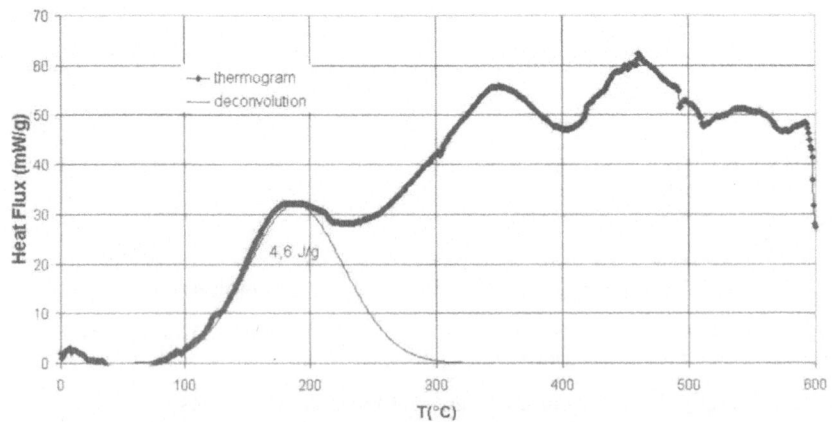

$Figure$ 14. DSC-spectra of eutectoid steel wire drawn to $\varepsilon = 3.74$ (heating rate = 40K/s).

Two different stages, named stage 1 and 2, are distinguished in the first temperature range, corresponding to different mechanisms.

4.1.1 Stage 1:Role of Interstitial Carbon in Ferrite

The first stage of strain ageing is almost classical and not specific of heavily drawn pearlitic steels: ferrite interstitial atoms (essentially carbon atoms) diffuse in the crystal and migrate to dislocations to pin them because carbon-dislocation binding energy in ferrite is greater than the interstitial activation energy [35]. The pinning of intralamellar ferrite dislocations by the so-called Cottrell atmospheres modifies some properties such as yield strength, electrical resistivity or anelastic behaviour. This pinning mechanism is linked to diffusion and therefore depends mainly on time and temperature.

Concerning the kinetic, the number of atoms arriving on single dislocation per time unit (t) is proportional to $t^{2/3}$ [36]. Many studies show evidence of this first stage by analyzing the evolution during isothermal ageing of a material's property $P(t)$ with the empirical Johnson-Mehl-Avrami equation:

$$\frac{P(t) - P(t=0)}{P(t=\infty) - P(t=0)} = 1 - \exp(-(kt)^n) \qquad (4)$$

$$\text{and } k = k_0 \exp\left(-\frac{E_a}{RT}\right) \qquad (5)$$

where E_a is the apparent activation energy of the thermally activated mechanism, R the universal gas constant and T the temperature. The main results are listed in table 2.

TABLE 2: Kinetic parameters related to the first stage of ageing in drawn pearlitic steel wires.

Réf.	Property P(t)	C % wt.	ε	n	E_a (kJ/mol)	T(°C)
[37]	Resistivity ρ	0.76	0.65	-	82.9	70-100
[38]	Yield strength $\sigma_{0.1}$	0.78	1.7	1/3	59 to 145.3	80
[39]	Yield strength $\sigma_{0.1}$				87.9	
	Relaxation strength Q^{-1}	0.82	1.96	1/3	82.8	60-100
	Resistivity ρ				89.5	

The time exponent n measured between room temperature and 100°C is equal to 1/3 and not 2/3 [38,39]. Lement et al. [40] showed that the value of 1/3 can be attributed to the diffusion of interstitial atoms to cells or walls of dislocations. Then, the first stage of strain is explained by the unidirectional diffusion of interstitial carbon atoms to the interface dislocations [39]. The activation energy measured is estimated between 80 and 90 kJ/mol [37,39] which corresponds to the carbon diffusion activation energy in ferrite estimated to 84 kJ/mol. [41]. As mentioned by Watté et al. [34], interstitial carbon diffusion in pearlitic ferrite is unlikely to yield the high heat fluxes shown in figure 14.

The influence of this first stage on the mechanical properties is not so significant. The reason of this is often attributed to the weak solubility of carbon in ferrite at room temperature ($<10^{-6}$ wt.%) which is too small to induce a sufficient pinning of dislocations. Nevertheless, as already mentioned in the previous parts, the real amount of carbon in ferrite is so important and the scale of diffusion so small that we can wonder whether this kind of kinetic analyses are pertinent or not [14].

4.1.2 *Stage 2*

This stage of ageing corresponding to a significant rise of the tensile strength is more specific to heavily deformed high carbon steels. The temperature range of this stage lies between 100 and 250°C. Previous studies mentioned in the last section (§ 4.3.1) provide similar analyses of this temperature range, listed in table 3.

The time exponent n is measured to ½. Based on the studies from Aaron *et al.* [42] who show that the dissolution of a planar precipitate leads to a kinetic exponent $n=1/2$, some authors interpret this stage as a thermally activated dissolution of cementite [37,38,42]. Carbon atoms issued from this dissolution would then pin the dislocations at the interfaces. The quantity of cementite dissolved and the dislocation density, both supposed to be high, would then account for the significance of this ageing stage.

TABLE 3: Kinetic parameters related to the stage 2 of ageing in drawn pearlitic steel wires.

Réf.	Property P(t)	C % wt.	ε	n	E_a (kJ/mol)	T(°C)
[37]	Resistivity ρ	0.76	0.65	-	117	150-250
[38]	Yield strength $\sigma_{0.1}$	0.78	1.7	1/2	97 to 151.4	100-120
	Yield strength $\sigma_{0.1}$				114.6	120-200
[39]	Relaxation strength Q^{-1}	0.82	1.96	1/2	-	130
	Resistivity ρ				125.2	130-200

Nevertheless, Araujo *et al.* [12] showed by mean of Mössbauer spectroscopy that the cementite dissolution occurs during the drawing process and not during further heat treatment (1h at 180°C). The second stage is then explained as an indirect consequence of cementite decomposition during drawing [12,39]. The latter would induce the formation of fine carbon enriched domains in the interfaces dislocation walls. As the interaction energy between carbon atoms and intralamellar dislocations is supposed to be higher than the one that links these atoms to the interfacial zones, these zones would dissolve by thermal activation. The final step would be the same as stage 1 but enhanced by the amount of carbon atoms.

Direct observations of the microstructure seems to infirm the proposed mechanisms of thermally dissolved cementite mentioned above.

Comparing by MET and X-ray diffraction the microstructure of a drawn wire ($\varepsilon = 3.5$) in the as drawn condition and heat treated for 1h at 200°C, Languillaume *et al.* [14] reported no evolution of the mean interlamellar spacing and of the ferrite microstrains $\langle\varepsilon^2\rangle^{1/2}$. They observed, between adjacent ferrite lamellae, an increase of the presence of small carbides precipitates that they identified as cementite with neutron diffraction.

Hong *et al.* [15] investigated wires ($\varepsilon = 4.22$) in the same conditions of heat treatment by means of TEM and APFIM. They detect no significant evolution of the microstructure and especially of the carbon distribution in the lamellae, as shown in figure 15. The observed increase of the yield strength is explained by the locking of mobile dislocations in ferrite. This explanation is close to the mechanism of first stage of ageing, but the cementite dissolution is now taken into account.

As a conclusion, no significant and visible evolution of the nanostructure that could account for the high heat fluxes measured has yet been identified in this temperature range.

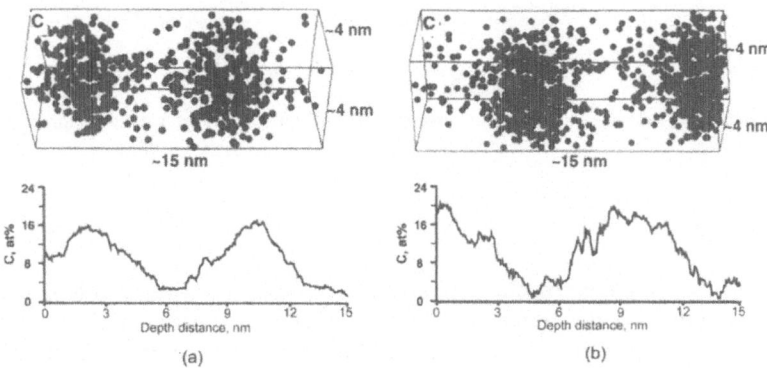

Figure 15. 3-DAP map of carbon in a pearlitic wire ($\varepsilon = 4.22$) in the as drawn condition (a) and annealed at 200°C for 1h (b) [15].

4.2 ANNEALING ABOVE 250°C

In this part, we briefly expose the main characteristics of the mechanisms occurring in the temperature ranges above 250°C.

Languillaume *et al.* [14] report, in the (200-400°C) temperature range, a significant decrease of ferrite microstrains $<\varepsilon^2>^{1/2}$ and an increase of the mean interlamellar spacing. They also observe an increase of the size of the cementite crystallites. This point is confirmed by APFIM studies [15]. The softening in this temperature range could be linked to the microstructure's progressive recovery that decreases the obstacle to mobile dislocations slip.

The authors mentioned above observed a recrystallization of ferrite and a spheroidization of cementite between 400 and 600°C, as shown in the figure 16. Languillaume *et al.* [33] suggest that damage events could be expected to explain the sharp decrease of ductility.

268

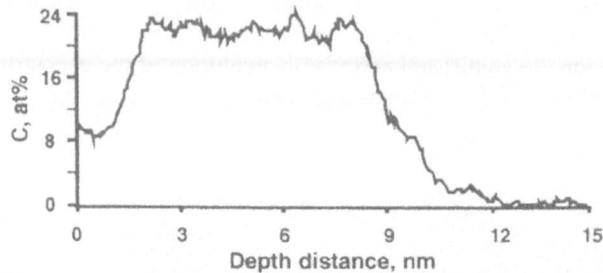

Figure 16. 3-DAP profile of carbon in a eutectoid steel wire (ε = 4.22) aged at 400°C for 1h [15].

5. Conclusion

The origin of the very high mechanical strength of heavily drawn eutectoid steel wires is usually attributed to the refinement of the lamellar microstructure down to a few nanometers. The cementite is supposed to limit the dislocation motion inside the lamellae. Nevertheless, the modelling of cementite acting as a non penetrable barrier to dislocations is not consistent with the observation of slip bands, which length is in the order of one micrometer and also with the cementite dissolution, at high drawing strains. The high level of residual stresses measured, at different scales, must certainly be taken into account for a better understanding, even though the links with the mechanical properties are not clearly evidenced. The cementite dissolution leads to an out-of-equilibrium nanostructure which effects on ageing evolution are observed even though the related mechanisms, compatible with the high heat fluxes encountered, are not yet characterized, especially in the 100-250°C temperature range. A better comprehension of the mechanisms involved will probably rise with the use of the recent tools of nanosciences such as molecular dynamic simulations.

6. Acknowledgements

The authors would like to thank E. Depraetere and P. Lescœurs of the Michelin Company providing materials and many fruitful discussions.

7. References

1. Embury, J.D. and Fisher, R.M. (1966) The structure and properties of drawn pearlite, *Acta Metallurgica* **14**, 147-159.
2. Langford, G. (1970) A study of the deformation of patented steel wire, *Metallurgical Transactions* **1**, 465-477.
3. Gonzales, B.M., Buono, V.C.T.L., Paula E Silva, E.M., Lima, T.M. and Andrade, M.S. (2000) Atomic force microscopy study of the behaviour of pearlite during drawing of a eutectoid steel, *Proceedings of the 70th annual convention of the Wire Association International*, 152-156.
4. Nam, W.J. and Bae, C.M.(1995) Void initiation and microstructural changes during wire drawing of pearlitic steels, *Materials Science & Engineering A* **203**, 278-285.

269

5.   Langford, G. (1977) Deformation of pearlite, *Metallurgical Transactions A* **8A**, 861-875.
6.   Danoix, F., Julien, D., Sauvage, X. and Copreaux, J. (1998) Direct evidence of cementite dissolution in drawn peerlitic steels observed by tomographic atom probe, *Materials Science & Engineering A* **250**, 8-13.
7.   Hosford, W.F. Jr. (1964) Microstructural changes during deformation of [011] fiber-textured Metals, *Transactions of the Metallurgical Society of AIME* **230**, 12-15.
8.   François, M. (1991) Détermination de contraintes résiduelles sur des fils d'acier eutectoïde de faible diamètre par diffraction des rayons X, Ph.D Thesis (ENSAM)
9.   Heizmann, J.J., Montesin, T. And Vadon, A. (1994) Circular texture in thin wires, *Materials Science Forum* **157-162**, 701-708.
10. Heizmann, J.J., Tidu, A., Bolle, B. and Peeters, L. (July 1999), Influence of the crystallographic texture on the torsional behavior of steel cord, *Wire Journal International,* 150-158.
11. Gridnev, V.N. and Gavrilyuk, V.G. (1982) Cementite decomposition in steel under plastic deformation (a review), *Phys. Metals* **4(3)**, 531-551.
12. Araujo, F.G.S., Gonzales, B.M., Cetlin, P.R., Coelho, Á.R.Z. and Mansur, R.A. (Feb. 1993) Cementite decomposition and the second stage static strain aging of pearlitic steel wires. *Wire Journal International,* 191-194.
13. Korznikov, A.V., Ivanisenko, Y.V., Laptionok, D.V., Safarov I.M., Pilyugin, V.P. and Valiev R.Z. (1994) Influence of severe plastic deformation on structure and phase composition of carbon steel, *Nanostructured Materials* **4**, 159-167.
14. Languillaume, J., Kapelski, G. and Baudelet, B. (1997) Cementite dissolution in heavily cold drawn pearlitic steel wires, *Acta Materialia* **45(3)**, 1201-1212.
15. Hong, M.H., Reynods, W.T., Tarui, T., and Hono, K. (1999) Atom probe and Transmission Electron Microscopy investigations of heavily drawn pearlitic steel wire, *Metallurgical and Materials Transactions A* **30**, 717-727.
16. Read, H.G., Reynolds, W.T., Hono, K. and Tarui, T. (1997) APFIM and TEM studies of drawn pearlite wire, *Scripta Materialia* **37(8)**, 1221-1230.
17. Sauvage, X., Copreaux, J., Danoix, F. and Blavette, D. (2000) Atomic-scale observation and modeling of cementite dissolution in heavily deformed pearlitic steel wires, *Philosophical Magazine A* **80(4)**, 781-796.
18. Nam, W.J., Bae, C.M., Oh, S.J. and Kwon, S. (2000) Effect of interlamellar spacing on cementite dissolution during wire drawing of pearlitic steel wire, *Scripta Materialia* **42**, 457.
19. Hono, K., Ohnuma, M., Murayama, M., Nishida, S., Yoshie, A. and Takahashi T. (2001) Cementite decomposition in heavily drawn pearlite steel wire, *Scripta Materialia* **44**, 977-983.
20. Fasika, E.J. And Wagenblast, H. (1967) Dilation of alpha iron by carbon, *Transactions of the Metallurgical society of AIME* **239**, 1818-1820.
21. Sauvage, X., Thilly, L. and Blavette, D. (200) Microstructure evolutions in pearlitic steels and Cu/Nb wires resulting from svere plastic deformation during drawing, in E. Aeby-Gautier, M. Clavel and F. Dunne (eds), *Euromech-Mecamat'2000,* EDP Sciences, Les Ulis, 4-27.
22. Gleiter, H. and Fichtner, M. (2002) Is the enhanced solubility in nanocomposites an electronic effect?, *Scripta Materialia* **46**, 497-500.
23. Nishida, S., Yoshie, A. and Imagumbai, M. (1998) Work hardening of hypereutectoid and eutectoid steels during drawing, *ISIJ International* **38(2)**, 177-186.
24. Langford, G. and Cohen, M. (1969) Strain hardening of iron by severe plastic deformation, *Transactions of the ASM* **62**, 623-638.
25. Gil Sevillano, J. (1991) Substructure and strengthening of heavily deformed single and two-phase metallic materials, *J. Phys III* **6**, 967-988.
26. Dollar, M., Bernstein, I.M. and Thomson A.W. (1988) Influence of deformation substructure in flow and fracture of fully pearlitic steel, *Acta Metallurgica* **36(2)**, 311-320.
27. Embury, J.D. and Hirth, J.P. (1994) On dislocation storage and the mechanical response of fine scale microstructure, *Acta Metall. Mater.* **42(6)**, 2051-2056.
28. Janecek, M., Louchet, F., Doisneau-Cottignies, B., Bréchet, Y. and Guelton, N. (2000) Specific dislocation multiplication mechanisms and mechanical properties in nanoscaled multillayers: the example of pearlite, *Philosophical Magazine A* **80(7)**, 1605-1619.
29. Withers, P.J. and Bhadeshia, H.K.D.H. (2001) Residual stress: Part 1 – measurement techniques (overview), *Materials Science & Technology* **17**, 355-365.
30. Zolotorevsky, N. Yu. and Krivonosova, N. Yu (1996) Effect of ferrite crystals' plastic anisotropy on residual stresses on cold-drawn steel wire, *Materials Science & Engineering* **A205**, 239-246.

31. Van Acker, K., Root, J., Van Houtte, P. and Aernoudt, E. (1996) Neutron diffraction measurement of the residual stress in the cementite and ferrite phases of cold-drawn steel wires, *Acta Materialia.* **44(10)**, 4039-4049.
32. Lukáš, P., Tomota, Y., Harjo, S., Neov, D., Strunz, P. and Mikula, P. (2001) *In situ* neutron diffraction study of drawn pearlitic steel wires upon tensile deformation, *Journal of neutron research* **9**, 415-421.
33. Languillaume, J., Kapelski, G. and Baudelet, B. (1997) Evolution of the tensile strength in heavily cold drawn and annealed pearlitic steel wires, *Materials Letters* **33**, 241-245.
34. Watté, P., Van Humbeeck, J., Aernoudt, E. and Lefever, I. (1996) Strain ageing in heavily drawn eutectoid steel wires, *Scripta Materialia* **34(1)**, 89-95.
35. Cottrell, A.H. and Bilby, B.A. (1949) Dislocation theory of yielding and strain ageing of iron, *Proc. Phys. Soc.* **62**, 49-62.
36. Friedel, J. (1956) *Les dislocations*, Éditions Gauthier-Villars.
37. Yamada, Y. (1976) Static strain aging of eutectoid carbon steel wires, *Transactions ISIJ* **16**, 417-426.
38. Kemp, I.P., Pollard, G. and Bramley, A.N. (1990) Static strain aging in high carbon steel wire, *Materials Science & Technology* **6**, 331-337.
39. Buono, V.T.L., Andrade, M.S. and Gonzales B.M. (1998) Kinetics of strain aging in drawn pearlitic steels, *Metallurgical and Material Transactions A* **29A**, 1415-1423.
40. Lement, B.S. and Cohen, M. (1956) A dislocation attraction model for the first stage of tempering, *Acta Metallurgica* **4**, 469-476.
41. Fast, J.D. (1961) Frottement intérieur des métaux, *Métaux, Corrosion, Industries* , 311-320
42. Aaron, H.B. and Kotler, G.R. (1971) Second phase dissolution, *Metallurgical Transactions A* **2**, 393-408.

CLUSTER STRUCTURE OF THE AMORPHOUS STATE AND (NANO)CRYSTALLIZATION OF RAPIDLY QUENCHED IRON AND COBALT BASED SYSTEMS

PETER ŠVEC, KATARINA KRIŠTIAKOVÁ, and MARIAN DEANKO

Institute of Physics, Slovak Academy of SciencesDúbravská cesta 9, 842 28 Bratislava, Slovakia

Abstract. Micromechanisms and energetics of transitions from metastable to more stable state were investigated in complex metastable disordered systems prepared by rapid quenching from the melt from the viewpoint of spatially (structurally) correlated distribution of transformation rates of individual microprocesses controlling the transition process. Using a novel, model-independent method for determination of continuous distributions of process rates it was possible to obtain information on distributions of true activation energies of these microprocesses. Detailed analysis of subdistributions of microprocesses active at each stage a of transition yielded also the information on temperature dependence of the activation energies.

   We have analyzed different nanocrystal-forming iron and cobalt based systems with the focus on the origin of the clustered amorphous state. New information was obtained with respect to the original local ordering of atoms in the amorphous state and its influence on the formation of nanostructures. Additional information was extracted which allowed comparison of the processes in the early stages of nanocrystallization with those activated at the end of this transformation. The origin of distributions of microprocess rates or, alternatively, of activation energies, i. e. dynamically heterogeneous behaviour, is discussed and correlated with the expected clustered structure of the amorphous state, i. e. spatial heterogeneities having distinct ordering within the disordered matrix.

1. Introduction

A typical feature of modern materials with complex structure such as metallic glasses or glassy polymers is their frozen-in metastable state. Non-crystalline structures, in particular amorphous solid structures in metastable state, inherently possess high degree of freedom for local topological and chemical arrangements, which persists in amorphous state. Such structures often function as promising precursors for controlling nano-scale structure through the control of the local atomic environments during transformation to a more stable state.

T. Tsakalakos, et al. (eds.) pgs. 271 - 294
Nanostructures: Synthesis, Functional Properties and Applications;
© *2003 Kluwer Academic Publishers.*

The area of phase transitions related to non-crystalline systems, especially concerning their formation and further transformations, is not fully understood and is still highly controversial, in spite of several decades of extensive studies. This area is connected with the effect of quenched random disorder – transformation from undercooled liquid or melt to amorphous solid. It includes phenomena related to the structure and dynamics of complex disordered metastable systems of different origin. The mechanism of formation of these systems is similar for conventional metallic glasses, bulk metallic glasses, supercooled glass-forming molecular liquids, polymeric systems and miscellaneous amorphous systems such as biomolecules, proteins, colloids, etc. As the systems are cooled down below their melting or freezing point, the dynamics slows down drastically due to cooling and even more drastically in the proximity of the glass transition. Supercooled system falls out of equilibrium and exhibits behavior different from that of a homogeneous liquid, attaining a metastable state characterized by heterogeneous behavior, i. e. by varying dynamics in different regions lying close to each other. The principal and still open question is whether such heterogeneous dynamics can be correlated with the frozen-in spatial heterogeneity and disorder [1].

The relationship between the thermodynamic state of a complex system (e. g. metallic glass, etc.) and the role of microstructural processes controlling transformation rates [2] is not unambiguously clear, especially with respect to the metastability of the initial amorphous state. It is therefore very desirable to provide a model-free distribution of transformation rates and eventually of activation energies from experimental data, which can then be more reliably coupled to theoretical calculations and objectively interpreted. The analysis of experimental results of transformation rates combined with detailed knowledge of microstructure on scales as close to the atomic-level resolution as possible becomes therefore the most important area for further theoretical considerations related to the nature of the metastable state and the corresponding concepts of the reaction rate theory [3].

The focus of this work is on phenomena related to the local atomic-scale structure and transformation behavior of amorphous metastable alloy systems prepared by rapid quenching from the melt. Both subjects, short-range ordering, i. e. cluster structure, in amorphous state and its manifestations through dynamic behavior of these systems, belong to the key factors for understanding the nature of amorphous metastable state and the mechanisms of formation of (nano)crystalline phases.

2. Heterogeneities in Complex Disordered Metastable Systems

The concept of heterogeneity in complex disordered metastable systems [1] is quite broad, involving structures and phenomena in noncrystalline systems of different origin such as high molecular weight liquids, polymers, undercooled liquids or melts and metallic glasses. The interest in metallic glasses lies in the fact, that they represent systems with strongly interacting constituents and provide possibility to investigate phase transitions from amorphous or undercooled states in the vicinity of the glass transitions.

Amorphous alloys exhibit inherent properties of complex systems. Local heterogeneity is present as a consequence of glassy structure which contains high degree of disorder due to rapid quenching of the precursor liquid (or melt) and, as a paradox, atomic disorder exists along with a certain specific degree of local ordering (clusters, polyatomic aggregates, etc.) [4]. The metastable and disordered state of these systems imposes questions about the genesis, type, number, structure and stability of clusters consisting of several, usually metal and metalloid, atoms and about the structure of the matrix surrounding such clusters. Another important point is that the complexity of such metastable system consisting of several subsystems must have influence on the dynamics of its transition to a more stable state.

In spite of the progress in theoretical approaches, computing facilities and advances in numerical methods, the first-principle models for amorphous metallic systems are rare, mostly due to the inapplicability of a periodic potential to disordered structures; in most cases the existence and structure of any low-energy structural unit or cluster has to be postulated *a-priori*. Further complications for modeling are caused by the presence of $3d$ atoms, by electron localization and by combination of metals and metalloids and their respective bonding in majority of real metallic glassy systems. Recently, however, it was shown that a similar effect related to the special type of bonding is present also in certain metal-metal systems [5].

From the viewpoint of dynamics, transformations in metallic glasses are typically heterogeneous. They are nonexponential on time and non-Arrhenius on temperature scales, especially in a wider interval of these variables. This is a common phenomenon for complex metastable disordered systems, observed in true dynamic relaxations (after a suitable excitation), in structural relaxations with preservation of the amorphous state as well as in phase transitions from amorphous to a more stable state. It is of principal interest to verify the correlation between such dynamically heterogeneous behavior and microstructural (spatial) heterogeneities in amorphous structure manifested through specific micromechanisms controlling transition thermodynamics [6].

In simple processes (corresponding to relaxations or phase transitions in simple systems) a single microprocess rate λ given in classical Arrhenius form as $\lambda = \lambda_0 \exp(-E_a/RT)$ with the corresponding constant activation energy E_a and constant preexponential (or frequency) factor λ_0 may be expected. If spatial regions with distinguishable dynamic properties exist in the investigated, more complex systems, then a more exact treatment accounting for the nonexponential behavior is by using several discrete microscopic rates λ_i with specific discrete weights or a continuous distribution (probability density function) of process rates $pdf(\lambda)$, which is an alternative to the use of the empirical KWW equation [7]. Such approach, however, even in the case of existence of local heterogeneities, is valid strictly only for parallel processes running independently of each other, i. e. where each degree of freedom or spatial region "relaxes" independently with its own λ. Under certain, not very limiting assumptions, however, it may be applied to processes on the path to equilibrium through series of serial (sequential) processes as well.

An alternative "homogeneous" approach may in principle be adopted assuming intrinsically nonexponential behavior of a microscopic response function $f(t, \lambda)$ which would replace the respective exponential term $\exp[-(t.\lambda)]$ from the previous cases. Then pdf(λ) would correspond to a generalized distribution function; however, simultaneous determination of both pdf(λ) and the functional dependence of f(t,λ) would be problematic without prior assumptions. A certain estimate of the situation may be obtained by introducing a degree of dynamic heterogeneity for complex systems [8] ranging from zero for the homogeneous to unity for fully heterogeneous behavior [9].

The distribution of processes reflected through their rate distribution pdf(λ) should have direct correlation with microscopic sources of such distributions - with intrinsic microstructural heterogeneities.

3. Spatial Heterogeneity – Cluster Structure of Metastable Amorphous Systems

While the origin of spatial heterogeneity of the dynamics in supercooled molecular liquids or polymer systems above the glass transition may be unclear, in the glassy state heterogeneities are more plausible. The glassy system may be expected to behave more solid-like. The same can hold for supercooled liquids below a certain critical temperature, yet still above the glass transition [10]. This picture is suggestive of a convenient potential energy landscape, which in solid-like systems immediately provides a distribution of initial thermodynamic states [11, 12] and transitions are thermally activated from this energy landscape with different potential energy minima corresponding to different local packings of the constituents. However, the same effect may also be due to fluctuations of local density or entropy [13] related to local atomic or orientational ordering. Another appealing approach [14] correlates the distribution of dynamics with spatial distribution of frustration-limited domains.

Of all questions concerning structural heterogeneities in complex metastable amorphous systems such as their size, lifetime and dynamics, the question about their spatial and thermodynamic origin is the most important one. The answer must be related to the chemical identity of the glass formers and especially to their microstructure on atomic scales viewed from the point of existing interatomic or structurally correlated cluster bonding [15]. This effect may be different for supercooled molecular liquids and for metallic glasses due to the nature of constituent atoms. In the former case the size of "molecules" makes the notion of larger clusters or cooperatively rearranging regions [16] immediately plausible; the latter case is more difficult to evidence, especially due to a much smaller size of the expected locally ordered region [17]. Such reasoning may be also convenient in detailed treatment of the degree of fragility of glass formers and the origin of the glass-forming ability [18, 19]. It has to be noted that the vast majority of experiments treats heterogeneity in dynamics; a correlation with structural heterogeneity has to be proved or disproved individually for each case by suitable local probes, e. g. small angle neutron scattering in supercooled molecular liquids [20] and in bulk metallic glasses [21].

At least two very plausible causes for the genesis of spatial heterogeneities may be envisaged. One of the causes is common to all mentioned amorphous systems and their supercooled liquid precursors. It resides in the nature of the formation of amorphous rapidly quenched systems due to the thermodynamics of formation of these systems, either by deep quench or by rapid solidification of strongly interacting multi-component melts, leading to nonequilibrium, yet long living states, enhanced further by the state from which the quenching was realized: a dynamically stable structure of complex liquids or melts, possibly heterogeneous on atomic scales [22], is prevented from equilibration, leading to quenched-in heterogeneities with the structure of the melt and with spatial relaxation times increased by tens of orders of magnitude as the temperature decreases from above the melting point down towards the glass transition.

It is worthwhile to mention that in reality the preparation of a fully amorphous metallic glass is usually a non-trivial two-stage process where the system passes through the melting temperature to the liquid state twice, inducing effects that ought to be of primary importance for the formation of the glass structure and its local ordering. In the first stage the components in correct amounts with respect to the desired final composition of the glassy system are heated and melted together to produce a polycrystalline master alloy after conventional cooling. Subsequently the alloy is melted again and rapidly quenched with the quenching rates ranging up to several millions of Kelvins per second in order to obtain amorphous structure. The procedure is schematically depicted in Fig. 1. As was shown in [23], melt history and physical casting conditions have definite influence on the degree of clustering; especially the history of the master alloy in molten state during the first preparation stage is crucial in determining the importance of this effect. Depending on the temperature-time regime in molten state, different metastable equilibria are established between the locally more ordered clusters originating from melting of heterogeneous constituents (often badly miscible in solid state) and the surrounding melt [24]. When quenching such melt, these microstructures are quenched into the resulting amorphous, yet locally heterogeneous structure on atomic scales with the corresponding potential relief, denoted as "potential energy landscape" [25]. Therefore, the process or rapid quenching itself can create a distribution of initial thermodynamic states in the amorphous system and thus a distribution of activation energies for the same type of "relaxation" process on the path to a more stable state. Such qualitative picture, where the potential energy landscape is projected into one dimension [26], is shown on Fig. 1. The initial states correspond to a number of inherently structured local potential minima separated by saddle points. Transformation behavior of the system depends on the relationship between the depth of these minima and the temperature. For higher temperatures (well above the glass transition) the system has sufficient kinetic energy to sample the full energy landscape and it was shown that in such case it exhibits temperature-independent activation energy similar to that of free diffusion [27]. With decreasing temperature the activation energy increases relatively to the available kinetic energy and the behavior becomes progressively landscape-influenced and landscape-dominated. It is worthwhile to mention that the typical values of potential energies are of the order of few eV below zero and the different minima are spread over a few percents of this value, while the differences between deep local (metastable amorphous states) and absolute energetic minima (crystalline state) generally do not exceed values of the order of 0.1eV.

The initial metastable state of metallic glass reflects the atomic structure and the heterogeneities of the melt precursor, preferential local ordering between different types of atoms and possibly also the dynamics of uneven distribution of temperatures in the course of rapid quenching. The initial state of even the simplest amorphous system should thus be distributed over a range of possible local orderings and energetic states around a certain mean system-specific value, which ought to be reproducible, reflecting

276

also the reproducibility of the rapid quenching process. However, the formation of more complex, relatively stable structural units – clusters – in the melt has been predicted and experimentally established [28, 29]. Such clusters are expected to be relatively stable and the transformation process to a more stable (crystalline) state should include their transition to crystalline state without a total rearrangement of atoms constituting the cluster by a process combining suitable collective motion across the interface with amorphous matrix and slight rearrangement of atoms to equilibrium positions.

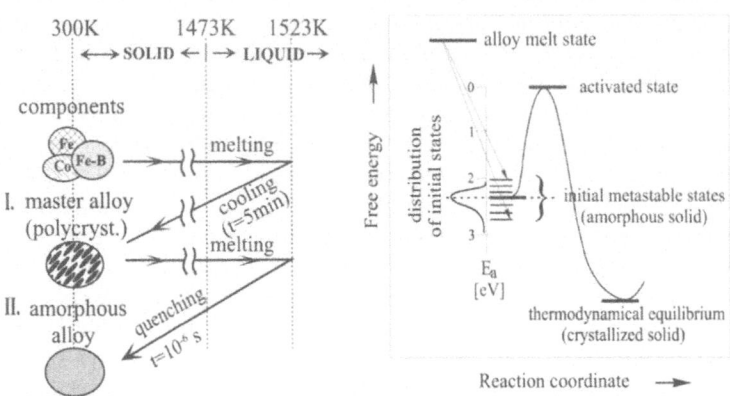

Figure 1. Physical scheme of the two-stage process of preparation of amorphous metallic systems (left) and qualitative scheme (right) showing the origin of the distribution of thermodynamic processes controlling transition from amorphous to a more stable state.

Another natural and thermodynamically easily acceptable source of spatial heterogeneities related to dynamic heterogeneities is the dynamically stable distribution of embryos [30] in any undercooled liquid, melt or amorphous solid on the path towards a more stable state – prior to eventual crystallization, yet without sufficient mobility to realize the transition via e. g. homogeneous nucleation. Such a mechanism leads to formation of additional SRO, possibly also competing with the one induced previously by the melt dynamics. Further quantification of cluster formation in such systems is based on statistical mechanics formalism [31], including both energetic and geometrical criteria and the concept of connectivity, allowing to define separately physical clusters and chemical clusters with respect to the connectivity lifetime and the stability of bonding between intra-cluster particles [32]. Especially in liquids with strong heteroatomic bonding where regular solution and Becker's model of nucleation cannot be applied, this kind of clustering with sufficiently long lifetimes and concentration fluctuations can act as additional, and thermodynamically important, sources of spatial heterogeneities [33] with pronounced influence on the transition dynamics [34, 35, 36].

4. Energetics of Transformations from (Clustered) Amorphous State

Intense research effort has been devoted over the last decades to the transitions from the metastable state in amorphous alloys using various models for the interpretation of the experimental observations. Among these are e. g. adapted models developed originally for transformations in polycrystalline solids or rather complicated single-atom diffusion models, which suppose concentration gradients being established also in the amorphous phase with proceeding transformation [37]. However, the assumption about the concentration gradient was not verified in the majority of cases [38]. Many other experiments indicate complications in using classical transformation theory [39], especially with respect to the anomalous temperature dependence of the Arrhenius factor in diffusion measurements.

In amorphous metastable metallic systems most transformation processes to more stable states are thermally activated and are usually approached through the notion of the activated state. Dynamics of such transitions can be studied by the analysis of isothermal time evolution of transformed volume fraction x(t) using either a simple Avrami equation $x(t)=1-\exp[-(t\lambda)^n]$, where n is the Avrami parameter, for (rare) simple cases with single process rate λ or using an Avrami-like analogue of the KWW equation for complex systems

$$x(t,T) = 1 - \int_0^\infty pdf(\lambda)\exp[-(t.\lambda)^n]d\lambda. \tag{1}$$

In the case of amorphous metallic systems the transformation to a less metastable (crystalline) state is usually a two-stage process. The first crystallization stage leads to the formation of one single, usually metastable crystalline phase, which occupies about half of the entire sample volume; its subsequent transformation takes place in the crystalline state [40]. The morphology of the crystalline grains is typically three-dimensional, most frequently spherulitic. The grains are mutually not connected, indicating low level of impingement and thus also a special micromechanism controlling the process of their formation, growth and their maximum size, which is a phenomenon of importance especially in the formation of nanocrystalline structures [41]. Such a transformation may be plausibly and simply described by one single type of reaction without invoking the necessity for a concentration gradient and with the same particle morphology (with constant morphology parameter). This implies the same transformation micromechanism throughout the whole transformation stage and does not exclude different initial potentials, which are reflected in different activation energies distributed over the entire assembly and lead to a distribution of reaction rates.

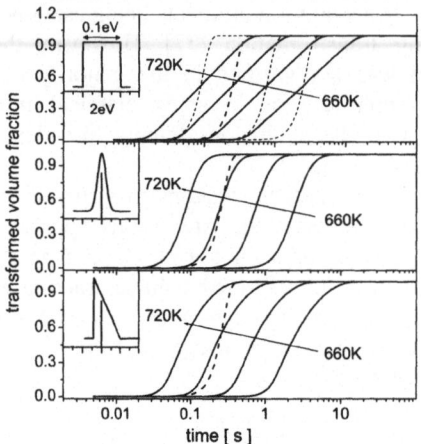

Figure 2. Time evolution of crystalline volume fractions x(t) for transformations controlled by three types of activation energy distributions (in insets) compared with x(t) for single-rate transformation at 700K (solid and dashed lines, respectively) with the same mean E_a=2.00eV, λ_0=10^{16}s^{-1} and the Avrami parameter n=4.

The importance of even a very narrow distribution of activation energies E_a for the "relaxation" or transformation dynamics and especially for its heterogeneous behavior is shown in the following consideration. Let us assume a distribution of activation energies pdf(E) normalized to unity around a mean arbitrary value of 2 eV, commonly observed for transformations in metallic glasses. In relaxation processes a box-like distribution of ±0.1eV is plausible. Fully disordered systems may be expected to have a Gaussian distribution; let its width in half-maximum be 0.07eV. The last case with triangular distribution decreasing from ~1.97 to 2.05eV may be considered a reasonable approximation for nucleation and growth processes. In all three cases shown as insets in Fig. 2 the value of E_a~2eV ± 5%, far below the accuracy of most experimental methods. Assuming transformation kinetics given by Eq. (1) for complex processes with rate distributions pdf(λ)=pdf(E_a)/RT, as derived in [42], the transformed volume fractions x(t) in the temperature range from 650 to 720K are shown in Fig. 2. The box distribution yields x(t) with linear part in log t ranging over several decades on the time scale, much wider than that for simple Avrami relation with only a single mean transformation rate λ (dashed line for 700K) and very similar to the usually observed structure relaxation behavior. The Gaussian and triangular distributions lead to narrower, yet still quite broad x(t) with specific smooth behavior both at the onset and at the end of the process. In all cases the presence of the distributions leads to pronounced deviations from the simple single-rate transformation behavior [43] where both the shape and the (small) width of the distributions play crucial role. The results demonstrate the importance of including distribution of initial metastable states and their temperature dependencies into considerations about stability, kinetics and transition thermodynamics of amorphous metastable systems.

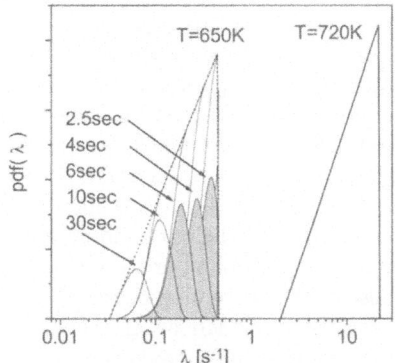

Figure 3. Rate distributions for triangular activation energy distributions from Figure 2 and the active process windows at different times of transformation, computed using Eq. (3).

However, even such model manifestations of dynamic heterogeneity, cannot be analyzed, either qualitatively or quantitatively, by the classical kinetic analysis methods which provide single values of kinetic parameters only. It is, however, possible to introduce the notion of processes annealed out till time t. Their distribution, $P_{out}(t, T)$ would be given by the subintegral function in Eq. (1) expressed in log λ. Then we can define processes active at this time as a difference of processes annealed out between the time t and $t+dt$ (in logarithm of time) as

$$P_{act} = \partial P_{out}/\partial t \sim (\lambda t)^n \exp[-(\lambda t)^n] \, pdf(\lambda), \tag{2}$$

which represents a product of steeply increasing power function and steeply decreasing power exponential function representing "observation" window for the rate distribution $pdf(\lambda)$ with a maximum at $t \sim 1/\lambda$ and a width of about half order of magnitude in log λ. Thus at any time t only a relatively narrow portion of the entire rate distribution can be active and "visible" (Fig. 3). The experimentally observed transformation rate at this time,

$$dx/dt = \int_0^\infty P_{act}(t,T)d\lambda, \tag{3}$$

however, provides only integral information about the active process rates. Processes with $\lambda < t^{-1}$ are "sleeping" processes, to be activated as the observation window is shifted with increasing time. This phenomenon is closely related also to the information content of the transformation curves and capability of predicting transformation behavior [44]: unless the process obeys a simple law similar to the Avrami equation in its simplest form, the knowledge of the entire process rate (or activation energy) distribution is necessary. Otherwise, predictability ranges only to times within one order of magnitude of the longest observation time or the slowest process rate. In potential energy lansdscape terminology this is equivalent to the necessity of probing the entire clustered, energy-distributed landscape for thermodynamic behavior of all cluster types and sizes.

The initially amorphous clustered structure (and, consequently, with initial energy distribution) can lead to enhanced temperature dependence of transformation rates compared to that given by the Arrhenius-type relation valid for simple transitions involving the interaction or motion of single atoms. It was shown that even a random potential ensemble of Arrhenius-type processes, as may be expected in amorphous matter, with suitable energy distribution leads to a non-Arrhenius behavior of process rates [45]. A more complex atomic environment may induce an effective or apparent dependence of the activation energy on temperature or may lead to a non-Arrhenius temperature dependence of process rates [46]. Another reason for this may lie in the fact that atomic jumps may influence the surrounding matrix and the thermodynamic states of other atoms therein or by enabling processes with thermodynamic parameters slightly different from those of the "parent" process [47]. Activation of one process in an array of processes distributed in activation energies may change the resulting structure and the density of the remaining processes. All these factors lead either to a real or apparent dependence of the activation energy on temperature or to a commonly observed apparent dependence of the preexponential factor λ_0 on the activation energy.

In real transformation processes a true temperature dependence of activation energy of process rates can be expected. While in simple processes involving single atoms only (diffusion) the effective activation energy is a constant, in processes controlled by viscous flow where cooperative motion of larger number of (clustered) atoms takes place, the activation energy decreases with temperature; in nucleation-controlled processes the activation energy increases with temperature [30]. Furthermore, activation energy of crystallization from amorphous or undercooled states $E(T)$ ought to be considered as a weighted sum of activation energies of nucleation and of growth, $E(T)=n_n.E_n(T)+n_g.E_g(T)$, where $n_n+n_g=n$ and n_n and n_g represent the character of nucleation and dimensionality of grain growth, respectively [48].

In a selected interval of annealing temperatures the general $E(T)$ dependence may be approximated [42, 43] by a straight line as $E(T)\sim E_0+E'(T-T_0)$ where $E'=dE(T)/dT$ is the temperature coefficient of activation energy and $E_0=E(T_0)$ at a mean annealing temperature T_0. Then the correct form for microprocess rates can be written as $\lambda(T) = \lambda_{00} \exp[-E(T)/RT]$, where λ_{00} is now the true constant preexponential factor. It can be shown easily that $\log \lambda_{00}= \log \lambda_0 - E'/RT_0$ and the "classical" Arrhenius $E_a=E_0-E'T_0$. Evidently, the term E' is closely related to the changes in (configurational) entropy; this field of interpretation is under progress in combination with the evolution of new methods for characterization of disordered structure on atomic-scale resolution and will not be treated herein. It is, however, obvious that important new information may be obtained from the sign and magnitude of $E'(E)$ and from the distributions of activation energies $pdf(E(T))$ controlling transformation, which can be determined from $pdf(\lambda)$. Simple processes controlled by single-atom diffusion may be expected to possess $E'=0$. Processes influenced in major part by viscous flow of atoms or groups of atoms exhibit effective activation energy with $E'<0$, while nucleation-like processes, where the energy of formation of critical-size embryos has to be considered, would have $E'>0$. However, except in specific cases or under specific (usually limiting) assumptions, it was, until recently, difficult to obtain access to this kind of information.

5. Cluster Structure Visualization

The possibilities for investigation of the cluster structure depend on the effect that the changes of the cluster structure have on the overall atomic structure, properties and stability of the amorphous matter. An ideal case is when the amorphous matter is composed of phases with different cluster type and selective heat treatment is capable of bringing about changes in one type of cluster only. This is the case of systems where undercooled liquid and amorphous solid coexist, such as Ni-Zr-Al based alloys [49, 50, 51, 52]. In these systems is was possible to enhance the stability of the amorphous structure against crystallization by a suitable sequential thermal treatment in undercooled liquid state. Another, more difficult, possibility in the case when a single type of cluster is expected (e. g. in metal-metalloid amorphous systems), is either to induce changes in the cluster structure in the course of its formation (e. g. while still in molten state [23]) or to investigate the changes in the amorphous matter and interpret the stability and evolution of the cluster structure upon transition from amorphous state. In such special cases positron annihilation lifetime spectroscopy was used, yielding not only single values of positron lifetimes but also positron annihilation rate distributions, together with special annealing sequences for enhanced visualization of the evolution of the cluster structure. These experiments were performed alongside with the investigation of time and temperature dependencies of different physical properties, e. g. electrical resistivity, dilatation, magnetostriction.

A progressive and more accurate continuous distribution approach (K-S approach) to the analysis of transition dynamics represented by Eq. (1) was developed [26]. This model-free approach makes use of isothermal time dependences of suitable physical property P(t), selected to reflect the transformation process in complex structures. The set of transformation isotherms for all investigated temperatures (Eq. 1) is considered as ill-posed problem solved as a system of Fredholm integral equations of the first kind. In ideal case the solution yields directly the distribution of transformation rates, pdf(λ). The effect of measuring instrumentation needed to obtain the the time dependence of the transformed volume fraction x(t) and experimental noise were eliminated by the use of deconvolution procedure and inverse Laplace transformation. From rate distributions, using moment analysis and by introducing the notion of subdistributions of transformation processes active at selected times the distributions pdf(E) of the true activation energy E are obtained. The dependencies of the preexponential factor λ_{00} and temperature coefficient of activation energy E' are obtained as functions of E_0. Simple recalculation yields the values of pdf(E(T)) at arbitrary T and thus allows to predict kinetic behavior even in a complicated thermal treatment sequence. The information about the effects of temperature dependence of activation energies, E(T), not accessible till now, was obtained from isothermal measurements.

The results allow to consider each transformation stage as controlled by a single type of microprocess with identical micromechanism (i. e. identical preexponential factors and constant morphology parameter n) yet coming from different initial energetic states, as witnessed by the obtained distributions of activation energies. As the processes take place in a viscous amorphous medium, their temperature dependence is more complex than Arrhenius relation. The results provide the sign and magnitude of the deviation of process rates from the Arrhenius behavior and allow to justify the choice of the controlling mechanism (nucleation, growth, diffusion, viscous flow, etc.) [53] without any need for prior postulation of the presence of these mechanisms and based only on fundamental thermodynamic presumptions and the nature of the amorphous state.

As noted briefly in the initial sections, simple single-rate exponential dependencies or classical Avrami-type equations are valid for ideal transformations characterized by a single process only. Any distribution of processes with different activation energies, however simple, would lead to the time dependences of the property P(t), used to monitor the transformation, which would deviate from isokinetic behavior. The above-described approach easily allows the analysis of such curves using advantageously the notion of moments and subdistributions of processes. It also provides a powerful tool for prediction and extrapolation of dynamic behavior in conditions which are experimentally difficult to access, e. g. long-time low-temperature stability [53] or high temperature behavior due to high transformation rates and thus short reaction times. It is also possible to predict easily the evolution of dynamics in step-annealing regime, which is an important asset for the basic physical research as well as for technological purposes.

The formulation of Eq. 1 is based on the assumption that the i-th process, defined by its own λ_i, is predetermined to take place on several suitable larger-sized regions, ending in formation of final crystallinity produced by this type of process, $x_i(t\rightarrow\infty)$ where $x(t\rightarrow\infty)=\Sigma_i x_i(t\rightarrow\infty)=1$ and $x(t)=\Sigma_i x_i(t)$. Thus the transformation of amorphous structure to a more stable state takes place as a weighted sum of all possible subtransformations. The complexity and heterogeneity of the metastable structure on different size scales (short and medium-range ordering of amorphous structure) make such an approach plausible.

The K-S approach represents also a contribution to the principal philosophy about the cluster structure of the amorphous state, which is in accordance with recent microstructural observations, especially the notion of the transformation of entire, relatively stable, amorphous clusters into a metastable crystalline phase. This transformation mechanism may be thought to take place by preservation of the local ordering of the cluster structure in the first crystalline phase formed in the transformation process via small motion of entire clusters and slight atomic rearrangement taking place upon crossing the interface. Careful use of the approach provides the possibility of obtaining information on early stages of the transformation (nucleation processes), their temperature dependence and their influence on the transformation mechanism. From this viewpoint the concept of primary crystallization controlled by long-range diffusion, resulting from the use of a classical single-atom approach to transformation rates, in clustered amorphous media becomes quite inapplicable. Short-range ordering and cluster structure override the contribution from the motion of single atoms over (long) distance due to a small portion of "free" unclustered atoms in the matter.

6. Cluster Structure and Cluster Stability – Positron Annihilation in Fe-Co-B

The origin, structure and stability of the clusters was investigated in detail in [23] and especially in [54] on amorphous $Fe_{64}Co_{21}B_{15}$ by positron annihilation lifetime spectroscopy and electrical resistivity. The results presented in [23] show a definite correlation between the expected melt precursor structure induced by special thermal history of the melt and the presence of specific type of clusters in the amorphous matrix. The changes in the cluster structure, which take place at high melt temperatures, are preserved in the corresponding amorphous phase by rapid quenching. These changes are visible especially in amorphous remains, which are left over after the first-stage crystallization and which contain mostly clustered atoms. Differences in local ordering have been found for samples from master alloys with different thermal histories, pointing to relatively high stability of the clusters formed in the melt and preserved in the amorphous matter by rapid quenching.

Positrons in disordered matter image vacant spaces where atoms are missing or where the density of atoms (or electrons) is reduced; by investigation of positron capture sites it was possible to obtain information about the corresponding local atomic structure surrounding these sites and to follow their stability against structural transformation. The results indicated the presence of at least two rather similar positron capture sites within the atomic arrangement of the rapidly quenched Fe-Co-B amorphous structure throughout the whole devitrification process. By a special treatment of the data it was possible to determine the positron lifetime distributions for these two sites.

A local short range ordering model between metal and metalloid atoms in form of "bridging complexes", presented in [17], was used in [54] for the interpretation of the results obtained from electron-positron annihilation data. The disordered amorphous structure has been approximated for convenience by its immediate crystalline successor, bcc-(FeCo) structure. Such model allows for two positron capture sites with identical atomic surroundings but differing in the surrounding electron density due to the presence of different kinds of interatomic bonding in the bridging complex. Both positron capture sites are located in the centers of the faces of the bcc cell (in fcc position), the difference, however, lies in the electrochemical neighborhood of the sites. The first site is located in the intermediate space between the complexes. It is surrounded by metallic bonds without proximity of the metal-metalloid pairs and its size and occurrence is affected mainly by the presence of additional metal atoms. The second site is located close to the center of the complex and in the proximity of the metal-metalloid bonds in the bridging complex, where the electron density is reduced due to the strength of the semicovalent character of the (FeCo)-B bonds inside the bridging complex. This site is far from equivalent to the former site and should remain unaffected by crystallization from the amorphous state, because the complexes are structurally similar to the arrangements in the Fe_3B borides. Similar effects were recently observed also in some Al-V based intermetallic phases, where the strength of bonds leads to the formation of a gap in the electron density around the Fermi level [5].

The overall behavior of the lifetime distributions (Fig. 4) for both sites indicates high stability of the complexes (clusters) against crystallization. Even the second-stage crystallization, which leads to complete devitrification, does not lead to the disappearance of this kind of short-range order which takes place only after recrystallization in fully crystalline state into bcc-(FeCo) and (FeCo) diborides, where a complete atomic rearrangement takes place. The presence of such stable ordered microstructures implies that all structure evolution micromechanisms active during transition from amorphous to the crystalline state are controlled by cooperative motion and slight rearrangements of entire complexes (or atoms within these complexes) into equilibrium positions. Only complete recrystallization at high temperatures (~1000K) leads to their disruption, as shown in the right part of Fig. 4. by the disappearance of the corresponding lifetime distribution. Since the high quenching rates achieved during sample preparation make the idea of cluster formation during quenching improbable, the high thermal stability of these clusters and measurements of magnetic properties of similar melts [55] suggest that this kind of local ordering is present already in the melt and is preserved by rapid quenching.

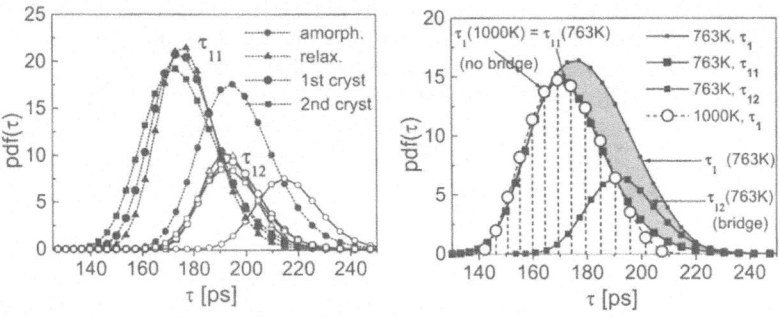

Figure 4. Left: Decomposition of the main peak of the positron annihilation lifetime distributions of $Fe_{64}Co_{21}B_{15}$, pdf(τ_1) into two components, pdf(τ_{11}) and pdf(τ_{12}) (full lines), which represent two kinds of capture sites for four types of structure states – as-quenched amorphous, relaxed at 573K, after the first crystallization at 693K and after the second crystallization at 763K. Right: Existence and disappearance of the bridging complex as seen by pdf(τ_{12}) for $Fe_{64}Co_{21}B_{15}$. The distribution pdf(τ_1) at 1000K is identical with pdf(τ_{11}) at 763K (after the second crystallization).

7. K-S Approach and Step Annealing

One of the experimental procedures convenient for evaluation of special transformation effects is step annealing where the sample is annealed for the same time intervals sequentially at two different temperatures where the investigated transformation takes place. Using the isothermal electrical resistivity dependencies of $Fe_{64}Co_{21}B_{15}$ metallic glass upon crystallization measured in a wide range of temperatures from 650 to 720 K, the distributions of activation energies, constant preexponential factor and temperature dependence of activation energies have been determined [26]. Full resistivity isotherms for two selected temperatures 650 and 680 K) are shown by dashed lines in Fig. 5a. Using this information, fresh glassy sample has been isothermally annealed for 10 minutes in a special fast and highly stable isothermal furnace at the temperature of 650 K. After removing the sample the temperature was increased to 680 K and the sample has again been annealed for another 10 minutes; the procedure was repeated again, exposing the sample to prolonged annealing (600 min) at the initial temperature of 650 K. Resistivity measurements performed in all three annealing steps are shown in Fig. 5a as thick lines, the transitions between steps indicated by arrows. Using the pdf(E(T))s for the two temperatures, portions of activation energies annealed out in the three steps can be computed (shaded areas in Fig. 5b) and subtracted from the original distributions, allowing thus to compute the transformation curves x(t) and time dependences of relative resistivity. The results represented by open circles in Fig. 5a, coincide extremely well with the measured resistivity curves. Thick dotted lines show the measured step anneals and the simulated isotherms shifted in time by the value corresponding to the respective crystallized fraction on the original annealing curves. The prediction potential for this type of treatment is shown by excellent agreement between the thick lines and open circles and by agreement between the dashed and thick-dotted lines. The procedure can be used for optimization of long-term low-temperature annealing experiments, e. g. replacing conventional annealing by step annealing to obtain the same crystalline state of the sample, however, in much shorter annealing time. As can be seen, the third step annealing even in its initial stage already corresponds to the annealing time of ~200 minutes for simple 650K treatment.

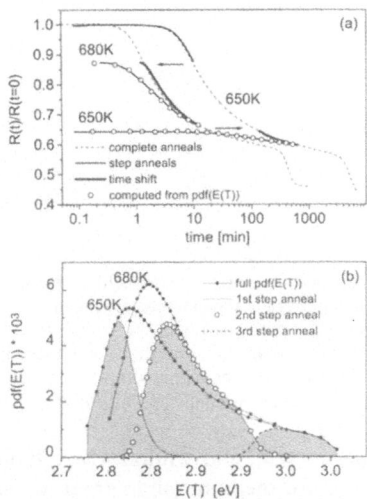

Figure 5. (a) Electrical resistivity isotherms for Fe64Co21B15 measured in the step-annealing experiment. Dashed lines - complete transformation curves, thick lines - step anneals, thick dotted lines – step anneals shifted in time, open circles - values computed from pdf(E(T)), see text. (b) Complete distributions of activation energies at 650 and 680 K and portions of pdf(E)s annealed out in the consecutive annealing steps (shaded areas).

8. Nanocrystallization of Fe-based and Fe-Co-based Systems

Using the newly developed K-S approach together with novel information about the atomic structure and local ordering on atomic scales, the dynamics of transitions from amorphous state in $Fe_{73.5}Cu_1Nb_3Si_{13.5}B_9$ Finemet [56] and $Fe_{87}Zr_7B_6$ and $Fe_{86}Zr_7B_6Cu_1$ Nanoperm [57] systems was revisited. The aim was to determine and to quantify the relationship between the observed phenomena and the microprocesses controlling them. Different stages of nanocrystallization were identified by means of the distributions of activation energies and of their temperature dependences, especially the effect of Cu precipitation in the initial stages of transformation and the influence of the cluster structure of amorphous state on the energetics of the observed stages.

The processes of Cu clustering, formation of nanograins from Fe-Si clusters and their extremely slow subsequent growth [56] are well identified on electrical resistivity isotherms and on the pdf(E) and E'(E) dependences (Fig. 6); the clustering of Cu by a nucleation-like process exhibits high positive values of E' and temperature dependent shifts of the corresponding part of pdf(E). Nanocrystalization is characterized by lower E' decreasing to zero for the process of nanograin growth which terminates by inclusions of single Fe atoms from the surrounding amorphous remains corresponding to very slight grain size increase from ~12 to ~15 nm in a slow process observed only for prolonged annealing times.

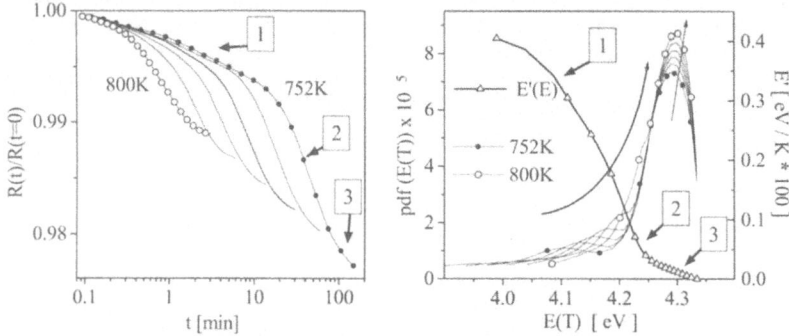

Figure 6. Experimental isothermal time dependences of electrical resistivity of $Fe_{73.5}Cu_1Nb_3Si_{13.5}B_9$ in the course of nanocrystallization together with the temperature dependence of activation energy E' and with the activation energy distributions pdf(E(T)) correlated with the processes and phenomena observed during the transformation. The numbers in boxes indicate the process of Cu-clustering prior to nanocrystallization, the nucleation-and-growth process of formation of nanograins from Fe-Si clusters and their slow size increase vias single-atom motion, respectively. Full arrows indicate shifts of pdf(E) with increasing temperature.

The structure of Nanoperms with and without Cu, although amorphous, was found [58] to exhibit a certain degree of medium-range ordering containing mainly Fe atoms. Negative values of E' observed in this case [57] indicate low degree of rearrangement needed for such kind of clusters to form critical nuclei, with E'<0, being dominated by viscous-flow type temperature dependence which describes well the incorporation of coordinated units – clusters – into the growing nanocrystals. The abundance of such clusters decreases with proceeding transformation and the process including regions with low Zr content terminates by involving clusters with lower number of Fe atoms (down to single atom process) and, consequently, E' attaining zero. The transition then proceeds by involvement of more stable Fe-Zr rich regions, yet with very low rates and E' becoming again negative. In Cu-containing Nanoperm this effect is overridden by Cu clustering [59] in later stages of nanocrystallization and increase E' from zero to positive values for this process which needs higher degree of rearrangement with behavior characteristic for nucleation.

The analysis included also a new progressive system based on FeCoZrB [60], denoted as Hitperm [61] where again two cluster types with short range ordering in the amorphous state dissimilar to bcc arrangement were identified and the involvement of each of them in the nanocrystallization process was identified by the behavior of E'>0 throughout the nanocrystallization process, with transition through slow, single-atom processes in the intermediate stages with E' reaching zero after using up all available clusters of the first type (rich in Fe-Co with low Zr content). Activation of the second type of Fe-Co clusters rich in Zr leads to increase of E' again, although with low probability density.

In this way it was possible to obtain quite a broad and more general systematic notion of the spatial and dynamic heterogeneities in modern nanocrystal-forming systems [57]. The microprocesses controlling the transition were shown to involve clusters with different local ordering and different sizes. After depletion of the amorphous matrix of the previously active clusters the process proceeds by involvement of clusters from the remaining matrix, although with significantly decreased rates, reflecting their higher thermodynamic stability. The diversity of the cluster sizes and types is the reason for the complicated activation energy distributions and the temperature dependences of the activation energies observed in dynamic experiments.

9. Formation of Nanocrystalline Structure by Simulation of Nucleation and Growth

As mentioned above, crystallization and especially nanocrystallization from amorphous state exhibits certain general features. One of them is the presence of two crystallization reactions where the first one involves mainly one type of clusters and covers ~50% of the entire volume filled by three-dimensional spherulitic or polyhedral grains embedded in amorphous matrix and formed, presumably, by incorporation of entire clusters and without need for their breakup or long-range diffusion. Stopping of grain growth of two neighboring grains may again be considered as being due to exhaustion of clusters available for further growth in the vicinity of the grains, which exhibit only very low degree of impingement.

A simple simulation of disordered solid with random homogeneous nucleation and spherical growth (n=4) was performed in a cube with the side of 1000 arb. units, 1 arb. unit corresponding to 1 nanometer and in time scale of steps corresponding to seconds; grain growth was ceased for grains coming into contact with another grain. Thermodynamic quantities used in simulation were similar to those calculated for the case of Fe-Co-B [26]. The transformation curve was found to behave as $x(t) \sim t^4$ for the initial stages (Fig. 7a). For later stages, however, a sharp deviation from the theoretical Avrami behavior $x(t)=1-\exp[-(t.\lambda)^4]$ was observed (Fig. 7b). The transformed volume in simulated transformation attained ~50 vol.% only after very long times and the values of the apparent Avrami parameter decreased from ideal n=4 sharply to almost zero (similar as in experimental observations). These effects are due to the fact that although the number of all grains present increases linearly with time, the number of grains still growing decreases due to the impingement effect. Thus the grain size distributions (Fig. 7c) for early crystallization stages differ radically from those for later stages, where

288

practically no larger grains are growing any more and a high number of very small grains is observed. The reason for this phenomenon lies in the fact that only small-sized volumes for nucleation are available which, however, make up ~50% of the entire remaining "amorphous" volume. This is again in agreement with experimental observations, e. g. in Finemet after nanocrystallization (Fig. 7d). While the main contribution to the overall crystallinity, which is typically about half of the entire volume, comes from 10-12 nm grains, the presence of large number of very small grains can be observed.

The model may easily be adapted to include softer impingement conditions such as stopping the growth even prior to actual impingement effect. Also the effects of heterogeneous nucleation with specific content of heterogeneous nucleation centers on the final grain size can easily be investigated [41]. Results of such transformation process where both homogeneous and heterogeneous (instantaneous) nucleation processes take place show that the final or mean (characteristic) grain sizes become substantially affected only by the presence of heterogeneity content above ~1% of the total number of nucleation events, which is a useful estimate of practical procedure for grain size refinement.

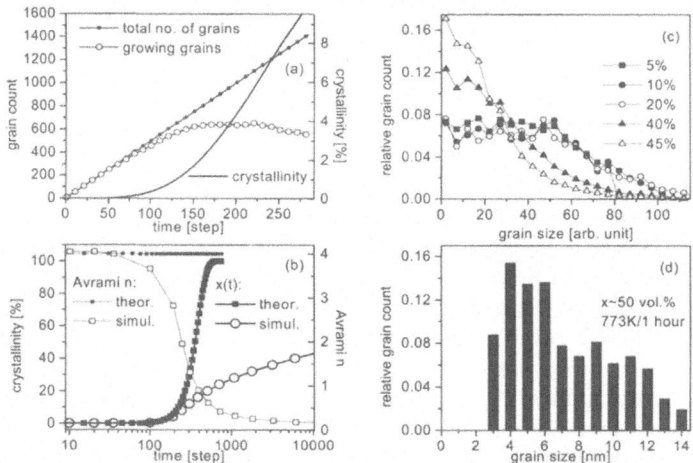

Figure 7. Time dependence of total and growing grain count and crystallinity for early stages of simulated transformation process (a); comparison of simulated and theoretical time evolution of crystallinity together with the behavior of the Avrami parameter n (b); grain size distribution for different crystalline volume content (c) and experimental grain size distribution in Finemet alloy after nanocrystallization (d).

10. Cluster Structure and Phase Selection

Addition of Ni into the Finemet-based system of the $Fe_{73.5-x}Ni_xCu_1Nb_3Si_{13.5}B_9$ type (x=0, 10, 20, 30, 40, [62]) leads to a specific behaviour both in amorphous state as well as during crystallization. Alloying Fe with Ni brings about the possibility of specific transformation of the amorphous structure with martensitic-like character which was formed and preserved during the solidification path in the undercooled melt prior to amorphisation by melt quenching [63]; the structure retains the high-temperature fcc-like (γ-FeNi) short-range atomic arrangement within the Fe-(Ni)-Cu-Si clusters discussed for the classical Finemet [64, 60]. During isothermal treatment at sufficiently high temperatures (730-800K) although still in amorphous state the structure of Finemet with 10 at.% of Ni exhibits a tendency to transform to a more bcc-like (α-FeNi) short-range ordering in a sluggish athermal reaction, well monitored by resistivity decrease which is not influenced by the annealing temperature; the isotherms at temperatures from 770 up to 800K are practically identical (Fig. 8). Amorphous state is fully retained for times up to ~20 minutes, as checked by both X-ray and TEM analyses. Longer annealing in this temperature range leads to formation of bcc-FeNi and $(FeNi)_3Si$ nanocrystals.

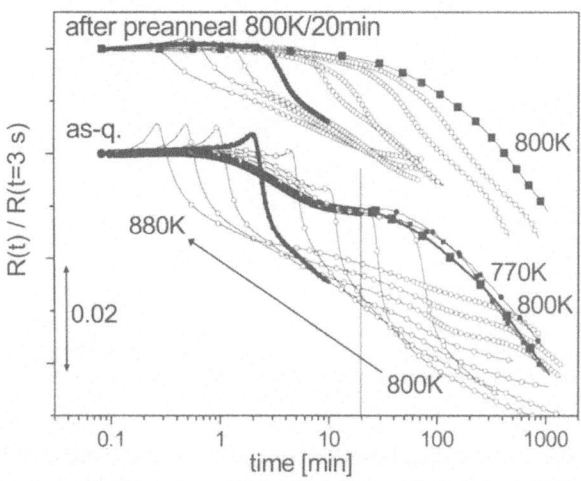

Figure 8. Time dependence of relative electrical resistivity of Finemet with 10 at.% Ni annealed from as-quenched state (bottom set of curves) and after preannealing at 800K for 20 min (top set of curves) in the temperature range from 800 to 880K in steps of 10K. Isotherm at 770K (close to that at 800K) is shown to illustrate sluggish temperature dependence of the curves in this temperature range. Thick lines indicate annealing for 10 min at 850K for both cases.

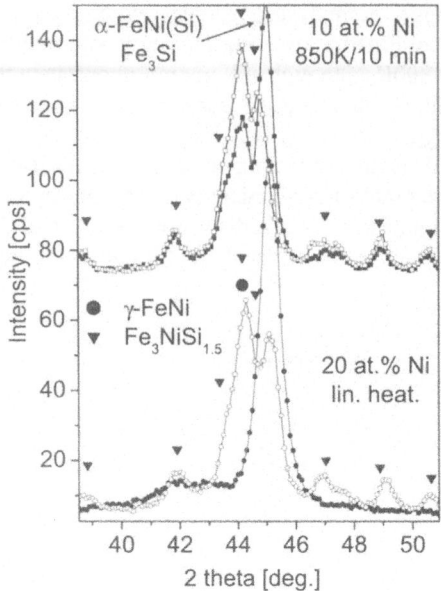

Figure 9. X-ray patterns for Finemet with 10 at.% Ni annealed isothermally at 850K for 10 min from as-quenched amorphous state (open symbols) and after preannealing at 800K for 20 min (full symbols) and for Finemet with 20 at.% Ni linearly heated with fast (40K/min, open symbols) and slow (2K/min, full symbols) heating rates.

At temperatures above 800K, however, the phases formed are mainly bct-$Fe_3NiSi_{1.5}$, especially in the first stages of annealing indicated by the slight resistivity increase, and bcc-FeNi. The content of bcc-FeNi and especially $(FeNi)_3Si$ is much lower. This behavior may be reversed by preannealing at 800K for 20 minutes to induce the fcc-to-bcc like change of short range ordering in the amorphous state.

Subsequent step annealing of preannealed samples at temperatures equal or less than 800K leads to resistivity dependencies which follow the same pattern and shape as samples without preannealing. The thick line for 800K in the bottom right part of Fig. 8 starting at ~20 minutes is the resistivity time dependence for sample preannealed at 800K for 20 minutes (full squares), cooled and annealed further at the same temperature. The plot is shifted to the respective position on the 800K isotherm to show very close shape resemblance. This behavior is similar to the case of step annealing for Fe-Co-B presented above and is an indication that the transformation processes are controlled by the same mechanism. However, such behavior cannot be obtained for subsequent annealing of preannealed samples at temperatures above 800K. The isotherms for preannealed samples annealed at the same temperatures are plotted in the top part of Fig. 8. It can be seen that the shape of the isotherms for preannealed samples differs drastically from those annealed from as-quenched state.

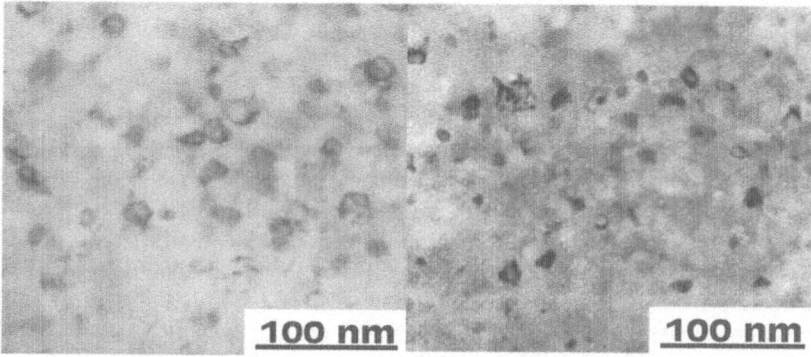

Figure 10. TEM view of the Finemet with 10 at.% Ni after sequential annealing at 800K for 20 min and 850K for 10 minutes (left) and after simple annealing of as-quenched amorphous sample at 850K for 10 min (right).

Subsequent step annealing of preannealed samples at temperatures above 800K (under identical conditions as in the previous case) leads to nano-crystallization of bcc-FeNi and $Fe_3NiSi_{1.5}$, however, the content of $Fe_3NiSi_{1.5}$ is substantially lower than for the case of samples annealed from as-quenched state (Fig. 9). The final crystalline phases present after very prolonged annealing are bcc-FeNi and $Fe_3NiSi_{1.5}$ with very low traces of $(FeNi)_3Si$ and fcc-FeNi. Prolonged annealing leads to increase of the intensities of the bcc-FeNi lines; the peaks corresponding to $Fe_3NiSi_{1.5}$ do not increase in magnitude. Quantitative X-ray and TEM analyses for the case of samples from Fig. 9 annealed from as-quenched state show that the mean size of grains is ~18 nm and the volume content of crystallinity is ~30%. Preannealed samples have typical grain sizes larger, ~21 nm, however, they exhibit lower crystallinity content of ~25%. The grain morphology for the preannealed samples is much more regular than that for the samples annealed from as-quenched state (Fig. 10), where twinning and surface kinks may be observed [65], probably due to fcc-bcc martensitic transformation within the nanocrystalline grains.

Applying the KS-analysis to the set of preannealed isotherms yields a behavior quite similar to that of the classical Finemet (without Ni) except for a slightly lower and positive value of E' in the initial stages. This may be connected with increased solubility of Cu in Ni at higher temperatures, leading to suppression of the Cu-clustering effect on the initial nucleation stages and also to the increase of mean grain size; however, the nucleation-like character of the nanocrystallization is quite evident. Straightforward annealing yields E'<0. Assuming that the sluggish martensitic-like transformation of fcc-like SRO in the as-quenched amorphous phase into bcc-like amorphous SRO within the Fe-(Ni)-Cu-Si clusters takes place at temperatures below 800 K, where the thermodynamic conditions for nanocrystallization are yet not fulfilled, the transformation mechanism of preannealed samples may be easily understood. The first crystalline phase formed, as stated e. g. in [40], would preserve the local ordering of the original cluster structure, which is of the bcc type in this case. The same holds for samples annealed from as-quenched state, which however still possesses the original fcc-like SRO, thus $bct-Fe_3NiSi_{1.5}$ and small amount of fcc-FeNi are formed in the initial stages with lower need for nucleation-like rearrangement and thus E'<0. Further annealing at temperatures between 800 and 900K leads to at least partial transformation of these phases into bcc-FeNi. The observed irregularity of nanograins may be ascribed to this mechanism [66, 67, 68].

The temperature of the proposed change of SRO may be expected to decrease with increasing Ni content [68]. An effect related to the one demonstrated in Fig. 8 can be seen for the Finemet-based alloy with 20 at. % of Ni on the temperature dependencies of electrical resistivity R(T) during linear heating with very different heating rates (Fig. 11). Slow heating rates may be expected to correspond to preannealing at lower temperatures and lead to a transformation behavior evidently different from that observed for high heating rates. This is shown also by the X-ray analysis (Fig. 9). Slow heating rates provide sufficient time for the change of fcc-to-bcc like SRO and the nanocrystalline phase formed till 880 K contains practically only bcc-FeNi with no traces of other phases. Fast heating rates (40 K/min) up to the same temperature lead to formation of substantial amount of bct-$Fe_3NiSi_{1.5}$ and much less of the bcc phase. The same phenomenon is observed for isothermal annealing at 880 K – samples annealed from as-quenched state exhibit increased formation of bct-$Fe_3NiSi_{1.5}$ with prolonged annealing, while in preannealed samples the content of bct-$Fe_3NiSi_{1.5}$ does not increase and only further formation the bcc-phase takes place. It is worthwhile to note that the effective Ni to Fe content in the two cases investigated here, 10 and 20 at.% of Ni, is about 14 and 28 at.% of Ni, respectively.

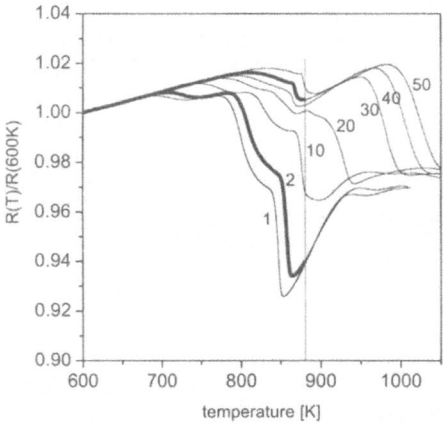

Figure 11. Temperature dependence of relative electrical resistivity of the Finemet with 20 at.% Ni using different heating rates (in K/min). Thick lines indicate treatment of the samples with rates of 2 and 40 K/min to 880K.

11. Conclusions

The initial thermodynamic states, energetics, structures and micromechanisms active in transitions of complex metastable metallic systems to a more stable state were investigated from the viewpoint of spatially correlated distribution of rates of microprocesses controlling the transformation processes. Using the novel activation energy distribution approach we have obtained results, which point to the conclusion that transitions from clustered amorphous state prepared by rapid quenching take place by transformation of entire clusters into metastable crystalline phases with preservation of the local atomic ordering within the cluster. Attention was focused on the origin of the amorphous structure, on the formation and stability of the clusters and on formation and phase selection of subsequently created (nano)crystalline structures. The use of the potential of the novel method for interpretation, prediction and control of transformation processes was demonstrated on specific cases of modern amorphous and nanocrystal-forming systems.

While concentrating on rapidly quenched amorphous metallic alloy systems, the considerations presented, concerning the role of cluster structure, its origin and stability, ought to be applicable in several instances also to structure and transition rate theory of disordered systems in general. The knowledge about the processes controlling thermal and time stability and the rate of transformation from metastable (amorphous) to a more stable (crystalline) state yields vital information for detailed determination of the thermodynamic nature of microprocesses controlling the transformation. Its correlation with the original local atomic structure and with the character of microscopic properties provides understanding and possibilities to master the control of macroscopic properties in metastable disordered condensed matter.

12. Acknowledgement

Support of the the the Grant Agency for Science of Slovakia and NATO Science for Peace (Grants No. 2/2038/22 and SfP-973649) is acknowledged.

13. References

1. M. D. Ediger, Annu. Rev. Phys. Chem. **51** (2000) 99.
2. P. Hänggi, P. Talkner, M. Borovec, Rev. Mod. Phys. **62** (1990) 251.
3. R. Bohmer, Curr Opin. Solid St. & Mat. Sci. **3** (1998) 378.
4. T. Egami, Mat. Res. Bull. **13** (1978) 557.
5. M. Krajci, J. Hafner, J. Phys.: Condens. Matter **14** (2002) 1865.
6. J. Qian, A. Heuer, Eur. Phys. J. B **18** (2000) 501.
7. J. C. Phillips, Rep. Prog. Phys. **59** (1996) 1133.
8. R. Böhmer, R. V. Chamberlin, G. Diezemann, B. Geil, A. Heuer, G. Hinze, S. C. Kuebler, R. Richert, B. Schiener, H. Sillescu, H. W. Spiess, U. Tracht and M. Wilhelm, J. Non-Cryst. Sol. **235-237** (1998) 1.
9. H. Wendt, R. Richert, Phys. Rev. E **61** (2000) 1722.
10. A. P. Sokolov, J. Non-Cryst. Sol. **235-237** (1998) 190.
11. M. Goldstein, J. Chem. Phys. **51** (1969) 3728.
12. F. H. Stillinger, Science **267** (1995) 1935.
13. F. Sciortino, W. Kob, P. Tartaglia, Phys. Rev. Lett. **83** (1999) 3214.
14. D. Kivelson, S. A. Kivelson, X. L. Zhao, Z. Nussinov, G. Tarjus, Physica A **219** (1995) 27.
15. M. Oguni, J. Non-Cryst. Solids **210** (1997) 171.
16. Y. Hiwatari, T. Muranaka, J. Non-Cryst. Sol. **235-237** (1998) 19.
17. P. Duhaj, P. Hanic, phys. stat. sol. (a) **76** (1983) 476.
18. J. C. Dyre, J. Non-Cryst. Sol. **235-237** (1998) 142.
19. C. A. Angell, Science **267** (1995) 1924.
20. R. Leheny, D. Menon, S. R. Nagel, K. Volin, D. L. Price, P. Thiyagarjan, J. Chem. Phys. **105** (1996) 7783.
21. J. F. Löffler, W. L. Johnson, Mat. Sci. Eng. A **304-306** (2001) 670.
22. V. Sidorov, P. Popel, M. Calvo-Dahlborg, U. Dahlborg, V. Manov, Mat. Sci. Eng. A **304-306** (2001) 480.

294

23. K. Krištiaková, P. Švec, J. Krištiak, P. Duhaj, O. Šauša, Mat. Sci. Eng. A **226-228** (1997) 321.
24. P. S. Popel, O. A. Chikova, V. M. Matveev, High Temp. Mater. Process **4** (1995) 219.
25. F. H. Stillinger, Science **267** (1995) 1935.
26. K. Krištiaková, P. Švec, Phys. Rev. **B64** (2001) 184202.
27. P. G. Debedenetti, F. Stillinger, Nature **410** (2001) 259.
28. B. Predel, Physica B 103 (1981) 113.
29. A. I. Zaitsev, N. E. Shelkova, Z. Metallkd. **91** (2000) 992.
30. J. W. Christian, *The Theory of Transformations in Metals and Alloys*, Pt. I, Pergamon Press, Oxford, 1975.
31. T. L. Hill, J. Chem. Phys. **23** (1955) 617.
32. L. A. Pugnaloni, F. Vericat, J. Chem. Phys. 116 (2002) 1097.
33. E. Cini, B. Vinet, P. J. Desré, Philos. Mag. A **80** (2000) 955.
34. P. Duhaj, P. Švec, Key Eng. Mat. **40-41** (1990) 69.
35. P. Duhaj, P. Švec, Mat. Sci. Eng. A **226-228** (1997) 245.
36. W. Swiatkowski, J. Non-Cryst. Sol. **262** (2000) 162.
37. D. Crespo, T. Pradell, M. T. Clavaguera-Mora, N. Clavaguera, Phys. Rev. B **55** (1997) 3435.
38. K. Hono, D. H. Ping, M. Ohnuma, H. Onodera, Acta Mater. **47** (1999) 997.
39. H. Kronmüller, W. Frank, A. Hörner, Mat. Sci. Eng. A **133** (1991) 410.
40. P. Duhaj, P. Švec, T. Zemčík, Materials Lett. **9** (1990) 235.
41. M. Deanko, P. Švec, Proc. APCOM 2002 ed. J. Mudroň, Military Acad, Liptovský Mikuláš, 2002 p. 27.
42. K. Krištiaková, P. Švec, Mat. Sci. Forum **360-362** (2001) 467 / J. Metastab. Nanocryst. Mater. **10** (2001) 467.
43. K. Krištiaková, P. Švec, Czech J. Phys. **52** (2002) Suppl. A 133.
44. K. Krištiaková, P. Švec, Mat. Sci. Eng. **A304-306** (2001) 343.
45. T. A. Vilgis, J. Phys. C **21** (1988) L299.
46. H. S. Chen, J. Non-Cryst. Solids **22** (1976) 135.
47. M. R. Gibbs, J. E. Evetts, J. A. Leake, J. Mater. Sci. **18** (1983) 278.
48. S. Ranganathan, M. von Heimendahl, J. Mater. Sci. **16** (1981) 2401.
49. K. Krištiaková, J. Krištiak, P. Švec, P. Duhaj, O. Šauša, NanoStructured Materials **6** (1995) 505.
50. K. Krištiaková, P. Švec, J. Krištiak, O. Šauša, P. Duhaj, J. Non-Cryst. Solids **192** (1995) 277.
51. K. Krištiaková, J. Krištiak, P. Švec, O. Šauša, P. Duhaj, Mat. Sci. Eng. B **39** (1996) 15.
52. G. Vlasák, P. Švec, P. Duhaj, Mat. Sci. Eng. A **304-306** (2001) 472.
53. P. Švec, K. Krištiaková, Mat. Sci. Forum **360-362** (2001) 475-480.
54. K. Krištiaková, P. Švec, Phys. Rev. **B64** (2001) 014204.
55. W. Weiss, H. Alexander, J. Phys. F: Metal Phys. **17** (1983) 1987.
56. K. Krištiaková, P. Švec, Scripta Materialia **44** (2001) 1275.
57. P. Švec, K. Krištiaková, P. Duhaj, D. Janičkovič, Czech J. Phys. **52** (2002) 145.
58. Y. Zhang, K. Hono, A. Inoue, A. Makino, T. Sakurai, Acta Mater. **44** (1996) 1497.
59. D. Ohkubo, H. Kai, D. H. Ping, K. Hono, Y. Hirotsu, Scripta Mater. **44** (2001.
60. K. Krištiaková, P. Švec, D. Janičkovič, Mater. Transaction JIM **42** (2001) 1523.
61. M. E. McHenry, M. A. Willard, D. E. Laughlin, Prog. Mater. Sci. **44** (1999) 291.
62. P. Duhaj, P. Švec, J. Sitek, D. Janičkovič, Mat. Sci. Eng. **A304-306** (2001) 178-186.
63. J. J. Rayment, O. Ashira, B. Cantor, Proc. Int. Conf. TMS-AIME, Warrendale, 1982, pp. 1385-1389.
64. I. Maťko, P. Duhaj, P. Švec, D. Janičkovič, Mat. Sci. Eng. **A179/180** (1994) 557.
65. K. Kadau, P. Entel, T. C. Germann, P. S. Lomsdahl, B. L. Holian, J. de Phys. IV **11** (2001) 17.
66. A. Gilbert, W. S. Owen, Acta Metallurgica **10** (1962) 45.
67. T. Suzuki, M. Shimono, S. Takeno, Phys. Rev. Lett. **82** (1999) 1474.
68. T. Suzuki, M. Shimono, M. Wuttig, Scripta Mater. **44** (2001) 1979.

THE INFLUENCE OF THE GRAIN BOUNDARY PHASE TRANSITIONS ON THE PROPERTIES OF NANOSTRUCTURED MATERIALS

B.B. STRAUMAL
Institute of Solid State Physics, Russian Academy of Sciences
Chernogolovka, Moscow distr., 142432 Russia

Abstract Grain boundary (GB) phase transitions can change drastically the properties of nanograined polycrystals, leading to enhanced plasticity or brittleness, increasing diffusion permeability. They influence also liquid-phase and activated sintering, soldering, processing of semi-solid materials. The GB wetting phase transition can occur in the two-phase area of the bulk phase diagram where the liquid (L) and solid (S) phases are in equlibrium. The GB wetting tie line appears in the $L+S$ area. Above the temperature of the GB wetting phase transition a GB cannot exist in equlibrium contact with the liquid phase. The liquid phase has to substitute the GB and to separate both grains. The GB wetting tie-line can continue in the one-phase area of the bulk phase diagram as a GB solidus line. This line represents the GB premelting or prewetting phase transitions. The GB properties change drastically when GB solidus line is crossed by a change in the temperature or concentration. In case if two solid phase are in equilibrium, the GB "solid state wetting" (or covering) can occur. In this case the layer of the solid phase *2* has to substitute GBs in the solid phase *1*. Such covering GB phase transition occurs if the energy of two interphase boundaries between phase *1* and *2* is lower than the GB energy in the phase *1*.

1. Introduction

The properties of modern materials, especially those of nanocrystalline, superplastic or composite materials, depend critically on the properties of internal interfaces such as grain boundaries (GBs) and interphase boundaries (IBs). All processes which can change the properties of GBs and IBs affect drastically the behaviour of polycrystalline metals and ceramics [1]. GB phase transitions are one of the important examples of such processes [2]. Recently, the lines of GB phase transitions began to appear in the traditional bulk phase diagrams [2–7]. The addition of these equilibrium lines to the bulk phase diagrams ensures an adequate description of polycrystalline materials, particularly their diffusion permeability, deformation behavior and the evolution of the

T. Tsakalakos, et al. (eds.) pgs. 295 - 312
Nanostructures: Synthesis, Functional Properties and Applications;
© *2003 Kluwer Academic Publishers.*

microstructure. In this work the following GB phase transitions are discussed: (a) GB wetting, (b) GB prewetting (or premelting) and (c) GB «wetting» (covering) by second solid phase. The recently obtained experimental data are discussed. Using these data, the new GB lines in the conventional bulk phase diagrams are constructed.

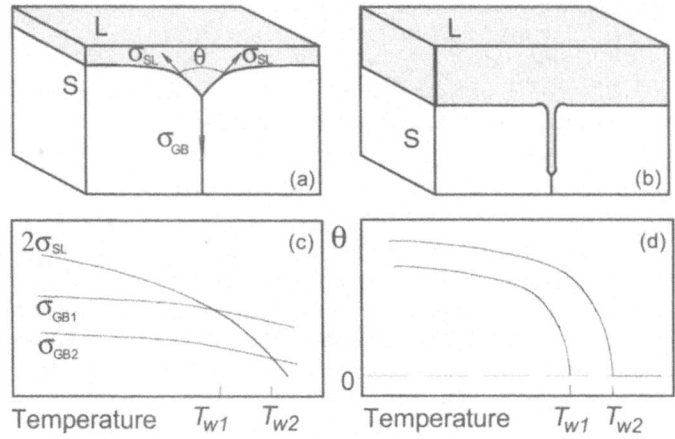

Figure 1. (a) Scheme of the equilibrium contact between the grain boundary in the solid phase S and the liquid phase L (incomplete wetting). (b) Complete GB wetting. (c) Scheme of the temperature dependence for the GB energy σ_{GB} (for two different GBs) and the energy of the solid-liquid interface boundary σ_{SL}. (d) Scheme of the temperature dependence of the contact angle θ for two grain boundaries with energies σ_{GB1} and σ_{GB2}. T_{w1} and T_{w2} are the temperatures of the GB wetting phase transition.

2. Grain Boundary Wetting Phase Transitions

In this work GB wetting, prewetting and premelting phase transitions are considered. The GB melting, GB faceting transition and the "special GB – random GB phase transitions" are analyzed elsewhere [8–10]. One of the most important GB phase transitions is the *GB wetting transition*. Since their prediction by Cahn [11] the study of wetting phase transitions has been of great experimental and theoretical interest, primarily for planar solid substrates and fluid mixtures [12–14]. Particularly, it was experimentally shown that the wetting transition is of first order, namely the discontinuity of the surface energy was measured and the hysteresis of the wetting behavior was observed [15, 16]. The important difference is that in case of GB wetting only two phases coexist, namely the liquid (melt) phase and the solid one containing the boundary between the misoriented grains. Therefore, the contact angle θ also depends only on two different surface energies (the GB energy σ_{GB} and the energy of the solid/liquid interphase boundary σ_{SL}) instead of three ones in the

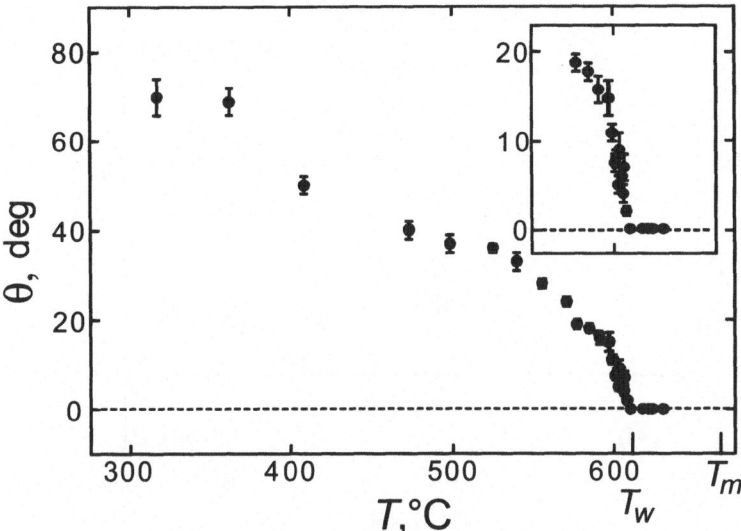

Figure 2. Temperature dependences of the contact angle between Sn-rich melt and tilt Al GB 32°<011>{001}. (T_w = 604±1°C). The insert shows the in details the $\square$(T) dependence close to T_w.

usual experiments: σ_{GB} = 2 σ_{SL} cos (θ/2). If σ_{GB} < 2σ_{SL}, the GB is incompletely wetted and the contact angle θ>0 (Fig. 1a). At the temperature T_w of the *GB wetting phase transition* σ_{GB} = 2σ_{SL} and at $T \geq T_w$ the GB is completely wetted by the liquid phase and θ = 0 (Fig. 1b). If two GBs have different energies the temperatures of their GB wetting transitions will also differ: the lower σ_{GB}, the higher T_w (Figs. 1c and 1d). If the GB wetting phase transition is of first order, there is a discontinuity in the temperature derivative of the GB energy at T_w which is equal to [$\partial\sigma_{GB}/\partial T - \partial(2\sigma_{SL})/\partial T$] [11, 16]. If the GB wetting phase transition is of second order, $\partial\sigma_{GB}/\partial T = \partial(2\sigma_{SL})/\partial T$ at T_w. The theory predicts also the shape of the temperature dependence θ (T) at $T \rightarrow T_w$: it must be convex for a first order wetting transition [$\theta \sim \tau^{1/2}$ where $\tau = (T_w-T)/T_w$] and concave for a second order wetting transition: $\theta \sim \tau^{3/2}$ [12].

Nowdays, the GB phase transitions of the second order were not observed experimentally. All observed temperature dependences $\theta(T)$ have a discontinuity in the temperature derivative of the GB energy at T_w. The $\theta(T)$ dependences are convex (like those shown in Fig. 2 for the Al–Sn system) and follow the $\theta \sim \tau^{1/2}$ law [17].

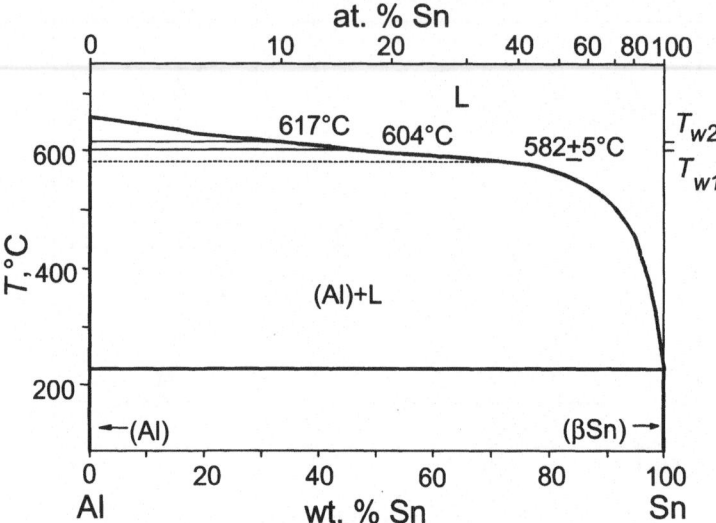

Figure 3. The Al–Sn phase diagram. Thick solid lines represent the bulk phase transitions. Thin solid lines are the tie lines of the GB wetting phase transitions. Thin dotted line represents the estimation for the GB wetting phase transiton for the GB with highest possible energy.

Consider the contact between a bicrystal and a liquid phase L. If the GB energy σ_{GB} is lower than the energy of two solid/liquid interfaces $2\sigma_{SL}$, the GB is not completely wetted and the contact angle $\theta > 0$ (Fig. 1a). If $\sigma_{GB} > 2\sigma_{SL}$ the GB is wetted completely by the liquid phase and $\theta = 0$ (Fig. 1b). If the temperature dependences σ_{GB} (T) and $2\sigma_{SL}$ (T) intersect, then the GB wetting phase transition proceeds at the temperature T_w of their intersection (Fig. 1c). The contact angle θ decreases gradually with increasing temperature down to zero at T_w. (see also the Fig. 2 for the system Cu–In) At $T > T_w$ the contact angle $\theta = 0$ (Figs. 1d and 2). The *tie-line of the GB wetting phase transition* appears at T_w in the two-phase region $(S+L)$ of the bulk phase diagram (Fig. 3). Above this tie line GBs with an energy σ_{GB} cannot exist in equilibrium with the liquid phase. The liquid phase forms a layer separating the crystals. In Fig. 3 two GB wetting tie lines are shown for two GBs with different energies obtained by measurements of $\theta(T)$ dependences (Fig. 2). In polycrystals the whole spectrum of GBs exist with varuious energies. Therefore, in polycrystals the maximal T_{wmax} and minimal T_{wmin} can be found for high-angle GBs with minimal and

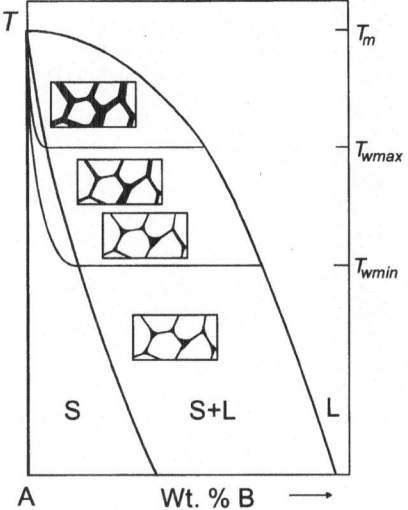

Figure 4. Scheme of the phase diagram with lines of bulk and GB phase transitions. Thick lines represent the bulk phase transitions. Thin lines represent the tie-lines of the GB wetting phase transition in the $S + L$ area for the high angle GBs having maximal and minimal possible energy and the GB premelting phase transition in the solid solution area S.

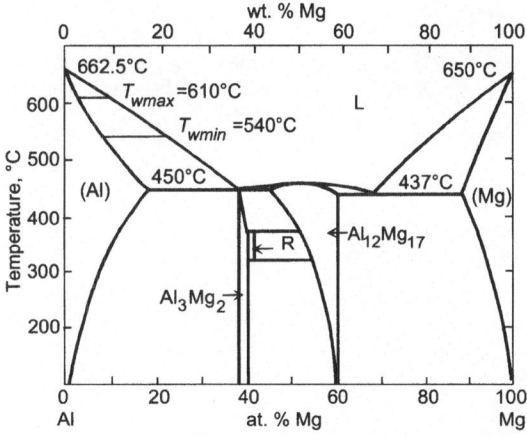

Figure 5. The Al–Mg phase diagram. Thick lines represent the bulk phase transitions. Thin lines are the tie-lines of the GB wetting phase transitions (T_{wmax} = 610°C and T_{wmin} = 540°C)

300

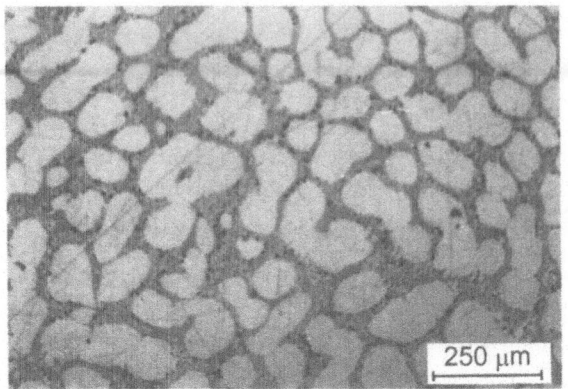

Figure 6.
Microstructure of
Al – 10 wt.% Mg
polycrystals in two phase
S+L area of
Al – Mg phase diagram

(a) 610°C, all GBs are
wetted by the melt

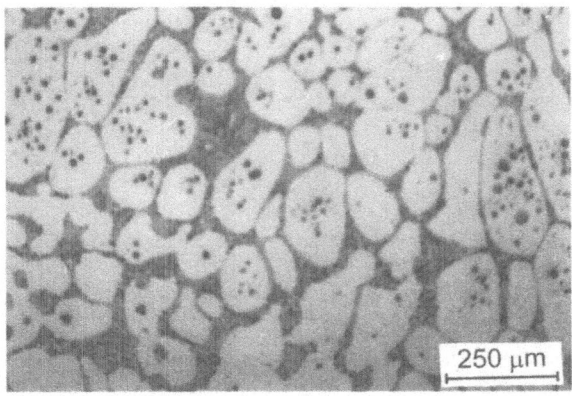

(b) 581°C, some GBs are
wetted by the melt,
another GBs are not
wetted

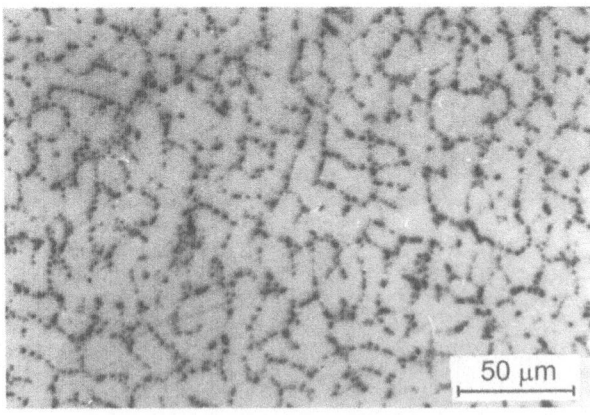

(c) 490°C, no wetted
GBs in polycrystal

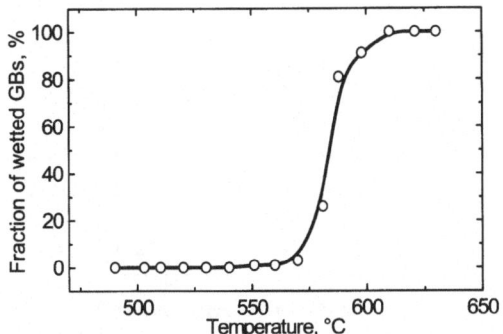

Figure 7...Temperature depen-dence of the fraction of wetted GBs in two-phase Al–Mg poly-crystals.

maximal energy σ_{GBmin} and σ_{GBmax}, respectively (Fig. 4). The tie-lines at T_{wmax} and T_{wmin} are shown also in the Al–Mg phase diagram (Fig. 5). Above T_{wmax} all GBs are completely wetted (see Fig. 6a). At the temperature between T_{wmax} and T_{wmin} some GBs are wetted by the liquid phase and other GBs are not wetted (Fig. 6b). Below T_{wmin} all GBs are not wetted, and the melt has a shape of separated inclusions (Fig. 6c). With increasing temperature between and the fraction of the wetted GBs increases from 0 at T_{wmin} to 100% at T_{wmax} (Fig. 7).

First indications of the GB wetting phase transitions were found by measuring of the contact angles in polycrystals [17]. Correct measurements were later perfomed using metallic bicrystals with inndividual tilt GBs in the Al–Sn (Figs. 2 and 3), Cu–In [4], Al–Pb–Sn [3,18,19], Al–Ga, Al–Sn–Ga [20, 21], Cu–Bi [5, 22, 23, 30], Fe–Si–Zn [24–27], Mo–Ni [28], W–Ni [29] and Zn–Sn [7] systems. The tie-lines of the GB wetting phase transition were constructed basing on the experimental data [3, 4, 7, 18–29]. The difference in the GB wetting phase transition temperature was experimentally revealed for GBs with different energies [4, 18]. The precize measurements of the temperature dependence of the contact angle revealed also that the GB wetting phase transition is of the first order [18]. The indications of presence of the liquid-like phase along the dislocation lines were also found [23].

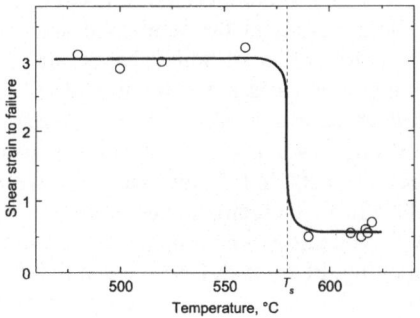

Figure 8..Temperature depen-dence of shear strain to failure for Al–5 wt. % Mg alloy in solid and semi-solid state [37].

302

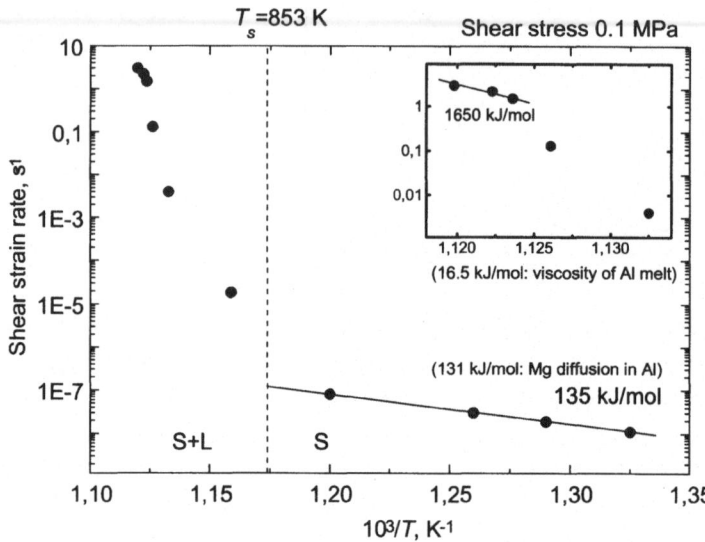

Figure 9. Temperature dependence of shear strain rate for Al–5 wt. % Mg alloy in solid and semi-solid state [37].

The deformation behavior of metals in the semi solid state has been extensively investigated from the viewpoint of rheological flow [30–32]. These studies have shown that the viscosity of the semi-solid metals depends on the volume fraction and morphology of the solid phase and the shear strain rate. In addition, the deformation behavior in a semi-solid state at the early stages of melting has been investigated by compressive creep tests [33–36]. Vaandrager and Pharr [24] showed that the deformation mechanism in a semi-solid state at the early stages of melting is grain boundary sliding accommodated by cavitation in a liquid phase for the copper containing a liquid bismuth. This deformation mechanism in the semi-solid state at the early stages of melting appears to be different from that in the semi-solid state during solidification. The presence of a liquid phase gives rise to complicated effects on the deformation behavior in the semi-solid state. Deformation in the semi-solid state is phenomenologically divided as follows: plastic deformation of solid phases, sliding between solid phases, flow of liquid incorporating solid phases and liquid flow [22]. For compressive deformation, because the liquid phase is squeezed out of boundaries experiencing compressive stresses in a very short time [24], it is difficult to investigate deformation related to the liquid flow by compressive tests. In [37] the shear tests were carried out over a wide temperature range of 480–620°C, including temperatures below and above the solidus temperature, for Al – 5 wt.% Mg alloy to investigate deformation behavior in semi-solid states at early stages of melting. Pharr et al. [36] showed that the liquid phase significantly affects creep behavior of alloys when a significant portion of the grain boundary area, in excess of 70%, is wet. This revealed that the volume fraction of the liquid phase is an important factor in the deformation characteristics in the semi-solid state. The same trend has been reported in a semi-solid state at solidification [34]. However, deformation in the semi-solid state is very complicated and cannot be characterized only by the volume fraction of a liquid phase. In [37] the pure shear of Al–5 wt% Mg alloy was investigated. This method permits to exclude the squeezing of the liquid phase from the sample. The shear strain to failure drops drastically at the solidus temperature (Fig. 8) [37]. In the semi-solid phase it is about 6 times lower than in the solid solution. Using the micrographs of the structure of polycrystals in the semi-solid state from [37] we calculated the fraction of the fully wetted GBs in dependence on the temperature. This dependence is shown in the Fig. 7. The continuous change of the fraction of wetted GBs influences strongly the mechanism of the deformation. In Fig. 9 the temperature dependence of the shear strain rate is shown recalculated from the data [37]. In the solid solution the of the shear strain rate increases moderately with increasing temperature, and the activation energy (135 kJ/mol) is very close to the activation energy of Mg diffusion in Al (131 kJ/mol). In the semi-solid state the shear strain rate increases drastically. Close to T_{wmax} the (formally calculated) activation energy is about 1650 kJ/mol, i.e. ten times higher than the activation energy for the viscosity of Al melt. It means that in the semi-solid state no unique thermally activated mechanism is working. Due to the temperature increase of the fraction of wetted GBs, the structure of the solid skeleton changes continuously. It becomes more and more cutted with increasing temperature, therefore, making the shear easier in addition to the pure temperature activation.

3. Grain Boundary Prewetting (Premelting) Phase Transitions

It was pointed out by Cahn [38] that, when the critical consolution point of two phases is approached, GBs of one critical phase should be wetted by a layer of another critical phase, and in the one-phase region of a phase diagram there should be a singularity connected with an abrupt transition to a microscopic wetting layer. We distinguish two possible situations: the first one, when a layer of the new phase is formed on the GB (*prewetting transition*), and the second one, when the GB is replaced by a layer of the new phase (*premelting phase transition*). At the prewetting transition the difference between two phases must be small, while at the premelting transition the wetting phase may differ from that of the bulk dramatically. The lines of the GB prewetting or premelting phase transitions appear in the one-phase areas of the bulk phase diagrams where only one bulk phase can exist in the thermodynamic equilibrium (e.g. solid solution S, see Fig. 4). These lines continue the tie-lines of the GB wetting phase transitions and represent the GB solidus (Fig. 4). The thin liquid-like layer of the GB phase exists on the GBs between the bulk solidus and GB solidus in the phase diagram. During the GB premelting phase transition this layer appears abruptly on the GB by the intersection of GB solidus. As a result, the GB properties (diffusivity, mobility, segregation) change dramatically.

In other words, above the GB wetting tie line T_W in the $S+L$ area of the bulk phase diagram $\sigma_{GB} > 2\sigma_{SL}$. This is true also if we intersect the bulk solidus at T = const and move into the one-phase area S of the bulk phase diagram. The GB energy σ_{GB} in this part of the one-phase region is still higher that the energy $2\sigma_{SL}$ of two solid-liquid

304

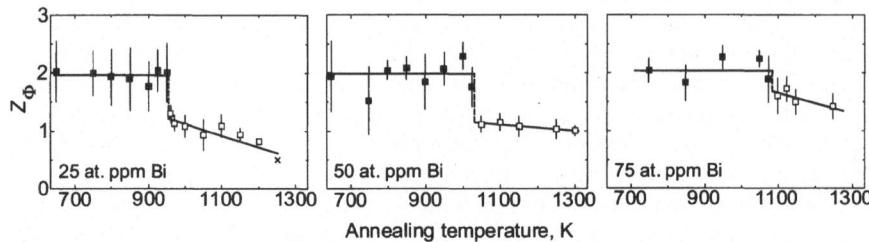

Figure 10. Temperature dependence of the GB Gibbsian excess of Bi in Cu(Bi) polycrystals of various compositions, measured by AES. The sudden change of the GB segregation corresponds to the intersection of GB solidus line.

interphase boundaries. Therefore, the GB still can be substituted by two solid-liquid interfaces, and the energy gain $G = \sigma_{GB} - 2\sigma_{SL}$ appears by this substitution. G permits to stabilize the GB layer of the liquid-like phase. The appearance of the liquid-like phase (otherwise unstable in the bulk) between two S/L interfaces instead of GB leads to the energy loss $\Box g$ per unit thickness and unit square. Therefore, the GB layer of the liquid-like phase has the thickness l defined by the equation $\sigma_{GB} - 2\sigma_{SL} = \Delta g l$. Thickness l depends on the concentration and temperature and becomes $l = 0$ at the line of GB premelting (or prewetting) phase transition.

The premelting transition has been revealed in the ternary Fe–Si–Zn system by measurements of Zn GB *diffusivity* along tilt GBs in the Fe–Si alloys [24–27]. It was found that the penetration profiles of Zn along GBs consist of two sections, one with a small slope (high GB diffusivity $D_b\delta$) at high Zn concentrations and one with a large slope (low GB diffusivity) at low Zn concentrations. The transition from one type of behavior to the other was found to occur at a definite Zn concentration c_{bt} at the GB, which is an equilibrium characteristic of a GB and depends on the temperature and pressure. The GB diffusivity increases about two orders of magnitude which is an indication of a quasi-liquid layer present in the GBs at high Zn concentration. The line of GB premelting phase transition in the one-phase area of the bulk phase diagram continues the line of the GB wetting phase transition in the two-phase $L+S$ area: by pressure increase both the GB wetting and the GB enhanced diffusivity disappear together at the same pressure value [27].

The *GB mobility* was studied for two tilt GBs in bicrystals grown of high purity 99.999 wt.% Al and of the same material doped by 50 wt. ppm Ga [21]. The GB mobility increased about 10 times by addition of the Ga content for the both GBs studied. Normally, the addition of a second component can only decrease the GB mobility due to the solution drag [32]. The increase of the GB mobility can only be explained by the formation of the liquid-like Ga-rich layer on the GBs as a result of a premelting phase transition

The *GB segregation* of Bi in Cu was studied in the broad temperature and concentration interval [5, 22, 23, 39, 40]. It was shown that at a fixed Bi concentration

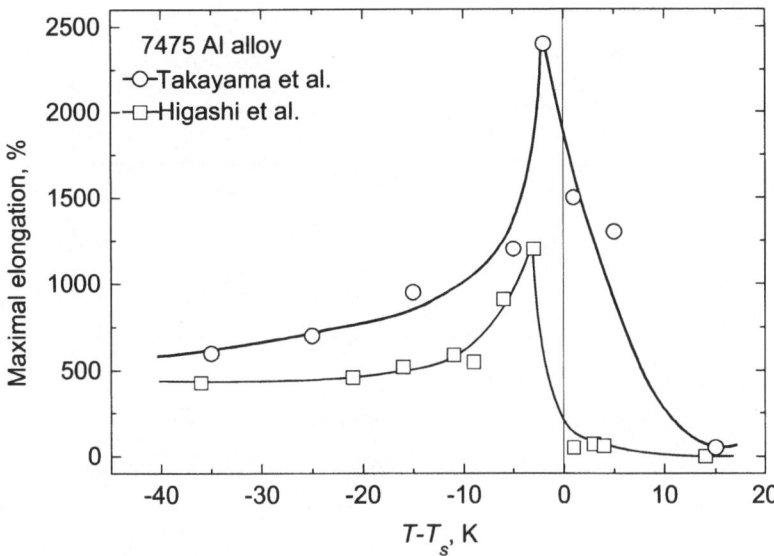

Figure 11. The temperature dependence of the maximal elongation of the 7475 Al–Zn–Mg alloy samples. T_s is the solidus temperature. Circles represent the data [50] and squares are taken from [51].

the GB segregation Z_Φ changes abruptly at a certain temperature (Fig. 10). Below this temperature the GB Bi concentration is constant and corresponds to a thin layer of pure Bi (GB phase). Above this temperature the GB segregation is lower than one monolayer of Bi and decreases gradually with increasing temperature according to the usual laws. These features indicate also the formation of a thin layer of a GB phase in the one-phase area of the bulk Cu–Bi phase diagram. The points of the abrupt change of the GB segregation form the GB solidus line in the bulk Cu–Bi phase diagram [22, 23, 39, 40]. GB segregation was measured with the aid of Auger electron spectroscopy (AES) on the GB fracture surfaces in samples broken *in situ* in the AES instrument. In other words, the multilayer GB segregation in Cu–Bi alloys leads to the increased *GB brittleness*. In [41] the GB energy was measured in Cu–Bi alloys using individual $\Sigma 19$ GB in bicrystals with the aid of the GB thermal grooves. The thermal groove profile was obtained witt the aid of atomic force microscopy. The GB Bi segregation was measured simultaneously in the same conditions. The abrupt change of the segregation coincides with the *discontinuity of GB energy*. This fact demonstrates that the GB premelting (or prewetting) phase transition is of first order. The low-temperature measurements of resistivity temperature coefficient $d\rho/dT$ and residual resistivity ρ_0 at 4 K were performed in [40] using the Cu–Bi polycrystals annealed at high temperature and subsequently quenched. Both $d\rho/dT$ and ρ_0 demonstrate well pronounced break exactly at the same position where the sudden change of GB segregation was observed. In other words, the formation of GB layers of liquid-like phase leads to the measurable changes of *resistivity*.

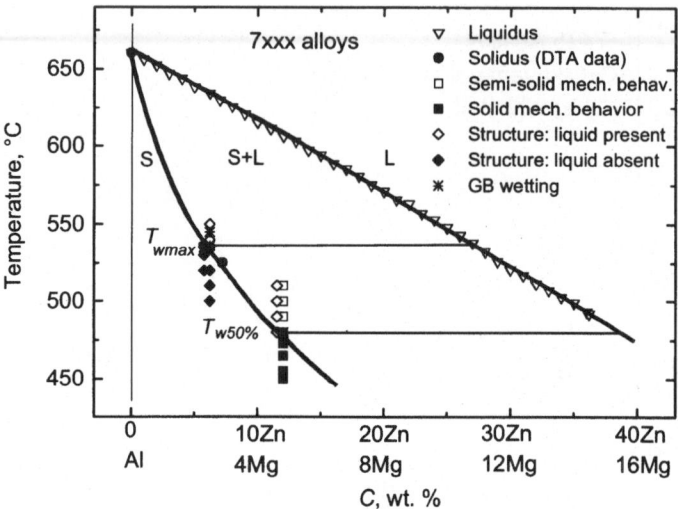

Figure 12. The phase diagram containing the GB wetting phase transition tie-lines constructed for the 7xxx Al–Zn–Mg alloys.

Superplastic forming of micrograined and nanostructured materials is a commercial, viable, manufacturing technology. One of the major drawbacks of conventional superplastic forming is that the phenomenon is only found at relatively low strain rates, typically about 10^{-4} to 10^{-3} s^{-1}. Recently, a number of studies have indicated that superplasticity of nanostructured materials can sometimes occur at extremely high strain rates (greater than 10^{-3} s^{-1} and up to 10^2 s^{-1}). A specific example is a tensile elongation of over 1250% recorded at a strain rate of 10^2 s^{-1} [42]. Thus far, this phenomenon, denoted as high-strain-rate superplasticity (HSRS), has been reported in several classes of materials, including metal alloys [42], metal-matrix composites [43–46] and mechanically-alloyed materials [47–49]. Despite these extensive experimental observations, the fundamental understanding of the factors leading to HSRS has not yet been arrived at. One very pertinent fact is that all of the materials that exhibit HSRS have a very fine grain size (~ 1 μm and less). Another is that the phenomenon is observed at rather high homologous matrix temperatures and very close to the matrix solidus temperature. In Fig. 11 the example is shown of HSRS for the 7475 Al–Zn–Mg alloy. The data are taken from independent works [50, 51] and reveal the very good reproducibility of the effect. Both temperature dependences have rather narrow maximum few degrees below the solidus temperature T_s. It is important to mention that the soludus temperature was measured by the differential thermal analysis (DTA) in the same works [50, 51]. The maximum elongation to failure reaches 1250 %. Below T_s the maximal elongation is about 500 %, above T_s the maximal elongation drops very quickly down to almost 0%.

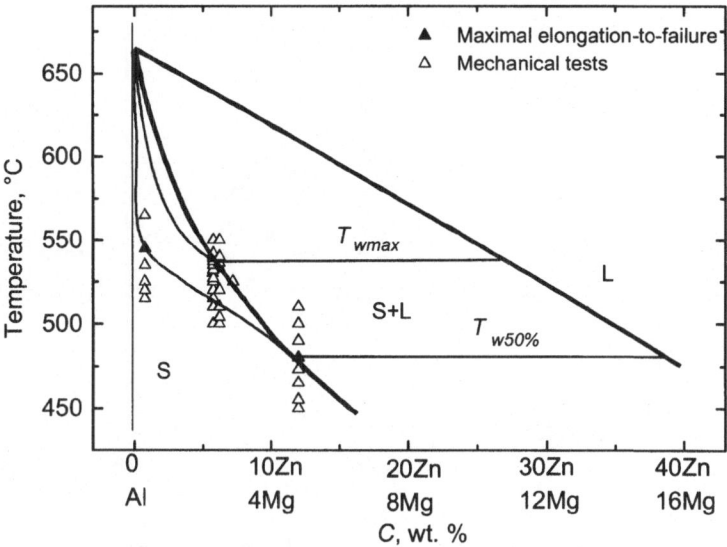

Figure 13. The phase diagram containing the GB wetting phase transition tie-lines and GB solidus lines constructed for the 7xxx Al–Zn–Mg alloys.

We suppose that the HSRS phenomenon can be explained using the ideas on the GB phase transitions in the two-phase $S+L$ area and the solid solution area of the bulk phase diagrams. Using the data published in the literature, we constructed the lines of the GB wetting phase transition for the 7xxx Al–Zn–Mg alloys (Fig. 12). The liquidis line (thick solid line, open down triangles) have been cosntructed using the linear interpolation of liquidus lines for the binary Al–Mg and Al–Zn phase diagrams [52]. Solidus line (thick solid line, full circles) has been drawn through the melting point for Al [52] and experimental points obtained using DTA for the 7xxx alloys [35, 50, 51, 53]. Open and full squares mark the solid and semi-solid mechanical behaviour, respectively [35]. Open and full diamonds mark the samples where the microstructural observations revealed the presence or absence of the liquid phase, respectively [35, 50, 51]. The analysis of the microstructures published in [35, 50, 51] permitted us to estimate the fraction of completely and partially wetted GBs (data marked by stars). These estimations allow to construct the GB wetting transition tie-lines (thin solid lines) for the T_{wmax} (above T_{wmax} all high-angle GBs in the polycrystal are completely wetted) [50, 51] and $T_{w50\%}$ (above $T_{w50\%}$ about 50% of the high-angle GBs in the polycrystal are completely wetted) [35].

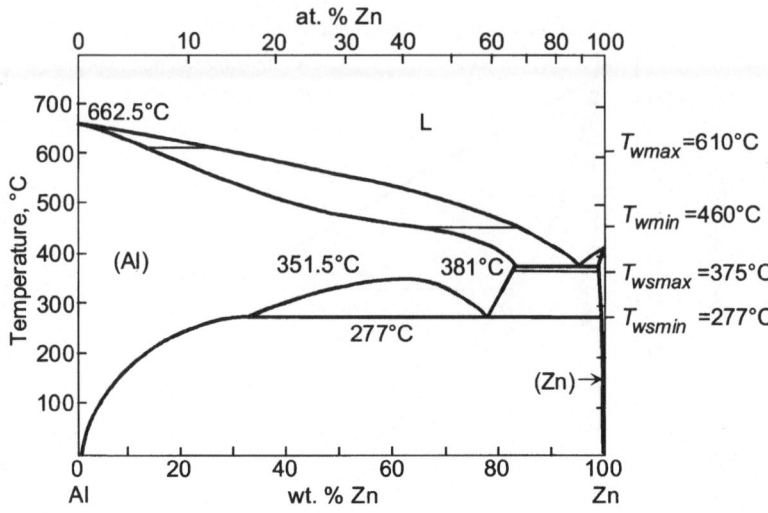

Figure 14. The Al–Zn phase diagram. Thick lines represent the bulk phase transitions. Thin lines are the tie-lines of the GB wetting phase transitions (T_{wmax} = 610°C and T_{wmin} = 460°C) and tie-lines of the GB covering («solid state wetting») phase transitions (T_{wsmax} = 375°C and T_{wsmin} = 277°C).

In the Fig. 12 the liquidus, solidus lines and GB wetting tie lines obtained in the Fig. 11 are repeated without experimental points. The data on mechanical tests (full and open up-triangles) are added [50, 53–55]. Full triangles mark the maximal elongation-to-failure obtained in the tests performed at different temperatures [50, 51, 54]. Full triangles lye either below the bulk solidus line or coincide with it. The temperature difference between temperature of the maximal elongation-to-failure and T_s decreases with increasing concentration of Mg and Zn. According to the thermodynamics, the tie lines of the GB wetting phase transition cannot finish at the intersection with the bulk solidus. They have to form the GB solidus line which continue in the solid solution area of the bulk phase diagram and finish in the melting point of the pure component. In the limiting case the degenerated GB solidus coincide with the bulk solidus. But in some systems it can extend into the solid solution area like it is shown in Fig. 4. In that case the layer of the liquid-like GB phase exist in the GBs between the GB and bulk solidus lines. We have shown above that the presence of such liquid-like layer in GB leads to the enhanced GB diffusivity, mobility and segregation of the second component [5, 11–17]. Such GB solidus lines are drawn also in Fig. 13 (thin solid lines). They continue the T_{wmax} and $T_{w50\%}$ GB wetting tie lines and finish in the melting point of Al. The GB solidus lines are drawn in such a way that the points of the maximal elongation-to-failure are between the GB and bulk solidus. Therefore, the enhanced plasticity of the nanograined polycrystals can be explained by the GB phase transition leading to the formation of the liquid-like layer on the GB in the narrow band of the solid solution area, just below the bulk solidus line. The phase diagrams similar to those shown in Figs 12 and 13 can be constructed using the published data on HSRS, DTA and microscopy also for the 2xxx (Al–Zn–Mg) [45, 47–59, 51, 56–58], 5xxx (Al–Mg) [47, 51, 59, 60] and 6xxx (Al–Si–Mg) [59, 61–64] alloys. All authors studied the HSRS phenomenon mention that the physical reason of such a huge and reproducible increase of the palsticity is unknown. We suppose for the first time that the HSRS phenomenon can be explained by the existence of the equilibrium GB liquid-like layer close to the bulk solidus.

4. Grain Boundary Wetting (Covering) by Solid Phase

The situation illustrated in Fig. 1 can repeat in case if second phase (β) is not liquid but also solid. In other words, if in the phase α the GB energy $\sigma_{\alpha\alpha}$ is lower that the energy of two α/β solid/solid interfaces, the GB αα has to be substituted by the layer of the second solid phase β. Such process can be called the GB wetting (or covering) by solid phase. It is clear, that the kinetics of the equlibration processes in case GB wetting (or covering) by solid phase is much slower than in case of wetting by liquid phase. Our preliminary experiments with Al–95 wt.% Zn alloy demonstrate that after about 1 month of annealing the difference in the morphology of Al-rich phase precipitates (Al) at the (Zn)/(Zn) GBs in Zn-rich phase can be observed (Fig. 14). Namely, at high tempreratures just below the eutectic temperature in the Al–Zn system, more than 50% of (Zn)/(Zn) GBs are covered by continuous layer of the Al-rich phase. With decreasing temperature the portion of the (Zn)/(Zn) GBs covered by the (Al) layer decreased, and at the temperatures just below the eutectoid point all (Al) precipitates at the (Zn)/(Zn) GBs have the shape of isolated particles. Another examples of the GB covering phase transitions can be found by the analysis of the data published in the literature, for example, for Zr–Nb [65] or W–Cu systems [66–71].

5. Conclusions

The GB phase transitions can be observed both in two-phase and one-phase areas of bulk phase diagrams. In the two-phase $S+L$ area where solid and liquid phases are in equilibrium the GB wetting phase transition can take place at T_w. Above T_w the GB disappears being substituted by two solid/liquid interfaces and the (macroscopically thick) layer of the liquid phase. The tie-lines of the GB wetting phase transition must have a continuation (GB solidus) in the one-phase S area of the bulk phase diagram. By intersection of GB solidus line the GB prewetting or premelting phase transition proceeds. Between the lines of GB and bulk solidus the grain boundary is substituted by two solid/liquid interfaces and the thin layer of the liquid-like phase. This liquid-like phase is stable in the GB and unstable in the bulk. The liquid-like phase is stabilized in GB due to the energy gain which appears as a result of

310

substitution of GB by two solid/liquid interfaces. The GB wetting and prewetting (premelting) phase transitions observed up-to-date are of first order. If the GB energy is higher than the energy of two solid/solid interfaces, the GB solid state wetting (covering) phase transition can occur in a two-phase S_1+S_2 area of the phase diagram.

6. Acknowledgements

The financial support of Russian Foundation for Basic Research and the Government of the Moscow district (contracts 01-02-97039 and 01-02-16473), Deutsche Forschungsgemein-schaft (contracts Gu 258/12-1 and Ba-1768/1-2), Copernicus program of EU (contract ICA2-CT-2001-10008) and the German Federal Ministry for Education and Research (contract WTZ RUS 00/209) is acknowledged.

7. References

1. Langdon, T.G., Watanabe, T., Wadsworth, J., Mayo, M.J., Nutt, S.R., and Kassner, M. E. (1993) Future research directions for interface engineering in high-temperature plasticity, *Mater. Sci. Eng. A* 166, 237.
2. Straumal, B.B. and Gust, W. (1996) Lines of grain boundary phase transitions on the bulk phase diagrams, *Mater. Sci. Forum* 207–209, 59-68.
3. Straumal, B., Molodov, D., and Gust, W. (1994) Tie lines of the grain boundary wetting phase transition in the Al–Sn system, *J. Phase Equilibria* 15, 386-391.
4. Straumal, B., Muschik, T., Gust, W., and Predel, B. (1992) The wetting transition in high and low energy grain boundaries in the Cu(In) system, *Acta metall. mater.* 40, 939-945.
5. Chang, L.-S., Rabkin, E., Straumal, B.B., Hofmann, S., Baretzky, B., and Gust, W. (1998) Grain boundary segregation in the Cu–Bi system, *Defect Diff. Forum* 156, 135-146.
6. Straumal, B., Semenov, V., Glebovsky, V., and Gust, W. (1997) Grain boundary wetting phase transitions in the Mo–Ni system. *Defect Diff. Forum* 143–147, 1517-1522.
7. Straumal, B.B., Gust, W., and Watanabe, T. (1999) Tie lines of the grain boundary wetting phase transition in the Zn-rich part of the Zn–Sn phase diagram, *Mater. Sci. Forum* 294–296, 411-414.
8. Ernst, F., Finnis, M.W., Koch, A., Schmidt, C., Straumal, B., and Gust, W. (1996) Structure and energy of twin boundaries in copper, *Z. Metallk.* 87, 911-922.
9. Straumal, B.B. and Shvindlerman, L.S. (1985) Regions of existence of special and non-special grain boundaries, *Acta metall.* 33, 1735-1749.
10. Maksimova, E.L., Shvindlerman, L.S., and Straumal, B.B. (1988) Transformation of □17 special tilt boundaries to general boundaries in tin, *Acta metall.* 36, 1573-1583.
11. Cahn, J.W. (1977) Wetting transitions on surface, *J. Chem. Phys.* 66, 3667-3679.
12. Dietrich, S. (1988) Wetting transitions in interfaces, in C. Domb and J.H. Lebowitz (eds.), *Phase Transitions and Critical Phenomena*, 12, Academic Press, London, pp. 2-218.
13. Jasnov, D. (1984) Phase transitions on surfaces, *Rep. Prog. Phys.* 47, 1059-1070.
14. de Gennes, G. (1985) Wetting: statics and dynamics, *Rev. Mod. Phys.* 57, 827-863.
15. Kellay, H., Bonn, D., and Meunier, J. (1993) Prewetting in a binary liquid mixture, *Phys. Rev. Lett.* 71, 2607-2610.
16. Schmidt, J.W. and Moldover, M.R. (1983) First-order wetting transition at a liquid-vapor interface, *J. Chem. Phys.* 79, 379-387.
17. Eustatopoulos, N., Coudurier, L., Joud, J.C., and Desre, P. (1976) Solid-liquid interface tension of Al–Sn, Al–In and Al–Sn–In systems, *J. Crystal Growth* 33, 105-115.
18. Straumal, B., Gust, W., and Molodov, D. (1995) Wetting transition on the grain boundaries in Al contacting with Sn-rich melt, *Interface Sci.* 3, 127-132.
19. Straumal, B., Molodov, D., and Gust, W. (1996) Grain boundary wetting phase transitions in the Al–Sn

311

and Al–Pb–Sn systems, *Mater. Sci. Forum* **207–209**, 437-440.

20. Straumal, B., Risser, S., Sursaeva, V., Chenal, B., and Gust, W. (1995) Grain growth and grain boundary wetting transitions in the Al–Ga and Al–Sn–Ga alloys of high purity, *J. Physique* **IV 5-C7**, 233-241.

21. Molodov, D.A., Czubayko, U., Gottstein, G., Shvindlerman, L.S., Straumal, B.B., and Gust, W. (1995) Acceleration of grain boundary motion in Al by small additions of Ga, *Phil. Mag. Lett.* **72**, 361-368.

22. Chang, L.-S., Straumal, B.B., Rabkin, E., Gust, W., and Sommer, F. (1997) The solidus line of the Cu–Bi phase diagram, *J. Phase Equilibria* **18**, 128-135.

23. Chang, L.-S., Rabkin, E., Straumal, B., Lejcek, P., Hofmann, S., and Gust, W. (1997) Temperature dependence of the grain boundary segregation of Bi in Cu polycrystals, *Scripta mater.* **37**, 729-735.

24. Rabkin E.I., Semenov, V.N., Shvindlerman, L.S., and Straumal, B.B. (1991) Penetration of tin and zinc along tilt grain boundaries 43°[100] in Fe–5at.%Si alloy: Premelting phase transition? *Acta metall. mater.* **39**, 627-639.

25. Noskovich, O.I., Rabkin, E.I., Semenov, V.N., Shvindlerman, L.S., and Straumal, B.B. (1991) Wetting and premelting phase transitions in 38°[100] tilt grain boundaries in (Fe–12 at. % Si)–Zn alloy in the vicinity of the A2–B2 bulk ordering in Fe–12 at. % Si alloy, *Acta metal. mater.* **39**, 3091-3098.

26. Straumal, B.B., Noskovich, O.I., Semenov, V.N., Shvindlerman, L.S., Gust, W., and Predel, B. (1992) Premelting transition on 38°<100> tilt grain boundaries in (Fe–10 at. % Si)–Zn alloys, *Acta metall. mater.* **40**, 795-801.

27. Straumal, B., Rabkin, E., Lojkowski, W., Gust, W., and Shvindlerman, L.S. (1997) Pressure influence on the grain boundary wetting phase transition in Fe–Si alloys, *Acta mater.* **45**, 1931-1940.

28. Rabkin, E., Weygand, D., Straumal, B., Semenov, V., Gust, W., and Brèchet, Y. (1996) Liquid film migration in a Mo(Ni) bicrystal, *Phil. Mag. Lett.* **73**, 187-193.

29. Glebovsky, V.G., Straumal, B.B., Semenov, V.N., Sursaeva, V.G., and Gust, W. (1994) Grain boundary penetration of a Ni-rich melt in tungsten polycrystals, *High Temp. Mater. Proc.* **13**, 67-73.

30. Flemings, M.C. (1991) Behavior of metal alloys in the semisolid state, *Metall. Trans. A* **22**, 957-981.

31. Kumar, P., Martin. C.L., and Brown, S. (1993) Shear strain rate thickening flow behaviour of semisolid slurries, *Metall. Trans. A* **24**, 1107-1116.

32. Chen, C.P. and Tsao, C.-Y.A. (1997) Semi-solid deformation of non-dendritic structures .1. Phenomenological behavior, *Acta mater.* **45**, 1955-1968.

33. Roth, M.C., Weatherly, G.C., and Miller, W.A. (1980) The temperature dependence of the mechanical properties of aluminum alloys containing low-melting-point inclusions, *Acta metall.* **28**, 841-853.

34. Vaandrager, B.L. and Pharr, G.M. (1989) Compressive creep of copper containing a liquid bismuth intergranular phase, *Acta metall.* **37**, 1057-1066.

35. Baudelet, B., Dang, M.C. and, Bordeaux, F. (1995) Mechanical behaviour of an aluminium alloy with fusible grain boundaries, *Scripta Metall. Mater.* **32**, 707-712.

36. Iwasaki, H., Mori, T., Mabuchi, M., and Higashi, K. (1998) Shear deformation behavior of Al–5% Mg in a semi-solid state, *Acta mater.* **46**, 6351-6360.

37. Pharr, G.M., Godavarti, P.S., and Vaandrager, B.L. (1989) Effects of wetting on the compression creep-behavior of metals containing low melting intergranular phases, *J. Mater. Sci.* **24**, 784-792.

38. Cahn, J.W. (1982) Transitions and phase equilibria among grain boundary structures, *J. Phys.Colloq.* **43-C6**, 199-213.

39. Chang, L.-S., Rabkin, E., Straumal, B.B., Baretzky, B., and Gust, W. (1999) Thermodynamic aspects of the grain boundary segregation in Cu(Bi) alloys, *Acta mater.* **47**, 4041-4046.

40. Straumal, B., Sluchanko, N.E., and Gust, W. (2001) Influence of the grain boundary phase transitions on the properties of Cu–Bi polycrystals, *Def. Diff. Forum* **188–190**, 185-194.

41. Schölhammer, J., Baretzky, B., Gust, W., Mittemeijer, E., and Straumal, B. (2001) Grain boundary grooving as an indicator of grain boundary phase transformations, *Interf. Sci.* **9**, 43-53.

42. Higashi, K., Tanimura, S., and Ito, T. (1990) Superplastic behaviour at high-strain rates in a particulate 6061 aluminium composites, *MRS Proc.* **196**, 385-390.

43. Imai, T., Mabuchi, M., Tozawa, Y., and Yamada, M. (1990) Superplasticity in beta-silicon nitride whisker reinforced 2124 aluminum composite, *J. Mater. Sci. Lett.* **2**, 255-257.

44. Mabuchi, M. and Imai, T. (1990) Superplasticity of Si_3N_4 whisker reinforced 6061 aluminum at high strain rate, *J. Mater. Sci. Lett.* **9**, 761-762.

45. Nieh, T.G., Henshall, C.A., and Wadsworth, J. (1984) Superplasticity at high strain rates in a SiC whisker reinforced Al alloy, *Scripta Metall.* **18**, 1405-1408.

46. Mabuchi, M., Higashi, K., Okada, Y., Tanimura, S., Imai, T., and Kubo, K. (1991) Superplastic behaviour at high-strain rates in a particulate Si_3N_4 6061 aluminium composite, *Scripta Metall.* **25**, 2003.

312

47. Nieh, T.G., Gilman, P.S., and Wadsworth, J. (1985) Extended ductility at high strain rates in a mechanically alloyed aluminum alloy, *Scripta Metall.* **19**, 1375-1378.
48. Bieler, T.R., Nieh, T.G., Wadsworth, J., and Mukherjee, A.K. (1988) Superplastic-like behaviour at high strain rates in mechanically alloyed aluminum, *Scripta Metall.* **22**, 81-86.
49. Higashi, K., Okada, Y., Mukai, T., and Tanimura, S. (1991) Positive exponent strain-rate superplasticity in mechanically alloyed aluminum IN9021, *Scripta Metall.* **25**, 2053-2057.
50. Takayama, Y., Tozawa, T., and Kato, H. (1999) Superplasticity and thickness of liquid phase in the vicinity of solidus temperature in a 7475 aluminum alloy, *Acta mater.* **47**, 1263-1270.
51. Higashi, K., Nieh, T.G., Mabuchi, M., and Wadsworth, J. (1995) Effect of liquid phases on the tensile elongation of superplastic aluminum alloys and composites, *Scripta metall. mater.* **32**, 1079-1084.
52. Apykhtina, I., Bokstein, B., Khusnutdinova, A., Peteline, A., and Rakov, S. (2001) Kinetics of diffusion-controlled grooving in solid-liquid systems, *Def. Diff. Forum* **194–199**, 1331-1336.
53. Imai, T., Mabuchi, M., Tozawa, Y., Murase, Y., and Kusul, J. (1990) in R.B. Bhagat. et al. (eds.), *Metal & Ceramic Matrix Composites: Processing, Modeling & Mechanical Behavior*, TMS-AIME, Warrendale, Pennsylvania, pp. 235-239.
54. Mabuchi, M., Higashi, K., Imai, T., and Kubo, K. (1991) Superplastic-like behavior in as-extruded Al–Zn–Mg alloy matrix composites reinforced with Si_3N_4 whiskers, *Scripta Metall.* **25**, 1675-1680.
55. Furushiro, N., Hori, S., and Miyake, Y. (1991) in S. Hori et al., (eds.) *Proc. Int. Conf. Superplast. Adv. Mats (ICSAM-91)*, Jap. Soc. Res. Superplast., Tokyo, pp. 557-562.
56. Mabuchi, M., Higashi, K., and Langdon, T. (1994) An investigation of the role of a liquid-phase in Al–Cu–Mg metal-matrix composites exhibiting high-strain rate suparplasticity, *Acta metall. mater.* **42**, 1739-1745.
57. Mabuchi, M., Higashi, K., Wada, S., and Tanimura, S. (1992) Superplastic behavior in as-extruded Al–Cu–Mg alloy matrix composite reinforced with 20 vol. % Si_3N_4 partciculates, *Scripta Metall.* **26**, 1269-1274.
58. Nieh, T.G. and Wadsworth, J. (1993) *Scripta Metall.* **28**, 1119.
59. Mabuchi, M., Higashi, K., Inoue, K., and Tanimura, S. (1992) Experimental investigation of superplastic behavior in a 20 vol. % $Si_3N_4P/5052$ aluminium composite, *Scripta Metall.* **26**, 1839.
60. Koike, J., Mabuchi, M., and Higashi, K. (1995) In-situ observation of partial melting in superplastic aluminium-alloy composites at high-temparatures, *Acta metall. mater.* **43**, 199-206.
61. Mabuchi, M., Higashi, K., Okada, Y., Tanimura, S., Imai, T., and Kubo, K. (1991) Very high strain-rate superplasticity in a particulate $Si_3N_4/6061$ aluminium composite, *Scripta Metall.* **25**, 2517-2522.
62. Hikosaka, T., Imai, T., Nieh, T.G., and Wadsworth, J. (1994) High-strain rate superplasticity of a SiC particulate-reinforced aluminium alloy composite by a vortex method, *Scripta Metall.* **31**, 1181-1186.
63. Grishaber, R.B., Mishra, R.S., and Mukherjee, A.K. (1996) Effect of testing environment on intergranular microsuperplasticity in an aluminum MMC, *Mat. Sci. & Eng. A* **220**, 78-84.
64. Nieh, T.G., Lesuer, D.R., and Syn, C.K. (1995) Tensile and fatigue properties of a 25 vol. % SiC SIC particulate-reinforced 6090-Al composite at 300°C, *Scripta Metall. Mater.* **32**, 707-712.
65. Iribarren, M.J., Agüero, O.E., and Dyment, F. (2001) Co-diffusion along the alpha/beta interphase boundaries of a Zr–2.5% Nb alloy, *Def. Diff. Forum.* **194–199**, 1211-1216.
66. Geguzin, Ya.E. (1984) *Physics of Sintering*, 2nd edition. Nauka, Moscow (in Russian).
67. Eremenko, V.N., Naidich, Yu.V., and Lavrinenko, I.A. (1968) *Sintering in the Presence of Liquid Phase*, Naukova dumka, Kiev (in Russian).
68. Panichkina, V.V., Sirotjuk, M.M., and Skorokhod, V.V. (1982) Liquid-phase sintering of higly disperced W–Cu mixtures, *Poroshk. Metall.* **6**, 21 (in Russian).
69. Skorokhod, V.V., Panichkina, V.V., and Prokushev, N.K. (1986) Microstructural inhomogeneity and localization of densification during liquid-phase sintering of W–Cu powder mixtures *Poroshk. Metall.* **8**, 14 (in Russian).
70. Skorokhod, V.V., Solonin, Yu.M., Filippov, N.I., and Poshin, A.N. (1983) Sintering of W–Cu mixtures *Poroshk. Metall.* **9**, 9 (in Russian).
71. Huppmann, W.J. and Riegger, H. (1975) Modeling of rearrangement processes in liquid phase sintering, *Acta metall.* **23**, 965-971.

ATOMIC ORDERING AS A NEW METHOD
OF PRODUCING A NANOSTRUCTURE

A. I. GUSEV and A. A.REMPEL
Institute of Solid State Chemistry, Ural Division of the Russian Academy of Sciences Pervomaiskaya 91, Yekaterinburg 620219, Russia

1. Introduction

The problem of manufacturing of nanocrystalline powders of metals, alloys, and compounds and nanostructured metallic and ceramic materials for various areas of technology has been discussed for a long time. An interest in this area of research is caused by considerable effect of the nanocrystalline state on the physical, chemical, mechanical, thermal, magnetic and other properties of metals and solid-phase compounds [1-5].

There are many methods for producing a nanostructured state in solids. Present paper is devoted to a new method for preparing bulk and powdered nanostructured substances. Nonstoichiometry and atomic ordering are the physical basis of this method.

What is a nonstoichiometry? According to known definition of Dalton, compound is named stoichiometric if its composition can be presented with the use of small integer coefficients (subscripts), for example, MgO, CH_4, $K_3Fe(CN)_6$. However, there are many solid-phase compounds that contain large quantity of atomic defects. These compounds are normally said to be nonstoichiometric. The composition of nonstoichiometric compound can not be written with the use of only integer coefficients. Vacancies that are unfilled sites of the crystal lattice and interstitial atoms are the atomic defects in nonstoichiometric compounds. Probably, the most familiar among these compounds is wustite FeO which is not available in the stoichiometric form. It always contains excess oxygen, which is due to the presence of vacancies in the iron sublattice. For example, at 1300 K the wustite composition is $Fe_{0.88}O$. At high temperatures wustite has the homogeneity interval from $Fe_{0.84}O$ to $Fe_{0.96}O$ and it has no stoichiometric composition [6]. A considerable deviation from stoichiometry, which is accompanied by the formation of vacancies in the metal sublattice, is observed in iron $Fe_{0.85}S$ and copper $Cu_{1.73}S$ sulfides with the $B8$ (NiAs-type) structure and other compounds.

Nonstoichiometry in solids can be realized by point defect called structural or constitutional vacancy □ [6, 7]. Introduction of structural vacancy in solids leads to a change of a number of atoms in crystal. Therefore structural vacancy is different from Frenkel, Schottky and anti-Schottky defects which are also point defects in a crystal but do not change the stoichiometry of a crystal. Structural vacancies can be formed not only in binary, but also in ternary and more complex compounds. From crystallographical point of view the presence of structural vacancies is due to the mismatch between the chemical composition, i. e. the relative number of atoms of different species, and the relative number of sites in different crystal sublattices occupied by these atoms. From physical or chemical point of view the reason of structural vacancy formation is following. If the chemical composition do not corresponds to the crystal structure and atoms from one species sublattice can not occupy the sites of other species sublattice, i. e. formation of antisite defects is energetically not possible, then structural vacancies are formed in a crystal.

Nonstoichiometry takes place in crystals with two or more kind of atoms and appears more distinctly in oxides, nitrides and carbides of the Group IV and V transition metals. Nonstoichiometry leads to appearance of unfilled crystal lattice sites, i. e. structural vacancies □. The concentration of structural vacancies in above compounds can be varied very precisely and may reach about 50 at.%. The crystals with concentration of structural vacancies of 10 at.% and more, i. e. with a large deviation from stoichiometry are the strongly nonstoichiometric crystals.

By strongly nonstoichiometric interstitial compounds one currently understands transition-metal carbides, nitrides, and oxides with a related cubic $B1$ (NaCl-type) or hexagonal $L'3$ (W_2C-type) structure that arises when carbon, nitrogen, or oxygen atoms intrude into the octahedral interstices of the face-centered cubic or hexagonal close-packed crystal lattice formed by transition-metal atoms [6-9]. The nonstoichiometric interstitial compounds MX_y (or $MX_y\square_{1-y}$) possess wide homogeneity intervals. In the nonmetallic sublattice of a disordered nonstoichiometric compound $MX_y\square_{1-y}$ the interstitial atoms X and structural vacancies □ form a substitutional solution. The presence of structural vacancies may give rise to ordering. For strongly nonstoichiometric interstitial compounds, the disordered state is a state of thermodynamic equilibrium only at a temperature above 1000 to 1300 K, while at lower temperatures (below the order-disorder transition temperature T_{trans}) the state of thermodynamic equilibrium is an ordered state; a disordered state is metastable below T_{trans}.

Experimental investigations of atomic ordering in nonstoichiometric carbides and nitrides MC_y and MN_y (M = Ti, Zr, Hf, V, Nb, and Ta) have been conducted rather intensively over the past twenty years. By the present time more than 30 ordered carbide, nitride and oxide phases have been found [6, 7]. Experimental results indicate that superstructures of the types M_2X, M_3X_2, M_4X_3, M_6X_5, and M_8X_7 are apt to form in cubic nonstoichiometric compounds MX_y. Common formulae of these superstructures is $M_{2t}X_{2t-1}$ where t is equal to 1, 1.5, 2, 3, or 4 [6, 10, 11].

Transition-metal carbides MC_y ($MC_y\square_{1-y}$) are strongly nonstoichiometric compounds [6, 7]. Carbides $MC_y\square_{1-y}$ have a $B1$ cubic structure in the disordered state and their nonmetal sublattice may contain up to 30 or 50 at.% of structural vacancies □

[6-11]. When $T < 1300$ K, the $B1$ structure becomes unstable and disorder-order phase transformations take place in nonstoichiometric carbides. Ordering leads to appearance of ordered phases with complicated superstructures . The order-disorder transformations in the carbides represent transitions of the 1st kind [6-11], which are accompanied by an abrupt change of the volume. Ordering is realized by diffusion and, consequently, it takes a relatively long time. The carbides are synthesized at a temperature from 1400 to 1800 K, which is higher than the temperature of disorder-order phase transformations, T_{trans}. When a nonstoichiometric carbide is cooled from the synthesis temperature to room temperature, it passes the ordering temperature T_{trans} and tends to the ordered state. If a specimen is cooled quickly, the ordering process has not time to complete and the nonstoichiometric carbide remains in a metastable disordered state. Since the lattice parameters of the disordered and ordered phases are different, stresses arise in specimens, leading with time to cracking of crystallites along interfaces between the disordered and ordered phases. By controlling the formation of ordered phase, it is possible to prepare nanostructured nonstoichiometric carbides in dispersed and bulk states.

The present study deals with the structure and properties of nanocrystalline nonstoichiometric vanadium carbide. A nanostructured state is produced by a fundamentally new method, namely by ordering of a nonstoichiometric vanadium carbide. The ordering phenomenon has not been used so far for creation of a nanostructure in solids. The vanadium carbide is chosen as the object of study, because ordering is pronounced most in this compound [12-15]. Both volume effects (connected with the size of particles) and surface effects (connected with the state and the structure of interfaces) are significant for nanocrystalline solids and nanopowders [3-5]. Therefore the study of physical and chemical properties is focused, in particular, on the state of the vanadium carbide surface.

2. The Structure and the Composition

2.1. POWDERED CARBIDE

The starting powdered $VC_{0.875}$ vanadium carbide with particles 1 to 2 μm in size is prepared by carbothermal reduction of the V_2O_5 oxide and then is aged for a long time at the ambient temperature in a stopped vessel preventing the ingress of water vapor from the atmosphere. The aged powder of the vanadium carbide proved to be very hygroscopic. Immediately after the powder was removed from the vessel, it contained not more than 0.2 mass % of physically adsorbed water; after the powder was kept in the ambient atmosphere for a few months, the water concentration was 2.0 mass % and reached the saturation level. A usual coarse-grained VC_y powder does not possess such hygroscopic property. Therefore a high hygroscopicity indicates indirectly to a highly developed surface of the aged vanadium-carbide powder.

A chemical analysis showed that the aged vanadium carbide has the formula $VC_{0.875}$ corresponding to the upper boundary of the homogeneity interval of the $B1$ cubic phase.

The total concentration of the Ti, Nb and Ta metal impurities, which are detected only on the surface of specimen by the EDX method, is about 1 mass %.

TABLE. Chemical composition of the aged vanadium carbide powder (mass %)

V	C_{bond}	C_{free}	H_2O	$O_{chemisorp}$	$O_{lattice}$	N
76.8±0.1	15.9±0.1	0.9±0.1	2.0±0.2	3.1±0.1	0.1	0.2±0.02

The total concentration of oxygen in the powder was as high as 5 mass %. In this connection, a detailed analysis was performed in order to determine the forms of oxygen in the specimen. According to the data of the quantitative gas chromatography, most of the oxygen (3.1 mass %) was present in a chemisorbed form. Another 1.8 mass % of oxygen was contained in adsorbed water (the water content was determined by calcination in a vacuum and using the methods of gas chromatography and thermogravimetry). A small quantity of oxygen (0.1 mass %) is dissolved in the vanadium-carbide lattice. As a result, a small fraction (about 0.4 %) of structural sites in the carbon sublattice is occupied by oxygen atoms. Finally, a very small quantity of oxygen forms a surface oxide film whose thickness is 2 to 3 atomic monolayers. The film contains the V_2O_3 oxide, in the main. The nitrogen concentration is only 0.2 mass % and could not affect considerably the powder properties.

A microscopic examination of the vanadium-carbide powder led to the following findings. A magnification of 100 times revealed separate irregular-shaped agglomerates 5 to 50 μm in size comprising particles nearly 1 μm in size. However, a larger magnification shows that the particles have a complicated structure and represent a set of a large number of nanocrystallites (nanoparticles). At a magnification of 10000 times (Fig. 1), it is readily seen that each of the object about 1 μm in size looks like an open rosebud, green salad or a loose cabbage-head and is made up of nanocrystallites.

Figure 1. Microstructure of vanadium carbide powder $VC_{0.875}$ subjected to prolonged aging at ambient temperature in ambient atmosphere. Particles about 1 μm in size have a complicated structure and contain large number of nanocrystallites (magnification of 10000 times).

Figure 2. Morphology of nanoparticles of aged powder of $VC_{0.875}$ (V_8C_7) carbide at a magnification of 50000 times. Each of the particles, about 1 μm in size, is seen as an open rosebud and consists of nanocrystallites which are shaped as bent petals.

The Figure 2 is taken at a magnification of 50000 times in a DSM 982 Gemini high-resolution scanning electron microscope. It is seen that the nanocrystallites are shaped as bent petals, leaves or flakes. They joined together to form a structure resembling corals or a very loose cabbage-head. This vanadium carbide nanostructure unique and has never been observed before. As the first approximation, the nanocrystallites may be modeled by a disk 400 to 600 nm in diameter and about 15 to 20 nm thick. Although the volume of such disk corresponds to a coarse spherical particle 150 to 220 nm in diameter, the ratio between the surface area S and the volume V is $S/V = 0.107$-0.143 nm^{-1} thanks to a small thickness of the disk. This value corresponds to a specific surface of the powder from 19 to 26 $m^2 g^{-1}$. An electron microscopic examination also reveals a small dispersion of size of nanocrystallites in the powder.

The crystal structure is studied by the method of X-ray diffraction analysis in the $CuK\alpha_{1,2}$ radiation. Figure 3 presents X-ray diffraction spectra of powders of a disordered coarse-grained $VC_{0.875}$ carbide, an annealed coarse-grained ordered $VC_{0.875}$ (V_8C_7) carbide, and a nanostructured $VC_{0.875}$ carbide.

318

In addition to structure reflections of the $B1$ basic phase, only superstructure reflections of a cubic ordered V_8C_7 phase with the $P4_332$ space group are observed in the X-ray diffraction pattern of the $VC_{0.875}$ powder (Fig. 3c).

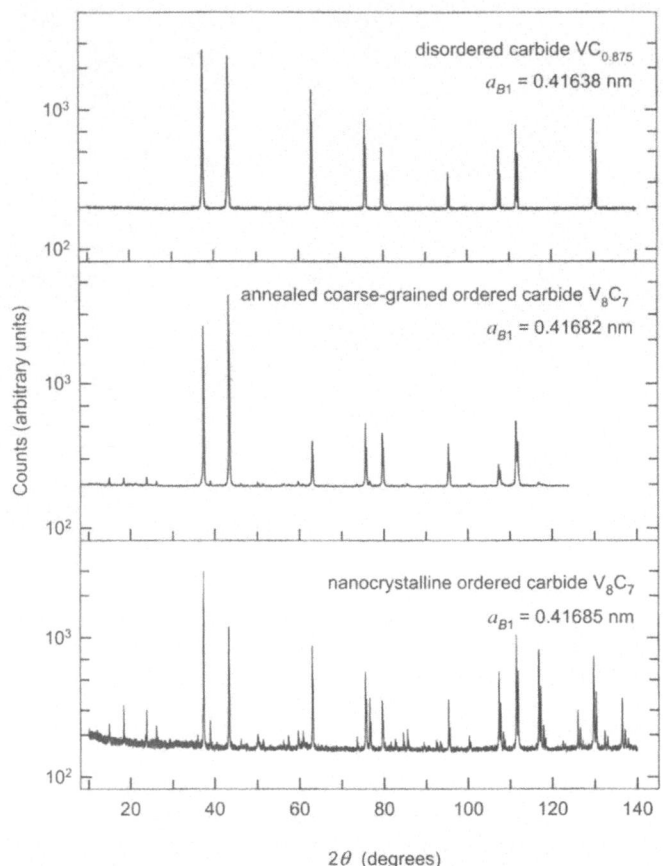

Figure 3. X-ray diffraction patterns of powders of disordered coarse-grained $VC_{0.875}$ carbide (*a*), annealed coarse-grained ordered $VC_{0.875}$ (V_8C_7) carbide (*b*), and nanostructured $VC_{0.875}$ carbide (*c*). $CuK\alpha_{1,2}$ radiation, intensity is shown in logarithmic scale, structure reflections of the $B1$ basic phase are marked. Superstructure reflections which are observed in the X-ray diffraction patterns (*b*) and (*c*) correspond to ordered cubic (space group $P4_332$) V_8C_7 phase.

The lattice constant of the ordered phase is 0.8337 ± 0.0001 nm. An ideal cubic superstructure of the M_8C_7 type with the $P4_332$ space group has a double lattice constant compared to the lattice constant of a disordered $B1$ basic phase [6, 7]. Therefore the lattice constant of the basic phase in the vanadium carbide is $a_{B1} = 0.41685$ nm. This value is much larger (by 0.47 pm) than the lattice constant of the disordered $VC_{0.875}$ carbide. Such large difference of the lattice constants of the ordered and disordered $VC_{0.875}$ carbides is possible only if the degree of order is a maximum or approaches a maximum. Indeed, the intensity ratio of structure and superstructure reflections suggests that the degree of the long-range order in the vanadium carbide approaches a maximum. Moreover, from this ratio it follows that the ordered phase occupies the whole volume of the material, i. e. the powder contains one phase.

Remarkably, the intensity of superstructure reflections I_{super} of coarse-grained ordered $VC_{0.875}$ carbide diminishes with increasing diffraction angle 2θ (Fig. 3b). At 2θ > 100° the intensity of superstructure reflections of the vanadium carbide nanopowder does not decrease, but, on the contrary, grows (Fig. 3c). This is caused probably by large relaxation static displacements of vanadium atoms near carbon vacancies. Such displacements were detected in an ordered V_8C_7 carbide [14].

Although thickness of the nanocrystallites is about 20 nm, the diffraction reflections does not reveal considerable deviations from the instrumental width. Since all atoms in a crystallite were scattered coherently, the absence of broadening of the diffraction lines agrees with the presence of a large number of atoms in the nanocrystallites because of their large diameter.

The nanostructured $VC_{0.875}$ powder has a density of 5.15 g cm^{-3} as measured by the pycnometric method. This value is much smaller than the theoretical density. The carbide, which is annealed in a vacuum at 900 K, has a density $\rho = 5.62$ g/cm^3, which coincides with the theoretical density. This means that a small starting density is due to adsorbed impurities of water and oxygen.

2.2. BULK SPECIMENS

To prepare specimens of a bulk nonstoichiometric $VC_{0.875}$ vanadium carbide, a powder of a disordered $VC_{0.875}$ carbide is pressed at a temperature of 2000 K and a pressure of 20 to 25 MPa in a flow of extra-pure argon. The bulk specimens are treated under three thermal routes. First route is annealing at a temperature of 1370 K for 2 h and slow cooling to 300 K at a rate of 100 K h^{-1}. Second and third routes are quenching from 1420 K to a room temperature and quenching from 1500 K to a room temperature. For quenching the bulk specimens are put in evacuated quartz ampoules and are annealed at 1420 and 1500 K for 15 minutes. Then ampoules with specimens are dropped in water. The quenching rate is 100 K sec^{-1}. The grain size of quenched bulk vanadium carbide is 10 to 60 μm.

An examination of the structure of the annealed and quenched specimens of the $VC_{0.875}$ carbide shows that along with structure reflections, the X-ray diffraction spectra contain additional weak lines after annealing or quenching (Fig. 4). Judging by their position, the additional lines are superstructure reflections and correspond to an ordered cubic V_8C_7 phase with the $P4_332$ space group. The superstructure reflections of the annealed and quenched specimens have nearly equal integral intensities, but their width is different. The widest superstructure reflections are observed in the X-ray diffraction pattern of the specimen quenched from 1500 K (Fig. 4c).

In accordance with the phase equilibrium diagram of the V − C system [6, 7, 16], an ordered V_8C_7 phase is formed thanks to the disorder–order transformation at a temperature T_{trans} = 1380 K. The experimental phase transformation temperature is 1413 ± 20 K [12]. This evidence suggests that quick cooling from 1420 K or 1500 K

320

must lead to saving of the disordered state of nonstoichiometric $VC_{0.875}$ vanadium carbide as a metastable state. However, even if the specimen is quenched from 1500 K, an ordered V_8C_7 phase appears and the relative intensity of the superstructure reflections is nearly equal to the relative intensity in the specimens quenched from 1420 K or annealed at 1370 K (Fig. 4).

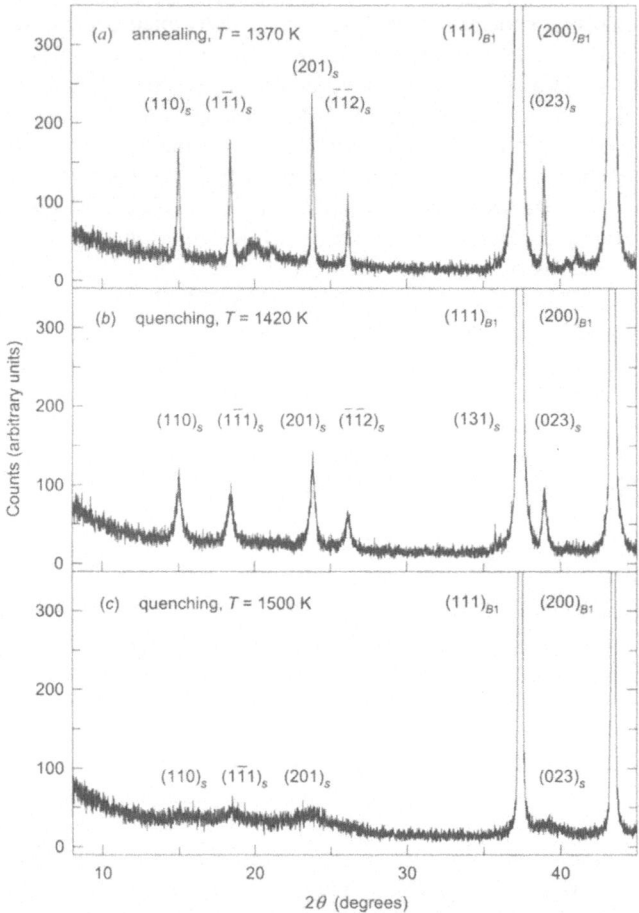

Figure 4. X-ray diffraction patterns of bulk specimens of $VC_{0.875}$ carbide which are produced by hot pressing and are subjected to additional thermal treatment: (*a*) annealing at 1370 K for 2 h and subsequent slow cooling to 300 K at rate of 100 K h^{-1}, (*b*) quenching from 1420 to 300 K at rate of 100 K s^{-1}, (*c*) quenching from 1500 to 300 K at rate of 100 K s^{-1}. Cu$K\alpha_{1,2}$ radiation, intensity is shown in logarithmic scale, structure reflections of the $B1$ basic phase and superstructure reflections of the ordered V_8C_7 phase are marked.

As a result of ordering, each grain of the disordered basis phase breaks down into domains of the ordered phase. A domain has a high degree of order, while the mutual spatial location of the domains is chaotic and depends on the ratio between the structures of the ordered phase and the disordered matrix. Optical microscopic examination at magnification of 200 times shows that formation of the ordered phase started on the boundaries of grains of the disordered phase. The grains of basic phase in the disordered carbide have a clear strength boundaries (Fig. 5a); after annealing and ordering these boundaries become curved (Fig. 5b). It means that the domains of ordered phase grows in grains of disordered basic phase in the direction from boundary to center of grain. Unfortunately, small size of domains and the same cubic symmetry of the disordered and ordered phases do not allow to observe the domain boundaries by optical microscopy.

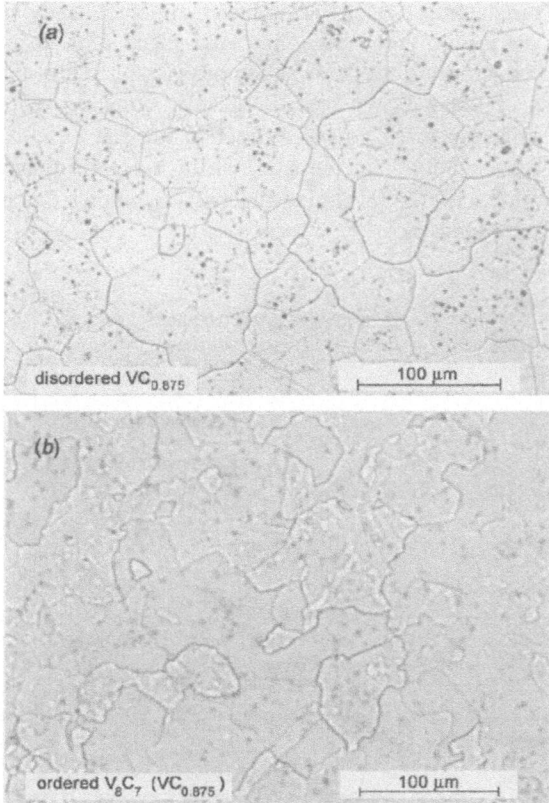

Figure 5. Microstructure change of compacted nonstoichiometric vanadium carbide $VC_{0.875}$ as a result of ordering: (*a*) grains of disordered $VC_{0.875}$ carbide have clear strength boundaries, (*b*) boundaries of grains of basic phase in ordered V_8C_7 ($VC_{0.875}$) carbide are curved because of formation of domains of the ordered phase.

According to X-ray investigation, the width of the structure reflections is independent of the thermal treatment conditions of the bulk $VC_{0.875}$ specimens. Therefore one may think that the size of grains of the basic phase remains unchanged during ordering. On the contrary, the superstructure lines are broadened considerably. This broadening may be due to a small size of the domains of the ordered phase, which is formed under different thermal treatment conditions. Does a nanostructure appear in bulk specimens of thermally treated $VC_{0.875}$ carbide containing an ordered V_8C_7 phase? How small are domains of the ordered phase? These questions may be answered if one measures the width of experimental diffraction reflections and compares it with the resolution function of the diffractometer, θ_R.

The resolution function of a Siemens-D500 autodiffractometer is determined in a special diffraction experiment using an annealed specimen of a stoichiometric tungsten carbide with grains 10 to 20 μm in size. The tungsten carbide has not a homogeneity interval therefore broadening which is induced inhomogeneity is absent. Size broadening of the diffraction lines does not take place with these grains either. Annealing of the specimen rules out deformation broadening. Thus, the width of some diffraction reflection of the tungsten carbide actually coincides with the resolution function θ_R of the diffractometer at a given diffraction angle θ.

A comparison of the width of superstructure diffraction reflections of specimens of a bulk vanadium carbide with the resolution function, θ_R, shows that the experimental reflections are broadened (Fig. 6).

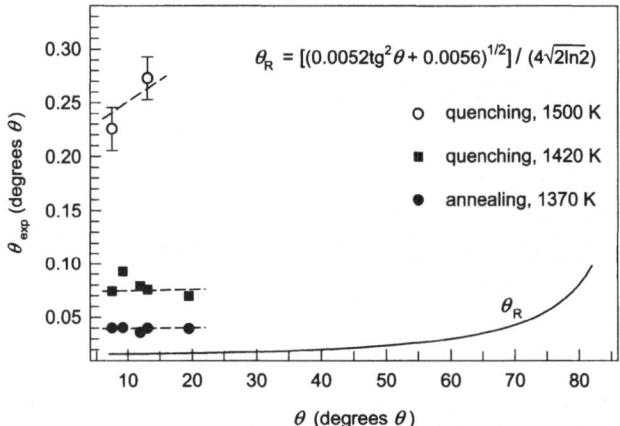

Figure 6. Broadening of X-ray diffraction reflections of ordered phase V_8C_7 for quenched and annealed bulk specimens of vanadium carbide $VC_{0.875}$

Since the width of these reflections depends on the domain size and the instrumental resolution, the size of domains can be determined from the measured broadening $\beta = \sqrt{\theta_{exp}^2 - \theta_R^2}$. We shall assume as the first approximation that deformation broadening is absent and the observed broadening β is due only to a small size of domains, therefore $\beta = \beta_s$. In turn, size broadening $\beta_s(2\theta) \equiv 2\beta_s(\theta)$ measured in radians is related to an average domain size D as

$$D = \frac{K_{hkl}\lambda}{\beta_s(2\theta)\cos\theta} = \frac{K_{hkl}\lambda}{2\beta_s(\theta)\cos\theta}, \tag{1}$$

where $K_{hkl} \approx 1$ is the coefficient, which depends on a shape of a particle (crystallite, domain) and on Miller indices of diffraction reflection (hkl), λ is the radiation wavelength.

From Figs. 4 and 6 it is seen that broadening of the superstructure reflections is the greatest for the specimen quenched from 1500 K. The least broadening of the superstructure reflections is observed for the vanadium carbide specimen annealed at 1370 K. This means that domains of the ordered phase are the smallest in the specimen quenched from 1500 K. Domains are the largest in the annealed vanadium carbide specimen, because annealing and subsequent slow cooling favor the domain growth. The size of domains in the annealed specimens (Fig. 4a) is equal to 127 ± 10 nm; in the specimens quenched from 1420 K (Fig. 4b) and 1500 K (Fig. 4c) the size of domains is 60 ± 9 and 18 ± 12 nm, respectively, with the probability confidence of 95 %.

Thus, annealing of bulk specimens of a nonstoichiometric $VC_{0.875}$ vanadium carbide and quenching of these specimens from $T_{trans} \pm 100$ K causes appearance of a nanostructure representing a set of domains of the ordered phase. The least domain size corresponds to the largest quenching temperature. Depending on the thermal treatment conditions of the bulk specimens of nonstoichiometric carbide, the domain size can be varied from 10 to 100 nm.

3. Electron-Positron Annihilation

Electron-positron annihilation is the most efficient and sensitive method for the study of defects at interfaces and on the surface of nanoparticles. The capture of positrons in such defects as vacancies or nanovoids prolongs the lifetime of positrons as compared to their lifetime in defect-free materials [17]. The defect type may be determined from the positron lifetime.

The positron lifetime τ is measured using powdered nanocrystalline vanadium carbide, which was calcined beforehand at 400 K to remove water. For comparison, the positron lifetime in a sintered coarse-grained specimen of the $VC_{0.875}$ vanadium carbide is measured. The obtained spectra of the positron lifetime are shown in Fig. 7.

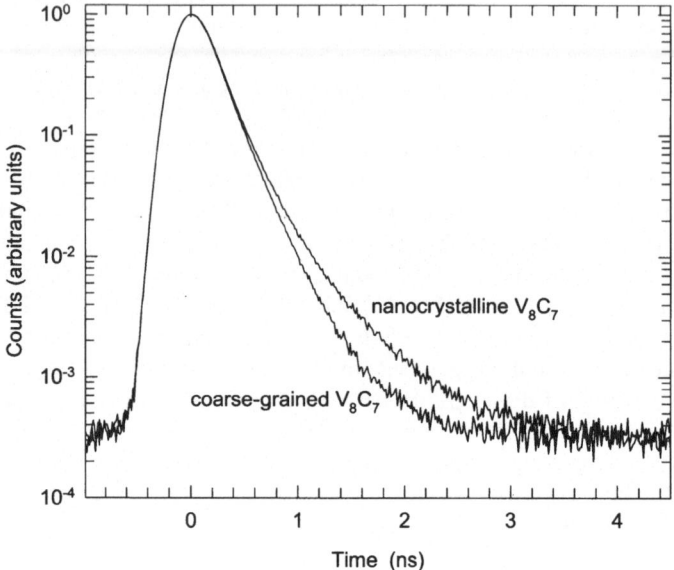

Figure 7. Positron lifetime spectra of nanocrystalline and coarse-grained V_8C_7 carbide

These spectra suggest that the mean positron lifetime in the nanopowder is much longer than in the polycrystal. The spectrum of the coarse-grained vanadium carbide contains only a short component of 157 ± 2 ps, which corresponds to annihilation of positrons at a structural vacancy of the carbon sublattice [18-20]. In addition to the short component, the spectrum of the nanostructured powder has a long component of 500 ps with a relative intensity $I_2 = 7\%$. The long component is due to annihilation of positrons at defects on the particle surface. The intensity of the components is proportional to the volume fraction of the phases containing different types of defects. Therefore the relative intensity of the long component, I_2, coincides with the volume fraction of the surface $\Delta V_{surf} = \Delta D \times S/V$ in the vanadium carbide nanopowder. Calculation shows that thickness of the surface layer is $\Delta D = 0.5$–0.7 nm, which corresponds to 3–4 atomic monolayers.

Thus, measurements of the positron lifetime shows that nanocrystallites of the powdered vanadium carbide only contain nonmetallic structural vacancies in the bulk and defects like vacancy agglomerates on the surface.

4. Sintering of Nanopowder

Nanostructured vanadium carbide V_8C_7 ($VC_{0.875}$) powder was compacted at a room temperature and pressure of 10 MPa. The density of compact was equal to 68 % of theoretical density; this value is much larger than bulk density of nanopowder which is equal to 36 %. After that compacted specimen was sintered in vacuum of 10^{-3} Pa in temperature interval from 400 to 2000 K with a step of 100 K. The holding time at each temperature was 2 hours. Considerable change of the density of sintered specimen as compared with the density of compact was not detected.

In spite of large porosity (~30 %) of the specimen, sintering allowed to measure Vickers microhardness. The sintered specimen was ground and Vickers microhardness was measured at load of 200 and 500 g and loading time of 10 s. Microhardness does not exhibit any dependence versus the load within the measurement error. Microhardness H_V was equal to 60-80 GPa. For a comparison, a measured microhardness of coarse-grained vanadium carbide $VC_{0.875}$ specimen, which was prepared by hot pressing, is 21 GPa at load of 100 g. According to the data [21], a microhardness of coarse-grained disordered vanadium carbide is equal to 29 GPa under load of 100 and 200 g. Thus, microhardness of a specimen prepared by sintering of vanadium carbide nanopowder is 2 to 3 times higher than microhardness of coarse-grained disordered vanadium carbide and approaches microhardness of diamond. It is known that at 300 K microhardness of nanomaterials is 2 to 7 times higher than H_V of polycrystalline materials [3, 4]. High microhardness of the carbide $VC_{0.875}$ specimen prepared from a nanopowder may be explained by the Hall–Petch law $H_V \approx H_0 + kD^{-1/2}$, i. e. may be due to the decrease in grain size. Considering the results obtained, it may be assumed that ordering also contributes to the increase in microhardness of nanocrystalline carbide $VC_{0.875}$. Generally, formation of an ordered nanostructure in vanadium carbide $VC_{0.875}$ causes microhardness to increase by a factor of 2 to 3.

It should be noted that nanostructure is not found in sintered vanadium carbide specimen. The study of sintered specimen in ISI-DS 130 scanning electron microscope shows that its microstructure represents a set of well sintered agglomerates with free space between them (Fig. 8). It agrees with a large porosity of sintered specimen. In order to determine, whether sintered agglomerates have a nanostructure, additional high-resolution transmission electron microscopic studies are necessary in a future.

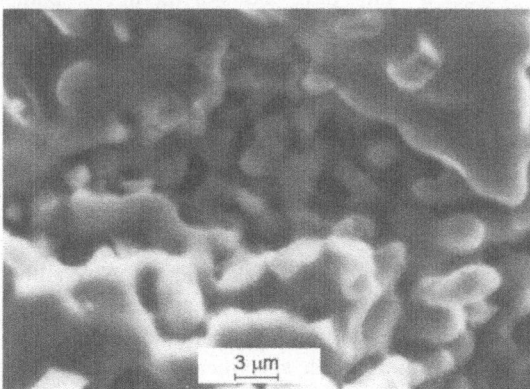

Figure 8. Microstructure of sintered vanadium carbide nanopowder observed by scanning electron microscopy. Very densely sintered agglomerates and free space (about of some tens μm in size) between them are seen.

5. Conclusion

Powdered and bulk specimens of a nonstoichiometric $VC_{0.875}$ vanadium carbide form a nanostructure thanks to a disorder–order $VC_{0.875} \to V_8C_7$ phase transformation in the carbide. Nanocrystallites of a powdered nonstoichiometric vanadium carbide represent strongly bent disk plates 400 to 600 nm in diameter and 15 to 20 nm thick. The bulk of the nanocrystallites are an ordered V_8C_7 carbide with a high degree of long-range order and a negligibly small concentration of dissolved oxygen. The surface layer of the crystallites contains about 3 mass % of chemisorbed oxygen and a considerable number of vacancy agglomerates. This points to a loose structure of the surface layer. Thickness of the surface phase does not exceed 0.7 nm or four atomic monolayers.

The observed morphology of the nanopowders of a nonstoichiometric vanadium carbide may be explained by cracking of grains at the interface between the disordered and ordered phases. Indeed, high-temperature X-ray measurements [12] showed that at a temperature of 1413 ± 20 K the lattice constant of the face-centered cubic sublattice increases abruptly by 0.4 pm as a result of a disorder–order $VC_{0.875} \to V_8C_7$ phase transformation. The size of domains of the ordered phase is about 20 nm. According to [6, 7, 15, 16], $VC_{0.875} \to V_8C_7$ ordering is realized by the mechanism of the phase transition of the 1st kind at 1368 ± 12 K. At 300 K the basic lattice constant a_{B1} of a quenched disordered $VC_{0.875}$ carbide is smaller than that of the ordered carbide with a similar carbon concentration [13, 15]. Different volumes of the disordered and ordered phases lead to stresses and cracking along interfaces between the phases.

A nanostructure is formed by another mechanism in bulk specimens of a nonstoichiometric $VC_{0.875}$ vanadium carbide. Domains of the ordered phase appear after the disorder–order transformation. The least domain size corresponds to the largest temperature of the thermal treatment (quenching or annealing) and the largest cooling rate.

The study shows that ordering represents an efficient method for creation of a nanostructure in powdered and bulk nonstoichiometric compounds. Probably, disorder–order transformations, which are accompanied by a sudden change of the molar volume, may be used for producing a nanostructured state in such substances as strongly nonstoichiometric oxides and nitrides and substitutional solid solutions including some alloys (especially, intermetallic alloys with large brittleness).

6. Acknowledgement

The authors would like to thank Mr. H. Labitzke and Ms. A. Weißhardt for the electron microscopic examination of the vanadium carbide specimens, Dr. V. N. Lipatnikov for the microscopic investigation of ordering of vanadium carbide and Ms. O. V. Makarova for her help in certification of the carbide specimens.

7. References

1. Gleiter, H. (1992) Materials with ultrafine microstructure: retrospectives and perspectives, *Nanostruct. Mater.* **1**, 1-19.
2. Gleiter, H. (1995) Nanostructured materials: state of art and perspectives, *Nanostruct. Mater.* **6**, 3-14.
3. Gusev, A. I. (1998) Effects of the nanocrystalline state in solids, *Uspekhi Fiz. Nauk* **168**, 55-83 (in Russian). (Engl. transl.: (1998) *Physics - Uspekhi* **41**, 49-76)
4. Gusev, A. I. (1998) *Nanocrystalline Materials:Preparation and Properties*, Ural Division of the Russian Academy of Sciences, Yekaterinburg (in Russian).
5. Gusev, A. I. and Rempel, A. A. (2000) *Nanocrystalline Materials*, Nauka–Fizmatlit, Moscow (in Russian).
6. Gusev, A. I., Rempel, A. A. and Magerl, J. (2001) *Disorder and Order in Strongly Nonstoichiometric Compounds: Transition Metal Carbides, Nitrides and Oxides*, Springer, Berlin-Heidelberg.
7. Gusev, A. I. and Rempel, A. A. (2000) *Nonstoichiometry, Disorder and Order in Solids*, Ural Division of the Russian Academy of Sciences, Yekaterinburg (in Russian).
8. Gusev, A. I. and Rempel, A. A. (1988) *Structural Phase Transitions in Nonstoichiometric Compounds*, Nauka, Moscow (in Russian).
9. Gusev, A. I. (1991*) Physical Chemistry of Nonstoichiometric Refractory Compounds*, Nauka, Moscow (in Russian).
10. Gusev, A. I. (1991) Disorder and long-range order in nonstoichiometric interstitial compounds: transition metal carbides, nitrides and oxides, *Phys. Stat. Sol.* (b) **163**, 17-54.
11. Gusev, A. I. and Rempel, A. A. (1993) Superstructures of non-stoichiometric interstitial compounds and the distribution functions of interstitial atoms, *Phys. Stat. Sol.* (a) **135**, 15-58.
12. Athanassiadis, T., Lorenzelli, N. and de Novion, C. H. (1987) Diffraction studies of the order-disorder transformation in V_8C_7, *Ann. Chim. France* **12**, 129-142.
13. Lipatnikov, V. N., Lengauer, W., Ettmayer, P., Keil, E., Groboth, G. and Kny E. J. (1997) Effects of vacancy ordering on structure and properties of vanadium carbide, *J. Alloys Comp.* **261**, 192-197.
14. Rafaja, D., Lengauer, W., Ettmayer, P. and Lipatnikov, V. N. (1998) Rietveld analysis of the ordering in V_8C_7, *J. Alloys Comp.* **269**, 60-62.
15. Lipatnikov, V. N., Gusev, A. I., Ettmayer, P., Lengauer, W. (1999) Phase transformations in non-stoichiometric vanadium carbide, *J. Phys.: Condens. Matter* **11**, 163-184.
16. Gusev, A. I. (2000) A phase diagram of the vanadium–carbon system taking into account ordering in nonstoichiometric vanadium carbide, *Zh. Fiz. Khimii* **74**, 600-606 (in Russian). (Engl. transl.: (2000) *Russ. J. Phys. Chem.* **74**, 510-516).
17. Würschum, R. and Schaefer, H.-E. (1996) Interfacial free volumes and atomic diffusion in nanostructured solids, in A. S. Edelstein and R. C. Cammarata (eds.), *Nanomaterials: Synthesis, Properties and Applications*, Institute of Physics Publishing, Bristol, pp.277-301.
18. Rempel, A. A., Forster, M. and Schaefer, H.-E. (1992) Positron lifetime in carbides with *B*1 structure, *Doklady Akad. Nauk SSSR* **326**, 91-97 (in Russian). (Engl. transl.: (1992) *Sov. Physics Doklady* **37**, 484-487)
19. Rempel, A. A., Forster, M. and Schaefer, H.-E. (1993) Positron lifetime in non-stoichiometric carbides with a *B*1 (NaCl) structure, *J. Phys.: Condens. Matter* **5**, 261-266.
20. Rempel, A. A., Zueva, L. V., Lipatnikov, V. N. and Schaefer, H.-E. (1998) Positron lifetime in the atomic vacancies of nonstoichiometric titanium and vanadium carbides, *Phys. Stat. Sol.* (a) **169**, R9-R10.
21. Ramqvist, L. (1968) Variation of hardness, resistivity and lattice parameter with carbon content of group 5b metal carbides, *Jernkontors Annaler.* **152**, 467-475.

PARTICULARITIES OF MARTENSITIC TRANSFORMATIONS IN NANOSRUCTURED FE-MN SYSTEM OBTAINED BY MECHANICAL ALLOYING

L.YU.PUSTOV[1], S.D. KALOSHKIN[1], E.I. ESTRIN[2], G. PRINCIPI[3]
V.V. TCHERDYNTSEV[1], E.V. SHELEKHOV[1]
[1]*Moscow State Institute of Steel and Alloys, Leninsky prosp.4, Moscow, 119991, Russia*
[2]*Central Research Inst. for Ferrous Metalurgy, 2rd Baumanskaya st, 9/23, Moscow, 107005, Russia*
[3]*Settore Materiali and INFM, DIM, via Marsolo 9, 135131 Padova, Italy*

1. Introduction

Martensitic transformation (MT) in the Fe-rich alloys always attracts much attention. Iron-rich alloys of Fe-Mn system belong to martensitic class. Depending on the Mn content two types of martensite structures b.c.c and/or h.c.p., may be formed through non-diffusional martensitic mechanism upon cooling from the temperature range of f.c.c. phase stability [1]. Only b.c.c. type appears when Mn content is less than 10 %, at higher Mn concentrations h.c.p. martensite exists.

Austenite to martensite transformation temperature (M_s) occur sensitive to parameters of alloy structure determined by the method of its preparation and treatment, including deformation, and these questions always were the subject of metal science research [2]. It has been well established by some previous studies that the austenite grain size has a considerable influence on M_s temperature in Fe alloys [3-5]. For example, significant decrease of the M_s temperature was found for Fe-Ni alloys rapidly quenched from the melt [6]. Thus, the grain size is one of the important factors, which affects the behavior of MT.

It is well known that martensite nucleation is heterogeneous in nature [7] and the probability of finding a nucleation site for MT in powder decreases compared to that for bulk alloys [7]. Moreover, other important factors as different structure defects, specimens size, external stresses strongly effect on MT [3-5].

Mechanical alloying (MA) of elemental powders, rather simple method of the most severe mechanical treatment of materials resulting nonequilibrium and very defective, as well as nano- structure formation at relatively low temperature (0.3-0.4 T_m), causes the degree of material's deformation in this method is unlimited

T. Tsakalakos et al. (eds.) pgs. 329 - 337
Nanostructures: Synthesis, Functional Properties and Applications,
© *2003 Kluwer Academic Publishers.*

It is known that after nanostructuring even well-known alloys and phases may attain new properties. In a contrast to quenching especially significant influence on MT points can be achieved at using of MA method of alloy producing. Actually, using the MA technique for Fe-Ni alloys preparation resulted in significant decrease of MT points [9,10]. Recently the phase compositions of MA Fe-Mn [10] and Fe-Ni [11,12] alloys were studied in wide concentration ranges. We also reported about significant retardation of MT in $Fe_{100-y}Ni_y$ (y=20÷30), alloys obtained by MA from elemental powder mixtures, comparing with the equilibrium alloys [13,14].

The present study was undertaken to observe the MT behavior in nanostructured by MA Fe-Mn alloys.

2. Procedure

The MA of powder mixtures of carbonyl iron (99,95 %), electrolytic manganese (99,5 %) was carried out in the planetary ball mill AGO-2U for preparation of binary compositions $Fe_{100-x}Mn_x$ (x = 6,8,10,14,16,18,20,22,30 at.% Mn). 15 g of material and 150 g of steel balls with 7.8 mm diameter were loaded into steel hermetic vials of 160 cm^3. The process was conducted in argon atmosphere. The intensity of milling, determined both experimentally and theoretically [15-16], was about 30 watt per gram. In spite of water-cooling of the vials, the average temperature of the material, determined as described in [15-17], was about 300°C.

X-ray diffractometry (XRD) was used to determine the structure and phase composition of the samples. Temperatures of α→γ and γ→α transformations were determined by thermo-magnetic method using differential measuring transformer. Before heating the Fe-Mn samples were placed in the hermetic quarts ampoules, then absorption of oxygen and nitrogen was realized by preliminary heating of magnesium getter inside. Low temperature measurements were conducted by dipping of the samples into liquid nitrogen. Temperature was measured by thermocouple buried into the sample powder. Heating and cooling speed was about 20 °C/min. Annealing of the samples was carried out in an argon atmosphere. Slight deformation was performed by pounding the powder material in mortar with pestle.

3. Results and Discussion

3.1. STRUCTURE OF THE SAMPLES AFTER MECHANICAL ALLOYING

Phase compositions of the as-milled and annealed samples, determined by XRD are shown in Figure 1. Single b.c.c. phase was observed in as-milled $Fe_{100-x}Mn_x$ alloys at x ≤ 8 at.%, and mixture of b.c.c. and f.c.c. phases was obtained at 10 ≤ x ≤ 30 at.%. Figures 2a and 2d show XRD patterns of "boundary" $Fe_{92}Mn_8$ and $Fe_{90}Mn_{10}$. compositions

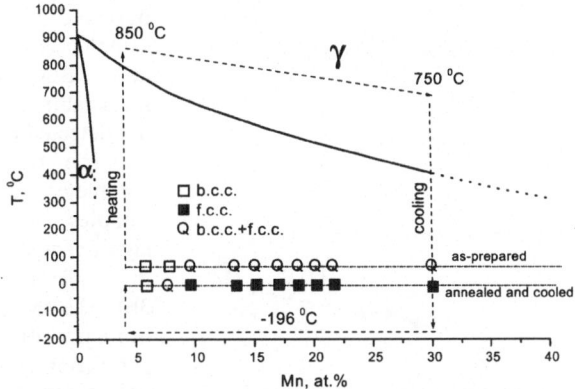

Figure 1. Phase composition of MA Fe-Mn samples, as-prepared and after thermal treatment, plotted on the part of Fe-Mn stable phase diagram.

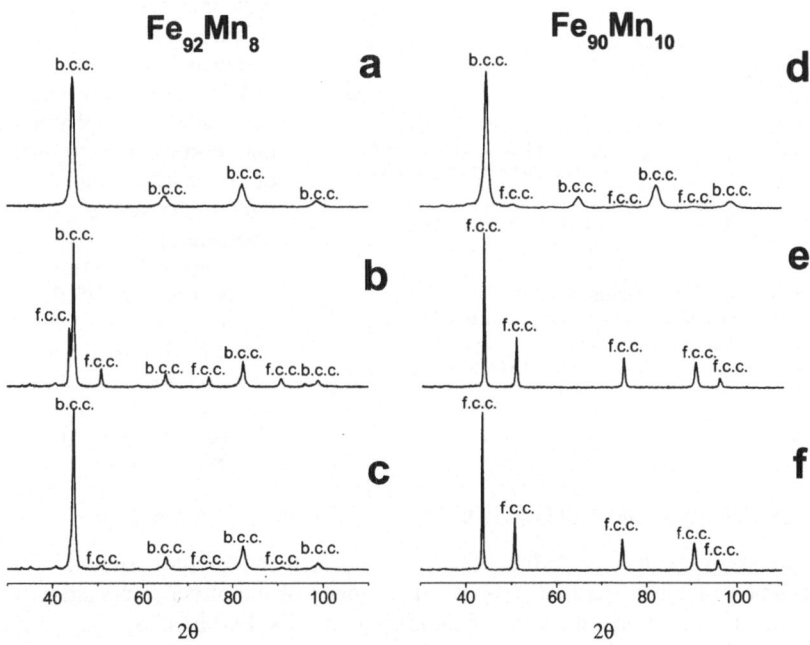

Figure 2. XRD patterns of MA Fe$_{92}$Mn$_8$ and Fe$_{90}$Mn$_{10}$ alloys (room temperature): (a, d) - as-milled, (b,e) – annealed at 850 $^{\circ}$C and 800 $^{\circ}$C, correspondingly, (c, f)- then cooled to -196 $^{\circ}$C.

The results of quantitative phase analysis are presented in Figure 3a. Good agreement of f.c.c. phase content with the data for conventionally prepared [18] was observed for as-milled samples containing up to 22 at.% Mn. However, the metastable h.c.p. phase was not detected in these alloys.

The f.c.c phase lattice parameter (a) of as-milled samples occured to be significantly higher than for the stable alloys [19] (Figure 3b). The same phenomena of influence of MA on lattice parameter was oserved previously for Fe-Mn [10] and Fe-Ni [20] systems.

The stacking faults concentration (P) decreases with increase of Mn content (Figure 3c), which is associated with increasing of difference in Gibbs energy of f.c.c. and h.c.p. phases [10], that corresponds to increase of the stacking faults energy with increasing of Mn concentration.

Figure 3d shows that the block size (D) of the b.c.c. phase determined from XRD was found to be from 8 to 13 nm, and about 10 nm for the f.c.c. one. This is an evidence of nanostructure formation by MA treatment.

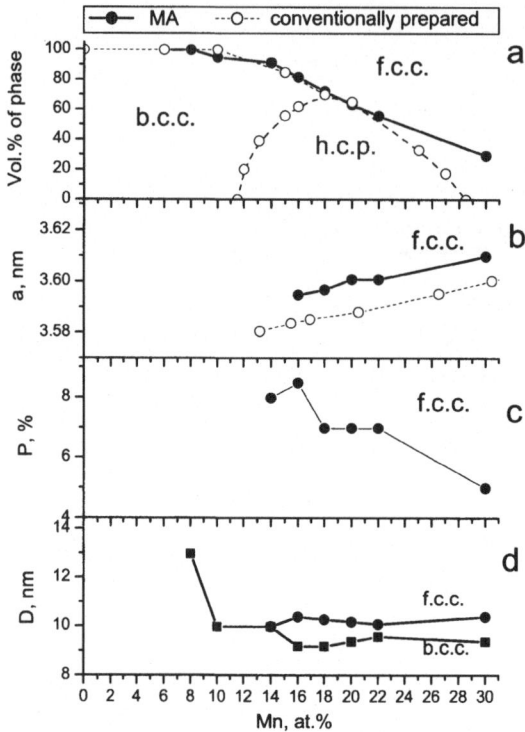

Figure 3. Concentration dependences of (a) – phase composition, (b) – lattice parameters for f.c.c. phase, (c) - stacking fault concentration in f.c.c., (d) – block sizes of b.c.c. and f.c.c. for as-prepared MA $Fe_{100-x}Mn_x$ alloys

3.2. PHASE TRANSFORMATIONS DURING HEATING AND COOLING

As might be expected annealing of severe deformed samples should lead to decomposition of supersaturated phases and approach of the phase composition to the stable state. However, in the case of annealing of MA Fe-Mn alloys the situation absolutely different. F.c.c. structure occurs much more stable in MA Fe-Mn alloys as compared with conventionally prepared ones. As it seen in Figure 4, heating of the as-milled samples resulted in the decrease of the magnetisation due to lessening of the b.c.c. phase content owing to start of $\alpha \rightarrow \gamma$ transformation. At the end of the transformation, samples become non-ferromagnetic, and there was only f.c.c. phase in the structure. Similar heating curves were obtained for all alloys underinvestigation.

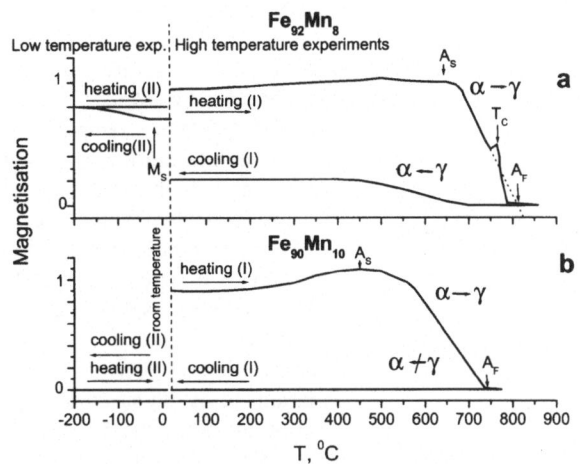

Figure 4. Heating-cooling curves of the as-milled $Fe_{92}Mn_8$ (a) and $Fe_{90}Mn_{10}$ (b) samples.

Two types of cooling curves were obtained depending on Mn content: with and without appearance of ferromagnetism.

In low manganese Fe-Mn MA alloys, in particular, in 8 at.% Mn sample, cooling results in partial $\gamma \rightarrow \alpha$ transformation. Cooling curve of $Fe_{92}Mn_8$ alloy is shown in Figure 4a. According to thermo-magnetic data, approximately 30 % of the b.c.c. phase appeared in the structure after cooling to 20 °C.

The two-phase structure was confirmed by XRD (Figure 2b), though according to XRD data there was 74 vol. % of b.c.c. martensite in the structure. The difference between the magnetization and XRD data, may be explained by continuation of isothermal MT while being kept at 20 °C after magnetization measurement until XRD experiment (about 800 hours).

In fact, isothermal MT was detected in $Fe_{92}Mn_{08}$ alloy upon cooling to temperatures lower than 20 °C - the second series of magnetization measurements. The isothermal character of transformation was established by the increasing the intensity of the magnetization of the sample during warming from -196 to 20 °C and at subsequent cooling cycle (Figure 4a, II). This experiment was performed about 1000 hours later the first one. Results of the intensity of magnetization agreed well with that for XRD analysis: b.c.c. phase content was increased in structure in comparison with that for the alloy cooled to 20 °C from 850 °C (Figure 2c). According to XRD analysis 90 vol. % of b.c.c. martensite was observed in this alloy after three cycles of cooling-warming from the 20 to -196 °C and back.

The block size of b.c.c. martensite obtained by XRD was found to be from 27 to 46 nm, i.e., martensite crystallites upon cooling were nanosized due to preliminary MA treatment.

In the alloys with ≥ 10 at.% Mn, f.c.c. phase, derived during heating, did not transform into b.c.c. while cooling to -196 °C. Typical cooling curve of the $Fe_{90}Mn_{10}$ sample of this range is shown in Figure 4b. According to XRD analysis, there were just f.c.c. phase peaks on spectra of these alloys (Figure 2e).

Thus, the obtained f.c.c. phase possessed the higher stability to MT into b.c.c. phase compared with that for alloys obtained by conventional techniques.

Summarising the data on phase composition of MA and subsequently annealed Fe-Mn alloys of this research(Figure 1), one can see that the concentration ranges of f.c.c.

phase existence became significantly wider when compared to that for conventionally prepared alloys. This phenomena was noticed previously for MA Fe-Ni system too [13, 14]

Annealing led to the increase of the block size of f.c.c. structure, though it remained nanostructured. Block sizes of f.c.c. structure for MA $Fe_{90}Mn_{10}$, $Fe_{84}Mn_{16}$ and $Fe_{82}Mn_{18}$ samples are illustrated in Figure 5. Long-term annealing of the MA samples results in further increase of the block size of f.c.c. phase (Figure 5). After 3 h annealing at 750 ^{0}C MT is still suppressed for $Fe_{84}Mn_{16}$ and $Fe_{82}Mn_{18}$ compositions, there was no martensite in the structure. After 8 h of annealing at 750 ^{0}C only 3 % of b.c.c. martensite was detected in structure of $Fe_{84}Mn_{16}$ alloy (Figure 3a), while for conventionally prepared alloy, there must be approximately 20% of b.c.c. and 62 % h.c.p. phases.

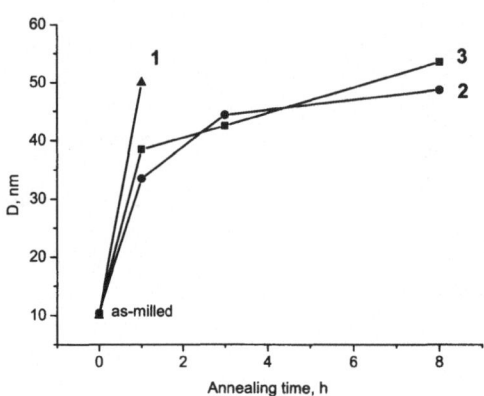

Figure 5. Dependence of f.c.c. structure block size on annealing regime for MA samples:
1 - $Fe_{90}Mn_{10}$ (annealing at 800 ^{0}C), 2 - $Fe_{84}Mn_{16}$ (750 ^{0}C) and 3- $Fe_{82}Mn_{18}$ (750 ^{0}C).

MA $Fe_{82}Mn_{18}$ alloy remained fully austenic after the same annealing regimes. Increasing the f.c.c. structure block size (Figure 5) did not lead to the development of MT. There is no b.c.c. martensite in the structure of the conventionally prepared alloy of the same composition, but there has to be 70 % of h.c.p. martensite (Figure 3a). However, appearance of h.c.p. martensite at cooling was not detected neither in this nor in other MA Fe-Mn alloys of studied concentration range.

Probably the main cause of suppression of MT is not only small block size of MA f.c.c. structure, but also the structure defects and their interactions.

3.3. MARTENSITIC TRANSFORMATIONS INDUCED BY DEFORMATION

Slight deformation of MA and subsequently annealed powder samples resulted in changing of phase composition: f.c.c. phase under deformation partially transforms to b.c.c. and/or h.c.p. depending on composition.

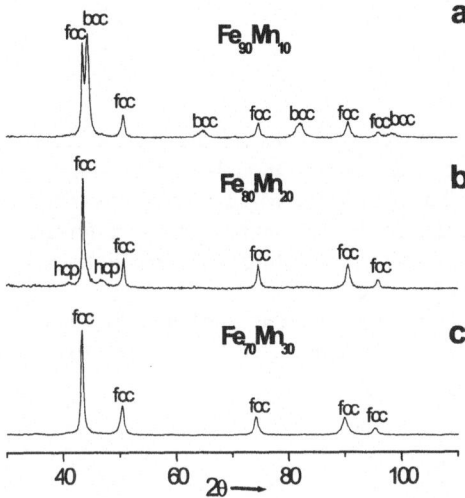

Figure 6. XRD patterns of MA $Fe_{100-x}Mn_x$ alloys after heating-cooling cycle, then deformed: x=10 (a), 20 (b) and 30 % (c).

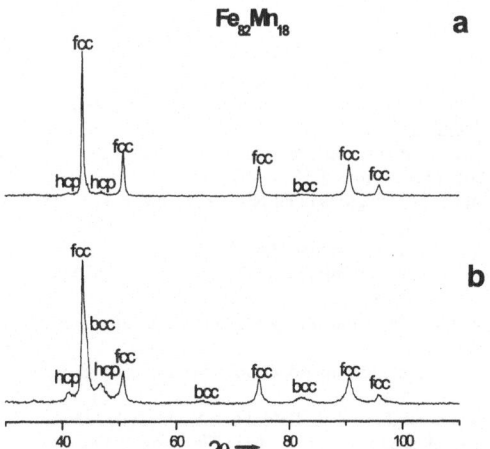

Figure 7. XRD patterns of MA $Fe_{82}Mn_{18}$ pounded in mortar after annealing at 750 ^{0}C for different times: a- 1 h and b- 3 h.

An example of structure obtained after mechanical deformation in MA austenic $Fe_{100-x}Mn_x$ at x=10, 20, 30 % samples is illustrated in Figure 6. As it is seen for $Fe_{90}Mn_{10}$ alloy deformation resulted in formation of b.c.c. phase (57 %), whereas for $Fe_{80}Mn_{20}$ - b.c.c. (5 %) and h.c.p. (25 %) phases appeared. There was no MT under deformation in $Fe_{70}Mn_{30}$ sample.

Also, it was noticed that the sensitivity of austenite structure to MT under deformation increased with time of annealing of the samples in the temperature range of austenite phase stability.

It is shown for the MA $Fe_{82}Mn_{18}$ alloy that deformation of the sample after annealing for 1 hour at 750 ^{0}C resulted in the formation of only 3 % b.c.c. and 3 % h.c.p. martensites correspondingly, while for the same alloy annealed for 3 h the same procedure led to the formation of 12 % b.c.c. and 20 % h.c.p. martensites (Figure 7). Block size of f.c.c. structure in $Fe_{82}Mn_{18}$ alloy altered insignificantly at increasing annealing time from 1 to 3 hours, nevertheless, MT developed more intensively.

At further annealing block size and amounts of maratensites continued to increase (Figure 5): after deformation of sample annealed for 8 hours there were 15 % b.c.c. and 23 % h.c.p. martensites in alloy structure.

These results agree with the assumption, that the suppression of MT in MA alloys is mostly caused by small block size of MA f.c.c. structure and presence of structure defects.

It is worth to be mentioned that the block size of b.c.c. martensite obtained by deformation of the austenic samples is lower than for thermally obtained martensite. Martensite grains formed under deformation were found extremely fine (from 14 to 22 nm).

4. Summary

The martensitic transformations in nanostructured by mechanical alloying Fe-Mn alloys was found to be significantly suppressed. The concentration border of single f.c.c. phase, formed at annealing, decreased to 10 at.% of Mn in MA Fe-Mn alloys. The nano f.c.c. phase may be partially transformed to b.c.c. and/or h.c.p. martensite by deformation. The block sizes of martensite obtained at cooling or deformation are from 15 to 46 nm and are smaller at deformation than at cooling.

The origin of martensite transformation suppression is attributed to block size refinement of MA f.c.c. structure as well as structure defects provided by severe deformation of alloy at mechanical alloying.

5. Acknowledgement

SDK thanks for financial support Italian Ministry of Foreign Affairs and Landau Network-Centro Volta for Fellowship of his work in Padova University

6. References

[1] Houdremont, E. (1956) *Handbuch der Soderstahlkunde*, Springer-Verlag, Berlin

[2] Thatnani, N.N., and Meyers, M.A. (1986) *Progr. in Mater.Sci.* **30**, p.1-20 (Review)

[3] Maki, T., Shimooka, S., and Tamura, I.(1971) The M_S temperature and morphology of martensite in Fe-31 Pct Ni-0.23 Pct C alloy, *Metallurgical Transactions* **2**, 2944-2945

[4] Takaki, S., Nakatsu, H., and Torunaga, Y. (1993) Effects of austenite grain size on ε martensitic transformation in Fe-15 mass%Mn alloy, *Materials Transactions JIM* **34**, 489-495

[5] Hayzelden, C., and Cantor, B. (1986) The martensite transformations in Fe-Ni-C, *Acta metal.* **34**, 2, 233-242

[6] Inokiti, Y., and Cantor., B. (1976) Splat-quenched Fe-Ni alloys, *Scripta Met.* **10**, 655-659

[7] Cech, R.E., and Turnbull, D., (1956) Heterogeneous nucleation of the martensite transformation J.Met **8**, 2, 124-132

[8] Kurt, C. and Schultz, L. (1993) Phase formation and martensitic transformation in mechanically alloyed nanocrystalline Fe-Ni, *J. Appl. Phys.* **73**, 1975-1980.

[9] Kurt, C. and Schultz, L. (1993) Phase formation and martensitic transformation in mechanically alloyed nanocrystalline Fe-Ni, *J. Appl. Phys.* **73**, 6588-6590

[10] Tcherdyntsev, V.V., Kaloshkin, S.D., Tomilin, I.A., Shelekhov, E.V., Baldokhin, Yu.V (1999) Phase Composition and Structure of Fe-Mn Alloys Prepared by Mechanical Alloying from Elemental Powders, *Z. Metallkde.* **90**, 747 - 752

[11] Hong L.B., and Fultz B. (1996) Two-Phase Coexistence in Fe-Ni Alloys Synthezed by Ball Milling, *J. Appl. Phys.* **79**, 3946 - 3954

[12] Scorzelli R.B. (1997) A Study of Phase Stability in Invar Fe-Ni Alloys Obtained by Non-Conventional Method, *Hyp. Int.* **110**, 143-150

[13] Kaloshkin, S.D., Tcherdyntsev, V.V., Baldokhin, Yu.V., Tomilin, I.A., and Shelekhov, E.V.(2001) Mechanically alloyed low-nickel austenite Fe-Ni phase: evidence of single-phase paramagnetic state, J. Non-Crystalline Solids **287**, 329-333

[14] Kaloshkin, S.D., Tcherdyntsev, V.V., Baldokhin, Yu.V., Tomilin, and I.A., Shelekhov, E.V.(2001) Phase Transformations in Fe-Ni System at Mechanical Alloying and Consequent Annealing of Elemental Powder Mixtures, *Physica B.* **299**, 236 – 241.

[15] Pustov, L.Yu., Kaloshkin, S.D., Tcherdyntsev, V.V., Tomilin, I.A., Shelekhov, E.V., and Salimon, A.I. (2001) Experimental Measurement and Theoretical Computation of Milling Intensity and Temperature for the Purpose of Mechanical Alloying Kinetics Description, *Mater. Sci. Forum.* **360-362**, 373 – 378.

[16] Shelekhov, E.V., Tcherdyntsev, V.V., Pustov, L.Yu., Kaloshkin, S.D., and Tomilin, I.A. (2000) Computer simulation of mechanoactivation process in the planetary ball mill: determination of the energy parameters of milling, *Mater. Sci. Forum* **343-346** , p.603-609

[17] Tcherdyntsev, V.V., Pustov, L.Yu., Kaloshkin, S.D., Tomilin, I.A., and Shelekhov, E.V. (2000) Calculation of energy intensity and temperature of mechanoactivation process in planetary ball mill by computer simulation, *NATO Science Partnership Sub-series: 3 High Technology* **80**, p.139-146

[18] Balychev, Yu, M., and Tkachenko, F.K. (1979), *Izv. AN SSSR. Metally* **3**, 169-171

[19] Vol, A.E. (1962*) Stroenie i svoistva dvoinykh splavov (Structure and Properties of Binary Systems)*, Vol. 2, Nauka, Moscow

[20] Tcherdyntsev V.V., Kaloshkin S.D., Tomilin I.A., and Shelekhov, E.V. (1999) Formation of iron-nickel nanocrystalline alloys by mechanical alloying, *Nanostr. Mater.* **12**, 139 - 142.

MISFIT DEFECTS IN NANOSTRUCTURED FILMS

A. E. ROMANOV
*Ioffe Physico-Technical Institute, Russian Academy of Sciences,
Polytechnicheskaya 26, St.Petersburg 194021, Russia*

Abstract. Titanium aluminides based on TiAl and Ti_3Al are potential materials for high temperature aerospace application. Their low density, high temperature creep resistance and strength, high oxidation resistance make them excellent potential engine materials. It is reasonable to develop processing strategies for protective and high temperature coating based on them. On magnetron sputtering the Ti-48% Al alloy the films, the composition of which corresponds to the target's one, have been obtained. When annealed, the aluminum atoms have been discovered to diffuse into the substrate. The redistribution and the leveling of the films' compositions in the volume take place. After annealing the film contains a higher titanium concentration. On sputtering the formation of the Ti_3Al-based metastable phase in the films has been observed. The phase is the main in the films and has a different morphology on different substrates. It is stable and doesn't decompose at annealing. During annealing the new intermediate metastable phases, which are stable only within the definite temperature and concentration range, have developed. The structure, kinetics of phase transformation and evolution of the film microstructures depend on the structure and morphology of the substrate.

Keywords: Thin films transport phenomena; Phase transitions; Intermetallics compound; Sputtering.

1. Introduction

Thin micro- and nanoscale films for various physical applications are usually grown on commercially avalible substrates or substrates, which are suitable for modern integrated technologies. In many cases the material of the film has substantial crystal lattice mismatch with respect to the material of substrate. Even the materials with equivalent (for example) cubic crystal lattice structure possess mutual mismatch, which can be characterized by misfit parameter (misfit strain)

T. Tsakalakos et al. (eds.) pgs. 339 - 352
Nanostructures: Synthesis, Functional Properties and Applications;
© 2003 Kluwer Academic Publishers.

340

$$\varepsilon_m = \frac{a_s - a_f}{a_f} \quad , \tag{1}$$

where a_s and a_f are lattice parameters for substrate and film respectively (see fig. 1a). The possible reasons for mismatched film structure generation are illustrated in fig. 1.

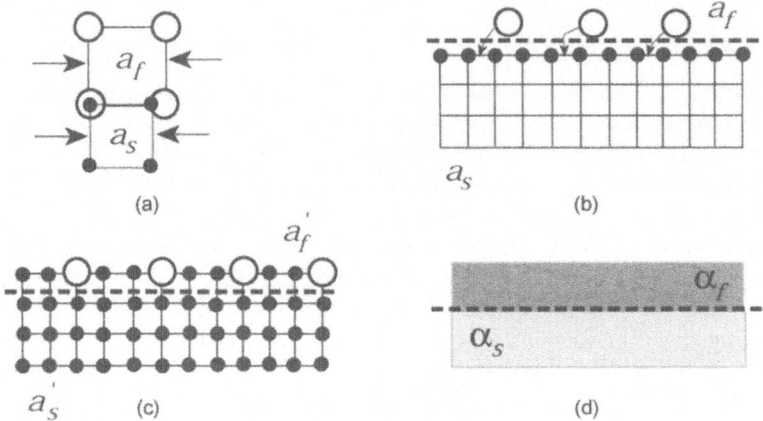

Figure 1. Crystal lattice mismatch at film/substrate interface: (a) two contacting crystals with different lattice parameters a_s and a_f; (b) epitaxial film of different chemical composition $a_s \neq a_f$; (c) impurity mismatch $a'_s \neq a'_f$; (d) difference in thermal expansions $\alpha_s \neq \alpha_f$; dashed lines indicate interface.

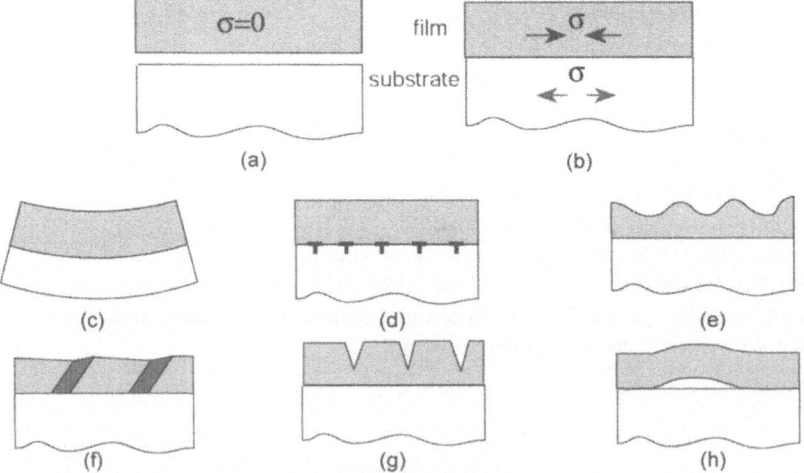

Figure 2. Relaxation of crystal lattice mismatch in film/substrate systems: (a) film and substrate in unconstrained state; (b) constrained state resulting in elastic strains and stresses; (c) elastic bending; (d) misfit dislocation (MD) generation; (e) roughening of the free surface of a strained film; (f) domain pattern formation; (g) cracking due to tensile stresses; (h) loss of coherency.

The presence of mismatching interfaces in film/substrate structures leads to the generation of elastic strains and stresses in contacting layers [1]. When the substrate is thick enough practically all elastic distortions are localized in the film (fig. 2b). When the thickness of mismatching layers are comparable there exists a possibility for elastic bending (fig. 2c), which leads to the redistribution of elastic strains and to the diminishing of stored elastic energy. Alternative mechanisms of stored energy relaxation appeal to non-elastic deformation or fracture of materials under consideration. In the most relevant technological case of thin micro- and nanoscale films all relaxation processes take place in the film. The most common relaxation mechanism is associated with misfit dislocation (MD) formation at film/substrate interface (fig. 2d) [2]. In the present article we discuss some typical features related to misfit defects, which also include threading dislocations (TDs), in growing films. In alternative to MD formation the following ralaxation modes can operate in growing nanoscale films: roughening of the film surfaces (fig. 2e) [3] and the formation of the domain patterns (fig. 2f) [4]. If the material of the film is brittle or cohesion at film/substrate interface is low two fracture modes [5]: normal to interface crack opening (fig. 2g) and loss of coherency (fig. 2h) at the interface operate in the system depending on the sign of misfit stress (tension or compression). We may note that various relaxation modes may accompany each other. For example surface morphology changes as a result of MD formation giving characteristic cross-hatch surface relief [6]. We discuss this last relation in details below.

2. Background

In lattice mismatch heteroepitaxy, the film is uniformly stressed when it grows coherently on a thick substrate. In the case of a one-dimensional (1D) lattice mismatch, the non-zero component of stresses in the film are given as [3]:

$$\sigma_{xx}^{1D} = \frac{2G}{1-\nu}\varepsilon_m; \qquad \sigma_{zz}^{1D} = \frac{2\nu G}{1-\nu}\varepsilon_m \qquad (2)$$

where the x-axis is chosen along the direction of lattice mismatch, which is characterized by the misfit parameter (misfit strain) ε_m, and the y-axis is perpendicular to the film/substrate interface. The film material is assumed to be elastically isotropic with shear modulus G and Poisson ratio ν. In the case of equi-biaxial mismatch, the stress state in the film is given as:

$$\sigma_{xx}^{2D} = \sigma_{zz}^{2D} = 2G\frac{1+\nu}{1-\nu}\varepsilon_m. \qquad (3)$$

The stored elastic energy w per unit area of the film/substrate interface associated with the stresses given by Eqs. (1) or (2) is proportional to the film thickness h:

$$w^{2D} = 2(1+\nu)w^{1D} = 2G\frac{1+\nu}{1-\nu}\varepsilon_m^2 h. \qquad (4)$$

Increasing the stored energy with increasing film thickness h will eventually lead to the possibility for the manifestation of a variety of relaxation processes (schematically shown in fig. 2) in the elastically stressed film.

Misfit dislocation generation at the interface between the film and substrate has been shown to be the most common mechanism for the relaxation of elastic stress [2,7,8]. In the majority of cases the MDs are associated with threading dislocations (TDs), which are concomitant to MDs but have their lines going through the film to the free surface [8,9]. The appearance of high density of TDs is dictated by the mechanisms of MD nucleation in mismatched films, as explained in fig. 3. MD dislocations can be formed by trapping of lattice dislocations generated by surface source at the film/substrate interface as shown in fig. 3 a. In this case each MD segment produces a pair of TDs. On the other hand the existing TD (for example growth dislocation) can be bent and moved (see fig. 3b) under the action of the of misfit stresses given by Eqs. (2) or (3). As a result of such motion a new portion of MD appears at the interface to form typical TD/MD configuration.

The Matthews-Blakeslee [10] critical thickness h_c for MD generation may be derived [11] by considering the energetics of a combined TD/MD configuration (see fig. 3b) in a stressed film:

$$h_c = \frac{b}{\varepsilon_m (1+v) 8\pi \cos \lambda} (1 - \cos^2 \beta) \ln\left(\frac{\alpha_o h_c}{b}\right) \approx \frac{b}{\varepsilon_m}, \qquad (5)$$

where $b = |\vec{b}|$ is the magnitude of the dislocation Burgers vector $\vec{b}$, λ is the angle between the Burgers vector and a line that lies in the film/substrate interface normal to the MD line, β is the angle between the MD line and $\vec{b}$, and α_o is the dislocation core cutoff parameter.

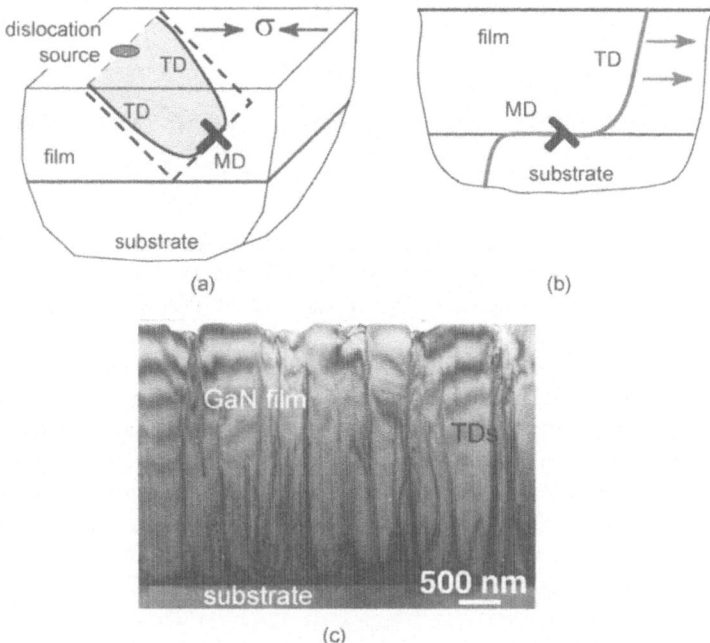

Figure 3. Threading dislocations: (a) nucleation of a dislocation half-loop by the surface source resulting in generation of a MD at the film/surface interface and a pair of TDs; (b) bending of the existing TD; (c) an example of high density ($10^9 cm^{-2}$) TDs in GaN film [12].

For a completely relaxed film (when the film thickness $h \gg h_c$), the linear density of MDs $\rho_{MD,relaxed}$ multiplied by the edge component of the MD Burgers vector parallel to the interface $b_{\parallel} = b \cos \lambda$, is equal to the misfit strain ε_m, *i.e.*,

$$\rho_{MD,relaxed} b_{\parallel} = \varepsilon_m .$$ (6)

For films of finite thickness that are grown on semi-infinite substrates, the equilibrium linear MD density $\rho_{MD,equil}$ may be shown to depend on the film thickness as follows [1]:

$$\rho_{MD,equil} = \rho_{MD,relaxed} (1 - \frac{h_c}{h}).$$ (7)

The actual dislocation density realized at a particular stage of the film relaxation is almost always less those given by Eqs. (6) and (7) $\rho_{MD,actual} \leq \rho_{MD,equil} < \rho_{MD,relaxed}$.

The extent of strain relaxation R can be therefore defined with the help of $\rho_{MD,actual}$:

$$R = \frac{\rho_{MD,actual}}{\rho_{MD,relaxed}} = \frac{\varepsilon_m - \varepsilon^*}{\varepsilon_m} = \frac{a_s - a_f^*}{a_s - a_f}, \tag{8}$$

where we take into account that the residual elastic strain ε^* in the film during relaxation is $\varepsilon^* = \varepsilon_m - \rho_{MD,actual} b_\parallel$. Here a_f^* is the measured in-plane lattice spacing of the film.

It is useful to note that MDs are equilibrium misfit defects in the film/substrate system with the equilibrium density given by Eq. (7). On the other hand TDs are essentially nonequilibrium misfit defects. The appearance of TDs has the origin in the kinetics of MD nucleation. The equilibrium TD density is zero but in real films can be as high as 10^{10} cm^{-2} (see for example micrograph of TD structure in GaN film grown on sapphire and shown in fig. 3c). For a wide variety of electronic and opto-electronic device applications, particularly for minority carrier devices, TDs are deleterious for physical performance [13]. In recent years there has been a substantial experimental effort to reduce TD densities. Despite the large literature to theoretically and experimentally understand critical thickness for MD generation, there have been relatively few theoretical efforts to understand the mechanisms by which TDs are eliminated in thin films. We report here on the recent results on TD density reduction in growing films.

3. Misfit Dislocations and Surface Morphology of Growing Films

In lattice mismatched semiconductor systems, large undulations in the surface height profile appear as a characteristic cross-hatch pattern, which can be revealed by atomic force microscopy (AFM) and other experimental techniques [6,14,15]. Cross-hatch morphology is a common feature for low mismatch (<2%) systems, *i.e.* InGaAs on GaAs and SiGe on Si, which grow in a planar 2D layer-by-layer mode [16,17]. Figure 4 illustrates the most common crystallography for dislocation assisted strain relaxation during (001) epitaxial film growth of face centered cubic (fcc) materials. Relaxation occurs via threading dislocation (TD) motion and misfit dislocation (MD) formation on the inclined {111} glide planes [11]. The orthogonal <110> directions, along which the cross-hatch pattern develops, are the same directions as the intersections of the glide planes with the film surface.

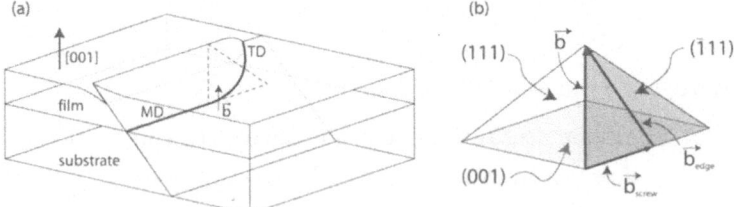

Figure 4. Dislocation geometry for a strained heteroepitaxial film with an inclined slip plane. (a) gliding threading dislocation segment with trailing misfit dislocation; (b) dislocation Burgers vector is decomposed into the $\vec{b}_{edge}$ and $\vec{b}_{screw}$ components.

The articles [15,18] report on details of cross-hatch surface morphology observation in epitaxial films of $In_xGa_{1-x}As$ grown by MBE technique on GaAs (001) semi-insulating substrates. The films were grown at 520 °C with compositions of 15% In and 25% In, which correspond to 1.1 and 1.8% mismatch and a critical thickness approximately 100 and 50 Å, where h_c is the equilibrium critical thickness for misfit dislocation nucleation at film/substrate interface (see Eq. (5)). The film thicknesses h were 10, 20, 30, and 60 times h_c. The degree of relaxation and film composition were determined by {115} off-axis high-resolution x-ray diffraction (XRD) measurements [15]. Atomic force microscopy was used to determine the surface height profile [15,18]. All films were partially relaxed. Figure 5 presents an AFM image of the surface of a 1000 Å (20 h_c) $In_{0.25}Ga_{0.75}As$ film that is 70% strain relaxed. The experimentally observed cross-hatch patterns for III-V semiconductors exhibit an anisotropy in the initial relaxation between the two orthogonal <110> directions, which is measurable by XRD and visible in the cross-hatch pattern. The films in this study, the degrees of strain relaxation were found to differ by up to 20%. The peak-to-valley amplitude in the observed films increases with relaxation of the initial film strain. The relaxation of approximately 5, 20, 50, 70, and 100% resulted in maximum cross-hatch amplitudes of approximately 15, 25, 45, 60, and 100 Å, respectively (results for ~ 70% relaxed film are shown in Fig. 5). It is important to note that the film relaxation increases with increasing film thickness.

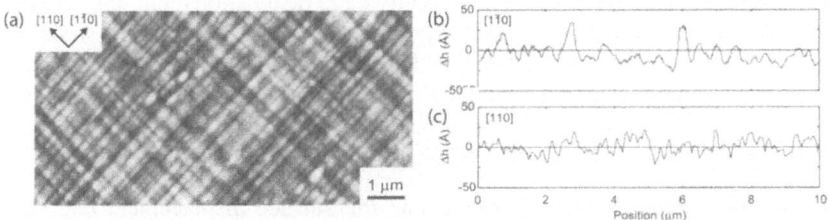

Figure5. Experimental observation of cross-hatch morphology in ~70% strain relaxed $In_{0.25}Ga_{0.75}As$ film (h = 20 h_c) [15]. (a) AFM image of the film surface; (b), (c) cross sections of two orthogonal <110> directions in the AFM cross-hatch pattern.

Below we propose a model for the surface height profile developement. Figure 6 illustrates the general idea for the evolution of the film surface undulations. Starting from a fully coherent strained epitaxial film (Fig. 6a), the film partially relaxes, via dislocation nucleation and motion, forming surface steps (Fig. 6b). Steps are then eliminated by their lateral flow resulting in the surface profile shown in Fig. 6c. The described sequence of the events can be also described with the help of a hypothetical Eshelby-like procedure [18,19] where the strained film is removed from the substrate and relaxed by slip (Fig. 6b') [18]. The surface is then smoothed by film growth or lateral mass transport (Fig. 6b''). Finally the film is reattached to the substrate. The underlying slip steps can be directly affiliated to misfit dislocations at film substrate interface.

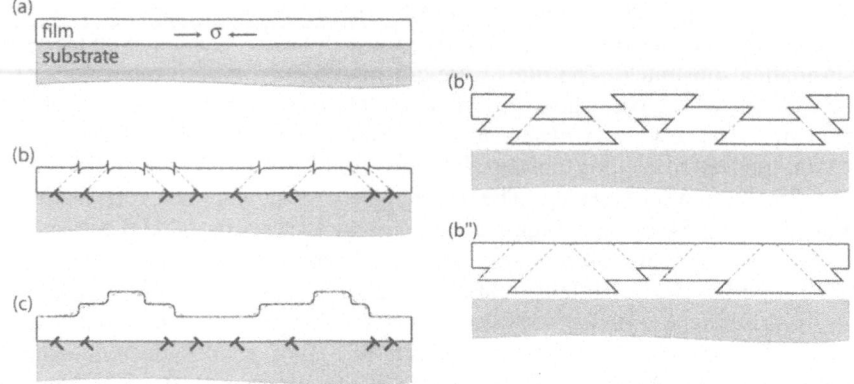

Figure 6. Schematic representation of the cross-hatch morphology development in strained films. (a) coherent compressively strained film; (b) strain relaxation in the film by dislocation glide resulting in the formation of surface steps and misfit dislocations; (c) final film state resulting from subsequent film growth via surface step elimination. The transition from (a) to (c) can be represented as Eshelby-like process where the unconstrained film plastically relaxes (b') then the steps are eliminated by subsequent growth (b'') and then the film is reattached to substrate (c).

The details required to model the cross-hatch pattern can be separated into geometry, strain relaxation, and growth. The geometry for the (001) oriented surface with {111} slip systems is shown in Fig. 1. Only edge dislocations with the component of Burgers vector parallel to the film/substrate interface relieve the misfit strain in the film. That leaves two possible dislocation Burgers vectors per slip plane with both screw and edge components that create a surface step. The edge Burgers vector component $\vec{b}_{edge}$ can be decomposed again into the edge component parallel $\vec{b}_{\parallel}$ and perpendicular $\vec{b}_{\perp}$ to the surface. To calculate the surface displacement and stresses from the underlying misfit dislocations we use fully analytical elasticity solutions for edge dislocations with Burgers vectors oriented parallel or perpendicular to the surface [18]. For example, the component of stress $\sigma_{yy}^{b_{\perp}}$, resulting from a misfit dislocation with the Burgers vector perpendicular to the free surface is:

$$\sigma_{yy}^{b_{\perp}} = \frac{Gb_{\perp}}{2\pi(1-v)}\left[\frac{-x\left(x^2+3(y+h)^2\right)}{\left(x^2+(y+h)^2\right)^2} - \frac{x\left(x^2+3(y-h)^2\right)}{\left(x^2+(y-h)^2\right)^2} + \frac{2x\left(x^4+4x^2(y+h)^2-2x^2yh+3(y+h)^4+6yh(y+h)^2\right)}{\left(x^2+(y+h)^2\right)^3}\right] \qquad (9)$$

The surface displacement expressions for perpendicular and parallel misfit dislocations Burger vectors are given in Eqs. (2):

$$\left. u_y^{b_\parallel} \right|_{y=0} = \frac{b_\parallel}{\pi} \frac{h^2}{x^2 + h^2} \, , \qquad \left. u_y^{b_\perp} \right|_{y=0} = \frac{b_\perp}{\pi} \left(-\frac{xh}{x^2 + h^2} - \tan^{-1}\left[\frac{x}{h}\right] \right). \tag{10}$$

x, y are spatial coordinates where the y-axis is normal to the surface, h is the film thickness and G and v are shear modulus and Poisson ratio of the material of the film. In our description, the surface step created during dislocation nucleation and motion was introduced by adding a Heaviside step function to the surface displacement caused by the subsurface misfit dislocation. This step is located at the intersection of the dislocation glide plane and the surface. To account for the surface profile after complete step elimination we use the displacement from Eqs. (10), which produces no slip step for the subsurface misfit dislocation.

A Monte Carlo algorithm was then used to simulate the misfit dislocation array generation and surface steps. Figure 7 illustrates the application of the algorithm for modeling a ~70% strain relaxed In$_{0.25}$Ga$_{0.75}$As film. Each possible surface nucleation site was included in a one-dimensional model, Fig. 7a, and the dislocation nucleation site was randomly selected. Then the glide plane was randomly selected and the resulting surface stresses and displacements were calculated. The nucleation of the next dislocation was modeled in the same way and the resulting total stress and displacements were calculated by linear superposition of surface displacement from the individual misfit dislocations and the related surface steps. The introduction of dislocations continued until a final value of relaxation was realized. The final stress profile at the surface (σ_{xx}), average film stress after relaxation (σ_f) and the original stress (σ_0) of the fully coherent film are shown in Fig. 7b. The surface height profile without film growth and lateral mass transport results is a locally rough film (Fig. 7c). Elimination of the surface steps (Fig. 7d) results in a locally smooth surface with large height amplitude undulations. The experimental height profile for the same film composition is shown in Fig. 7d.

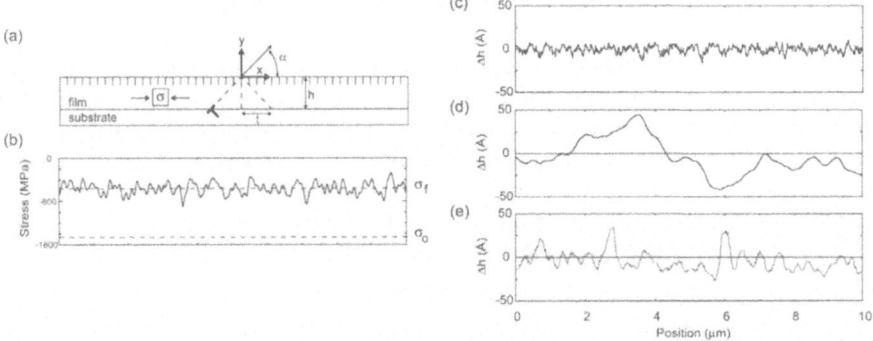

Figure 7. Modeling surface stresses and displacement. (a) schematic representation of the 1 D model; (b) resulting stress (σ_{xx}) at the surface after 70% strain relaxation; (c) surface profile created during film relaxation without step elimination; (d) surface profile after step elimination; (e) experimentally observed surface profile for strain relaxed In$_{0.25}$Ga$_{0.75}$As film.

For comparison with experimental data, the material properties of $In_xGa_{1-x}As$ on GaAs (001) substrate were used in the model. The results of the model do depend on material properties. The surface stress and surface height profiles were calculated for various extents of strain relaxation and for different film thickness and full results are given in [18]. The results of the model share many important trends with the experimental data. Fully strained and only slightly relaxed films are significantly smoother than the more relaxed films. The average amplitude of the surface undulations and their apparent period both increase with increasing film relaxation. A film surface with only the slip steps is mesoscopically smooth (~1000 Å length scale) but locally very rough. While the film surface with eliminated surface steps is locally smooth, it is a mesoscopically rough surface with large amplitude undulations.

The surface profiles are thickness dependent. For very thin films, the displacement from misfit dislocations given by Eqs. (10) is laterally confined. Thus the surface appears to have many sharp transitions in the profile. As the film thickness increases and dislocation screening is more prevalent, the surface height transitions are more gradual. The surface of thicker films develops larger undulating amplitude and a locally smoother appearance.

The qualitative agreement of the experimentally observed surface cross-hatch with the modeled surface profile is good. The influence of thickness and strain relaxation on surface morphology is very similar. The quantitative discrepancies in the overall height amplitude and frequency of undulations that are seen in Fig. 7d and 7e can be attributed to many factors not currently included in the model. For example, stress induced surface diffusion is neglected. The incomplete elimination of steps due to surface diffusion would lead to a rougher surface than for the complete step elimination. The simulation algorithm applied here introduces only a single dislocation and slip step during each iteration, so multiple steps do not interact and do not form step bunchings. Real films can in addition have dislocation multiplication sources in the bulk.

4. Theoretical Models for Threading Dislocation Density Reduction

TDs are nonequilibrium defects that raise the free energy of the film (on the contrary to MDs which are defects with the equilibrium density given by Eq. (7)). Therefore, there is a thermodynamic driving force to diminish the TD density. The high densities of TDs facilitate the development of a kinetic approach for TD reduction [9,20]. The kinetic approach considers the reactions between TDs in relation to their densities and relative motion; this is effectively is a "reaction kinetics" treatment. The TD motion r is defined as the lateral movement of the intersection point of the TD with the free surface. It depends on the parameters of the problem: film thickness h and misfit ε_m, but also additional internal and external factors (for example, the concentration of point defects). Changes in film thickness h give rise to lateral TD motion (Fig. 8a). A vacancy or interstitial supersaturation may lead to TD climb (Fig. 8b). Misfit strain may lead to MD generation and concurrent TD motion (Fig. 8c). This effect has special features in superlattices where TDs can have a set of successive displacements (Fig. 8c).

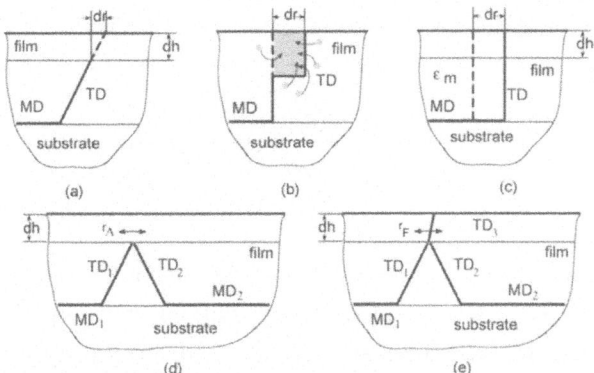

Figure 8. Two fundamentals of treading dislocation reduction: lateral TD motion and mutual TD reactions: (a) effective TD motion due to the film growth,; (b) TD climb; TD glide caused y misfit strain; (d) reaction of TD annihilation; (e) fusion of two TDs and appearance of a new TD.

Together with relatively slow TD motion, the other more rapid process can be defined in a TD ensemble, namely, dislocation reactions (as shown Fig. 8 d,e). One characterizes the reactions by the radius r_1. The reaction radius represents the distance at which the interaction between TDs is sufficient to overcome the Peierls barrier σ_p for TD glide or climb. Once initiated, this movement results in reaction product which depends on the reacting dislocation Burgers vectors. The possible reactions between TDs include annihilation, fusion, and scattering. In the annihilation reaction, TDs with opposite Burgers vectors that fall in annihilation radius r_A react and stop the propagation of both TDs to the film surface. In a fusion reaction with a characteristic reaction radius r_F, two TDs react to produce a new TD with Burgers vector that is the sum of Burgers vectors of the reacting TDs. The possible reactions between TDs in the case of (001) fcc and (0001) hcp epitaxial film growth have been analysed in details in [12,21,22].

The above outlined approach was applied for the first time [9] for the analysis of the TD reduction in homogeneous buffer layers to explain experimentally observed dependence of TD density on the buffer layer thickness h. As an example, consider the possibility of growing a relaxed (i.e. free from misfit strain) layer with finite lateral dimension λ as it might be realized in selective are growth over a relaxed film with a high TD density. For this case, the change in TD density with film thickness may be written as [20,23]

$$\frac{d\rho_{TD}}{dh} = \frac{\rho_{TD}}{\Lambda} - K\rho_{TD}^2 ,$$ (11)

where $\Lambda = \lambda / J$, $K = 2Jr_1$ and $J = \dfrac{dr}{dh}$ is a geometric factor that describes TD motion during film growth and commonly $J \approx 1$ for inclined TDs in (001) cubic semiconductors films [20]. Eq.(11) gives the following solution if ρ_0 is the initial density of TDs at $h = h_0$:

$$\rho_0 = \frac{\rho_0}{(1 + K\Lambda\rho_0)\exp[\frac{h - h_0}{\Lambda}] - K\Lambda\rho_0}. \tag{12}$$

This solution has two asymptotes that are relevant for TD reduction. In the first limit, the mesa size is large in comparison with film thickness, $\Lambda \gg h - h_0$, and the solution predicts the inverse proportional dependence of TD density with film thickness. The full crystallographic details of this problem have been included into consideration in [21,22]. Computer simulation approach has also been used [24] in addition to the continuum description of TD reduction. It has been found a very good agreement between results of continuum and computer simulation approaches. This is an agreement with experimental observations of TD density in homogeneous buffers. In the case $\Lambda \to 0$, which is a limiting case of small mesas, there is exponential decay of TD density with film thickness [23].

It has been demonstrated that the continuum approach is helpful in treating the problem of TD behaviour in stressed layers [25] and/or superlattices [26]. Under these conditions, the effect of TD bending is very important for cleaning dislocations from the working zones of semiconductor devices. Finally, so-called reaction enhancing interlayers were shown to be very effective in threading dislocation density reduction [27].

5. Summary and Conclusions

Cross-hatch surface morphology observed in nanoscale films is the result of misfit defect (misfit dislocation (MD)) assisted relaxation of elastic stresses, which originate from crystal lattice mismatch, and film growth. New model for the cross-hatch pattern formation considers the elimination of surface steps and explains experimentally observed film surface profiles. This model will enable the use of cross-hatch observations to learn about misfit strain relaxation pathways, particularly dislocation sources, blocking, multiplication and possible routes to more efficient relaxation in lattice mismatched heteroepitaxy.

Physical mechanisms responsible for misfit defect (threading dislocation (TD)) density reduction involve mutual motion and reactions between dislocations. For inclined TDs in growing buffer layers their motion is governed by only geometrical factors. In the strained layers TD motion is associated with MD generation and misfit stress relaxation. The developed approach predicts the simple 1/h scaling law for TD reduction in buffer layers of III-V compounds and shows the limitations of TD reduction in strained layers and in growing relaxed GaN films.

6. Acknowledgments

This work has been supported by the Program "Physics of Solid State Nanostructures" by the Ministry of Industry and Science of Russia.

7. References

1. Tsao, J.Y. (1993) *Materials Fundamentals of Molecular Beam Epitaxy*, Academic, New York.

2. Matthews, J.W. (1979) Misfit dislocations, in F.R.N.Nabarro (eds.), *Dislocations in Solids* **2**, North-Holland Publ. Co., Amsterdam, pp.461-500.

3. Gao, H. (1994) Some general properties of stress-driven surface evolution in a heteroepitaxial thin-film structure, *J. Mech. Phys. Solids* **42**, 741-772.

4. Speck, J.S., Daykin, A.C., Seifert, A., Romanov, A.E., Pompe, W. (1995) Domain configurations due to multiple misfit relaxation mechanisms in epitaxial ferroelectric thin films. III. Interfacial defects and domain misorientations, *J.Appl.Phys.* **78**, 1696-1706.

5. Ohring, M. (1992) *Material Science of thin films*, Academic, New York.

6. Burmeister, R.A., Pighini, G.P and Greene, P.E. (1969) Large area epitaxial growth of GaAs1-xPx for display applications, *Trans. TMS-AIME* **245**, 587-594.

7. Fitzgerald, E.A. (1991) Dislocations in strained-layer epitaxy: theory, experiment, and applications, *Mater. Sci. Rep.* **7**, 87-142.

8. Beanland, R., Dunstan, D.J.and Goodhew, P.J. (1996) Plastic relaxation and relaxed buffer layers for semiconductor epitaxy, *Adv. Phys.* **45**, 87-146.

9. Speck, J.S., Brewer, M.A., Beltz, G., Romanov, A.E.and Pompe, W. (1996) Scaling laws for the reduction of threading dislocation densities in homogeneous buffer layers, *J. Appl. Phys.* **80**, 3808-3816.

10. Matthews, J.W. and Blakeslee, A.E. (1974) Defects in epitaxial multilayers. 1. Misfit dislocations, *J. Cryst. Growth* **27**, 118-125.

11. Freund, L.B. (1992) Dislocation mechanisms of relaxation in strained epitaxial films, *MRS Bulletin* **17** (**7**), 52-60.

12. Mathis, S.K., Romanov, A.E., Chen, L.F., Beltz, G.E., Pompe, W., Speck, J.S. (2001) Modeling of threading dislocation reduction in growing GaN layers, *J. Cryst. Growth* **231**, 371-390.

13. Speck, J.S. (2001) The role of threading dislocations in the physical properties of GaN and its alloys, *Mater. Sci. Forum* **353-356**, 769-778.

14. Springholz, G. (1999) Observation of large-scale surface undulations due to inhomogeneous dislocation strain fields in lattice-mismatched epitaxial layers, *Appl. Phys. Lett.* **75**, 3099-3101.

15. Andrews, A.M., Romanov, A.E., Speck, J.S., Bobeth, M. and Pompe, W. (2000) Development of cross-hatch morphology during growth of lattice mismatched layers, *Appl. Phys. Lett.* **77**, 3740-3742.

16. Olsen, G.H. (1975) Inrefacial lattice mismatch effects in III-V compounds (device performance), *J. Cryst. Growth* **31**, 223-239.

17. Kishino, S., Ogirima, M. and Kurata, K. (1972) Cross-hatch pattern in GaAs1-xPx epitaxially grown on GaAs substrate, *J. Electrochem. Soc.* **119**, 617-622.

18. Andrews, M., Speck, J.S., Romanov, A.E., Bobeth, M. and Pompe, W. (2002) Modeling cross-hatch surface morphology in growing mismatched layers, *J. Appl. Phys.* **91**, 1933-1943.

19. Eshelby, J.D. (1957) The determination of the elastic field of an ellipsoidal inclusion, and related problems, *Proc. Roy. Soc.* **A241**, 376-396.

20. Romanov, A.E., Pompe, W., Beltz, G.E. and Speck, J.S. (1996) An approach to threading dislocation "reaction kinetics, *Appl. Phys. Lett.* **69**, 3342-3344.

21. Romanov, A.E., Pompe, W., Beltz, G.E. and Speck, J.S. (1996) Modeling of threading dislocation density reduction in heteroepitaxial layers. I. Geometry and crystallography, *Physica Status Solidi B* **198**, 599-613.

22. Romanov, A.E., Pompe, W., Beltz, G.E. and Speck, J.S. (1997) Modeling of threading dislocation density reduction in heteroepitaxial layers. II. Effective dislocation kinetics, *Physica Status Solidi B*, **199**, 33-49.

23. Beltz, G.E., Chang, M., Eardley, M.A., Pompe, W., Romanov, A.E. and Speck, J.S. (1997) A theoretical model for threading dislocation reduction during selective area growth, *Mat. Sci. and Eng. A* **234-236**, 794-797.

24. Beltz, G.E., Chang, M., Speck, J.S., Pompe, W. and Romanov, A.E. (1997) Computer simulation of threading dislocation density reduction in heteroepitaxial layers, *Phil. Mag. A* **76**, 807-835.

25. Romanov, A.E., Pompe, W., Mathis, S., Beltz, G.E. and Speck, J.S. (1999) Threading dislocation reduction in strained layers, J.Appl. Phys. **85**, 182-192.

26. Martisov, M.Yu and Romanov, A.E. (1990) Dislocation annihilation mechanism in stressed superlattices, *Soviet Physics Solid State* **32**, 1101-1102.

27. Romanov, A.E. and Speck, J.S. (2000) Relaxation enhancing interlayers (REIs) in threading dislocation reduction, *J. Electr. Mat.* **29**, 901-905.

MICROSTRUCTURAL DEVELOPMENT OF Ti-Al THIN FILMS DURING ANNEALING

S.E. ROMANKOV, B.N. MUKASHEV
*Institute of Physics & Technology, 480082,
Almaty82, Kazakstan*

Abstract. Titanium aluminides, based on TiAl and Ti₃Al, are potential materials for high temperature aerospace application. Their low density, high temperature creep resistance and strength, high oxidation resistance, make them excellent potential engine materials. It is reasonable to develop processing strategies for protective and high temperature coating based on them. On magnetron sputtering, the Ti- 48% Al alloy thin films, the composition of which corresponds to the target's one, have been obtained. When annealed, the aluminium atoms have been discovered to diffuse into the substrate and the redistribution, as well as the leveling of the films' compositions in the volume, take place. After annealing, the film contains a higher titanium concentration. On sputtering, the formation of the Ti₃Al-based metastable phase in the films has been observed. This phase is the main in the films and has a different morphology on different substrates. It is stable and doesn't decompose at annealing. During annealing, the new intermediate metastable phases, which are stable within a finite temperature and concentration range, have developed. The structure, kinetics of phase transformation and evolution of the film microstructures, depend on the structure and morphology of the substrate.

Keywords: Thin films transport phenomena; Phase transitions; Intermetallics compound; Sputtering.

1. Introduction

Intermetallic alloys, based on γ-TiAl which contain a small fraction of α_2- Ti₃Al phase, are now regarded as promising candidates for high-temperature structural materials because of their attractive properties, e.g. low density, high melting temperature, inverse temperature dependence of strength and reasonable oxidation resistant [1-3]. The properties of these alloys are known to be very sensitive to microstructure which, in turn, varies considerably with change in minor alloy composition and in processing parameters [4-6]. The various microstructural features can be obtained by a combination of different thermal treatments [7-10]. As titanium aluminides are excellent materials for high temperature application, it is reasonable to develop processing strategies for protective and high temperature coating based on them [11].

T. Tsakalakos et al. (eds.) pgs. 353 - 361
Nanostructures: Synthesis, Functional Properties and Applications;
© *2003 Kluwer Academic Publishers.*

The point is that the thin films can be produced under different non-equilibrium conditions and as a result many metastable states, which it is impossible to receive in bulk materials, can be formed in them [12,13]. In present work, we have developed the mode of sputtering Ti-Al alloys, created Ti-48Al thing films on different substrates, and investigated their microstructural evolution during annealing by SEM and X-ray diffraction.

2. Experimental Procedures

Thin films were produced by magnetron sputtering of the Ti-48Al alloy (all compositions are given in atomic percent). The Ti-48Al alloy were prepared by electric-art melting 99.999% Al and 99.999% Ti. The one was homogenized at 1150^0C for 160 h. Pressure during sputtering was 0.8 Pa. The temperature of the substrate during sputtering was 300^0C. The speed of sputtering was 0.11nm/c. The thickness of the received films was 6 μm. As the substrates glassceramics and polycore were used. The thin films were annealing in vacuum of 10^{-5} Pa at 650^0C for 0.5, 2 and 6 h and further cooled in a furnace. Thin films was studied by SEM and X-ray diffraction analysis. The resulting microstructures were observed using second electron imaging. Microanalysis using X- ray energy dispersive spectroscopy was carried out on JCXA 733 (Superprobe 733).

3. Experimental results and discussion

The target's chemical composition is listed in Table1. After sputtering the surface layers of the target contain more titanium and less aluminium than in the initial state. Figure 1 shows the XRD patterns of the target before and after sputtering. In the initial as-annealed state the target consists of two phases, i.e., the TiAl- based solid solution with an ordered tetragonal face- centered $L1_0$ structure and a small quantity of the Ti_3Al-based solid solution with the DO_{19} hcp structure. From Figure1 it is seen that during sputtering a phase of layering Ti_3Al and TiAl – based solid solutions takes place. According to the XRD patterns during this process in the alloy a wide range of solid solutions based on these compounds is formed, and they differ by the parameters of their lattices and, accordingly, by their compositions. That is, the process of target's sputtering is not uniform.

TABLE 1 Chemical composition of the target

State	Ti	Al
Before Sputtering	51.889	48.111
After Sputtering	55.578	44.422

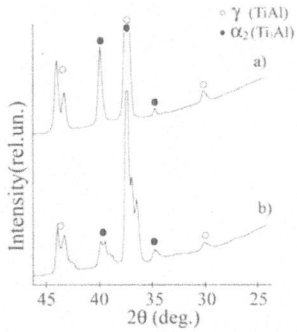

Figure1. X-ray diffraction patterns of the taget in various states: (a) before sputtering; (b) after sputtering

Figure 2a shows the microstructure of the thin film obtained on the glassceramic substrate. The surface of the film, on the whole, reflects the structure of the substrate itself. The film is porous and has surface inhomogeneities, such as canonical formations and cavities. There are also small inclusions, having a smooth and even surface. Surface cracks have not been observed. The chemical composition of different parts of the film is listed in Table 2. The average chemical composition of the film after sputtering is practically identical to that of the target (Table 1). It is possible to assume, that the sputtering coefficient for aluminum is higher than that for titanium, that is why at the very beginning of sputtering the target's surface loses aluminum. Soon on the target's surface there appears the titanium enriched domain. Some steady-state target composition is achieved, and on further sputtering the film with the same composition as the initial target's one is deposited. However, the film's rough and smooth areas are considerably different in their element composition and differ from the film's average composition.

TABLE 2. Chemical composition of the thin films on the glassceramic substrate

Area	Initial	Annealed, 0.5 h	Annealed, 2 h	Annealed, 6 h
Average Composition	Al- 49.553 Ti– 49.841 Substrate- 0.606	Al- 35.982 Ti- 62.871 Substrate- 1.147	Al- 37.512 Ti- 61.672 Substrate- 0.816	Al- 36.699 Ti- 62.733 Substrate- 0.568
Even Areas	Al- 63.238 Ti- 36.335 Substrate- 0.427	Al- 43.671 Ti- 55.463 Substrate- 0.866	Al- 33.678 Ti- 65.764 Substrate- 0.558	Al- 39.551 Ti- 59.897 Substrate- 0.552
Rough Areas	Al- 46.996 Ti- 52.709 Substrate- 0.295	Al- 23.542 Ti- 74.891 Substrate- 1.567	Al- 35.272 Ti- 63.915 Substrate- 0.813	Al- 37.691 Ti- 61.795 Substrate- 0.564

356

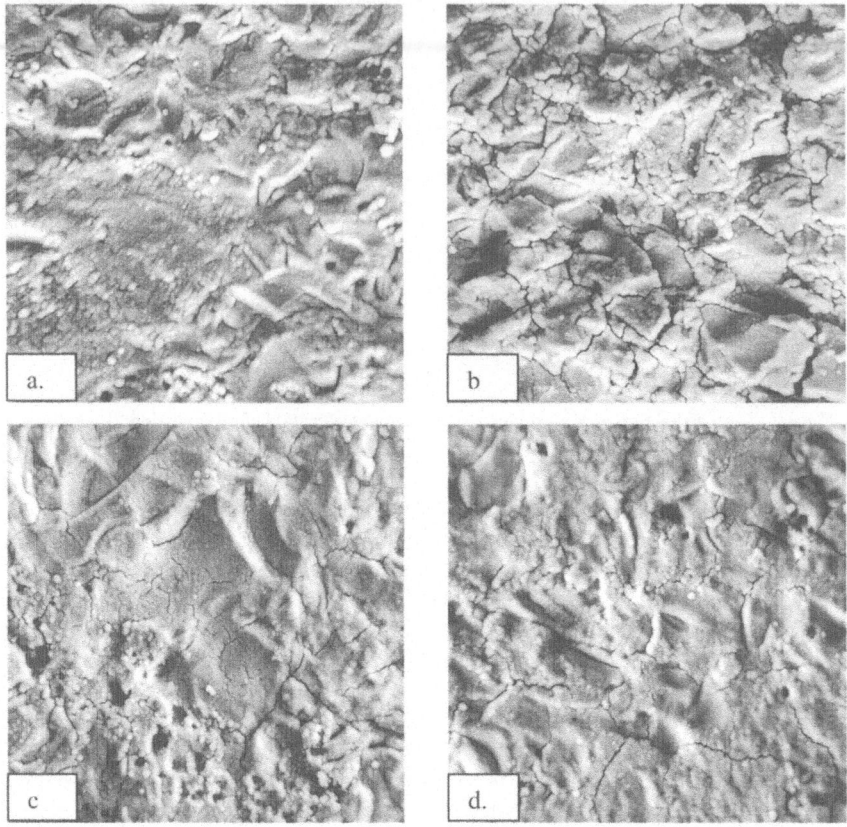

Figure 2. Microstructure of the thin films on the glassceramic substrate in various state: (a) initial state; (b) annealed for 0.5, (c) 2, (d) 6 h (T=650⁰C, for all cases). 1- even areas and 2- rough areas

Figure 3a shows the microstructure of the film obtained on the polycore. In contrast to the glassceramic substrate film, the film on polycore has a more finely sputtered structure and consists of individual non-closely spaced grains. The grains are not identical in size and have a spherical form. Some grains have a smooth and even surface, others- a rough one. The chemical composition of the polycore film is listed in Table 3. The composition of the film sputtered on the polycore is practically identical to that of the target. However, the rough and smooth grains differ greatly in their composition (Table 3). It should be emphasized that after sputtering the chemical composition of the films obtained on the polycore and glassceramicis substrates is different (Tables 2 and 3).

TABLE 3. Chemical composition of the thin films on the polycore substrate

Area	Initial	Annealed, 0.5 h	Annealed, 2 h	Annealed, 6 h
Average Composition	Al- 44.853 Ti– 54.774 Substrate- 0.373	Al- 39.154 Ti- 60.185 Substrate- 0.661	Al- 38.129 Ti- 61.416 Substrate- 0.455	Al- 45.495 Ti- 54.127 Substrate- 0.378
Even Grains	Al- 50.966 Ti- 48.646 Substrate- 0.388	Al- 37.179 Ti- 62.450 Substrate- 0.371	Al- 42.728 Ti- 56.859 Substrate- 0.413	Al- 57.468 Ti- 42.010 Substrate- 0.522
Rough Grains	Al- 38.654 Ti- 61.346	Al- 38.279 Ti- 61.226 Substrate- 0.495	Al- 39.247 Ti- 60.402 Substrate- 0.351	Al- 50.121 Ti- 49.246 Substrate- 0.633

Figures 4a and 5a shows the XRD patterns of the films obtained on glassceramicis and polycore substrates. After sputtering only the Ti_3Al- based phase is determined in the films by the XRD. Ti_3Al has an ordered lattice, which is characterized by the superstructure lines in the XRD patterns. However, the superstructure reflections have not been displayed despite the very intensive structural lines. It indicates that after sputtering the disordered Ti_3Al- based phase is formed in the films, or that antiphased domains are strongly disoriented relative to each other,

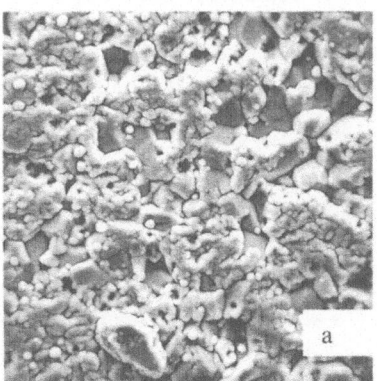

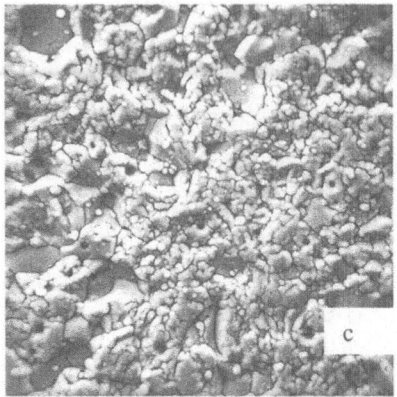 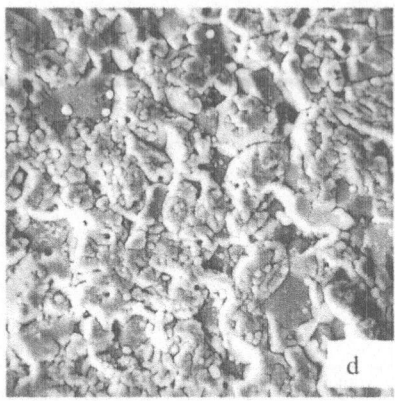

Figure 3. Microstructure of the thin films on the polycore substrate in various state: (a) initial state; (b) annealed for 0.5, (c) 2, (d) 6 h (T=650⁰C, for all cases). 1- even grains and 2- rough grains

and this is known to result in the considerable weakening of the superstructure lines. Besides, according to the Ti-Al equilibrium phase diagram for the films with given chemical composition TiAl- based phase must be main and stable. The X-ray analysis did not display such phase in the films. It proves that the Ti_3Al- based phase, obtained on the films after sputtering, is metastable. Metastability is maintained due to the deformation fields, which are always present in the films.

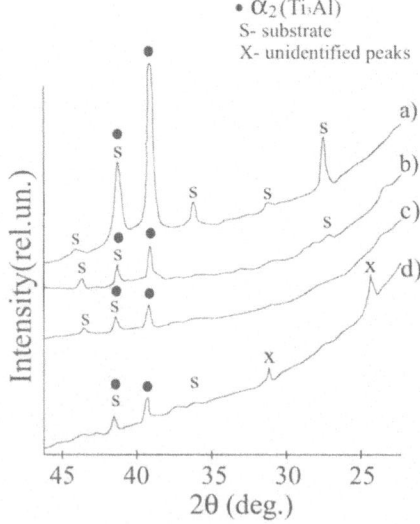

Figure 4. X-ray diffraction patterns of the thin films on the glassceramic substrate in various states: (a) initial state; (b) annealed for 0.5, (c) 2, (d) 6 h (T=650⁰C, for all cases)

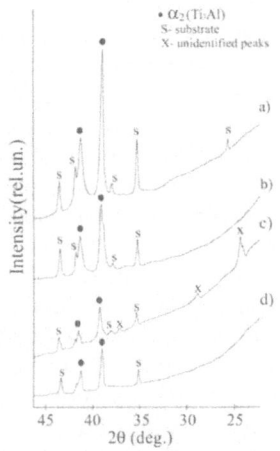

Figure 5. X-ray diffraction patterns of the thin films on the polycore substrate in various states: (a) initial state; (b) annealed for 0.5, (c) 2, (d) 6 h (T=650^{0}C, for all cases)

Figure 2b shows the microstructure of the film obtained on the glassceramicis substrate after annealing for 0.5 h. After annealing surface cracks appear. The crack network covers the whole surface of the film. It indicates that during annealing the strong deformation strains appear, either due to different volume changes of the film and substrate upon heating or due to structural rearranging of thin film. Besides, after annealing the films contain more titanium and less aluminium than in the initial state (Table 2). Taking into account that the mobility of aluminium atoms is higher than that of titanium and increasing concentration of substrate elements into the film it can be assumed that aluminium atoms diffuse into substrate during annealing. As a result the average chemical composition of the films changes dramatically.

Figure 3b shows the microstructure of the film obtained on polycore substrate after annealing for 0.5 h. In the case of the film on polycore substrate, after annealing for 0.5 h the crack density increase was observed, that process was accompanied by the increase in the titanium concentration and the decrease in the aluminium content in the film in comparison with the initial state (Table 3). At the same time the composition of different grains is becoming equal and approaching the average film composition. After annealing for 0.5 h the chemical composition of the films on glassceramicis and polycore is only slightly different. In addition, it should be noted that after annealing for 0.5 h the chemical composition of the films on glassceramicis and polycore is only slightly different.

After annealing for 0.5 h the integrated intensity of diffraction lines in the XRD patterns of substrates and films decreases dramatically in comparison with the initial state (Fig.4b and 5b). No new diffraction lines have been found. After annealing both polycore and glassceramicis the Ti$_3$Al– based phase lines become asymmetric and diffuse smeared near the small angles. It indicates that film structure is non-homogenous. Therefore, the above mentioned results show that during annealing the redistribution of atoms resulting in the formation of new bonds takes place in the film.

On increasing the annealing time to two hours "the healing" of film cracks on the glassceramicis substrate is observed (Fig.2c). The film becomes more morphologically homogenous and continuous than after annealing for 0.5 h (Fig.2b). The average chemical composition of the film does not really change relative to that obtained after annealing for 0.5 h (Table 2). The redistribution of the components between the different parts of the film and the process of their compositions becoming similar have been observed (Table 2). After annealing for 2 h no new phases have been observed. However, on increasing the time of annealing to 2 h the intensity of diffraction lines of the substrate and the film continues to decrease. (Fig.4c). The redistribution of the film's crystallites relative to the adjoining atomic plane of the substrate may take place. The crystallites start to be oriented in the definite direction, which results in the simultaneous intensity decrease of the diffraction lines of the substrate and the film.

Figure 3c shows the microstructure of the film obtained on polycore substrate after annealing for 2 h. On increasing the annealing time to 2 h the crack density on film obtained on polycore substrate goes on increasing. The average chemical composition of the film does not really change relative to that obtained after annealing for 0.5 hour (Table 2). At the same time individual grains change their composition, which does not already correspond to the average composition of the film. Besides, after annealing for 2 h the new phase lines have been displayed on the XRD patterns (Fig.5c). We failed to define this phase structure by three diffraction lines. With the increase of the annealing time to 2 h the simultaneous decrease of the diffraction lines intensities of the substrate and film have been observed (Fig.5c). It may be assumed that during annealing the structural rearranging of thin film begins and simultaneously the crystallites start to be oriented in the definite crystallographic direction in order to decrease the deformation strains.

With the increase of the annealing time to 6 h the structure of the film on the glassceramicis substrate is gradually becoming more homogenous (Fig.2d). The cracks go on "healing". The composition of different parts is becoming similar and approaching the average film composition (Table 2). The intensity of the diffraction lines of the substrate and the film on the XRD patterns continues to decrease with the increase of annealing time (Fig.3d), which again indicates that the film's crystallites are oriented in the definite crystallographic direction. Besides, after 6 h annealing the lines of the new phase have been displayed in the XRD patterns. By their angular location they are close to the lines' angles, displayed on the films obtained on the polycore substrate after annealing for 2 h (Fig.5c).

Figure 3d shows the microstructure of the film obtained on polycore substrate after annealing for 6 h. On increasing annealing time to 6 h the growth of individual big smooth grains of the film on polycore substrate (Fig.3d) is observed, and simultaneously the film cracks are gradually "healed". Besides, the change of the film average chemical composition takes place. After annealing for 6 h the increase of the aluminium concentration and the decrease of the titanium concentration take place in the film. It is interesting to note that after annealing for 6 h the average composition of the film on polycore substrate is the same as the composition of the initial film obtained after sputtering (Table 3). Besides, the new phase lines observed on the XRD patterns after annealing for 2 h (Fig.5c) completely disappear with the increase of the annealing time to 6 h (Fig.5d). It can be assumed that the new phase observed on the film on polycore substrate after annealing for 2 h is an intermediate metastable phase, which is formed within a very narrow temperature and concentration range, and which is dissolved when a more stable phase is formed.

4. Conclusion

1. On magnetron sputtering the Ti- 48% Al alloy the films, the composition of which corresponds to the target's one, have been obtained. When annealed, the aluminium atoms have been discovered to diffuse into the substrate. The redistribution and the leveling of the films' compositions in the volume take place. After annealing the film contains a higher titanium concentration.
2. On sputtering the formation of the Ti_3Al-based metastable phase in the films has been observed. This phase is the main in the films and has a different morphology on different substrates. It is stable and doesn't decompose at annealing. During annealing the new intermediate metastable phases, which are stable only within the definite temperature and concentration range, have developed.
3. The structure, kinetics of phase transformation and evolution of the film microstructures depend on the structure and morphology of the substrate.

5. References

1. F.H.Froes, C.Suryanarayana, and D.Elizier (1991) Production, characteristics and commercialization of titanium aluminides, *ISIJ International* **31,** 1235-1248.
2. Y.-W. Kim (1989) Intermetallic alloys based on gamma titanium aluminide, *JOM* **41,** 24-32.
3. -Y.-W.Kim and D.M.Dimiduk (1991) Progress in understanding of gamma titanium aluminides, *JOM* **43,** 40-47.
4. D. Hu, A. Godfrey, P.A. Blenkisop (1998) Prossesing–Property– Microstructure relationship in Ti-Al-based alloys, *Met.Trans.A.***29,** 919-925.
5. E.A. Ott, T.M. Pollock (1998) Microstructure development and creep deformation in equiaxed γ, γ+α and γ+α₂+B2 titanium aluminiders, *Met.Trans.A.***29,** 965-978.
6. P.J. Maziasz, C.T. Liu (1998) Development of ultrafine lammellar structures in two phase γ- Ti alloys, *Met.Trans.A.***29,** 105-117.
7. Y-W. Kim (1992) Microstructural evolution and mechanical properties of a forged gamma titanium aluminide alloy, *Acta metall. mater.***40,** 1121-1134.
8. T. Kumagai, E. Abe, M.Nakamura (1998) Microstructure evolution through the α→γ phase transformation in a Ti- 48 at. pct. Al alloy, *Met.Trans.A.***29** 19-26.
9. W.J. Zhang, G.L. Chen, E. Evangelista (1999) Formation of [alpha] phase in the massive and feathery [gamma]- TiAl alloys during aging in the single [alpha] field, *Met.Trans.A.***30** 2591-2598.
10. S.E. Romankov, T.V. Volkova, V.D. Melikhov (2002), Phase and structural transformations in the Ti-48%Al-2%Nb alloy upon aging, *Phys. Met. Metall.* **93,** 338-348.
11. J.Creus, H. Jdrissi, H.Marille, F.Sanchette, P.Jacquot (1999) Corrosion behaviour of Al/Ti coating elaborated by cathodic arc PVD process onto mild steel substrate, *Thin Solid Films* **346,** 150-154.
12. M.L. Ovecoglu, O.N.Senkov, N. Srisukhumbowornchai, F.H.Froes (1999) Microsturctural evolution of a nanocrystalline Ti-47Al-3Cr alloy during annealing in the α+γ–phase field, *Met.Trans.A.***30,** 751-761.
13. E. Abe, M. Ohnura, M. Nakamura (1999) The structure of a new ε- phase formed during the early stage of crystallization of Ti-48at.%Al amorphours film, *Acta Mater.* **47,** 3607-3616.
14. H.M. Flower, J. Christodoulou (1999) Phase equilibria and transformations in titanium aluminides, *Mater.Sci.and Technol.* **15,** 45-52.

NANOSTRUCTURED METAL OXIDE FILMS WITH ROOM TEMPERATURE GAS SENSING PROPERTIES

G. KIRIAKIDIS, H. OUACHA, and N. KATSARAKIS
Institute of Electronic Structure and Laser, Foundation for Research &
Technology-Hellas, PO Box 1527, Vasilika Vouton, 71110 Heraklion,
Crete, Greece

Abstract   The interest on nanocrystalline films, as a result of their outstanding properties associated with nanoscale and interface effects, has been recently extended to a range of metal oxides such as InOx and ZnO. These oxides are known for their fabrication simplicity, their high transparency in the visible and high reflectivity in the infrared regions, and have attracted a pronounced attention due to their remarkable room temperature sensing properties to reactive atmospheres such as Ozone. They exhibit changes in conductivity from five to eight orders of magnitude, after photoreduction with UV or laser irradiation and subsequent oxidation in reactive atmosphere, in a fully reversible manner. This combination of optical and electrical properties favors numerous applications as transparent conductive electrodes in flat panel devices and solar cells, coatings for architectural glasses and, more recently, as grating materials in optoelectronic devices.

A comparison of the room temperature ozone sensing properties of polycrystalline Zinc oxide and Indium oxide films to an ozone atmosphere has been carried out. Thin films, with a thickness of 10 nm to 1100 nm, have been produced by the DC magnetron sputtering technique. The initially high resistive as-grown films were brought to a high conducting state through a photoreduction process by UV light exposure and, subsequently, they were exposed to a controlled ozone atmosphere resulting in a strong increase of the resistivity of the films caused by re-oxidation. This treatment was shown to be fully reversible over many cycles of photoreduction and oxidation. The films exhibited resistivity changes of several orders of magnitude during the cycles. The sensor response, i.e. the ratio between the conductivity in the conductive state and the insulating state, has been studied for a variety of deposition parameters. Structural investigations, carried out by XRD and AFM, revealed that there is a strong correlation between crystallinity, surface topology, and ozone sensitivity for the materials under investigation. The highest sensor response of the films was always achieved at room temperature.

1. Introduction

The use of transparent conducting oxides (TCO), in the field of photonic and electronic devices, makes these thin film materials important for technological applications and the

T. Tsakalakos et al. (eds.) pgs. 363 - 382
Nanostructures: Synthesis, Functional Properties and Applications;
© *2003 Kluwer Academic Publishers.*

subject of extensive studies. The history of TCO goes back to 1907, when the first cadmium oxide (CdO) films were prepared by the thermal oxidation of sputtered cadmium [1]. With the advancement of technology, the development of TCO has progressed greatly and many TCO films, based on different metals, have appeared. The interest arises because their features and properties are high infrared reflectance, high luminous transmittance, and have good electrical properties (conductivity / resistivity). These characteristics make TCO attractive for many applications, such as; transparent electrodes for solar cells and flat panel displays, electrochromic window coatings capable of yielding indoor comfort and energy efficiency, films for organic light-emitting devices, hardness, excellent substrate adherence, and gas sensors [2-7].

The conductivity in semiconductor materials is given by $\sigma = Nq\mu$, where N is the carrier concentration, q is the elementary charge, and μ the carrier mobility. As can be seen from the above expression, any change in N and/or μ will cause changes in the conductivity. There are many physical mechanisms that have an impact on the conductivity of the material, such as, local random events which include scattering mechanisms undergone by the carriers that produce fluctuations of their number or/and their velocity. On the other hand, chemical mechanism, which occurs at the surface of the semiconductor, has shown to significantly induce changes in the magnitude of the conductivity/resistivity of the material. Such mechanism can be due to the adsorption of gas [5-7]. When a semiconductor surface is brought into contact with a gas the adsorption's phenomenon starts and successful effort has been made to make use of this behaviour for gas detection purposes [8-15]. The dangers of exposure to natural gas and the fast growth in motor traffic, which leads to increasing pollution of our environment, are of important concern. This is driving researchers to develop gas sensor devices for domestic, environmental, and industrial applications. A number of these devices are found to be based on semiconducting metal oxide thin films as the gas-sensitive materials. The choice of these films is based on their portability, simplicity, low cost, and reduced size. These characteristics make the gas sensors with a thin sensing layer interesting for the realization of these detectors on industrial scale. In general, a sensor consists of an active layer whose resistance is measured by two electrodes. The structure is completed with a resistor used to heat the device and a temperature sensor. Moreover, these oxide films exhibit a significant change in their conductivity when a change in the ambient occurs and this property forms the basis for a new generation of gas detection. The change in the conductivity can generally be explained by two mechanisms: the transfer of electrons from the adsorbed gas to the oxide semiconductor or the reaction of the adsorbed gas with previously chemisorbed surface oxygen. Nowadays, research on gas sensors is focused on the development of novel sensing layers and, the main goal, to find TCO films with good electrical and optical properties for specific sensing applications operating at room temperature. For this purpose, intensive studies and investigations on a variety of materials are conducted to produce different gas sensors that include: films such as indium oxide (InOx), zinc oxide (ZnO), tin dioxide (SnO2), tin-doped indium oxide (In2O3: Sn) symbolized by ITO, tantalum pentoxide (Ta2O5), and tungsten trioxide (WO3). Besides their application as gas sensors, the advantage of thin sensing films is that they can be economically fabricated by using the existing silicon microelectronics technology which offers the possibility of an array of large number of sensors on a small area.

Comini et al. [16] investigated the response of tin monoxide SnO silicon doped thin films toward two gases of great importance for environmental monitoring: nitrogen dioxide (NO2) and carbon monoxide (CO). The best result for the gases tested is obtained for a working temperature of 300 °C. The films were found to be capable of detecting concentrations as low as 15 ppm of CO and 200 ppm of NO2 with a response time of approximately 2-5 min. On the other hand, Sharma et al. [17] reported on a sensitive, selective, and stable SnO2 for CO sensing. The results indicated that the sensitivity of the thin film was observed between 270-320 °C and at 1000 ppm CO gas concentration while higher sensitivity was obtained as the operating temperature increases. The response time was found to be 5-7 s. The resistance changes of SnO showed an opposite sign to that of SnO2 [18] which is widely used as active layer in resistive type detectors. This property has been used to integrate different sensors in an array designed to identify different types of gases [19]. The source of ozone (O3), which is the main origin of smog, is the photolytic decomposition of NO2 by solar radiation. The characteristics of dc sputtered WO3 thin films as NOx gas sensor and its ozone sensing properties have been investigated, respectively [20-21]. It has been observed that WO3 films for sensing NOx gas at least require a post-annealing process at 600 °C. For the films deposited at 500 °C and annealed at 600 °C, the sensitivity value of about 179 was measured at the temperature of 200 °C. On the other hand, in their preliminary work, Aguir et al [21] studied the influence of the temperature over the response time of their sensors for the same ozone concentration. The conductivity of WO3 films was found to vary up to a factor higher than 1000 in the presence of ozone in a reproducible way, and the best answer was obtained between 200 and 300 °C. Tungsten trioxide is among the promising new metal oxides, and its sensitivity towards O3, NOx, H2S, CO has also been studied by other groups too [22-23]. Other TCO films, such as, tin-doped indium oxide, In2O3: Sn and Ta2O5 have been investigated, and their sensing characteristics are reported elsewhere [24-25].

The deposition conditions, such as, deposition temperature, oxygen pressure, substrate temperature, and film thickness, are usually reported to have an impact on the sensing properties of these films. The technology is now seeking to maximize the surface to volume ration because the sensing mechanism for TCO thin films is primarily a surface effect. Besides the above-mentioned studies, additional work by other groups concerning mainly high operating temperatures for gas sensors are reported [26-28]. However, for commercial applications there is a demand for gas sensors with low power consumption since the sensors work round the clock. Thus, there is a clear preference for sensors operating at room temperature (RT) suppressing power consumption. Existing semiconductor gas sensors operate at elevated temperatures to ensure appropriate response time and high sensitivity. In this work, we study the sensing properties of selected TCO materials, such as, indium oxide and zinc oxide, and demonstrate their sensing properties to reactive atmospheres, such as ozone at RT, by making use of the photoreduction behaviour of these thin films.

For state of the art gas sensors based on TCO, usually a planar structure is selected. Thus, our sensitive films are binary metal oxides which are supported by Corning 7059 glass substrates equipped with two electrodes. Our nanostructured films, InOx and ZnO, were prepared by magnetron sputtering, the technique found to be the most widely used for producing high quality TCO films with controlled properties. It provides good film uniformity, good adhesion, thickness control, surface smoothness, and less waste of expensive source material.

2. Experiments

Nanocrystalline indium oxide (InOx) and zinc oxide (ZnO) thin films were deposited by DC magnetron sputtering in an Alcatel sputtering system. They were grown in a mixture of argon-oxygen plasma at a total pressure of 8×10^{-3} mbar using 99.999% pure indium and zinc targets. The base pressure of the vacuum chamber was always 5×10^{-7} mbar. All films were deposited onto Corning 7059 glass substrates which had two thermally evaporated NiCr electrodes for two-probe conductivity measurements. During the same run, similar substrates without electrodes were coated for performing the structural and morphological characterization. Furthermore, InOx and ZnO films were deposited onto silica substrates for optical characterization. The substrate temperature, during deposition, was varied between RT and 400 °C. ZnO films were deposited with a thickness of 150 nm and 1000 nm while, for InOx films, the thickness varied between 10 and 1100 nm as measured in-situ by a Leybold XTM/2 quartz crystal deposition monitor and controlled after deposition with the Stylus method. All other parameters, i.e., total pressure, composition of the sputtering atmosphere, and sputtering rate, were kept constant for all the depositions discussed.

X-ray diffraction (XRD), using a Rigaku diffractometer with Cu Kα X-rays, was applied in order to determine the crystal structure of the deposited films. From the diffractograms, the lattice constant and the grain size in the growth direction were derived. The surface morphology (lateral grain sizes and surface roughness) was investigated with a Nanoscope II Atomic Force Microscope (AFM) in tapping mode. The optical transmittance of the as-deposited films was measured using a Perkin-Elmer UV/VIS photo-spectrometer. From the spectra, the optical energy gap was calculated.

The electrical characterization was performed in a special designed reactor described elsewhere [29]. The as-deposited films were all in an insulating state, for example, the as-deposited ZnO thin films showed a conductivity of less than 10^{-10} Ω-1cm-1 in vacuum. For photoreduction, the samples were directly irradiated in vacuum by the UV light of a mercury pencil lamp at a distance of approximately 3 cm for 20 minutes in order to achieve a steady state. For the subsequent oxidation, the chamber was backfilled with oxygen at a pressure of 560 Torr and the samples were shielded from the lamp which, in this case, served as a source for ozone production. This treatment lasted 40 min after which no further changes of the conductivity could be observed. Finally, the chamber was evacuated and the photoreduction-oxidation cycle described above was repeated for a few times. An electric field (1 or 10 V/cm) was applied during the whole cycling procedure to the samples and the electrical current was measured with an electrometer. The contact geometry for the two-probe measurement of the film's conductivity is described in detail in [30]. In this work, all conductivity measurements were carried out at room temperature. An I-V curve was recorded before the cycling started in order to ensure the Ohmic nature of the contacts. The photoreduction treatment results in an increase of the conductivity up to 10^{-4} Ω-1cm-1 for the ZnO films and 10^{-1}-10^2 Ω-1cm-1 for the InOx films while conductivity values, as low as

~10-9 Ω-1cm-1 for ZnO and 10-3-10-6 Ω-1cm-1 for InOx, are obtained by subsequent ozone oxidation. This behavior was completely reversible through many cycles of photoreduction and oxidation treatments as shown for ZnO in Figure 1.

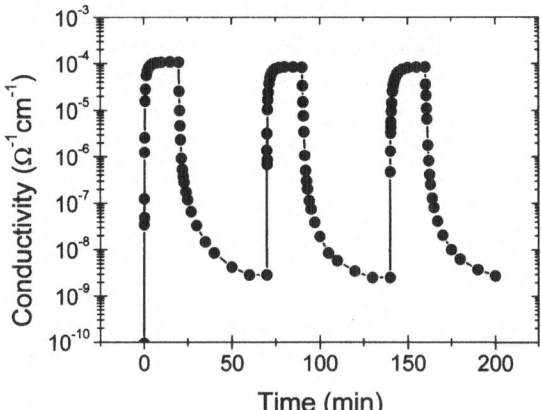

Figure 1: Typical photoreduction-oxidation cycle of a ZnO film.

In the following discussion, maximum conductivity (σmax) denotes the conductivity in the conducting state of the sample after the irradiation procedure while minimum conductivity (σmin) denotes the conductivity in the insulating state after re-oxidizing the sample. Finally, the sensor response of the films is defined as the ratio σmax/σmin.

3. Results and Discussion

3.1 OPTICAL CHARACTERIZATION

The as-deposited InOx and ZnO thin films were found to be highly transparent in the visible wavelength region with an average transmittance of 80% and showed an absorption edge in the UV depending on deposition temperature. The optical energy gap Egap was derived assuming a direct transition between the edges of the valence and the conduction band for which the variation, in the absorption coefficient α with photon energy hv, is given by:

$$\alpha(hv) = A(hv - E_{gap})^{1/2} \tag{1}$$

In equation (1), Egap denotes the optical energy gap between the valence and the conduction band. By plotting $\alpha 2$ vs. hv and extrapolating the linear region of the resulting curve, Egap was obtained. A direct optical gap of 3.62 ± 0.05 eV was obtained for InOx which is in good agreement with a previous study (Eg = 3.70 ± 0.05 eV) [29]. Similar results have been reported by Naseem et al. [31] who measured energy gaps in the range from 3.67 eV to 3.92 eV for evaporated InOx films, depending on the oxygen partial pressure during deposition. The calculated values of the direct optical energy gap varied between 3.21 eV and 3.33 eV for ZnO thin films, depending on the substrate temperature during deposition. In Fig. 2, the dependence of the optical energy gap on substrate temperature is depicted for 150 nm and 1000 nm thick ZnO films. It can be observed, that the energy gap decreases with increasing substrate temperature for both series of samples. Furthermore, the calculated energy gap for the thicker films is systematically smaller which could be attributed to variations of the film density.

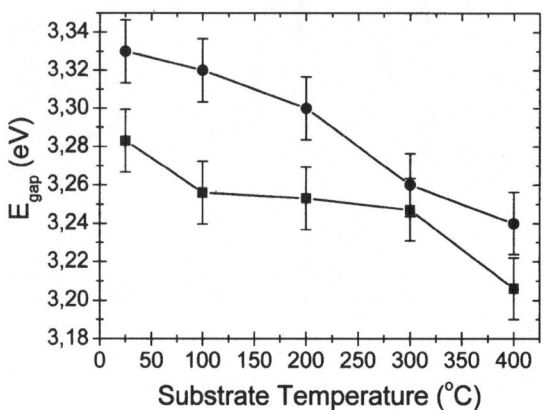

Figure 2: Dependence of the optical band gap of 150 nm (●) and 1000 nm (■) thick ZnO films on substrate temperature during deposition.

It has been suggested, that two concurring phenomena affect the optical band gap [32]. First, the well-known Burstein-Moss shift (BM shift) depends on the shape of the band edge and leads to an increase of the optical energy gap, due to the filling of the lowest levels of the conduction band with charge carriers in degenerate semiconductors. The observed energy gap is then,

$$E_{gap} = E_{gap,0} + \Delta E_{BM} \tag{2}$$

For parabolic band edges, ΔE_{BM} is given by:

$$\Delta E_{BM} = \frac{\hbar}{2m_{vc}^*} (3\pi^2 n)^{2/3} \tag{3}$$

The second phenomenon, which is responsible for the band gap shrinkage due to increased tailing of the absorption edge, is attributed to the merging of the donor states with the conduction band in the vicinity of the semiconductor-metal transition [33] and to electron-electron interactions [34]. The observation of a decreasing band gap, with increasing substrate temperature, leads to the assumption that the effect of band merging probably plays a dominant role in the films studied, presumably caused by building up a donor band below the conduction band edge for films deposited at elevated temperatures. A detailed analysis of the transport properties which is, however, beyond the scope of the work presented, could provide further insight to the observed behavior.

3.2 INFLUENCE OF SUBSTRATE TEMPERATURE ON ELECTRICAL AND STRUCTURAL PROPERTIES

At a fixed thickness of 150 nm, the substrate temperature during the deposition was varied between RT and 300 °C for InOx and between RT and 400 °C for ZnO dc-sputtered thin films. All other deposition parameters were kept constant.

3.2.1. *InOx dc –sputtered Films*
Figure 3 shows a sequence of AFM images for InOx films deposited at different substrate temperatures. The surface exhibited a granular structure with a mean grain size between 150 and 350 Å for all temperatures.

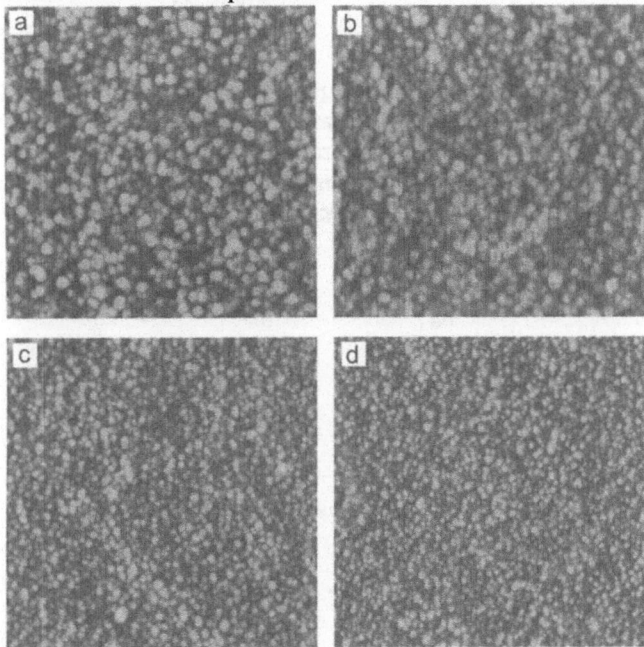

Figure 3: AFM images of InOx films deposited at different substrate temperatures: RT (a), 100 °C (b), 200 °C (c) and 300 °C (d). The area shown is (1x1) µm2. The z-scale is 10 nm.

The XRD measurements revealed that all InOx films were polycrystalline (Fig. 4), even those deposited at room temperature, in contrast to the amorphous growth observed during reactive evaporation [35] or laser ablation [36], indicating an unintended heating of the growing surface due to the plasma.

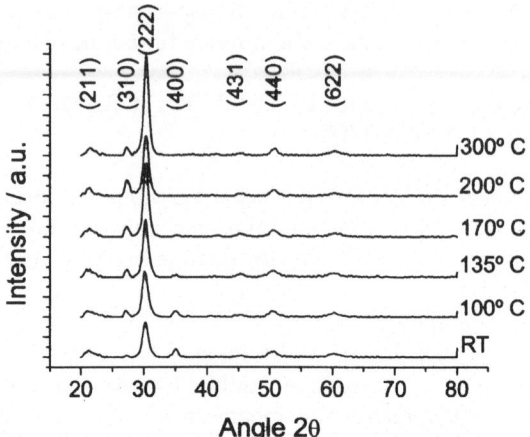

Figure 4: X-ray-diffractograms of InOx films prepared at different substrate temperatures.

No evidence was found for phases other than the cubic bixbiyte structure of In2O3. All films showed a preferred orientation along the [222] axis and an improved crystallinity when the substrate temperature during deposition was increased. Mean grain sizes and lattice constants were derived from the full width at half maximum and the position of the (222) diffraction peak, respectively (Fig. 5).

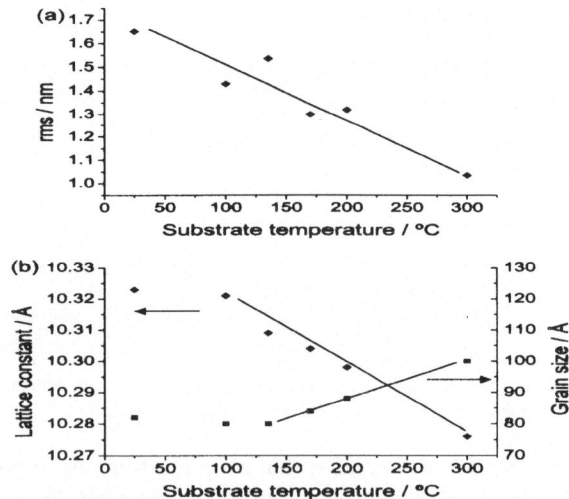

Figure5: (a) rms roughness derived from AFM measurements, (b) lattice constant and grain size derived from XRD measurements of InOx films as a function of substrate temperature during deposition.

The grain size of the samples was found to be approximately 80 Å for films deposited at substrate temperatures up to 135 C, and increased to ~100 Å for films deposited at higher temperatures. Thus, a transition in the growth mechanism occurs from pure grain re-nucleation to grain growth according to the structure zone model for thin films [37] at ~170 C. At this temperature, Thilakan et al. [38] also found a crystallization of evaporated InOx films during post growth annealing experiments.

On the contrary, the lateral grain size at the surface as estimated by AFM was decreasing with increasing temperature (Fig.3). In addition, the surface roughness was found to decrease (Fig. 5). This unusual behavior was already observed by Muranaka et al. [35] for thin films deposited by reactive evaporation. The growth of their InOx films was initiated by the formation of a quasi-amorphous film with a thickness up to 10 nm. This phase crystallized during the continuous growth. With increasing substrate temperature, this amorphous-crystalline transition occurred at a lower thickness with a higher number of initial crystallites in the amorphous matrix. Thus, the resulting grain density increased and, for the fully crystallized film, the grain size decreased.

All derived lattice constants, of the InOx films, were larger than the literature value [39] of 10.118 Å for the cubic bixbiyte structure of In2O3. However, they decreased from 10.32 Å to 10.28 Å with increasing temperature indicating, an improved crystallinity with less defects.

The electrical measurements confirmed the strong change in the film properties around 170 °C (Fig. 6).

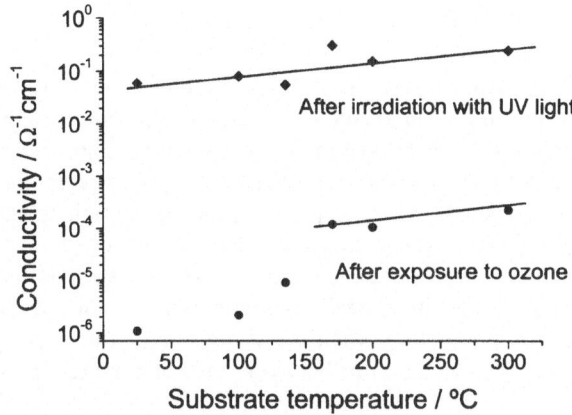

Figure 6: Maximum and minimum conductivities of InOx films prepared at different substrate temperatures.

While the maximum conductivity was increasing only slightly with the deposition temperature from $5.5\times10\text{-}2$ Ω-1cm-1 to $3\times10\text{-}1\Omega$-1cm-1, the minimum conductivity showed a very different behavior for the low and high temperature growth regime. Up to 170 C, the conductivity in the oxidized state increased more than two orders of magnitude from $1.1\times10\text{-}6\Omega$-1cm-1 to $1.2\times10\text{-}4\Omega$-1cm-1. Above 170 C, the slight increase to $2.3\times10\text{-}4$ Ω-1cm-1 followed, qualitatively, the behavior of the maximum conductivity.

3.2.2. *ZnO dc –sputtered Films*

The XRD measurements revealed that all 150 nm-thick ZnO films are polycrystalline, including those deposited at RT (see Fig. 7). No evidence for the existence of phases, other than the hexagonal wurtzite structure of ZnO, was found. All films showed a preferred growth orientation along the c-axis, i.e. (002) plane which is perpendicular to the glass substrate.

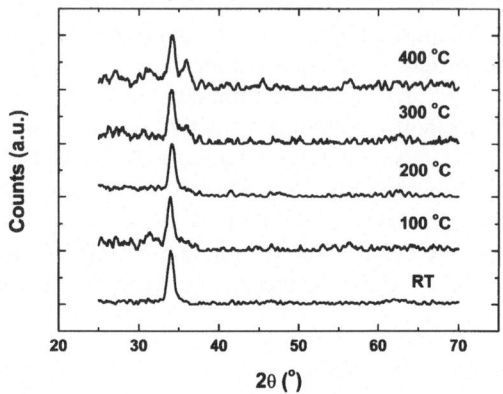

Figure 7: XRD patterns of 150 nm-thick ZnO films deposited at different substrate temperatures.

However, for ZnO films deposited at RT, only the (002) diffraction peak is observed while films deposited at higher temperature (T ≥100 °C) showed more reflections. This indicates a rather random orientation of the crystallites even though the (002) reflection always displayed the higher intensity. This effect was more pronounced for the 1000 nm-thick films and will be discussed in the next section. Thus, it can be concluded that the substrate temperature during deposition strongly affects the preferred growth orientation of our dc sputtered ZnO films. Low deposition temperature (T<100 °C) favors a crystallite growth along the (002) plane while higher deposition temperature (T ≥ 100 °C) results more in rather random oriented ZnO crystallites. Bao et al. have observed a similar change in crystallite orientation for sol-gel deposited ZnO films which lost the c-axis-orientation when the firing temperature was raised to more than 550 °C [40]. To our knowledge, this behavior has not been reported for sputtered films yet.

Mean grain sizes were derived from the full width at half maximum (FWHM) of the (002) diffraction peak for the 150 nm-thick ZnO films while the lattice constants, α and c, were calculated from the positions of the (002) and (110) diffraction peaks (Fig. 8). For all films, the derived lattice constants were in good agreement with the literature values of α=3.253 Å and c=5.209 Å reported for the hexagonal structure of ZnO [41].

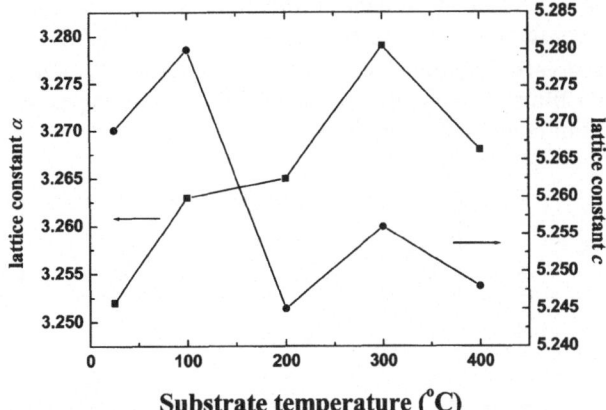

Figure 8: Dependence of the lattice constants α (●) and c (■) of 150 nm-thick ZnO films on the substrate deposition temperature.

It was observed, that α increased from 3.252 Å to 3.268 Å while c decreased from 5.269 Å to 5.248 Å with increasing temperature from RT to 400 °C. Again, this indicates a significant change in the crystallite structure of the films. The grain size of the 150nm-thick ZnO films slightly increased when the substrate temperature was raised from RT to 400°C (10.94 nm at RT and 11.59 nm at 400°C).

The electrical measurements are presented in Fig.9 for the films with a thickness of 150 nm. The maximum conductivity changed by about two orders of magnitude with increasing substrate temperature from RT to 400 °C. In detail, for T<200°C, σmax remained almost constant (~3×10-3Ω-1cm-1) and then increased to 3×10-1Ω-1cm-1 for 200°C<T≤400°C.

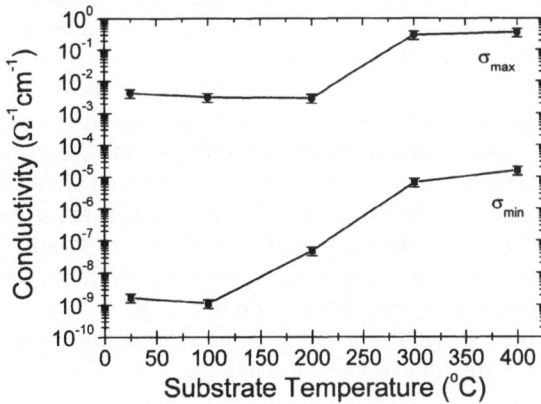

Figure 9: Dependence of the maximum conductivity after UV photoreduction (●) and the minimum conductivity after re-oxidation (■) on the deposition temperature of 150 nm-thick ZnO films.

The minimum conductivity, however, showed more pronounced changes and increased by four orders of magnitude between RT and 400°C. For T≤100°C, σmin remained almost constant (~2×10-9Ω-1cm-1) while, for T≥200 °C, it increased abruptly reaching a value of ~10-5 Ω-1cm-1 at 400°C. As a result, the sensor response of the 150nm-thick ZnO films to ozone decreased with increasing deposition temperature by almost two orders of magnitude (Fig.10).

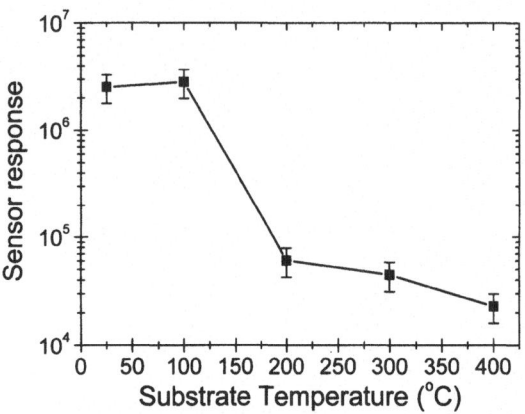

Figure 10: Sensor response of 150 nm-thick ZnO films in dependence on the deposition temperature.

The increase in σmin with deposition temperature is probably due to an enhanced density of donor states which are likely forming a donor band just below the conduction band. The existence of such a donor band could explain the increase in σmin as well as the observed decrease of the optical band gap with increasing substrate deposition temperature, as discussed in the previous section. The step observed in σmin, σmax, and the sensor response, versus substrate temperature curves in the range of 100<T<200°C, can then be related with the change in the preferred growth orientation at this temperature interval. The c-axis oriented ZnO films showed higher sensor response to ozone than the random oriented ones.

AFM images show that the surface morphology of the 150nm-thick ZnO films, deposited at room temperature 200°C, and 400°C, do not differ significantly. All films presented regularly shaped columnar grains with similar sizes. The rms roughness calculated from the images varies between 1.6nm and 5.6nm without a clear dependence on deposition temperature. The same holds for the lateral grain sizes estimated from the AFM images which all lie in the range between 55nm and 75nm. Therefore, tt can be assumed that the surface morphology of the sputtered ZnO films is similar and does not depend on the deposition temperature. Moreover, the structural changes observed, concerning the favored growth orientation of the ZnO films, do not correspond to any significant variation in the film surface morphology.

3.3. INFLUENCE OF FILM THICKNESS ON ELECTRICAL AND STRUCTURAL PROPERTIES

3.3.1. *InOx dc –sputtered Films*

The influence of the film thickness on the electrical and structural properties of InOx was investigated in the range from 10 to 1100nm comprising two orders of magnitude. The substrate temperature was set to 200 °C which is above the crystallization temperature of amorphous InOx films [39]. As described above, at this temperature the electrical properties are only slightly affected by the structural properties of the films and, therefore, a more unambiguous dependence on the film thickness can be expected.

The XRD measurements revealed that all films but the thinnest, with 10nm and 17nm thickness, showed a polycrystalline structure. Again, only the cubic bixbiyte structure of In2O3 was found with a preferred orientation along the [222] axis independent of the film thickness. The mean grain size increased slightly from 69 Å to 103 Å with increasing film thickness and the lattice constant decreased from 10.26Å for the 34nm-thick film to 10.17 Å for the 1100nm-thick film, indicating that thicker films exhibit less strain and/or less defects (Fig.11).

The AFM images presented in Fig.12 show the surface morphology of InOx films with a thickness between 10nm and 725nm. The thinnest film with 10nm thickness (Fig. 12a) exhibited only small features at the surface which can be assigned to very small crystallites in an amorphous matrix according to the model of Muranaka, et al. [35].

With increasing film thickness, the surface appeared to be fully crystallized (Fig. 12b) and, for a thickness above 200nm, an aggregation of the crystallites to clusters was observed (Figs. 12c and 12d). Both the average lateral grain size and the surface roughness (Fig.11) increased when the film thickness became larger. Thus, with increasing film thickness small grains tend to be overgrown by larger neighbors. Therefore, the lateral grain sizes increased more strongly than the ones derived from XRD data.

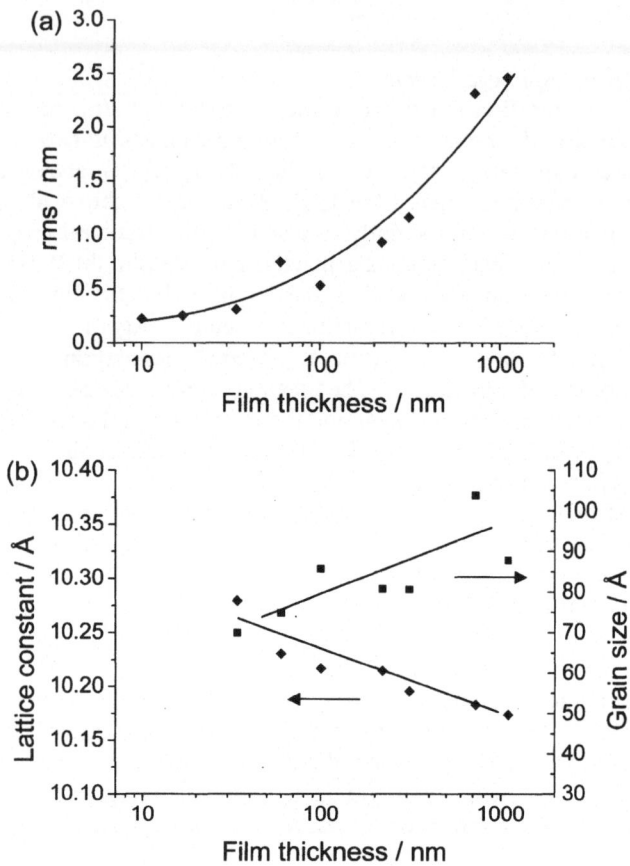

Figure 11: (a) rms roughness derived from AFM measurements, (b) lattice constant and grain size derived from XRD measurements of InOx films as a function of film thickness.

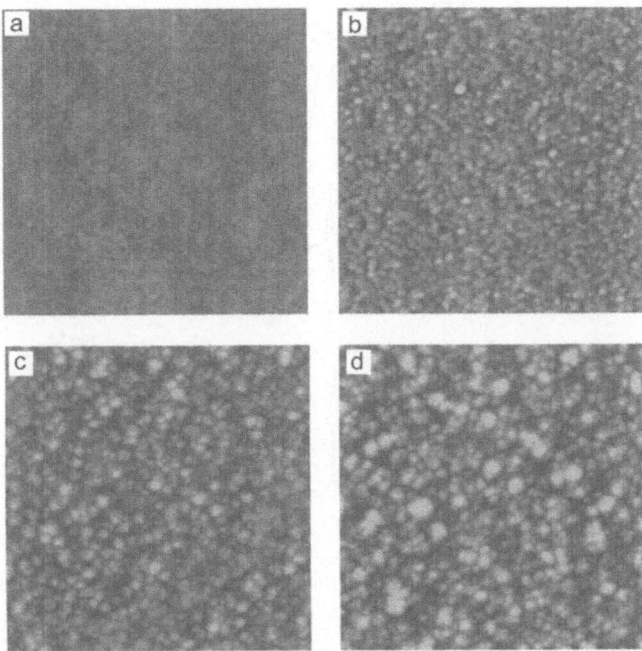

Figure 12: AFM images of InOx films with a thickness of (a) 10 nm, (b) 60 nm, (c) 310 nm and (d) 725 nm. The area shown is (1x1) μm2. The z-scale is 20 nm.

The film thickness mainly affected the maximum conductivity after irradiating the InOx thin films with UV light (Fig.13). The highest conductivity values were measured for the thinnest films. When the thickness of the films was increased about two orders of magnitude from 10 nm to 1100nm, the maximum conductivity dropped by two orders of magnitude from approximately $100\Omega\text{-}1cm\text{-}1$ to $1\Omega\text{-}1cm\text{-}1$. The minimum conductivity, however, was almost constant in the range of $10\text{-}3\Omega\text{-}1cm\text{-}1$.

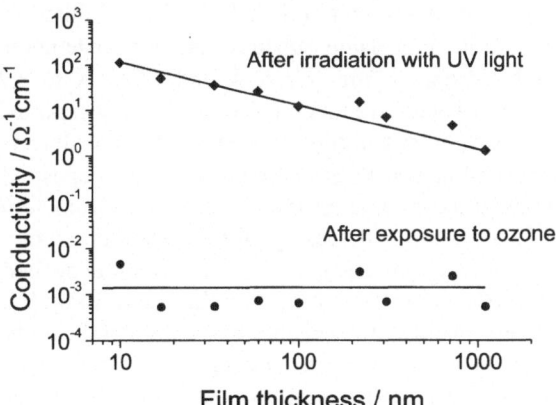

Figure 13: Maximum and minimum conductivities of InOx films as a function of film thickness.

3.3.2. *ZnO dc –sputtered Films*

The XRD measurements revealed that the 1000nm-thick ZnO films were polycrystalline for all deposition temperatures from RT up to 400°C (Fig. 14) like the 150nm-thick ones.

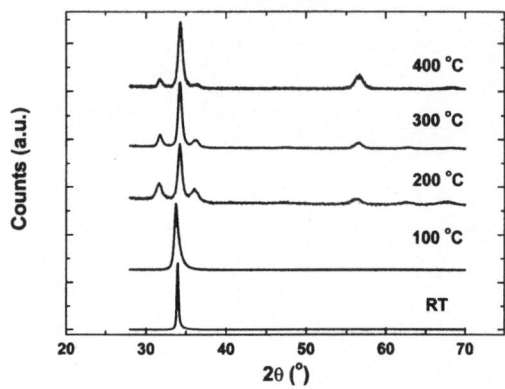

Figure 14: XRD patterns of 1000 nm-thick ZnO films deposited at different substrate temperatures.

However, the preferred growth orientation along the c-axis, i.e. (002) plane, at T≤100°C was even more pronounced for the thicker ZnO films. Only the (002) diffraction peak was observed for films deposited at RT and T=100°C. Moreover, the FWHM of the (002) diffraction peak for films deposited at RT was 0.24 attesting to a high degree of crystallite orientation. For films deposited at T>100°C more diffraction peaks were observed pointing rather to a random crystallite orientation, although the (002) direction still remained the favorite growth orientation.

Mean grain sizes were derived from the full width at half maximum (FWHM) of the (002) diffraction peak while the lattice constants α and c of the 1000 nm-thick ZnO films were calculated from the positions of the (002) and (110) diffraction peaks for films deposited at T≥200°C. The absence of any other diffraction peaks, except of the (002) reflection for films deposited at RT and 100°C, prohibited the calculation of their lattice parameters. With increasing substrate deposition temperature from 200°C to 400°C, both α and c dropped from 3.267 Å and 5.231 Å to 3.247 Å and 5.222 Å, respectively. The latter values are much closer to the ones reported in the literature and demonstrate the improved crystallinity with less structural defects and/or strain of the 1000nm-films deposited at 400°C compared to the ones deposited at 200°C. One can also remark, that the lattice parameters α and c decreased from 3.268nm and 5.248nm to 3.248nm and 5.222nm with increasing film thickness from 150nm to 1000nm, respectively, at 400°C. The mean grain size for films deposited at 400°C increased from 11.59nm to 15.46nm with increasing film thickness from 150nm to 1000nm.

The film thickness mainly affected the maximum conductivity after irradiating the ZnO thin films with UV light (Fig. 15). The highest conductivity values were measured for the 150nm-thick films ($3\times10-3\Omega$-1cm-1 at RT). With increasing thickness to

1000nm, the maximum conductivity dropped at RT by about three orders of magnitude from approximately 3×10-3Ω-1cm-1 to ~10-6Ω-1cm-1. The minimum conductivity, however, remained almost constant in the range of ~10-9Ω-1cm-1. Both, the maximum and minimum conductivity of the 1000nm-thick ZnO films, remained almost constant with increasing deposition temperature from RT to 100°C while they increased with increasing temperature from 200°C to 400°C. This trend could be linked to the increased crystallinity of the films deposited at 400°C compared to the ones deposited at 200°C and has also been observed, e.g., by Zhang et al. [42] in the case of aluminum doped ZnO films and by Liu et al. [43] for undoped ZnO.

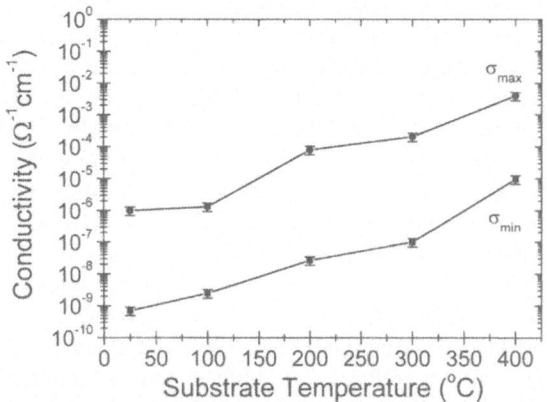

Figure 15: Dependence of the maximum conductivity after UV photoreduction and the minimum conductivity after re-oxidation on the deposition temperature of 1000 nm-thick ZnO films.

3.4. CONDUCTION MECHANISM

The mechanism responsible for the large conductivity changes in InOx and ZnO films is the formation and annihilation of oxygen vacancies. UV irradiation of the sample with energies above the bonding energy between In (Zn) and O leads to a transformation of an oxygen atom from a bound state to the gaseous state. In this process, the two electrons of the oxygen ion are left in the vacant site. If both of these two electrons are localized at the vacancy, charge neutrality is preserved and the vacancy has zero effective charge. If one or both of the localized electrons are excited and transferred away from the vacancy, the vacancy is left with an effective positive charge. The charged oxygen vacancy (singly or doubly ionized) becomes an electron-trapping site but in this process one or two electrons are available for conduction. The subsequent oxidation in ozone leads to annihilation of the charged oxygen vacancies by incorporation of oxygen into the film. Thus, the charge carrier concentration decreases drastically.

The electrical measurements carried out for InOx and ZnO films showed that the maximum conductivity after photoreduction is about three orders of magnitude higher for the thinner films compared to the series of thicker films (see Figs. 13 and 15). The minimum conductivities after oxidation, however, are comparable. This is to be expected according to the mechanism described above, where the conductivity can be assumed to be composed of a contribution of the bulk of the film and a contribution of the surface of the film [45]. After photoreduction, the "surface-near-layer" contributes with a high conductivity and, therefore, the conductivity is thickness-dependent. When the film is re-oxidized, the surface is depleted from free carriers due to adsorbed oxidizing species and merely the bulk conductivity is observed. Thus, the conductivity in this state is independent of film thickness.

4. Conclusions

In conclusion, we have reviewed our recent work on nano-structured InOx and presented new data on ZnO dc sputtered thin films. A comparison of the room temperature sensing properties of polycrystalline Zinc oxide and Indium oxide films to an ozone atmosphere has been carried out. All but the thinnest InOx films were found to be polycrystalline, as revealed by XRD analysis. The initially high resistive as-grown films were brought to a high conducting state through a photoreduction process by UV light exposure and subsequently they were exposed to a controlled ozone atmosphere resulting in a strong increase of the resistivity of the films caused by re-oxidation. This treatment was shown to be fully reversible over many cycles of photoreduction and oxidation. The films exhibited resistivity changes of several orders of magnitude during the cycles. The sensor response, i.e., the ratio between the conductivity in the conductive state and the insulating state, has been studied for a variety of thicknesses and substrate deposition temperatures. The highest sensor response has been achieved for the thinnest InOx and ZnO films deposited at room temperature (~106). The conduction mechanism has been discussed in terms of an ultra-thin surface layer, which is mainly contributing to the photoconductivity in polycrystalline InOx and ZnO films. On the other hand, the minimum conductivity represents the bulk conductivity of the films and is increasing with improved crystal structure after deposition at elevated substrate temperatures.

5. References

1. Frederick Ojo Adurodija (2002) Laser applications in transparent conducting oxide thin films processing, in Hari Singh Nalwa (eds.), Handbook of Thin Film Materials 1, 161.
2. D. G. Lim, D. M. Jang, and J. Yi (2002) A novel multicrystalline silicon solar cell using grain boundary etching treatment and transparent conducting oxide, Solar Energy Materials & Solar cells 72, 571.
3. C. G. Granqvist, and A. Hultaker (2002) Transparent and conducting ITO films: new developments and applications, Thin Solid Films 411, 1.
4. H. Kim, A. Pique, J. S. Horwitz, H. Mattoussi, H. Murata, Z. H. Kafafi, and D. B. Chrisey (1999) Indium tin oxide thin films for organic light-emitting devices, Appl. Phys. Lett. 74, 3444.
5. H. Meixner, J. Gerblinger, U. Lampe, and M. Fleisher (1995) Thin-film gas sensors based on semiconducting metal oxides, Sens. Actuators B 23, 119.
6. G. Kiriakidis, M. Bender, N. Katsarakis, E. Gagaoudakis, E. Hourdakis, E. Douloufakis, and V. Cimalla (2001) Ozone Sensing Properties of Polycrystalline Indium Oxide Films at Room Temperature Phys. Stat. Sol. (a) 185, 27.
7. E. Gagaoudakis, M. Bender, E. Douloufakis, N. Katsarakis, E. Natsakou, V. Cimalla, and G. Kiriakidis (2001) The influence of deposition parameters on room temperature ozone sensing properties of InOx films, Sensors and Actuators B 80, 155.
8. M. Radecka, J. Przewoznik, and K. Zakrzewska (2001) Microstructure and gas-sensing properties of (Sn,Ti)O2 thin films deposited by RGTO technique, Thin Solid Films 391, 247.

9. T. Miyata, T. Hikosaka, and T. Minami (2000) High sensitivity chlorine gas sensors using multicomponent transparent conducting oxide thin films, Sensors and Actuators B 69, 16.

10. E. Comini, M. Ferroni, V. Guidi, G. Faglia, G. Martinelli, and G. Sberveglieri (2002) Nanostructured mixed oxides compounds for gas sensing applications, Sensors and Actuators B 84, 26.

11. K. Zakrzewska (2001) Mixed oxides as gas sensors, Thin Solid Films 391, 229.

12. M. I. Baraton (2000) Surface characterization of nanostructured coating: Study of nanocrystalline SnO2 gas sensors, in GaN-Moog Chow, Ilya A. Ovidko and Thomas Tsakalakos (eds.), Nanostructured Films and Coatings 78, 187.

13. M. C. Horrillo, P. Serrini, J. Santos, and L. Manes (1997) Influence of the deposition conditions of SnO2 thin films by reactive sputtering on the sensitivity to urban pollutants, Sensors and Actuators B 45, 193.

14. P. Serrini, V. Briois, M. C. Horrillo, A. Traverse, and L. Manes (1997) Chemical composition and crystalline structure of SnO2 thin films used as gas sensors, Thin Solid Films 304, 113.

15. D. S. Vlachos, C. A. Papadopoulos, and J. N. Avaritsiotis (1995) The effect of film oxygen content on SnOx gas-sensor selectivity, Sensors and Actuators B 24-25, 883.

16. E. Comini, G. Faglia, and G. Sberveglieri (2001) CO and NO2 response of tin oxide silicon doped thin films, Sensors and Actuators B 76, 270.

17. R. K. Sharma, P. C. H. Chan, Z. Tang, G. Yang, I. M. Hsing, and J. K. O. Sin (2001) Sensitive, selective and stable tin dioxide thin-films for carbon monoxide and hydrogen sensing in integrated gas sensor array applications, Sensors and Actuators B 72, 160.

18. P. Ciureanu, and S. Middlehoek (1992) Thin film gas sensors, p. 456.

19. J. Calderer, P. Molinas, J. Sueiras, E. Llobet, X. Vilanova, X. Correig, F. Masana, and A. Rodrigues (2000) Synthesis and characterization of metal suboxides for gas sensors, Microelectronics Reliability 40, 807.

20. T. S. Kim, Y. B. Kim, K. S. Yoo, G. S. Sung, and H. J. Jung (2000) Sensing characteristics of dc reactive sputtered WO3 thin films as a NOx gas sensor, Sensors and Actuators B 62, 102.

21. K. Aguir, C. Lemire, and D. B. B. Lollman (2002) Electrical properties of reactively sputtered WO3 thin films as ozone gas sensor, Sensors and Actuators B 84, 11.

22. C. Cantalini, M. Pelino, H. T. Sun, S. Santucci, M. Passacantando, L. Lozzi, and M. Faccio (1996) Cross sensitivity and stability of NO2 sensors from WO3 thin film, Sensors and Actuators B 35, 112.

23. W. Qu, and W. Wlodarski (2000) A thin-film sensing element for ozone, humidity and temperature, Sensors and Actuators B 64, 42.

24. T. Sako, A. Ohmi, H. Yumoto, and K. Nishiyama (2001) ITO-film gas sensor for measuring photodecomposition of NO2 gas Surface and Coatings Technology 142-144, 781.

25. C. Chaneliere, J. L. Autran, R. A. B. Devine, and B. Balland (1998) Tantalum pentoxide (Ta2O5) thin films for advanced dielectric applications, Mater. Sci. Eng. R22, 269.

26. T. Miyata and T. Minami (1999) Stability and sensing mechanism of high sensitivity chlorine gas sensors using transparent conducting oxide thin films, Thin Solid Films 355-356, 35.

27. M. Fleischer, and H. Meixner (1999) Thin-film gas sensors based on high-temperature-operated metal oxides, J. Vac. Sci. Technol. A 17, 1866.

28. Q. Pan, J. Xu, X. Dong, and J. Zhang (2000) Gas-sensitive properties of nanometer-sized SnO2, Sensors and Actuators B 66, 237.

29. C. Xirouchaki, G. Kiriakidis, T. F. Pedersen, and H. Fritzsche, Photoreduction and oxidation of as-deposited microcrystalline indium oxide (1996) J. Appl. Phys. 79, 9349.

30. C. Xirouchaki (1998) Ph.D. Thesis, Physics Department, University of Crete, Greece, p.93.

31. S. Naseem, I. A. Rauf, and K. Hussain (1988) Effects of oxygen partial pressure on the properties of reactively evaporated thin films of Indium oxide, Thin Solid Films 156, 161.

32. I. Hamberg, and C. G. Granqvist (1986) Evaporated Sn-doped In2O3 films: Basic optical properties and applications to energy-efficient windows, J. Appl. Phys. 60, R123.

33. A. Roth, J. B. Webb, and D. F. Williams (1982) Band-gap narrowing in heavily doped ZnO, Phys. Rev. B 25, 7836.

34. L. Gupta, A. Mansingh, and P. K. Srivashava, Thin Solid Films 176 (1989) 33.

35. S. Muranaka, Y. Bando, T. Takada (1987) Influence of substrate temperature and film thickness on the structure of reactively evaporated In2O3 films, Thin Solid Films 151, 353.

36. F. O. Adurodija, H. Izumi, T. Ishihara, H. Yoshioka, M. Motoyama, and K. Murai (2000) Influence of substrate temperature on the properties of indium oxide thin films, J. Vac. Sci. Technol. A 18, 814.

37. J. A. Thornton (1977), Ann. Rev. Mater. Sci. 7, 239.

38. P. Thilakan, and J. Kumar (1998) Mat. Sci. Eng. B 55, 195.

39. R. W. G. Wyckoff (1964) Crystal Structures, Wiley, London.

382

40. D. Bao, W. Gu, and A. Kwan (1998) Sol⁻gel-derived c-axis oriented ZnO thin films, Thin Solid Films 312, 37.
41. D. R. Lide (1992) Handbook of Chemistry and Physics, CRC Press, Florida.
42. D. H. Zhang, T. L. Yang, J. Ma, Q. P. Wang, R. W. Gao, and H. L. Ma (2000) Preparation of transparent conducting ZnO: Al films on polymer substrates by r.f. magnetron sputtering, Appl. Surf. Sci. 158, 43.
43. J. M. Liu, and C. K. Ong (1998) Pulsed laser deposition of ZnO as conductive buffer layer of (001)-LiNbO3 thin films, Appl. Phys. A 67, 493.
44. P. Kofstad (1972), Non-stoichiometry, Diffusion and Electrical Conductivity in Binary Metal Oxides, Wiley, New York.
45. V. D. Das, S. Kirupavathy, L. Damodare, N. Lakshminarayan (1996) Optical and electrical investigations of indium oxide thin films prepared by thermal oxidation of indium thin films, J. Appl. Phys. 79, 8521.

ACHIEVEMENT OF NANOCRYSTALLINE STRUCTURES AFTER THERMAL ANNEALING AMORPHOUS W-SI-N SPUTTERED FILMS

A. CAVALEIRO AND P. MARQUES

*ICEMS, Dep. de Engenharia Mecânica, Universidade de Coimbra, 3030
Coimbra, Portugal.
Tel: 351 239 790745, Fax: 351 239 790701,
e-mail:albano.cavaleiro@dem.uc.pt*

1. Introduction

An increasing interest in nanocrystalline materials (NCM) has been revealed in the last decade due to their potential to replace traditional materials in many applications by the possibility to control the structure at an atomic level. Such a fine control allows tailoring the material with the optimum desired properties for many applications in the areas of magnetics, optics, and catalysis. In the particular case of mechanical field, very interesting properties can be reached, such as, superplastic deformation behaviour and super high hardness (over 50GPa) (see e.g. [1,2], special issues of MRS Bulletin dedicated to this subject). One of the most interesting points in the super-hard materials, is the possibility to simultaneously obtain very high fracture toughness values if a mixture of nanocrystalline grains, embedded in an amorphous matrix, can be produced [3].

Among the different ways that have been used to synthesize NCMs, their deposition in the form of thin films has been one of the most utilized. Many deposition techniques allow to deposit materials with large ranges of either structures from equilibrium to metastable (including amorphous), or grain sizes (1-1000nm), if suitable conditions are used during the thin films formation. Thus, this is an efficient way to accomplish the above mentioned tailoring of the materials properties.

Many authors have been trying to correlate the very high hardness of nanocrystalline films with the very small size of the grains, following the well known Hall-Petch relationship, which explains this phenomenon on the basis of dislocation pileups at grain boundaries [2]. However, this theory must break down. for this size of grains that cannot support a pileup, making the dislocation theory not appropriate in this size regime. Molecular-dynamics simulation proved that for NCMs, at very small grain sizes, intergranular deformation mechanisms supplant intragranular mechanisms [2]. This means that the mechanical strength of NCM materials will be a competition between the grain size (the large volume fration of the grain boundaries leads to a softening of the material) and the type of grain boundary. Thus, to reach very high

T. Tsakalakos et al. (eds.) pgs. 383 - 411
Nanostructures: Synthesis, Functional Properties and Applications;
© *2003 Kluwer Academic Publishers.*

hardness in NCMs, is is imperative not only to have a close control of the grain size but also of the type of the grain boundary.

Recently, many research studies were carried on the development of coatings with a multi-phase structure consisting of a nanocrystalline hard phase surrounded by a very thin layer (<1nm) of an amorphous/crystalline phase [3-5]. In the case of Ti-Si-N system the films structure consists of nanocrystals of TiN embedded in an amorphous (Si_3N_4)/crystalline $(TiSi_2)$ matrix.

Multiphase nanocrystalline materials can also be produced by the thermal annealing of amorphous phases. This is one of the procedures used in the development of magnetic materials with improved properties (see e.g. [6]). This procedure allows both the precise control of the grain size of the crystalline phase being formed and the fraction of the crystallised material. However, at our knowledge, this technique was never been applied for the development of coatings for mechanical applications.

Sputtering is a deposition technique very suitable to produce amorphous structures. On W-based nitrides and carbides, amorphous structures can be easily deposited by doping the films with many different elements [7] including Si [8]. Thus, W-Si-N seems to be a good system to be analyzed concerning the possibility of forming nanocomposite structures after thermal annealing. In this work, we report on the deposition and characterization of sputtered coatings from the W-Si-N system, keeping in mind, firstly, the relationship between the deposition and annealing conditions with the structural characteristics and, finally, their influence on the mechanical properties particularly on the hardness and Young's modulus

2. Chemical Composition

For this study, amorphous films of the W-Si-N system were deposited by d.c. reactive magnetron sputtering In order to reach different Si contents in the films two targets (150x150mm) were used: one of W which was incrusted with 8, 12, 16 and 20 round ($\varnothing$=12mm) pieces of Si and another of WSi_2. Also, by using N_2/Ar partial pressure ratio in the range [0-1], it was possible to reach coatings with different N contents. Figure 1 presents the chemical composition results (Electron Probe Microanalysis – EPMA) of the as-deposited W-Si-N films (only the films with amorphous structures are presented), expressed by the Si/W ratio plotted as a function of the N content. The deposition of the amorphous structure can be reached with a large range of Si and N contents. As a general trend, it can be concluded that both the addition of Si and N to W films contributes for the instability of the crystalline structure and the formation of low range order phases. The higher the Si content the lower can be the N content for promoting the deposition of amorphous phases.

The film deposited from the WSi_2 target without N_2 has a much lower Si/W ratio ($\approx$1) than that of the target ($\approx$2). However, in N-containing films the Si/W ratio increases in relation to W-Si film approaching the target Si/W ratio. Similar trend is also registered for the films deposited from the other targets. This behaviour can be explained by the preferential re-sputtering of silicon during the film growth [8], induced by the Ar^+ ion bombardment, due to the negative bias applied to the substrate during the deposition. When depositing in the $Ar+N_2$ reactive atmosphere, the formation of strong

Si-N bonds makes more difficult the preferential sputtering of Si leading to higher Si/W ratios in the films.

A research study on the chemical bonding in this type of films by X-ray Photoelectron Spectroscopy (XPS) [9] showed that during the deposition N and Si would be preferentially bonded each other, in such a way that either the whole Si available would be combined to N, or vice-versa, being the remaining content of N or Si then bonded to W. In the crystalline as-deposited samples, a nanocomposite arrangement was suggested having been formed, consisting in nanograins (3-10nm) of a W-based phase (α-W and/or β-W_2N) embedded in an amorphous Si-N phase. Even for amorphous films, particular arrangement among the atoms could be expected since similar features to those observed in crystalline films could be extracted from XPS analysis. Figures 2 a) and b) present the XPS W4f and Si2p peaks, respectively, as well as the peaks deconvolution after fitting analysis, obtained for the amorphous ($W_{26}Si_{30}N_{44}$) film. Taking into account the preferential bonding sequence above referred to, there is a good agreement between XPS and EPMA chemical composition results. The 30at.% of Si is bonded to 40at.% of N, leaving 4at.% of this element to be combined with W as demonstrated by the W-N bond contribution (33.1 and 35.4eV) found in the W4f peak. The presence of sub-peaks placed at 31.6 and 33.8eV can be attributed to W-W bonds [10], as should be expected taking into account that only a part of the W is consumed in the bonding with N. In the same way, only vestiges of sub-peaks corresponding to the Si-Si and/or Si-W (should appear at 99.6eV [10]) were detected in the Si2p signal, indicating that almost the entire Si is combined with N, as suggested above. Besides the Si-O peaks, related to the surface oxidation, only the sub-peak corresponding to the Si-N bond is clearly detected (101.8eV).

Amorphous films were crystallized by thermal annealing at temperatures up to 1000°C in an Ar+H_2 atmosphere to avoid oxidation problems. After crystallization, no significant changes in the chemical composition could be found (e.g. $W_{31}Si_{36}N_{33}$ and $W_{28}Si_{45}N_{27}$ films in Table 1) with the exception of particular cases, as follows:

- In the films with low Si and/or very high N contents, loss of N is detected after annealing in relation to the as-deposited chemical composition ($W_{63}Si_8N_{29}$ in Table 1). For this very high N/Si ratio, a part of N will be available to establish W-N bonds. However, due to the very low affinity of W for N, the stability of the W-N phases is very low, being observed their decomposition at high temperatures with the consequent liberation of nitrogen as other researchers [11,12] have found.

- Among the substrates used in these research studies, those containing Ni (310 (AISI) steel and Invar®), seemed not be suitable for the deposition of W-Si-N films for subsequent thermal annealing. In fact, for temperatures higher than 900°C, interdiffusion between the film and the substrate could be observed, changing the measured chemical composition (see examples in table 1) [13-15]. Figure 3 a) shows, for a particular case, the evolution of the chemical composition across the film thickness of the $W_{34}Si_{33}N_{33}$ film deposited onto a 310 steel substrate after have been annealed at 1000°C [14]. The points correspond to EPMA analyses performed along a line in the border of a crater carried out in the coated sample by ball abrasion (only Fe is presented, although similar behaviour was also found for Ni and Cr). As can be observed, severe outwards diffusion of Fe is observed from the substrate to the film. In the middle of the film, N content almost vanishes, being diffusing outwards or inwards. Depending on the loss of N for the annealing atmosphere and/or its accumulation in the surface layers of the coatings, the content measured by EPMA can change significantly. This behaviour explains the changes presented in table 1 for other coatings deposited on Ni-containing substrates and annealed at 1000°C. The same film deposited on other substrates (Fecralloy® and Mo-alloy materials) did not present such an interdiffusion (figure 3 b)).

2.1 STRUCTURE

Figures 4 a) and b) show XRD patterns of films with similar N at.% but different Si/W ratio and similar Si/W and different N content, respectively. In all cases, only a broad peak, centered in the 2θ range from 40° to 50°, is detected confirming the amorphous character of the structure. With increasing N content, a small shift in the peak to lower diffraction angles is observed, confirming the existence of some order arrangement already in the amorphous state. In crystalline films based on W, the addition of N also leaded to shift of the diffraction peaks for lower angles [8].

Figure 5 presents the crystallization temperatures (Tc) of the W-Si-N films as a function of their N content, for three different ranges of Si/W ratios. These values correspond to temperatures for which XRD patterns with well-defined diffraction peaks are obtained. As found by other authors [12], both N and Si contents are determinant on the thermal stability of the amorphous phase. The higher the content of these elements the higher the Tc value is. The thermal stability of metastable materials of metal-metalloid type, as for W-Si-N, is indicated in the literature [16] to be dependent of: (i) the difference in the atomic size of the elements of the system; (ii) the strength of the bonding established between the metal and the metalloid; (iii) the probability to form intermetallic compounds with different structures. In the system W-Si-N the increase in the Si and N contents contributes to enhance all these factors since, the heterogeneity in the size of the atoms increases, the formation of strong Si-N bonds is improved and the possibility to form tungsten silicides is higher.

The phases detected after crystallization confirms the combination above suggested between the chemical elements. The following sequence of phases would be expected as a function of the increase of the remaining Si that was not-bonded to N:

$$\alpha\text{-W} \rightarrow \alpha\text{-W}+\beta\text{-W (W}_3\text{Si [17])} \rightarrow \alpha\text{-W}+\text{W}_5\text{Si}_3 \rightarrow \text{W}_5\text{Si}_3 \rightarrow \text{W}_5\text{Si}_3+\text{WSi}_2 ,$$

i.e. with increasing Si content, silicides with increasing Si % would be obtained. Selected examples of XRD patterns illustrating these cases are presented in figure 6. Moreover, no crystalline nitrides phases were detected in spite of the formation of Si-N and W-N bonds in the as deposited conditions. Firstly, in the case of sputtered Si-N films, crystallization was only observed for temperatures higher than 1200°C [18]. Thus, silicon nitride should exist only in the amorphous state. Secondly, as referred to above, W nitrides have low stability at high temperatures and their decomposition and loss of N arise, leading to the formation of metal phase b.c.c. Only if the annealing is performed in N-containing atmospheres, the crystallization of W nitrides phases is possible [19].

In a general way, after crystallization, whatever is the crystalline phase, the grain sizes are in the range from 3 to 15nm. As can be observed in figure 7, the films containing high Si and N contents show after crystallization lower grain sizes. The presence of the amorphous Si-N phase should play an important role during the crystallization process working as stopper of the grain growing.

For the coated samples, on which the evolution of the grain size with temperature (those which have low Tc) was possible to be followed, no significant changes on the grain size were observed for N-containing films. Otherwise, an increase in the grain size with increasing temperatures was observed for W-Si films. The XRD patterns in figure 8 a) and b) for $W_{45}Si_{55}$ and $W_{38}Si_{52}N_{20}$ films, respectively, show that the diffraction peaks of film containing nitrogen have approximately the same width for all the annealing temperatures. On the other hand, $W_{45}Si_{55}$ film after crystallization presents broad diffraction lines which becomes progressively narrower with the increase in the annealing temperature. The presence of the amorphous Si-N phase in N-containing films should contribute to inhibit the grain growing. Veprek et al [20] found very high thermal stability in crystalline films of the Ti-Si-N system. In some cases, the grain growing was only registered for temperatures higher than 1100°C. Moreover, these authors also observed that the lower was the grain size the higher the temperature for grain growing. This result is against the Ostwald rippening theory [20] for grain growing..

The interdiffusion occurring between W-Si-N films and Ni-containing substrate materials, detected for annealing temperatures higher than 900°C, was confirmed by XRD patterns with the appearing of "new" phases ($Fe_2W/NiW/NiWSi$ or M_6C/M_6N with M=W,Fe,Cr,Ni) containing elements from the substrate [13-15]. Figure 9 shows an example of the occurrence of these phases. It was not possible to undoubtedly index these peaks to one of those phases, since no total match with standard positions is reached in any case, suggesting that probably a mixture of those phases can exist in the films.

2.2 MECHANICAL PROPERTIES

In figure 10, the hardness values measured in the W-Si-N films before and after annealing are presented. In this last case, the values refer to the highest hardness evaluated for each sample during the annealing process. The hardness of the samples was evaluated by depth sensing indentation, with an indentation load of 50mN, in order to assure that no influence of the substrate in the measurements occurred. The testing procedure, including the correction of the experimental results for geometrical defects of the indenter, thermal drift of the equipment and uncertainty in the initial contact, was described elsewhere [21]. The hardness of the as-deposited W-Si-N films are in the range (22-31GPa) of those of either Si-N sputtered amorphous films [22,23] or other amorphous W-based coatings such as W-Co-C or W-Ni-N [24,25]. After crystallization, in all the cases, significant improvements in the hardness were observed, having been obtained values (50GPa) higher than those measured in crystalline as-deposited films [8]. The increase in the hardness after thermal annealing was previously attributed to two main factors:

- The crystallization process; previous results have shown that for a large range of systems based on W [7,8], as-deposited sputtered crystalline films have higher hardness than amorphous ones. Films with similar chemical composition but having different structures (amorphous and crystalline), showed very different hardness (27 and 42 GPa for $W_{68}Si_{14}N_{18}$ and $W_{76}Si_{12}N_{12}$, respectively). Although some authors have referred having reached higher HV values in the amorphous than in crystalline structures (e.g. [26]), the inverse is commonly observed in the Materials Science community.

- Annealing at high temperatures has an effect on the stress relief and the structural relaxation of materials [27,28]. At high temperatures the existing stresses originated by the deposition process, can be partially or completely released. During the cooling from the annealing temperature, the difference in the thermal expansion coefficients of the film (αf) and the substrate (αs) can induce a "new" stress state in the film. Most of the substrates used for the deposition of the W-Si-N films, particularly those to which the results presented in figure 10 are concerned, have higher thermal expansion coefficients ($\alpha s > 12 \times 10^{-6} K^{-1}$) than the films (considering the thermal expansion coefficients of the bulk materials similar to the phases constituting the films, tungsten, silicides, nitrides, αs should be in the range 3-7 x $10^{-6} K^{-1}$ [29]). Thus, a compressive stress state is created in the films during the cooling of the coated samples down to the room temperature. As shown previously [30] and in agreement to the results of many authors on thin coatings (e.g [31,32]), compressive stresses gives rise to an increase of the hardness of materials.

The importance of the difference in the thermal expansion coefficients between the films and the substrates is clearly enhanced when the hardness results of the films deposited on substrates with different αs values are analysed (figure 11) [14]. As the films were deposited at 450°C, after deposition and due to the cooling down to room temperature, thermal stresses are induced, which are as high as the difference in αf and αs. The thermal expansion coefficient increases from Mo to 310 steel substrates and, as shown in figure 11, there is a tendency for higher hardness when the films are deposited on the substrates with high αs.

Although the low grain size of the films seems to be of crucial importance for the high hardness values found after crystallization, it should not be the only one. Figure 12 shows the evolution of the hardness of the W-Si film as a function of the grain size measured after annealing at different temperatures. There is a strong decrease in HV for the highest grain sizes. However, comparing films with similar structure after crystallization ($W_{64}Si_9N_{27}$ and $W_{68}Si_{14}N_{18}$ - b.c.c α-W) and approximately the same hardness (43GPa against 49GPa), there is a rather difference on their grain sizes (80nm against 160nm, respectively). Moreover, the film showing the higher H value has the highest grain size.

One possible explanation for the variation of the hardness among W-Si-N films after crystallization is their structural phase constitution. A good agreement was obtained between the hardness and the type of phase originated from the crystallization (figure 13). When the bcc α-W phase is presented the highest hardness values are reached. When it is mixed with silicide phases the hardness decreases as the type of silicide is increasing richer in Si. The hardness of silicides, even in the form of thin films (W_5Si_3 – 7.7GPa, WSi_2 in the range [11-14 GPa] [33]), is lower than the value indicated for α-W film [8], explaining the decrease observed when these phases are detected. It is also important to remark that the presence of the bcc α-W phase does not allow, only by itself, to explain the reached values. In previous studies it was shown that the hardness of single crystalline as-deposited W sputtered films was approximately 20GPa, value much lower than those obtained after crystallization of amorphous W-Si-N films. This value could be significantly increased (up to 42GPa) if small contents of N were added to W [8]. Taking into account the lattice parameters of the bcc α-W phase calculated from the position of the XRD diffraction lines for the annealed films and comparing the values with those of the as-deposited crystalline films containing the bcc α-W phase (table 2), it can be concluded that only small N and/or Si contents should be expected in the tungsten lattice. In fact, even considering the presence of compressive stresses, the calculated values are very close to the standard value of α-W phase (3.168Å [34]), and much lower than those of the crystalline as-deposited films containing N and/or Si and presenting the same structure. Moreover, in spite of the "lower" hardness found for films containing silicide phases, the measured values are over 30GPa, clearly above the literature values found for W-silicide phases. These results suggest that the nanocomposite structure consisting of nanograins of a crystalline phase mixed with amorphous silicon nitride resulting from the crystallization process should be responsible for these very high hardness values.

In spite of the importance of the hardness as an indicative way of the in-service behaviour of coated samples for mechanical applications, there is an increasing number of researchers who defend some sort of H/E ratio (e.g. H/E [35] or H^3/E^2 [36]) to better serve as the parameter to predict the tribological behaviour. In figure 14 the values of the Young's modulus as a function of the hardness are plotted for different film compositions. The results concern different films deposited on several substrates after having been annealed at increasing temperatures. Linear regression analysis was performed on the results of films containing the same amount of N. A very good correlation was obtained between the hardness and the elastic modulus in all the cases. A close look to the chemical compositon of the films allows to conclude that, for the same hardness, the increase in the N content leads to films with lower E values and, consequently, to higher H/E ratios. As above referred to, after crystallization only Si-nitrides are present in the films even if in an amorphous state. With increasing N content, higher amounts of Si-N phase are formed. Due to the lower E values of Si-N phase in relation to W-based materials [29], the coatings should also reach a lower elastic modulus. For visual aids, a line defining the ratio H/E=0.1 was plotted in the graphics. Matthews et al [35] give this value as an indication of the lower threshold to be reached for having a good tribological behaviour for a given material. As can be observed, there are several coatings that accomplish this condition showing H/E values higher than 0.1, envisaging promising high performance behaviour for future applications in the mechanical field.

3. Concluding Remarks

The use of thermal annealing for producing nanocrystalline films, from the crystallization of the amorphous state, can be an alternative whenever their direct deposition by sputtering is unfeasible. In the case of the W-Si-N system, the need to increase the Si content, to improve the oxidation behaviour of the films, leads to the deposition of amorphous films. The final properties, in particular the hardness and the elastic modulus of the nanocrystalline W-Si-N coatings achieved after thermal annealing, are better than of those obtained with crystalline structure after deposition.

Nevertheless, the application of this procedure for mechanical components should be careful : on one hand, only thermal stable substrates (at least up to 1000°C) can be used, since some of the amorphous W-Si-N coatings only crystallizes for temperatures higher than 950°C and on the other, some coated samples (films with high Si content deposited on Ni-containing substrates) are not chemically stable, showing elemental interdiffusion between the film and the substrate.

4. Acknowledgements

This research was sponsored by the Portuguese Foundation for Science and Technology (Fundação para a Ciência e Tecnologia, Portugal), as part of program POCTI in projects EME/33978/99 and CTM/33933/99.

5. References

[1] - "Nanoscale characterization of Materials", *MRS Bulletin*, Vol. 22, 1997.
[2] – "Mechanical Behaviour of Nanostructured Materials", *MRS Bulletin*, Vol. 24, 1999.
[3] - S. Veprek, A. Niederhofer, K. Moto, T. Bolom, H.-D. Mannling, P. Nesladek, G. Dollinger and A. Bergmaier (2000) Composition, nanostructure and origin of the ultrahardness in nc-TiN/a-Si$_3$N$_4$/nc-TiSi$_2$ nanocomposites with HV = 80 to > 105 GPa. *Surf. Coat. Technol.*, **133-134**, 152-159
[4] – C.Mitterer, P.H.Mayrhofer, M.Beschliesser, P.Losbichler, P.Warbichler, F.Hoffer, P.N.Gibson, W.Gissler, H.Hrubý, J.Musil and J.Vlček (1999) Microstructure and properties of nanocomposite Ti-B-N and Ti-B-C coatings. *Surf. Coat. Technol.*, **120-121**, 405-411
[5] - F. Vaz, L. Rebouta, P.Goudeau, J. Pacaud, H. Garem, J.P. Rivière, A. Cavaleiro, E. Alves (2000) Characterization of Ti$_{1-x}$Si$_x$N Nanocomposite Films. *Surf. Coat. Technol.*, **133-134**, 307-313.
[6] - *Nanocrystalline and Nanoscale Materials* (ed. J. Rivas and M.A. Lopez-Quintela), World Scientific Publishing Co, Singapore, 1998.
[7] – M.T. Vieira, A. Cavaleiro, B. Trindade (2002) The effects of a third element on structure and properties of W-C/N. *Surf. Coat. Technol.*, **151-152**, 495-504.
[8] - C. Louro, A. Cavaleiro (1999) Hardness versus Structure in W-Si-N Sputtered Coatings. *Surf. Coat. Technol.*, **116-119**, 74-80.
[9] - C. Louro e A. Cavaleiro (2001) How is the Chemical Bonding of W-Si-N Sputtered Films. *Surf. Coat. Technol.*, **142-144**, 964-970.
[10] - C.D. Wagner, W.H. Riggs, C.E. David, J.F. Moulder, G.E. Muilenberg, Handbook of X-Ray Photoelectron Spectroscopy, Perkin-Elmer Corporation, 1979
[11] – J.S. Reid, E. Kolawa, R.P. Ruiz, M.A. Nicolet (1993) Evaluation of amorphous (Mo, Ta, W)-Si-N diffusion barriers for Si/Cu metallizations. *Thin Solid Films*, **236**, 319-324.
[12] – J.S. Reid, E. Kolawa, G.M. Garland, M.A. Nicolet, F. Cardone, D. Gupta, R.P. Ruiz (1996) Amorphous (Mo, Ta, or W)-Si-N diffusion barriers for Al metallizations. *J. Appl. Phys.*, **79**, 1109-1115.
[13] – C. Louro, A.Cavaleiro, S.Dub, P.Smid, J.Musil and J.Vlcek (2003) The depth profile analysis of W-Si-N coatings after thermal annealing. *Surf. Coat. Technol.*, **161**, 111-119.
[14] – A. Cavaleiro, C. Louro, A.P. Marques (2002) The effect of the substrate on the annealing behaviour of W-Si-N sputtered films, presentation B3-1-3 at ICMCTF2002, San Diego, USA.
[15] – P. Marques and A. Cavaleiro (2002) Thermal and mechanical behaviour of sputtered coatings deposited from a WSi$_2$ target, *submitted to Thin Solid Films*.
[16] – M.G. Scott in *"Amorphous Metallic Alloys"*, (ed. F.E. Luborsky) Butterworths editions, London, Chap. 10, 1983.
[17] - H.J. Goldschmidt, *Interstitial Alloys*, ed. Butterworths, London, Chap. 7, 1967.

[18] – Y. Hirohata, N. Shimamoto, T. Hino, T. Yamashima, K. Yabe (1994) Properties of silicon nitride films prepared by magnetron sputtering. *Thin Solid Films*, **253**, 425-429.

[19] - L. Ferreira, C. Louro, A. Cavaleiro and B. Trindade (2002) Influence of heat treatment on the structure of (W,Si)N sputtered films, *Key Engineering Materials*, **230-232** 642-645.

[20] –A. Niederhofer, P. Nesladek, H.-D. Mannling, K. Moto, S. Veprek, M. Jilek (1999) Structural properties, internal stress and thermal stability of nc-TiN/a-Si3N4, nc-TiN/TiSix and nc-(Ti1-yAlySix)N superhard nanocomposite coatings reaching the hardness of diamond. *Surf. Coat. Technol.*, **120-121**, 173-178.

[21] - J.M. Antunes, A. Cavaleiro, L.F. Menezes, M.I. Simões and J.V. Fernandes (2002) Ulta-microhardness testing procedure with Vickers indenter. *Surf. Coat. Technol.*, **149**, 27-35.

[22] - A. Bendeddouche, R. Berjoan, E. Bêche, R. Hillel (1999) Hardness and stiffness of amorphous SiCxNy chemical vapor deposited coatings. *Surf. Coat. Technol.*, **111**, 184-190.

[23] - A. Grill, P.R. Aron (1983) RF-sputtered silicon and hafnium nitrides – properties and adhesion to 440C stainless steel. *Thin Solid Films*, **108**, 173-180

[24] – A. Cavaleiro, M.T. Vieira (1994) Evaluation of hardness of sputtered W-C-Co thin films. *Surface Engineering*, **10**, 147.

[25] - A. Cavaleiro, B. Trindade, M.T. Vieira (1999) Deposition and characterization of fine-grained W-Ni-C/N ternary films. *Surf. Coat. Technol.*, **116-119**, 944-948.

[26] – H. Leiste, U. Dambacher, S. Ulrich, H. Holleck (1999) Microstructure and properties of multilayer coatings with covalent bonded hard materials. *Surf. Coat. Technol.*, **116-119**, 313-320.

[27] - H. Kimura and T. Masumoto in *Amorphous Metallic Alloys* (ed. by F.E. Luborsky) Butterworths Monographs Materials, London, Chap. 12, 1983.

[28] - C.C. Chiu (1993) A method for measuring temperature-dependent stress and thermal expansion of coatings. *J. Mater. Sci.*, **28**, 5684-5692.

[29] - M.F. Ashby, *Materials Selection in Mechanical Design*, ed. Butterworth-Heinemann, London, 1992.

[30] - C. Louro and A. Cavaleiro (2000) Mechanical behaviour of amorphous W-Si-N sputtered films after thermal annealing at increasing temperatures, *Surf. Coat. Technol.*, **123** 192-198.

[31] - P.H. Mayrhofer, G. Tischler, C. Mitterer (2001) Microstructure and mechanical/ thermal properties of Cr-N coatings deposited by reactive unbalanced magnetron sputtering. *Surf. Coat. Technol.*, **142-144**, 78-84.

[32] - T. Mae, M. Nose, M. Zhou, T. Nagae, K. Shimamura (2001) The effects of Si addition on the structure and mechanical properties of ZrN thin films deposited by an r.f. reactive sputtering method. *Surf. Coat. Technol.*, **142-144**, 954-958.

[33] - *Properties of Metal Silicides* (ed. K. Maex, M. van Rossum), INSPEC publication, Leuven, Belgium, Chap. 1.2 , 1995.

[34] - International Centre for Diffraction Data, Swarthmore, PA, Card 04-0806.

[35] - A. Matthews and A. Leyland (2002) Developments in vapour deposited ceramic coatings for tribological applications. *Key Engineering Materials*, **206-213**, 459-466.

[36] - J. Musil, F. Kunc, H. Zeman, H. Polakova (2002) Relationships between hardness, Young´s modulus and elastic recovery in hard nanocomposite coatings, *Surf. Coat. Technol.*, **154**, 304.

Table 1 – Chemical composition and structure of W-Si-N films before and after annealing at increasing temperatures

Film	Substrate	Chemical Composition (at. %)			Ann. Temp (°C)
		W	Si	N	
$W_{31}Si_{36}N_{33}$	310	31.2	36.2	32.6	25
		31.3	37.3	31.4	700
		29.8	37.7	32.5	900
		33.7	37.6	28.7	950
		32.1	37.6	30.3	1000
$W_{28}Si_{45}N_{27}$	Fecralloy	28.4	44.7	26.9	25
		26.9	45.2	27.9	800
		27.5	45.4	27.1	900
		28.1	45.5	26.4	1000
$W_{63}Si_{8}N_{29}$	310	62.4	8.1	29.5	25
		73.3	10.2	16.6	700
		76.9	7.9	15.2	800
		76.5	8.6	14.9	850
		72.8	12.3	15.0	900
$W_{33}Si_{48}N_{19}$	Invar	33.2	47.6	19.2	25
		30.8	49.2	20.2	800
		30.6	51.0	18.4	900
		66.5	26.8	6.7	1000
$W_{30}Si_{46}N_{24}$	Invar	29.2	45.6	24.2	25
		27.8	45.3	26.7	800
		28.6	47.5	23.9	900
		42.5	21.9	35.6	1000
$W_{33}Si_{48}N_{19}$	310	32.9	47.8	19.3	25
		31.1	49.3	19.6	800
		32.5	47.6	19.9	900
		39.1	34.5	26.4	1000

Table 2 – Lattice parameter of the b.c.c. α-W phase indexed in sputtered W-Si-N films when deposited either in the crystalline state or after crystallization from the amorphous as-deposited condition.

As-deposited films	Lattice parameter (Å)
W	3.1981
$W_{95}N_5$	3.2137
$W_{89}N_{11}$	3.2145
$W_{72}N_{28}$	3.2385
$W_{99}Si_1$	3.1964
$W_{98}Si_2$	3.2056
$W_{95}Si_5$	3.2187
$W_{87}Si_2N_{11}$	3.2238
$W_{74}Si_2N_{24}$	3.2389
$W_{63}Si_1N_{36}$	3.2508
$W_{84}Si_9N_7$	3.2163
$W_{68}Si_{12}N_{12}$	3.2316
$W_{70}Si_{22}N_8$	3.1997

Annealed films	Lattice parameter (Å)
$W_{84}Si_{16}$	3.1748
$W_{69}Si_{23}N_8$	3.1867
$W_{48}Si_{36}N_{16}$	3.1721
$W_{68}Si_{14}N_{18}$	3.1927
$W_{64}Si_9N_{27}$	3.1790
$W_{34}Si_{35}N_{31}$	3.1748
$W_{30}Si_{36}N_{34}$	3.1839
$W_{32}Si_{27}N_{41}$	3.1688
$W_{31}Si_{23}N_{46}$	3.1845

394

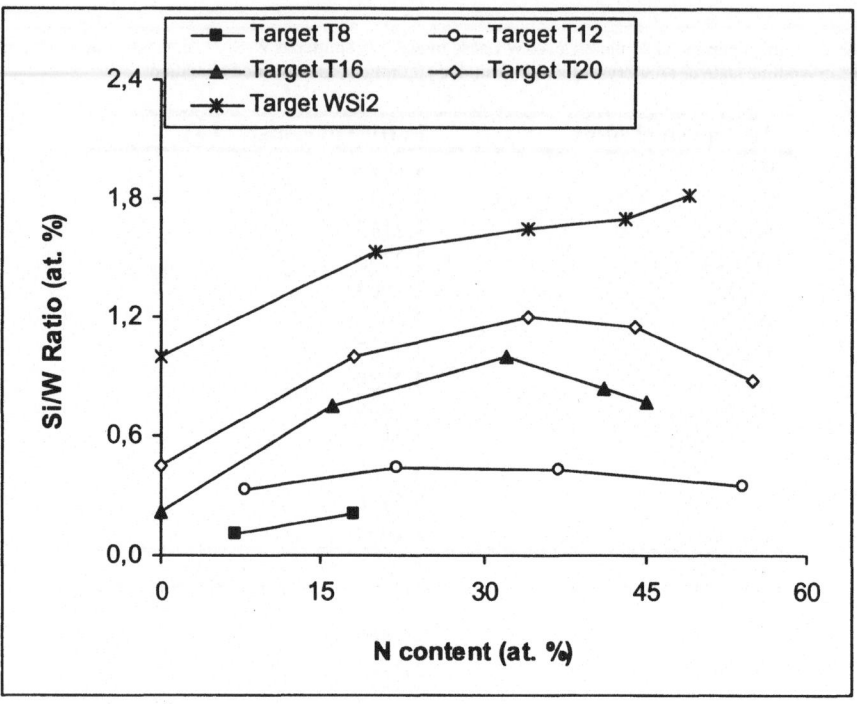

Figure 1 – EPMA results of the chemical composition of the as-deposited amorphous W-Si-N films expressed by the Si/W ratio as a function of the N content.

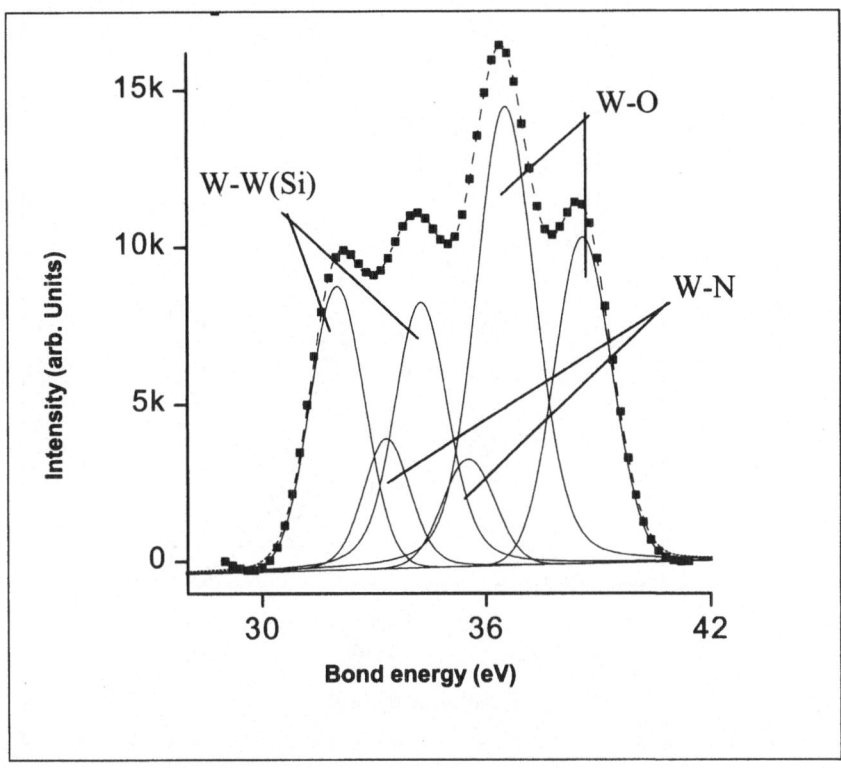

Figure 2 – XPS results of an amorphous W-Si-N sputtered film ($W_{26}Si_{30}N_{44}$); a) W4f and b) Si2p peaks.

396

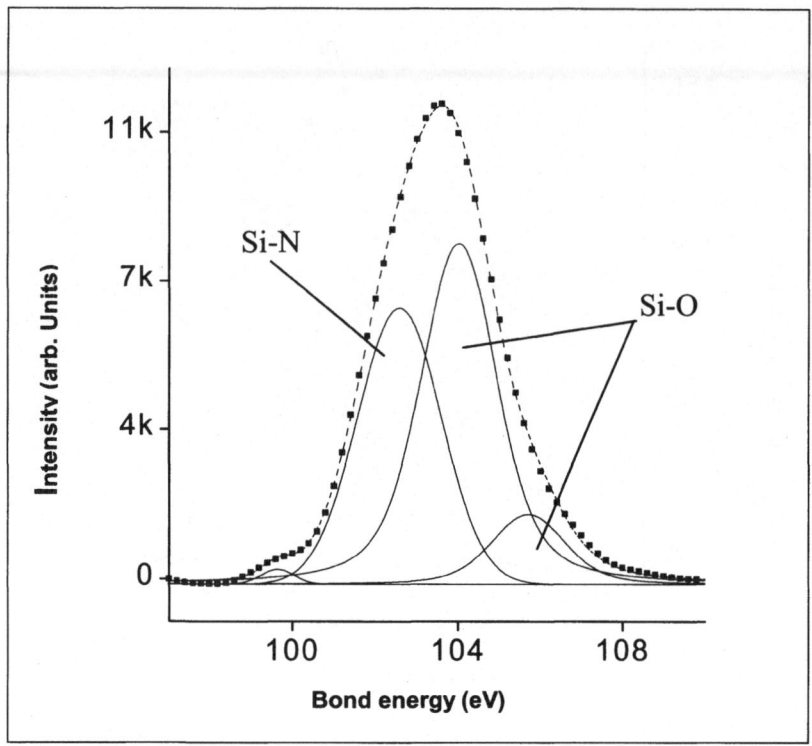

Figure 2b.

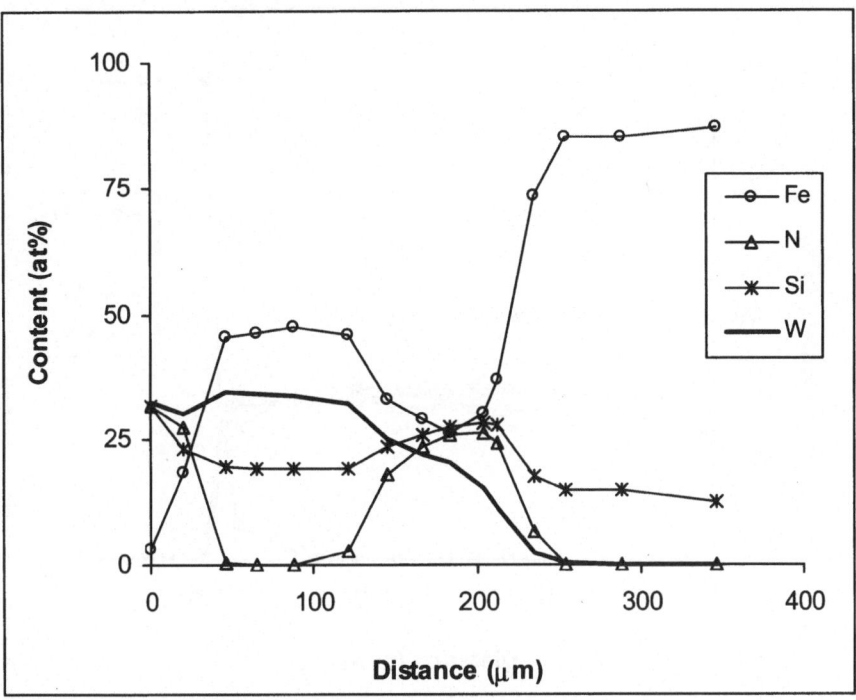

Figure 3 - Evolution of the chemical composition (EPMA results) across the thickness of the $W_{34}Si_{33}N_{33}$ film deposited onto a) 310 steel and b) Mo-alloy.

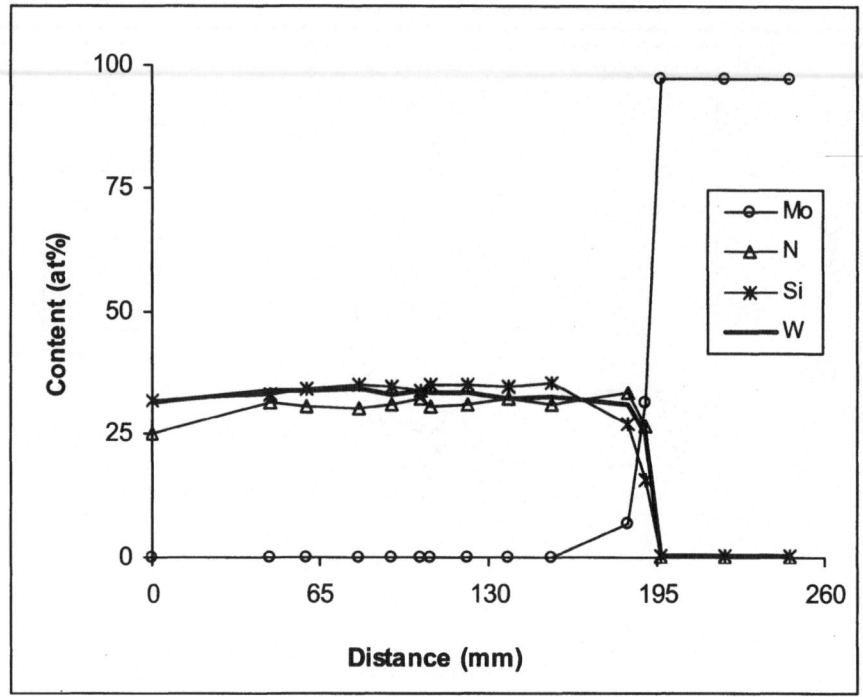

Figure 3b.

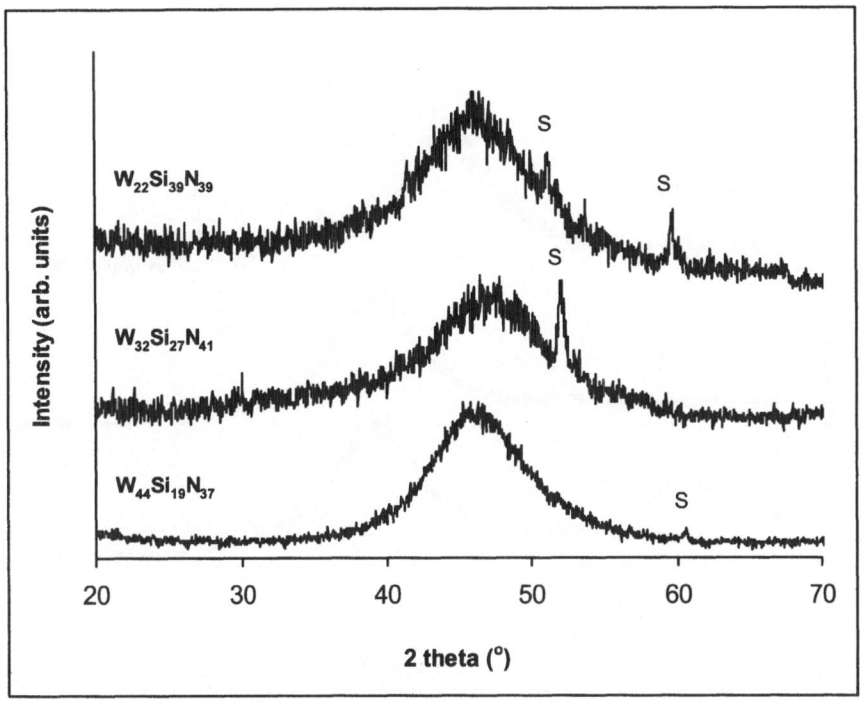

Figure 4 – XRD diffraction patterns of amorphous W-Si-N with a) similar Si/W ratios but increasing N at.% and b) similar N contents but increasing Si/W ratio.

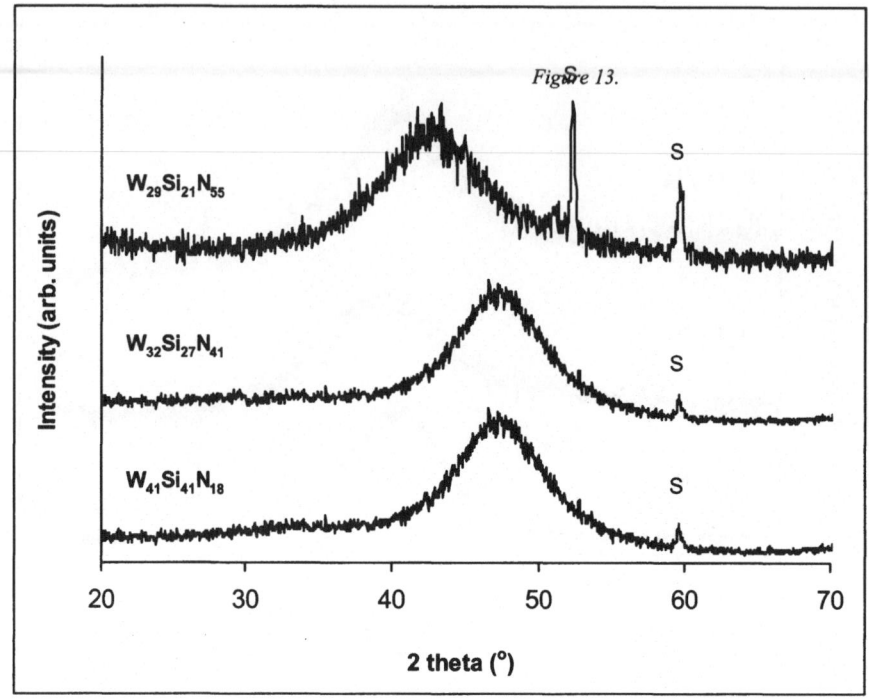

Figure 4b.

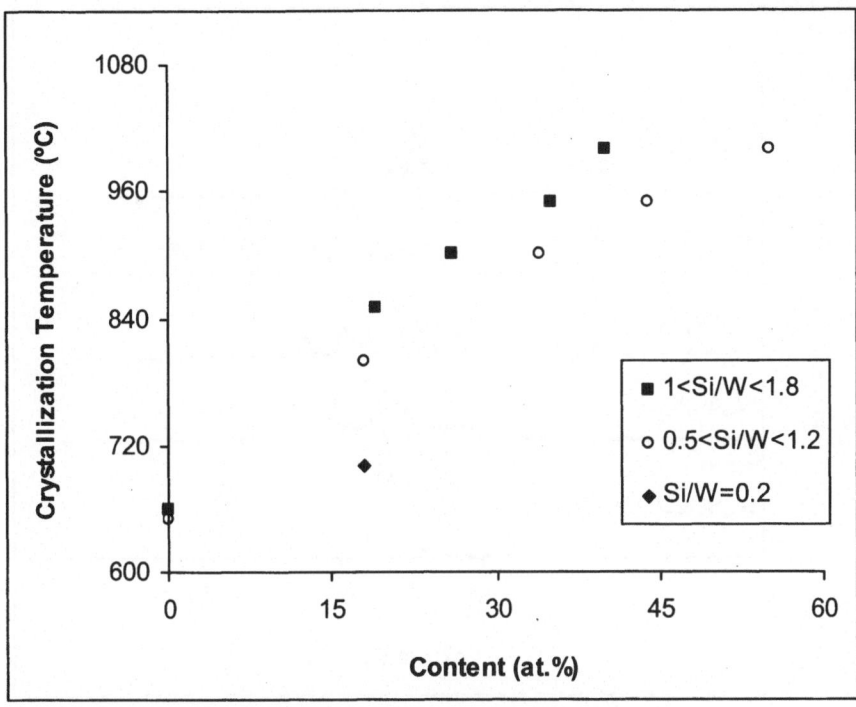

Figure 5 - Crystallization temperatures (Tc) of the sputtered amorphous W-Si-N films as a function of their N content.

402

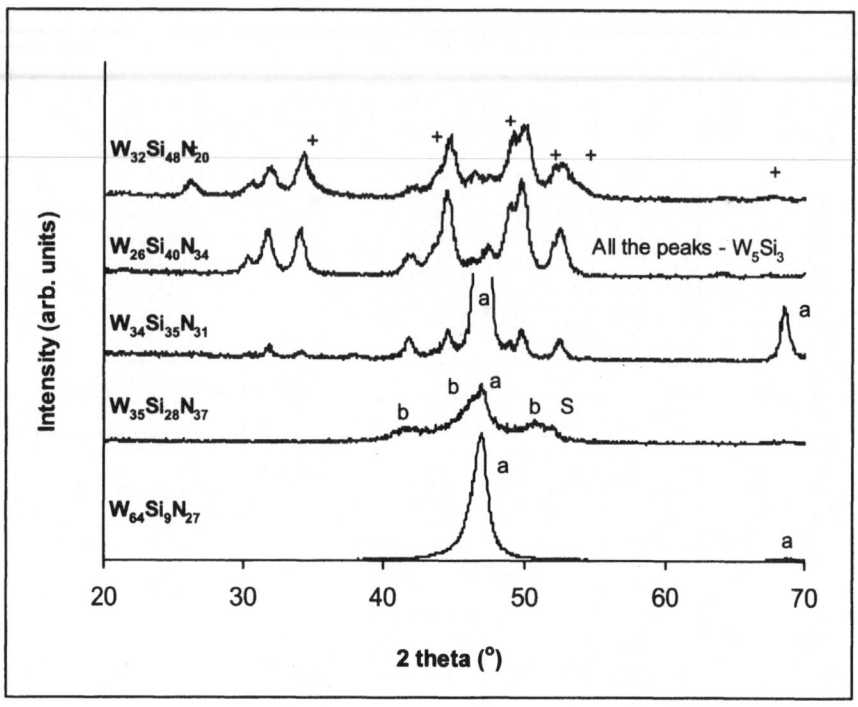

Figure 6 – XRD diffraction patterns of selected crystallised sputtered W-Si-N films **a** - α-W; **b** - β-W (W₃Si); + - WSi₂ ; **S** - substrate.

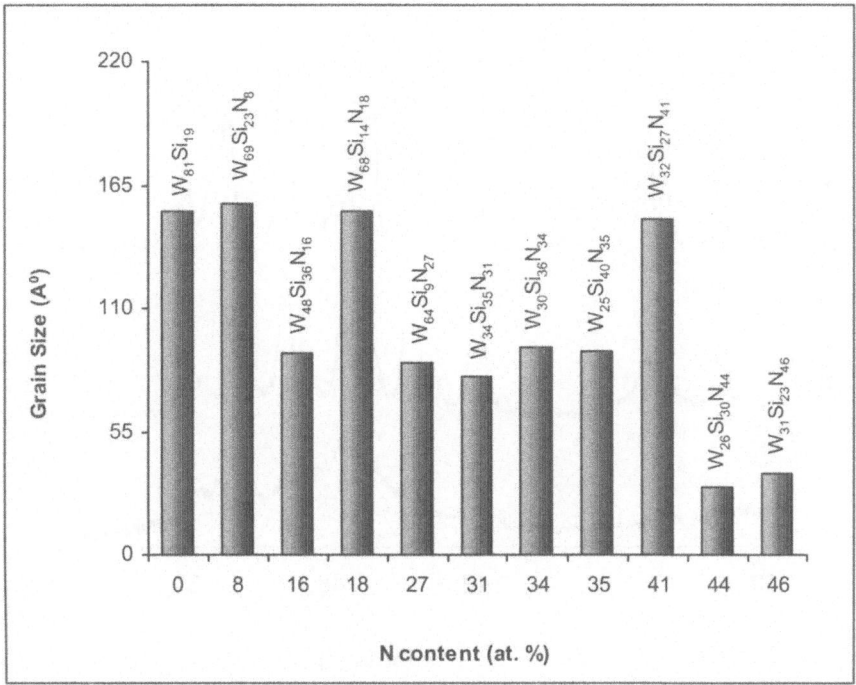

Figure 7 – Grain size determined by the application of the Sherrer's equation to the integral width of the main diffraction peak of W-Si-N films after crystallization.

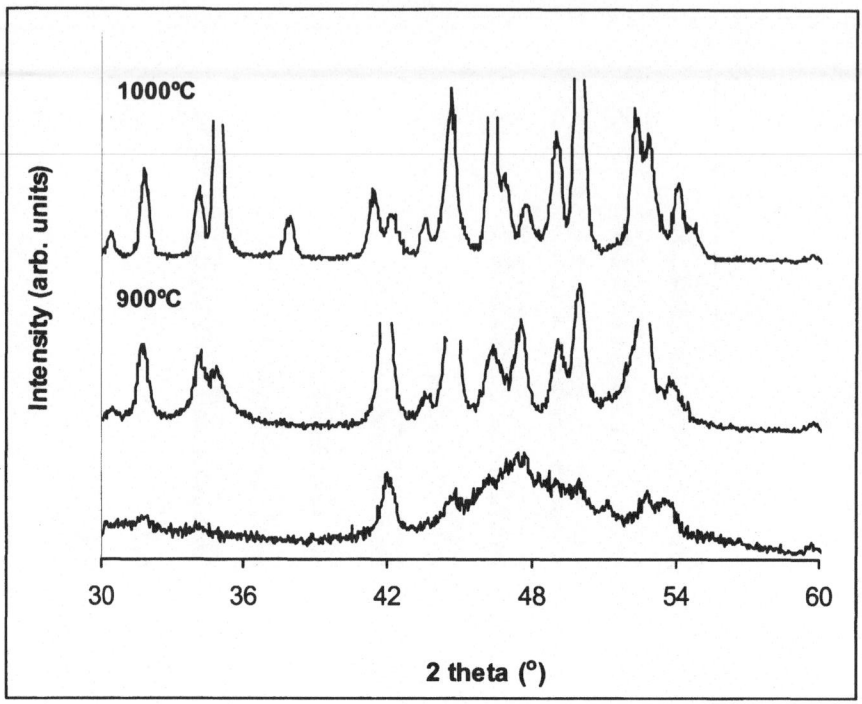

Figure 8 – XRD diffraction patterns of crystallised sputtered W-Si-N films at increasing temperatures a) $W_{45}Si_{55}$; b) $W_{38}Si_{52}N_{20}$.

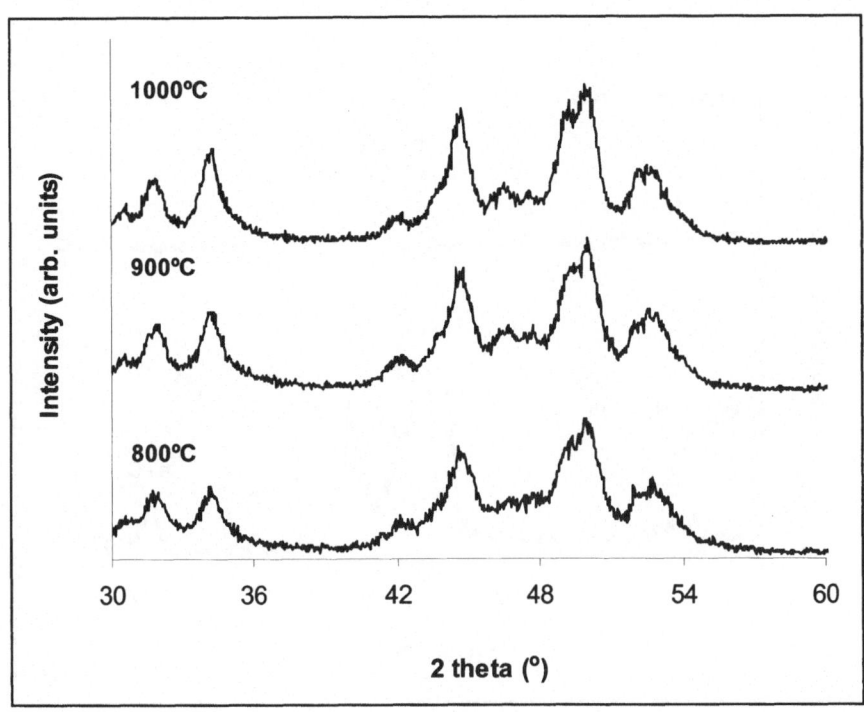

Figure 8b.

406

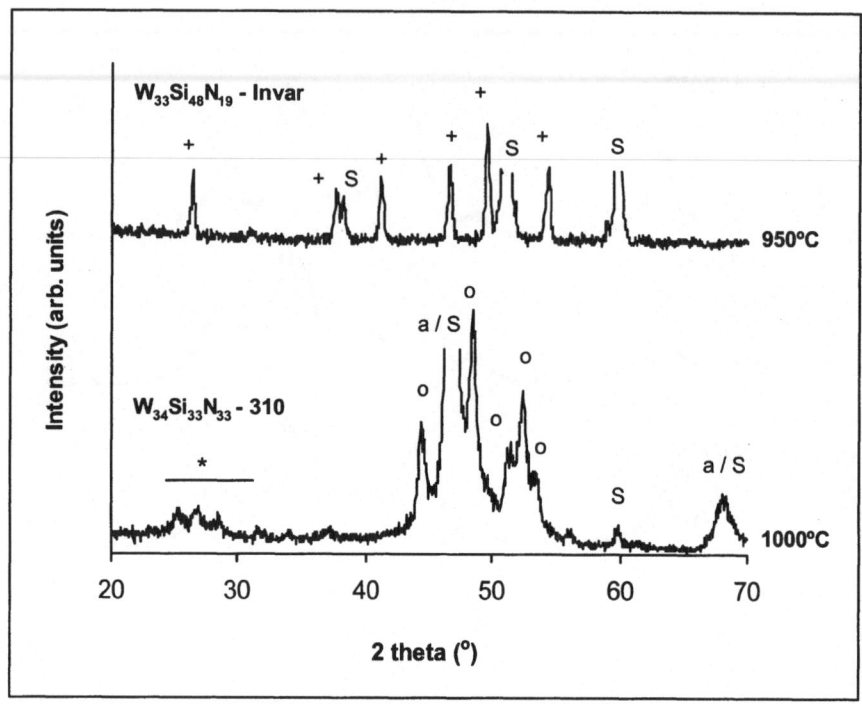

Figure 9 – XRD diffraction patterns of sputtered W-Si-N films deposited onto Invar alloy and 310 (AISI) steel showing crystalline phases containing elements from the substrate (+ - M_6C/M_6N, ° - $Fe_2W/NiW/NiWSi$, a - α-W, S – substrate).

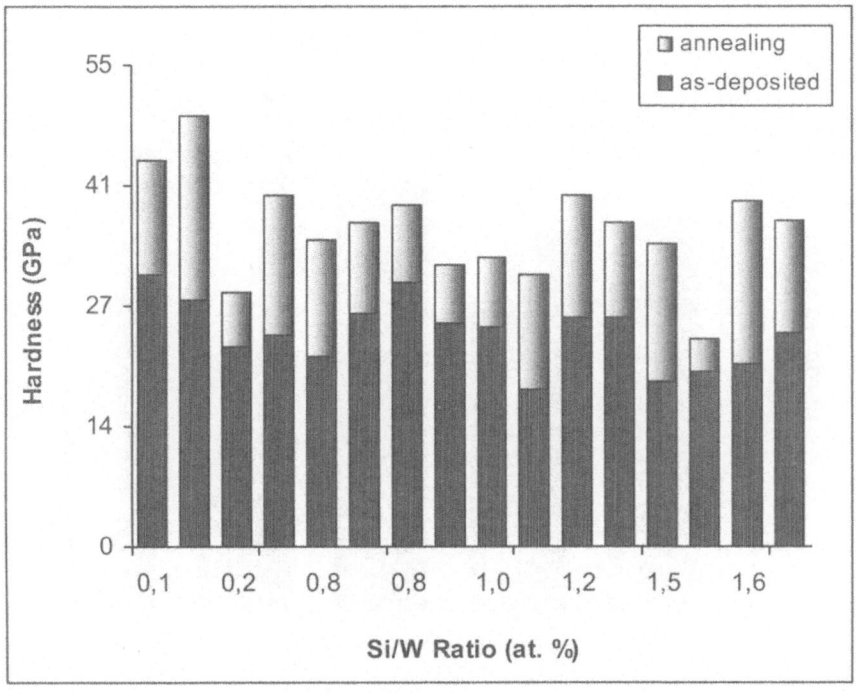

Figure 10 – Hardness results of sputtered amorphous W-Si-N coating: as-deposited and after crystallization upon thermal annealing.

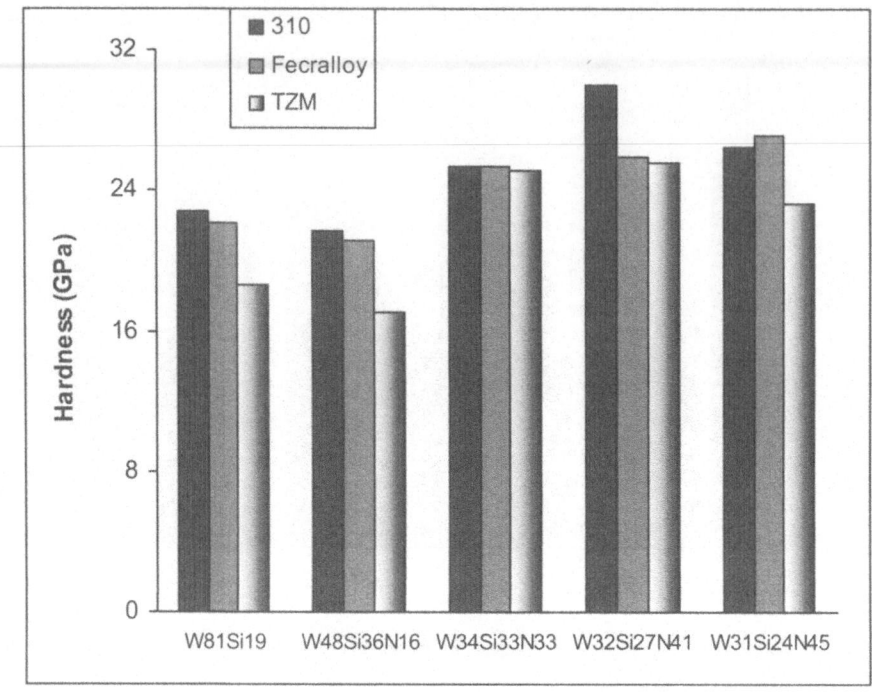

Figure 11 – Influence of the thermal expansion coefficient of the substrate on the hardness of sputtered amorphous W-Si-N coatings.

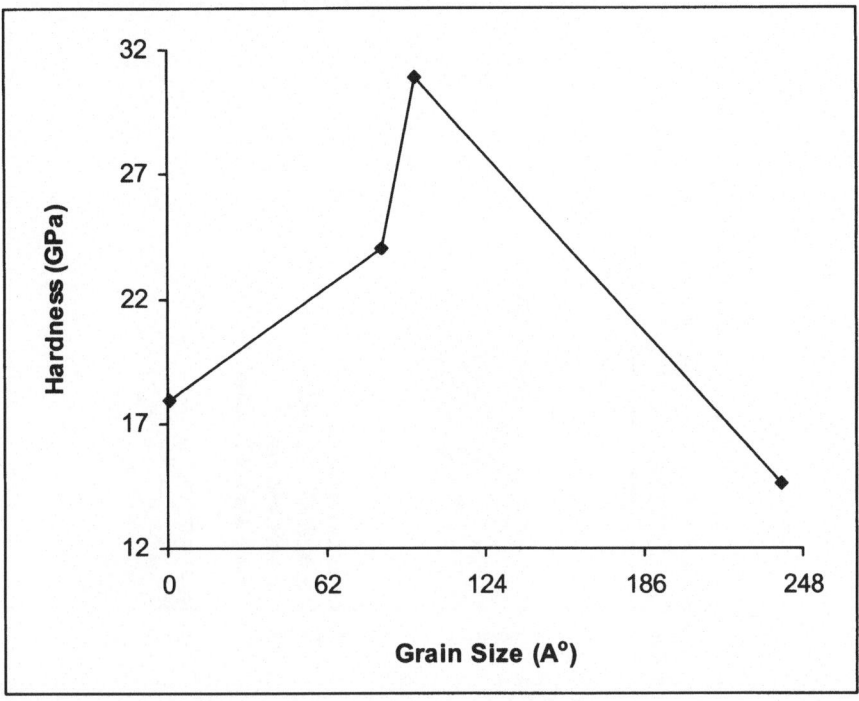

Figure 12 – Influence of the grain size on the hardness of a W-Si sputtered film ($W_{45}Si_{55}$) annealed at different temperatures.

410

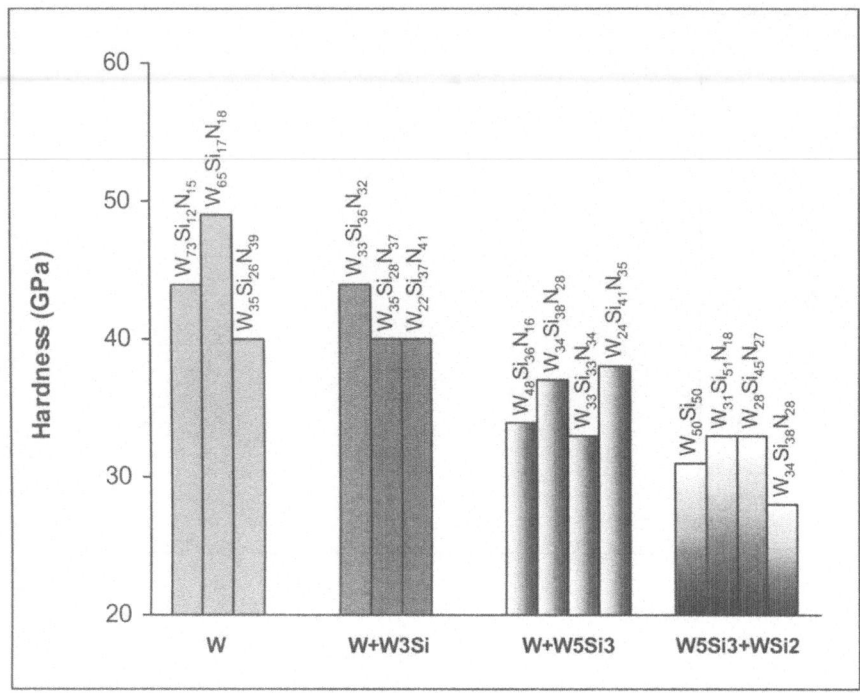

Figure 13 – Hardness of sputtered W-Si-N films as a function of the type of phases formed after thermal crystallization.

411

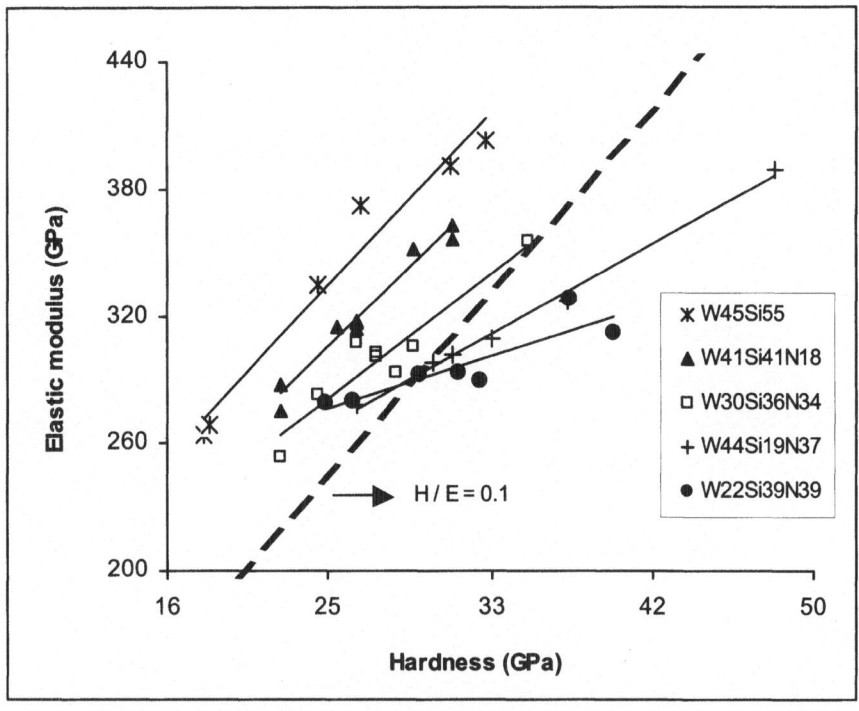

Figure 14 – Elastic modulus of W-Si-N sputtered films plotted as a function of the hardness values.

MICROWAVE JOINING OF ZrO$_2$ AND Al$_2$O$_3$ CERAMICS VIA NANOSTRUCTURED INTERLAYERS

Yu.V. BYKOV, S.V. EGOROV, A.G. EREMEEV, K.I. RYBAKOV,
N.A. ZHAROVA, M.A.. LOBAEV[*], A.W. FLIFLET[**], D. LEWIS III[**],
M.A. IMAM[**], A.I. RACHKOVSKII[***]
Institute of Applied Physics of the Russian Academy of Sciences, Nizhny Novgorod, Russia, [*]Nizhny Novgorod State University, Nizhny Novgorod, Russia, [**]Naval Research Laboratory, Washington D.C., USA, [***]Russian Federal Nuclear Center, Sarov, Russia

1. Introduction

In recent years considerable interest has been drawn to the development of functionally graded materials, and in particular, graded thermal barrier coatings (TBC) [1, 2]. One widely studied TBC system is ZrO$_2$ – Al$_2$O$_3$ – metal, in which zirconia is responsible for high-temperature stability of the coating, and alumina prevents oxygen diffusion to the metal and thereby enhances its resistance to corrosion. One of the most promising methods of creating thermal barrier structures on the basis of the ZrO$_2$ + Al$_2$O$_3$ ceramic composition is high-temperature diffusion joining of these materials. A major problem with the joining of dissimilar materials is high residual stresses in the contact zone and its vicinity. These stresses result from a mismatch of the coefficients of thermal expansion (CTEs) when the joint is cooled down after processing.

This work attempts to solve this problem by using an interlayer that matches the CTEs of the materials to be joined. Furthermore, it is proposed to prepare the interlayer from a nanostructured ceramic material with enhanced ductility. To retain fine microstructure of the interlayer, the process of joining requires fast heating of the joint zone, which is implemented by microwave heating. The use of microwave power of the millimeter-wave range ensures good absorption of the microwave energy in the oxide ceramic materials [3,4]. The suggested approach has been tested by experimental investigation, analytical and numerical modeling. The temperature and pressure of the process have been optimized, and different compositions of the interlayer material have been tested. As a result, the feasibility of producing ZrO$_2$ – Al$_2$O$_3$ graded composite structures by microwave joining is demonstrated.

T. Tsakalakos et al. (eds.) pgs. 413 - 426
Nanostructures: Synthesis, Functional Properties and Applications;
© *2003 Kluwer Academic Publishers.*

2. Theory

We will consider here the residual stresses which arise in the joined sample of dissimilar materials when it is cooled down from the temperature of joining. Due to the mismatch in the coefficients of thermal expansion (CTEs), the parts of the joined samples tend to shrink at different rates during cooling. Therefore, the material assumes a stressed state. Roughly speaking, if two initially stress-free materials are joined at a high temperature and then cooled down, in the material with the larger CTE there will be a tensile stress along the joint interface, and in the material with the smaller CTE there will be a compressive stress. In fact, due to finite dimensions of the samples all components of the stress tensor will be present. However, the distribution of all stress components in the configurations of practical interest can only be found numerically, whereas the principal stress components can be estimated on the basis of a simplified model, as shown below.

The free energy of a deformed, nonuniformly heated body is [5]

$$F = -K\alpha(T - T_0)\varepsilon_{ll} + \mu\left(\varepsilon_{ik} - \frac{1}{3}\delta_{ik}u_{ll}\right)^2 + \frac{K}{2}\varepsilon_{ll}^2, \tag{1}$$

where ε_{ik} is the strain tensor, K and μ are elastic moduli of compression and shear, respectively, α is the CTE, and $T - T_0$ is temperature difference.

The stress tensor is

$$\sigma_{ik} = \frac{\partial F}{\partial \varepsilon_{ik}} = -K\alpha(T - T_0)\delta_{ik} + K\varepsilon_{ll}\delta_{ik} + 2\mu\left(\varepsilon_{ik} - \frac{1}{3}\delta_{ik}\varepsilon_{ll}\right), \tag{2}$$

and the distribution of stresses and strains in the body is determined by solving the equations of equilibrium,

$$\frac{\partial \sigma_{ik}}{\partial x_k} = 0, \tag{3}$$

where x_k are components of the coordinate vector.

We will start our investigation of the residual thermoelastic stresses in a joined sample by considering the case of planar deformations, when the vector of displacement, $\mathbf{u}$, has only two components, u_x and u_y, which depend on two coordinates, x and y. The components of the strain tensor, ε_{xz}, ε_{yz}, and ε_{zz}, as well as those of the stress tensor, σ_{xz} and σ_{yz}, in this case turn to zero. Consider first an infinite plate of a uniform material at uniform temperature. The equations of equilibrium in this case take the form

$$\frac{\partial^2 u_x}{\partial y^2} + \frac{1}{1-2v}\frac{\partial^2 u_y}{\partial x \partial y} + \frac{2(1-v)}{1-2v}\frac{\partial^2 u_x}{\partial x^2} = 0;$$

$$\frac{\partial^2 u_y}{\partial y^2} + \frac{1}{2(1-v)}\frac{\partial^2 u_x}{\partial x \partial y} + \frac{1-2v}{2(1-v)}\frac{\partial^2 u_y}{\partial x^2} = 0, \tag{4}$$

where $v = \dfrac{3K - 2\mu}{2(3K + \mu)}$ is the Poisson ratio. Let one surface of the plate, $y = 0$, be free,

$$\sigma_{xy}(x, y = 0) = 0 \ ; \ \sigma_{yy}(x, y = 0) = 0 \ , \tag{5}$$

and the other, $y = h$, be deformed (Fig.1), which models the influence of the second plate with a different CTE in a joined material. It is convenient to set the deformation at the surface $y = h$ in the following form:

$$u_x(x, y = 0) = a \sin kx , \ u_y(x, y = 0) = 0 \tag{6}$$

(where a is some amplitude of deformation, $a \ll h$), which allows easy solving of the partial differential equations (4) via their reduction to the ordinary ones. Yet, any realistic distribution of tangential deformation over the surface can be expressed as a Fourier integral of sinusoidal spatial harmonics.

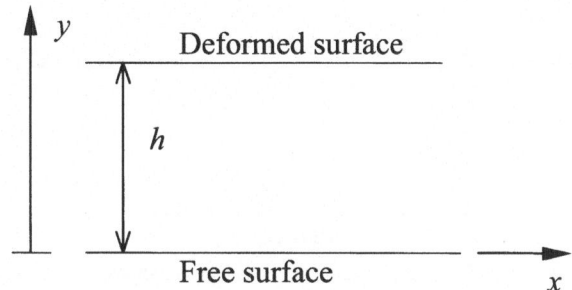

Figure 1. Infinite plate and the coordinate system.

The stress distribution in the problem set by Eqs. (4) with the boundary conditions (5) and (6) is

$$\sigma_{xx} = \tilde{E}ka\{(3 - 2v)\cosh[k(h + y)] + k(y - h)\sinh[k(h + y)] +$$
$$+ (2k^2hy + 5 - 6v)\cosh[k(h - y)] + [(4v - 3)ky - 3kh]\sinh[k(h - y)]\};$$
$$\sigma_{xy} = \tilde{E}ka\{k(y - h)\cosh[k(h + y)] + 2(1 - v)\sinh[k(h + y)] +$$
$$+ [kh - (4v - 3)ky]\cosh[k(h - y)] + 2(v - 1 - k^2hy)\sinh[k(h - y)]\};$$
$$\sigma_{yy} = \tilde{E}ka\{(2v - 1)\cosh[k(h + y)] + k(h - y)\sinh[k(h + y)] +$$
$$+ (1 - 2v - 2k^2hy)\cosh[k(h - y)] + [(4v - 3)ky - kh]\sinh[k(h - y)]\}, \tag{7}$$

where $\tilde{E} = E / \{2(1 + v)[2k^2h^2 + 4v(2v - 3) + 5 + (3 - 4v)\cosh(2kh)]\}$, $E = \dfrac{9K\mu}{3K + \mu}$ is the

Young modulus. Consider the case of smooth variations of deformation along the surface of the plate compared to the plate thickness:

$$kh \ll 1 . \tag{8}$$

The solution (7) in this case takes the form

$$\sigma_{xx} = Eka / (1 - v^2);$$
$$\sigma_{xy} = Ek^2 ay / (1 - v^2); \tag{9}$$
$$\sigma_{yy} = Ek^3 ay^2 / [2(v^2 - 1)]$$

416

It can be seen from Eqs. (9) that in the case of smooth variations of deformation the principal component of the stress tensor is σ_{xx}, i. e., the normal stress parallel to the surface. The other components are of the first or second order of smallness with respect to ky.

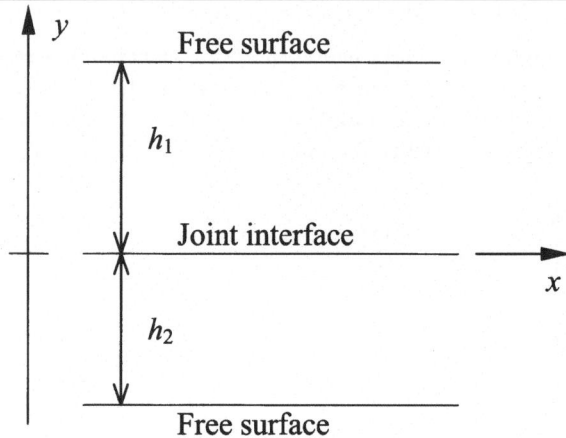

Figure 2. Two joined infinite plates and the coordinate system.

Consider now the case of two joined plates of dissimilar materials of thickness h_1 and h_2 which are cooled from the temperature $T(x)$ down to T_0 (Fig. 2). The equations of equilibrium in this case take (for each of the materials) the form

$$\frac{\partial^2 u_x}{\partial y^2} + \frac{1}{1-2v}\frac{\partial^2 u_y}{\partial x \partial y} + \frac{2(1-v)}{1-2v}\frac{\partial^2 u_x}{\partial x^2} = \frac{\alpha}{3}\frac{1+v}{1-2v}\frac{\partial T}{\partial x};$$

$$\frac{\partial^2 u_y}{\partial y^2} + \frac{1}{2(1-v)}\frac{\partial^2 u_x}{\partial x \partial y} + \frac{1-2v}{2(1-v)}\frac{\partial^2 u_y}{\partial x^2} = \frac{\alpha}{3}\frac{1+v}{2(1-v)}\frac{\partial T}{\partial y},$$

(10)

Let both outer surfaces of the joined plates, $y = h_1$ and $y = -h_2$, be free: $\sigma_{xy} = 0$, $\sigma_{yy} = 0$. The boundary conditions at the joint interface, $y = 0$, are continuity of the components u_x, u_y, σ_{xy}, and σ_{yy}. Let the temperature distribution be $T(x) = T_m \cos kx$, where T_m is a constant. Eqs. (10) with these boundary conditions can be solved analytically. The solutions are cumbersome, and we will present them only for the case of smooth variations of temperature, $kh_1 \ll 1$, $kh_2 \ll 1$, and equal Young's modulus and Poisson's ratio of the materials (i. e., only the CTEs of the materials are different):

$$\sigma_{xx} = \frac{E(\alpha_1 - \alpha_2)T_m}{3(1-v)}\frac{h_2\left(6h_1 y + h_1 h_2 - 4h_1^2 - h_2^2\right)}{\left(h_1 + h_2\right)^3}\cos kx;$$

$$\sigma_{xy} = \frac{E(\alpha_1 - \alpha_2)T_m}{3(1-v)} \frac{h_2(3h_1 y + h_1 h_2 - h_1^2 - h_2^2)}{(h_1 + h_2)^3} k(y - h_1)\sin kx; \qquad (11a)$$

$$\sigma_{yy} = \frac{E(\alpha_1 - \alpha_2)T_m}{6(1-v)} \frac{h_2(-2h_1 y - h_1 h_2 + h_2^2)}{(h_1 + h_2)^3} k^2 (y - h_1)^2 \cos kx$$

for the material 1;

$$\sigma_{xx} = \frac{E(\alpha_1 - \alpha_2)T_m}{3(1-v)} \frac{h_1(6h_2 y - h_1 h_2 + h_1^2 + 4h_2^2)}{(h_1 + h_2)^3} \cos kx;$$

$$\sigma_{xy} = \frac{E(\alpha_1 - \alpha_2)T_m}{3(1-v)} \frac{h_1(3h_2 y - h_1 h_2 + h_1^2 + h_2^2)}{(h_1 + h_2)^3} k(y + h_2)\sin kx; \qquad (11b)$$

$$\sigma_{yy} = \frac{E(\alpha_1 - \alpha_2)T_m}{6(1-v)} \frac{h_1(-2h_2 y + h_1 h_2 - h_2^2)}{(h_1 + h_2)^3} k^2 (y + h_1)^2 \cos kx$$

for the material 2.

Similar to the previously considered case, only the normal stress parallel to the joint interface, σ_{xx}, remains significant when the temperature variations are smooth enough. In this asymptotic case, σ_{xx} varies linearly with y within the plates and exhibits a discontinuity at the joint interface.

The deformation, ε_{xx}, obtained from the solution of Eqs. (10), is in this case

$$\varepsilon_{xx} = \frac{T_m(1+v)}{3(h_1 + h_2)^3} \left[\alpha_1 h_1^3 + (4\alpha_2 - \alpha_1)h_1^2 h_2 + (4\alpha_1 - \alpha_2)h_1 h_2^2 + \alpha_2 h_2^3 + 6(\alpha_1 - \alpha_2)h_1 h_2 y \right]\cos kx$$

$$(12)$$

in both materials. The deformation is also a linear function of y, with the same coefficient in both joined plates. Note that the following relationship between stress and deformation is valid for each of the plates:

$$\sigma_{xx} = \frac{E}{(1+v)(1-v)}(\varepsilon_{xx} - \alpha T). \qquad (13)$$

These considerations illustrate that the simplest, one-dimensional approximate analytical model for estimating stresses in layered structures can be constructed based on the following principles [6]: - the tangential (to the interlayer boundary) component of the deformation, ε_{xx}, varies linearly with the coordinate, y, within each layer. The deformation can be presented as

$$\varepsilon_{xx} = \varepsilon_i^0 + \frac{y}{R}, \qquad (14)$$

where R represents the radius of curvature of the bending plane of the system, i is the number of the layer;
- the deformation, ε_{xx}, is continuous at the boundaries between layers;
- the stress, σ_{xx}, is related to the deformation, ε_{xx}, via Eq. (12);

- the equilibrium of forces in the x direction and the equilibrium of moments about the y axis complete the system of algebraic equations from which the unknown ε_i^0 and R are found. Then the stresses are determined via Eq. (13).

However, in the realistic, not very thin samples the stress components other than σ_{xx} may reach significant values. The stress distribution in such samples can be obtained numerically. The numerical scheme can be constructed based on the variational principle. The solution minimizes the free energy, F (Eq. (1)). The minimum is found by the steepest descent method. In each grid point, the gradient of the free energy with the respect to the displacement vector, u, is determined. Over an iteration, the new displacement, $\mathbf{u}_{new}$, is found as

$$\mathbf{u}_{new} = \mathbf{u} + \tau \nabla_{\mathbf{u}} F ,\tag{15}$$

where τ is a small parameter. The convergence criterion is decrease of the free energy. This method has a clear physical analogy. The quantity $\nabla_{\mathbf{u}} F$ is in fact the volumetric force acting in the material. Considering τ as a time step, the iteration formula (15) can be viewed as a description of the relaxation of some initial stress state in a viscous medium.

The results of analytical estimates and numerical modeling of the stress state in the actual samples used in the experiments are presented in the "Results and discussion" section.

3. Experiment

The joining experiments were performed in a gyrotron system for microwave processing of materials [7]. The microwave source was a cw gyrotron with a power of 10 kW at a frequency of 30 GHz. The use of millimeter waves has allowed efficient heating of the low-loss ceramic materials and retaining fine microstructure of the interlayer material to preserve its ductility. The radiation from gyrotron was shaped as a Gaussian beam and fed into a supermultimode cylindrical cavity, 50 cm in diameter and 60 cm in height. The radiation was distributed uniformly over the cavity volume with the help of a spherical scatterer. The pressurizing fixture with a specially designed thermal insulation arrangement (Fig. 3) was positioned in the center of the cavity. The ceramic inserts in the pressurizing fixture contributed to thermally insulating the edges of the samples. The thermal insulation arrangement shaped the temperature profile so that the temperature in the joining zone was maximal and uniform. The maximum pressure was 1.5 MPa.

The microwave power fed into the cavity was controlled automatically by a computerized feedback loop control system using the temperature data. The temperature was measured by a B-type thermocouple which was brought in contact with the joining zone of the sample. The accuracy of temperature measurement was no worse than ± 5 °C. The heating rate in all experiments was 30 °C /min, the time of hold at the maximum temperature of joining was 20 min. The cooling rate in the interval from the maximum temperature of joining down to 1200 °C (in the first series of experiments) and 700 °C (in the second series of experiments) was kept equal to 10 °C /min by controlled decrease of the microwave power.

419

The samples used for joining were sintered cylinders of pure alumina and zirconia with a diameter of 8 mm and height of 6 mm (alumina) and 4 mm (zirconia). The density of the ceramics was about 98 % of the theoretical density, the grain size was 2–6 µm (alumina) and 0.3 µm (zirconia). The interlayers were sintered from nanosize powders (pure alumina powder or a mix of alumina and zirconia powder) in the shape of discs, 8 mm in diameter and 1 .. 2 mm in height. Nanosize powders were obtained by electrical explosion synthesis and the interlayers were compacted by magnetic pulsed compaction at the Institute of Electrophysics, Ekaterinburg, Russia [8].

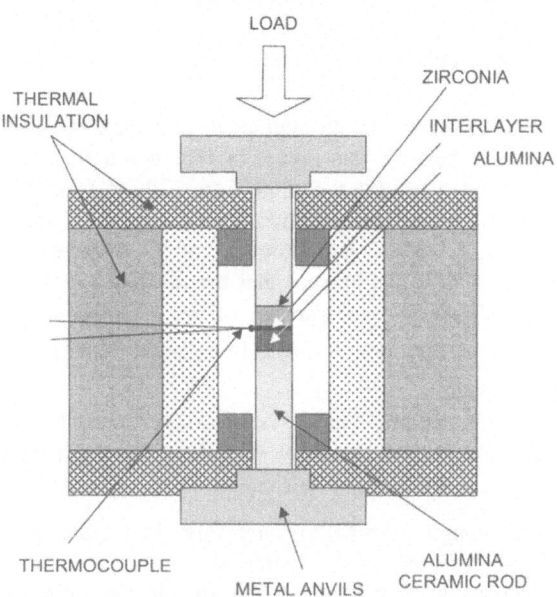

Figure 3. The pressurizing fixture with a thermal insulation arrangement

All sintering was done using microwave heating in the same system, except for the alumina cylinders which were sintered conventionally. An important condition for diffusion joining is preparation of the surfaces. The joining surfaces of samples were fine ground to roughness less than 3 µm and made parallel to the wedge angle less than $5 \cdot 10^{-2}$ °.

The joined samples were cut across the joint, polished, and subjected to the microstructural characterization. The electron microscopy of the joints was done using the electron probe microanalyzer JCMA-733 in the secondary electron imaging (SEI) and backscattered electron imaging (BEI) modes. Based on the results of microstructural analysis, the maximum temperature of joining and the pressure were optimized in the course of experiments.

4. Results and Discussion

Table 1 summarizes the properties of materials relevant to this work.

Table 1. Mechanical properties of zirconia and alumina

	ZrO$_2$	Al$_2$O$_3$
Density, ρ, g/cm^3	5.83	3.98
Coefficient of thermal expansion, α, 10^{-6} °C^{-1}	11.4	8.6
Young's modulus, E (at room temperature), GPa	169	374
Poisson's ratio, ν (at room temperature)	0.32	0.25
Tensile strength, MPa	145	258

At the first stage of experiments, an attempt to join the ZrO$_2$ and Al$_2$O$_3$ ceramics directly under microwave heating was made. In the search for the optimum joining regime, the temperature was varied in the range from 1200 to 1500 °C, and the pressure was varied from 0.4 to 1.5 MPa. The minimum temperature of joining was found to be 1430 °C at a pressure of 1.1 MPa. A photograph of the transverse cut of the joined sample is shown in Fig. 4, a. It can be seen that the residual tensile stresses parallel to the joint interface (σ_{rr}) have caused cracking of the ZrO$_2$ ceramic. In addition, it is seen that the axial stresses (σ_{zz}) have caused cracks in the Al$_2$O$_3$ ceramic directed parallel to the joint interface.

This experiment is well explained by the results of modeling (Fig. 4, b–d). It can be seen that the tensile stress σ_{rr} in the ZrO$_2$ ceramic at the joint interface exceeds its tensile strength (145 MPa). In addition, the tensile stress σ_{zz} in the Al$_2$O$_3$ ceramic is also close to its tensile strength (258 MPa). Therefore, the residual stresses are too high to permit a direct joining between the two materials.

In the next experiment an attempt was made to join the ZrO$_2$ and Al$_2$O$_3$ ceramic samples via an interlayer sintered from nanosize Al$_2$O$_3$ powder. The microstructure of the as-sintered interlayer material is shown in Fig. 5. The temperature and pressure needed for joining were found to be the same as in the previous case. It might be expected that the enhanced ductility of the nanostructured material would promote partial relaxation of the stress. However, the actual extent of stress relaxation occurred to be insufficient, and the joined sample cracked similar to the directly joined ZrO$_2$ – Al$_2$O$_3$ sample.

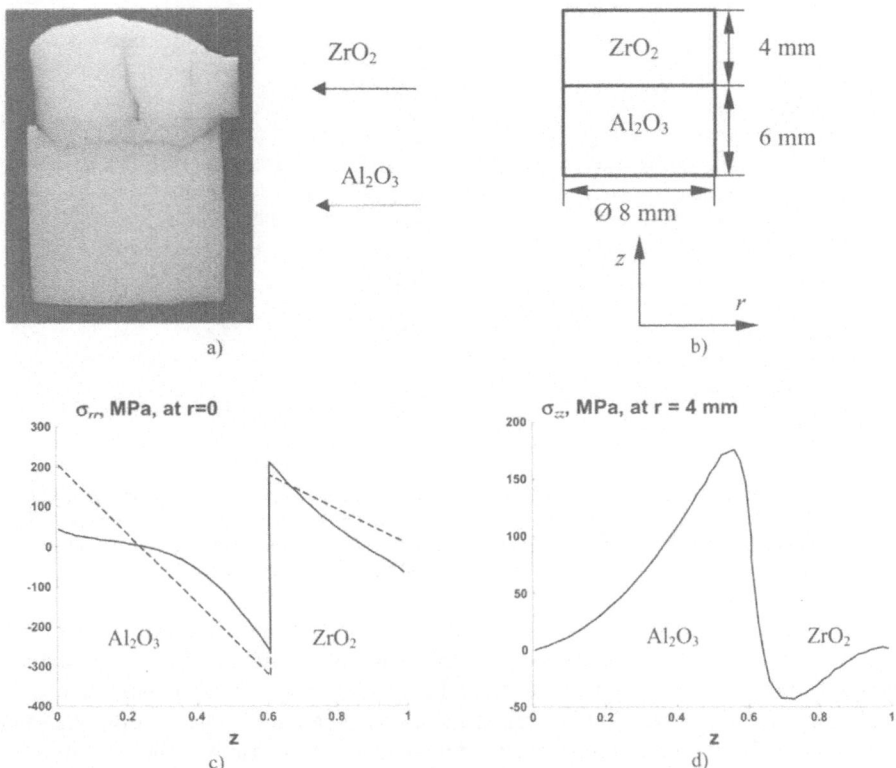

Figure 4. Direct joining of ZrO₂ and Al₂O₃ ceramics: a) photograph of a transverse section of the joined sample; b) dimensions for modeling; c, d) stress tensor components in the joined sample vs. the axial coordinate, z (solid lines – calculated numerically, dashed lines – obtained approximately within the analytical model).

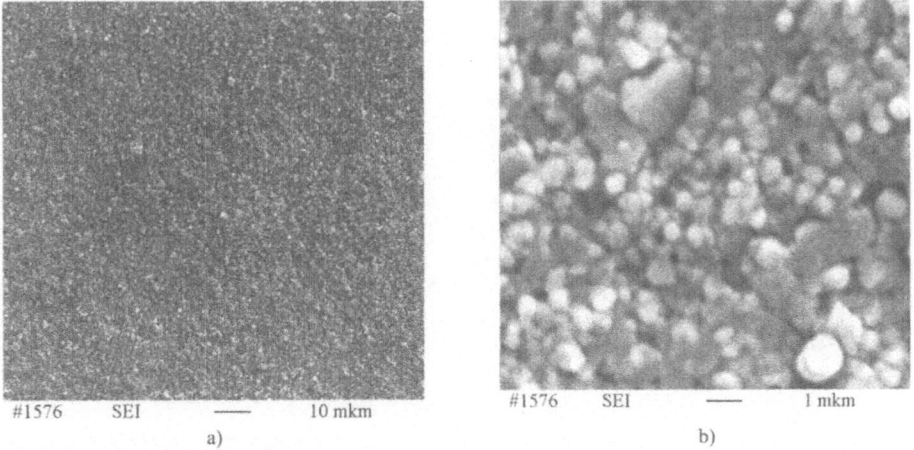

Figure 5. Microstructure of the Al₂O₃ ceramic sintered from nanosize powder for the interlayer material. Grains in the size range from 220 to 2000 nm are present in the material; average grain size is 570 nm.

422

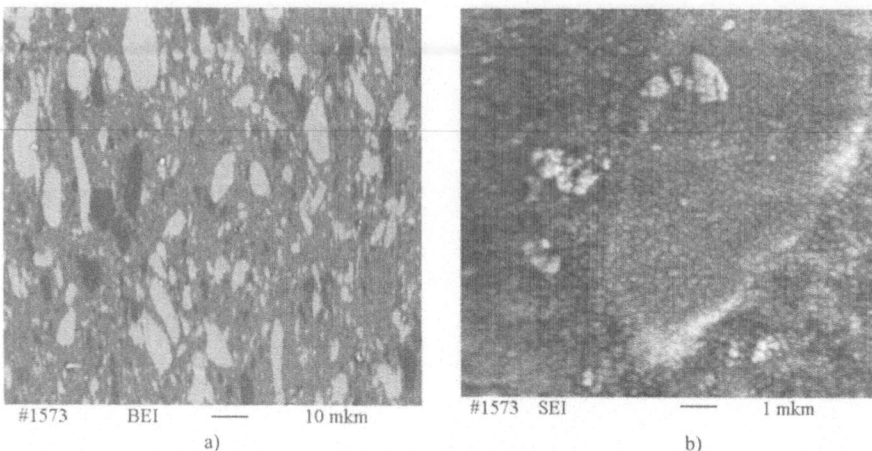

| #1573 | BEI | —— | 10 mkm | #1573 | SEI | —— | 1 mkm |

a)   b)

Figure 6. Microstructure of the 60 % ZrO_2 + 40 % Al_2O_3 ceramic sintered from nanopowders for the interlayer material. Zirconia grains in the size range from 80 to 250 nm are present; average size of zirconia grains is 160 nm. Alumina grains in the size range from 130 to 450 nm are present; average size of zirconia grains is 215 nm. Dark areas in Fig. 6, a are alumina-rich and bright areas are zirconia-rich.

A successful joining was achieved with the use of an interlayer sintered from a 60 % ZrO_2 + 40 % Al_2O_3 nanopowder mixture. The microstructure of the as-sintered interlayer material is shown in Fig. 6. Note that the presence of zirconia has reduced the final size of alumina grains, compared to the alumina ceramics sintered from the same powder (Fig. 5). The temperature needed for joining in this case was found to be 1490 °C at a pressure of 1.1 MPa. A photograph of the transverse cut of the joined sample is shown in Fig. 7, a. It can be seen that the insertion of the interlayer whose elastic properties and CTE are somewhat intermediate between the materials has apparently lowered the residual stresses and resulted in a crack-free joint.

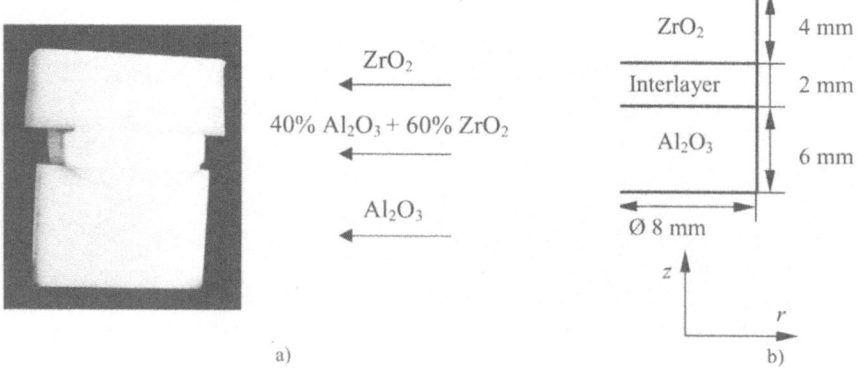

a)   b)

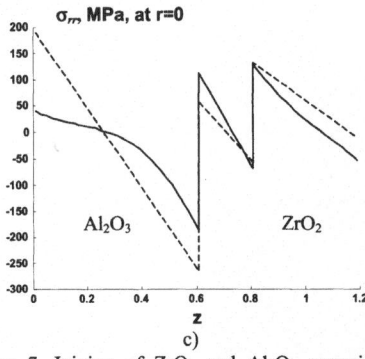

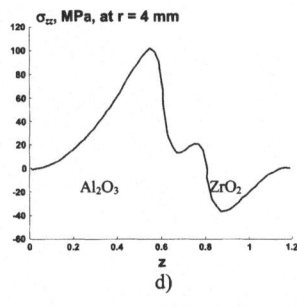

c)

d)

Figure 7. Joining of ZrO$_2$ and Al$_2$O$_3$ ceramics via an interlayer sintered from a nanopowder mix of composition 40%ZrO$_2$ + 60%Al$_2$O$_3$: a) photograph of a transverse section of the joined sample; b) dimensions for modeling; c, d) stress tensor components in the joined sample vs. the axial coordinate, z (solid lines – calculated numerically, dashed lines – obtained approximately within the analytical model). Note the absence of cracks in the joined sample.

This result is in agreement with the results of modeling. The stresses in a three-component system were estimated analytically and computed numerically in a manner similar to the two-component case. The effective elastic properties and the CTE of the 60 % ZrO$_2$ + 40 % Al$_2$O$_3$ composite materials were computed based on the mixing model proposed in [9]. The results, shown in Fig. 7, c, d, indicate that the tensile residual stresses do not exceed the tensile strengths anywhere in the sample.

The microstructure of the joint zone was investigated in all samples described above. The micrographs are shown in Figs. 8–10. It can be seen in the micrographs taken in the SEI mode that chains of pores are formed along the joint interface in all samples. These porous structures have the characteristic width of 1...2 μm, which is apparently associated with the roughness of the surfaces prior to joining.

At the final stage of experimentation, the possibility of elimination of the porosity in the joint zone by increasing the maximum temperature of joining was studied. The samples for joining were prepared from three types of 98 % dense alumina ceramic with the grain size 6 – 10 μm, 1 – 2 μm, and 0.4 – 0.6 μm. The zirconia samples and nanocomposite interlayers were the same as in the previous experiment. Experimentation has shown that a porosity-free joint is achieved at joining temperatures of 1600 °C and higher. Figure 11 shows the microstructures of the obtained joints.

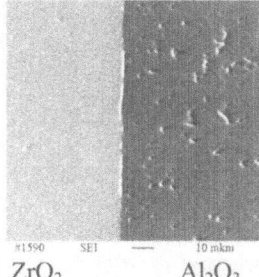

ZrO$_2$ Al$_2$O$_3$

Figure 8. Microstructure of the joining zone in the directly joined ZrO$_2$ – Al$_2$O$_3$ sample.

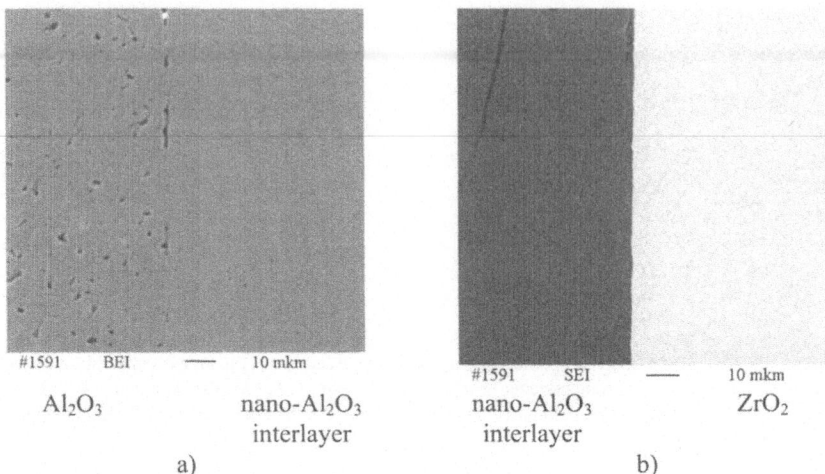

Al₂O₃   nano-Al₂O₃   nano-Al₂O₃   ZrO₂
          interlayer   interlayer

a)   b)

Figure 9. Microstructures of the joining zones in the Al₂O₃ – ZrO₂ sample joined via a nano-Al₂O₃ interlayer.

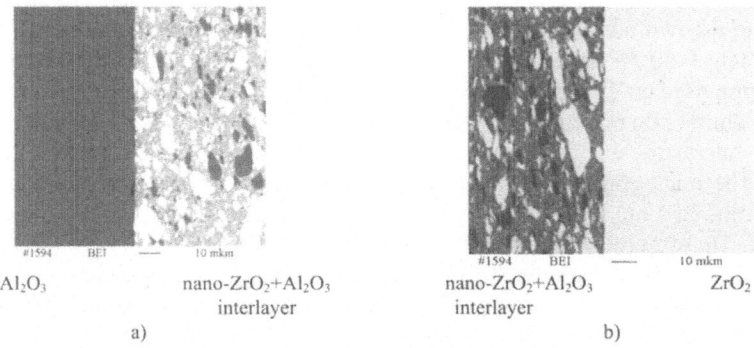

Al₂O₃   nano-ZrO₂+Al₂O₃   nano-ZrO₂+Al₂O₃   ZrO₂
           interlayer   interlayer

a)   b)

Figure 10. Microstructures of the joining zones in the Al₂O₃ – ZrO₂ sample joined via a nano-60 % ZrO₂ + 40 % Al₂O₃ interlayer.

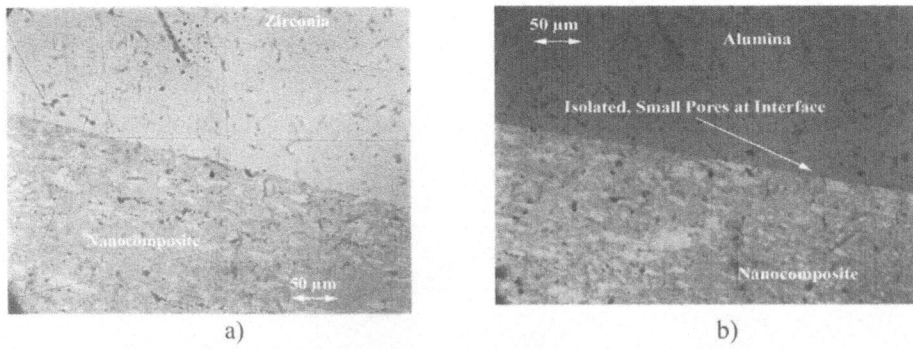

a)   b)

Figure 11. Microstructure of the joint interfaces in the samples joined via a 60 % ZrO₂ + 40 % Al₂O₃ nanocomposite interlayer at a temperature of 1600 °C: a) zirconia – nanocomposite interface; b) nanocomposite – alumina interface.

5. Summary and Conclusions

The possibility of creating graded thermal barrier coatings by joining the ZrO_2 and Al_2O_3 ceramics under microwave heating has been demonstrated. The minimum temperature needed for joining is 1430 °C. The residual stresses that develop in a direct joint between the two materials are prohibitively high. The use of a nanocomposite, 60 % ZrO_2 + 40 % Al_2O_3 interlayer reduces the stresses by matching the CTEs of zirconia and alumina and makes it possible to obtain a crack-free joined sample. The increase of the maximum temperature of joining to 1600 °C facilitates removal of the chains of pores from the joint interface.

These experimental results are in good agreement with the theoretical analysis of the residual stresses in the joints. The stresses have been estimated analytically and computed numerically. The analytical model has been demonstrated to be suitable for qualitative estimates of the stresses. The results of analysis suggest that normal stresses in the direction parallel to the interface between the joined plates of dissimilar materials are prevailing provided the temperature is uniform enough along the joint. The stresses at the joint interface are generally tensile in the material that has higher thermal expansion coefficient. The use of an interlayer with properties that are intermediate between the two materials to be joined reduces the stresses considerably.

Although not yet tested by modeling, we believe that the enhanced ductility of the ceramic interlayer sintered from nanosize powders also contributes to the reduction of stresses in the joint.

In conclusion, the results of this study suggest that microwave heating is a viable method for creating graded thermal barrier coatings.

6. Acknowledgement

This research is supported in part by the International Science and Technology Center under Project # 1607. International collaboration was supported by the Collaborative Linkage Grant CLG 974624 from NATO.

7. References

1. Functionally graded materials (1996), edited by I. Shiota and Y. Miyamoto, Elsevier, Amsterdam.
2. *Materials Research Society Bulletin*, **XX** [1] (1995).
3. Ho, W.W. (1988) High-temperature dielectric properties of polycrystalline ceramics, in W.H. Sutton, M.H. Brooks, I.J. Chabinsky (eds.) *Microwave Processing of Materials* (Materials Research Society Symp. Proc., Vol. 124), Materials Research Society, Pittsburgh, pp. 137–148.
4. Bykov, Yu.V., Rybakov, K.I., Semenov, V.E. (2001) High-temperature microwave processing of materials (topical review), *Journal of Physics D: Applied Physics*, **34**, R55 – R75
5. L.D. Landau and E.M. Lifshitz (1986), *Theory of Elasticity*, 3rd edition. Pergamon, Oxford.
6. O.T. Iancu, D. Munz (1990) Residual stress state of brazed ceramic-metal compounds, determined by analytical methods and X-ray residual stress measurements, *J. Am. Ceram. Soc.*, **73**, 1144–1149.
7. Bykov, Y., Eremeev, A., Flyagin, V., Kaurov, V., Kuftin, A., Luchinin, A., Malygin, O., Plotnikov, I., Zapevalov, V. (1995) The gyrotron system for ceramics sintering, in D.E. Clark, D.C. Folz, S.J. Oda, R.

Silberglitt (eds.) *Microwaves: theory and applications in material processing III* (Ceramic Transactions, vol 59). Amer. Ceram. Soc., Westerville, pp. 133–140

8. Ivanov, V., Kotov, Y. A., Samatov, O. H., Böhme, R., Karow, H. U., Schumacher, G. (1995) Synthesis and dynamic compaction of ceramic nano powders by techniques based on electric pulsed power, *NanoStructured Materials* **6**, 287–290.

9. Ravichandran, K.S. (1994) Elastic Properties of Two-Phase Composites, *J. Am. Ceram. Soc.*, **77**, 1178–84.

SURFACE CHEMISTRY AND FUNCTIONALIZATION OF SEMICONDUCTING NANOSIZED PARTICLES

MARIE-ISABELLE BARATON
SPCTS – UMR 6638 CNRS, University of Limoges (France)

1.   Introduction

The tremendous problem with surfaces is that their composition, structure, reactivity are dependent not only on the synthesis conditions but also on the surrounding environment of the material. Indeed, a surface is in constant evolution to balance the energy at the interface. Therefore, one can never define the surface composition of any material without describing the surrounding environment. The obvious result is that the surface properties depend on the environmental conditions. For nanosized particles which can be considered as surfaces in three dimensions, this means that the control of the overall properties cannot be achieved without the control of the surface in terms of structure, chemical composition and reactivity. This is an important drawback which can be turned into an advantage, since modifications of the surface chemical composition, for example, make it possible to tailor the nanosized particles properties. Indeed, it is now well recognized that the reproducibility of the properties of nanosized particles can only be obtained if the surface chemical composition and surface chemistry of the nanosized particles are perfectly controlled [1]. In addition, surface functionalization of nanosized particles has become an asset for the optimisation of targeted properties [2].

This chapter will essentially focus on the capabilities of Fourier transform infrared (FTIR) spectroscopy to analyze the surface of nanosized particles. The fundamentals of surface analysis and the experimental techniques are briefly reported and applications of Fourier transform infrared surface spectroscopy to semiconducting nanosized particles are presented, taking titanium oxide (TiO_2) or tin oxide (SnO_2) as examples. Both materials are used in the fabrication of resistive gas sensors. It is demonstrated how FTIR spectroscopy can be used to follow the electrical conductivity variations of the semiconductor when the surrounding atmosphere is varied. It is also explained how and why chemical modifications of the surface species result in changes in the gas sensing properties.

T. Tsakalakos et al. (eds.) pgs. 427-440
Nanostructures: Synthesis, Functional Properties and Applications;
© *2003 Kluwer Academic Publishers.*

2. Fundamentals of Infrared Spectroscopy

2.1. BACKGROUND

Infrared spectroscopy is based on the absorption by a sample of infrared wavelengths which excite interatomic vibrations in the molecule (for more details on infrared spectroscopy, see for example [3-5]). In other words, infrared spectra originate from transitions between discrete vibrational and rotational energy levels of molecules. The interatomic vibrations may be generally described as bond stretching, bond bending and eventually torsional modes. The wavelength (or frequency) of the absorbed infrared radiation essentially depends on the atomic weight of the vibrating atoms and on the force constant of the bond between these atoms. In general, it is possible to determine relatively narrow ranges of frequencies where defined vibrations of particular chemical groups will absorb regardless of their chemical environment. For example, the absorption ranges of the stretching vibrations of the CO bonds are clearly identified depending on the nature of the bond [6]: The stretching vibration of the C-O single bond in alcohols usually absorbs in the 1000-1300 cm^{-1} range, the stretching vibration of the C=O double bond in carbonyl groups absorbs in the 1550-1900 cm^{-1} range, whereas gaseous CO_2 and CO have their stretching vibrations at 2348 and 2143 cm^{-1} respectively. Therefore, by studying the frequency ranges where the absorption bands fall, it is theoretically possible to determine which kinds of atomic bonds constitute the molecule.

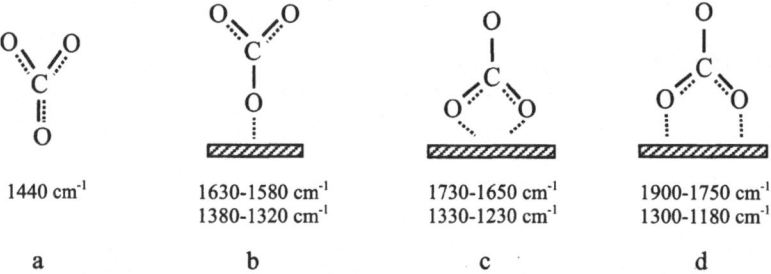

a	b	c	d
1440 cm^{-1}	1630-1580 cm^{-1} 1380-1320 cm^{-1}	1730-1650 cm^{-1} 1330-1230 cm^{-1}	1900-1750 cm^{-1} 1300-1180 cm^{-1}

Figure 1. $v(CO)$ absorption frequency ranges of the carbonate ion in various coordination: a) free ion; b) monodentate coordination; c) bidendate coordination; d) bridged ion.

Let us now consider the case of a particular interatomic bond, such as the CO bond in the carbon monoxide molecule or in a particular chemical group such as the carbonate ion CO_3^{2-}. The carbon monoxide molecule in the gas phase has a very well defined absorption frequency for the stretching vibration (2143 cm^{-1}). But, when the molecule interacts with or adsorbs on a solid surface, the electronic distribution is perturbed within the CO molecule. As a consequence, the force constant of the bond is changed and the stretching frequency is shifted downwards or upwards depending on the nature of the interaction. On the other hand, the free carbonate ion (Figure 1a) has a characteristic absorption frequency $v(CO)$ around 1440 cm^{-1} but, when this ion adsorbs onto a solid surface, this absorption frequency splits into two components due to symmetry changes[7]. The frequency shift between these two components is very dependent on the nature and geometry of the adsorbed species and on the force of the interaction between the ion and the surface. For example [7], the carbonate ion in monodentate coordination (Figure 1b) has its $\nu(CO)$ stretching absorptions in the 1630-1580 cm^{-1} and 1380-1320 cm^{-1} ranges, whereas the stretching frequencies of the CO_3^{2-} ion in bidentate coordination (Figure 1c) fall in the 1730-1650 cm^{-1} and 1330-1320 cm^{-1} ranges. But if the CO_3^{2-} ion is bridged on two adsorption sites (bridged carbonates, Figure 1d), the $\nu(CO)$ absorption frequencies can be found in the 1900-1750 cm^{-1} and 1300-1180 cm^{-1} ranges. To summarize, by observing the perturbation of a particular absorption frequency of a molecule, it is possible to obtain information on the nature and the strength of the interaction between the molecule and its environment.

2.2. SURFACE FTIR SPECTROSCOPY

It is well-known that a clean surface can only be maintained in this state under ultra high vacuum [8-9]. Most of the time, surfaces are exposed to gases, liquids or solids and become contaminated by atoms or molecules from the ambient. Indeed, the existence of a surface means that bonds are broken and that the atoms at the very surface are in low coordination number and therefore highly reactive. Molecules surrounding a clean surface will therefore react with these unsaturated surface atoms to balance the forces at the interface. Obviously, to maintain balanced forces, the nature of the contaminating atoms at the surface evolves with the surrounding environment and, as a consequence, one can never define the surface composition of any material without describing its surrounding environment.

One of the most probable reactions occurring at a surface is the dissociation of water to form hydroxyl groups which are known to have a very important role in the surface chemistry [10]. Additional contamination of the surface may originate from synthesis residues such as carbonate or nitrate groups. As a result, contaminating species on a material surface are likely to be of organic nature.

As a consequence of the surface contamination, the chemical composition of the surface is different from that of the bulk. Therefore, the surface of nanosized particles calls for specific characterization and, from the points described in section 2.1, infrared spectroscopy appears to be a quite appropriate tool to determine the chemical nature of these surface species and the kinds of interactions between these chemical groups and the material surface. Indeed, when an infrared beam crosses whatever particle, the transmitted beam contains information not only on the interatomic bonds constituting the bulk, but also on the chemical groups at the particle surface. Obviously, these surface groups are minority by far. But, when the size of the particle is decreased down to the nanometer scale, the concentration of the surface groups relatively to that of the bulk interatomic bonds increases and the contribution of these surface groups to the overall infrared absorption becomes no longer negligible. As a result, the absorption bands of the surface chemical species can be clearly observed in the infrared transmission spectrum in addition to the absorption bands due to the bulk modes.

However, it is necessary to discriminate the bands due to the surface species and the bands due to the bulk modes. The first obvious criterion for such discrimination is that the bands of the bulk modes are usually much more intense than the bands due to the surface groups. In addition, the bulk modes of inorganic materials (such as metal oxides, nitrides or carbides) fall in the low wavenumber region of the infrared spectra (usually below 1000 cm^{-1}) whereas the surface species which are organic-like groups have major absorption bands in the highest wavenumber region (usually above 1000 cm^{-1}). But the final step of the discrimination is performed by adsorbing selected molecules on the material surface [11-16]. These probe-molecules will perturb the surface groups easily accessible while leaving the bulk unchanged. So under probe-molecules, only the surface absorption bands will be modified.

2.3. DRUDE-ZENER THEORY. APPLICATIONS TO GAS SENSORS

In the case of semiconductors, the free carriers (electrons or holes) contribute to the infrared absorption [17-19]. The absorption of these free carriers is usually over the total infrared range and does not correspond to narrow bands with defined absorption maxima [19]. A variation of the free carrier density translates into a change of the background absorption level of the infrared spectrum [20]. In other words, materials such as metals with a high electrical conductivity, that is with a high density of free carriers are opaque to the infrared radiation. This relationship between free carrier density and absorbance is mathematically described by the Drude-Zener theory [18]. Under specific conditions, a direct correlation can be established between absorbance and electrical conductivity at each wavelength, as explained below.

According to the Beer's law, the absorbance A at a given wavelength λ is directly related to the absorption coefficient K:

$$A(\lambda) = K(\lambda)\,z$$

where z is the thickness of the sample.

When the sample is a semiconductor, part of the infrared absorption is due to the free carriers. From the Drude-Zener theory, it can be shown that the absorption coefficient is directly related to the electrical conductivity:

$$K(\lambda) = \sigma(\lambda)/\varepsilon_0\,c\,n = f(\lambda^2)$$

where $\sigma(\lambda)$ is the electrical conductivity which depends on the wavelength λ, ε_0 the permittivity, c the light velocity and n the refractive index.

Therefore, the absorbance of a semiconductor at a given wavelength is directly related to the electrical conductivity [20]. Obviously this equation is valid under specific conditions and, practically speaking, several factors must be taken into account to obtain a quantitative relationship between absorbance and electrical conductivity. But, if one assumes that only the free carrier density varies, the variation of the absorption of the semiconductor over the total infrared range, will reflect the variation of the free carrier density, that is the variation of the electrical conductivity. As a consequence, by observing the variations of the background infrared absorption, we can obtain information on the variations of the free carrier density. In the following, this property has been used to evaluate the gas sensing properties of semiconducting nanosized particles by FTIR spectroscopy.

The gas detection by chemical sensors based on semiconducting materials is due to electrical conductivity variations induced by adsorption of gases on the semiconductor surface. For example, when oxygen adsorbs on a semiconductor, negatively charged oxygen species (or ionosorbed species) are formed [21-22]:

$$O_2 + 2\,e^- \rightarrow 2\,O^-_{ads}$$
$$O_2 + e^- \rightarrow O^-_{2\,ads}$$

The oxygen ionosorption causes electron transfer from the surface of the grain toward the adsorbed species, thus leading to the formation of an electron-depleted surface layer. As a result, the electrical conductivity of a n-type semiconductor decreases.

On the contrary, when a reducing gas, such as CO, adsorbs, electrons are injected into the conduction band and the electrical conductivity of a n-type semiconductor increases. When CO adsorbs in presence of ionosorbed oxygen, the following reactions can possibly occur [22-25]:

$$2\,CO + O^-_{2\,ads} \rightarrow 2\,CO_2 + e^-$$
$$CO + O^-_{ads} \rightarrow CO_2 + e^-$$
$$CO + 2\,O^-_{ads} \rightarrow CO_3^{2-}{}_{ads}$$

The first two reactions which produce delocalized electrons are responsible for electrical conductivity changes.

2.4. SUMMARY

To summarize the above fundamental considerations, a thorough analysis of the infrared spectrum can theoretically provide information on:
- the interatomic bonds constituting the bulk
- the chemical nature of the surface bonds and surface groups
- the possible presence of contaminating species on the surface
- the nature of the adsorption sites
- the free carrier density, in the case of a semiconductor
- the surface reactions and interactions occurring in presence of whatever gaseous environment

When this technique is applied to nanoparticles, the contribution of the surface can be clearly identified in the infrared absorption spectrum because of the high surface-to-bulk ratio of these nanoparticles.

3. Experimental Approach

The experimental details of the FTIR surface analysis of nanosized particles have been described elsewhere [14-16]. To briefly summarize, the nanopowder is slightly pressed into a thin pellet. If needed, the pellet can be pre-treated by heating under dynamic vacuum (*activation*) in order to clear the nanoparticle surface from contaminating species. Monitoring of surface chemical modifications can be performed *in situ* and the stability of the surface functionalization can also be checked *in situ* under various conditions by varying parameters such as temperature, vacuum or gaseous environments.

When studying semiconducting materials, the nanoparticle pellets should undergo an oxidizing treatment under pure oxygen before analysis. Indeed the semiconducting nanoparticles can be strongly reduced by the thermal pretreatment under dynamic vacuum and consequently, as previously explained, n-type semiconductors become opaque to the IR radiation. To screen the gas sensing properties of the nanopowders, the pellets which simulate the gas sensors are subjected to oxygen and carbon monoxide at temperatures corresponding to the optimized operating temperatures of real sensors. The variations of the total infrared energy transmitted by the nanopowder pellet are measured during "gas addition/evacuation" sequences. Simultaneously, the infrared spectrum of the sample is recorded at each step of the experiment to follow the evolution of the surface reactions. All surface analyses are performed *in situ*.

The tin oxide nanopowder used for this study was synthesized by laser evaporation of commercial powders [26]. The average particle size is 15 nm and it is mainly in the rutile phase. Titanium oxide is a commercial product (Degussa P25). The primary particle size given by the manufacturer is 21 nm and the material is a mixture of anatase (70%) and rutile (30%) phases.

The infrared spectra were recorded in the 5000-450 cm^{-1} range (4 cm^{-1} resolution) by using a FTIR spectrometer (Perkin Elmer Spectrum 2000) equipped with a MCT cryodetector. All the gases were provided by Air Products Corp. (oxygen: 99.999% pure; carbon monoxide: 99.9% pure). Hexamethyldisilazane (Fluka, 99% pure) was used without further purification. Water was bi-distilled and deionised.

4. Study of Tin Oxide Nanoparticles

Figure 2a shows the infrared transmission spectrum of a pellet of tin oxide nanopowder at 623 K under dynamic vacuum. The main chemical species present on the surface of the tin oxide nanoparticles are hydroxyl groups either isolated or hydrogen-bonded [27]. All these OH groups absorb in the highest wavenumber region of the spectrum (3700-3200 cm^{-1}). Figure 2b corresponds to the spectrum of the same tin oxide pellet under oxygen (50 mbar). The decrease of the overall absorbance indicates a decrease of the free carrier density. Indeed, under oxygen adsorption, electrons are transferred from the surface toward adsorbed oxygen (cf. section 2.3), thus leading to the formation of ionosorbed oxygen species at the surface. As a consequence, the thickness of the depletion layer increases while the electrical conductivity decreases. When CO is adsorbed in presence of oxygen (Figure 2c), the increase of the background absorption is immediate corresponding to the expected increase of the electrical conductivity under this reducing gas. In addition, new infrared absorption bands are observed and are assigned to newly formed carbon dioxide and surface carbonate species. Both carbon dioxide and carbonate species are totally eliminated by evacuation (Figure 2d).

As previously mentioned, the oxidation reaction of CO into CO_2 is responsible for the electrical conductivity increase by generating delocalized electrons. Additionally, part of the newly formed carbon dioxide adsorbs on the O^{2-} basic surface sites to form carbonate groups according to the following reaction [25]:

$$CO_2 + O^{2-}_s \rightarrow CO_3^{2-}_{ads}$$

Indeed, additional experiments have proven that carbon dioxide is first formed and it then reacts with the basic surface sites leading to reversible surface carbonates but without any change of the electrical conductivity [28].

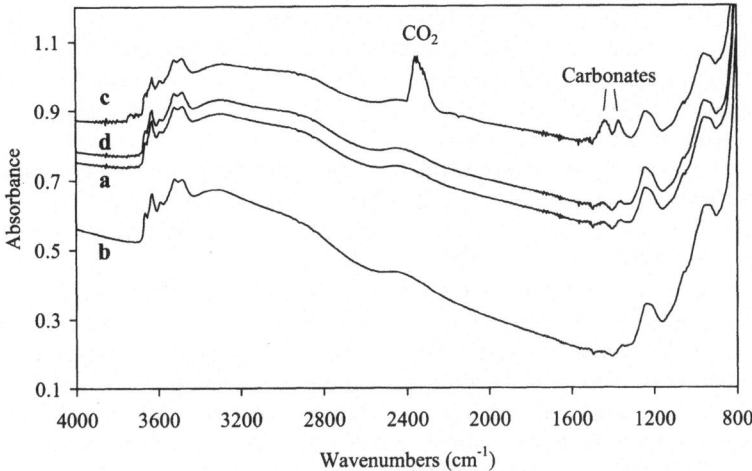

Figure 2. Infrared spectra of the SnO$_2$ nanopowder at 623 K: a) under vacuum; b) under 50 mbar oxygen; c) after CO addition (10 mbar) in presence of oxygen; d) after evacuation. (The spectra have NOT been shifted).

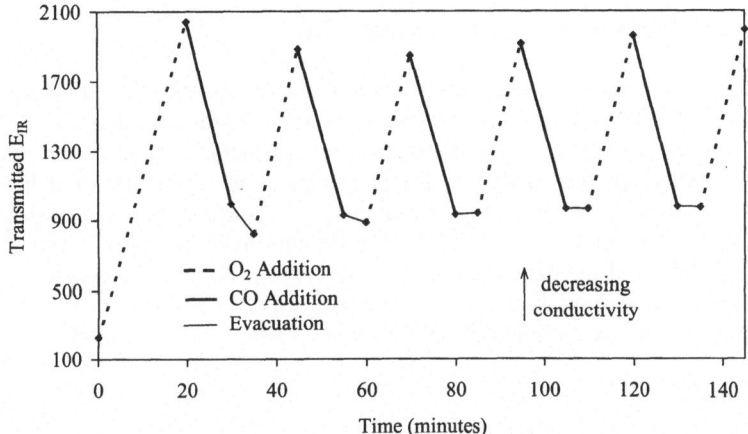

Figure 3. Variations of the infrared energy transmitted by the SnO$_2$ nanopowder at 623 K versus gas exposures.

Figure 3 shows the variations of the infrared energy transmitted by the tin oxide nanoparticles versus gas exposures at 623 K. Obviously, an increase of the transmitted infrared energy corresponds to a decrease of the overall absorption, that is to a decrease of the electrical conductivity. The origin of the curve corresponds to Figure 2a, that is to the sample under vacuum and totally reduced. When oxygen is in contact with the sample (corresponding to Figure 2b), the transmitted infrared energy rapidly increases, indicating a decrease of the electrical conductivity resulting from the oxidation of the tin oxide nanopowder. Then, carbon monoxide is adsorbed in presence of oxygen (corresponding to Figure 2c). The transmitted infrared energy decreases showing the increase of the electrical conductivity due to the reducing effect originating from the carbon dioxide formation. A quick evacuation followed by addition of a new dose of oxygen leads to a recovery of the oxidation state and quite reproducible variations of the infrared energy can be observed during the "CO addition-evacuation-oxygen addition" sequences.

In previous works, we have proved that this curve can be directly compared to the sensor electrical response curve [29]. The lapse of 10 minutes between two measurements has been chosen to ensure that the system has reached the chemical equilibrium and that the surface reactions are completed. It does not correspond to the response time of the real sensor which actually is less than 30 seconds [30].

It is important to note that, with this technique, one actually measures the variation of the free carrier density in each particle independently from the others. To perform such measurement, there is no need for necks between the particles and no need for an electron transfer from a particle to another, contrary to standard electrical measurements. Actually, this technique makes it possible to evaluate the amplitude of the electrical conductivity variations that a semiconducting nanopowder can eventually undergo in presence of oxidizing or reducing gases with no need for metallic electrodes that are known to possibly generate perturbations in the gas sensor response.

5. Surface Modification of TiO_2 Nanoparticles

An important drawback of the semiconductor-based sensors is their sensitivity to humidity which makes their response unreliable for outdoor operation. The effect of humidity on the electrical conductivity of semiconductor sensors is not perfectly understood. Although the surface hydroxyl groups seem to be involved in the gas detection mechanism in humid environment, the surface reactions which may eventually occur are not clear [22, 31]. Experiments have been performed to check whether a decrease of the density of surface hydroxyl groups could modify the sensor response in presence of humidity. To this end, the response of titanium oxide to carbon monoxide and humidity is compared before and after surface modification.

5.1. GRAFTING REACTIONS

The surface of the titanium oxide nanoparticles is modified by grafting hexamethyldisilazane (HMDS) on the surface hydroxyl groups. The reaction is expected to take place as follows [32-34]:

$$2\ Ti\text{-}OH\ +\ [Si(CH_3)_3]NH[Si(CH_3)_3]\ \rightarrow\ 2\ Ti\text{-}O\text{-}Si(CH_3)_3\ +\ NH_3$$

If the reaction is complete, no hydroxyl groups should be left on the surface and the surface is expected to become hydrophobic.

Figure 4 shows the evolution of the titania surface under HMDS grafting. The first spectrum (Figure 4a) has been recorded at room temperature after heating at 723 K under dynamic vacuum to clear the surface from adsorbed humidity and other physisorbed and weakly chemisorbed contaminating species. At the end of the thermal treatment, several bands can be observed in the 3780-3600 cm^{-1} range. All of them are assigned to v(OH) stretching vibrations [35-38]. The multiplicity of these bands indicates that several types of OH groups are present on the titania surface. The number and the vibrational frequencies of these OH groups depend on the extent of the dehydroxylation, on the crystalline phase and on the possible presence of impurities [38-41]. As a consequence, comparisons between different samples and between the results from different works become extremely difficult and perhaps even meaningless. Additionally, different kinds of OH groups simultaneously exist since an hydroxyl group can be bonded to one titanium atom, or linked to two titanium atoms or bridged to three titanium atoms. Despite of this complexity, precise assignments of all the v(OH) bands can be found in the literature for various samples and under various conditions [35-42], thus giving a good picture of the surface of any titania sample. The OH groups can eventually react with the chemical environment as either proton donor or proton acceptor [43].

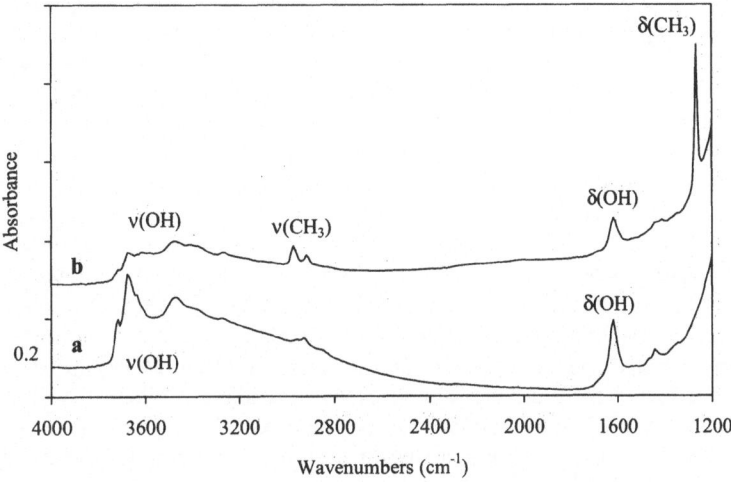

Figure 4. Infrared spectra of the TiO$_2$ nanopowder at room temperature: a) after activation at 723 K; b) after HMDS grafting followed by thermal desorption at 673 K. (The spectra have been shifted for clarity sake).

Then, this heat-treated surface is subjected to 7 mbar of HMDS liquid vapor at room temperature. After a thirty-minute contact, HMDS is steadily desorbed at increasing temperatures up to 673 K. During the HMDS contact with the surface at room temperature, most of the v(OH) bands disappear very rapidly (spectrum not shown). Although a slight recovery is observed after the thermal desorption, the surface OH groups are still less intense compared to the original titania surface (Figure 4b). HMDS is not eliminated, as indicated by the very intense band at 1267 cm^{-1} assigned to the δ(CH$_3$) bending vibration of Si-CH$_3$ groups [6, 33]. Moreover, new bands appear (in the 2900 cm^{-1} region and 1265 cm^{-1}), characteristic of the CH$_3$ group vibration. These bands are the obvious proof of the surface modification by Si(CH$_3$)$_3$ trimethylsilyl groups according to the above-mentioned reaction. Ammonia formed during the grafting reaction is eliminated by the thermal desorption. Therefore, by grafting HMDS on titania, the surface chemical species and the surface reactivity have been irreversibly modified.

5.2. MODIFICATION OF THE SENSING PROPERTIES

The sensing properties towards CO of these non-grafted and HMDS-grafted titanium oxide nanopowders have been checked in presence and in absence of humidity. To this end, series of experiments on titanium oxide were performed consisting in the introduction of four pure CO doses (6 mbar), referred to as "dry CO doses", followed by four doses of CO mixed with a small amount of water vapor (10%), referred to as "wet CO doses", and then followed by four new "dry CO doses". The operating temperature is set at 673 K. For sake of simplicity, these preliminary experiments have been performed in an oxygen-free environment.

Figures 5 and 6 show the variations of the transmitted infrared energy versus gas exposures at 673 K for non-grafted and HMDS-grafted samples,respectively. On the non-grafted sample (Figure 5), the first dose of "dry CO" is added for 10 minutes and causes the expected decrease of the transmitted infrared energy resulting from the increase of the free carrier density. On the infrared spectrum, the formation of CO_2 is observed (not shown). After evacuation of the first CO dose, the transmitted infrared energy does not get back to its original intensity. This is also expected because in absence of an oxidizing atmosphere the adsorbed oxygen species used for the CO_2 formation are not regenerated. Then, other "dry CO doses" are subsequently added. Each dose causes a decrease of the transmitted infrared energy, whereas each evacuation leads to an increase of the infrared energy. But the sample is steadily reduced by the subsequent CO additions. The addition of "wet CO doses" causes a strong decrease of the electrical conductivity although without any reproducibility. In parallel, the infrared spectra show an intensity increase of the $\nu(OH)$ absorption range. The overall effect of the four "wet CO doses" is an oxidation. When "dry CO" is added again, the energy evolution is similar to that observed during the addition of the first "dry CO doses". Moreover, the baseline drift shows the same downward trend, thus indicating an overall reducing effect.

The second curve (Figure 6) corresponds to the same experiment performed on the HMDS-grafted titanium oxide nanopowder. As in the previous case, the addition of "dry CO doses" has a reducing effect. Concomitantly, no perturbation can be observed on the IR spectrum and the formation of CO_2 is hardly visible. Like in the case of the non-grafted sample, the oxidation state is not restored by evacuation. But differently, the "addition-evacuation" sequences of "dry CO doses" lead to an oxidizing effect. This phenomenon could be tentatively explained by a rearrangement of the surface under CO because, in addition to the modification of the surface species, the HMDS grafting by itself has a reducing effect by producing ammonia (cf. section5.1).

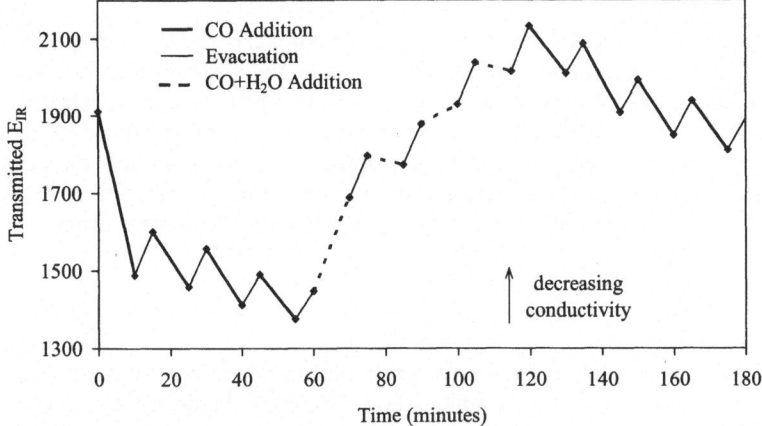

Figure 5. Evolution of the infrared energy transmitted by the non-grafted TiO₂ nanopowder versus gas exposures at 673 K.

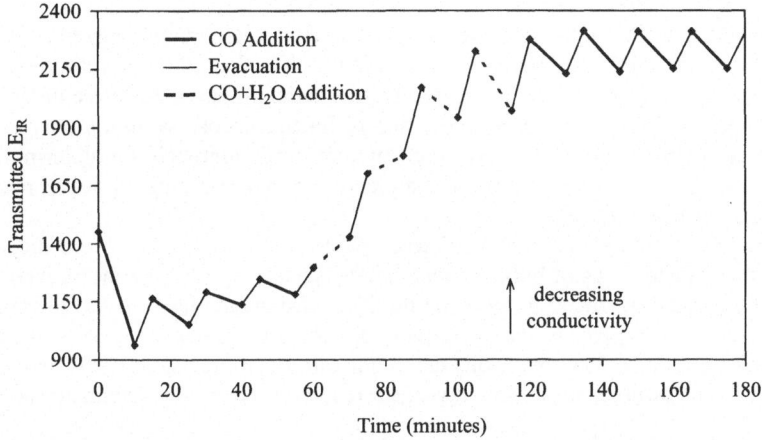

Figure 6. Evolution of the infrared energy transmitted by the HMDS-grafted TiO₂ nanopowder versus gas exposures at 673 K.

The addition of the first two "wet CO doses" amplifies the oxidizing effect. On the infrared spectra, we note the increase of a new band which is assigned to the formation of new Si-OH surface groups by reaction of water with the Si(CH₃)₃ grafted groups [34, 44]. Then, for the subsequent two "wet CO doses", the reducing action becomes preponderant and no further changes are noted in the IR spectra. When "dry CO" is added again, we only note a slight reversible shift of the new Si-OH band. But, surprisingly, the transmitted infrared energy evolution shows a perfectly reproducible response and no drift of the baseline, that is no change in the oxidation state after the

"CO addition-evacuation" sequences. The formation of CO_2 appears slightly more visible.

Although a thorough analysis and discussion of these results is out of scope of this chapter, the only observation of these curves clearly demonstrates that, by changing the chemical species on a nanopowder surface, one can affect the overall properties of the material like, in this particular case, the response of a semiconductor to gases [45]. Therefore, a specific property of a nanopowder can be modified on purpose by controlling the surface composition and the surface chemistry. The obvious counterpart is that a contamination of the surface may adversely affect the desired property.

6. Conclusions

In conclusion, it should be emphasized that Fourier transform infrared spectroscopy should not be seen as a routine technique useful only for characterizing chemical bonds in the bulk of materials. In addition to results specific to the surface composition and chemistry of nanosized particles, this technique can provide valuable information on the *in situ* monitoring of surface reactions, resistance of surface functionalization under various conditions, electrical conductivity of semiconductors without contamination by metallic contacts, among others.

However, while high-performance spectrometers are commercialized at relatively low cost compared to other equipment, the drawback is that all the accessories, such as cells, furnaces, gas feeders are not available on a standard basis. All these items have to be designed and built by the user according to his/her needs. Sampling may also be a problem as there is no other way than an empirical approach for determining the specific and optimum conditions to sample every material and, in some occasions, every batch of nanopowder.

In addition, the study of the infrared spectra is a complex problem most of the time as the contributions of bulk, surface bonds, surface species, surface contaminants, free carriers, surface states, gaseous or liquid environments have to be discriminated. This requires many precise and systematic experiments under different and perfectly controlled conditions. Moreover, this technique allows so precise analysis that different batches of nanoparticles may show divergent results if the surface chemistry is not well mastered.

But the outcome of all these experiments is definitely of tremendous value for understanding the physical and chemical behavior of nanosized particles, tailoring their properties, controlling their surface chemistry, and mastering the functionalization of their surface for targeted applications.

7. Acknowledgment

Part of this work has been financially supported by the Commission of the European Community under the BRITE-EURAM III program (contract N° BRPR-CT95-0002) and under the Information Society Technologies program (contract N° IST-1999-12615).

8. References

1. Gonsalves, K.E., Baraton, M.-I., Singh, R., Hofmann, H., Chen, J.X., and Akkara, J.A. (eds.) (1998) *Surface Controlled Nanoscale Materials for High-Added-Value Applications*, MRS Symp. Proc. Vol. 501, MRS Publisher, Warrendale (PA).
2. Baraton, M.-I. (ed.) (2002) *Synthesis, Functionalization and Surface Treatment of Nanoparticles*, American Scientific Publishers, Stevenson Ranch (CA).
3. Herzberg, G. (1962) *Molecular Spectra and Molecular Structure*, Van Nostrand, Princeton.
4. Wilson, E.B. Jr, Decius, J.C., and Cross, P.L. (1955) *Molecular Vibrations. The Theory of Infrared and Raman Vibrational Spectra*, Dover, NewYork.
5. Schrader, B. (ed.) (1995) *Infrared and Raman Spectroscopy. Methods and Applications*, VCH, Weinheim.
6. Colthup, N.B., Daly, L.H., and Wiberley, S.E. (1964) *Introduction to Infrared and Raman Spectroscopy*, Academic Press, New York.
7. Busca, G. and Lorenzelli, V. (1982), Infrared Spectroscopic Identification of Species Arising from Reactive Adsorption of Carbon Oxides on Metal Oxide Surfaces, *Mater. Chem.* 7, 89-126.
8. Somorjai, G.A. (1990) *Introduction to Surface Chemistry and Catalysis*, Wiley, New York.
9. Somorjai, G.A. (1998) From Surface Materials to Surface Analysis, *MRS Bulletin* 23(5), 11-29.
10. Boehm, H.-P. and Knözinger, H. (1983) Nature and Estimation of Functional Groups on Solid Surfaces, in J.R.A. Anderson and M. Boudart (eds.), *Catalysis* Vol. 4, Springer-Verlag, Berlin, pp. 39-207.
11. Hair, M.L. (1967) *Infrared Spectroscopy in Surface Chemistry*, M. Dekker, New York.
12. Knözinger, H. (1976) Specific Poisoning and Characterization of Catalytically Active Oxide Surfaces, *Advances in Catalysis* 25, 184-201.
13. Davydov, A.A. (1984) *Infrared Spectroscopy of Adsorbed Species on the Surface of Transition Metal Oxides*, John Wiley & Sons, New York.
14. Baraton, M.-I. (1994) IR and Raman Characterization of Nanophase Ceramic Materials, *J. High Temp. Chem. Processes* 3, 545-554.
15. Baraton, M.-I. (1998) The Surface Characterization of Nanosized Powders: Relevance of the FTIR Surface Spectrometry, in G.M. Chow *et al.* (eds.), *Nanostructured Materials*, NATO-ASI Series, Kluwer Academic Publishers, Dordrecht, pp. 303-317.
16. Baraton, M.-I. (1999) FTIR Surface Spectrometries of Nanosized Particles, in H.S. Nalwa (ed.), *Handbook of Nanostructured Materials and Nanotechnology*, Academic Press, San Diego, pp. 89-153.
17. Harrick, N.J. (1967) *Internal Reflection Spectroscopy*, Interscience, Wiley, New York (2nd printing by Harrick Scientific Corp., Ossining, N.Y., 1979).
18. Gibson, A.F. (1958) Infrared and Microwave Modulation using Free Carriers in Semiconductors, *J. Scientific Instruments* 35, 273-278.
19. Chabal, Y.J. (1988) Surface Infrared Spectroscopy, *Surf. Sc. Reports* 8, 211-357.
20. Harrick, N.J. (1962) Optical Spectrum of the Semiconductor Surface States from Frustrated Total Internal Reflection, *Phys. Rev.* 125(4), 1165-1170.
21. Morrison, S.R. (1990) *The Chemical Physics of Surfaces*, Plenum Press, New York.
22. Henrich, V.E. and Cox, P.A. (1994) *The Surface Science of Metal Oxides*, Cambridge University Press, Cambridge.
23. Clifford, P.K. (1981) *Mechanisms of Gas Detection by Metal Oxide Surfaces*, Ph.D. Thesis, Carnegie Mellon Univ., Pittsburg.
24. Harrison, P.G. and Willett, M.J. (1988) The Mechanism of Operation of Tin(IV) Oxide Carbon Monoxide Sensors, *Nature* 332(6162), 337-339.
25. Willett, M.J. (1991) Spectroscopy of Surface Reactions, in Mosley, P.T., Norris, J.W.O., and Williams D.E. (eds.), *Techniques and Mechanisms in Gas Sensing*, Adams Hilger, Bristol, pp. 61-107.
26. Riehemann, W. (1998) Synthesis of Nanoscaled Powders by Laser-Evaporation of Materials, in K.E. Gonsalves, M.-I. Baraton *et al.* (eds.), *MRS Symp. Proc. Vol. 501*, MRS Publisher, Warrendale (PA), pp. 3-13.
27. Thornton, E.W. and Harrison, P.G. (1975) Tin Oxide Surfaces. I. Surface Hydroxyl Groups and the Chemisorption of Carbon Dioxide and Carbon Monoxide on Tin(IV) Oxide, *J. Chem. Soc., Faraday Trans. I* 71, 461-472.
28. Baraton, M.-I., unpublished results.

440

29. Baraton M.-I. and Merhari, L. (2001) Determination of the Gas Sensing Potentiality of Nanosized Powders by FTIR Spectrometry, *Scripta Mater.* **44**, 1643-1648.
30. Williams, G. and Coles G.S.V. (1999) SMOGLESS Final Report (contract N°: BRPR-CT95-0002), unpublished results.
31. Morrison, S.R. (1994) Chemical Sensors, in S.M. Sze (ed.), *Semiconductors Sensors*, John Wiley & Sons, New York, pp. 383-413.
32. Hertl, W. and Hair, M.L. (1971) Reaction of Hexamethyldisilazane with Silica, *J. Phys. Chem.* **75**(14), 2181-2185.
33. Chancel, F., Tribout, J., and Baraton, M.-I. (1997) Modification of the Surface Properties of a Titania Nanopowder by Grafting: A Fourier Transform Infrared Analysis, *Key Eng. Mater.* **136**, 236-239.
34. Baraton, M.-I., Chancel, F., and Merhari, L. (1997) In Situ Determination of the Grafting Sites on Nanosized Ceramic Powders by FTIR Spectrometry, *Nanostructured Materials* **9**, 319-322.
35. Primet, M., Pichat, P., and Mathieu, M.-V. (1971) Infrared Study of the Surface of Titanium Dioxides. I. Hydroxyl Groups, *J. Phys. Chem.* **75**(9), 1216-1220.
36. Ho, S.-W. (1996) Surface Hydroxyls and Chemisorbed Hydrogen on Titania and Titania Supported Cobalt, *J. Chinese Chem. Soc.* **43**, 155-163.
37. Morterra, C. (1988) An Infrared Spectroscopic Study of Anatase Properties, *J. Chem. Soc., Faraday Trans. I* **84**(5), 1617-1637.
38. Morrow, B.A. (1990) Surface Groups on Oxides, in J.L.G. Fierro (ed.), *Spectroscopic Characterization of Heterogeneous Catalysis* (Part A), Elsevier, Amsterdam, pp. A161-A224.
39. Busca, G., Saussey, H., Saur, O., Lavalley, J.-C., and Lorenzelli, V. (1985) FT-IR Characterization of the Surface Acidity of Different Titanium Dioxide Anatase Preparations, *Applied Catal.* **14**, 245-260.
40. Tsyganenko, A.A. and Filimonov, V.N. (1972) Infrared Spectra of Surface Hydroxyl Groups and Crystalline Structure of Oxides, *Spectr. Letters* **5**(12), 477-487.
41. Primet, M., Pichat, P., and Mathieu, M.-V. (1968) Etude par Spectrométrie Infrarouge des Groupes Hydroxyles de l'Anatase et du Rutile, *C. R. Acad. Sc. Paris* **267B**, 799-802.
42. Yates, D.J.C. (1961) Infrared Studies of the Surface Hydroxyl Groups on Titanium Dioxide, and of the Chemisorption of Carbon Monoxide and Carbon Dioxide, *J. Phys. Chem.* **65**, 746-753.
43. Primet, M., Pichat, P., and Mathieu, M.-V. (1971) Infrared Study of the Surface of Titanium Dioxides. II. Acidic and Basic Properties, *J. Phys. Chem.* **75**(9), 1221-1226.
44. Baraton M.-I., Merhari L., Chancel F., and Tribout J. (1997) Chemical characterization by FT-IR spectrometry and modification of the very first atomic layer of a TiO_2 nanosized powder, *MRS Symp. Proc. Vol. 448*, MRS Publisher, Warrendale, pp. 81-86.
45. Baraton M.-I. (2002) Surface Functionalization of ceramic Nanoparticles: Applications to Ion-Sensing and Gas-Sensing Devices, *MRS Symp. Proc. Vol. 705*, MRS Publisher, Warrendale, pp. 159-170.

SPECTROSCOPIC CHARACTERIZATION OF NANOCRYSTALLINE V/Ce OXIDES FOR NOVEL COUNTER ELECTRODES

Z. CRNJAK OREL[1] and A. TURKOVIĆ[2]
[1]National Institute of Chemistry, Hajdrihova 19, P.O Box 3430, SI-1001, Ljubljana, Slovenia;
[2]Rudjer Bošković Institute, P.O. Box 118, Bijenička 54, HR-10002 Zagreb, Croatia

1. Introduction

The preparation of vanadium oxide and new mixed vanadium cerium (V/Ce) oxides thin films as intercalation compound for lithium ions started from aqueous inorganic precursors. Thin films of pure V and mixed (V/Ce) oxides at 78, 55, 38 and 32 at% of V were obtained via sol-gel process by dip-coated method which is a well known synthetic route for preparation of various oxide and mixed oxides materials. These films were prepared for counter electrode in electrochromic devices. All films were investigated by cyclic voltammetry CV, UV-VIS technique, FT-IR and Raman spectroscopy, atomic force microscopy (AFM), grazing-incidence X-ray reflectivity (GIXR), and grazing-incidence small-angle X-ray scattering (GISAXS) and grazing-incidence wide angle diffraction (GIWAXD). Improved electrochemical properties and cycling behaviour of vanadium oxide films were obtained after the addition of CeO_2. The intercalation of Li^+ ions in V/Ce films was followed by FT-IR spectroscopy in combinations with CV measurements. The transmittance did not change significantly after cycling. The fundamental absorption edge shifts after the intercalation of Li^+ ions. The best intercalation properties and cycling durability were obtained for films prepared at 55 and 38 at% of V. The amorphous grains between 50-100 nm were observed on the films surface at 55 at% of V, with AFM. No grain structure was observed at 38 at% of V but after the cycling increase the surface roughness and grain with grains size of 100 nm were observed. Layered structure in all V/Ce oxides was revealed by GIXR method. The average grain radius <R>, obtained by GISAXS, was correlated within layer thickness. Porosity of samples was obtained by GISAXS. The structure of V/Ce oxides was studied using IR and Raman spectroscopies.

Novel nanophase materials require new methods of characterisation. GISAXS, GIXR and GIWAXD, performed at synchrotron ELETTRA, are such advanced methods in use. In this paper, GISAXS method is supported by all above mentioned optical and electrical methods of measurements.

Observed sample which consists of nanometer grains, or have variations in electron density, will scatter incident electromagnetic wave at small angles. For x-rays, scattered intensity is measured as a function of scattering angle Θ. Such

T. Tsakalakos et al. (eds.) pgs. 441 - 447
Nanostructures: Synthesis, Functional Properties and Applications;
© 2003 Kluwer Academic Publishers.

experimental method is called SAXS (small-angle x-ray scattering). Wave vector is given by equation (1).

$$q = \frac{4\pi}{\lambda}\left[\sin\frac{\Theta}{2}\right]$$

(1)

where λ is wave vector of probe.

For crystalline materials, peaks of scattered intensity occur at such scattering angles, that wave vector obeys relation:

$$q = \frac{2\pi}{l}, \text{ - } BRAGGS \text{ } scattering/diffraction$$

(2)

l = distance between atomic planes

For nanocrystalline grains, which have well defined long range atomic ordering, nanometer grains are randomly oriented each to other, and Bragg's peaks will be known as well known peaks of polycrystalline material with significant broadening of diffraction lines.

2. Experimental

V/Ce mixed oxides on glass substrate were prepared by sol-gel dip-coating procedure [1,2]. Data that we obtain from scattering curve depend about the geometry of experimental set-up. Standard way of measurements is with perpendicular transmission of x-rays through the sample.

a) *incoming beam at right angle* – transmittance occurs if beam income at right angle to the surface of the sample. Incoming rays pass through the layer and substrate and scatter at the other side of sample. Substrate has to be as thin as possible and not too strong absorbant of x-rays and the path through the layer depends about the thickness of the layer.

Next relation represents outcoming intensity I, of röentgen ray which passes through material of thickness d:

$$I = I_0 e^{-\mu d}$$

(3)

where are;

μ = linear coefficient of absorption

d = thickness of material

I_0 = intensity of incoming beam

a) *incoming beam at small grazing-incidence angle* Beam incoming to sample is closing a small angle with its surface. With the change of beams incident angle, we choose part of sample that is irradiated and scatters the irradiation. We control the penetration depth of x-rays by changing the incidence angle and substrate background is relatively small. Our samples of V/Ce oxides were measured in geometry b) - incoming beam at small grazing-incidence angle. Such experimental method is called GISAXS (grazing-incidence small-angle x-ray scattering).

b) Experiment of scattering x-rays at small angle was performed at Austrian SAXS beamline of synchrotron ELETTRA in Trieste [4,5]. Wavelength of x-rays was λ=1,55 Å, and photon energy of x-rays equals KeV.

3. Results

3.1. METHODS OF ANALYSING DATA OBTAINED BY SCATTERING OF X-RAYS AT SMALL ANGLE

At synchrotron light source, by analysing scattering curves; if scattered intensities are big enough in a wide-angle range we can use Guinier approximation [5] and Porod's law [6,7,8].

• *Guinier approximation* enables us to calculate radius of gyration, from which we can calculate particle/grain size.

• Guinier approximation of starting part of the scattering curve is given by following relation:

$$I(q) = I_0 e^{-\frac{q^2 R^2}{3}} \qquad (4)$$

where are;

a) R = radius of gyration, the only parameter.

b) I_0 = intensity of incoming beam, constant in scattering function

Equation represents very good approximation in finite angle area, especially for spherical particles *Porods Law:*

• Porods approximation enables us to calculate specific surface S.

• Porod's law is described by following relation:

$$I(q) = (\Delta\rho)^2 \frac{2\pi}{q^4} S = K q^{-4} \qquad (5)$$

where are:

K Porod's constant

$(\Delta\rho)$ difference between electron density of particles and medium.

Porod's law states that scattering intensity for big wave vectors q is proportional to q^{-4}. It is valid not only for one particle but also for close packed systems. So, following is valid:

If there is a jump in electron density of particle, no matter if it is big or small, inside the particle or at its edge, exponent n=4, i.e. condition of Porod's law is fulfilled, which is valid also for homogenous particles.

444

Constant value of Porod's product $q^4 \langle I \rangle$ is proportional to surface of sphere at which border jump occurs and to depth of this jump. This is to consider at application of Porod 's law for determining of specific surface of nanostructured material.

3.2. V_2O_5 AND MIXED V/CE OXIDES

Vanadium oxide, such as V_2O_5, has been extensively studied because it tends to form a lamellar structure that allows the intercalation/de-intercalation of different ions between its layers. The use of V_2O_5 is closely connected with a way of preparing material for its final usage. It can be used as a catalyst, in electrochromic devices, in an advanced electrochemical cell concept, especially in lithium batteries. Our measurements (9) have revealed them be the nanophase materials. Nanostructured properties of these materials pose them as candidates for nanophase solar cell electrode in analogy with previously used TiO_2 and CeO_2 and as intercalation electrodes with a solid polymeric electrolyte [10-12].

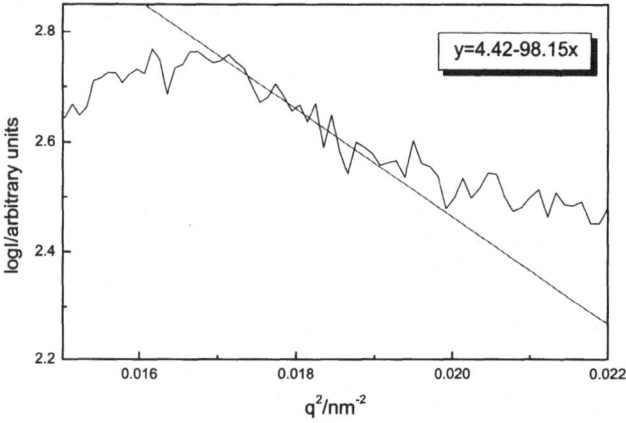

Figure 1. Linear fit to Log$I$=f(q^2) for SAXS at grazing angle 0.6^0 for V/Ce oxide at 55 of V.

Figure 1. represents Guinier plot: dependence of logarithm of intensity about square of wave vector i.e. angle Θ. From the coefficient of direction of line, k =109.96 we can determine radius of gyration <R_0>, and average diameter of grains, <D> = $2(5/3)^{1/2}$<R> = 10.68 nm.

Vanadium oxide, V_2O_5 and V/Ce oxides on glass substrate were prepared by sol-gel dip-coating process. Average grain radius <R> obtained by GISAXS method for pure V_2O_5 is (7.49±1.06) nm [9]. Average grain radius <R> of mixed oxides depends of atomic percent of V in sample. Our example of analyse of results of measurements performed at synchrotron ELETTRA Trieste at Austrian beamline SAXS, is by graphical method of calculation of grain sizes and specific surface, which determines porosity of material for V/Ce oxide with 55 at % of vanadium at grazing angle of 0.6^0. Average grain radius is <R> =(5.34±1.71) nm, which is different than value for pure V_2O_5. This material with 55 % of V shows one of the largest capacities for intercalation of Li ions regarding other mixed oxides [2].

Figure 2. shows dependence of logarithm of intensity I, about the logarithm of wave vector i.e. logarithm of angle Θ. 2Θ, is scattering angle in degrees proportional to wave vector. At larger angles, curve equals to line with coefficient of direction <-4 which means that the condition for application of Porod´s approximation is not quite fulfilled. The value of -2.3 indicates that some other interpretation of data should be attempted for $CeVO_4$ (55 at % V in V/Ce oxide). We had performed calculation [9] of fractal dimensions by formalism stated by Martin & Hurd [13] and obtained radius of pores as 3.4 nm.

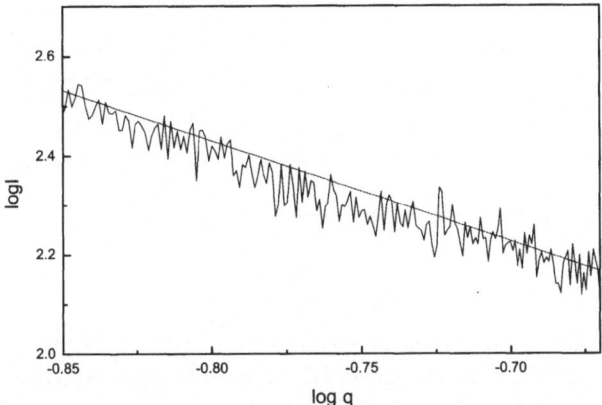

Figure 2. Representation of the data as log$I$=f(log$q$) for V/Ce oxide at 55% of V as a test whether we can apply the Porod formula.

Figure. 3 represents determination of limes of function [q4 I (q)] for values of wave vector q > 1.5 nm^{-1}.

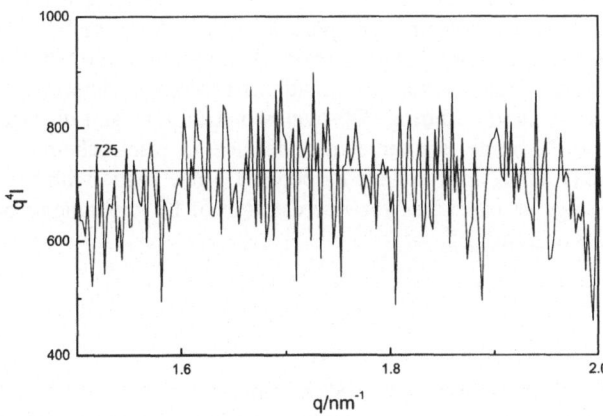

Figure 3. Determination of limiting value for function $[q^4I(q)]$ to higher values of q for V/Ce oxide at 55 % of V.

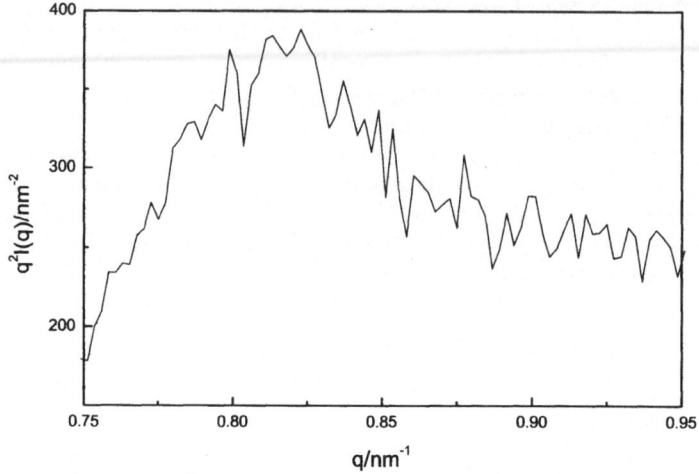

Figure 4. Calculation of invariant Q according to the equation [6] for V/Ce oxide at 55% of V.

Invariant Q is calculated according to equation:

[6]

$$Q = \int_{q_m}^{q_M} q^2 I(q) dq$$

Area A, under curve dependence of function $[q^2 I(q)]$ of q, is determined till value of vector $q_M = 1.05$ nm^{-1} and is equal to the integral in above equation.

Specific surface according Mittelbach-*Porod* formula:

$$S_s = \frac{S}{V} = \pi \frac{\lim_{q \to \infty} [q^4 I(q)]}{Q}$$

[7]

For sample of V/Ce oxides with 55 at % of V, is 0.65 nm^{-1}, which is somewhat bigger value than value of 0.40 nm^{-1} for pure V_2O_5 [9]. So, pure V_2O_5 for grazing angle 0.6 ^{0} has bigger grains and lower porosity than mixed oxide V/Ce with 55 at. % of V. Addition of cerium dioxide to vanadium pentoxide increased stability of layers of vanadium pentoxide during 700 cycles of charging and discharging with constant current. V/Ce oxide is intercalant electrode in electrochemical cell with lithium electrolyte (LiClO$_4$ in propylene carbonate). V/Ce oxide with 55 at % of V shows one of the biggest capacities for intercalation of Li ions among other V/Ce oxides, which is 20-mC/cm^2 [2].

4. Conclusion

In conclusion, this study shows that GISAXS measurements at synchrotron light source ELETTRA can be applied for determination of grain sizes of mixed, nanostructured *V/Ce* oxides. In comparison with other methods, as microscopy,

Raman spectroscopy and roentgen diffraction, GISAXS is suitable for analysis of porous nature of nanstructured materials. Beside grain sizes. porosity is a key parameter for application in efficient electrochemical and solar cells.

5. Acknowledgment

This work was supported by the Ministry of Education, Science and Sport of Republic of Slovenia (P-0512) and Ministry of Science and Technology of Croatia.

6. References

1. Crnjak Orel, Z. (1999) New counter electrode prepared as vanadium and V/Ce oxide films: preparation and characterisation, *Solid State Ionics* **116**, 105-116.
2. Crnjak Orel Z. and Mušević, I. (1999) Characterisation of vanadium oxide and new V/Ce oxide films prepared by sol-gel process, *NanoStructured Materials* **12**, 399-404.
3. Amenitsch, H., Bernstorff, S. and Laggner, P. (1995) High-flux beamline for small-angle X-ray scattering at ELETTRA, *Rev. Sci. Instrum,* **66**, 1624-1626.
4. Amenitsch, H., Rappolt, M., Kriechbaum, M, Mio, H., Laggner, P. and Bernstorff, S. (1997) Small Angle X-ray Scattering - Beamline at ELETTRA: A New Powerful Station for structural Investigations with Synchrotron Radiation, *Book of abstracts: Sixth Croatian-Slovenian Crystallographic Meeting* **56**, Umag, Croatia, June 19-21.
5. Guinier, A..(1939) Théorie et Technique de la Radiocristallographie, *Ann. Phys.* Paris **12**, (1939) 161-237.
6. Porod, G. (1951) Determination of General Parameters by Small-Angle X-Ray Scattering, *Kolloid-Zeitschrift und Zeitschrift fuer Polymere* **124**, 83.
7. Porod, G. (1952) Determination of General Parameters by Small-Angle X-Ray Scattering, *Kolloid-Zeitschrift und Zeitschrift fuer Polymere* **125**, 108.
8. Porod, G. (1953) Determination of General Parameters by Small-Angle X-Ray Scattering, *Kolloid-Zeitschrift und Zeitschrift fuer Polymere* **133**, 51.
9. Turković, A., Crnjak Orel, Z. and Dubček, P.(2001) Grazing-incidence small-angle X-ray scattering on nanosized vanadium oxide and V/Ce oxide films, Materials Science & Engineering **B79,** 11-15.
10. Turković, A., Drašner, A., Šokčević, D., Ritala, M., Asikainen, T., and Leskelä, M.(1995) Comparison Between CVD and ALE Produced TiO_2 Cathodes in $Zn/(PEO)_4ZnCl_2/TiO_2,SnO_2$ or ITO Galvanic Cells, Journal de Physique IV **5**, C5-1133-C5-1139.
11. Turković, A., and Crnjak Orel, Z. (1996) Electrical and Optical Properties of Thin Films $Zn/(PEO)_4ZnCl_2/CeO_2$, or $CeO_2/SnO_2(17\%)$,ITO Galvanic Cells, Solid State Ionics **89**, (3-4), 255-261.
12. Turković, A., and Crnjak Orel, Z. (1997) Dye-sensitized solar cell with CeO_2 and mixed CeO_2/SnO_2 photo anodes, Solar Energy Materials and Solar Cells **45/3**, 275-281.
13. Martin, J.E., Hurd, A.J. (1987) J.Appl.Cryst.**20**, 61-78.

CHARACTERIZATION OF NANOCRYSTALLINE ALLOYS BY MÖSSBAUER EFFECT TECHNIQUES

MARCEL MIGLIERINI
Department of Nuclear Physics and Technology, Slovak University of Technology, Ilkovičova 3, 812 19 Bratislava, Slovakia

Abstract. The use of a non-destructive nuclear-physical method, namely ^{57}Fe Mössbauer spectroscopy, is discussed for the investigation of magnetic and structural arrangement of Fe-based nanocrystalline alloys. Transmission Mössbauer spectroscopy (TMS) as well as conversion electron Mössbauer spectroscopy (CEMS) are reviewed using FINEMET- and NANOPERM-type nanocrystalline alloys as examples. They consist of nanocrystalline grains embedded within a residual amorphous matrix thus exhibiting a two-phase magnetic behaviour. Hyperfine field distributions derived from Mössbauer spectra provide information about the structure and magnetic states of atoms located in different structural positions. Prior to this, basic features of Mössbauer spectra are briefly summarised. Influence of composition, content of nanograins, and interactions among them are demonstrated as a function of annealing temperature and measuring temperature for bulk (TMS) and surface (CEMS) of the investigated nanocrystalline alloys.

1 Nanocrystalline Alloys

Nanocrystalline alloys result from an annealing of amorphous precursors under controlled atmosphere to prevent oxidation. Since their discovery by Yoshizawa *et al.* [1], the nanocrystalline alloys were widely studied. These new materials are very attractive because they exhibit unusual two-phase structural and magnetic behaviour and are promising for industrial applications due to their excellent soft magnetic properties combining high saturation magnetic flux density with high permeability [2]. The magnetic coupling between crystalline grains via the intergranular phase, the suppression of the effective magnetic anisotropy, and the decrease of the magnetostriction are sources of these soft magnetic properties [3].

Structural properties of nanocrystalline alloys are studied by X-ray diffraction, electron diffraction, and transmission electron microscopy along with differential scanning calorimetry [4]. These methods used to determine the average particle size and the distribution of the particle size or to identify the crystalline phases. Macroscopic magnetic parameters are derived from magnetic measurements. Mössbauer spectroscopy is a non-destructive nuclear-physical method which is able to provide information on structural arrangement and magnetic states of the resonant atoms at the

T. Tsakalakos et al. (eds.) pgs 449 – 462
Nanostructures: Synthesis, Functional Properties and Applications;
© 2003 Kluwer Academic Publishers.

same time. Because iron is one of the principal elements constituting nanocrystalline alloys, it is not surprising that Mössbauer spectroscopy plays a crucial role in the investigation of these materials.

2 Basic Features of Mössbauer Spectroscopy

Mössbauer spectrometry, that is complementary to diffraction techniques, is often proving decisive in the area of material science [5]. Indeed, its local behaviour enables to be an atomic scale sensitive tool while its time of measurement allows investigation of relaxation phenomena and dynamic effects. It is important to emphasise that ^{57}Fe is a valuable isotope that can be encountered in different kinds of materials including nanocrystalline alloys.

Recoilless nuclear gamma resonance - the Mössbauer effect [6, 7] - takes place between the same type of nuclei located in a source and an absorber (sample). Radiation from the source is modulated by a Doppler effect, which eliminates differences between energy levels of the source and the absorber. Mössbauer spectrum is a plot of intensity of gamma radiation accepted by the detector versus velocity. Differences in energy levels provide information on the character of nearest neighbourhoods of resonant atoms both from the point of view of structure and hyperfine interactions.

Spectral line positions represent intensities of hyperfine magnetic fields or electric field tensors, and/or electron densities at the nuclei. Linewidths give information on the structure (defects, order), and line intensities (line areas) correspond to the orientation of the sample's magnetization and/or determine the texture. Area under all lines of an individual subspectrum is directly proportional to the relative amount of resonant atoms located in that particular structural site.

In *transmission geometry,* we measure the attenuation of intensity of the source radiation due to resonance effects at the absorber (sample). Gamma photons are propagating through the sample completely, which puts forward some restrictions as far as geometrical dimensions of the absorber are concerned. Thick samples can be measured in the so-called *scattering geometry.* The detector is located out of the direction of a collimated primary beam from the source and only scattered radiation is detected. Counting rate of such *emission Mössbauer spectra* is rather low, which extends the time of experiment.

The de-excitation of the absorber (transition from the excited to the ground state) is accomplished via release of: (i) gamma photons, (ii) X-rays, and (iii) conversion electrons. X-rays and conversion electrons have smaller energy than the original gamma photon and that is why their penetration depth is short. On the other hand, we have an excellent tool for investigations of surface layers as demonstrated in Fig. 1.

2.1 MÖSSBAUER SPECTROSCOPY IN ORDERED (CRYSTALLINE) SYSTEMS

Spectral parameters of crystalline materials are discrete values and similar as fingerprints unambiguously identify different structural positions of the resonant atoms. Other, *i.e.* non-resonant atoms present in the lattice affect the nearest neighbourhoods of the resonant atoms, too. In such a way, the resonant atoms act as probes even in materials, which do not contain the respective resonant nuclei in their original composition. Figure 2 introduces model Mössbauer spectra showing electric quadrupole (non-magnetic material), and magnetic dipole interactions for ordered (crystalline) and disordered (amorphous) specimens together with the respective hyperfine parameters.

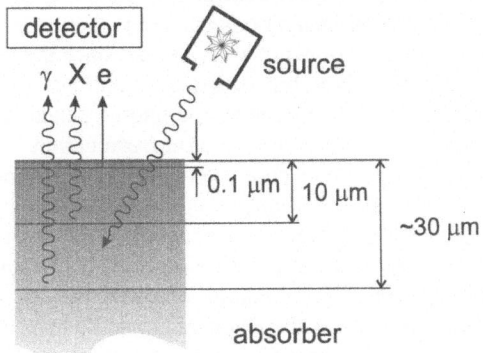

Figure 1 Escape depths of de-excitation radiation in Mössbauer effect experiments.

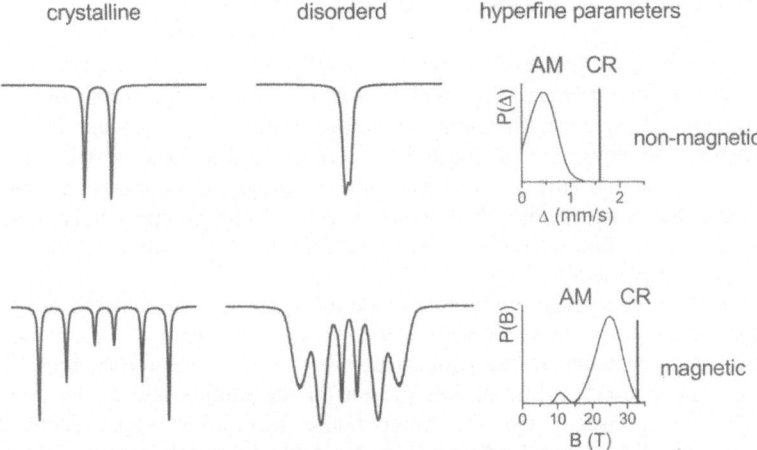

Figure 2 Model Mössbauer spectra of crystalline (ordered) and disordered systems with corresponding hyperfine parameters in non-magnetic and magnetic states.

2.2 MÖSSBAUER SPECTROSCOPY IN DISORDERED SYSTEMS

Disordered structures such as amorphous, metallic glasses or very small particles cannot be in general analysed by optical and/or diffraction techniques because of broad lines and halo effects. Mössbauer spectroscopy, on the other hand, provides signal from *any* structural position of resonant atoms including highly disordered states. Better structural identification than in classical diffraction methods is achieved via remarkable difference in spectral line widths between crystalline structure characterised by a long-range order (periodicity of the crystalline lattice), and an amorphous one. Although the resulting spectral lines are broad and often overlapped, they can be further analysed. Consequently, distributions of respective hyperfine parameters are provided as demonstrated on the right-hand side of Fig. 2.

The line broadening is a consequence of structurally non-equivalent nearest neighbourhoods of resonant atoms. Topological and/or chemical order is maintained only for a short distance over several nearest co-ordination shells - the so-called *short-range order*. Evaluation of such spectra is not a trivial task since the respective spectral parameters are no more discrete values but depict distributions in their values. The individual hyperfine interactions often influence one another, thus leading to spectral line asymmetry that further complicates the spectrum evaluation. Since only a few methods yield distributions of hyperfine parameters one surely benefits from rather laborious fitting procedures. Indeed, Mössbauer spectroscopy plays an unmatched role in the study of microstructure and magnetic arrangement of amorphous materials, metallic glasses, quasicrystals, and nanocrystalline materials. It is a suitable tool for the investigation of disordered structures on a microscopic (atomic and/or nuclear) level [8]. Scanning the nearest surroundings of the resonant atoms it is possible to answer the questions *where*, *when* and *what* ordering is found in the samples studied [9].

3 Mössbauer Spectroscopy and Nanocrystalline Alloys

Nanocrystalline alloys prepared by partial crystallization of metallic glasses essentially consist of crystalline grains of the order of 5-20 nm embedded in a residual amorphous phase. As a consequence, different structural positions of resonant atoms can be distinguished. Namely, atoms located in: (i) crystalline grains, (ii) the retained amorphous phase, and (iii) between the bulk of the nanocrystals and the amorphous phase. Thus, the corresponding Mössbauer spectra depict complex behaviour. Due to obvious variance in line widths associated with different structures, however, the latter can be readily classified.

Classical magnetic measurements provide information about macroscopic magnetic properties (coercivity, magnetization, *etc*). The data obtained from macroscopic magnetic measurements reflect simultaneously the contributions both from the crystalline and non-crystalline phases present in the sample and in this respect they cannot be distinguished. On the other hand, Mössbauer spectroscopy provides information on all structurally different components separately giving a possibility to study magnetic and/or structural arrangements.

Figure 2 illustrates how the structural distinctions between ordered (crystalline) and disordered (amorphous, interface) arrangements of resonant atoms are reflected in Mössbauer spectra. Apart from diffraction techniques (*e.g.*, X-ray diffraction), the Mössbauer spectra provide also simultaneous evidence on magnetic states of the atoms regardless their structural origin.

There are two main families of nanocrystalline alloys – FINEMET [1] with a general composition of FeCuNbSiB and NANOPERM [10], *i.e.* FeZr(Cu)B alloys. During crystallization of the FINEMET-type alloys, Fe-Si ultra-fine grains are created with Fe atoms occupying different crystallographic sites. Depending on the Si content, the FeSi phase exhibits a bcc and/or a DO_3 structure. Alternatively, Fe-Si solid solution is formed. On the other hand, NANOPERM-type nanocrystalline alloys show evidence of only one crystalline phase – bcc Fe.

3.1 FINEMET-TYPE NANOCRYSTALLINE ALLOYS

3.1.1 *Evaluation of Mössbauer Spectra*

Mössbauer spectra of Si-containing FINEMET-type nanocrystalline alloys are complex, showing multiple narrow lines superimposed on a broadened feature. The former represents structurally different crystallographic sites of the particular crystalline phase whereas the latter is ascribed to a residual amorphous phase. A typical Mössbauer spectrum of $Fe_{73.5}Cu_1Nb_3Si_{13.5}B_{16}$ nanocrystalline alloys is shown in Fig. 3.

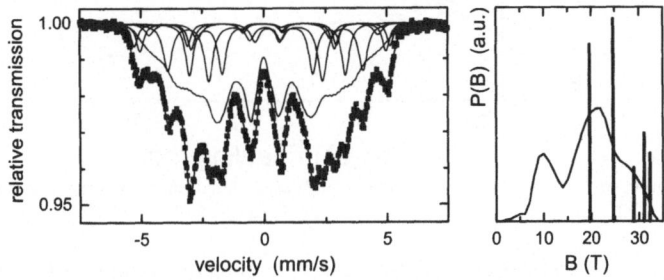

Figure 3 Room temperature Mössbauer spectrum (left) of the $Fe_{73.5}Cu_1Nb_3Si_{13.5}B_{16}$ nanocrystalline FINEMET with spectral components and distribution of hyperfine fields (right). Vertical lines represent hyperfine fields of the crystalline components.

As discussed in [11-13], the problem with the evaluation of FINEMET-type Mössbauer spectra is twofold: (i) number of lines representing the crystalline phase which usually varies between 4 and 7, and (ii) refinement of the amorphous phase. As for the latter, in some procedures a single sextet of broad Lorentz [14, 15], pseudo-Lorentz [16], or Voigt [17] profiles are applied. However, because of disordered nature of the intergranular remainder, distributions of hyperfine fields should be employed. Again, several approaches were proposed which make use of constrained profiles of Gaussian [18], double Gaussian asymmetrical [19], and "skew Gaussian" [20] functions or unconstrained profiles resulting from distributions of discrete hyperfine fields of Lorentzian sextets [11, 21-23]. Recently, hyperfine field distributions with a bimodal behaviour were described by Borrego *et al.* [24, 25] that consist of two Gaussian components.

3.1.2 *Examples of Analyses*

Mössbauer spectrometry in the FINEMET-type nanocrystalline alloys is usually used for the identification of crystalline phases containing Fe which are created in the process of thermal annealing [14-16, 19], for the study of the kinetics of crystallization [11, 18], or to investigate the effects of the Nb substitutions by other elements [23-26]. Mössbauer spectroscopy can be used to follow the temperature-dependent magnetization of the amorphous as well as the nanocrystalline grains [17].

Distributions of hyperfine fields were studied and conclusions regarding the

454

magnetic and structural arrangement of the retained amorphous phase were done for $Fe_{73.5-x}Cu_1Nb_{3+x}Si_{13.5}B_9$ nanocrystalline alloys with respect to varying composition [11]. Quantification of the decrease of the residual amorphous phase with the time of annealing is shown in Fig. 4. Attempts to characterise the amorphous rest from the point of view of structural and magnetic properties are presented in [27, 28].

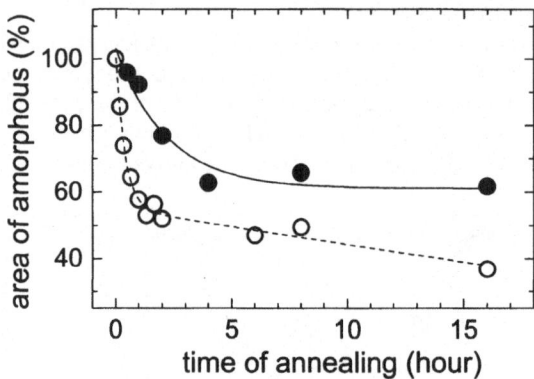

Figure 4 Evolution of the relative fraction of the amorphous residual phase as a function of the annealing time for $Fe_{73.5-x}Cu_1Nb_{3+x}Si_{13.5}B_9$ nanocrystalline alloy: $x = 0$ (O) and 1.5 (●).

Formation of crystalline phases after complete crystallization on the surface and in the bulk of the studied $Fe_{73.5}Cu_1Nb_3Si_{13.5}B_{16}$ alloy was carried out by Rixecker *et al.* [16] employing simultaneous measurement of gamma rays in transmission geometry (TMS), X-rays in scattering geometry (CXMS), and conversion electrons (CEMS) in addition to X-ray diffraction, scanning and transmission electron microscopy as well as microprobe analysis. Another way of precise identification of the formation of crystalline phases is the use of radio-frequency fields as demonstrated by Graf *et al.* [29] on a $Fe_{73.5}Cu_1Nb_3Si_{13.5}B_{16}$ FINEMET.

3.2 NANOPERM-TYPE NANOCRYSTALLINE ALLOYS

3.2.1 *Fitting Models of Mössbauer Spectra*
Only a bcc-Fe phase is present during the first crystallization of NANOPERM-type (*i.e.*, without Si) nanocrystalline alloys. Because iron is principally suited for Mössbauer spectrometry more detail insight into the local atomic arrangement than in case of FINEMET-type alloys was envisaged. Taking into consideration the nanosized dimensions of the crystallites, one can expect that the atoms localised on the surface of nanograins should demonstrate signs of the symmetry breaking, *i.e.* lack of ordering, by deviations in their respective spectral parameters when compared to those positioned in the bulk of the nanograins. Because of the nanometer size of the crystalline grains, the contribution of surface atoms to the resulting spectrum cannot be neglected and, consequently, the corresponding spectral component is anticipated. Its contribution, however, might differ because of such parameters as the size of the nanograins, their amount, shape, and potential clustering which are, in turn, governed by the thermal treatment applied as well as chemical composition of the as-quenched precursor. Indeed, along with a broadened spectral component, which characterises disordered

positions of resonant atoms in the retained amorphous matrix, and narrow six lines of bcc-Fe nanocrystals, a third spectral component is unambiguously detected. It presents broad lines positioned at the inner side of the narrow sextet. However, even though Mössbauer spectra of NANOPERM-type alloys are relatively simple as compared to those of FINEMETs the interpretation of the third spectral component is a point of controversy in a literature. This component is completely hidden in complex spectra of FINEMET alloys.

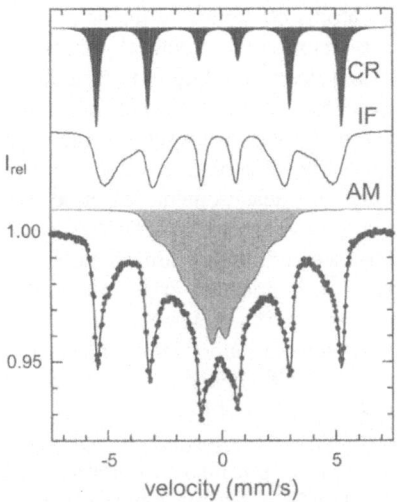

Figure 5 Mössbauer spectrum of a NANOPERM-type nanocrystalline alloy with its components representing the amorphous residual matrix (AM – grey), the crystalline phase (CR – black), and the third component (IF – white) positioned at the inner side of the narrow CR sextet.

As mentioned above, besides a broad spectral component of the amorphous matrix and a narrow one of the bcc-Fe nanocrystals, a third component must be used in order to obtain a satisfactory fit to the experimental data as demonstrated in Fig. 5. What is more important, its existence is justified from structural assumptions. Hyperfine field of this component is by about 10% smaller than that of the crystalline component. Today, its presence is not doubted and it is generally accepted. Nevertheless, the following fitting procedures are employed to refine the spectral parameters of NANOPERM alloys: (1) one continuous hyperfine field distribution (HFD) which extends over the whole range of hyperfine field values and comprises all three components [30, 31], (2) one HFD with a peak at around 30 T assigned to the amorphous-crystal interface and one sextet representing the bcc-Fe nanocrystals [32, 33], (3) one HFD and two sextets introduced by Suzuki *et al.* [34, 35] where one broadened sextet is ascribed to non-magnetic near neighbour atoms in the bcc lattice [36-41], (4) two HFDs and one sextet ascribing one independent HFD to the atoms located on the surface of crystalline nanograins and in their immediate vicinity – the so-called *interface zone* [42-45] and, the other distribution is eventually used as a distribution of quadrupole splitting and/or broadened doublet to emulate paramagnetic regions inside the amorphous residual phase [46, 47].

456

In addition to these "conventional" approaches, a completely different fitting strategy based on a large meaningful number of spectral lines was proposed by Chuev *et al.* [48] and applied in temperature-dependent studies of FeCuNbB alloys [49, 50] justifying the presence of the interface regions. The latter are thoroughly discussed also in the recent paper of Grenèche and Ślawska-Waniewska [51].

3.2.2 *Magnetic Structure and Microstructure of NANOPERM Nanocrystalline Alloys*

Applying Mössbauer spectroscopy, we can investigate the hyperfine interactions associated with all resonant atoms contained in the samples studied. It is important to note that in the case of a disordered structural arrangement, the information on particular atomic sites and their corresponding hyperfine interactions are in general not directly related to each other. They cannot be unambiguously correlated and the results from other techniques should be looked for to support the proposed interpretation of the Mössbauer spectra. Alternatively, Mössbauer effect measurements under different conditions (*e.g.*, temperature of measurement, external magnetic field) should be employed to reveal trends in the evolution of particular spectral parameters, which can be subsequently compared to those of similar (amorphous) systems [52].

The interface zone plays an important role in the magnetic behaviour of the nanocrystalline systems by transmitting ferromagnetic exchange coupling among adjacent crystalline grains through the amorphous matrix [53, 54]. This phenomenon can be precisely studied only in Mössbauer spectra with *one* crystalline site (bcc-Fe).

Because of relative simplicity of a FeM(Cu)B system, the Mössbauer spectrometry can be exhaustingly exploited and the information regarding not only the amorphous remainder and the crystalline phase but also the interface zone can be derived. Such information is provided (to the best of our knowledge) exclusively by Mössbauer spectrometry. Furthermore, an impact of the interface zone on the topology of hyperfine interactions in nanocrystalline alloys can be estimated [55].

Temperature of Annealing. By means of varying the temperature of annealing we are able to control the amount of nanocrystallites. Magnetic states of iron atoms are significantly affected by already small crystalline fraction. Originally paramagnetic samples show distributions of quadrupole splitting values and after some bcc-Fe nanocrystallites have emerged, magnetic interactions are observed in the amorphous residual phase. They are demonstrated by apparent broadening of the corresponding spectral component as documented in Fig. 6 and originate from ferromagnetic interactions among the grains which are penetrating into the amorphous residue [54]. Combined magnetic and Mössbauer effect studies are often employed to clarify the role of exchange coupling between the grains through the surrounding amorphous matrix [56]. Another possibility is to vary the composition of NANOPERM-type alloys [41, 57-59] and to perform the measurements at low temperatures [58, 60].

457

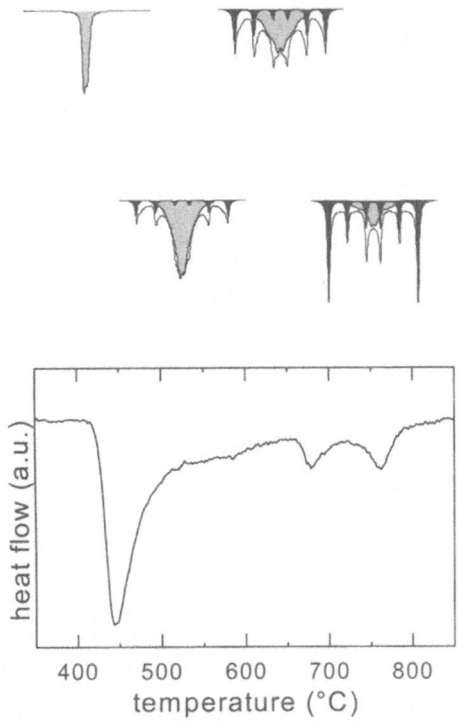

Figure 6 Mössbauer spectra of the $Fe_{80}Mo_7Cu_1B_{12}$ alloy (CR - black, AM - grey) taken at room temperature after the indicated heat treatment. The corresponding DSC curve is given below.

Temperature of Measurement. By varying temperature of measurement, it is possible to separate one from another the amorphous and interface contributions to the overall Mössbauer spectrum due to differences in their magnetic states [52, 61]. This both simplifies the fitting procedure, and opens a new approach to the investigation of hyperfine fields because both components are better resolved. When the measurement temperature increases and becomes higher than the expected Curie temperature of the residual amorphous phase, the hyperfine structure drastically changes: the broad line magnetic sextet tends to collapse to a paramagnetic doublet (Fig. 7). It is important to emphasise that the evolution of Mössbauer spectra strongly depends on the volumetric fraction of the crystalline phase, which governs the nature and the strength of intergrain magnetic interactions and of exchange magnetic interactions within the crystalline grains and the residual amorphous matrix. Again, magnetic measurements are often employed in addition to Mössbauer effect studies [36, 62]. Evolution of hyperfine magnetic fields of the crystalline component with temperature is investigated, too [41, 61-63].

458

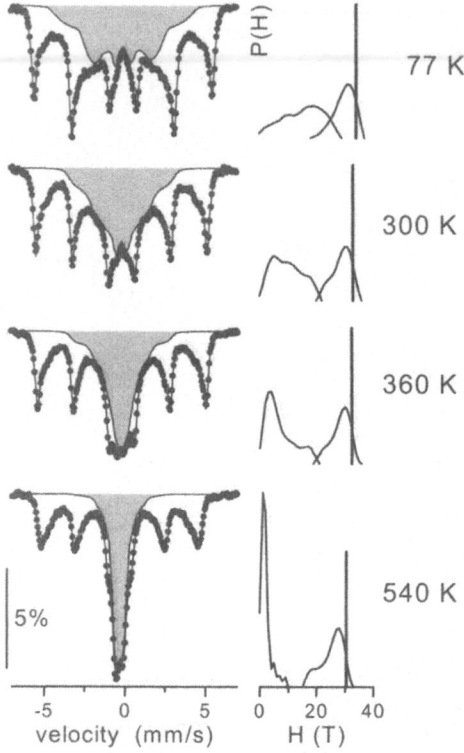

Figure 7 Mössbauer spectra (left, AM - grey) and hyperfine fields (right, thick vertical lines correspond to CR) of the $Fe_{80}Mo_7Cu_1B_{12}$ alloy annealed at 440°C/1h taken at the indicated temperatures.

Topography of hyperfine fields. During the crystallization Fe atoms segregate from the residual amorphous matrix into bcc-Fe crystalline grains. Consequently, chemical and/or topological short-range order of the amorphous matrix is changed which gives rise to regions depleted in Fe atoms. On the other hand, the nearest surroundings of these Fe atoms are enriched in other constituent elements. As a result, the magnetic moments of Fe atoms are lowered inside these regions and considerable fraction of Fe is located in non-magnetic sites. To allow a thorough view on the evolution of hyperfine interactions within the nanocrystalline material corresponding HFDs are derived from Mössbauer spectra.

Employing HFDs, we are able to distinguish among different structural arrangements in nanocrystalline alloys. At the same time, hyperfine interactions describe structural positions from the point of view of their magnetic states [52, 55].

Radio frequency field and surface studies. The unconventional radio-frequency (rf) Mössbauer technique, in which the rf collapse and rf sideband effects are employed, was used in the study of magnetic structure of Zr-containing FeMCuB nanocrystalline alloys. Kopcewicz *et al.* [64-66] have used rf fields which allowed them to reach conclusions regarding the size of the bcc–Fe grains as a function of annealing temperature. The rf Mössbauer experiments performed on $Fe_{93-x-y}Zr_7B_xCu_y$ alloys permit them to distinguish the soft nanocrystalline bcc-Fe phase from the magnetically harder microcrystalline α–Fe phase. The magnetic behaviour of the disordered spectral components is discussed in [64], where single-block HFD showed two distinct types of short-range order in the remaining amorphous phase toward higher temperatures of annealing.

Surface crystallization was studied by means of conversion electrons (CEMS) for FeZr(Cu)B [66-68] and for FeMCuB-type nanocrystalline alloys [69-71]. Grabias *et al.* [70, 71], employed also the radio-frequency Mössbauer technique. Both quantitative and qualitative distinctions between surface and bulk are apparent. The former are depicted by more rapid progress of the crystal formation on the surface in Fig. 8 whereas the latter are documented by deviations in the magnetic microstructure by the help of three-dimensional HFDs in Fig. 9.

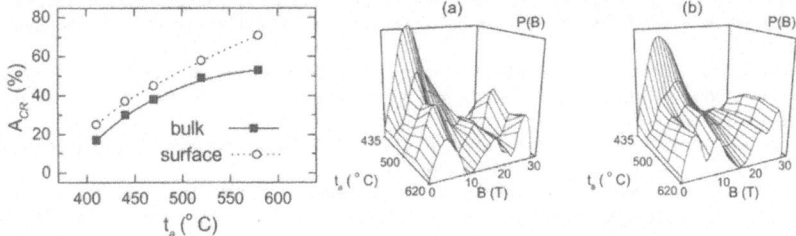

Figure 8 Content of nanocrystallites A_{CR} as a function of annealing temperature t_a for the surface and bulk of nanocrystalline corresponding $Fe_{80}Nb_7Cu_1B_{12}$ alloy.

Figure 9 Three-dimensional HFDs derived from (a) CEMS and (b) transmission Mössbauer spectra of nanocrystalline $Fe_{80}Nb_7Cu_1B_{12}$ alloys to surface and bulk amorphous matrix respectively.

4. Conclusions

The principal findings from Mössbauer effect investigations of FINEMET and NANOPERM nanocrystalline alloys can be summarised as follows: (i) crystallization starts on the surface at temperatures well below the temperature of primary crystallization determined from differential scanning calorimetry, (ii) significant relaxation is observed below the crystallization onset which leads to structural and/or magnetic rearrangement, (iii) amount of nanograins on the surface is higher than in the bulk, (iv) distributions of hyperfine fields point out differences between short range order of surface and bulk regions. Using the results from Mössbauer spectroscopy, the magnetic structure and/or microstructure of nanocrystalline alloys can be modelled.

5. Acknowledgement

This work was supported by the project SGA 1/8305/01.

460

6. References

1. Yoshizawa, Y., Oguma, S. and Yamauchi, K. (1988) New-Fe-based soft magnetic alloys composed of ultrafine grain structure, *J. Appl. Phys.* **64**, 6044-6046.

2. Hernando, A., Vázquez, M. and Páramo, D. (1998) Applications of amorphous and nanocrystalline magnetic materials as sensing elements, *Mater. Sci. Forum* **269-272**, 1033-1042.

3. Herzer, G. (1993) Nanocrystalline soft magnetic materials, *Phys. Scr.* **T49**, 307-314.

4. Miglierini, M., Kopcewicz, M., Idzikowski, B., Horváth, Z. E., Grabias, A., Škorvánek, I., Dłużewski, P. and Daróczi, Cs. S. (1999) Structure, hyperfine interactions and magnetic behavior of amorphous and nanocrystalline $Fe_{80}M_7B_{12}Cu_1$ (M = Mo, Nb, Ti) alloys, *J. Appl. Phys.* **85**, 1014-1025.

5. G. J. Long (ed.), (1984, 1987, 1989), in *Mössbauer Spectroscopy Applied to Inorganic Chemistry Vol 1, 2, and 3*, Plenum Press, New-York.
   G. J. Long and F Grandjean (eds.), (1993, 1996), in *Mössbauer Spectroscopy Applied to Magnetism and Materials Science Vol 1, 2*, Plenum Press, New-York.
   M. Miglierini and D. Petridis (eds.), (1999), in *Mössbauer Spectroscopy in Materials Science*, Kluwer Academic Publishers, Dordrecht.

6. Mössbauer, R. L. (1958) Kernresonanzfluoreszenz von Gammastrahlung in Ir^{191}, *Z. Phys.* **151**, 124-143.

7. Mössbauer, R. L. (1958) Kernresonanzeabsorption von Gammastrahlung in Ir^{191}, *Naturwissenschaften* **45**, 538-539.

8. Campbell, S. J., Aubertin, F. (1989) Evaluation of distributed hyperfine parameters, in G. J. Long and F Grandjean (eds.), *Mössbauer Spectroscopy Applied to Inorganic Chemistry Vol 3*, Plenum Press, New-York, pp. 183-242.

9. Gonser, U., Limbach, C. T. and Aubertin, F. (1988) Mössbauer spctroscopy in amorphous metals: failures and successes, *J. Non-Cryst. Solids* **106**, 395-398.

10. Suzuki, K., Kataoka, N., Inoue, A., Makino, A. and Masumoto, T. (1990) High saturation magnetization and soft magnetic properties of FeZrB alloys with ultrafine structure, *Mater. Trans. JIM* **31**, 743-746.

11. Miglierini, M. (1994) Mössbauer-effect study of the hyperfine field distributions in the residual amorphous phase of Fe-Cu-Nb-Si-B nanocrystalline alloys, *J. Phys.: Condens. Matter* **6**, 1431-1438.

12. Grenèche, J. M. (1997) Nanocrystalline iron-based alloys investigated by Mössbauer spectrometry, *Hyperfine Interactions* **110**, 81-91.

13. Miglierini, M. and Grenèche, J. M. (1999) Mössbauer spectrometry applied to iron-based nanocrystalline alloys II. Hyperfine fields of amorphous and interfacial regions, in M. Miglierini and D. Petridis (eds.), *Mössbauer Spectroscopy in Materials Science*, Kluwer Academic Publishers, Dordrecht, pp. 257-272.

14. Jiang, J., Aubertin, F., Gonser, U. and Hilzinger, H. R. (1991) Mössbauer spectroscopy and X-ray diffraction studies of the crystallization in the amorphous $Fe_{73.5}Cu_1Nb_3Si_{13.5}B_9$, *Z Metallk.* **82**, 698-702.

15. Hampel, G., Pundt, A. and Hesse, J. (1992) Crystallization of $Fe_{73.5}Cu_1Nb_3Si_{13.5}B_9$: structure and kinetics examined by x-ray diffraction and Mössbauer effect spectroscopy, *J. Phys.: Condens. Matter* **4**, 3195-3214.

16. Rixecker, G., Schaaf, P. and Gonser, U. (1992) Crystallization behaviour of amorphous $Fe_{73.5}Cu_1Nb_3Si_{13.5}B_9$, *J. Phys.: Condens. Matter* **4**, 10295-10310.

17. Gupta, A., Bhagat, N. and Principi, G. (1995) Mössbauer study of magnetic ineractions in nanocrystalline $Fe_{73.5}Cu_1Nb_3Si_{16.5}B_6$, *J. Phys.: Condens. Matter* **7**, 2237-2248.

18. Pradell, T., Clavaguera, N. Zhu, J. and Clavaguera-Mora, M. T. (1995) A Mössbauer study of the nanocrystallization process in $Fe_{73.5}CuNb_3Si_{17.5}B_5$ alloy, *J. Phys.: Condens. Matter* **7**, 4129-4143.

19. Zemčík, T. (1993) Phase analysis of amorphous and nanocrystalline FeCuNbSiB alloys by ^{57}Fe Mössbauer spectroscopy, *Key. Eng. Mater.* **81-3**, 261-266.

20. Knobel, M., Sato Turtelli, R. and Rechenberg, H. R. (1992) Compositional evolution and magnetic properties of nanocrystalline $Fe_{73.5}CuNb_3Si_{17.5}B_5$, *J. Appl. Phys.* **71**, 6008-6012.

21. Randrianantoandro, N., Grenèche, J. M., Jędrika, E., Ślawska-Waniewska, A. and Lachowicz, H. K. (1995) Nanocrystallized Fe-based metglasses investigated by Mössbauer spectrometry, *Mater Sci Forum* **179-181**, 545-550.

22. Randrianantoandro, N., Ślawska-Waniewska, A. and Grenèche, J. M. (1997) Magnetic interactions of nanocrystallized Fe-Cr amorphous alloys, *Phys. Rev. B* **56**, 10797-10800.

23. Borrego, J. M., Peña-Rodríguez, V. A. and Conde, A. (1997) Mössbauer study of the nanocrystallization of the amorphous system $Fe_{73.5}Si_{13.5}B_9Cu_1Nb_1X_2$ with X = Nb, Mo, V and Zr, *Hyperfine Interactions* **110**, 1-6.

24. Borrego, J. M., Conde, A., Peña-Rodríguez, V. A. and Grenèche, J. M. (2000) A fitting procedure to describe Mössbauer spectra of FINEMET-type nanocrystalline alloys, *Hyperfine Interactions* **131**, 67-82.

461

25. Borrego, J. M., Conde, C. F., Conde, A., Peña-Rodríguez, V. A. and Grenèche, J. M. (2000) Devitrification process of FeSiBCuNbX nanocrystalline alloys: Mössbauer study of the intergranular phase, *J. Phys.: Condens. Matter* **12**, 8089-8100.

26. Brovko, I., Petrovič, P., Zatroch, M. and Konč, M. (1993) Crystallization of the FeNbCu(Au,Ag,Pd,Pt,Mn)SiB alloys investigated by X-ray and by Mössbauer spectroscopy, *Key. Eng. Mater.* **81-82**, 183-188.

27. Gorría, P., Orúe, I., Plazaola, F., Fernández-Gubieda, M. L. and Barandiarán, J. M. (1993) Magnetic and Mössbauer study of amorphous and nanocrystalline $Fe_{86}Zr_7Cu_1B_6$ alloys *IEEE Trans. Magn.* **29**, 2682-2684.

28. Navarro, I., Hernando, A., Vázquez, M. and Yu Seong-Cho (1995) Mössbauer spectroscopy in nanocrystalline $Fe_{88}Zr_7B_4Cu_1$, *J. Magn. Magn. Mater.* **145**, 313318.

29. Graf, T., Kopcewicz, M. and Hesse, J. (1995) Mössbauer studies of radio frequency induced effects in nano- and microcrystalline FINEMET, *NanoStructured Mater* **6**, 937-940.

30. Ciurzyńska, W. H., Varga, L. K., Olsewski, J., Zbroszczyk, J. and Hasiak, M. (2000) Mössbauer studies and some magnetic properties of amorphous and nanocrystalline $Fe_{87-x}Zr_7B_6Cu_x$ alloys, *J. Magn. Magn. Mat.* **208**, 61–68.

31. Zbroszczyk, J., Fukunaga, H., Olsewski, J., Ciurzyńska, W. H., Hasiak, M. and Błachowicz, A. (2000) Mössbauer and magnetic studies of amorphous and nanocrystalline $Fe_{85.4}Zr_{6.8-x}Nb_xB_{6.8}Cu_1$ (x = 0, 1, 2) alloys, *J. Magn. Magn. Mat.* **215-216**, 419 – 421.

32. Kopcewicz, M. (1999) Mössbauer study of nanocrystalline alloys, *Acta Phys. Pol. A* **96**, 49–68.

33. Kopcewicz, M., Grabias, A. and Idzikowski, B. (1999) Influence of the alloy composition on the magnetic properties of nanocrystalline $Fe_{80}M_7B_{12}Cu_1$ (M: Ti, Ta, Nb, Mo), in: *Mat. Res. Soc. Symp. Proc., Vol. 577*, Materials Research Society, pp. 487-492.

34. Suzuki, K., Cadogan, J. M., Sahajwalla, V., Inoue, A. and Masumoto, T. (1996) Mössbauer study of amorphous and nanocrystalline Fe-Nb-B alloys, *Mater. Sci. Forum* **225-227**, 707 – 712.

35. Suzuki, K., Cadogan, J. M., Sahajwalla, V., Inoue, A. and Masumoto, T. (1997) The role of alloying elements in Cu-free nanocrystalline Fe-Nb-B soft magnetic alloys, *Mater. Sci. Eng. A* **226-228**, 554–558.

36. Suzuki, K. and Cadogan, J. M. (1998) Random magnetocrystalline anisotropy in two-phase nanocrystalline systems, *Phys. Rev. B* **58**, 2730–2739.

37. Suzuki, K. and Cadogan, J. M. (1999) The effect of the spontaneous magnetization in the grain boundary region on the magnetic softness of nanocrystalline alloys, *J. Appl. Phys.* **85**, 4400–4402.

38. Suzuki, K. and Cadogan, J. M. (2000) Effect of Fe-exchange-field penetration on the residual amorphous phase in nanocrystalline $Fe_{92}Zr_8$, *J. Appl. Phys.* **87**, 7097–7099.

39. Kemény, T., Kaptás, D., Balogh, J., Kiss, L. F., Pusztai, T. and Vincze, I. (1999) Microscopic study of the magnetic coupling in a nanocrystalline soft magnet, *J. Phys.: Condens. Matter* **11**, 2841–2847.

40. Kaptás, D., Kemény, T., Balogh, J., Bujdosó, L., Kiss, L. F., Pusztai, T. and Vincze, I. (1999) *J. Phys.: Condens. Matter* **11** L179-L185.

41. Balogh, J., Bujdosó, L., Kaptás, D., Kemény, T., Vincze, I., Szabó, S. and Beke, D. L. (2000) *Phys. Rev. B* **61**, 4109–4116.

42. Miglierini, M. and Grenèche, J. M. (1997) Mössbauer spectrometry of Fe(Cu)MB-type nanocrystalline alloys: I. The fitting model for the Mössbauer spectra, *J. Phys.: Condens. Matter* **9**, 2303-2319.

43. Grabias, A. and Kopcewicz, M. (1998) Crystallization of the amorphous $Fe_{81}Zr_7B_{12}$ alloy inuced by short time annealing, *Mater. Sci. Forum* **269-272**, 725–730.

44. Grabias, A. and Kopcewicz, M. (2000) Transmission and conversion electron Mössbauer study of crystallization of amorphous FeZrB alloys, *Acta Phys. Pol. A* **96**, 123–130.

45. Garitaonandia, J. S., Schmoll, D. S. and Barandiarán, J. M. (1998) Model of exchange-field penetration in nanocrystalline $Fe_{87}Zr_6B_6Cu$ alloys from magnetic and Mössbauer studies, *Phys. Rev.* **B58**, 12147-12158.

46. Miglierini, M., Škorvánek, I. and Grenèche, J. M. (1998) Microstructure and Hyperfine Interactions of the $Fe_{73.5}Nb_{4.5}Cr_5CuB_{16}$ Nanocrystalline Alloys: Mössbauer Effect Temperature Measurements, *J. Phys.: Condens. Matter* **10**, 3159–3176.

47. Miglierini, M. and Grenèche, J. M. (1998) Methodology of Interfacial Regions in FeMCuB-type Nanocrystals, *Hyperfine Inter.* **113**, 375–382.

48. Chuev, M., Hupe, O., Bremers, H., Hesse, J. and Afanas'ev, A. M. (2000) A novel method for evaluation of complex Mössbauer spectra demonstrated on nanostructures ferromagnetic FeCuNbB alloys, *Hyperfine Ineractions* **126**, 407–410.

462

49. Hupe, O., Bremers, H., Hesse, J., Afanas'ev, A. M. and Chuev, M. A. (1999) Structural and magnetic information about a nanostructured ferromagnetic FeCuNbB alloy by novel model independent evaluation of Mössbauer spectra, *NanoStruct. Mat.* **12**, 581–584.

50. Hupe, O., Chuev, M. A., Bremers, H., Hesse, J. and Afanas'ev, A. M. (1999) Magnetic properties of nanostructured ferromagnetic FeCuNbB alloys revealed by a novel method for evaluating complex Mössbauer spectra, *J. Phys.: Condens. Matter* **11**, 10545–10556.

51. Grenèche, J. M. and Ślawska-Waniewska, A. (2000) About the interfacial zone in nanocrystalline alloys, *J. Magn. Magn. Mat.* **215-216**, 264–267.

52. Grenèche, J. M. and Miglierini, M. (1999) Mössbauer spectrometry applied to iron-based nanocrystalline alloys I. High temperature studies, in M. Miglierini and D. Petridis (eds.), *Mössbauer Spectroscopy in Materials Science*, Kluwer Academic Publishers, Dordrecht, pp. 243–256.

53. Hernando, A. and Kulik, T. (1994) Exchange interactions through amorphous paramagnetic layers in ferromagnetic nanocrystals, *Phys. Rev.* **B49**, 7064–7067.

54. Navarro, I., Ortuño, M. and Hernando, A. (1996) Ferromagnetic interactions in nanostructured systems with two different Curie temperatures, *Phys. Rev.* **B53**, 11656–11660.

55. Miglierini, M. and Grenèche, J. M. (1997) Mössbauer spectrometry of Fe(Cu)MB-type nanocrystalline alloys: II. The topography of hyperfine ineractions in Fe(Cu)ZrB alloys, *J. Phys.: Condens. Matter* **9**, 2321–2347.

56. Garitaonandia, J. S., Gorria, P., Fernández Barquín, L. and Barandiarán, J. M. (2000) Low temperature magnetic properties of Fe nanograins in an amorphous Fe-Zr-B matrix, *Phys. Rev.* **B61**, 6150–6155.

57. Olszewski, J., Varga, L. K., Zbroszczyk, J., Ciurzyńska, W. H., Hasiak, M. and Blachowicz, A. (2000) Magnetic behaviour of amorphous and nanocrystalline $Fe_{92-x}Zr_7Cu_1B_x$ (x = 2 or 6) alloys, *J. Magn. Magn. Mat.* **215-216**, 416–418.

58. Kemény, T., Kaptás, D., Kiss, L. F., Bujdosó, L., Gubicza, J., Ungár, T., and Vincze, I. (2000) Structure and magnetic properties of nanocrystalline $(Fe_{1-x}Co_x)_{90}Zr_7B_2Cu_1$ ($0 \leq x \leq 0.6$), *Appl. Phys. Lett.* **76**, 2110–2112.

59. Hasiak, M., Fukunaga, H., Ciurzyńska, W. H. and Yamashiro, Y. (2001) Effect of Co addition on microstructure and magnetic properties of $(Fe_{86-x}Co_x)$-Zr-B alloys, *Scipta mater.* **44**, 1465–1469.

60. Miglierini, M. and Grenèche, J. M. (1999) Hyperfine fields of amorphous residual and interface phases in FeMCuB nanocrystalline alloys: a Mössbauer effect study, *Hyperfine Interactions* **120/121**, 297–301.

61. Miglierini, M. and Grenèche, J. M. (1999) Temperature dependence of amorphous and interface phases in the $Fe_{80}Nb_7Cu_1B_{12}$ nanocrystalloiine alloys, *Hyperfine Interactions* **122**, 121–128.

62. Škorvánek, I., Kováč, J. and Grenèche, J. M. (2000) Structural an magnetic properties of the intergranular amorphous phase in FeNbB nanocrystalline alloys, *J. Phys.: Condens. Matter* **12**, 9085–9093.

63. Miglierini, M., Grenèche, J. M. and Idzikowski, B. (2001) Temperature Mössbauer effect study of nanocrystalline FeMCuB alloys, *Mater. Sci. Eng.* **A304-306**, 937–940.

64. Kopcewicz, M., Grabias, A., Nowicki, P. and Williamson, D. L. (1996) Mössbauer and X-ray study of the structure and magnetic properties of amorphous and nanocrystalline $Fe_{81}Zr_7B_{12}$ and $Fe_{79}Zr_7B_{12}Cu_2$ alloys, *J. Appl. Phys.* **79**, 993–1003.

65. Kopcewicz, M., Grabias, A. and Williamson, D. L. (1997) Magnetism and nanostructure of $Fe_{93-x-y}Zr_7B_xCu_y$ alloys, *J. Appl. Phys.* **82**, 1747–1758.

66. Kopcewicz, M. and Grabias, A. (1996) Mössbauer study of the surface crystallization of the amorphous and nanocrystalline $Fe_{81}Zr_7B_{12}$ alloy, *J. Appl. Phys.* **80**, 3422–3425.

67. Grabias, A., Kopcewicz, M. and Idzikowski, B. (1999) Surface analysis of the nanocrystalline Fe-based alloys by conversion electron Mössbauer spectroscopy, in: *Mat. Res. Soc. Symp. Proc., Vol. 577*, Materials Research Society, pp. 543–548.

68. Bibicu, I., Garitaonandia, J. S., Plazaola, F. and Apinaniz, E. (2001) X-ray diffraction, transmission Mössbauer spectrometry and conversion electron Mössbauer spectrometry studies of the $Fe_{87}Zr_6B_6Cu_1$ nanocrystallization process, *J. Non-Crystal. Solids* **287**, 277–281.

69. Miglierini, M. and Seberíni, M. (2002) Magnetic microstructue of nanocrystalline $Fe_{80}Nb_7Cu_1B_{12}$ investigated by Mössbauer spectrometry, *phys. stat. sol. (a)* **189**, 351–355.

70. Grabias, A., Kopcewicz, M. and Idzikowski, B. (1999) Mössbauer study of the nanocrystalline $Fe_{80}Ti_7B_{12}Cu_1$ alloy, *NanoStruct. Mater.* **12**, 899–902.

71. Grabias, A., Kopcewicz, M. and Idzikowski, B. (2000) On the formation of the nanostructure in $Fe_{80}M_7B_{12}Cu_1$ (M: Ti, Ta, Nb, Mo) alloys, *Hyperfine Interactions* **126**, 21–25.

NANO-STRUCTURED MAGNETIC FILMS INVESTIGATED WITH LORENTZ TRANSMISSION ELECTRON MICROSCOPY AND ELECTRON HOLOGRAPHY

JEFF DE HOSSON, NICOLAI G. CHECHENIN, and TOMAS VYSTAVEL
Department of Applied Physics, Materials Science Center and Netherlands Institute for Metals Research, University of Groningen, Nijenborgh 4, 9747 AG Groningen, The Netherlands.(e-mail: hossonj@phys.rug.nl)

1. Introduction

In general microscopy is devoted to link microstructural observations to physical properties. Although the microstructure-property relationship is in itself a truism, the actual linkage between structural aspects of defects in a material studied by microscopy on one hand and its physical property on the other is in most cases almost elusive. The reason is that various physical properties are actually determined by the collective behavior of defects. Even the behavior of one singular defect is often irrelevant. There are at least two reasons that prevent a straightforward correlation between microscopic structural information to the physical properties of materials: one fundamental and one practical reason. First, in the field of dislocations and interfaces we are faced with highly non-linear and non-equilibrium effects. This is a fundamental problem because there doesn't exist an appropriate analysis of these effects. Secondly, one should realize that a quantitative electron microscopy evaluation of the structure-property relationship is also hampered by a practical reason, namely the limited statistics. In particular, in situations where there is only a small volume fraction of defects present or a very inhomogeneous distribution statistical sampling may be a problem. Nevertheless, the situation is not hopeless and nowadays *in-situ dynamic* studies are possible to link microscopy information to functional and structural properties. In addition electron microscopy can be employed not only *to observe* but also as an instrument *to measure* functional properties

Along these lines this contribution concentrates on the application of transmission electron microscopy to functional materials, such as ultra-soft magnetic films for high-frequency inductors, to reveal the structure-property relationship. There exists an increasing demand for further miniaturization in portable appliances (e.g. mobile phones, palmtops), that is to say in communication tools. To this end the use of high-frequencies (e.g. 10-1000 MHz) in combination with thin magnetic materials is desirable. The use of magnetic films allows the integration of transformers and inductors into silicon IC circuitry. Soft-magnetic films are also widely used in modern electromagnetic devices as a high-frequency (>100 MHz) field-amplifying component,

T. Tsakalakos et al. (eds.) pgs 463 – 480
Nanostructures: Synthesis, Functional Properties and Applications;
© *2003 Kluwer Academic Publishers.*

e.g. in read-write heads for magnetic disk memories for computers and as a magnetic shielding material, e.g. in turners. The main requirements for the film material are: a high saturation magnetization, combined with a low coercivity and a small but finite anisotropy field. In addition the material should have a reasonably high specific electrical resistivity to reduce eddy currents, and also appropriate mechanical properties. To obtain the desired properties (low coercivity, little strain and very small magnetostriction) the use materials with grain size of the order of 10 nm nanometers, like nano-crystalline iron based materials, becomes attractive.

Knowledge of local magnetic properties is essential for the development of new magnetic nano-sized materials. One of technique that is suitable for the measurement of local magnetic structures is Lorentz-Fresnel (or defocused) imaging mode of transmission electron microscopy. This rather classical TEM technique [1][2] has several outstanding advantages: uncomplicated application to various parts of thin foil, possibility of dynamical studies and good spatial resolution. Nevertheless, to obtain quantitative information from Lorentz micrographs is relatively difficult due to in-direct link between image contrast and spatial variation of magnetic induction, which is problematic in regions of abrupt magnetization changes [3].

In this paper the possibility of a quantitative analysis of the magnetic properties of nano-crystalline iron using transmission electron microscopy is presented. The goal is to delineate a more quantitative way to obtain information of the magnetic induction and local magnetization. In particular the latter physical quantities affect the functional properties of ultra-soft magnetic materials for high-frequency inductors. One of the magnetic features that can provide quantitative magnetic information are the so-called magnetization or magnetic ripples, caused by local variation of magnetic induction that deviate from the mean magnetization direction [4]. Lorentz Fresnel (LTEM) and (off-axis) electron holographic modes are used to analyze the magnetic structures.

2. Experimental Results

Nano-crystalline Fe-Zr-N films have been prepared by DC magnetron reactive sputtering with a thickness between 50 and 1000 nm. The presence of zirconium is to catch the nitrogen in the iron matrix. Iron was chosen because it is easy to prepare and cheaper than other soft magnetic materials, e.g. cobalt. The nitrogen is added to get a small (nano-sized) grain size. Pure (99.96 at%) Fe sheets partially covered with Zr wires were used as targets. The N and Zr contents were controlled by varying the sputter power and/or the Ar/N2 gas mixture. An 800 Oe magnetic field was applied in the plane of the samples during deposition. More details on the film deposition can be found in [5].

The films have been deposited on a glass or silicon substrates at several temperatures between room temperature and 200 °C. The DC-sputtered samples were deposited on either a silicon substrate covered by a polymer, which was removed in acetone after sputtering, or on a silicon substrate covered by a Si_3N_4 layer. The former samples were extracted on copper TEM-grids for support, while the latter samples kept their substrate because the layers were very thin. Because all the films are sputtered or electrochemically deposited, they are of uniform thickness. The deposition conditions were chosen to obtain a composition $(Fe_{99}Zr_1)_{1-x}N_x$, where the concentration of nitrogen was in the range $x \leq 25$ at%. The best films as far as nano-size dimensions are concerned have been obtained for 8 at%<x<20 at%. The nitrogen concentration has been measured with Elastic Recoil Detection technique, Neutron Depth Profiling methods and compared with shifts in the XRD pattern. Standard $\theta - 2\theta$ XRD scans showed that up to x=10 at% as sputtered films were BCC single-phase materials with a strong (110) fiber texture. The grain size, estimated from the width of the (110) peak decreased monotonically with N content from typically 100 nm in the case of N-free films to less than 10 nm for films containing 8 at%.

The specimens were examined with a JEOL 2010F 200 kV transmission electron microscope equipped with a post column energy filter (GIF 2000 Gatan Imaging Filter, with a resolution of 1024x1024 pixels), which provides an additional magnification around 20 at the plane of CCD camera with respect to the maximal attainable magnification by using the objective minilens (magnification: 6 10^3). For holography the microscope is mounted with a biprism (JEOL biprism with a 0.6 μm diameter platinum wire). Grain size determination was done with several tilting experiments using an ACT: Automated Crystallography for the TEM, from TSL/EDAX and a Gatan dual-view CCD camera (resolution 1300x1030 pixels). A single-tilt specimen holder and a double-tilt heating specimen holder were used. The images are acquired and edited using DigitalMicrograph (DM) 3.3 and 3.4 on both Apple Macintosh and Microsoft Windows PC's. Furthermore an additional script package for DM was used: the NCEM Package Image that was developed at the national center for electron microscopy.

XRD, as well as conventional TEM and selected area diffraction (SAD), reveal a 2-30 nm size of crystallites after various tilting experiments for most of the investigated sputter deposited films (see Figure 1) and in the diffraction pattern of Figure 1b the allowed reflections for the bcc structure were found (110, 200, 211). No signs of Fe_4N or $Fe_{16}N_2$ were detected in XRD. A very weak ring within the first bright ring can be observed that originates from an oxide layer on the surface. The average measured grain size is approximately 34 nm as shown in Figure2. The grain size distribution is rather broad, even for sputtered films. Indeed, one should be careful with this analysis when the film thickness is much larger than the grain size. The information of more grains on top of each other may cause the ACT to measure an incorrect grain size and a different orientation distribution. It should be stressed that the grain size distribution depends on the exact condition of the deposition process, i.e. on the nitrogen concentration. There are many defects in individual grains. With in-situ annealing till 250^0 C the grain size distribution becomes sharper and this phenomenon is related to the crystallization of the inter-granular amorphous-like phase.

Figure 3 shows three domain walls and the magnetic ripples within the different grains as observed in LTEM (objective lens turned down) and Figure 4 shows a cross-tie wall. Because the magnetization is perpendicular to the magnetic ripple structure within the magnetic domains, magnetic induction vectors can be drawn in the image(s) (figure 3 and 4). Figure 3 shows that there are two 90°-degree Néel domain walls, which is to be expected in thin films like these. The magnetic ripples are characterized by the mean wavelength and the mean angle of deviation of the local magnetization variation. These could be scrutinized by a careful analysis of the Lorentz images in Fourier space. A typical magnetic contribution in frequency domain of a ripple structure is represented by a "bow-tie" like feature (see figure 5). To find the ripple wavelength the dominant frequency in frequency domain has to be found whereas the deviation angle can be

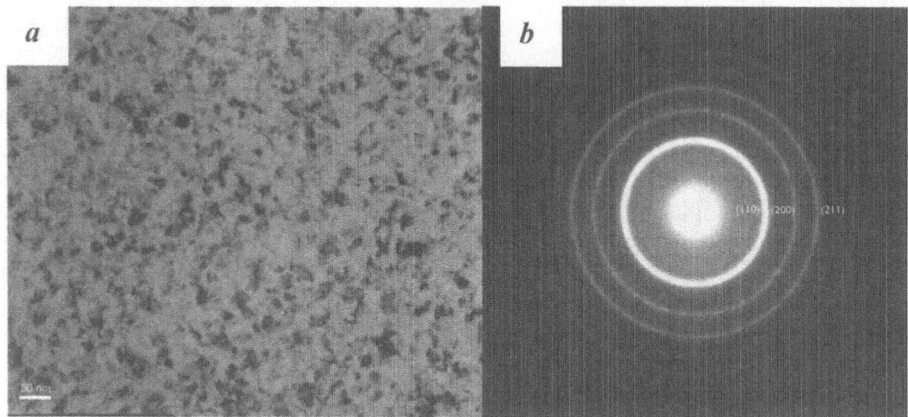

Figure 1 (a) Nano-crystalline iron and (b)SA diffraction pattern of nano-crystalline $Fe_{94}N_5Zr_1$

found by determining the (half) angle of the bow-tie (angle θ in figure 5). The procedure of image analysis is as follows and displayed in figure 6. First an image containing homogeneous ripples is acquired. To determine the ripple wavelength the rotational average of the modulus of the fast Fourier transform (FFT) is calculated (figure 6 B and 6 C). If a line profile from the center of the image is taken the wavelength can be determined. This is done by looking at the first order maximum (indicated by a vertical line in figure 6 D). Using the distance between this peak and the central peak the average ripple wavelength can be calculated [6].

Figure 2 Grain size and grain size distribution from ACT300 analysis (.Automated Crystallography for the TEM, from TSL/EDAX)

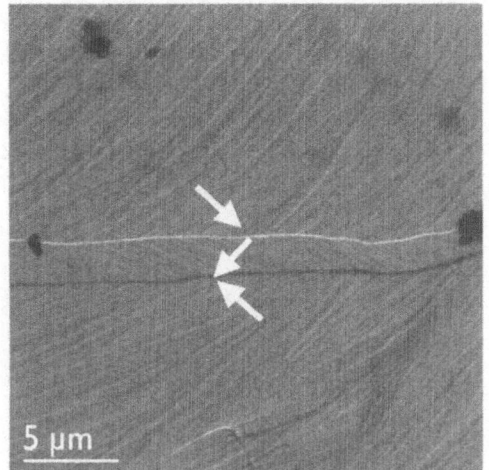

Figure 3 Magnetization direction within magnetic domains

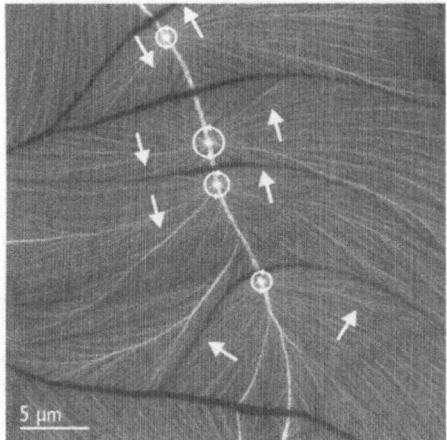

Figure 4 Magnetization direction in a cross-tie wall

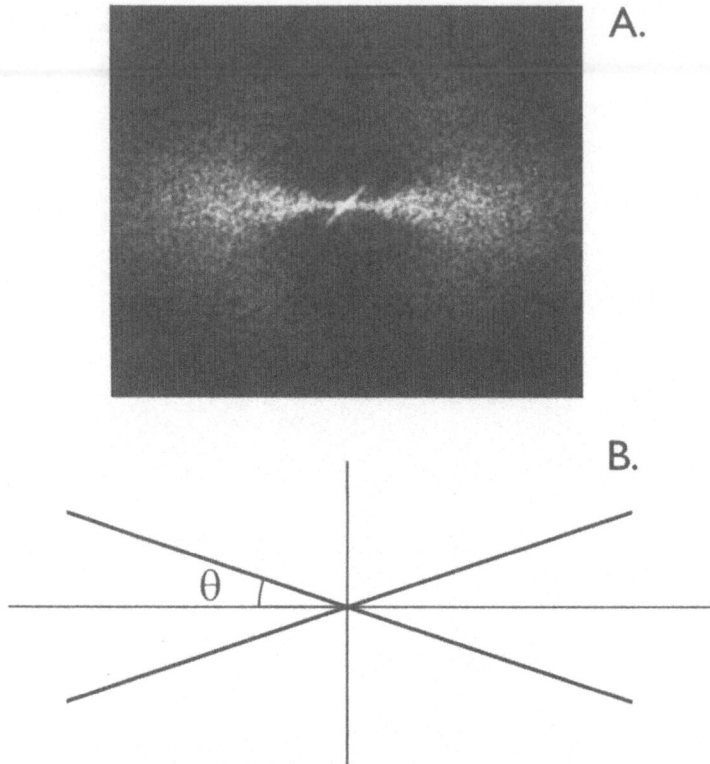

Figure 5 Typical diffractogram of the magnetic ripple structure

The deviation angle of the ripple spectrum can also be found by calculating the modulus of the FFT [7]. The FFT contains two triangular shapes, from which the mean deviation angle can be obtained (figure 6). A band pass mask is applied on the FFT at the distance from the origin where the first order maximum lies. Then a rotational profile is taken within this filtered FFT. Then, the ratio between the peak width (indicated by the rectangular shape in figure 6 F) and the entire profile determines the angle. The peak width is determined by looking at the maximum intensity and by finding the points where the intensity falls to $\frac{1}{2}\sqrt{2}$ ($\approx 70\%$) of the maximum intensity.

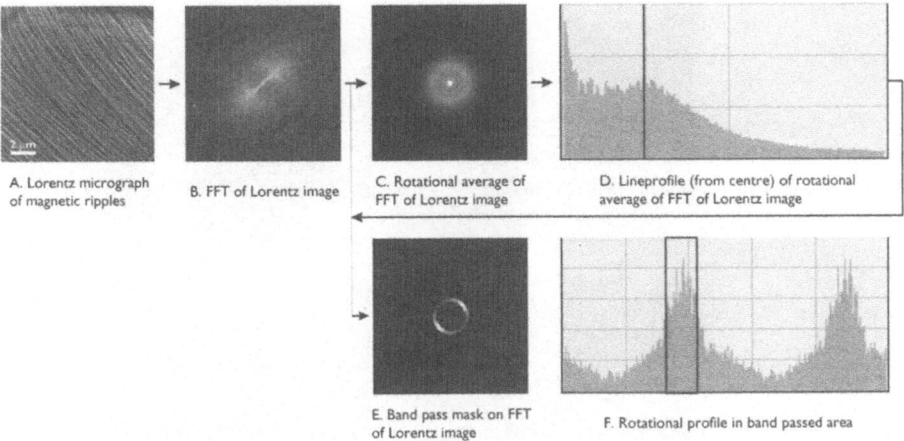

A. Lorentz micrograph of magnetic ripples

B. FFT of Lorentz image

C. Rotational average of FFT of Lorentz image

D. Lineprofile (from centre) of rotational average of FFT of Lorentz image

E. Band pass mask on FFT of Lorentz image

F. Rotational profile in band passed area

Figure 6 Schematic description of the quantification method

An alternative transmission electron microscopy technique for obtaining information of the magnetization is electron holography, which is based on recording an interference pattern from which both the amplitude and phase of an object can be reconstructed [8][9]. Magnetic thin films are strong phase objects and the phase shift of the electrons passing through the specimen is proportional to the magnetic flux enclosed by the electron paths. There are various electron holographic techniques [10], but one of the most popular is off-axis electron holography. The early holograms were limited to the brightness and with that to the coherence of the filament sources. The development of the field emission gun (FEG), which gives a coherent beam with a high intensity, contributed greatly to the implementation of electron holography in practice. For off-axis holography a specimen is chosen that does not completely fill the image plane (for example a small magnetic element or the edge of an extended film) so that only part of the electron beam passes through the specimen. An electrostatic bi-prism, a thin (< 1μm) metallic wire or quartz fiber coated with platinum (or gold), is then used to recombine the specimen beam and the reference beam so that they interfere and form a hologram. This can be digitized and digital image-processing techniques can be applied to reconstruct an image of the magnetic domain structure. The intensity of the recorded hologram can be written as:

$$I(x,y) = |\Psi_1(x,y)|^2 + |\Psi_2(x,y)|^2 + |\Psi_1(x,y)||\Psi_2(x,y)|\left[e^{i(\phi_1-\phi_2)} + e^{-i(\phi_1-\phi_2)} \right]$$
$$= A_1^2 + A_2^2 + 2A_1A_2 \cos \Delta\phi$$

(1)

where Ψ is an electron wave function, φ represents the phase, and the subscripts refer to the reference and object waves, which travel through the opening and the specimen, respectively.

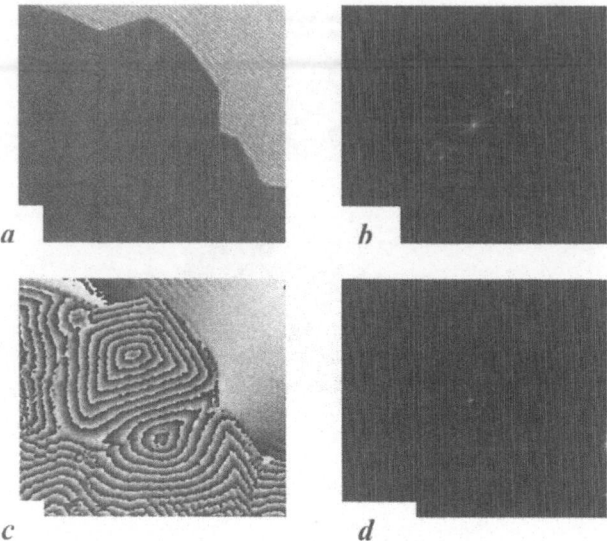

Figure 7 Step by step procedure to numerical reconstruct the phase from a hologram; A: Hologram; B: FFT of A, power spectrum with side bands; C: cut and center B; D: inverse FFT of C to obtain the phase map.

Equation (1) shows that the phase and the amplitude information are enclosed in the holographic image [11]. The phase can be numerically reconstructed according to the following procedure. First the fast Fourier transformation (FFT) of the holographic image (figure 7 A) is taken (figure 7 B). In frequency domain two sidebands can be detected. If one of the two side bands of the FFT is cut out and centered (figure 7 C) and the inverse FFT of this centered sideband is taken the phase will appear (figure 7 D). Because one of the off-axis sidebands is centered to obtain the phase information the method is called off-axis holography.

After a holographic image has been produced and the phase has been extracted it has to be processed before the in-plane magnetic information perpendicular to the electron beam can be extracted in an induction map. First, the phase of the image has to be known to get magnetic information [12][2] If neither $\vec{B}$ or V vary with the penetration depth z and neglecting magnetic and electric fields outside the sample, then the phase becomes

$$\varphi(x,y) = C_E V(x,y)t(x,y) - \frac{e}{\hbar} \int\int \vec{B}_\perp(x,y)t(x,y)dxdy \quad (2)$$

In one dimension Eq.(2) reads:

$$\frac{d\varphi(x)}{dx} = C_E \frac{d}{dx}\{V(x)t(x)\} - \frac{e}{\hbar}\vec{B}_\perp(x)t(x) \quad (3)$$

where V(x) is the mean crystal potential, t(x) represents the thickness as a function of the place, $B_\perp$ is the component of the magnetic induction perpendicular to both x and z. Assuming that the thickness is constant over the whole image (because sputtered films are used this assumption can be made) and the composition is homogeneous, the first right-hand side term drops and only the second term remains. With Maxwell's equation $\nabla \cdot \vec{B} = 0$, $B_\perp$ can be split and written as:

$$B_\perp(x,y) = B_x(x,y) + B_y(x,y) \begin{cases} B_x(x,y) = \dfrac{h}{2\pi et} \dfrac{d\varphi(x,y)}{dy} \\[2mm] B_y(x,y) = -\dfrac{h}{2\pi et} \dfrac{d\varphi(x,y)}{dx} \end{cases} \qquad (4)$$

In summary, provided the phase is fully recovered the derivative of the phase multiplied by a constant will give the magnetic induction in the x- and y-direction as a function of the position in the holographic image. To obtain the correct values for the magnetic induction the unwrapped phase of the hologram is needed. In order to convert the phase to the unwrapped phase the 2π steps have to be removed. This can be done by looking at the phase steps along a row or column of pixels in the image and by adding or subtracting 2π where appropriate.

In Figure 8 an electron hologram and its (unwrapped) reconstructed phase of the nano-crystalline Fe film is shown. When the phase and the Lorentz Fresnel image of this cross-tie wall are compared in Figure 8, it can be seen that the cross-ties inside the cross-tie wall in both Lorentz Fresnel and holography modes are clearly visible. To calculate the in-plane magnetic induction the derivative of the phase has to be calculated and multiplied by the proper constants. The procedure is delineated in Figure 9. To calculate the magnetization from the phase the unwrapped phase has to be calculated first [12]. Consequently the image has to be free of singularities. Because a part of the images is free of singularities only a smaller area can be observed and analyzed. Depending on the frame of reference chosen the light or the dark areas show a higher value for the local magnetic induction. In Figure 9a a higher intensity value represents a higher

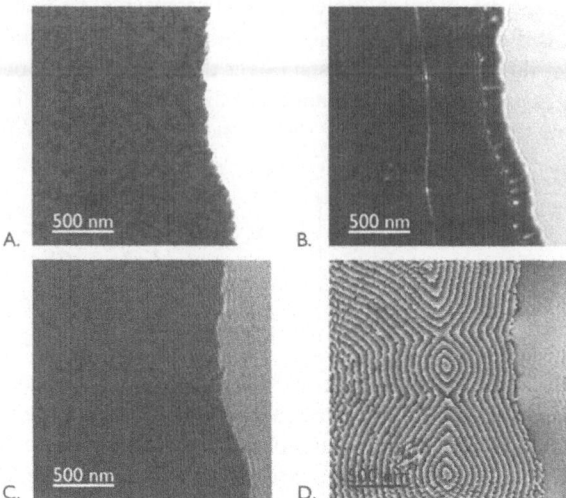

Figure 8 A structure image (A), a Lorentz Fresnel image (B), a hologram (C) and the reconstructed phase (D) of nano-crystalline $Fe_{94}N_5Zr_1$

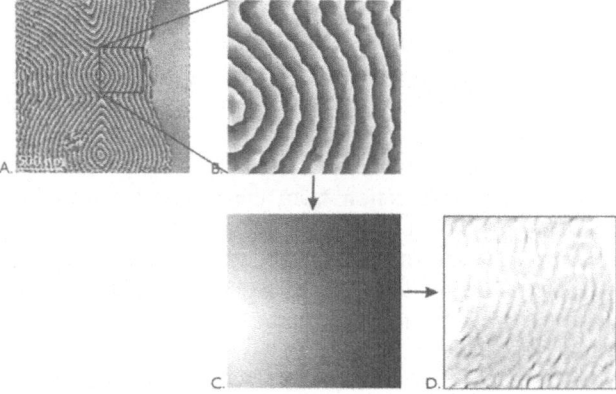

Figure 9 Example of determining the magnetic induction (D) from a hologram (A). A singularity free section is selected (B) then the unwrapped phase is calculated (C), after which the derivative multiplied with a constant gives the magnetic induction (D).

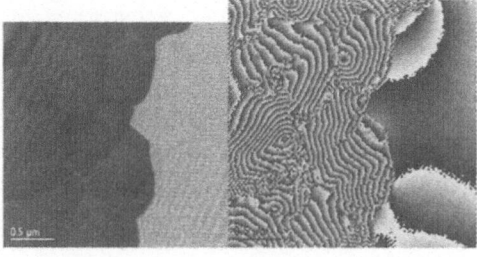

Figure 10 Hologram and phase image of nano-crystalline $Fe_{94}N_5Zr_1$

value for the magnetic induction. In Fig 10 another example of the reconstructed phase of nano-crystalline iron is shown. In this case there is also magnetic contrast outside the sample (light area), which means that there is also a magnetic flux present outside the sample.

3. Discussion

To obtain the ripple wavelength in the LTEM observation of the nano-crystalline $Fe_{94}N_5Zr_1$ film (thickness 55 nm) the values of the ripple wavelength at different defocus values are measured and interpolated at zero defocus. In this particular case this value gives a ripple wavelength of 400 ± 10 nm and deviation angle $\theta=6\pm1.5^0$. The experimental ripple wavelength can be compared with the theoretical value (λ_{th}) [4] using the following theoretical formula for the ripple wavelength,

$$\lambda_{th} = 2\pi \left(A/K_u \right)^{1/2} \left(h \mp 1 \right)^{-1/2} \tag{5}$$

where A is the exchange constant, $K_u = M H_k/2$ the uni-axial anisotropy constant and h $= H/H_k = H M/2K_u$ is the reduced field. Also the angle of deviation (θ_{th}) can be compared with the theoretical prediction, according to:

$$\theta_{th} = \frac{1}{4\pi^{1/2}} \frac{DK}{(2t)^{1/4} M^{1/2}} \frac{1}{\left\{ AK_u(h+1) \right\}^{3/8}} \tag{6}$$

where D is the mean grain size, t the thickness of the film, K the crystal anisotropy constant, M the saturation magnetization, K_u the uni-axial anisotropy constant and h = H/H_k is the reduced field. The difficulty when comparing the experimental data with theory stems from the fact that not all the physical parameters are accurately known. This is due to two reasons: only values for pure iron are known and not for nano-crystalline iron and secondly, even the values that are measured are often just estimates. For calculating λ_{th} the following parameters are chosen: the exchange constant of (pure) iron is $1.49 \ 10^{-11}$ J/m, the magnetization ranges from 1 up to 1.5 Tesla, H_k is sample dependent and varies between 5 and 18 Oe. For calculating θ_{th}, besides the thickness and mean grain size of the sample, the crystal anisotropy constant for (pure) iron 4.7 10^4 J/m^3 is used. These values give a theoretical values of $\lambda_{th}=1400$ nm and $\theta_{th}=49^0$, that is to say both predicted values are much larger than the experimental findings. Besides the fact that the magnetic parameters are not precisely known, the principal reason of this discrepancy lies in the circumstance that the classical theory is probably not applicable to the case of nano-structured materials. One of critical assumptions is that the effective crystalline anisotropy constant is much smaller than the uni-axial anisotropy constant ($K_{eff} << K_u$) but in the nano-crystalline material both K_{eff} and K_u are of the same order of magnitude. Therefore we suggest the following novel approach.

Clearly a quasi-periodic oscillation of the transversal component of the local magnetization is observed by Lorentz transmission electron microscopy (LTEM). A set of almost parallel fringes in under- or over-focused image is generated. In the classical theory of the LTEM imaging of micromagnetic ripple [2] and in a diffraction approach [13] the oscillations are one-dimensional and the absolute value of the magnetization

474

has been assumed to be constant. These assumptions are rather questionable in nano-structured materials and LTEM provides the possibility to elucidate the magnetic properties to a greater detail. It should be emphasized that we do not discuss any time dependence and consequently fluctuations are related only to space modulations either of magnetization in the film or of intensity in the LTEM image.

Due to the random orientations of the grains in nano-crystalline material, the magnetocrystalline anisotropy is averaged out to a large extent. A uni-axial anisotropy is induced during film deposition due to an applied magnetic field. The exchange interaction reduces the angular spread of the magnetization vector due to the local residual crystalline anisotropy, leading to a correlated wiggling of the magnetization around the easy axis (EA) producing the ripples observed in Lorentz microscopy. This wiggling can be characterized by an amplitude of the transversal component of the magnetization, ΔM_y , which is a periodic function in x-direction. In the simplest approximation we have

$$\Delta M_y = M \beta_0 \sin(2\pi x / \lambda_x) \qquad (7)$$

where the EA is oriented along x-axis. The amplitude of the angle of the wiggling, β_0, is of the order of one degree and depends on the grain size and on the applied magnetic field [15] A transversal oscillations $\Delta M_y(y)$ is energetically unfavorable [2] and the oscillation of ΔM_y (x) is called the longitudinal one. The component normal to the plane, ΔM_z, is suppressed due to the demagnetizing factor. The wiggling induces the internal stray field oriented parallel and antiparallel to the main magnetization direction. The magnitude of the stray field can easily be estimated by

$$\Delta B_{str,x} \approx -\pi M \beta_0^2 \cos(4\pi x / \lambda_x) \qquad (8)$$

The presence of this mode induces an oscillating Lorentz force that moves the electrons perpendicular to the magnetization direction. However, the magnitude of this force is of one to two orders of magnitude less than the one induced by the transversal component of Eq.(7). Moreover, the period of the oscillation in Eq.(8) is half of that in Eq.(7), meaning that the mode of Eq.(8) will not produce any phase shift at the points of maximum phase shift due to the mode of Eq.(7). That leads to the conclusion that in principle the mode described by Eq.(8) only leads to a certain distortion of the phase contrast due to the mode given by Eq.(7). So far, the direction of magnetization vector oscillates but it is constant in magnitude. That is a traditional point of view. Now, we assume that the magnetization has a certain spread due to the non-averaged crystalline anisotropy. In the case of random orientation of the crystallites the residual crystalline component, ΔM_r, has a uniform orientation. In the presence of exchange interaction the orientation is forced to correlate. However, the magnitude of the magnetization is free to vary and the major effect is expected for the transverse variation of the x-component. For simplicity, we again restrict ourselves to a simple harmonic oscillations, i.e.

$$\Delta M_x = \Delta M_r \sin(2\pi y / \lambda_y) \qquad (9)$$

Next, let us estimate the phase contrast in LTEM as produced by these magnetic features. The phase shift between two points of the image is determined by the magnetic flux through the area surrounded by lines connecting the corresponding points on the upper and lower surfaces of the films and the trajectories of electrons passing these points.Ifone of the points is at the origin of the coordinates we write for the phase shift:

$$\varphi(\vec{r}) = \frac{\pi}{\Phi_0} \int_0^t dz \int_0^{r(x,y)} \vec{B}(\vec{r},t).d\vec{S}$$

(10)

where $\Phi_0 = 2e/\hbar = 2.06 \cdot 10^{-15}$ Wb, e is the electron charge, $\hbar$ is Planck's constant, t is the film thickness, $\vec{B}(\vec{r},t)$ is the local magnetic induction and $d\vec{S}$ is an element of the area represented by the vector oriented along the normal to the area. For a uniform thickness and a uniform distribution of magnetization all through t Eq.(10) can be rewritten as

$$\varphi(\vec{r}(x,y)) = \frac{\pi}{\Phi_0} t \int_0^{r(x,y)} \left[B_y(x',y')dx' - B_x(x',y')dy' \right]$$

(11)

where

$$B_y = \Delta M_y \text{ and } B_x = M + \Delta M_x + \Delta B_{strx}$$

(12)

According to Eqs. (7-9) the phase can be written as:

$$\varphi(x,y) = \frac{\pi t}{\Phi_0} \left[\frac{M\beta_0 \lambda_x}{2\pi} (1 - \cos\frac{2\pi x}{\lambda_x}) - M_y - \frac{\Delta M_r \lambda_y}{2\pi}(1 - \cos\frac{2\pi x}{\lambda_x}) \right]$$

(13)

where the contribution due to the mode described by Eq.(8) was neglected. The intensity variation of the electron image of LTEM can be evaluated using the Fourier transforms technique, i.e. [14]:

$$I(x',y') = \left| \int \int F(k_x,k_y)T(k_x,k_y)e^{2\pi i(k_x x' + k_y y')} dk_x dk_y \right|^2$$

(14)

where

$$F(k_x,k_y) = \int \int f(x,y)\exp\left[-2\pi i(k_x x + k_y y) \right] dxdy,$$

(15)

is the Fourier transform of the exit electron wave function. Assuming that a plane electron wave enters the film $exp(2\pi i k_0 z)$ where $k_0 = 1/\lambda_0$ and λ_0 is the electron wave length ($\lambda_0 = 2.5$ pm for the 200 kV-electrons) and only the phase is modified by the magnetic film the z-dependent part of the $f(x,y)$ can be omitted in Eqs.(14,15), leading to:

$$f(x,y) = \exp\left[i\varphi(x,y) \right]$$

(16)

Neglecting effects of the aperture and of the spherical aberration in the Fresnel mode of LTEM the transfer function in Eq.(14) reads:

$$T(k_x,k_y) = \exp\left[i\pi\Delta z \lambda_0 (k_x^2 + k_y^2) \right],$$

(17)

where Δz is the defocus value (- for overfocus, + for underfocus). A nonzero value of Δz makes the ripple image observable. Next, Eq.(14) can be represented in the form of a convolution between $f(x,y)$ in Eq.16 and the Fourier transform of $T(k_x,k_y)$, which yields:

$$I(x',y') = \left| \int_{-\infty}^{\infty} dx \int_{-\infty}^{\infty} dy \exp[i\varphi(x,y)] \exp\left[-\pi i \frac{(x-x')^2+(y-y')^2}{\Delta z \lambda_0}\right] \right|^2 \quad (18),.$$

which can be easily integrated. Neglecting constant multipliers we find
$I(x,y) = I_x(x)I_y(y)$, where

$$I_y(y) = 1 - 2S_y \cos[(2\pi y - S_{y^2}\Delta z \lambda_0)/\lambda_x]$$

$$\sin[\Delta z \lambda_0 /(2\pi\lambda_y^2)] + 2S_y^2 \cos^2[(2\pi y - S_{y^2}\Delta z \lambda_0)/\lambda_y] \quad (19)$$

where $S_x = \Delta M_y t \lambda_x/2\Phi_0$, $S_y = \Delta M_t t \lambda_y/2\Phi_0$ and $S_{y2} = \pi M t /\Phi_0$ are factors determining the contrast of the magnetic fluctuations in LTEM image.

From the analysis presented in Eqs. (7) to (19) follows that the intensity of the LTEM image will vary in both, x- and y-directions, responding to different modes of magnetic oscillations (see Eqs. (7) to (9)). The longitudinal oscillations of the transversal component of the magnetization described by Eq. (7) gives a phase contrast of the fringes that is oriented perpendicular to the main magnetization vector as can be concluded from from Eqs. (7), (13), (18) and (19). The period of these fringes corresponds to λ_x. From the experimental observations follows that this period for different films may vary between 0.1 to 1.5 μm. If $S_x \ll 1$, then from Eq.(19) the contrast of the ripple pictures reads

$$C_x = \frac{I_x(0) - I_x(\lambda_x/2)}{I_x(\lambda_x/4)} = 2S_x \sin\frac{\lambda_0 \Delta z}{2\pi\lambda_x^2} \quad (20)$$

The maximum contrast is reached when the defocusing conditions correspond to the sin-term in $I_x(x)$ equals unity, leading to

$$C_x^{max} = \Delta M_y \lambda_x t /\Phi_0 \quad (21)$$

This relation is similar to the one derived in [15] but in Eq.(20) the defocusing Δz has been taken into account. The contrast depends linearly on Δz at small defocus values. The mode described by Eq.(9) produces a contrast that varies in the y-direction, i.e. along the ripple fringes. The period of this oscillation λ_y can be determined from the experiment. If $\Delta \xi_x = \Delta z \lambda_0 /(2\pi\lambda_x^2)$ is the defocus-argument of the sinusoidal function in $I_x(x)$ or C_x then the corresponding argument of the sinusoidal function in $I_y(y)$ is $\Delta \xi_y = \Delta \xi_x (\lambda_x /\lambda_y)^2$. The contrast of $I_y(y)$ oscillations represented by

$$C_y = 2S_y \sin(\Delta \xi_y) = \Delta M_t t \lambda_y \sin(\Delta \xi_y)/(2\Phi_0) \quad (22)$$

is in general not optimal when the contrast of x-oscillations is maximal. The magnitude of the variation of the magnetization vector can be estimated from the contrast described by Eq.(22).

The 2D variation of the intensity in the image can now be calculated based on Eq. 19. To illustrate the outcome of this analysis an example of the LTEM image with the ripple structure for the nano-crystalline Fe-Zr-N film is depicted in Figure 11 where a linear scan perpendicular to the ripple fringes (T-profile) shows a quasi-periodic

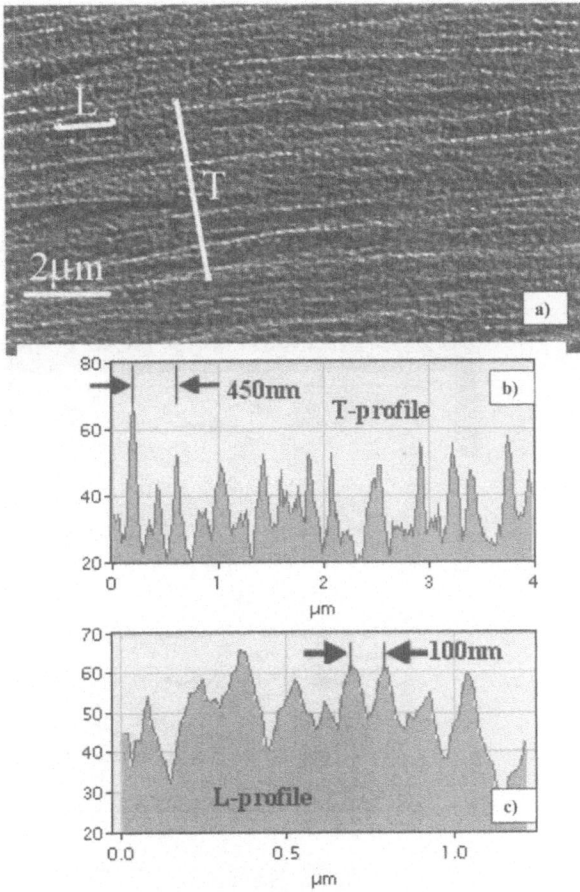

Figure 11 Analysis of the ripple structure in nano-crystalline $Fe_{94}N_5Zr_1$

arrangement of the fringes, Figure 11b. The analysis reveals the presence of two waves in the T-scan with almost the same wavelength $\lambda_x \approx 0.45\mu m$, which are shifted relative

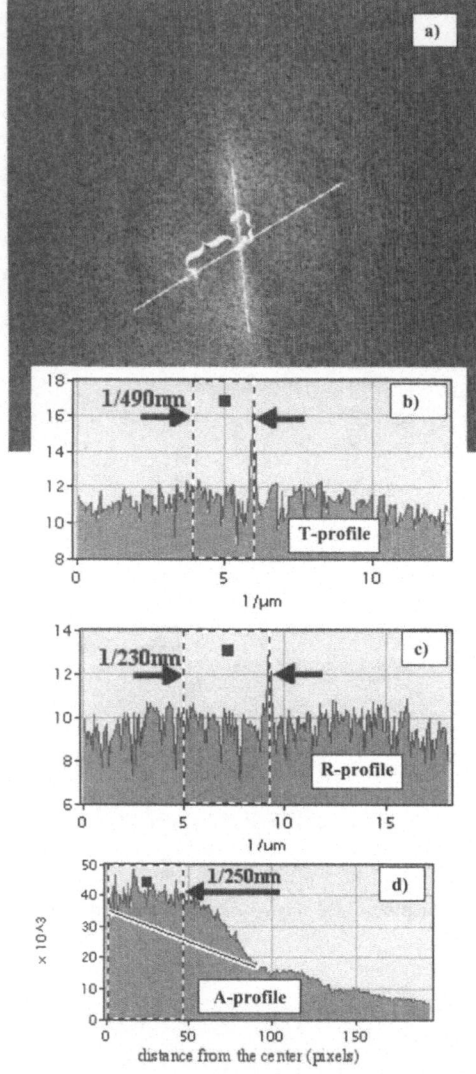

Figure 12 FFT analysis of the ripple structure in nano-crystalline $Fe_{94}N_5Zr_1$

to each other by a distance slightly more than the half of the period. This type of fringes corresponds to a wiggling of the magnetization vector perpendicular to the easy axis, i.e. the mode of Eq.(9) discussed above. The contrast is of the order of $C \approx 0.6$. This leads to an amplitude of the transversal component $\Delta M_y \approx 23$ mT, or an amplitude of the wiggling angle $\beta_0 \approx 0.9^0$.

As expected from our theoretical consideration the intensity of the image also is inhomogeneous along the ripple direction. This is demonstrated in the longitudinal scan (L-profile) in Figure 11c. This variation corresponds to the mode of Eq.(9) discussed in the previous section. The periodicity lies in between $\lambda_y \approx 0.1\text{-}0.15$ nm and the contrast $C \approx 0.2$. That gives the longitudinal fluctuation of the magnetization of the order of $\Delta M_x \approx 30$ mT, which is only slightly larger than the amplitude of the transversal component.

The Fourier transform (FFT) of the image is depicted in Figure 12. The enhanced intensity of the FFT in the direction perpendicular to the fringes is clearly seen. However, the angular spread around this direction is significantly larger than $\beta_0 \approx 0.9^0$. In addition there is a significant background almost over the complete azimuth of the FFT. The radial intensity variations along the maximum of the intensity, corresponding to the T-scan in Figure 12 and in an arbitrary direction (R-scan) through the center of the FFT picture, are plotted in Figs. 12b and 12c, respectively. Both profiles show a smooth and wide bump. In the T-scan the averaged radius corresponds to a periodicity of ~0.49 μm and in the R-scan to ~0.23 μm, i.e. of the same order of magnitude as that of the image in Figure 11. In addition in Figure 12d the radial profile is shown for azimuthally averaged FFT, which also shows a wide bump, positioned at an average radius corresponding to a periodicity of ~250 nm. From an experimental viewpoint, it should be stressed that for the delineated analysis special attention should be paid to the illumination convergence angle as well as to the value of the defocus. The ripples have to be nicely visible and this is accomplished at considerable defocus values. However, if the defocus is too extended the measured ripple length will deviate largely. In this particular case reproducible values of the ripple period were obtained with a defocus of ± 1000 to 4000 nm. From a determination of the ripple wavelength at different defocus values the effect of the defocus on the measured ripple wavelength can be estimated. This produces the following results: within a defocus bandwidth the ripple wavelength is (approximately) constant. If the defocus is bigger than a certain value the ripple wavelength rapidly increases, probably due the change of the condenser lens current. The demand for a good ripple contrast is in the case of the ripple deviation angle even more compelling then in the case the ripple wavelength.

4. Conclusions

This paper delineates a new possibility of LTEM to study the structure-property of nanostructured functional materials by concentrating on the magnetic properties of the soft magnetic films. We have shown that in contrast to the traditional point of view not only the direction of the magnetization vector in nano-crystalline $(Fe_{99}Zr_1)_{1\text{-}x}N_x$ films makes a correlated small-angle wiggling but also the magnitude of the magnetization modulus fluctuates. The latter produces a rapid modulation in the LTEM image. Analysis of the ripple structure corresponds to an amplitude of the transversal component of the magnetization ΔM_y of 23 mT and a longitudinal fluctuation of the magnetization of the order of $\Delta M_x = 30$ mT. Besides LTEM off-axis electron

480

holography technique was explored as an alternative method for obtaining information of the magnetization. It turned out that electron holographic methods could also be used on these $(Fe_{99}Zr_1)_{1-x}N_x$ films to recover the phase. The derivative of the phase with respect to the position provided qualitative information of the magnetic induction in the x- and y-direction as a function of the position in the holographic image. For further quantification of the results obtained with the off-axis electron holographic method a relation with the quantitative information of the LTEM technique will be explored in the future.

5. Acknowledgement

The work is part of the research program of the Priority Program on Materials of the Netherlands Organization of Research (NWO- The Hague) and supported by the Netherlands Foundation for Technical Sciences and the Netherlands Institute for Metals Research.

6. References

[1] Grundy, P.J., Tebble, R.S. (1968). Lorentz Electron Microscopy. *Advances in Physics,* **17**: p. 153-242.

[2] Fuller, H.W., Hale, M.E. (1960). Determination of magnetization distribution in thin films using electron microscopy. *J. Appl. Phys.*, **31**, 238-248.

[3] De Graef, M. (2000). Lorentz microscopy, in Experimental Methods in the Physical Sciences, Volume 36: Magnetic Imaging and its Applications to Materials, De Graef, M & Zhu,Y.(Eds), Chapter 2, pp.27, Academic Press, N.Y.

[4] Hoffmann, H. (1964). Quantitative calculation of the magnetic ripple of uniaxial thin permalloy films. *J. Appl. Phys.*, **35**,1790-1798.

[5] Chezan , A.R., Craus, C.B., Chechenin, N.G., Niesen, L., Boerma, D.O. (2002). Structure and soft magnetic properties of Fe-Zr-N films. *Physica Status Solidi (a)*, 189,833-841.

[6] Herrmann, M., Zweck, J., Hoffmann, H.(1994). Observation and Characterization of Micromagnetic Structures. *ICEM 13*, p.245.

[7] Gillies, M.F., Chapman, J.N., Kools,.J.C.S. (1995). Micromagnetic characteristics of single layer permalloy films in nanometre range. *J. Magn. and Magnetic Materials*, **140-144**:,721-722.

[8] Gabor, D. (1949). Microscopy by reconstructed wave-fronts,*Proceedings of the Royal Society A*, **197**, 454-487.

[9] Tonomura, A. (1999). *Electron Holography*. Berlin, Germany: Springer Series in Optical Sciences.

[10] Cowley, J. (1992). Twenty forms of electron holography.*Ultramicroscopy*, 41,335-348.

[11] McCartney, M.R., Dunin-Borkowski, R.E., Smith,D.J. (2000).Electron holography and its application to magnetic materials, in: Experimental Methods in the Physical Sciences, Volume 36: Magnetic Imaging and its Applications to Materials, De Graef,M. & Zhu,Y. (Eds), Chapter 4, pp.111, Academic Press, N.Y..

[12] Midgley, P.A.(2001).An introduction to off-axis electron holography. *Micron*, 32,167-184.

[13] Wohlleben, D. (1967). Diffraction effects in Lorentz microscopy. *J.Appl.Phys.* **38** , 3341-3352.

[14]Chapman,J.N. (1984). The investigation of magnetic domain structures in thin foils by electron microscopy. *J. Phys. D: Appl. Phys.* **17** , 623-647.

[15] Chechenin, N.G., Chezan, A.R., Craus, C.B., Vystavel, T., Boerma, D.O., De Hosson, J.Th.M., Niesen, L. (2002). Microstructure of nanocrystalline FeZr(N)-films and their softmagnetic properties. *J. Magn. Magn. Mater.*,in press.

PHASE SEPARATION IN CMR MATERIALS: THE ROLE OF SPIN NANOCLUSTERS.

A. SIMOPOULOS, M. PISSAS, G. KALLIAS AND E. DEVLIN
Institute of Material Sciences, NCSR Demokritos, 153 10 Aghia Paraskevi, Athens, Greece

ABSTRACT: Manganese perovskites, with the general formula $REMnO_3$ (RE=Rare Earth), display a wealth of new phenomena on substitution of part of the trivalent RE by a divalention of the alcaline earth series (Ca, Sr etc.). This doping results in the formation of a mixed valence system (Mn^{3+}, Mn^{4+}) with a rich phase diagram with ferromagnetic insulators at low doping, ferromagnetic conductors, at higher doping and antiferromagnetic insulators at doping higher than 50%. The system displays magnetic phase separation in the phase boundaries while application of a magnetic field brings the system in its ground state which is usually the metallic ferromagnetic state. The result of this process is the appearance of colossal negative magnetoresistance (CMR) with many potential applications. There has been an intensive study of these systems during the past 5 years in order to elucidate this new phenomenon at the atomic level. The double electron exchange (Zener) and the Jahn-Teller effect are the basic principles governing the CMR effect, although there are basic questions that are still open. It has now been established, within the relevant scientific community, that these phenomena arise from an interplay of the charge, magnetic moment (spin and orbital) and lattice degrees of freedom. Among the many experimental techniques that have been employed for these studies, the spectroscopic techniques for hyperfine interaction studies are of prime interest since they provide information at the atomic level. In this work, we present the results of Moessbauer studies in $La_{0.5}Ca_{0.5}MnO_3$ doped with 1% ^{57}Fe and 1% ^{119}Sn. These results show the coexistence of phases, in particular close to the phase boundaries, the dynamic transformation of one phase to the other and the role of the formation of spin-nanoclusters during this transformation.

1. Introduction

The discovery of colossal magnetoresistance (CMR) in manganese based perovskites has stimulated intense research on the physical bases of the CMR effect [1]. The perovskite manganites $L_{1-x}A_xMnO_3$ (L=lanthanides and A=alkaline earths) present a remarkably complicated phase diagram as the relative concentration of Mn^{3+} and Mn^{4+} changes [2,3]. For example, the ground state of the undoped compound $LaMnO_3$ is an antiferromagnetic (AFM) insulator. Doping the system with holes, that is substituting Ca^{2+} for La^{3+}, the AFM state is suppressed for a ferromagnetic (FM) metallic state

T. Tsakalakos et al. (eds.) pgs. 481 - 500
Nanostructures: Synthesis, Functional Properties and Applications;
© *2003 Kluwer Academic Publishers.*

$(0.2 < x < 0.5)$ that shows colossal negative magnetoresistance. The spin of the e_g ($S=1/2$) extra electron in Mn^{3+} ($3d^4$) is ferromagnetically coupled to the t_{2g} ($S=3/2$) local spin according to Hund's rule, thus making it energetically favorable for the e_g charge carriers to hop from one ion to the next without changing their spin direction. For $x > 0.5$, the AFM superexchange interaction in the $Mn^{3+(4+)} O^{-2}$-$Mn^{3+(4+)}$ bonds dominates over the FM double exchange in the Mn^{3+}-O^{-2}-Mn^{4+}. At half-doping ($x=1/2$), the system undergoes a paramagnetic (insulating) to FM (metallic) transition at $T_c = 225$ K and then to an AFM (insulating) state at $T_N = 155$ K. The low temperature AFM state presents an unusual charge and spin ordered structure, called the CE type [2], where real space ordering of Mn^{3+} and Mn^{4+} takes place.

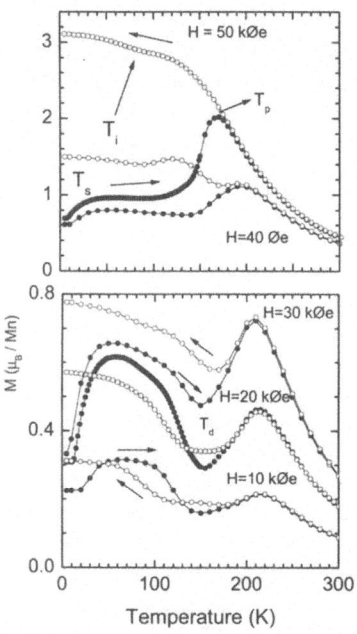

Figure 1 Magnetic moment per Mn ion versus temperature (ZFC and FC) of $La_{0.5}Ca_{0.5}Mn_{0.99}Fe_{0.01}O_3$ for magnetic fields up to 50 kOe. The filled circles represent the ZFC curves, while the open ones are the FC ones. The relevance of the various temperatures is explained in the text.

The co-existence of FM and AF phases (or phase separation) in this doping regime has been predicted, theoretically, [4,5] and witnessed by a number of techniques. This co-existence can be expected since the energies of the two relevant interactions (double-exchange and superexchange) are comparable. This energetic argument is supported by the fact that application of an external magnetic field changes the energy balance enhancing the FM phase through a melting transition of the charge-ordered lattice [6]. In bulk magnetization measurements, the high value of the magnetic moment observed at low temperatures implies the existence of a ferromagnetic component (its amount varies with the sample preparation route) [3,6,7]. Electron and neutron diffraction studies [8,9,10] have shown that, between 95 and 135 K, $La_{0.5}Ca_{0.5}MnO_3$ is an inhomogeneous spatial mixture of incommensurate charge-ordered and ferromagnetic charge-disordered microdomains with a size of 20-30 nm (chemical inhomogeneities have been ruled out with electron microprobe analysis with a spatial resolution of 20 nm).

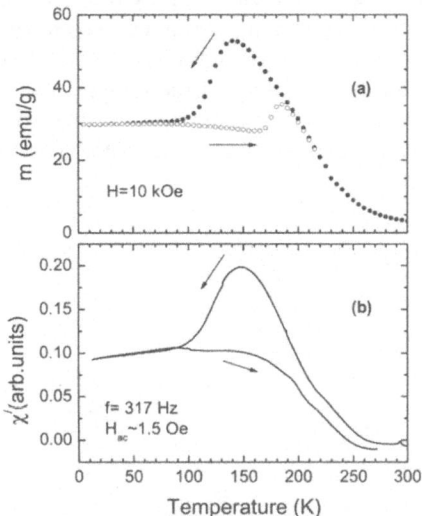

Figure 2 (a) The dc magnetic moment upon field-cooling and field-cooling warming in H=10 kOe and (b) the real part of the ac-susceptibility in zero dc magnetic field for $La_{0.5}Ca_{0.5}Mn_{0.99}Sn_{0.01}O_3$.

1.1 HYPERFINE INTERACTIONS.

Among the various experimental techniques, those utilizing the hyperfine interactions through local probes are of particular importance since they give a microscopic picture at the atomic level. The basic Hamiltonian describing these interactions is $H=IAS$, where A is the hyperfine tensor, I is the nuclear spin of the probe atom and S is the electronic spin. Extensive investigations of CMR materials with hyperfine interaction techniques have been performed with Nuclear Magnetic Resonance (NMR), Moessbauer Spectroscopy (MS) and muon spin rotation (μ-SR) [1]. In the NMR technique the local probes are the isotopes [139]La, [141]Pr, [55]Mn. In μ-SR the hyperfine interactions of muons embedded in the crystal with its neighboring atoms are detected [11].

Appropriate impurities for Moessbauer Spectroscopy are introduced in the lattice for this technique. So far [57]Fe and [119]Sn have been used for absorption experiments. The former substitutes for Mn^{3+} and the latter for Mn^{4+} according to their ionic radii. In emission experiments [57]Co in ppm quantities has been used [12].

We present in this work an extensive investigation of the system $La_{0.5}Ca_{0.5}MnO_3$ focusing our attention to the determination of the various magnetic phases and their evolution with temperature (Part of this work has appeared in Ref. 13 and 14). For this purpose, we employed Moessbauer Spectroscopy assisted with bulk measurements (magnetization and resistivity). In order to investigate the local environment, as seen by Mn^{3+} and Mn^{4+} ions, we have studied two samples, one doped with 1% ^{57}Fe (Fe sample) and one with 1% ^{119}Sn (Sn sample) which substitute respectively the corresponding ions. The results show phase separation at all temperatures below the charge ordering (CO) temperature as witnessed by the FM and AF components of the Moessbauer spectra. A third component, attributed to spin nanoclusters, seems to play a decisive role in the transformation of one phase to the other and the formation or disintegration of the CO state according to the temperature trip (lowering or raising respectively). The relative intensities of these components display thermal and magnetic hysteresis indicating that the FM-AF transition is of first order.

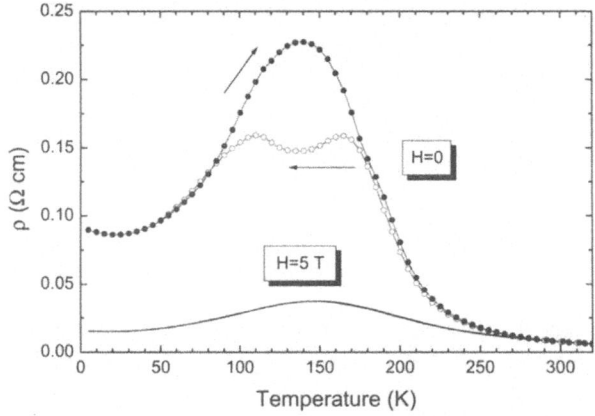

Figure 3 The temperature dependence of resistivity at H=0 (open and filled circles) and H=50 kOe (solid line).

2. Experimental Techniques

2.1 SAMPLE PREPARATION

Samples with nominal composition $La_{0.5}Ca_{0.5}Mn_{1-x}Fe_xO_3$ (x=0 and x=0.01) were prepared by thoroughly mixing high purity stoichiometric amounts of La_2O_3, $CaCO_3$, MnO_2 and Fe_2O_3 (90% enriched with ^{57}Fe). The mixed powders were pelletized and annealed in air at 1375°C for approximately 300 h with intermediate grindings and reformation into pellets each time. Finally, the samples were slowly cooled in the furnace. The x-ray powder diffraction data revealed single phase materials. The lattice parameters, obtained from Rietveld analysis, assuming the orthorhombic Pnma space group are a=0.5411 nm, b=0.7668 nm, c=0.5408 nm for the x=0.01 sample.

We also prepared a sample with nominal composition $La_{0.5}Ca_{0.5}Mn_{0.99}Sn_{0.01}O_3$ by thoroughly mixing high purity stoichiometric amounts of La_2O_3, $CaCO_3$, MnO_2 and SnO_2 (90% enriched in ^{119}Sn). The mixed powders were pelletized and annealed in air at 1390°C and 1410°C for approximately 300 h with intermediate grindings and reformation into pellets each time. Finally, the sample was slowly cooled in the furnace. The x-ray powdered diffraction data revealed a single phase material. The lattice parameters obtained from Rietvel analysis assuming the orthorhombic Pnma space group are a=0.54175(2) nm, b=0.76401(2) nm and c=0.5427(3) nm, very close to those of the undoped sample with x=0.5.

2.2 MAGNETIC MOMENT MEASUREMENTS

DC magnetization measurements were performed in a SQUID magnetometer (Quantum Design) for fields up to 50 kOe. ac susceptibility measurements were performed at zero dc field by means of a laboratory constructed probe at a frequency of 317 Hz and with an ac external field of amplitude h_{ac}=1.5 Oe. Four-probe resistivity measurements with and without a magnetic field were performed on a sintered bar in the temperature range 4.2-320K.

2.3 MOESSBAUER SPECTROSCOPY

MS is a nuclear gamma-resonance technique. It consists of a radioactive source emitting γ-rays of a specific energy and the absorber which contains the under study system. The energy of the emitted γ-rays is modulated by the Doppler effect by moving the source in a constant acceleration mode. The spectrum of the γ-rays consist of absorption lines which are characteristic of the hyperfine interactions of the system under study. The following physical parameters are determined from the analysis of the Moessbauer spectra:

(a) *Isomer shift*. It gives the energy shift (in velocity units) due to the difference of the s-electron density at the nucleus between the different chemical environments of the nucleus at the source and the absorber. It determines accurately the valence of the respective ion.

(b) *Quadrupole interaction.* It gives the interaction between the electric field gradient at the nucleus induced by the charges around it and the quadrupole moment of the nucleus. It allows the accurate determination of the site symmetry of the respective ion. In the absence of magnetic field (internal or external) the corresponding spectrum appears as a doublet for the Moessbauer nuclei ^{57}Fe and ^{119}Sn employed in the present study.

(c) *Magnetic hyperfine interaction.* It describes the hyperfine interaction of the nuclear and electronic spins given by the Hamiltonian H=I·A·S. The quantity A·S corresponds to the magnetic hyperfine field H_{hyp} acting on the nucleus. The absorption spectra consist of six lines and their analysis allows the determination of all three parameters discussed above.

Dynamic effects, due to the electronic spin relaxation frequency as compared to the nuclear Larmor frequency, modulate the spectra allowing the determination of the relevant physical parameters. An extreme case of such phenomena is the superparamagnetism. In this case, the magnetic moment of magnetic small clusters oscillates fast with respect to the nuclear Larmor precession and the magnetic hyperfine field averages to zero. The corresponding spectrum of this magnetic nanoparticle becomes a single or double absorption line.

The Moessbauer spectra of the present study were recorded with a conventional constant acceleration spectrometer with a 50 mCi ^{57}Co source (Fe sample), or a 5mCi Ca^{119}SnO$_3$ (Sn sample) source moving at room temperature (RT), while the absorber was kept fixed in a variable temperature cryostat equipped with a superconducting magnet (65 kOe) in a geometry perpendicular to the γ-ray beam. The calibration of the spectrometer was made with α-Fe and α-Fe$_2$O$_3$ absorbers and isomer shifts are quoted with respect to α-Fe (Fe sample) or CaSnO$_3$ (Sn sample).

3. Results and Discussion

3.1 MAGNETIC MOMENT MEASUREMNTS

3.1.1 Fe sample.
Figure 1 shows magnetic moment m vs T measurements in both zero field (ZFC) and field cooling (FC) modes for magnetic fields $0 \leq H_{ext} \leq 50$ kOe. At $H_{ext}=50$ kOe, for which the picture is simpler than for other fields, the observed behavior is strongly hysteretic. The FC branch displays a large magnetic moment of 3.1 μ_B/ion at 4.2 K. Field cooling at 50 kOe drives most of the system in a FM state (>80%, inferred from the theoretical value of 3.51 μ_B for an ideal ferromagnetic state in a 1% Fe-doped sample). In the undoped system, a 50 kOe field cooling lowers the FM-AFM transition point by 40 K. To ascertain the magnetic state of the sample we measured an m vs H loop after the field cooling to 4.2 K, which was characteristic of a soft ferromagnetic sample with a saturation magnetization of 3.1 μ_B. So not only is the system driven into an FM state after field cooling but it is trapped in this state, i.e. it remains ferromagnetic even after removing the field. At 40 kOe, the corresponding FM phase percentage is around 40, while for $H \leq 30$ kOe it is less than 20%, showing that field

cooling in higher fields dramatically increases the amount of the FM phase. For H≥40 kOe hysteresis is observed between the ZFC and FC branches close to $T_p(H_{ext})$, which means that the AFM-FM transition is of first order. For fields up to 30 kOe there is practically no difference between m(T) close to $T_p(H_{ext})$ measured on warming and cooling, suggesting that the transitions to and from the AFM are second order. Indeed, in the FC mode (10≤H_{ext}≤30 kOe) the decrease observed below $T_p(H_{ext})$ implies the existence of the AFM phase. So, the iron substitution for Mn^{3+} in the compound under study expands the AFM region to higher temperatures, in agreement with previous works in $La_{0.47}Ca_{0.53}MnO_3$ (ref. [15]) and $Pr_{0.5}Sr_{0.5}MnO_3$ (ref. [16]). An explanation for the enhancement of the FM phase after field cooling can be based on the H-T phase diagram of $La_{0.5}Ca_{0.5}MnO_3$ [6]. Decreasing the temperature in the presence of an applied magnetic field will cause supercooling of the FM phase, which will persist down to 4.2 K together with the AFM phase.

i.   Sn sample.

Figure 2a shows bulk magnetic moment m vs T measurements in field-cooling (FC) and field-cooling warming (FCW) modes in an external magnetic field of 10 kOe. The sample undergoes a rather broad paramagnet-to-ferromagnet transition at ~245 K (defined in a measurement at 50 Oe).

On cooling, m increases smoothly with a maximum magnetic moment around 140 K. Below this temperature the bulk magnetic moment drops to a value that is approximately 55% of the maximum value and then remains constant down to 4.2 K. The value of m at 4.2 K is rather large and certainly cannot be explained by an antiferromagnetic or canted antiferromagnetic phase. The presence of a ferromagnetic phase is needed. Warming from 4.2 K shows a strong hysteresis in the region 100-185 K, indicating the first-order nature of this transition. Under an applied field of 5 T the transition is smeared out and the bulk magnetic moment reaches 80% of its full FM saturation value at low temperatures (for an ideal ferromagnetic state in the sample it is 101.5 emu/g or 3.5 μ_B). The ac susceptibility measurement shown in fig.2b is in agreement with the dc data.

Resistivity measurements in zero and 50 kOe fields performed following the same sequence as the magnetic measurements are shown in fig.3. At zero field, on cooling, the resistivity increases below 250 K and shows two small maxima at T~160 and ~100 K, whereas on warming a broad peak appears at T~140 K, showing in strong hysteresis. In a 50 kOe measurement the hysteresis disappears and only a broad maximum centered around 150 K is observed. The magnetoresistance effect ($\rho(0)-\rho(50$ kOe))/ $\rho(0)$ is ≈ 85% at the maximum of the resistivity curve (around 140 K).

The maximum value of the resistivity observed at T~140 K indicates that above this temperature "activated" conduction occurs and below this temperature, conduction takes place probably through a percolation path created by FM regions in the sample, in agreement with the magnetization data

In a recent study Roy et al. [7] have investigated the region around x=0.5 and found differences between the x value and the Mn^{4+} fraction which reflect on the magnetization and resistivity behavior of the corresponding samples. Comparing our

488

bulk measurements with their detailed data we can conclude that the Sn-doped sample is slightly below x~0.5 (e.g. x=0.49) in the magnetic (T vs x) phase diagram of the $La_{1-x}Ca_xMnO_3$ manganites.

3.2 MOESSBAUER SPECTRA

3.2.1 Fe sample.

Figure 4 shows the temperature evolution of MS. The characteristic of these spectra is that although at 4.2K they display a normal 6-line pattern, as the temperature increases an asymmetric line broadening appears which increases gradually with temperature until

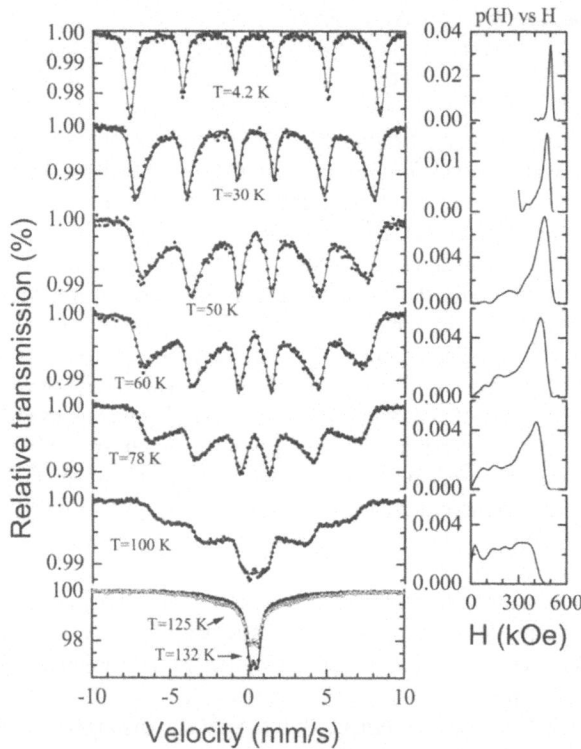

Figure 4 Moessbauer spectra of $La_{0.5}Ca_{0.5}Mn_{0.99}Fe_{0.01}O_3$ in the temperature range 4.2-132 K. The solid line through the experimental points corresponds to spectrum calculated using the hyperfine field distribution displayed on the right of the spectrum. At 125 and 132 K a spin relaxation model was employed as explained in the text.

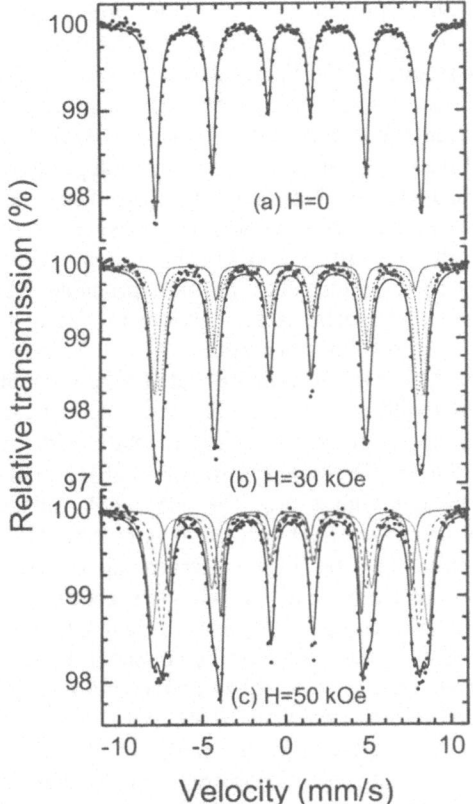

Figure 5 Moessbauer spectra of $La_{0.5}Ca_{0.5}Mn_{0.99}Fe_{0.01}O_3$ at 4.2 K after ZFC at H=0, 30 and 50 kOe.

the spectrum collapses to a paramagnetic doublet at the transition temperature T_N (139K). Furthermore, close to this temperature, a paramagnetic doublet coexists with the magnetic component. By using a hyperfine field continuous distribution program [17] we were able to fit the spectra in the whole temperature range. The result of this fit is shown in figure 4. We should note here that this fitting procedure was not satisfactory close to T_N where the paramagnetic component appears. The gradual disappearance of the resolved Zeeman hyperfine structure upon raising the temperature has been observed in all Moessbauer studies of the doped manganites reported so far [13, 18-23]. As we have mentioned in previous work for the x=0.33 sample [23], its origin may have various reasons of static or dynamic origin. For systems doped with the ^{57}Fe Moessbauer probe the static case is well represented by the modification of the spin excitation spectral density arising from the magnetic interaction of the Fe^{3+} impurity (S=5/2) with the neighboring magnetic moments of the Mn ions (S=2,3/2). Near T_N however, the static distribution model fails to reproduce the prominent doublet feature. Its coexistence with the magnetic component and its increase with temperature as we approach T_N is reminiscent of superparamagnetic behavior. Such a behavior could be attributed to the formation of spin clusters with sizes of ~100 Angstroms as we raise the temperature and approach T_N. This behavior is demonstrated in a clear way in the corresponding spectra of the Sn sample (*vide infra*). We have examined this possibility of spectral broadening from dynamic reasons by employing a spin relaxation model developed by van der Woude and Dekker [24] with free parameters the spin relaxation rate Ω and the order parameter η while we kept the experimental linewidth and the hyperfine magnetic field constant to their 4.2K values. The results are satisfactory and they improve a lot if we use two relaxing components instead of one, implying thus that we have a distribution of spin-cluster sizes (see the spectra at 125 and 132K). Thus we may conclude that the most probable reason for the lineshape broadening of the magnetic spectra is the mixture of FM and AF phases which at higher temperatures break down to spin clusters.

Figure 5 shows the MS at 4.2 K after ZFC at H=0, 30 and 50 kOe. The zero field spectrum could be fitted with two sextets in order to account for the small asymmetry between lines 1 and 6, but we preferred a simpler model with a narrow distribution of hyperfine fields (ΔH=5 kOe). The isomer shift is δ=0.463(1) mm/s indicating that Fe is high spin 3+; i.e. an S-state with S=5/2. The quadrupole interaction is small (-0.015 mm/s) as expected for a nearly perfect octahedral symmetry. The H_{hf}=498 kOe value is considerably lower than the 564 kOe observed in $LaFeO_3$ [25], where purely superexchange interactions are present. The 50 kOe spectrum (fig.5) displays a six line pattern with broadened lines (compared to the zero field and 30 kOe spectra). It is clear that in lines 2 and 5 there is a narrow spectral component. Lines 1 and 6 are broadened and have an unresolved structure. In order to understand this spectrum we outline the MS expected for an antiferromagnet, a canted AF and a mixture of AF and FM phases. First, as pointed out by Wertheim et al. [26], for the case of a polycrystalline antiferromagnet with large magnetic anisotropy in an external field, the MS are modulated by a distribution of hyperfine fields. The spin system remains aligned in the easy direction and the net field at the nucleus ranges from H_0-H to H_0+H (H_0 is the internal and H is the external field). The resulting MS are broad but not similar to the present spectra. For a canted antiferromagnet the situation is similar. For a mixture of an AF phase with large anisotropy and a FM phase with small anisotropy, we expect the ferromagnetic component to be narrow, the field reduced by an amount equal to the applied field, and a 3:4:1 line intensity ratio (since the magnetic field is perpendicular to γ-rays) and the antiferromagnetic component to be as described above. Thus, the spectrum shows the behavior of a mixture of AF and FM phases. Following this conclusion, we modeled the spectrum with one component with $\Gamma/2$=0.2 mm/s and intensity ratio 3:4:1 and two components (H_{hf}=483 and 517 kOe) modulated by a distribution of hyperfine fields (ΔH=7 kOe) with intensity ratio 3:x:1 (x=1.7). This intensity ratio is in contradiction with a random distribution of spin axes with respect to the external field. A logical explanation for this discrepancy can be given by supposing that we have rotation of the moments out of the plane. The FM component represents 20% of the total spectrum with H_{hf}=458 kOe.

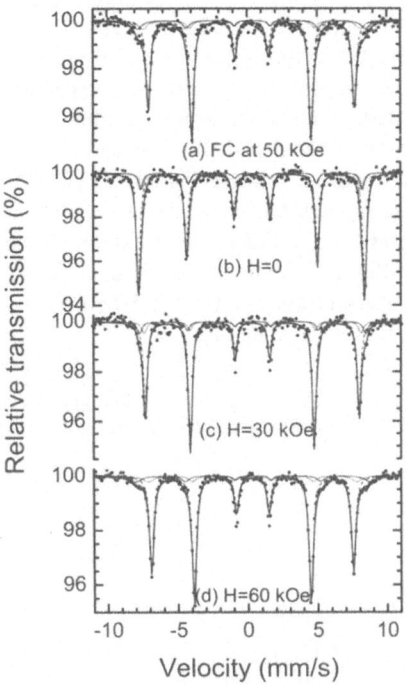

Figure 6 Moessbauer spectra of $La_{0.5}Ca_{0.5}Mn_{0.99}Fe_{0.01}O_3$ at 4.2 K (a) after FC at H=50 kOe, (b) after removing the 50 kOe magnetic field, (c) increasing the field to 30 kOe and (d) further increasing the field to 60 kOe.

Figure 6 shows the MS at 4.2 K after cooling the sample in a 50 kOe field. It is a narrow 6-line pattern whose absorption line intensities have a 3:4:1 ratio, showing that the magnetic moments have been oriented parallel to the external magnetic field. The hyperfine field is H_{hf}=458 kOe indicating that the spin of Fe is ferromagnetically coupled to its Mn neighbors. This can be accomplished only if the Mn ions themselves are ferromagnetically coupled. It is interesting to note that the average $\varepsilon=(1/8)e^2qQ(3\cos^2\theta-1)$ is exactly zero in this case because while the hyperfine field is parallel to the applied field, the z-axis of the electric field gradient is randomly distributed in the surface of a sphere. To model the MS we used two extra subspectra, consistent with the spectrum and fit presented in fig. 5c, to fit the AF component. These subspectra have broad lines (H_{hf}=483 and 507 kOe, ΔH=13 kOe) and 3:2:1 line intensity ratios. The AFM component represents approximately 20% of the spectrum. The above picture agrees with the H-T phase diagram for the undoped compound proposed by Gang Xiao et al. [6]. In that work it was shown that there is a line $H_m(T)$ representing a first order transition between the AF charge ordered phase and a FM melted (charge disordered) phase. Due to the first order nature of the transition, supercooling leads to the stabilization of the metastable FM phase at 4.2 K, even

492

though the thermodynamically stable phase is the AF. Fragments of the AF phase can also exist, resembling the situation in $Nd_{0.5}Sr_{0.5}MnO_3$ [27].

Removing the external field (Fig.6b) results in a narrow 6-line pattern also fitted with three subspectra. The major component, approximately 80% of the spectrum ($H_{hf}=503$ kOe), is attributed to the FM phase since the hyperfine field is increased by almost exactly by the removed field and the line width is narrow (0.165 mm/s). This means that the sample has locked into the FM phase, i.e. the amount of the FM phase remains unchanged on removing the magnetic field. Further, without warming the sample we applied a 30 and 60 kOe field, which did not change the relative areas of the two phases. Moreover, after FC down to 4.2 K in a field of 40 kOe, reducing the magnetic field to 8 kOe also did not change the 50:50 ratio of the two phases. This is additional evidence for a first order transition.

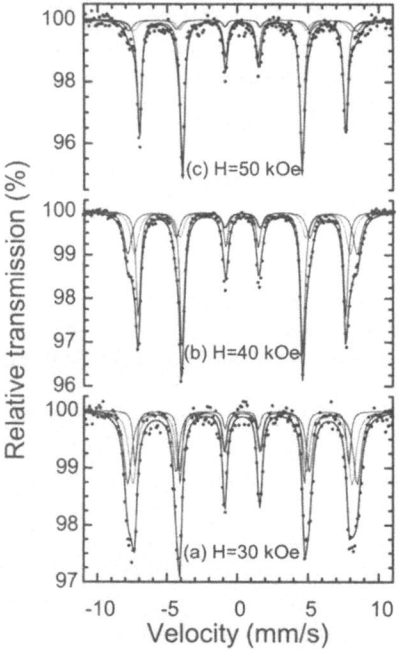

Figure 7 Moessbauer spectra of $La_{0.5}Ca_{0.5}Mn_{0.99}Fe_{0.01}O_3$ at 4.2 K after field cooling in H=30 kOe,H=40 kOe, and 50 kOe.

Figure 7 shows the MS after field cooling at H=30, 40 and 50 kOe (the 50 kOe spectrum is shown again for comparison). All three spectra were analyzed using a three component fit, as in the ZFC spectrum at H=50 kOe (Fig. 5c). The 30 kOe spectrum can be analyzed as a mixture of AF and FM components in a ratio 90:10, at 40 kOe a 50:50 ratio and at 50 kOe the ratio becomes 20:80. These results also support a first order phase transition. After field cooling in magnetic fields H≤30 kOe, there is no supercooling of the high temperature FM phase. Summarizing, in the FC mode we have the behavior predicted by a first order transition, while ZFC reveals a mixture of AF and FM components. The overall picture from the MS at 4.2 K is consistent with the H-T diagram proposed by Gang Xiao et al. [6]. When the sample is cooled down to 4.2 K in zero field, it consists mainly of an AF phase. After field cooling in a sufficiently high magnetic field (H≥40 kOe), a considerable part of the sample (which increases with increasing magnetic field) becomes ferromagnetic due to supercooling of the high temperature FM phase. In this case, keeping the temperature at 4.2 K, the amounts of the FM/AFM phase remain unchanged whether we reduce or increase the magnetic field.

3.2.2 Sn sample

Figure 8 shows Moessbauer spectra at 300 K, 4.2 K and 4.2 K after field-cooling in the presence of an external field of 60 kOe. The RT spectrum consists of a single absorption line with linewidth $\Gamma=0.78$ mm/s and isomer shift $\delta=0.10$ mm/s. This latter value corresponds to a 4+ valence state of Sn, indicating that Sn ions substitute for Mn^{4+}. The 4.2 K spectrum is fitted with three magnetic components, one with a large hyperfine field ($H_{hf}=249$ kOe) and the other two with smaller hyperfine fields (93 and 47 kOe).

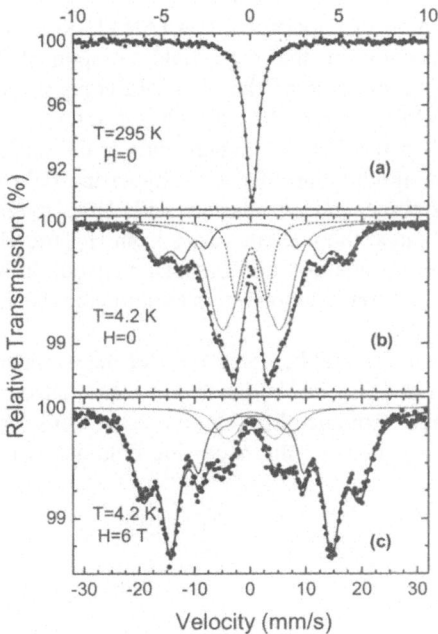

Figure 8 Moessbauer spectra (a) at room temperature, (b) at T=4.2 K in zero applied field, and (c) at T=4.2 K after field-cooling in an applied field of 60 kOe.

The FC spectrum shows a dramatic increase of the high field component at the expense of the low field components. An analogous increase (actually a reversal of the FM and AF fractions) was observed in the Fe-sample spectra and, as discussed above, it was attributed to the "melting" of the CO-AF state to a FM state.

The fact that the Sn nucleus senses transferred hyperfine field from the overlapping of the 3d orbitals of the Mn neighbors in the surrounding Mn octahedron with its 5s orbitals, leads us to the following assignment of the three components in the 4.2 K Sn spectrum. The large field component (H_{hf}=249 kOe) arises from a ferromagnetic environment where all the Mn moments are parallel and add up to a large transferred hyperfine field at the Sn nucleus (*FM component*). The 93 kOe component arises from an antiferromagnetic environment where the antiparallel moments contribute transferred hyperfine fields with opposite sign and which tend to cancel. (*AF component*). Since in the CE structure [7,28] there are four Mn^{3+} ions with antiparallel moments and two Mn^{4+} ions with parallel moments in the Mn octahedron surrounding the Sn^{4+} probe, the cancellation is not total and a hyperfine field value of ≈1/3 of the field of the FM component would be expected for the AF component. As mentioned above, this component transforms to the FM component under the influence of cooling in a 6 T field. We will discuss the nature of the third component (CL) with a hyperfine field of 43 kOe at a later stage.

The FC spectrum at 4.2 K (Fig. 8c) was modeled with one component (H_{hf}=288 kOe) with line intensity ratio 3:4:1 and two components (H_{hf}=120 and 80 kOe) modulated by a distribution of hyperfine fields (ΔH=21 kOe) with line intensity ratio 3:2:1. The relative intensity of the high field component (FM phase) increases considerably (80%) at the expense of the low-field components. The AF component present in the zero field spectrum has a reduced intensity of 20%, and the CL component is absorbed in the FM component due to the applied field. The two AF components describe the modulation of the net hyperfine field seen by the nucleus in the case of a polycrystalline antiferromagnet with large magnetic anisotropy in an external field. The net field at the nucleus varies from H_0-H to H_0+H (H_0 is the internal and H is the external field), causing the Moessbauer spectra to be broadened [26]. The overall effect of the applied field to the system (and thus to the spectrum) is the same as that observed in Fe-sample.

An important difference should be noted here for the hyperfine fields probed by the [119]Sn and [57]Fe nuclei. The [119]Sn ion is diamagnetic and detects the vector sum of the hyperfine fields transferred from neighboring moments. Thus opposite moments have a canceling effect. However, for [57]Fe the hyperfine field arises from the iron ion's own moment.

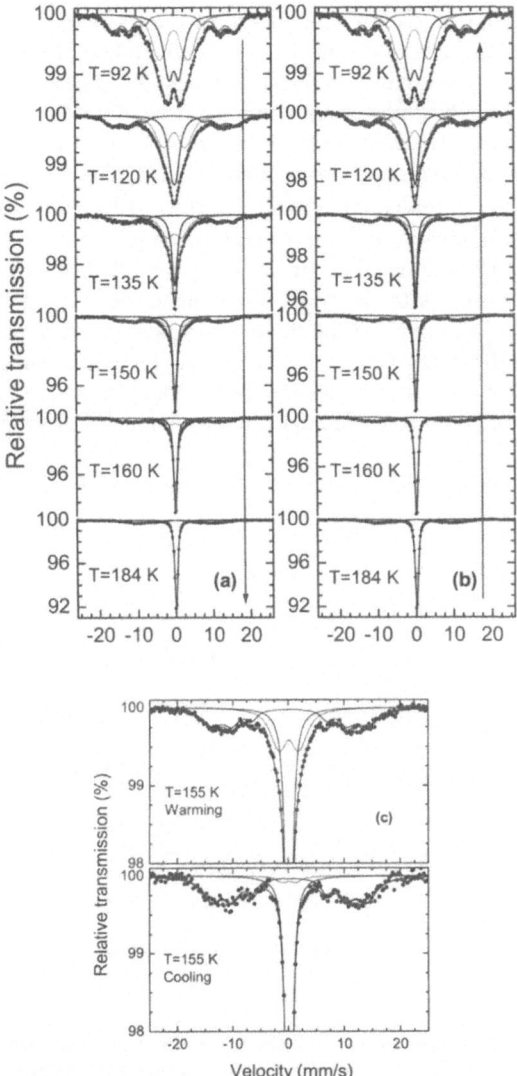

Figure 9 Representative Moessbauer spectra of $La_{0.5}Ca_{0.5}Mn_{0.99}Sn_{0.01}O_3$ in the temperature range 92 to 184 K upon (a) warming and (b)cooling. (c) Magnified view of the spectra at 155 K both upon warming and cooling.

The polarisation of the iron moment arises from the (super)exchange interactions of the Mn and Fe spins. The superexchange interaction may result in ferromagnetic or antiferromagnetic alignment, but the iron moment remains the same. Thus, parallel neighboring spins are equivalent to antiparallel spins, i.e. FM and AF environments result in the same hyperfine field.

We turn now to the temperature dependence of these three components. Figure 9 (upper panel) shows some characteristic spectra upon warming (fig.9a) and cooling (fig. 9b) in zero field. The general features of the spectra do not change up to ~120 K. Above this temperature a single peak emerges from the central region of the spectra whose intensity increases quickly as the temperature is raised. The FM component persists up to ~230 K and above this temperature the spectra consist of a single line only. Hysteresis is evident in the central part of these spectra. The wings which are present in the warming mode in the temperature range ~120 to ~180 K are reduced considerably at the corresponding temperature in the cooling mode (compare the spectra at 155 K in the two modes in the lower panel of fig. 9. We have fitted the spectra throughout the whole temperature range with the three components of the 4.2 K spectrum. In our fitting procedure the linewidths of the three components were kept constant to the RT value, the isomer shift values were the same for the three components allowing a small temperature variation due to the second order Doppler shift, and the quadrupole interaction was kept equal to zero due to the cubic symmetry of the Sn site.

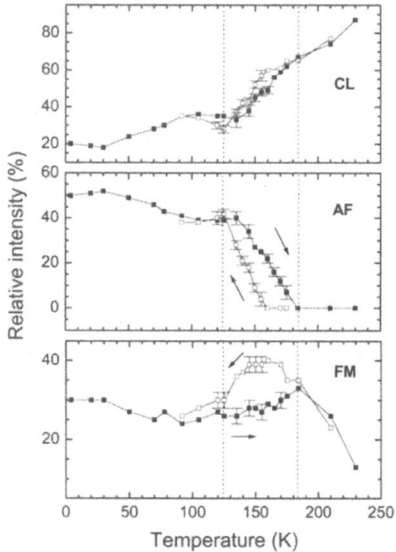

Figure 10 The temperature variation of the relative intensities of the components which appear in the Moessbauer spectra of $La_{0.5}Ca_{0.5}Mn_{0.99}Sn_{0.01}O_3$. The dashed lines mark the temperature range where hysteresis is observed.

The free parameters were the hyperfine field, the intensity of each spectral component, and a distribution of the hyperfine fields ΔH. The latter arises from small deviations of the Sn positions to which the transferred hyperfine field at the Sn nuclei is very sensitive.

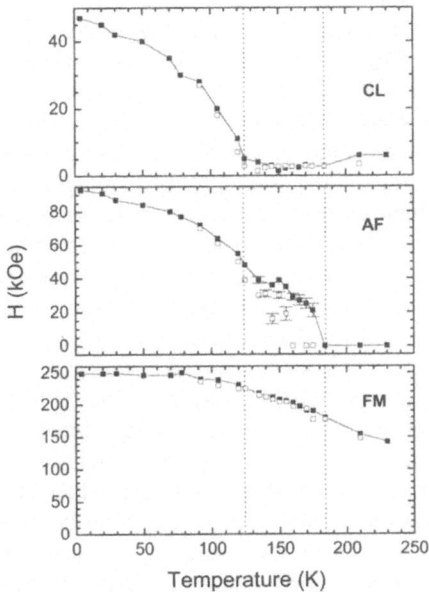

Figure 11 The temperature variation of the hyperfine fields of the components which appear in the Moessbauer spectra of $La_{0.5}Ca_{0.5}Mn_{0.99}Sn_{0.01}O_3$. The dashed lines mark the temperature range where hysteresis is observed.

Figure 10 shows the main result of this work: the temperature variation of the relative intensity of each component. The intensities of the FM and AF phases show hysteresis in the temperature range 120-185 K, as observed in the magnetic moment and resistivity measurements described in the previous section. In particular, on warming, the FM spectral area remains constant up to ~185 K and then drops, while on cooling it increases to 40% at ~150 K, then drops to ~25% at 100 K, and remains constant down to 4.2 K, in agreement with the magnetization data (Fig 2a), and with the La NMR results of Allodi et al. [29] obtained in the cooling mode. The opposite behavior is observed for the AF spectral area, which disappears at ~180 K on warming and reappears at ~150 K on cooling, thereby marking the charge-ordering temperature. It should be noted that this phase is absent in the zero field La NMR data due to the complete cancellation of the transferred hyperfine interactions from the neighbor Mn ions [29]. Also, it cannot be discerned in the [57]Fe Moessbauer spectra of the Fe sample since, as mentioned above, it coincides with the FM component. Thus, the [119]Sn probe is ideal to detect the AF phase. Neutron diffraction studies on the crystallographic and magnetic structure of $La_{1-x}Ca_xMnO_3$ for x=0.47, 0.50 and 0.53 have been published very recently by Huang et al. [30]. The results of this work are in excellent agreement with the present work regarding the temperature evolution of the phase separation and the hysteresis observed for the three doping regimes.

Hysteresis is also observed in the temperature variation of the hyperfine field of the AF phase (fig.11). The temperature variation of H_{hf} for the FM phase follows a typical meanfield-like curve, as for the case of a ferromagnet, up to 230 K, above which this component disappears. It is interesting to note that at this temperature the H_{hf} value is still considerable (142 kOe). This discontinuity has also been observed in La NMR experiments and it has been attributed to a first-order transition [29,31]. This zeroing of the volume of the FM phase and not of the hyperfine field (which is contrary to ordinary ferromagnets) implies the existence of a mixed state of FM and paramagnetic phases near T_c, thus leading to the possibility of ferromagnetic clusters [32].

Examining the temperature behavior of the third component (CL component) we notice that its hyperfine field decreases smoothly with temperature and drops to zero at ~125 K (fig. 11). Above this temperature this component appears as a single paramagnetic peak with some broadening, with its intensity increasing with temperature (see fig. 10 and the Moessbauer spectra in fig. 9). The zeroing of the hyperfine field of the CL component coincides with the temperature at which the charge-ordering is completed on cooling and is beginning to breakdown on warming. The small saturation field (47 kOe) of this component does not allow its assignment to any AF spin configuration of the Mn nearest neighbors. Thus, we ascribe this component to ferromagnetic clusters which exhibit a reduced hyperfine field either due to relaxation phenomena or to a reduced moment within the clusters. These clusters are present at 4.2 K and their spectral area remains practically constant up to ~125 K. Above this temperature they exhibit superparamagnetic behavior and their intensity increases rapidly at the expense of the AF and FM phases. The existence of ferromagnetic clusters in manganites has been used by Moreo et al. [33] to account for the density of states in the framework of one- and two-orbital models. Recently, Allodi et al. [34] have shown by [55]Mn NMR that the field-induced FM phase in $Pr_{0.5}Sr_{0.5}MnO_3$ develops by the nucleation of microscopic ferromagnetic domains, which may be similar in nature to the ferromagnetic clusters proposed here.

It should be noted that a paramagnetic-like component appears for $T<T_c$ in all the reported Moessbauer experiments of La manganites doped with Fe [13,18,19,20,22], Co [12,23] or Sn, [21] for values of Ca(Sr) doping between 0.17 and 0.60, which covers the phase diagram range where the ground state is FM ($x\leq0.45$) and AF ($x\geq0.55$). In addition, neutron scattering experiments for x=0.33, [35] and x=0.17, [36] have detected a quasielastic component for $T<T_c$ which becomes dominant as the temperature approaches T_c. We speculate that this component is associated with ferromagnetic clusters in all these cases. It appears that both Moessbauer and neutron scattering techniques have the sensitivity to directly detect the formation of ferromagnetic clusters. The coincidence of the zeroing of the hyperfine field of the CL component with the completion and breakdown (on cooling and warming respectively) of the charge ordering indicates that the cluster development and growth is directly associated with charge ordering or charge localization.

In summarizing the Sn Moessbauer data, we notice that the hysteretic range of the system is characterized by two temperatures, T_1=120 K and T_2= 185 K, in agreement with the bulk measurements. In the temperature interval 4.2<T<120 K the system comprises a FM phase (~30%), an AF phase (~50%) and FM clusters (~20%). The FM and AF phases display hysteresis in the range 120<T<185 K. The disintegration (formation) of the AF phase on warming (cooling) is clearly manifested through Moessbauer spectroscopy in this temperature interval. For T>125 K, ferromagnetic clusters with superparamagnetic behavior grow rapidly in number and dominate up to the ferromagnetic to paramagnetic transition at T_c=225 K. This picture for the $La_{0.5}Ca_{0.5}MnO_3$ system is in agreement with the phase separation models [4,5] and supplements the electron diffraction investigations [8,9] and the La NMR data [29,32].

4. Conclusions

Moessbauer spectroscopy reveals that both ^{57}Fe and ^{119}Sn probes monitor the phase evolution during the charge-ordering transition in $La_{0.5}Ca_{0.5}MnO_3$. Complementary bulk magnetic moment and resistivity measurements are in agreement with the Moessbauer findings. Below the charge-ordering transition, phase separation occurs into FM and AF phases down to 4.2 K. The hysteresis observed shows that the FM-AF transition is of first order. The phase separation is associated with the existence of FM clusters, which dominate in the system above the CO transition.

5. References

1. For recent reviews see: Dagotto, E., Hotta, T., Moreo, A. (2001) Colossal Magnetoresistant materials: the key role of phase separation, *Physics Reports* **344**, 1-153. Nagaev, E. L. (2001) Colossal-magnetoresistance materials: manganites and conventional ferromagnetic semiconductors, *Physics Reports* **346**, 387-531.

2. Wollan, E. O., Koehler, W. C. (1955) Neutron diffraction study of the magnetic properties of the series of perovskite-type compounds [(1-x) La, xCa]MnO₃, *Phys. Rev.* **100**, 545-563.

3. Schiffer, P., Ramirez, A. P., Bao, W., Cheong, S.-W. (1995) Low temperature magnetoresistance and the magnetic phase diagram of $La_{1-x}Ca_xMnO_3$, *Phys. Rev. Let.* **75**, 3336-3339.

4. Nagaev, E. L. (1998) Underdoped manganites: canted antiferromagnetic ordering or two-phase ferro-antiferromagnetic state?, *J. Exp Theor. Phys.* **87**, 1214-1220.

5. Moreo, A., Yunoki, S., Dagotto, E. (1999) Phase separation scenario for manganese oxides and related materials, *Science* **283**, 2034-2040.

6. Xiao, G., et al (1997) Magnetic field induced properties of manganite perovskites with colossal magnetoresistance, *J. Appl. Phys.*, **81** (8) 5324-5329.

7. Roy, M., Mitchell, J. F., Ramirez, A. P., Schiffer, P. (1999) A study of the magnetic and electrical crossover region of $La_{0.5}Ca_{0.5}MnO_3$, *J. Phys. Cond. Matt.*, **11**(25) 4843-4859;

8. Mori, S., Chen, C. H., Cheong, S-W. (1998) Paired and unpaired charge stripes in the ferromagnetic phase of $La_{0.5}Ca_{0.5}MnO_3$, *Phys. Rev. Lett.*, **76**, 3972-3275; Pissas, M., Kallias, G. (2002), Phase diagram of the $La_{1-x}Ca_xMnO_3$ compound for $0.5 \leq x \leq 0.9$ cond-matt. 0205410.

9. Chen, C. H. Cheong, S-W. (1996) Commensurate to Incommensurate charge ordering and its real -space images in $La_{0.5}Ca_{0.5}MnO_3$, *Phys. Rev. Lett.* **76**, 4042-4045.

10. Kallias, G., Pissas, M., Hoser, A. (2000) Neutron diffraction study of $La_{0.5}Ca_{0.5}MnO_3$ under an external magnetic field, Physica B **276-278**, 778.

11. Heffner, R. H. et al (1996) Ferromagnetic ordering and unusual magnetic ion dynamics in $La_{0.67}Ca_{0.33}MnO_3$, *Phys. Rev. Lett.* **77**, 1869-1972.

12. Chechersky, V. et al (1999) Comparative studies of the behavior of $La_{0.8}Ca_{0.2}MnO_3$ with different oxygen isotope substitutions, *J. Phys. Condens. Matter*, **11**, 8921-8931.

500

13. Kallias, G., Pissas, M., Devlin, E., Simopoulos, A., Niarchos, D. (1999) Moessbauer study of ^{57}Fe-doped $La_{0.5}Ca_{0.5}MnO_3$, *Phys. Rev. B,* **59,** 1272-1276.

14. Simopoulos, A., Kallias, G., Devlin, E. Pissas, M. (2000) Phase separation in $La_{0.5}Ca_{0.5}MnO_3$ doped with 1% ^{119}Sn detected by Moessbauer spectroscopy, *Phys. Rev. B,* **63** 054403-1-054403-7

15. Ahn, K. H., Wu, X. W., Liu, K., Chien, C. L. (1996) Magnetic properties and colossal magnetoresistance of La(Ca)MnO3 materials doped with Fe, *Phys. Rev. B,* **54,** 15299.

16. Maignan, A. Martin, C., Raveau, B. (1997) *Z. Phys. B* **102,** 19.

17. Le Caer, G., Dubois, J. M. (1979) *J. Phys. E* **12,** 1083.

18. Pissas, M., Kallias, G., Devlin, E., Simopoulos, A., Niarchos, D. (1997) Moessbauer study of $La_{0.75}Ca_{0.25}Mn_{0.98}Fe_{0.02}O_3$ compound, *J. Appl. Phys.* **81,** (8) 5770-2.

19. Ogale, B. et al (1998) Transport properties, magnetic ordering, and hyperfine interactions in Fe doped $La_{0.75}Ca_{0.25}MnO_3$: Localization-delocalization transition. *Pys. Rev. B* **57,** 7841-7845.

20. Tkatchuk, A. et al (1998) Dynamics of phase stability and magnetic order in magnetoresistive $La_{0.83}Sr_{0.17}Mn_{0.98}{}^{57}Fe_{0.02}O_3$, *Phys. Rev.* **57,** 8509-8517.

21. Simopoulos A., Kallias, G., Devlin, E., Panayotopoulos. I., Pissas, M. (1998) Spin fluctuations in $La_{2/3}Ca_{1/3}MnO_3$ probed by ^{57}Fe and ^{119}Sn Moessbauer spectroscopy, *J. Magn. Magn. Mater.* **177-181,** 860-861

22. Simopoulos, A., Kallias, G., Devlin, E., Panayotopoulos, I., Niarchos, D., Christides, C., Sonntag, R. (1999) Study of Fe-doped $La_{1-x}Ca_xMnO_3$ (x=1/3) using Moessbauer spectroscopy and neutron diffraction. *Phys. Rev. B* **59,** 1263-1271.

23. Chechersky, V. et al (1999) Evidence for breakdown of ferromagnetic order below T_c in the manganite $La_{0.8}Ca_{0.2}MnO_3$, *Phys. Rev. B* **59,** 497-502.

24. van der Woude, F. Dekker, A. J. (1965) The relation between magnetic properties and the shape of Moessbauer spectra, *phys. stat.solidi* **9,** 775.

25. Eibschutz, M., Shtriktman, S., Treves, D. (1967) Mössbauer Studies of ^{57}Fe in Orthoferrites, *Phys. Rev.* **156,** 562-577.

26. Wertheim, G. K., Buchanan, D.N.E., Wernick, J.,H. (1970) *Solid State Com.* **8, 2173.**

27. Kuwahara, H., Tomioka, Y., Asamitsu, A., Moritomo, Y., Tokura, Y. (1995) A first order phase transition induced by a magnetic field, *Science* **270,** 961-963.

28. Radaelli, P.G., Cox, D. E., Marezio, M., Cheong, S.-W. (1997) Charge, orbital, and magnetic ordering in $La_{0.5}Ca_{0.5}MnO_3$, *Phys. Rev. B* **55,** 3015-3023.

29. Allodi, G., De Renzi, R., Licci, F., Pieper, M. (1998) First order nucleation of charge -ordered domains in $La_{0.5}Ca_{0.5}MnO_3$ detected by ^{139}La and ^{55}Mn NMR, *Phys. Rev. Lett.* **81,** 4736-4739.

30. Huang, Q. et al (2000) Temperature and field dependence of the phase separation, structure, and magnetic ordering in $La_{1-x}Ca_xMnO_3$ (x=0.47, 0.50, and 0.53), *Phys. Rev. B* **61,** 8895-8905.

31. Papavassiliou, G. et al (1998) NMR in manganese perovskites: Detection of spatially varying electron states in domain walls, *Phys. Rev. B* **58,** 12237-12241.

32. Dho, J., Kim, I., Lee, S. (1999) Phase separation in $La_{0.5}Ca_{0.5}MnO_3$ observed by ^{55}Mn and ^{139}La NMR, *Phys. Rev. B* **60,** 14545-14548.

33. Moreo, A., Yunoki, S. Dagotto, E. (1999) Pseudogap formation in models for manganites, *Phys. Rev. Lett.,* **83,** 2773-2776.

34. Allodi, G., De Renzi, R., Solzi, M. Kamenev, G., Balakrishnan, G., Pieper, M. W. (2000) Field-induced segregation of ferromagnetic nanodomains in $Pr_{0.5}Sr_{0.5}MnO_3$ detected by ^{55}Mn NMR, *Phys. Rev. B* **61,** 5924-5927.

35. Lynn, J. W. et al (1996) *Phys. Rev. Lett.* **76,** 4046-4049.

36. Vasiliu-Doloc, L., et al (1997) Neutron scattering investigation of the structure and spin dynamics in $La_{0.85}Sr_{0.15}MnO_3$, *J. Appl. Phys.* **81,** 5491-5493.

EFFECTS OF INTERFACES ON THE PROPERTIES
OF NANOSTRUCTURED Ni-[Cu(II)-C-O] and CoCrPt FILMS

G.M. CHOW
Department of Materials Science
National University of Singapore
Kent Ridge, Singapore 119260, Republic of Singapore

Abstract. The control of the composition, structure (long range and short range orders), texture, and interfaces is important in order to control the magnetic properties of nanostructured films. In this paper an overview of some of our recent work on the effects of interphase interface and composition of textured long range order on the properties of nanostructured Ni-[Cu(II)-C-O] and CoCrPt films is presented.

1. Introduction

Nanostructured polycrystalline films with grain size less than 100 nm have unique properties [1]. These films have a significant amount of grain boundaries and interfaces. For many advanced materials such as those used in magnetic applications, careful control of the grain size, global composition, composition of a specific textured Bragg peak, long range and short range structural orders and interphase interfaces, is of essential importance in order to achieve the desirable properties.

Nanostructured Ni-[Cu(II)-C-O] and CoCrPt films were prepared using non-aqueous electroless polyol method and sputtering, respectively. The films were investigated using x-ray scattering, anomalous x-ray scattering (AXS), extended x-ray absorption fine structure spectroscopy (EXAFS), transmission electron microscopy (TEM), x-ray photoelectron spectroscopy (XPS) and vibrating sample magnetometry (VSM). Details of experimental methods can be found in the references. The effects of interphase interface and composition of a textured long range order on the magnetic properties are discussed.

T. Tsakalakos et al. (eds.) pgs. 501 - 509
Nanostructures: Synthesis, Functional Properties and Applications;
© 2003 Kluwer Academic Publishers.

502

2. Results and Discussions

2.1 ENHANCED MAGNETIZATION OF POLYOL DEPOSITED Ni/[Cu(II)-   C-O] FILMS

Nanostructured magnetic granular composite films with interesting properties have received much research interest [1,2,3]. These granular composite films consist of nanostructured magnetic particles embedded in a non-magnetic matrix. For magnetic alloys, the properties depend on the type, number, distance and symmetry of the neighbors of an atom in question. The magnetic moment and the Curie temperature (T_c) of a ferromagnetic solid solution such as Ni-Cu decrease with increasing concentration of non-magnetic Cu atoms.

The synthesis of micron-size and nano-size powders by the non-aqueous polyol process has been investigated [4,5,6,7]. This process involves the dissolution of solid precursor, reduction of metal ions and precipitation of metal powders in refluxing ethylene glycol at 194 °C. Ethylene glycol serves as both a solvent and a reducing agent. The polyol method has also been exploited for preparation of nanostructured films [8]. Film deposition occurs on suitable conductor or insulator substrates in a simple single-step process, without any pre-deposition surface treatment such as the catalyzation of insulator surface as required in traditional aqueous electroless plating. This method has been used to deposit nanostructured films of Cu [9,10], Ni_xCo_{100-x} [11,12,13,14,15,16] and Ni_xFe_{100-x} [17,18].

Recently it has been reported that Ni/[Cu(II)-C-O] films, deposited using the polyol process, showed unique magnetic properties such as anomalous enhancement of magnetization, high anisotropy field and reduced Curie temperature [19]. Film deposition was carried out on polished polycrystalline Cu substrates that were immersed in a mixture of nickel acetate and refluxing ethylene glycol for 5 min after the refluxing began.

The film thickness was about 200 nm as estimated from cross-sectioned samples using SEM. The magnetization (emu/g) was obtained by normalizing the measured magnetic moment with the deposited film weight. The grazing-incidence XRD of as-deposited films showed only Ni (111) peak without any crystalline oxide peaks. The EXAFS results showed that the average local atomic environment of Ni was disordered as compared to the reference bulk Ni. The features at the near-absorption edge indicated that Ni was not oxidized. The XPS chemical depth profile of deposited films showed the existence of Cu and its concentration decreased with increasing Ni film thickness from the substrate-film interface. Consistent with the XRD and EXAFS data, the XPS results showed that Ni existed as Ni^0 without any sign of oxidation. However, Cu existed as Cu^0 and Cu (II). The film composition, based on XPS depth profiling data and analysis, was estimated as ~ 82 and 18 weight % for Ni and the [Cu(II)-C-O] complex, respectively. It was suggested that chemical reaction of the solvent with the Cu substrate at the initial stage of film deposition leach Cu ions into the growing Ni film.

The bright field TEM of deposited film is shown in Fig. 1. The Ni particles (dark contrast) were dispersed in a matrix (light contrast), which was presumably the Cu (II) complex. Electron diffraction and dark field imaging showed that Ni particles consisted of polycrystalline crystallites. High resolution TEM lattice imaging further confirmed the polycrystalline Ni particles in the amorphous matrix.

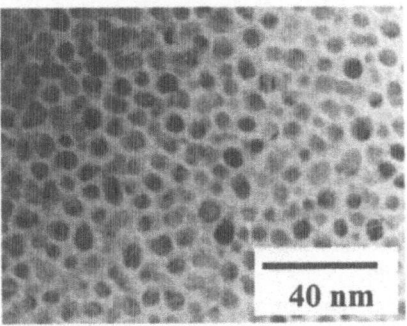

Figure 1. Bright field TEM image of polyol deposited Ni-[Cu{(II)-C-O] film

Figure 2 shows that the saturation magnetization M_s of Ni/Cu(II) complex film at room temperature was 112 emu/g, which was about 2 times that of bulk Ni (54.4 emu/g). The film showed in-plane (//) easy magnetization. The maximum theoretical anisotropy field H_k ($H_k = 4\pi M_s$) for thin film shape anisotropy should be about 6.1 kOe for Ni. The anisotropy field H_k of the Ni/Cu(II) complex film was about 8.1 kOe, which was significantly higher than that of the theoretical value. The high anisotropy cannot be explained by that Ni particles embedded in a non-magnetic matrix, because it exceeded the theoretical value of shape anisotropy for Ni. This indicated that the Ni/Cu(II) complex films required a much higher magnetic field to reach saturation. Figure 3 shows the dependence of the M_s of deposited films as a function of temperature. The M_s of deposited films increased by 25 % from 112 emu/g to 140 emu/g with decreasing temperature from 290 K to 0 K. However the increase of bulk Ni with decreasing temperature in the same temperature range is only about 6.6 %. The film showed a single Curie temperature T_c of 631K, which was 19 K lower than the bulk Ni foil.

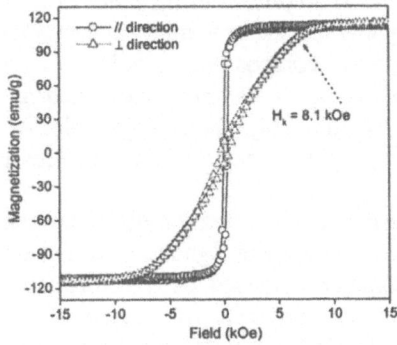

Figure 2. Magnetization of deposited film in parallel and perpendicular directions

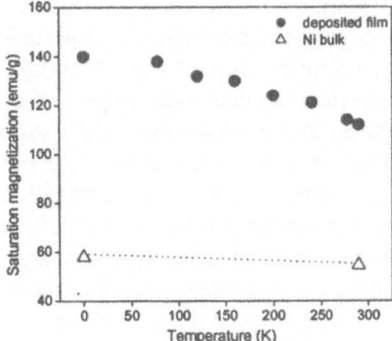

Figure 3. Magnetization of deposited film as a function of temperature from 0 K to 300 K

Assuming that the magnetization was solely attributed to Ni (82 wt % of film mass), the room temperature magnetic moment of such deposited Ni would be 1.78 μ_B. This unacceptably high value clearly indicated that deposited Ni alone could not be responsible for the observed magnetization since bulk Ni only has a magnetic moment of 0.568 μ_B. Although Ni and Cu are expected to form alloys at the reaction temperature, the observed enhancement instead of reduction of moment clearly ruled out that alloying had occurred. The enhanced magnetization was therefore proposed to arise from the Cu complex and possible magnetic interaction with the deposited Ni. The Ni^{2+} ion in oxides or other salts has a calculated moment of 2.83 μ_B based on the configuration of $3d^8$ electrons [20]. The absence of crystalline or amorphous oxide in the films did not support that the observed enhanced moment was caused by the Ni ions in the configuration of oxide.

As a result of chemical reaction of Cu surface with the solvent, Cu(II) ions were leached and subsequently incorporated in growing Ni films. The Cu ions formed an amorphous complex with C and O, separating the deposited Ni particles. The Cu (II) complex is probably magnetically ordered. Many Cu (II) complexes are magnetic [21] with a magnetic transition temperature as high as 435K [22]. The Cu (II) ion has a moment of 1.9 μ_B [21]. Calculation based on the weight % of Cu ions in the film would yield a moment of above 3 μ_B, which is higher than expected. Based on these results, it is therefore suggested that there existed magnetic interactions between the Ni and the Cu (II) complex, resulting in enhanced magnetization. Strong interactions may lead to ferromagnetic coupling of Ni and Cu (II) complex and enhanced moment of Ni in the interphase boundary region. Note that only a single Curie temperature was observed for the sample and it was significantly lower than that for bulk Ni. The result supported the possible strong coupling of Ni and Cu (II) complex. The high value of anisotropy also supported the existence of strong interactions of these magnetic components.

2.2 CORRELATION OF ELEMENTAL CHEMISTRY OF X-RAY DIFFRACTION PEAK OF CoCrPt MAGENTIC MEDIA

The magnetic properties of sputtered CoCr based magnetic media, for longitudinal and perpendicular recording applications, are controlled by the factors such as alloy composition, degree of phase separation of Co and Cr, alignment of magnetic columnar grains, degree of Cr segregation at grain boundary and mosaic of texture. It is crucial to achieve a proper crystallographic texture with optimal composition in order to control the magnetocrystalline anisotropy. The segregation of Cr has been explained as Cr segregation at grain boundary [23], formation of Co_3Cr [24], short-range order of Cr [25], formation of Cr oxide at grain boundaries [26], segregated microstructures and two-phase compositional separation [27,28].

The Cr segregation has been studied using various characterization methods such as transmission electron microscopy, x-ray microanalysis, thermomagnetic analysis, atom-probe field ion microscopy, nuclear magnetic resonance, extended x-ray absorption fine structure and ultra-high vacuum scanning tunneling microscopy. These methods however do not provide direct elemental correlation with the specific long range order in question.

X-ray scattering is a well-established technique to investigate the structure of alloys and composites. The appearance of a single set of diffraction peaks and the disappearance of elemental peaks are commonly accepted as evidence of alloy formation. However, when the crystallite size is reduced to below a critical length scale, a nanostructured solid solution cannot be unambiguously differentiated from a nanocomposite using conventional x-ray diffraction [29]. In a nanocomposite where the two phases have close lattice parameters and x-ray structural coherence, the Bragg peak of one phase has some degree of overlap with that of the other phase. Because of the effects of size and strain broadening and the contribution to diffraction amplitude by structural coherence of the two phases, a single peak will appear for a particular Bragg reflection when the size is below a certain limit. This single Bragg peak however has an average lattice spacing that has no correspondence in real space, and can be easily mistaken as evidence of formation of a solid solution. Using anomalous x-ray scattering to investigate the composition of a Bragg peak in question, it has been demonstrated that nanostructured NiCo films did not necessarily form solid solution as expected from their phase diagram or suggested by the results of conventional x-ray diffraction [30].

The absence of an element in a particular long-range order may be possibly due to segregation of the element to the grain boundary or the amorphous structure of the element in a nanocomposite. For coarse-grained, polycrystalline alloys and composites, the properties mainly depend on the bulk properties of the constituent grains (particles). The effects of any compositional surface segregation on material properties are generally not taken into consideration, as the total interface area is not significant. The large ratio of surface atoms to volume atoms in nanostructured materials renders the effects of surface segregation no longer insignificant. The prediction of phase miscibility in nanostructures is complicated by the unreliability of conventional phase diagrams that ignores the surface and interface effects. Any size-dependent surface segregation complicates the prediction of phase separation at grain boundaries and interphase interface, and may possibly lead to complete phase separation. A thermodynamic analysis of a NiCo alloy nanoparticle has been carried out to study the effects of particle effect on surface compositional segregation [31]. The surface compositions of NiCo nanoparticles were calculated based on the regular solution model, which took into account of the size effect, and compared with that of bulk alloy particles with a larger size. The particle size affects the compositional surface segregations due to the change in surface tensions caused by the variations in the number of broken bonds at the surface. It was observed that the segregated surface compositions in the NiCo binary alloy nanoparticles (with particle diameter of 10-100 nm) were few times larger than that of the corresponding bulk single crystal (with the particle diameter > 100 nm).

It is therefore possible that below a certain particle size of nanostructures, surface segregation in a solid solution would lead to complete phase separation. Structural information of compositional segregation cannot be obtained using conventional X-ray scattering or transmission electron microscopy, since Co and Cr have similar atomic numbers and scattering factors. It has been commonly but wrongly accepted that a single XRD Bragg peak without detectable elemental separation in these films supports that compositional segregation does not cause any perturbation in the crystalline order.

Recently a study of determining the elemental concentrations of the textured Bragg peak and the averaged local atomic environment of CoCrPt films using AXS and EXAFS has been reported [32]. The magnetic $Co_{66}Cr_{18}Pt_{16}$ layer (30 and 60 nm thick) and Ti underlayer (25 nm thick) were deposited on glass substrates at 250 °C using dc magnetron sputtering. The VSM hysteresis loops showed these films had perpendicular magnetization orientation. The magnetic properties of these films can be found in reference 33.

The x-ray powder (θ-2θ) scans showed both magnetic film and Ti underlayer were (002) textured (Fig. 4). Phase separation at the (002) reflection was not detected. For AXS study, powder θ-2θ scan was first performed to determine the position of the CoCrPt (002) reflection (Q_z = 2.984 Å^{-1}) at x-ray energies of 7.509 and 5.789 keV (below the respective K-edges of Co and Cr). In AXS, the momentum transfer was fixed to the position of (002) reflection at each photon energy and the scattering intensity was monitored as the x-ray energy was scanned through Co and Cr K-absorption edges, respectively. If the element in question is associated with the Bragg peak, then the elemental absorption causes a decrease in the Bragg intensity, and a cusp appears at its absorption edge [33]. Figure 5 shows the AXS spectra of the (002) Bragg peak taken in the vicinity of Co absorption K-edge of the $Co_{66}Cr_{18}Pt_{16}$ films with two different film thickness. The measured AXS data was fitted using a film-thickness independent simulation, based on the kinematic approximation of intensity with random mixing of elements. The simulation only accounted for the concentration of elements in the specific structural order. The experimental data was fitted against simulated spectra using different concentrations. The Co concentrations in the (002) peak for both films were ~62%, with an estimated uncertainty of ± 5 %. The Cr concentration in this Bragg peak was ~25 ± 6 % for the 60 nm film, and ~ 22 ± 6 % for the 30 nm film. Compared to the average global film composition, i.e. $Co_{66}Cr_{18}Pt_{16}$, the concentrations of Co decreased whereas that of Cr increased in the (002) peak.

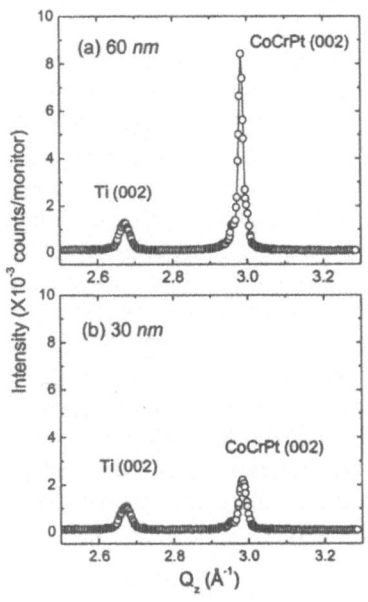

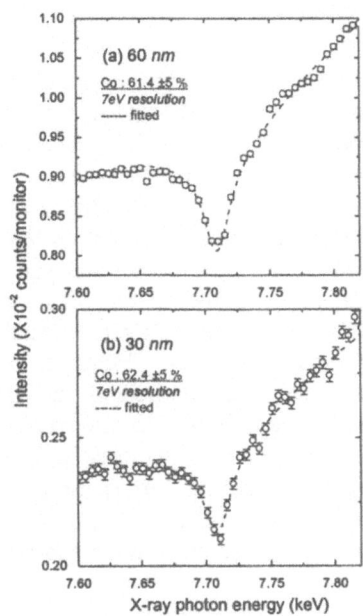

Figure 4. X-ray powder θ-2θ scans of (a) 60 nm film, and (b) 30 nm film

Figure 5. AXS spectra of the (002) peak in the vicinity of Co absorption *K*-edge. (a) 60 nm film, and (b) 30 nm film

The AXS method only probed the elemental concentrations in the (002) textured Bragg peak, it did not provide structure information besides this order. On the other hand, EXAFS measurements provided averaged global information of the local atomic environment of the element in question. The AXS results showed that Co maintained similar concentration in the (002) peak for both films. The EXAFS results indicated that both films had similar Co coordination number that was lower than the reference bulk Co foil. A certain degree of Co alloying with Cr and perhaps Pt had therefore occurred. Dilution of Co by Cr (> 13 %) addition stabilizes perpendicular magnetization and lowers M_s. Compared to the global film composition, the higher Cr concentration detected in the (002) Bragg peak for both films showed that the crystallinity of Cr was well established. Since grain boundaries are not expected to exhibit any long-range order, the observation of a higher Cr content in the textured peak ruled out the common assumption that a significant amount of Cr segregate to the grain boundary. If Cr segregation resulted in two crystalline phases, such phases must have the same *d*-spacing as the AXS measurements were fixed at the (002) momentum transfer. The segregation to two crystalline phases with the same *d*-spacing is equivalent to a single phase inhomogeneous alloy. Since Cr did not preferentially segregate to grain boundary, a resultant higher magnetic exchange coupling and reduced magnetocrystalline anisotropy constant reduced the coercivity. The low sputtering temperature and pressure used in this study could be responsible for inefficient Cr segregation in the grain boundary.

508

3. Summary

The magnetic properties of nanostructured films depend on many factors such as composition, structure, texture, and microstructures. The effects of interphase interface in granular Ni/[Cu(II)-C-O] films were discussed. Enhanced magnetization and anisotropy field were observed in the Ni/[Cu(II)-O-C] complex nanocomposite films deposited by the polyol process. It was suggested that the enhancement was caused by the magnetic interaction between the Ni particles and the amorphous Cu(II) complex. The enhanced magnetization was stable up to the Curie temperature around 400 °C. The correlation of elemental chemistry with a particular long-range order in question is essential to understand the structural contribution to magnetic properties in textured films such as CoCrPt. However this information cannot be obtained by common conventional characterization techniques. Using anomalous x-ray scattering and extended x-ray absorption fine structures it was shown that the elemental compositions of the textured peak in these polycrystalline nanostructured films differed from the averaged global film composition. The higher Cr concentration in the textured peak showed that a significant amount of Cr did not segregate to the grain boundary, due to the low sputtering temperature and pressure used in deposition.

4. Acknowledgment

The support of this research by the Academic Research Fund of the National University of Singapore and the grant from the Office of Naval Research (USA) is gratefully acknowledged. The synchrotron experiments were performed in Pohang Light Source, S. Korea, and Synchrotron Radiation Research Center, Taiwan.

5. References

[1] Chien, C.L. (1995) Magnetism and giant magnetotransport properties in granular solids, *Ann. Rev. Mater. Sci.* 25, 129-160

[2] Liou, S.H., Malhotra, S., Shan, Z., Sellmyer, D.J., Nafis, S., Woolam, J.A., Reed, C.P., DeAngelis, R. J. and Chow, G.M. (1991), The process-controlled magnetic properties of nanostructured Co/Ag composite films, *J. Appl. Phys.* 70, 5882-5884.

[3] Childress, J. R. and Chien, C.L. (1991) Granular cobalt in a metallic matrix, *J. Appl. Phys.* 70, 5885-5887.

[4] Fiévet, F., Lagier, J.P., and Figlarz, M. (1989) Preparing mndisperse metal powders in micrometer and submicrometer sizes by the plyl process, *Mater. Res. Sc. Bull.* December, p.29-34.

[5] Viau, G., Ravel, F., Acher., Fiévet -Vincent, F., and Fiévet, F. (1994) Preparation and microwave characterization f spherical and monodisperse $C_{20}Ni_{80}$ particles, *J. Appl. Phys.* 76, 6570-6572.

[6] Chw, G.M., Kurihara, L.K., Kemner, K.M., Schen, P.E., Elam, W.T., Ervin, A., Keller, S., Zhang, Y.D., Budnick, J., and Ambrse, T. (1995) Structural, morphological and magnetic study f nanocrystalline cobalt-copper powders synthesized by the plyl method, *J. Mater. Res.* 10, 1546-1554.

[7] Viau G., Fiévet-Vincent, F., and Fiévet, F. (1996) Nucleation and growth f bimetallic CNi and FeNi monodisperse particles prepared in plyls, *Slid State Inics* 84, 259-270.

[8] Kurihara, L.K, Chw, G.M., and Schen, P.E. (1995) Nanocrystalline metallic powders and films produced by the plyl method, *Nanstruc. Mater.* 5, 607-613.

[9] Chw, G.M., Kurihara, L.K., Ma, D., Feng, C.R., Schen, P.E., and Martinez-Miranda, L.J. (1997) Alternative approach t electroless Cu metallization f AlN by a nonaqueous plyl prcess, *Appl. Phys. Lett.* 70, 2315-2317.

[10] Martinez-Miranda, L.J., Li, Y., Chw, G.M., Kurihara, L.K. (1999) A depth study of the structure and strain distribution in chemically grown Cu films n AlN, *Nanstruc. Mater.* 12, 653-656.

[11] Chw, G.M., Ding. J., Zhang, J., Lee, K.Y. and Surani, D. (1999) Magnetic and hardness properties f nanostructured Ni-C films deposited by a non-aqueous electroless methd, *Appl. Phys. Lett.* **74**, 1889-1891.

[12] Zhang, J., Chw, G.M., Lawrence, S.H. and Feng, C.R. (2000) Nanstructured Ni films by plyl electroless depsitin, *Mater. Phys. Mech.* **1**, 11-14.

[13] Zhang, J. and Chw, G.M. (2000) Electrless plyl depsitin and magnetic prperties f nanstructured $Ni_{50}C_{50}$ films, *J. Appl. Phys.* **88**, 2125-2129.

[14] Chw, G.M., Li, Y.Y. and Hwu, Y.K, (2000) Mechanical and magnetic prperties f plyl electrdepsited NiC films, *Mater. Phys. Mech.* **1**, 67-72.

[15] Chw, G.M., Zhang, J., Li, Y.Y., Ding, J. and Gh, W.C., (2001) Electrless plyl synthesis and prperties f nanostructured $Ni_x C_{100-x}$, *Mater. Sci. Eng.* **A304–306**, 194–199.

[16] Blackwd, D. J., Li, Y. Y. and Chw, G. M. (2002) Plyl electroless and electrodepositin f nanostructured Ni-C films and pwders, *J. Electrchem. Sc.* **149**, D27-D34.

[17] Yin, H, Chan, H.S.. and Chow, G.M. (2001) Nanostructured iron-nickel thin films synthesized by electroless plyl deposition, *Mater. Phys. Mech.* **4**, 56-61.

[18] Yin, H. and Chw. G.M. (2002) Anmalus electrless plyl depsitin f FeNi pwders and films, *J. Electrchem. Sc.* **149**, C68-C73.

[19] Chw, G.M., Ding, J. and Zhang, J, (2002) Enhanced magnetizatin f nanstructured granular Ni/[Cu (II)-C-] films, *Appl. Phys. Lett.* **80**, 1028 -1030.

[20] Kittel, C. (1996) *Intrductin t Slid State Physics*, 7th ed., Jhn Wiley & Sns, USA, p. 426.

[21] Kahn, . (1993) *Mlecular Magnetism*, VCH Publisher, New Yrk.

[22] Yang, X.D., Si, L., Ding, J., Ranfrd, J.D. and Vittal, J. J. (2001) Cpper cmplex with a magnetic rdering temperature abve 400 K, *Appl. Phys. Lett.* **78**, 3502-3504 .

[23] uchi, K. and Iwasaki, S. (1985) Prperties f high rate sputtered perpendicular recrding media, *J. Appl. Phys.* **57**, 4013-4015.

[24] Chen, T., Charlan, G.B. and Yamashita, T. (1983) A cmparisn f the uniaxial anistrpy in sputtered C-Re and C-Cr perpendicular recrding media, *J. Appl. Phys.* **54**, 5103-5111.

[25] Haines, W. G. (1984) Effect of atomic distribution n the saturation magnetization f cobalt-Chromium films, *J. Appl. Phys.* **55**, 2263 -2265.

[26] Smits, J.W., Luitjens, S.B., and den Breder, F.J.A. (1984) Evidence fr micorstructural inhomgeneity in sputtered C-Cr thin films, *J. Appl. Phys.* **55**, 2260-2262.

[27] Maeda, Y. and Takahashi, M. (1989) Direct observation f the segregated microstructures within C-Cr film grains, *Jap. J. Appl. Phys.* **28**, L248 –L251.

[28] Hirayama, Y., Futamt, M., Kimt, K. and Usami, K. (1996) Compositional microstructures f C-Cr ally perpendicular magnetic recording media, *IEEE. Trans. Magn.* **32**, 3807-3809.

[29] Michaelsen, C. (1995) n the structure and homgeneity f slid solutions: the limits f conventional x-ray diffraction, *Phil. Mag. A.* **72**, 813-828.

[30] Chw, G.M., Gh, W.C., Hwu, Y.K., Ch, T.S., Je, J.H., Lee, H.H., Kang, H.C., Nh, D.Y., Lin, C.K., and Chang, W.D. (1999) Structure determinatin f nanostructured Ni-C films by anomalous x-ray scattering, *Appl. Phys. Lett.* **75**, 2503-2505.

[31] Jayaganthan, R. and Chow, G.M. (2002) Thermodynamics of surface compositional segregation in Ni-Co nanoparticles, *Mater. Sci. Eng.* **B95**, 116-123.

[32] Chow, G.M., Sun, C.J, Soo, E.W., Wang, J.P., Lee, H.H. Noh, D.Y. Cho, T.S., Je, J.H., Hwu, Y.K., (2002) Structural study of CoCrPt films by anomalous x-ray scattering and extended x-ray absorption fine structure, *Appl. Phys. Lett.* **80**, 1607-1609.

[33] Stragier, H., Cross, J.O., Rehr, J.J., Sorensen, L.B., Bouldin, C.E. and Woicik, J.C. (1992) Diffraction anomalous fine structure: a new x-ray structural technique, *Phy. Rev. Lett.* **69**, 3064-3067.

ELECTROPLATING AND ELECTROLESS DEPOSITION OF NANOSTRUCTURED MAGNETIC THIN FILMS

NICOLAE SULITANU
Department of Solid State Physics,
Faculty of Physics, "Al. I. Cuza" University,
6600 Iasi, Romania

1. Effect of Size Constrains and Nanoscale Magnetism

The dawn of nanoscale science can be traced to a now classic talk that Richard Feynman gave on December 29th, 1959 at the annual meeting of the American Physical Society at the California Institute of Technology. In this lecture, Feynman suggested that there exists no fundamental reason to prevent the controlled manipulation of matter at the scale of individual atoms and molecules. Twenty one years later, Eigler and co-workers [1] constructed the first man-made object atom-by-atom with the aid of a scanning tunneling microscope. This was just 7000 years after Democritus postulated atoms to be the fundamental building blocks of the visible world. A nanometer is thus the space occupied by 3-4 atoms placed end-to-end. Advances in the field have been accelerated following the invention by Binnig and Rohrer in the early 1980s of the scanning tunneling microscope [2]. This microscope, and its derivates, allows us to image and manipulate atoms, molecules and clusters in a controlled manner. It is this tool, which allows us, in a nano-workshop, to create and characterize individual structures whose dimensions are of the order of nanometers. It is forecast that many practical applications of nanotechnology will utilize massive arrays of such fabrication tools, combined with self-assembly techniques borrowed from nature and the biosciences, to create large numbers of nanoscale objects and structures. As opposed to the microscale, the nanoscale is not just another step towards miniaturization, but is a *qualitatively new* scale. Here quantum and size phenomena are allowed to manifest themselves either at a purely quantum level or in a certain "admixture" of quantum and classical components. At the foundation of nanosystems lie the quantum manifestations of matter that become relevant and measurable. Consequently, instead of being a limitation or an elusive frontier, quantum phenomena have become the crucial enabling tool for nanotechnology [3].

Nanoscale science and technology enables controlled component design and fabrication on atomic and molecular scales. Nano-related research and development unites findings and processes from biotechnology and genetic engineering with chemistry, physics, electronics and materials science with the aim of manufacturing cost-effective innovative products. Nanotechnology has been recognized by leading

T. Tsakalakos, et al. (eds.) pgs 511 - 532
Nanostructures: Synthesis, Functional Properties, and Applications;
© *2003 Kluwer Academic Publishers.*

512

industrialized countries to be of potential key economic significance in the 21st century [1, 3]. From the scientific point of view, nanotechnology is an interdisciplinary field of science and implies to exploit the techniques and processes of the microelectronic sector, while depending on new tools, fabrication and assembly techniques borrowed from the bio-, engineering-, chemical- and physics-communities. These techniques increasingly demand the controlled manipulation of matter on the atomic and molecular scale. Increasing miniaturization is accompanied by an irrevocable increase in the important of mastering and reliably implementing extreme nanoscale technologies in a mass production manufacturing environment.

Magnetism as a cooperative phenomenon lends itself to manipulation in small structures, where neighbor atoms can be replaced systematically by species with stronger or weaker magnetism. In fact, a class of magnetic/non-magnetic multilayer termed "spin valves" has been introduced into magnetic storage devices. The question is: how fine does a film need to be structured to have an impact on its magnetic properties? The wave function of electrons is going to change when they are confined to dimensions comparable with their wavelength (Fig.1). Confining electrons to small structures causes the continuous bulk bands to split up into discrete levels, for example, quantum well states in a slab [1, 3, 4]. For a coarse estimate of the corresponding slab thickness, one may set the energy E of the lowest level equal to kT. For room temperature, E = kT = 0.026 eV, one obtains a de Broglie wavelength $\lambda = h/p = h/(2mE)^{1/2} \approx 1.23$ nm/(E/eV)$^{1/2} \approx 8$ nm, which is comparable with spatial extent of the lowest quantum state. Thus, both the high electron density in magnetic metals and the requirement of room-temperature operation for quantum devices point to dimensions of a few nanometers.

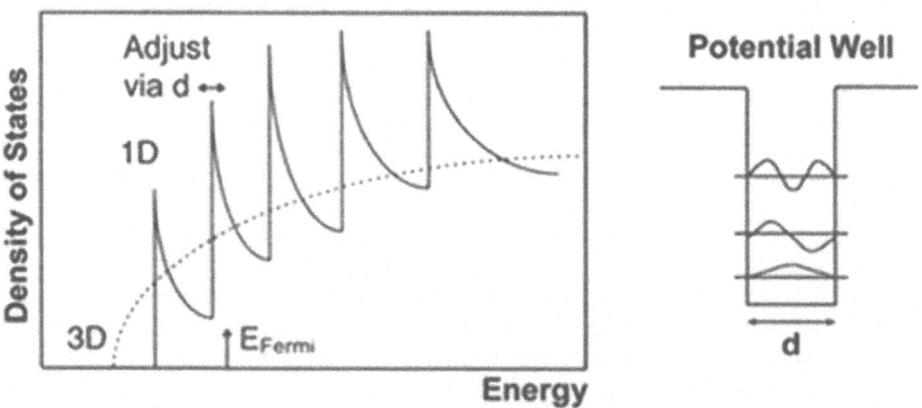

Figure 1. Tailoring electronic properties of materials by nanostructuring. Electron confined to nanostructure give rise to low-dimensional quantum well states, which modify the density state. States at the Fermi level trigger electronic phase transitions, such as magnetism and superconductivity. They are responsible for magnetic coupling and conductivity.

Therefore, nanostructured magnetic films must to contain fine grains (crystallites of nanometers size) which generally belonging to the 3d-transition metal series and are obtained by different methods calling for various skills [4, 5]. Nanostructured films posses unique properties due to both size and interface effects. They find many applications in area as data storage, magnetic or biological sensors, magneto-electronics devices and catalysis. Nanostructured or nanocrysalline magnetic films are typically make by sputtering, evaporation, chemical vapor deposition, and other highly sophisticated deposition techniques based on ion beam, plasma and vacuum technology, but they can also be made by electrodeposition. The study of electrodeposited metal nanostructures is still at a very early stage, but the combination of electrochemistry and patterning on the nanometer scale seems highly likely to lead to further exciting developments.

This work will present the advances being made in the field of electrodeposition of magnetic materials. First, the basic conditions for nanoscale magnetic film deposition will be discussed. Secondly two examples of nanostructured materials and their potential applications will be given. Finally an outlook to the future of electrodeposition within the microelectronics field will be given.

2. Electroplating and Electroless Plating of Nanostructured Magnetic Thin Films

Magnetic thin films are deposited with electroplating (electrodeposition) and electroless processes by electrolysis: magnetic metallic ions in an aqueous solution (the electrolyte) are reduced to metal atoms.

A usually electroplating cell consists of two electrodes, an electrolyte and a rectifier providing the external current.

In an electroless process, the metallic ions are also reduced to metal atoms but the electrons for the reduction are delivered by an oxidation reaction in the same electrolyte. There is no need for an external current supply in an electrode-less or electroless process and the oxidation and reduction reaction take place at one electrode [6-8].

Table 1. The periodic table of elements.
Indicated are the elements that can be electrodeposited from aqueous solutions.

1 H 1.0079																		2 He 4.0026
3 Li 6.941	4 Be 9.0122											5 B 10.811	6 C 12.011	7 N 14.007	8 O 15.999	9 F 18.998	10 Ne 20.18
11 Na 22.99	12 Mg 24.305											13 Al 26.982	14 Si 28.086	15 P 30.974	16 S 32.066	17 Cl 35.463	18 Ar 39.948
19 K 39.098	20 Ca 40.078	21 Sc 44.956	22 Ti 47.88	23 V 50.941	24 Cr 51.996	25 Mn 54.938	26 Fe 55.847	27 Co 58.933	28 Ni 58.693	29 Cu 63.546	30 Zn 65.39	31 Ga 69.723	32 Ge 72.61	33 As 74.922	34 Se 78.96	35 Br 79.904	36 Kr 83.8
37 Rb 85.468	38 Sr 97.62	39 Y 88.906	40 Zr 91.224	41 Nb 92.906	42 Mo 95.94	43 Tc (97.91)	44 Ru 101.07	45 Rh 102.91	46 Pd 106.42	47 Ag 107.87	48 Cd 112.41	49 In 114.82	50 Sn 118.71	51 Sb 121.76	52 Te 127.6	53 I 126.9	54 Xe 131.29
55 Cs 132.91	56 Ba 137.33	57 La 138.91	72 Hf 178.49	73 Ta 180.95	74 W 183.84	75 Re 186.21	76 Os 190.23	77 Ir 192.22	78 Pt 195.08	79 Au 196.97	80 Hg 200.59	81 Tl 204.38	82 Pb 207.2	83 Bi 208.98	84 Po (209)	85 At (210)	86 Rn (222)
87 Fr (223)	88 Ra (226)	89 Ac (227)	104 Rf (261.1)	105 Db (262.1)	106 Sg (263.1)	107 Bh (262.1)	108 Hs (265.1)	109 Mt (266.1)	110 Uun (269)	111 Uuu (272)	112 Uub (277)						

	58 Ce 140.12	59 Pr 140.91	60 Nd 144.24	61 Pm (144.9)	62 Sm 150.36	63 Eu 151.97	64 Gd 157.25	65 Tb 158.93	66 Dy 162.5	67 Ho 164.93	68 Er 167.26	69 Tm 168.93	70 Yb 173.04	71 Lu 174.97
Lanthanide Series														
Actinide Series	90 Th 232.04	91 Pa 231.04	92 U 238.03	93 Np (237)	94 Pu (244.1)	95 Am (243.1)	96 Cm (247.1)	97 Bk (247.1)	98 Cf (251.1)	99 Es (252.1)	100 Fm (257.1)	101 Md (258.1)	102 No (259.1)	103 Lr (262.1)

Table 1 is the periodic table of elements. Indicated are the elements that can be electrodeposited from aqueous solutions. By the simultaneous deposition of electroplated metals, an even larger number of thin films can be deposited as binary or ternary alloys. Other metals such as Mo, Ti, Mg, Te, W or non-metals such as P, S, B, C, N can be incorporated in thin films in a large concentration range by an induced deposition process after a surface diffusion.

In the last decade electrochemical fabrication techniques are becoming increasingly attractive for the deposition of new thin films alloys and novel nanostructures, since these open up a new possibility in controlling material properties [6-11]. Moreover, electrochemically deposition offers real advantages to automatically prepare identical thin films with large area surface and a great variety of structures and properties of obtained thin films [12-16]. Electrodeposition is an alternative method that, in many cases, is preferred to sputtering because of its low capital cost and because it generally yields better soft or hard magnetic properties at corresponding compositions, easily allows orientation of film anisotropy and most importantly enables the production of pole tips of very high quality by means of the plating-through-mask patterning process [17]. In addition, electrodeposition offers a unique feature compared to other deposition techniques: reversibility [18]. This means that films previously deposited onto a substrate can be fully or even partially removed by a simple potential change of the substrate without modification of the substrate surface. It is interesting to note that the scientific community has initially ignored electrodeposition as a means of producing nanostructured thin films, despite the fact that this approach is probably one of the oldest to synthesize such structures. There are numerous early reports in the literature on electrodeposition with ultrafine structures [19, 20]. Finally electrodeposition has been recognized now as a new, competitive way of preparing nanostructured thin films, multilayers and spin valve structures [4, 5, 21-26]. However, no systematic studies on the synthesis of nanocrystalline thin films by electrochemically methods to optimize certain properties by deliberately controlling the structure-property relationships in electrodeposited nanostructured thin films were published prior to the late 1990's. For all that, thanks to the constant effort of some researchers we can make some remarks on requirements for improved nanostructured thin films electrochemically obtained. Further on we try to summarize these requirements.

In general, results of dc electrodeposition suggest that the grain size of the film could be adjusted not only by the composition but also by the hydrogen and/or oxygen evolution that taking place during the thin film alloy deposition [27]. For deposition of thin films can be used a standard electrochemical cell consisting of three electrodes (Fig.2). A working electrode (cathode)-WE, a secondary electrode (anode)-SE and a saturated calomel reference electrode-RE. The purpose of the reference electrode (RE) is to allow the voltage to be monitored at one of the electrodes, thus allowing potentiostatic conditions to be maintained.

Standard electrochemical cell consisting of three electrods

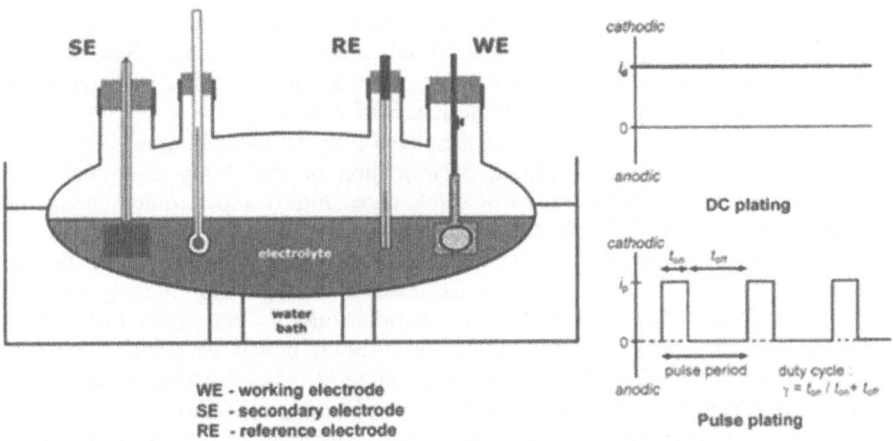

Figure 2. Standard electrochemical cell consisting of three electrodes.

On the other hand, pulsed electrodepositing (PEC) or pulse plating is a feasible method for controlling the composition and structure of nanocrystalline films [28, 29]. A current waveform for the pulsed electrodepositing of nanocrystalline Ni soft magnetic thin films is illustrated in Fig. 1 [29].

516

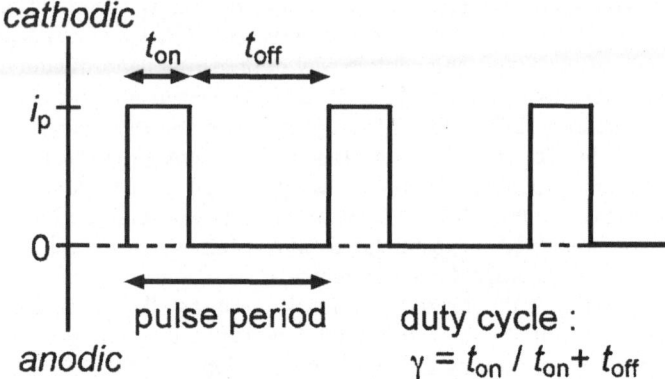

Figure 3. Current waveform for pulsed electrodepositing (PEC).

The duty cycle is defined as $t_{on}/(t_{on} + t_{off})$, where t_{on} is the pulse-on period and t_{off} is the relaxation period.

The thickness of films can be exactly tailored to achieve desired physical and particularly magnetic properties. This point seems to be most interesting in view of preparation processes which might automatically prepare films of identical magnetic properties controlled by appropriate feedback circuits. The structure of electrodeposited films are quite different depending on the electrolyte composition and the plating conditions such as current density, over potential, temperature and agitation, pH value or chemical complexes of the bath. However, most of the researches and studies are gradually beginning to reveal that the structure of a plated film is determinates in principal by the chemical composition of the plated film and the plating temperature [8, 19, 20]. Because plating is a method of formation of thin films at around room temperature this means the formation of solids under highly super cooled conditions. Therefore, various metastable phases are formed by the plating method and thus the mechanism of formation of thin films can be discussed in relation to the thermal equilibrium diagram. In the other words, researches regarding plating must be conducted not from the standpoint of electrochemistry, but from that of the crystallographic structure of plated films. As a result, it was established from the viewpoint of the plated films, first of all the structure of the plated film is determinate by the chemical composition of the plated film and plating temperature.

In many electroplating applications, small amounts of additives (such as surfactants, brighteners and leveling agents) are added in order to improve the brightness, the corrosion resistance etc. [30, 31]. The classification of the additives used in the DC electrodepositing of metallic thin films can be found in references 5-7 and 27. More recently a renewed interest in the effect of the additives was raised for the electroplating thin film alloys of transition metals in order to improve their magnetic properties [32-41]. The effects of these additives on the deposition kinetics depend on the nature of the metal or alloy, inhibiting or accelerating effects on the deposition process have been observed. The addition of brighteners and wetting agents improves the smoothness not only the external surface but also the planarity of the interfaces between the layers in multilayer systems. As a consequence such additions have a beneficial influence on the corrosion resistance of magnetic thin films and on the antiwearing properties. Some additives were shown to decrease the internal stress. Other additives are often used as antipitting agent in watts bath since they remove hydrogen bubbles formed at the cathode and reduce the number of pits. The additives can be used separately or together, since synergetic effects often occur. Moreover, some additives may also change the morphology and microstructure of magnetic thin films. Magnetic fields can be used during electroplating to affect the film growth, morphology, phase formation, texture, pore filling and the magnetic properties [42, 43]. Recently Gleiter [11] draw the attention on the possibility of tuning the electronic structure of electrodeposited nanostructured materials by means the deviation from charge neutrality. He suggests that nanostructured thin films may open the way to generate materials with an excess of a deficit of electrons or holes of up to 0.3 electrons/holes per atom. Such deviation from charge neutrality may be achieved either by means of an extremely applied voltage or by space charges at interfaces between materials with different chemical compositions (or combinations of both). The physical properties (ferromagnetic, electric, optical etc.) may be tuned by tuning the applied voltage. One way of generating thin films deviating in their entire volume significantly from charge neutrality for which the dimensions become comparable to the screening length is indicated in Fig. 4.

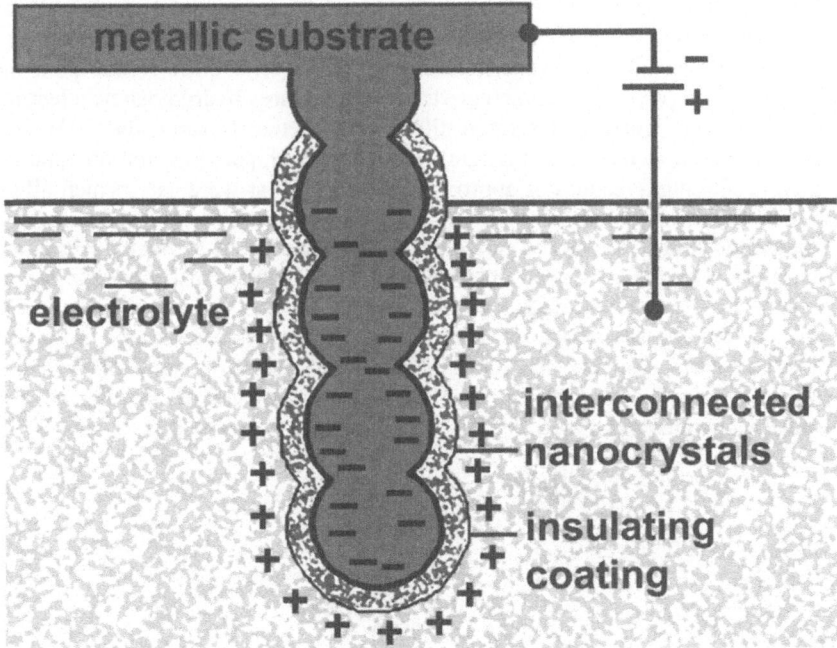

Figure 4. Chain of interconnected nanocrystals the free surface of which is coated with insulating layer. The chain is immersed into an electrolyte. Nanometer-sized Al crystals coated with a thin (e.g. 1 nm) layer of Al_2O_3 are an example of this kind of a structure [11].

Over the paste decade, we have prepared a number of binary alloy thin films (e.g. Ni-S [41], Ni-W, Fe-S, Co-S), ternary alloy thin films (e.g. NiFe-S, NiFe-W), quaternary alloy thin films (e.g. NiCoWS, CoNiFeS, CoNiFeW) and NiP_x/NiP_y multilayer system. A variety of methods such as dc plating, electroless plating and pulse plating have been used for this purpose. It should be noted that the main conditions for nanostructure formation are over potential phenomena (high nucleation rate) and adsorption/desorption processes of inhibiting molecules (slow grain growth). The two key mechanisms have been identified, as the major rate-determining steps are charge transfer at electrode surface and surface diffusion of adions on the crystals surface. In the following we report on the electrodepositing of nanocrystalline Ni-W thin films that exhibit high perpendicular magnetic anisotropy (PMA) and the electroless deposition of Co-S films.

3. Electrodeposition and Structure-Property Relationships of Nanocrystalline Ni-W Thin Films

The Ni-W films were deposited at a constant current density using a dc highly stabilized source. We have started from a sulfate bath with the same composition as in previous work [44] whose deposition parameters were strongly modified so that the conditions for the nanocrystal formation to be provide: high nucleation rate (over potential phenomena) and slow crystallite growth (adsorption/desorption processes of inhibiting molecules). The relevant experimental parameters those it was paid more attention were the physical parameters (bath temperature, hydrodynamic electrolyte conditions, current density, deposition time) and chemical parameters (pH value, addition of complex formers or inhibitors). Therefore, high pure preparation conditions equivalent to ultrahigh-vacuum of approximately 10^{-10} mbar were used which allowed the reproducible deposition of Ni-W films. Figure 5 shows the used typical cell for thin film electrodeposition. The plating cell deaeration was achieved by *in situ* saturation of the electrolyte with pure Ar gas. During alloy film preparation the plating cell was tightly capped in order to minimize the effect of air inflow. A self-valve mounted in the body cap allowed the electrolysis gas evacuation. The plating cell was thermostatically controlled in the temperature range 306 ± 0.5 K. The pH of the plating bath was fixed at either 2.0 using a 35 % NH_4OH solution and was controlled by means of the electronically operated pH-meter.

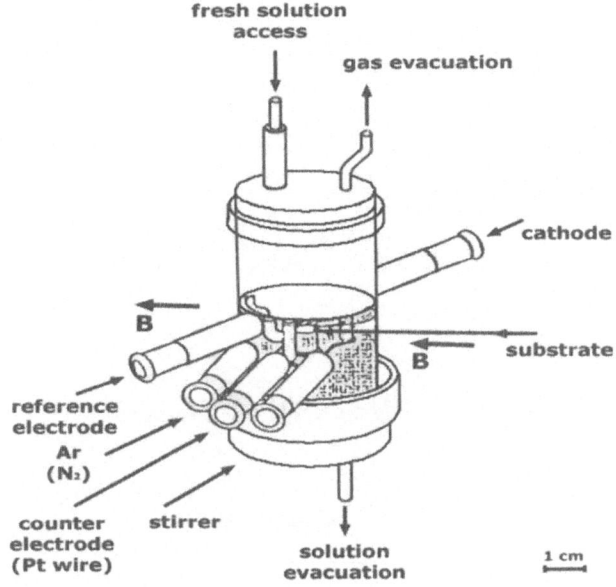

Figure 5. Typical cell for pure and reversible electrodeposition of nanocrystalline magnetic thin film.

While the electrolyte was being stirred (30-60 cycle/min) the dc current density reached an optimal value of 300-600 A/m², depending on W^{2+} ions concentration in electrolyte solution. These physical and chemical electrodeposition conditions give us the possibility that to intentionally adjust the film microstructure (crystallite size, distribution and shape, microstress), which in the last determines the physical, and chemical film properties (conductivity, magnetism, chemical stability, hardness). The plating bath contained as a base electrolyte nickel sulfate, boric acid, sodium and potassium ditartrate, sodium citrate and, respectively, sodium tungstate. Moreover, in contrast to the previous electrolyte [44], 0-7 mol m⁻³ of cumarine was added in order to avoid dendrite deposition. The Ni anode and copper cathode (substrate), situated at a distance of 3 cm, was disc-shaped and films with 1.8 cm in diameter were obtained. The film thickness, D, was currently ranging from 120-150 nm. The investigated films contain up to 18 wt. % in composition.

X-ray diffractometry clarified that the electrodeposited Ni-W films consisted of very fine fcc Ni grains whose (111) axes are preferentially oriented in the direction perpendicular to the film plane. Figure 1 shows XRD patterns for a pure Ni film and for two Ni-W films containing 6 wt. % W and, respectively, 13 wt. % W. The XRD patterns of all films indicate one higher peak from the {111} planes and other one very weak peak from the {220} planes only for pure Ni film.

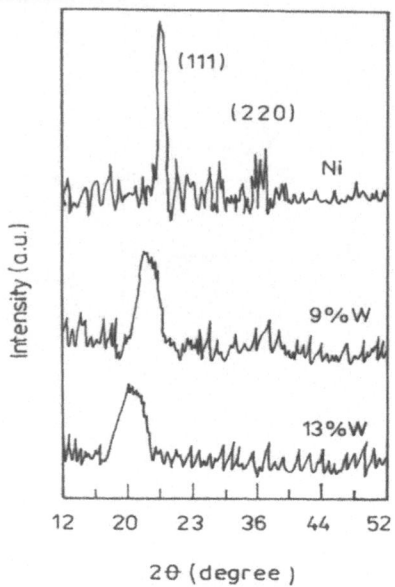

Figure 6. X-ray diffraction patterns of nanostructured films as a function of W content: for pure Ni film and for Ni-W films with 6 wt. % W and, respectively, 13 wt. % W (typical film) in composition.

It is concluded that the nickel grains have mainly the [111] preferred crystallographic growth orientation. One side, this suggests that the nickel atom diffusivity in the direction normal to the substrate was different from that parallel to the substrate. On the other side, this means that there is the anisotropy of surface diffusion, i.e., moving atoms on the surface has a tendency to ascend crystallites rather than to descend crystallites (the Schwoebel effect [6, 7]). The (111) peak intensity monotonically decreases while its width increases with an increase in W content up to 7 wt. %. This suggests the refinement of the grains. This kind of films consist of single fcc Ni phase. It was observed that the diffraction peak corresponding to Ni (111) steeply decreased in intensity and monotonically shifted to the side of the lower diffraction angle 2θ with an increase of W content over 7 wt. %. This means that the spacing between the adjacent Ni (111) planes was expanded due to the large internal stress resulting from high number of W atoms segregated at the boundaries of very fine Ni grains. These film structures are in a transition region from nanocrystalline fcc to amorphous phase. Therefore, depending on film composition, two types of nanostructures were observed: (a) single phase nanostructured films (< 7 wt. % W) which consist of nanocrystalline Ni grains cores (highly magnetic phase) separated by interfaces or "interphases", namely W enriched grain boundaries (14-27 nm), and (b) two-phase nanostructured films (7-18 wt. %) in which a second Ni-W amorphous phase or even amorphous-disordered mixture separates the magnetic Ni nanograins (6-14 nm).

Figure 7a shows TEM micrograph for a typical film which contains 13 wt. % W in composition. TEM micrograph reveals that the Ni-W film consists of a Ni-W amorphous matrix and single or here and there agglomerated Ni grains.

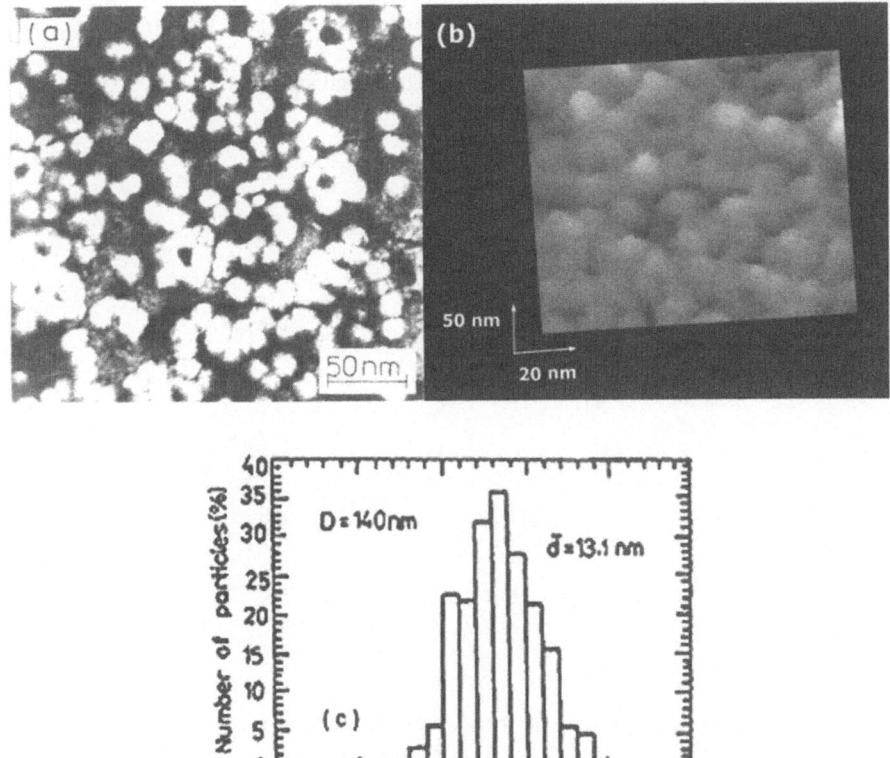

Figure 7. TEM micrograph (a) and AFM image (b) for a typical nanostructured film with 13 wt. % W in composition. The AFM image size is 100 x 80 nm. (c) Ni nanograins distribution (histogram) on film surface.

The grains are quite equiaxial in circular shape, having sizes in the nanometer range with a rather narrow size distribution. The AFM image of the Ni-W (13 wt. % W) typical film surface is shown in Fig. 7b. The surface morphology appears to be a set of continuous columnar grains which are well separated each from other. The grains average diameter was estimated at about 13 nm and this value is in well agreement with average size of about 12.5 nm obtained from Scherrer peak broadening analysis. Since the films have nanograined structure, there is a close correlation between the (111) film texture and the grain columnar shape [45]. Otherwise, a columnar structure was observed in the course of SEM investigation on the Ni-W film cross-section. The grains isolation is low in single-phase films (< 7 wt. % W), only by boundary interphase, but is sufficiently higher in two-phase nanostructured films (> 7 wt. % W) when a mixed nanocrystalline-amorphous Ni-W phase appeared at the boundary between grains. Therefore, the nanocrystalline-amorphous Ni-W films (7-18 wt. % W) behave as an assembly of Ni columnar nanograins (6-14 nm) embedded in a weak-magnetic Ni-W metallic matrix.

Figure 8 shows the change of crystallite size, d and saturation magnetization, M_s as a function of W content.

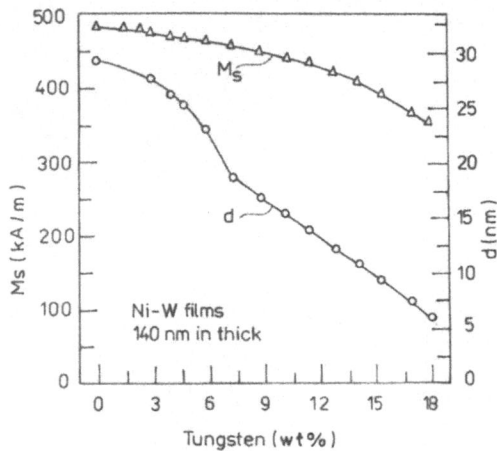

Figure 8. The diameter average size of Ni columnar particles, d and saturation magnetization M_s of Ni-W films as a function of tungsten content.

The grain size non-linearly decreases and displays a kink for approximately 7 wt. % W, which corresponds to the amorphous phase development beginning in the film. For two-phase nanostructured films the grain size, d, linearly decreases with W content increase. The amorphous volume fraction of the inter-grains regions increased with W content increase and the grain size decreased. As a consequence, by varying the nanocrystalline/amorphous ratio through W content change, the relative volume fraction of the grains and the size of consolidated grains can be manipulated. For pure Ni film with 29 nm grains size was found M_s = 482 kA/m. When the grains size decreases M_s also decreases reaching a minimum of about 350 kA/m for Ni-W film with columnar grains of 6 nm in diameter. The weak ferromagnetism of the mixed phase correlated with long-range magnetic interaction between the crystallites contributes to the low saturation magnetization of the nanograined Ni-W films. The reduction of magnetization in the smaller grains has been also attributed mainly to the surface-interface effects on the grain boundary [46]. The boundary region differs structurally as well as chemically from the grain core because the W atoms preferentially segregated at the grain surface. As a result, magnetic properties in the boundary region strongly differ in comparison with the grain core. For a typical film (13 wt. % W) with columnar crystallites of 12.5 nm in size it was found M_s = 420 kA/m.

Figure 9 shows the in-plane and transversal-to-plane magnetization curves (hysteresis loops) for three Ni-W films with different W content, i.e.: 6, 13 (typical film) and 18 wt. %. The hysteresis properties were measured at room temperature by using a vibrating sample magnetometer (VSM) in dc fields up to 720 kA/m.

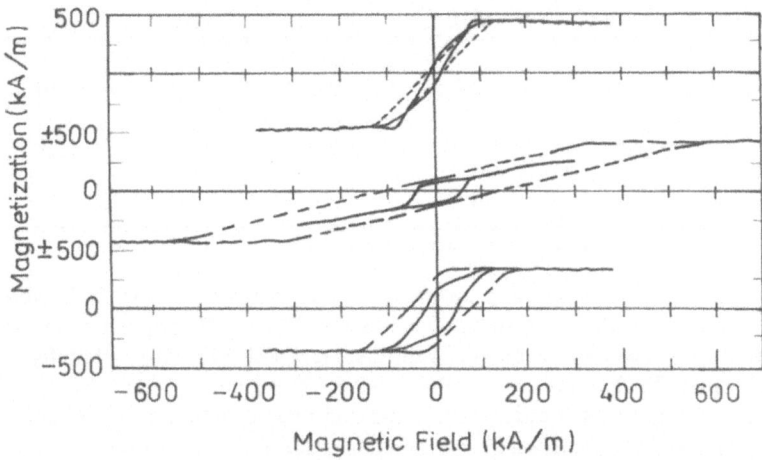

Figure 9. Magnetization curves (hysteresis loops) measured while applying the in-plane and, respectively, transversal-to-plane magnetic field: (a) 6 wt % W; (b) 13 wt % W (typical film); (c) 18 wt % W.

The in-plane magnetization curves have almost the same shape, i.e., the hysteresis loops undergo a small rapid change near the coercive field, $H_{c//}$, and a gradual (linear) change up to magnetization saturation. The higher rate of rapid change near $H_{c//}$ is reached for typical Ni-W film (13 wt. % W) when hysteresis loop clear has a rhomboidal shape. In the case of transversal-to-plane magnetization curves, since the magnetic field is applied perpendicular to the film plane, the influence of the demagnetizing field, $H_d = - N_\perp M_s = - M_s$ (the demagnetizing factor, $N_\perp = 1$ for thin films), must be considered. For these curves in Fig. 9, shearing correction has been carried out by subtracting H_d from the applied magnetic field. All the hysteresis loops are rhomboidal that is characteristic of films which exhibit perpendicular magnetic anisotropy (PMA), magnetization reversal by magnetic wall displacement and stripe-like magnetic structure [48]. The magnetic field decrease from saturation magnetization state results in a rapid decrease of magnetization at the beginning of the magnetization reversal region and then magnetization decreases almost linearly.

Figure 10 shows normalized magnetization ratio $M^* = M_s$ (pure Ni film)/M_s (Ni-film) and the in-plane $H_{c//}$ and, respectively, normal-to-plane $H_{c\perp}$ coercivity as a function of the average particle diameter of the Ni-W films. The saturation magnetization, in-plane $H_{c//}$ and transversal-to-plane $H_{c\perp}$ corcivities, anisotropy field H_k and PMA energy $K_\perp$ for each of three Ni-W films are summarized in Table II. The saturation magnetization and effective perpendicular anisotropy of films were measured at room temperature with an automatic torque-magnetometer (ATQM) in dc fields up to 480 kA/m [47].

Table II. Magnetic properties of Ni-W films for different tungsten content from in-plane and transversal-to-plane magnetization curves and torque-magnetometry measurements.

Tungsten wt. %	M_s kA/m	$H_{c//}$ kA/m	$H_{c\perp}$ kA/m	H_k kA/m	$K_\perp$ kJ/m^3
6	460	12.5	20	107	30
13*	420	50	120	455	120
18	350	32	69	91	20

* Typical Ni-W films so were denominated in the work presentation

The transversal hysteresis loop for Ni-W typical film with 13 wt. % W shown in Fig. 9b is quite similar to that of films which exhibit PMA with stripe domain structure and magnetic reversal by magnetic wall displacement [49, 50]. Therefore, it is estimated to have PMA in typical Ni-W (13 wt. %) film. This is supported by the fact that the field strength of saturation in the in-plane magnetization curve is higher than that in the transversal-to-plane magnetization curve and $H_{c\perp}$ is larger than $H_{c//}$. For this typical film $H_{c\perp}$ is twice that of others.

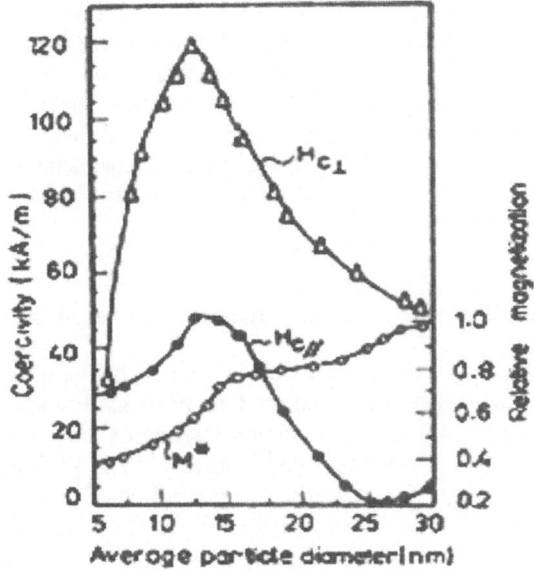

Figure 10. Normalized magnetization ratio M* = M$_s$ (pure Ni film)/M$_s$ (Ni-W film) and in-plane H$_{c//}$ and, respectively, perpendicular-to-plane H$_{c\perp}$, coercivity as a function of the average particle diameter of the Ni-W films.

The PMA energy K$_\perp$ = 120 kJ/m^3, M$_s$ = 420 kA/m and H$_{c\perp}$ = 120 kA/m are almost equal to those of Co-Cr thin films [5, 51]. The squareness ratio, S = M$_r$/M$_s$ (M$_r$ is remanent magnetization) of the in-plane M-H hysteresis loop was as low as 0.2. These results indicate that there are regions with strong uniaxial anisotropy in the direction perpendicular to the film plane. The origin of this large PMA in Ni-W typical films is mainly attributed to the magnetoelastic anisotropy associated with in-plane internal stress and positive magnetostriction [12, 45, 49]. The secondary source of PMA is believed to be the magnetocristalline anisotropy of <111> columnar crystallites and its shape magnetic anisotropy. The enhanced coercivity must be the result of the high uniaxial magnetic anisotropy of the isolated columnar Ni crystallites. The large K$_\perp$ and high H$_{c\perp}$ are sufficient to apply these alloy thin films to the perpendicular magnetic recording [50]. The low temperature of the electrodeposition (306 K) is also an advantage for deposition of Ni-W thin films on polymer tape substrates such as polyethylene terephtalate (PET) and polyethylene naphtalate (PEN), of which the glass transition temperatures are about 343 and 393 K, respectively, for production of flexible media such as floppy disks and recording tape. The large values of H$_{c\perp}$ and K$_\perp$ obtained for typical Ni-W film has not been attained in the sputtered Co-Cr thin film deposited at substrate temperature as low as such a temperature.

Summarize up, Ni-W electrodeposited thin films with W content of about 13 wt % consisted of fine columnar crystallites, d = 12.5 nm and exhibited large perpendicular magnetic anisotropy (PMA). Such Ni-W typical films are characterized by the fowling magnetic parameters: relatively high saturation magnetization, $M_s = 420$ kA/m, sufficiently large perpendicular coercivity, $H_{c\perp} = 120$ kA/m and large perpendicular magnetic energy $K_\perp = 120$ kJ/m^3. From this point of view, the Ni-W films with PMA had a critical composition for which it was established an equilibrium between magnetocrystalline anisotropy energy of <111> columnar grainss and magnetoelastic energy of the in–plane internal stress whatever their nature was. We consider that the appearance and development of the interface between magnetic core (Ni) and weak–magnetic boundary layer (Ni–W) for films with W content higher than 7 wt % is the driving mechanism of PMA arising.

4. Electroless Deposition and Unusual Magnetic Behavior of Co-S Films

Film samples of Co_xS_{1-x} were electrolessly deposited onto copper substrates, which were mechanically and chemically polished to have hard smooth surface that has previously described in [41]. The Co volume fraction of the samples were evaluated from the relation $x = V^{Co}/V^{Film} = M_s^{Film}/M_s^{Co}$ where V^{Co} and V^{Film} are the volume of Co in the film, M_s^{Film} and M_s^{Co} the saturation magnetization of the film and the pure Co, respectively. The selected films for investigation had a Co volume fraction $x = 0.36$ and the thickness D=50 μm. Annealing was performed in a vacuum chamber of 10^{-5} mbar, for an hour (1h) at 550 K and 600 K, respectively.

Figure 11 shows X-ray diffraction (XRD) patterns as a parameter of annealing temperature (T_a). For as-prepared film, despite larger content of Co, no diffraction peaks appear. The Co-S alloy film is apparently amorphous or heavily structurally disordered. For all that, the transmission electron microscopy (TEM) image (Fig. 12a) shows that film contain small volume fractions of precipitation (very fine grains) with diameters less than 10 nm that cannot be detected by large angle XRD.

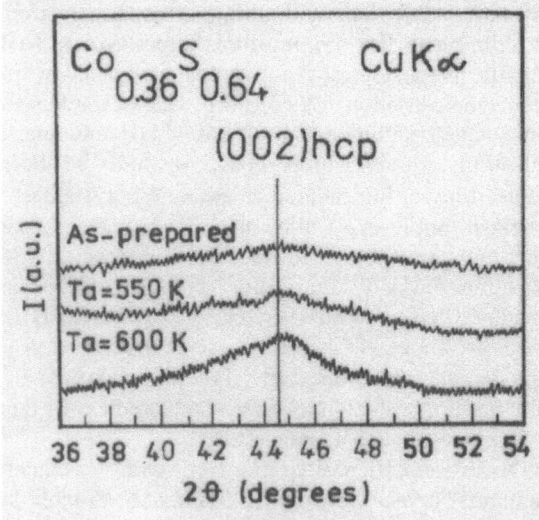

Figure 11. X-ray diffraction patterns of Co-S films as a parameter of annealing temperature (T_a).

Therefore, the as-prepared alloy films consist of ultrafine grains embedded in a Co-S amorphous matrix. Figure 12a shows the TEM micrograph of an as-prepared film.

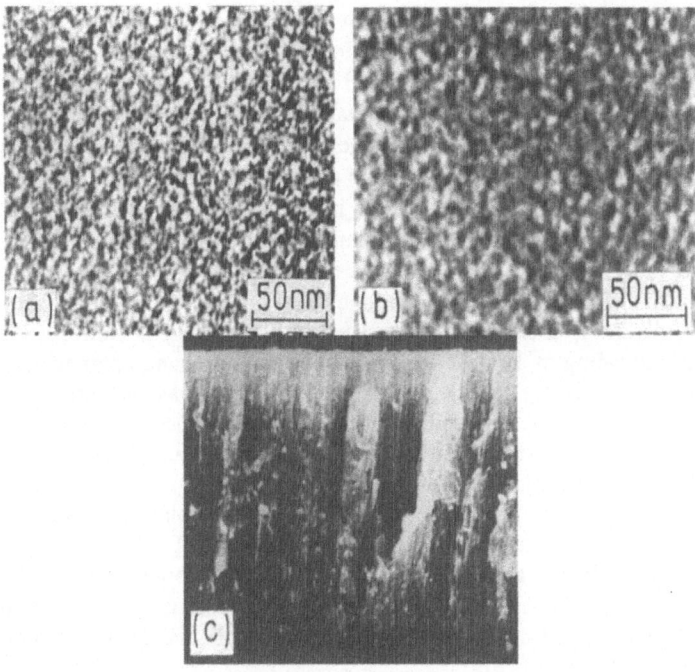

Figure 12. (a) TEM micrograph from as-prepared Co-S film; (b) TEM micrograph from annealed Co-S film at T_a = 600 K; (c) SEM image of cross-section of the as-deposited Co-S film.

From the figure 12a we can observe that the film consist of randomly oriented ultrafine grains, quite uniform size and shape with diameter in the range of 3 nm to 4.5 nm, and well separated from each other. The morphology in Fig. 12a is common to many nanostructured materials, and is controllable by changing of the electroless parameters [41]. Clearly, the inter-grains regions in Fig. 12a are of a lower electron density compared with grains. This is due to the establishment of a form of film growth in which columnar or needle-like (filiform) nanoscale units are separated by less dense, cobalt-sulfur amorphous or heavily disordered inter-grain regions. Figure 12c is a scanning electron microscopy (SEM) image of cross section of the as-deposited Co-S film. The nanostructured films have rough surface, where columnar (filiform) nanograins normal to the film plane are separated by amorphous and/or highly disordered regions. After films annealing for 1h at T_a = 550 K and, respectively at 600 K, the appearance of single α-Co (hcp) peak is observed on Fig. 11. Annealing treatment of the films promoted larger nanograins growth and, therefore, we clearly observe XRD patterns of hcp Co with only one broader and small peak. There is no morphological significant difference in the TEM images of the annealed samples, e.g. at T_a = 600 K (Fig. 12b) compared with those of as-prepared samples (Fig. 2a). The annealed samples also contain an amorphous phase that embedded the hcp Co nanograins with larger diameter (6-9 nm).

Figure 13 a-c shows the dc magnetic hysteresis loops at room temperature for as-prepared (a) and for the annealed films at T_a = 550 K (b) and at T_a = 600 K (c) measured by VSM. The dc hysteresis loops were measured, in plane (//) and perpendicular ($\perp$) to the film plane, respectively. From fig 13 a-c, we can note several outstanding features. The film magnetization easily saturates in plane than in perpendicular direction, but in a much larger field than expected. The saturation field $H_{s//}$ of about 3×10^5 A/m is required to align the magnetization in the film plane. The dc hysteresis loop with parallel field is linear in $H_{//}$, exhibiting very low remanence and coercivity. The saturation field in perpendicular direction to the film plane ($H_{s\perp}$) is even larger, about 5×10^5 A/m. Also, we note that magnetization is linear in $H_\perp$. These features, totally different from those homogeneous or polycrystalline Co films, suggest the existence of PMA, in addition to the usual in-plane shape anisotropy. As shown in Fig. 3b, after annealing at T_a = 550K, the results do not change appreciably. Only for the samples annealed at T_a = 600 K, the dc hysteresis loop has a square form for the in-plane measurement, and the perpendicular loop extends to the saturation magnetization. Evidently, the subtle nanostructure that causes PMA has been altered by the annealing at T_a = 600 K. A systematic study of the nanostructure and magnetic property changes induced by annealing is in progress. Figure 13d shows in-plane ac hysteresis loops of the as-prepared film measured with an ac inductive CAHLT at two different frequencies, i.e. at 50 Hz and 10 Hz. The ac hysteresis loops of Co-S films have a rhombic character ($H_{c//}$ = 3.6×10^3 A/m) that are characteristic loops for films which exhibit PMA, magnetization reversal by magnetic wall displacement and stripe-like magnetic domain structure [48].

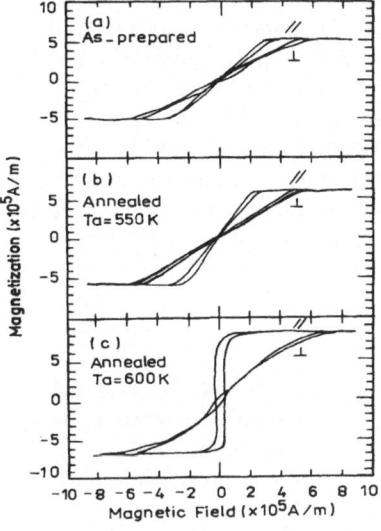

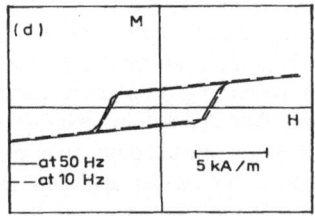

Figure 13. Magnetic hysteresis loops, at room temperature and measured with an vibrating sample magnetometer (VSM) from (a) as-prepared Co-S film, (b) annealed Co-S film at T_a = 550 K, and (c) annealed Co-S film at T_a = 600 K; (d) In-plane hysteresis loops, at room temperature and measured with a computer-controlled ac inductive hysteresis loop tracer (CAHLT) at two different frequencies, i.e. at 50 Hz and 10 Hz, for an as-prepared Co-S film.

The M-H linear dependence is characteristic for ultrafine magnetic particle assembly that exhibits dipolar interaction [52, 53]. The as-prepared Co-S alloy films behave as a magnetically disordered system (Fig. 3a) meanwhile, annealed Co-S films behave as a system near to the transition point between a magnetically disordered and ordered state (Figs. 13b and 13c). The wasp-like shape of the dc perpendicular hysteresis loops (Figs 13a-13c) indicates that Co-S film is not uniformly magnetized, but can divided into magnetic domains with up and down magnetization components. This fact was confirmed by the preliminary domain structure investigations undertaken on Co-S films. Moreover, the peculiarity of domain structures observed is that the domains embody a great number of Co grains, i.e. magnetic percolation takes place. This behavior may be understood as a result of the interplay between (i) PMA, which tends to place the magnetization perpendicular to the film plane, (ii) dipolar

530

interactions, which favor flux closure, and (iii) matrix mediated exchange interactions, which tend to align the magnetic moments parallel [52, 54].

In conclusion, we have found unusual histeresis loops in magnetic $Co_{0.36}S_{64}$ alloy films. The hysteresis loops are characterized by large saturation fields and a linear dependence of the applied field with nearly zero coercivity and remanence. The preliminary measurement pointed out a striped-like magnetic domain structure and the magnetic percolation. These unusual results are due to the perpendicular anisotropy present in the phase-separated films with certain nanostructure. The perpendicular anisotropy is related to the structure and not to the material.

5. Summary

Electrochemical methods can be used to produce nanostructured films and other interesting structure with a wide variety of interesting and potentially useful magnetic properties. These techniques open up new possibilities in controlling properties through the structure control. Moreover, electrochemical deposition offers real advantages to automatically prepare identical thin films with large area surface and a great variety of structures and properties. Electrodeposition is an alternative method that, in many cases (e.g. ultrathin films), is preferred to molecular beam epitaxy or sputtering, since of its low capital cost and because it generally yields better soft or hard magnetic properties at corresponding compositions, easily allows orientation of film anisotropy and most importantly enables the production of magnetic pole tips of very high quality by means of the plating-through-mask patterning process. In addition, electrodeposition offers a unique feature compared to other deposition techniques: reversibility. By combining electrodeposition and semiconductor substrate, it is possible to fabricate spin-valve, which shows promising results for sensor applications. The recent researches have pointed out that metal/oxide nanostructures can be fabricated by electrochemical methods and have a wide variety of interesting properties relating to single electron devices. Nanostructured films with high perpendicular magnetic anisotropy can be used for fabricated extremely high density recording heads and media.

6. References

1. Timp, G. (1999) *Nanotechnology*, American Institute of Physics Press, New York.
2. Binning, G., Rohrer, H., (1982) *Scanning tunneling microscopy*, Helvetica Physica Acta, Basel **55**, S. 726-735.
3. Noid, D.N., R. F. Tuzun, R.F., Sumpter, B.G. (1997) *On the Importance of Quantum Mechanics for Nanotechnology*, Nanotechnology **8**, 119-126.
4. Himpsel, F.J., Ortega, J.E., Mankey, G.J., Willis, R.F. (1998) *Magnetic nanostructures*, Adv. Phys. **47**, 511-597.
5. Wang, L.W., Liu, Y., Zhang, Z. (2002) *Handbook of Nanophase and Nanostructured Materials*, Kluwers Academic Publishers, Dordrecht.
6. Paanovic, M., Schesinger, M. (1998) *Fundamentals of Electrochemical Depositions*, John Wiley, New York.
7. Mallory, G., Hadju, J.B. (1990) *Electroless Plating: Fundaments and Applications*, AESF Orlando, Florida.
8. Watanable, T. (1994) *Formation of metastable phases by the plating method*, Mater Sci. Eng.

A179/A180, 193-197.

9. Kazeminezhad, I., Blythe, H.J., Schwarzacher, W. (2001) *Alloys by precision electrodeposition*, Appl. Phys. Lett. **78**, 1014-1016.

10. Schindler, W., Hofmann, D., Kirshner, J. (2000) *Nanoscale electrodeposition: a new route to magnetic nanostructures?*, J. Appl. Phys. **87**, 7007-7009.

11. Gleiter, H. (2001) *Tuning the electronic structure of solids by means of nanometer-sized microstructures*, Scripta Mater. **44**, 1161-1168.

12. Osaka, T. (2000) *Electrodeposition of highly functional thin films for magnetic recording devices of the next century*, Electrochimica Acta **45**, 3311-3321.

13. Sulitanu, N. (2000) *Microstructure and stripe domains in Ni-S ferromagnetic thin films*, J. Magn. Magn. Mater. **214**, 176-184.

14. Myung, N.V., Nobe, K. (2001) *Electrodeposited iron group thin-film alloys*, J. Electrochem. Soc. **148**, C136-C144.

15. Attenborough, K. (2001) J.P. Celis (2001) *Properties and applications of electrodeposited magnetic materials*, Galvanotechnik **92**, 488-494.

16. Sulitanu, N. (2002) *Electrochemical deposition of novel nanostructured magnetic thin films for advanced applications*, Mater. Sci. Eng. **B95**, 230-235.

17. Tumanski, S. (2001) *Thin Film Magnetoresistive Sensors*, Instiute of Physics Publishers, London.

18. Schindler, W., Koop, Th., Hofmann, D., Kirshner, J. (1998) *Reversible electrodeposition of ultrathin magnetic Co films*, IEEE Trans. Mag. **34**, 963-967.

19. Cziraki, A., Fogarassy, B., Gerocs, I., Toth-Kadar, E. and Bakonyi. I (1994) *Microstructure and growth of electrodeposited nanocrystalline nickel foils*, J. Mater. Sci. **29**, 4771-4777.

20. Erb, U., Palumbo, G., Zugic, R., Aust, K.T (1996) Structure-property relationships for electrodeposited nanocrystals, in C. Suryanaryana, J. Singh and F.H. Froes (eds.), *Processing and Properties of Nanocrystalline Materials*, The Minerals, Metals & Materials Society Press, Warrendale, Pennsylvania pp. 93-122.

21. Brenner, A. (1963) *Electrodeposition of Alloys. Principle and Practice*, Academic Press, New York.

22. Gawrilov, G.G. (1979) *Chemical (Electroless) Nickel-Plating*, Portcullis Press, Redhill, UK.

23. Schwartzacher, W. (1999) *Metal nanostructures. A new class of electronic devices*, Electrochem. Soc. *Interface* **8**, 18-22.

24. Searson, P.C., Cammarata, R.C., Chien, C.L. (1995) Electrochemical processing of nanostructured materials, in H. Merchant (ed.) *Defect Structure, Morphology and Properties of Deposits*, The Minerals, Metals & Materials Society Press, Warrendale, Pennsylvania pp. 345-357.

25. Cavallotti, P.L., Lecis, N., Fauser, H., Zielonka, A., Celis, J.P., Wouters, G., Machado da Silva, J., Brochado Oliviera, J.M., Sa, M.A. (1998) *Electrodeposition of magnetic multilaers*, Surf. Coat. Techn. **105**, 232-239.

26. Jansen, R. Van't Erve, O.M.J., Kim, S.D., Vlutters, R., Anil Kumar, P.S. (2001) *The spin-valve transistor: fabrication, characterization, and physics*, J. Appl. Phys. **89**, 7431-7436.

27. Budevski, E., Staikov, G., Lorenz, W.J. (1996) *Electrochemical Phase Formation and Growth*, VHC, Weinheim.

28. Perez, L., Attenborough, K., De Boek, J., Celis, J.P., Aroca, C., Sanchez, P., Lopez, E., Sanchez, M.C. (2002) *Magnetic properties of CoNiFe alloys electrodeposited under potential and current control conditions*, J. Magn. Magn. Mater. **242-245, Part I**, 163-165.

29. Saitou, M., Oshikawa, W., Mori, M., Makabe, A. (2001) *Surface roughening in the growth of direct current or pulse current electrodeposited nickel thin films*, J. Electrochem. Soc. **148**, C780-C783.

30. Natter, H., M., Hempelmann, R. (1996) *Nanocrystalline copper by pulsed electrodeposition: the effect of organic additives, bath temperature, and pH*, J.Phys. Chem. **100**, 19525-19532.

31. Chassaing, E. (2001) *Effect of organic additives on the electrocrystallization and the magnetoresistance of Cu-Co multilayers*, J. Electrochem. Soc. **148**, C690-C694.

32. Osaka, T., Sawaguchi, T., Mizutani, F., Yokoshima, T., Takai, M., Okinaka, Y., (1999) *Effects of saccharin and thiourea on sulfur inclusion and coercivity of electroplated soft magnetic CoNiFe film*, J. Electrochem. Soc. **146**, 3295-3299.

33. Peter, L., Kupay, Z., Cziraki, A., Padar, I., Toth, I., Bakonyi, I. (2001) *Additive effects in multilayer electrodeposition: properties of Co-Cu/Cu multilayers deposited with NaCl additive*, J. Phys. Chem. B **105**, 10867-10873.

34. Takai, M., Kondo, A., Mera, F., Kaseda, M., Osaka, T. (1998) *Electrodeposition of soft magnetic*

Ni-Fe-based film with high resistivity, J. Surf. Finish. Soc. Jpn. **49**, 292-296.

35. Tabakovic, I., Inturi, V., Riemer, S. (2002) *Composition, structure, stress, and coercivity of electrodeposited soft magnetic CoNeFe films. Thickness and substrate dependence*, J. Electrochem. Soc. **149**, C18-C20.

36. Sulitanu, N. (2000) *Electrolessly deposited NiCoWS alloy films for perpendicular recording media*, Bull. Polytechn. Inst. Jassy **46 (fasc. 3-4)**, 53-56.

37. Zhang, J., Chow, G.M. (2000) *Electroless polyoldeposition and magnetic properties of nanostructured $Ni_{50}Co_{50}$ films*, J. Appl. Phys. **88**, 2125-2129.

38. Kakuno, E.M., Da Silva, R.C., Mattoso, N., Schreiner, W.H., Mosca, D.H., Teixeira, S.R. (1999) *Giant magnetoresistance in electrodeposited $Co_{87}Fe_{13}/Cu$ compositionally modulated alloys*, J. Phys. D: Appl. Phys. **32**, 1209-1213.

39. Fenineche, N., Chaze, A.M., Coddet, C. (1996) *Effect of pH and current density on the magnetic properties of electrodeposited Co-Ni-P alloys*, Surf. Coat. Technol. **88**, 264-268.

40. Alper, M. (1995) *Electrodeposited Magnetic Superlattices*, Thesis, University of Bristol, UK.

41. Sulitanu, N. (2000) *Microstructure and magnetic properties of electrolessly deposited Co-S thin films*, Mater. Sci. Eng. **B77**, 27-32.

42. Coey, J.M., Hinds, G. (2001) *Magnetic electrodeposition*, J. Alloy. Comp. **326**, 238-245.

43. Coey, J.M., Hinds, G., O'Reilly, C., Ni Mhiochain, T.R. (2001) *Magnetic field effects on electrodeposition*, Mater. Sci. Forum **373-376**, 1-8.

44. Sulitanu, N.D. (1992) *A suitable method for obtaining Ni-W thin magnetic films*, Mater. Lett. **14**, 295-297.

45. Homma, T., Osaka, T., Yamazaki, Y., Namikawa, T. (1995) *Correlation between magnetic properties and phase-separated microstructure of electroless CoNiP perpendicular magnetic recording media*, Script. Metall. Mater. **33**, 1569-1573.

46. Krishnan, K. (1999) *Magnetism and microstructure: the role of interfaces*, Acta Mater. **47**, 4233-4244.

47. Sulitanu N (1992) *Automatic torque magnetometer for thin ferromagnetic*, Stud. Res. Phys. **44** 699-709.

48. Chikazumi, S. (1997) Physics of Ferromagnetism, Clarendon Press, Oxford, p.450 & p. 509.

49. Sulitanu, N. (2001) *Structural origin of perpendicular magnetic anisotropy in Ni-W thin films*, J. Magn. Magn. Mater. **231**, 85-93.

50. Grundy, P.J. (1998) *Thin film magnetic recording media*, J. Phys. D: Appl. Phys. **31**, 2975-2990.

51. Maeda, Y., Rogers, D.J., Song, O., Takei, K., Okhubo, T., Hirono, S., Suzuki, J., Morii, Y. (1997) *Magnetic microstructures produced by compositional separation in Co-Cr based alloy thin films*, IEEE Trans. Magn. **33**, 879-884.

52. Kechrakos, D., Trohidou, K.N. (1998) *Effects of dipolar interactions on the magnetic properties of granular solids*, J. Magn. Magn. Mater. **177-181**, 943-944.

53. Cowburn, R.P., Adeyeye, A.O., Welland, M.E. (1999) *Controlling magnetic ordering in coupled nanomagnet arrays*, New J. Phys. **1**, 16.1-16.9.

54. Franco, V., Battle, X., Labarta, A., O'Grady, K., (2000) *The nature of magnetic interactions in CoFe-Ag (Cu) granular thin films*, J. Phys. D: Appl. Phys. **33**, 609.

DISCLINATIONS IN SEVERE DEFORMED MATERIALS: A CASE FOR TEM CHARACTERIZATION

A.L. KOLESNIKOVA[1,2)], V. KLEMM[2)], P. KLIMANEK[2)], and
A. E. ROMANOV[2,3)]
[1)] Institute for Problems of Mechanical Engineering, Russian Academy of
Sciences, Bolshoj 61, Vas.Ostrov, St.Petersburg 199178, Russia
[2)] Freiberg University of Mining and Technology, Institute of Physical
Metallurgy, Gustav-Zeuner Str. 5, D-09596 Freiberg, Germany
[3)] Ioffe Physico-Technical Institute, Russian Academy of Sciences,
Polytechnicheskaya 26, St.Petersburg 194021, Russia

1. Introduction

The deformation of metallic materials up to large strains leads to the formation of fine grain structures (in the limit nanocrystalline structures) with high densities of grain boundaries and grain boundary junctions [1]. Such structures demonstrate significant lattice rotations and the presence of disclination defects at grain boundary junctions [1,2]. By definition partial disclinations are associated with terminated boundaries of misorientations in otherwise perfect crystals [3-5].

Wedge disclinations can be considered as the mesoscopic model for terminated tilt boundaries. Figure 1 presents the basic disclination description of defect structures, which can be often observed in severe deformed materials. These structures include low angle dislocation walls (Fig. 1a,b) with angle of misorientation φ_{DW}, twin boundaries (Fig. 1c) with misorientation φ_{TB}, high angle grain boundaries (Fig. 3d-f) with misorientation φ_{GB}.

The geometry and morphology of terminated boundaries of misorientation also can be different. For single terminated tilt boundaries the disclination strength (magnitude of Frank vector ω) is exactly equal to the angle of misorientation [5]. In Fig. 1a the so-called negative wedge disclination $-\omega$ with the strength

$$\omega = \varphi_{DW} \qquad (1)$$

is shown. In case of the dislocation wall terminated from two sides shown in Fig. 1b, the wall is equivalent to wedge disclination dipole with disclination strength $+\omega$ and $-\omega$ correspondingly. Disclination dipole configurations can be also associated with terminated lamellae with misoriented crystal lattice: misorientation and kink bands [3,5] or twins [6] (as shown in Fig. 1c). In this last case two terminated boundaries of misorientation (for example twin boundaries) are peculiar for the dipole. In the junction

T. Tsakalakos, et al. (eds) pgs. 533 - 541
Nanostructures: Synthesis, Functional Properties and Applications;
© 2003 Kluwer Academic Publishers.

534

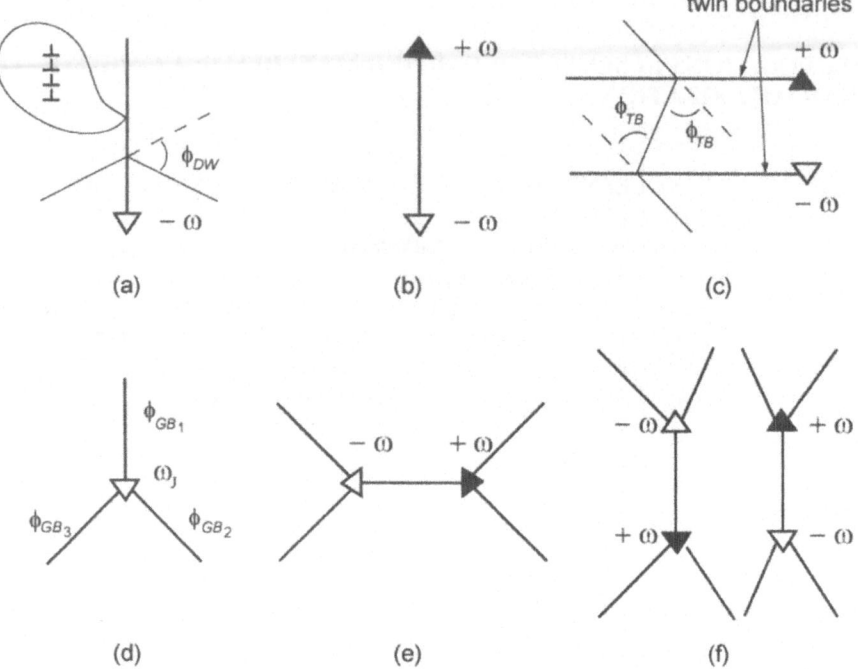

Figure 1. Disclination models for terminated tilt boundaries: (a) individual terminated dislocation wall with misorientation φ_{DW} ; (b) wedge disclination dipole – two side terminated wall; (c) twin lamella with twinning angle φ_{TB}; (d) triple junction of high angle grain boundaries; φ_{GB_i} $(i = 1,2,3)$ are misorientations at individual grain boundaries; (e) dipole of triple junctions with disclinations $\pm\omega$; (f) quadrupole of triple junctions.

of several boundaries (Fig.1d) the resulting disclination with the strength ω_J can appear under the condition

$$\omega_J = \sum_i \omega_i = \sum_i \varphi_{GB_i} \neq 0 \quad (i = 1,2,3), \qquad (2)$$

i.e. in the case of so-called non-compensated junctions. Figures 1e,f give the schematics for typical configurations (dipole and quadrupole) of non-compensated grain boundary junctions, which are peculiar for the ultra fine grain structure of severe deformed materials.

2. Observation of disclinations in the structure of deformed materials

At present one cannot say that the techniques for disclination identification for in severe deformed metals are developed as well as those for dislocations, inclusions and stacking faults in weakly deformed metals.

High defect densities and the influence of strong long-range elastic fields complicate the identification of disclination configuration in severe deformed metals. Possible examples of disclination defect observation in the structure of deformed metals are given in Fig.2.

One proposed technique for disclination identification is based on the calculation of the strength of the junction disclination from the misorientation at the associated boundaries. The description of the local misorientation between two regions A and B in the specimen can be easily done by the misorientation matrix. In its turn, the misorientation matrix can be calculated when the orientation of the crystal lattice in both regions is measured with respect to the laboratory coordinate system. The orientation of crystal regions can be directly found from electronograms [3]. However the accuracy of this technique (more than 1 degree [3,7]) is not enough for analysis low strength junction disclinations and disclinations related to low angle misorienation boundaries.

Recently proposed modification of the above technique, which is based on the analysis of Kikuchi diffraction pattern [8,9], can overcome these difficulties. It is demonstrated that the accuracy of misorientation determination in this case reaches 0.2 degrees. For small misorientations between neighboring grains the misorientation matrix follows directly from Kikuchi pattern of selected regions by applying special mathematical technique [9]. Product of all local misorientation matrixes across all dense dislocation walls around selected triple junction gives the resulting mismatch of misorientations at the junction. The Frank vector ω of the disclination can be calculated from the mismatch, but it is impossible to determine directly the line vector l of the disclination from Kikuchi patterns. Due to the high defect density in severe deformed materials with disclinations it is extremely difficult to reproduce a three-dimensional configuration of disclination line from two (or more) tilted TEM micrographs with nearly the same imaging conditions.

TEM micrographs of the material regions with non-compensated junctions of dense dislocation walls (Fig.2) show a characteristic black-white non-uniform contrast across and near the dense dislocation wall and a significant change in the run of the bending fringes caused by the long range distortion field. The tilt of a specimen changes the contrast distinctly. Such diffraction contrast is clearly related to elastic distortions present associated with terminated dislocation walls or grain boundary junctions. While elastic distortions have disclination origin we may call this contrast as disclination one.

It is well known that TEM diffraction contrast related to different defects (*i.e.* dislocations and stacking faults) can be well understood in the framework of the dynamical theory of electron diffraction [10]. The contrast analysis is useful, for example, for the determination of the dislocation Burgers vector [7,10]. In the case of two-beams (direct and diffracted beams) model by Howie and Whelan the influence of dislocation elastic fields can be relatively simple included in the model equations [7]. We propose to use the same two-beams model for the qualitative and quantitative analysis of the disclination contrast.

536

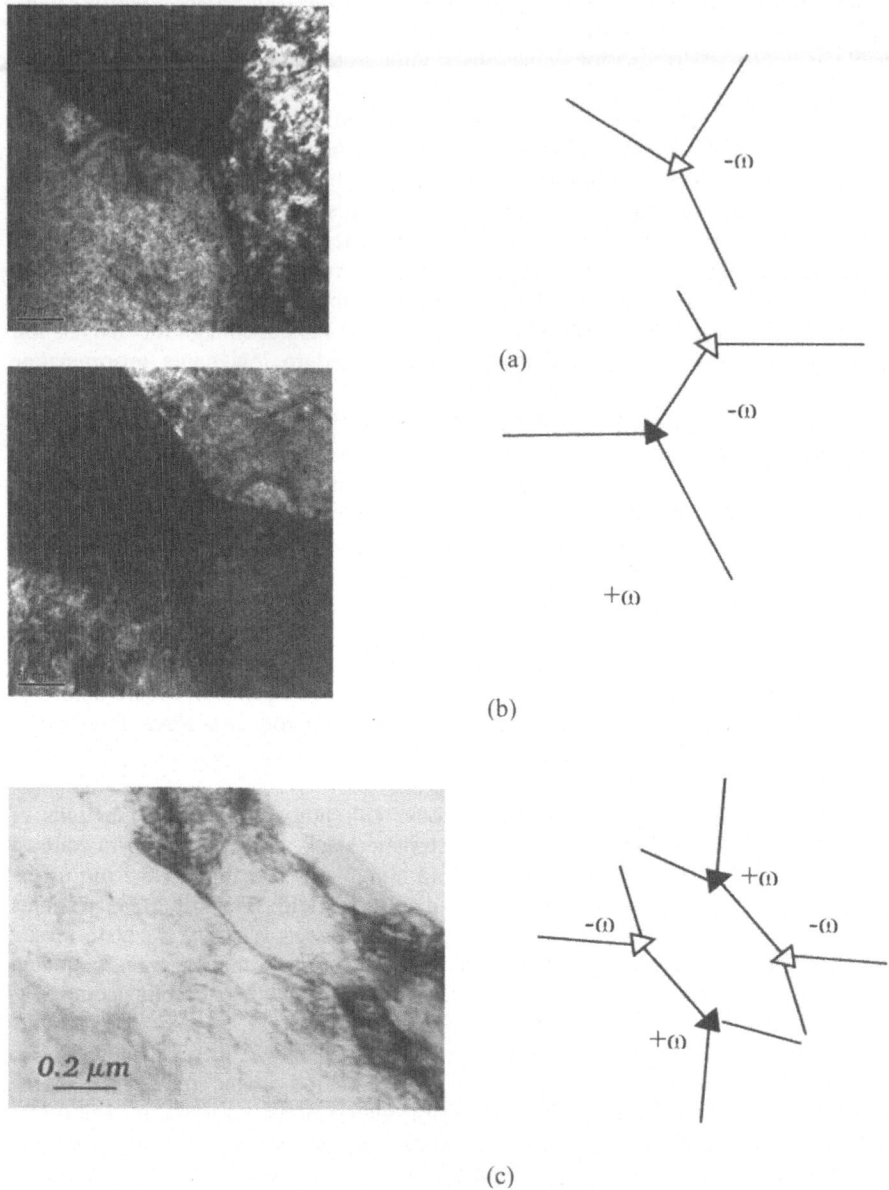

Figure 2. Examples and schematics for disclination configurations observed in strongly deformed metals: (a) non-compensated triple junction as a partial disclination and (b) disclination dipole in *W* rolled down to 70 % thickness reduction at 600°C; (c) disclination quadrupole in *Cu* rolled down to 70% thickness reduction at room temperatute.

3. Computer simulation of TEM Images of Disclination Defects

In the framework of Howie-Whelan two-beams model [10] the intensities of direct I_0 and diffracted I_g electron waves propagating throw the sample foil (Fig.3) containing disclination defects may be obtained by solving the following differential equations:

$$\begin{cases} \dfrac{d\Phi_0}{dz} = -\pi \dfrac{\xi_g}{\xi_0'} \Phi_0 + \pi(i - \dfrac{\xi_g}{\xi_g'})\Phi_g \\ \dfrac{d\Phi_g}{dz} = \pi(i - \dfrac{\xi_g}{\xi_g'})\Phi_0 + \left[-\pi \dfrac{\xi_g}{\xi_0'} + 2\pi i \xi_g (s + s_1) \right]\Phi_g \end{cases} \tag{3}$$

Here Φ_0 and Φ_g are the amplitudes of direct and diffracted electron waves respectively ($I_0 = |\Phi_0(z=t)|^2$ (bright field), $I_g = |\Phi_g(z=t)|^2$ (dark field)), $\dfrac{\xi_g}{\xi_0'}$

and $\dfrac{\xi_g}{\xi_g'}$ are parameters of normal and anomalous absorption respectively, ξ_g is the

extinction length, s is z-component of deviation vector due to non-defect nature,

$s_1 = \mathbf{g} \cdot \dfrac{d}{dz} \mathbf{u}$, $\mathbf{g}$ is diffraction vector, $\mathbf{u}$ is the field of the displacements of disclination defects.

Figure 4 presents the calculated bright field images of wedge disclinations with line perpendicular (Fig.4a) and parallel (Fig.4b) to the film surfaces. In these two cases the special method of virtual surface defects was applied to calculate the displacement field $\mathbf{u}$ of disclinations perpendicular and parallel to the plate surfaces [11-13]. Figure 5 shows the calculated TEM-images for compensated (no disclination) and non-compensated (with disclination) triple junctions of grain boundaries.

4. Conclusion

In the presented research we explored the technique of TEM diffraction contrast calculation for disclinations. We propose to use the developed technique for disclination identification in the structure of severe deformed materials.

From the calculations the following dependence of the TEM contrast on the mutual orientation of the disclination line, the rotational vector, and the diffraction vector becomes evident: (i) for wedge disclination line parallel to the foil surfaces there is the

538

condition of the contrast disappearance (when diffraction vector is parallel to the disclination line), (ii) for wedge disclinations normal to foil surfaces no such condition

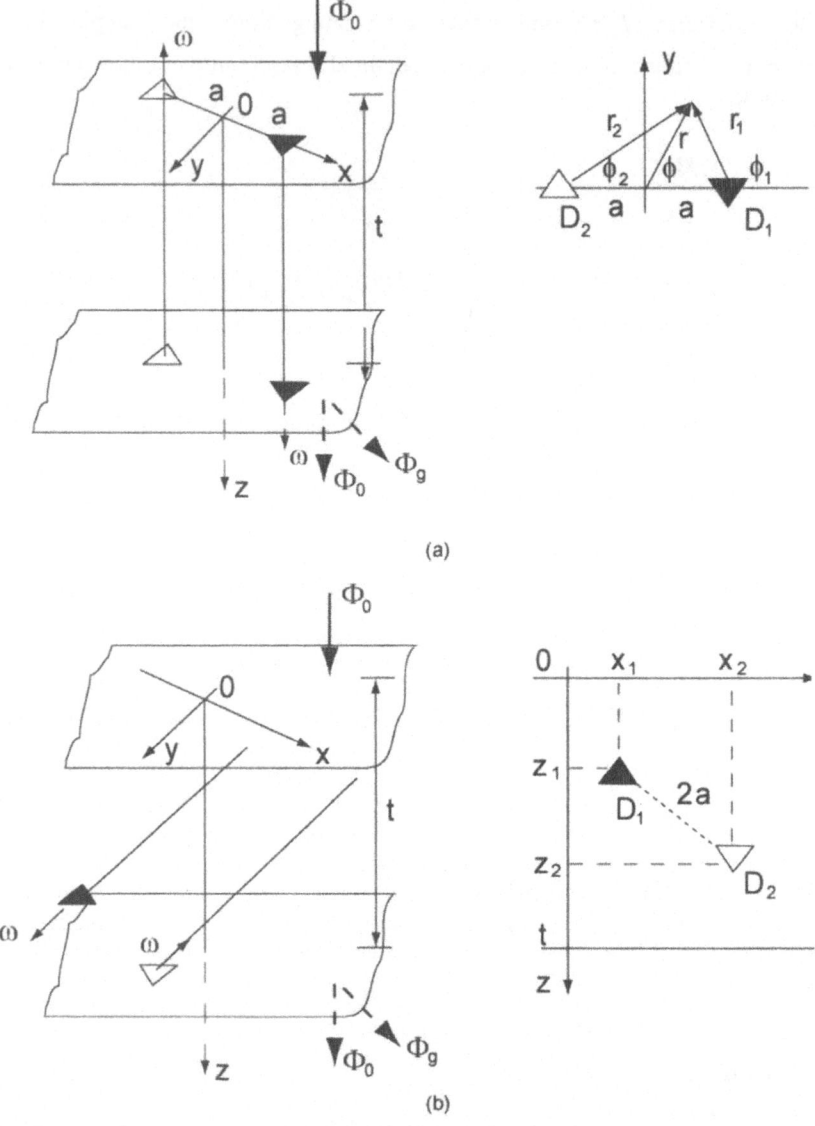

(a)

(b)

Figure 3. Wedge disclination dipole in a sample foil of finite thickness. (a) disclination lines D1 and D2 are perpendicular to the film surfaces ; (b) disclination lines are parallel to the foil surfaces; Φ_0, Φ_g are the amplitudes of the direct and diffracted electron waves.

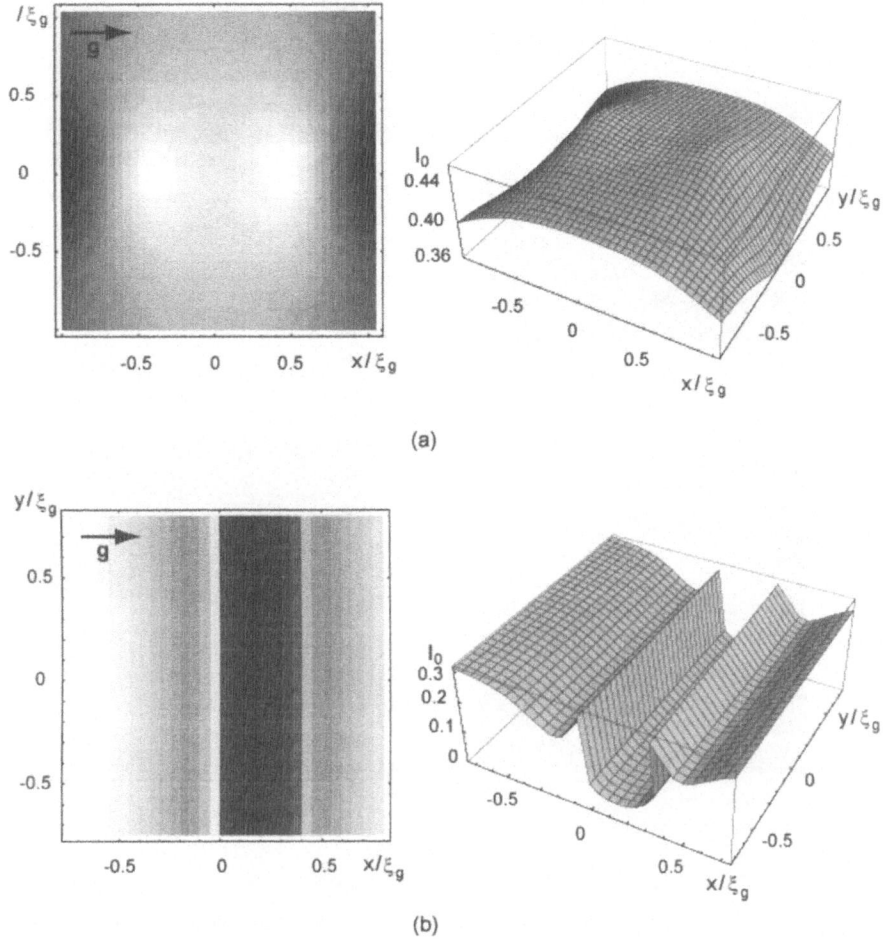

(a)

(b)

Figure 4. The bright field image for wedge disclination dipole in a thin foil. (a) disclination lines are perpendicular to the foil surfaces, dipole arm is $2a = 1.4\xi_g$; (b) disclination lines are parallel to the foil surfaces, dipole arm is $2a = 1.4\xi_g$; the depth co-ordinate for dipole is $100\,nm$ .The disclination strength is taken as $\omega = 3^0$. Foil thickness $t = 200$ *nm*, the diffraction vector **g** = (1,0,1) lies parallel to X-axis. Extinction length $\xi_g = 40.8746\ nm$ (Fe) . Bright field background is 0.32.

540

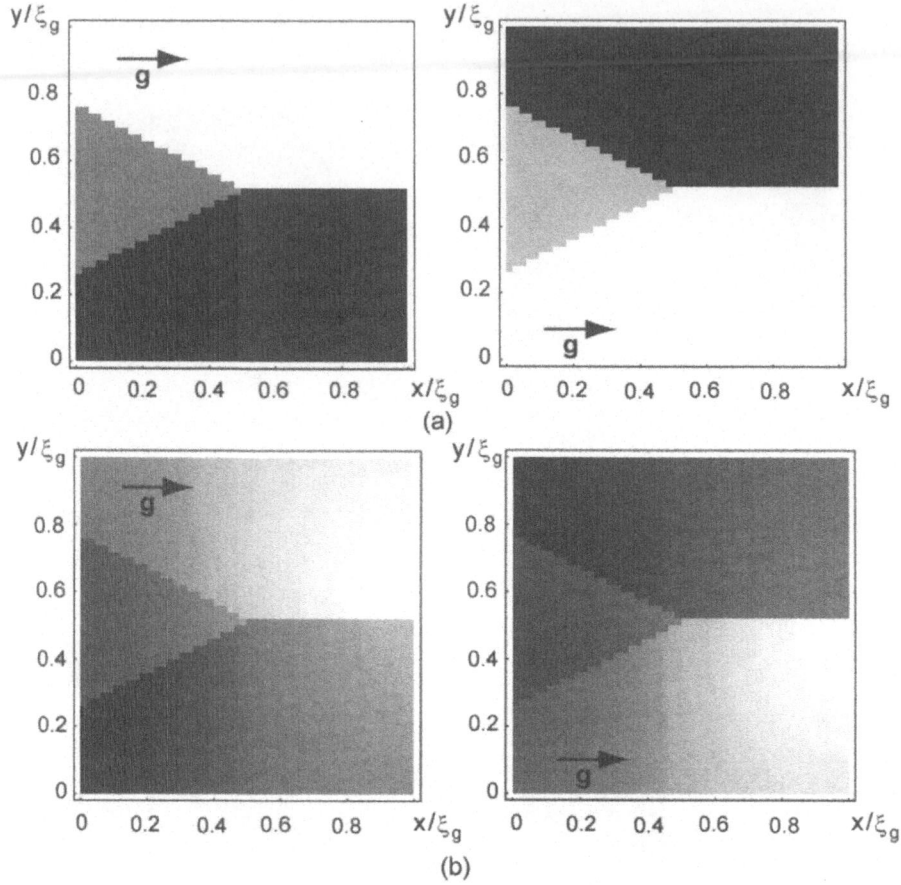

(a)

(b)

Figure 5. Compensated (a) and non-compensated (disclinated) (b) triple junctions. Left column – bright field images , right column – dark field images. Disclination strength is taken as $\omega = 1.83^0$. Foil thickness t = 200 nm, the diffraction vector g = (1,1,1) lies parallel to X-axis. Extinction length ξ_g = 41.8642 nm (Cu) . exists, but in this case the TEM image is very weak (for partial disclinations with strength $\omega \leq 10^0$ (compare Fig.4a and Fig.4b)).

TEM images for disclinations parallel to the foil surfaces have specific spatial features. The disclination line divides the image into two parts: bright and dark areas. The appearance of these features is due to rotational nature of disclination defects. The selection of the "jump" surface (at which the discontinuity in the displacement field takes place) may change the TEM images of wedge disclinations for all defect line orientations except the case when disclination line is perpendicular to the foil surfaces (at least in the framework of Howie-Whelan approach). Grain boundaries and low angle dislocation boundaries are natural physical sources for the discontinuities related to disclinations.

The difficulties in the experimental identification of disclination defects may be related to high density of defects and corresponding influence of another defect elastic fields and to unknown in advance orientation of disclination lines with respect to the surfaces of a sample. On the other hand the interference of the additional long-range elastic field with the elastic field of disclination origin may give rise to the specific picture of bending fringes near non-compensated grain boundary junctions. The modeling of the bending fringes run near disclinated grain boundary junctions is the subject of our outgoing research.

5. Acknowledgement

This work has been supported by Volkswagen Foundation (Research Project I/74645).

6. References

1. Valiev, R.Z., Islamgaliev, R.K., Alexandrov, I.V. (2000) Bulk nanostructured materials from severe plastic deformation, *Progr. Mater. Sci.* **45**, 103-189.
2. Klimanek, P., Klemm, V., Romanov, A.E., Seefeldt, M. (2001) Disclinations in plastically deformed metallic materials, *Adv. Eng. Mat.* **3**, 877-884.
3. Romanov, A.E., Vladimirov, V.I. (1983) Disclinations in solids, *Phys. Stat. Sol. (a)* **78**, 11-34.
4. Rybin,V.V. (1986) *Large Plastic Deformation and the Fracture of Metals,* Metallurgiya, Moscow (in Russian).
5. Romanov, A.E., Vladimirov, V.I. (1992) Disclinations in crystalline solids, in F.R.N.Nabarro (eds.), *Dislocations in Solids* **9**, North Holland Publ. Co., Amsterdam , pp.191-402.
6. Müllner, P. and Romanov, A.E. (1994) Between dislocation and disclination models for twins, *Scripta Met. Mat.* **31**, 1657-1662.
7. Williams, D.B., Carter, C.B. (1996) *Transmission Electron Microscopy. A Textbook for Materials Science*, Plenum Press, New York.
8. Klemm, V., Klimanek, P., Seefeldt, M. (1999) A microdiffraction method for the characterization of partial disclinations in plastically deformed method by TEM, *Phys. Stat. Sol. (a)* **175**, 569-576.
9. Klemm, V., Klimanek, P., Motylenko, M. (2002) TEM identification of disclinations in plastically deformed crystals, in P. Klimanek, A.E. Romanov, M. Seefeldt (eds.) *Local Lattice Rotations and Disclinations in Microstructures of Distorted Crystalline Materials*, Scitec Publications Ltd, Switzerland, pp.57-71
10. Hirsch, P.B., Howie, A., Nicholson, R.B., Pashly, D.W., Whelan, M.J. (1977) *Electron Microscopy of thin crystals*, Krieger, Huntington, New York.
11. Kolesnikova, A.L., Romanov, A.E. (1986) *Circular dislocation-disclination loops and their application to the solution of boundary problems in the theory of defects*, Preprint FTI, Leningrad (in Russian).
12. Vladimirov, V.I., Kolesnikova, A.L., Romanov, A.E. (1985) Wedge disclinations in an elastic plate, *Fizika Metallov i Metallovedenie* **60**, 1106-1115 (in Russian).
13. Kolesnikova, A.L., Klemm, V., Klimanek, P., Romanov, A.E. (2002) Transmission electron microscopy image contrast of disclination defects in crystals (computer simulation), *Phys. Stat. Sol. (a)* **191**, 467-481.

QUANTUM DOT SEMICONDUCTOR LASERS

V. M. USTINOV
A. F. Ioffe Physico-Technical Institute
Politekhnicheskaya 26, 194021 St. Petersburg, Russia

1. Introduction

Development of advanced active regions for semiconductor diode lasers was the main direction which gave the largest contribution to enormous progress of diode lasers in various applications. With each new approach in device design, and fabrication technology, the properties of lasers greatly improved and, in turn, gave a strong push to the development of new systems and, sometimes, new directions and branches of industry. The first step was the proposal of current-injection lasers and their realization. The decisive step for the beginning of the use of diode lasers in real industrial applications was the concept of double heterostructures which offered a possibility to fabricate devices, with low threshold, current density allowing continuous wave operation at room temperature. Further progress was associated with the use of effects of size quantization in semiconductor heterostructures. Fig.1 illustrates the progress of semiconductor diode lasers with reducing the dimensionality of the active region when the threshold current density is taken as the Figure of merit [1]. By threshold current density we mean the minimum current density of a semiconductor diode laser necessary to reach the population inversion and to overcome internal and external losses and to achieve modal gain.

The original idea to "exploit quantum effects in heterostructure semiconductor lasers to produce wavelength tunability" and achieve "lower lasing thresholds" belongs to Dingle and Henry [2]. The principal advantage of using the size-quantized heterostructures in lasers comes from the increase in the density of states for charge carriers near the band edges. When used as an active medium of a laser, this results in a concentration of most of the injected nonequilibrium carriers in an increasingly narrow energy range near the bottom of the conduction band and the top of the valence band. This enhances the maximum material gain and reduces the influence of temperature on device performance. The ultimate example of size quantization in solids is realized in a quantum dot (QD) which represents a semiconductor crystal with a size of only several nanometers coherently inserted in a larger bandgap semiconductor matrix [3].

T. Tsakalakos et al. (eds.) pgs. 543-559
Nanostructures: Synthesis, Functional Properties and Applications;
© 2003 Kluwer Academic Publishers.

544

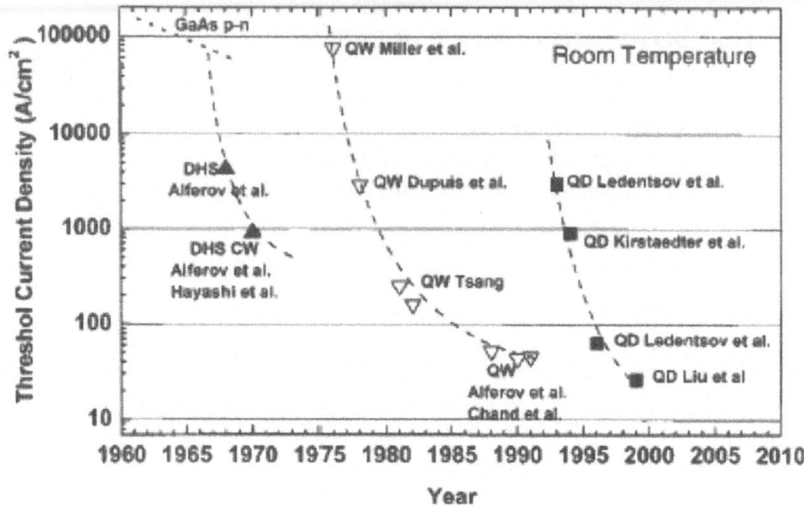

Fig. 1. Development of heterostructure lasers

It should be stressed, however, that real demonstration of the advantages of diode lasers based on low-dimensional structures is associated with the progress in the

epitaxial growth technology. The breakthrough is double heterostructure lasers occured when it was found that the AlGaAs/GaAs lattice matched system is stable under ambient conditions [4]. The drastic decrease in threshold current density of quantum well (QW) lasers became possible only after molecular beam epitaxy (MBE) and metal organic chemical vapor deposition (MOCVD) became sufficiently mature to fabricate device quality QW structures. The potential advantages of QD lasers were realized only when coherent self-organized quantum dot heterostructures were synthesized.

2. Self-Organized Quantum Dots and Control of their Characteristics

Semiconductor quantum dots exhibiting properties sufficient for laser applications are currently fabricated using the effect of spontaneous transformation of the growth surface at initial stages of strained layer heteroepitaxy. It was found that the growth of a strained material on a lattice-mismatched substrate first proceeds in a planar mode and so-called "wetting layer" is formed. However, at a certain critical thickness, this planar growth front is transformed into three-dimensional nanoscale islands on top of the thin wetting layer as was demonstrated for the first time for InAs/GaAs in [5]. When these InAs islands are covered with GaAs, a dense array of coherent nanoscale insertions in a GaAs matrix are formed. Since InAs has a bandgap, which is much less than GaAs, an array of InAs QDs is formed. This InAs QD ensemble usually demonstrates broad photoluminescence band around 1.2 eV at 77 K, i.e., at suprisingly longer wavelengths given the small average amount of InAs deposited (~ 2 monolayers). Extensive transmission electron microscopy (TEM) studies have shown that InAs/GaAs ensemble usually has surface island density around 4-5×10^{10} cm^{-2}, individual island shape is a pyramid with square base [6]. The lateral base size is about 100-150 Å, the height is 30-50 Å.

However, for various device applications, it is important to be able to control such parameters of a QD ensemble and individual islands as island surface density, island uniformity, island size and shape. These parameters affect electronic spectrum of quantum dots and significantly influence laser characteristics.

In the present work, we show that successive deposition of several planes of QDs separated by GaAs matrix spacers results in the formation of vertically coupled quantum dots which are characterized by increased height-to-base ratio. The use of the InAlAs seeding QD layer, with subsequent deposition of InGaAs QDs, leads to considerable increase in the InGaAs surface density. "Submonolayer" quantum dots formed as the result of deposition of alternate InAs and GaAs layers with less than one monolayer effective thickness demonstrate improved uniformity which manifests itself in considerably reduced PL peak width. And, finally, overgrowing InAs QDs with InGaAs ternary, rather than GaAs, results in the increase in the QD volume. These technological methods gave a possibility of realization of QD lasers with advanced properties.

2.1. VERTICALLY COUPLED QUANTUM DOTS

It has been found, that successive deposition of the InAs island sheets, and thin GaAs spacers, leads to the formation of the islands of the subsequent sheet just above the islands of the previous sheet if the spacer thickness is less than about 100 Å. The reason for this phenomenon is that the growth of the second InAs layer proceeds under the influence of the strain fields due to the presence of the previous QD layer. This leads to the preferential migration of the In atoms to places just above the location of the island of the previous sheet. If the spacer thickness is less than, or equal to, the height of the islands, the neighboring islands in vertical direction are characterized by the inified system of energy levels [7]. This means that varying the spacer thickness results in the shift of the PL emission line. The effect of vertical ordering is the basis for the increase of the surface density of quantum dots by the use of composite InAlAs/InAs QDs.

2.2. COMPOSITE VERTICALLY COUPLED InAlAs/InAs QUANTUM DOTS

Quantum dot density does not depend on the effective thickness of deposited InAs. It can be increased by stacking QDs but the number of QD layers can not be arbitrarily large and is limited by plastic strain relaxation and lateral association of neighboring islands. We proposed [8] to use InAlAs QDs whose surface density (~ 1.5×10^{11} cm^{-2}) has been found to be much larger than that of InGaAs QDs as nucleation centers for subsequent InAs QD formation.

Plan-view and cross-section TEM images of the structures containing three layers of stacked InAs QDs (#1), and one layer of InAlAs, followed by three layers of InAs stacked QDs (#2), are shown in Fig. 2. Cross section images show that vertical alignment between QDs in each row takes place for both structures. The surface densities extracted from the plan-view images are equal to 5×10^{10} cm^{-2} for structure #1 and 1×10^{11} cm^{-2} for structure #2. The average lateral sizes were estimated to be about 20 and 15 nm for the structures #1 and #2, respectively. Thus, the array of composite vertically alligned InAlAs/InAs QDs demonstrates higher density than that of InAs QDs. The increase in surface density leads to the decrease in lateral size of the islands since the amount of InAs deposited was the same for both structures studied.

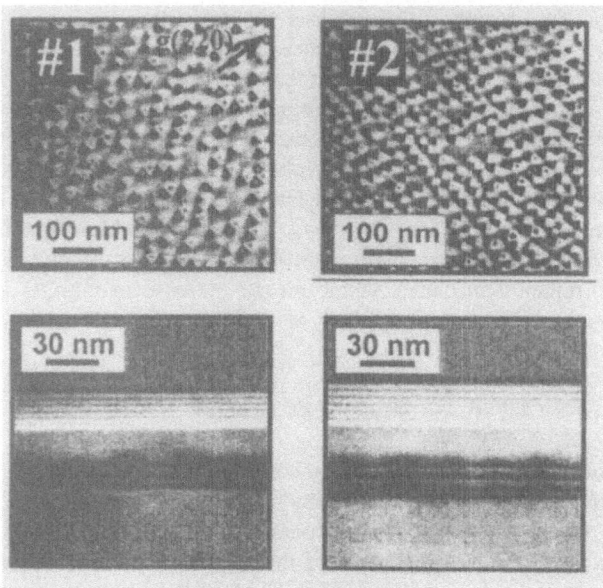

Fig. 2. Plan-view and cross-section TEM images of structures with composite InAlAs/InGaAs quantumdots

PL spectra taken from the samples #1 and #2 are shown in Fig. 3. One can see that the pre-deposition of InAlAs islands shifts the PL line toward higher energy which is in agreement with the decrease in the island size observed on TEM images.

Thus, InAlAs islands force InGaAs QDs to be transormed into the denser array as compared to the pure InGaAs case. The density of vertically aligned InAlAs/InGaAs QDs is set by the density of InAlAs QDs, whereas the energy of optical transition is determined by InGaAs QDs.

2.3. SUBMONOLAYER InAs/GaAs QUANTUM DOTS

Submonolayer quantum dots (SML QDs) are formed as the result of alternate deposition of InAs (0.5 ML) and GaAs (2.5 ML), repeated 10 times [9]. The sequence of SML quantum dots formation is schematically shown in Figs. 4(*a*)–(*d*).

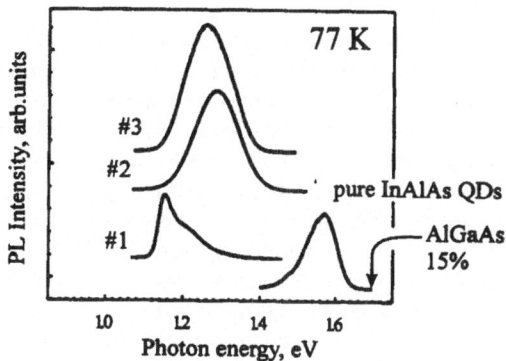

Fig. 3. PL spectra taken from structures with composite and conventional self-organized quantum dots.

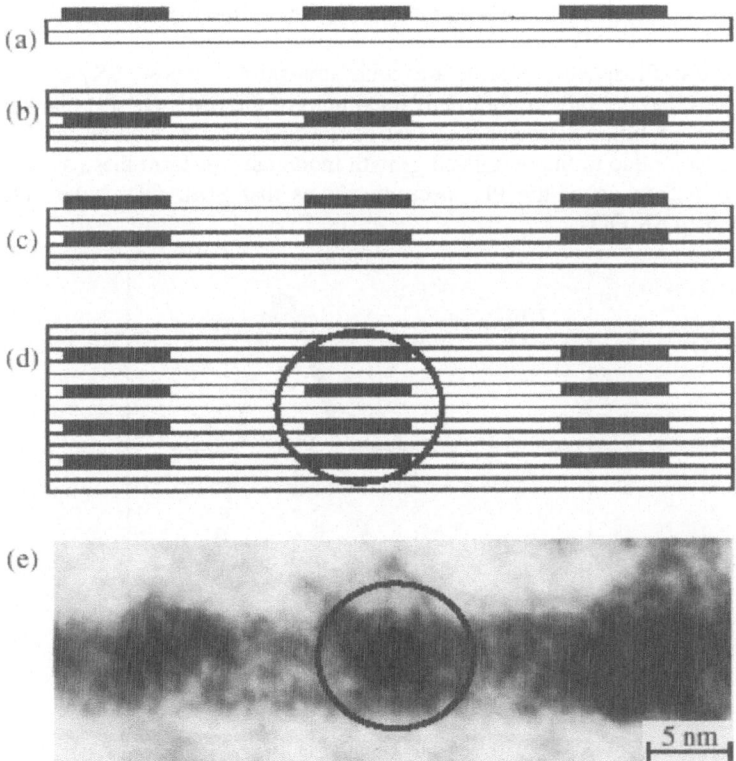

Fig. 4. Scheme of SML quantum dot formation: (*a*) deposition on InAs (<1 ML) on GaAs to give 1 ML high islands, (*b*) flat surface after deposition of several ML thick GaAs cap, (*c*) vertical correlation of InAs islands, (*d*) InGaAs QDs formed by SML deposition and (*e*) cross-sectional TEM image of an SML QD structure.

It was found previously that, when the InAs coverage is less then 1 ML, and optimized MBE growth conditions are used, thin InAs film transforms into an array of 1 ML high islands partly covering the surface. The islands of the second InAs layer, separated from the first layer by a thin GaAs spacer, spatially correlate with those of the preceding layer. As a result of multiple sub-monolayer deposition, In-rich QD like clusters consisting of several 1 ML high islands are formed, located one above another. A cross-sectional TEM image of the sub-monolayer InAs (0.5 ML)/GaAs (2.5 ML) superlattice is shown in Fig. 4(e). The QDs are visible as dark contrasting regions. The vertical correlation of QDs, separated by sufficiently thin spacers, has been previously observed in Stranski–Krastanow systems and explained by the effect of non-uniform strain fields. In the case of the sub-monolayer growth mode, the essential point is that all the constituent islands are characterized by the same height of 1 ML. Also, we can assume that the pyramidal SK QD shape, with usually large base-to-height aspect ratio, results in a strong effect of small deviations of the QD height and side facet angle on the quantization energy. Contrastingly, SML QDs are characterized by better stability with respect to small deviations in size and shape, owing to the symmetrical shape and aspect ratio close to unity (see Fig. 4). Thus, we expect much better uniformity of the SML QD array as compared to the QDs formed in the three-dimensional Stranski–Krastanow mode.

Fig. 5 shows PL spectra for structures with Stranski-Krastanow (SK) and SML QDs. It can be seen that the full-width-at-half-maximum of the SML QD PL line is about 19 nm, which is 3.5 times narrower than that of the SKQD, estimated to be 67 nm. We believe this to be due to the optimized growth mode used to form these sub-monolayer QDs. This narrowing of the PL spectrum shows that SML QDs have much better uniformity than SK QDs.

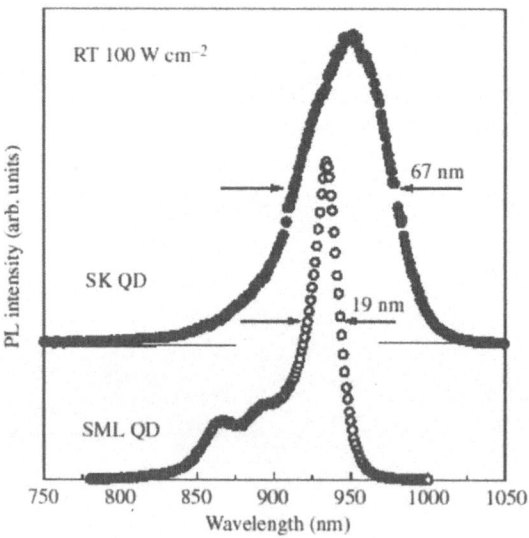

Fig. 5. Room temperature PL spectra taken from SK QD (•) and SML QD (°) structures.

2.4.   FORMATION OF THE InAs QUANTUM DOTS EMITTING AT 1.3 µm

We have shown earlier that the maximum emission wavelength which can be observed from the InAs/GaAs quantum dot ensemble, upon increasing the effective deposition thickness of InAs, is 1.24 µm at 300 K [10]. In this work we use the overgrowth of the InAs quantum dot ensemble by a thin $In_xGa_{1-x}As$ layer [11]. The TEM image shows that, in this case, the surface density of the quantum dots is about 5×10^{10} cm^{-2}. Both the increase in the effective thickness of deposited InAs (Q InAs) and the increase in the InAs mole fraction x in the In_xGa_{1-x} As ternary leads to a gradual increase in the emission wavelength which can achieve 1.3 µm at certain values of Q InAs and x [12]. Fig. 6 shows characteristic photoluminescence (PL) spectra of InAs/In_xGa_{1-x}As quantum dots. This figure also shows that we observe the red shift of the PL line even when we overgrow InAs quantum dots with the $In_xAl_yGa_{1-x-y}$ As quaternary whose bandgap is approximately equal to that of GaAs.

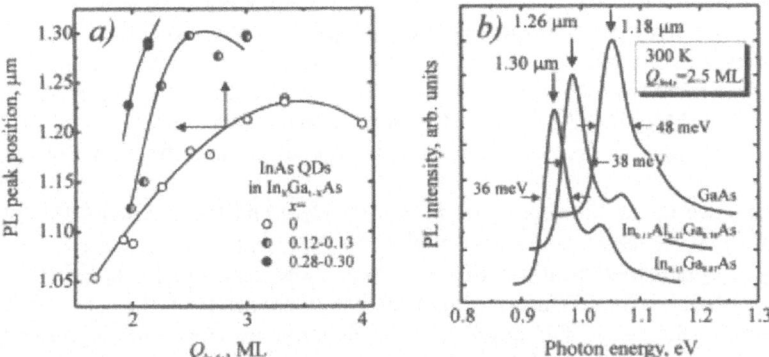

Fig. 6. (a) PL peak position at 300 K of InAs QD as a function of Q InAs for different QW compositions, (b) PL spectra of InAs QDs with different capping materials.

This unexpected result was explained after studying the cross-section TEM images of the structures which show that the effective volume of the InAs islands is increased which leads to the long-wavelength shift of the PL line. The characteristic PL spectrum (Fig. 6) consists of two features: the main peak is due to the ground state emission and the second (shorter-wavelength peak) is due to the excited state quantum dot emission. We should note that the InAs/InGaAs quantum dot structures are characterized by high indium content, which could potentially lead to strain relaxation with the formation of misfit dislocations. Therefore, we have minimized the overall In content in the structure, keeping 1.3 µm emission. We have found that in this case the structures demonstrate maximum PL intensity, indicating the lowest defect concentration [12]. The problem of the gain saturation characteristic for quantum dot lasers [13] leads to the use of multi-plane quantum dot structures in the active area of quantum dot lasers. We have found that, in the case of 1.3 µm InAs quantum dots, the use of thick (>20 nm) GaAs spacers between quantum dot sheets allows us to avoid misfit dislocations and to keep a high luminescence intensity.

3. Characteristics of Quantum Dot Lasers

The laser structures in this work were grown by elemental source molecular beam epitaxy (MBE) in a Riber 32 apparatus on an n + GaAs(100) substrate. Standard GRINSCH AlGaAs/GaAs laser structure design was used. The substrate temperature was 485, 700, and 600° C during the growth of the active region, cladding layers, and waveguide, respectively. The whole structure was grown under standard MBE As-rich conditions. The broad-area ridge devices with 0.1 mm-wide uncoated aperture were fabricated by shallow mesa etching. The p- and n- side metal contacts were Au/CrAu/ZnAu and GeAu/Au, respectively. Lasers of various lengths were tested under continuous wave (CW) and pulsed excitation (50ns pulse duration, 1kHz repetition frequency) conditions. For CW measurements, the devices were mounted p- side down on a Cu holder by In solder.

3.1. LASERS BASED ON COMPOSITE InAlAs/InGaAs QUANTUM DOTS

The active region was inserted into the middle of the 0.5 µm $Al_{0.15}Ga_{0.85}As$ waveguide and consisted of an array of composite vertically coupled InAlAs-InAs QDs [14]. First, three QD layers were formed by the successive deposition of 5.3 monolayers (ML) of InAlAs and then three rows of InAs QDs (2.5 ML) were grown. The QD rows were separated by 5nm $Al_{0.15}Ga_{0.85}As$ layers. QD formation was monitored *in situ* by the characteristic transition in the reflection high energy electron diffraction (RHEED) pattern.

The dependence of the threshold current density (J_{TH}) and the external differential efficiency (η_{DIF}) on the reciprocal cavity length ($1/L$) is presented in Fig. 7(a). In the 5–70 cm^{-1} range, J_{TH} increases linearly with mirror loss. Note, that superlinear behavior of J_{TH} is typically observed for QD lasers with the conventional design of the active region. This indicates that the denser QD array used in this structure overcomes significantly the problem of gain saturation. Moreover, the higher density of states in the QD array leads to a lower population in the wetting layer and the AlGaAs matrix resulting in a lower contribution of free carrier absorption to the internal loss. It can be seen that η_{DIF} is practically independent of the cavity length for the 5–60 cm^{-1} range of $1/L$. The internal loss for diodes with such L was estimated to be as low as 1.3 ±0.3cm^{-1}. The internal quantum efficiency was estimated to be ~75%.

The light output power per both facets (P_{OUT}) as a function of the drive current (I) for the 920 µm-long diode in pulsed and CW regimes are presented in Fig. 7(b). The heat sink temperature was 10° C. The CW lasing spectra for several values of I are shown on the inset. The 0.4A spectrum was taken just above the threshold. The threshold current (I_{TH}) was 380mA, which corresponds to a J_{TH} of ~400A/cm^2. Under pulsed operation the slope efficiency is nearly constant and equals 1.05 W/A (η_D = 73%) up to the sudden failure due to catastrophic optical mirror damage (COMD). The CW power-current dependence becomes slightly sublinear beyond 2.5A due to the heating of the active region, which is accompanied by a slight red-shift of the lasing line. The maximum output power recorded in CW and pulsed regimes is 3.5 and 4.8W at I = 4 and 5A, respectively. The effective (transverse) mode size was calculated to be 0.41 µm. The internal density of the optical power at COMD per facet is ~8MW/cm^2, which is in agreement with the COMD level for InGaAs-based lasers with uncoated facets. The device reaches COMD under pulsed operation at an internal optical power density of 11MW/cm², which is 40% higher than the COMD level under CW operation.

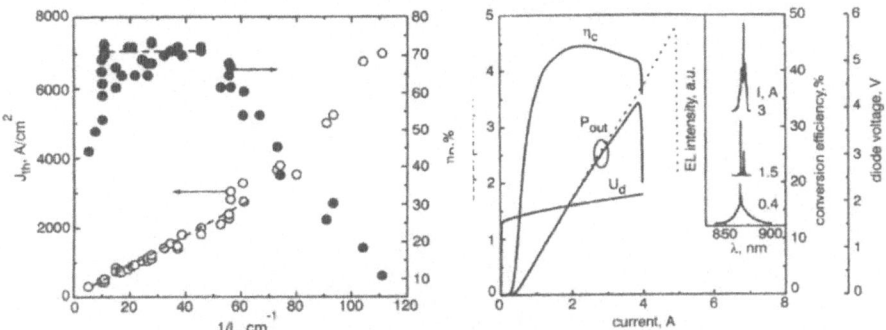

Fig. 7. Dependences of threshold current density and external differential efficiency on reciprocal cavity length (1/L) taken under pulsed excitation at 10°C (a) and light output power, diode voltage and conversion efficiency against drive current for 920mm-long stripe ———— CW operation, pulsed operation (1ms/1kHz)

Fig. 8 plots the threshold current density (J_{th}) versus the reciprocal cavity length (1/L) for SML and SK QD lasers [9]. The dependence of the lasing wavelength on 1/L is also shown. With decreasing diode length, the lasing wavelength is blue-shifted and the threshold current density increases to match the cavity loss. However, it is clearly seen that the SML QD laser demonstrates a weaker threshold current density rise and a better wavelength stability with increasing mirror loss. The slope of the J_{th}(1/L) dependence is inversely proportional to the differential gain, which is, thus, about two times higher in the SML QD laser as compared with the SK QD device. We believe that both these facts are direct consequences of the weaker inhomogeneous broadening of the gain spectrum, associated with the better uniformity of the SML QD array. It should be noted, that the densities of the SK and SML QD arrays are nearly the same of the order of 10^{11} cm². To compare these results with those obtained for InGaAs QW, we also plotted J_{th}(1/L) for our QW laser structure. The QW laser exhibits a much slower increase in J_{th} due to a much higher gain of QW as compared with QDs of both types. The reciprocal differential efficiency is plotted against the cavity length in Fig. 9.

It can be seen that, over the whole range of cavity lengths studied, the differential efficiency (external) of the SML QD laser is higher than that of SK QD. From this plot, the internal loss (α_i) and the internal quantum efficiency (η_i), were evaluated. The internal quantum efficiencies are comparable, being 100% and 96% for the SML QD and SK QD lasers, respectively. This fact reflects the high material quality and similar device designs. However, the internal losses in these cases are quite different. While the SML QD laser shows a reasonably low α_i of 2.3 cm⁻¹, the internal loss in the SK QD laser is much higher, 10 cm⁻¹. It has been shown, that α_i, of a self-organized QD laser, is mostly governed by free carrier absorption in the waveguide layer, the effective volume of the QD region being negligibly small.

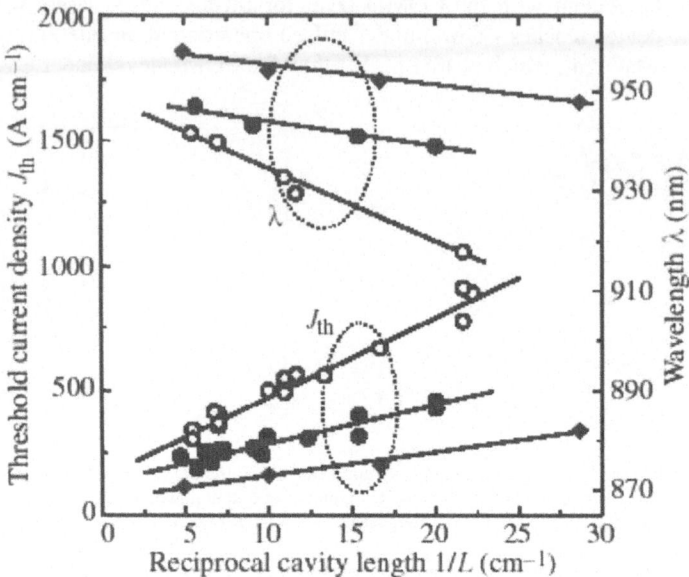

Fig. 8. Threshold current density and lasing wavelength versus reciprocal cavity length for 100 μm wide SML QD (•), SK QD (°) and QW (♦) laser diodes tested in pulsed mode 5 kHz / 1 μs.

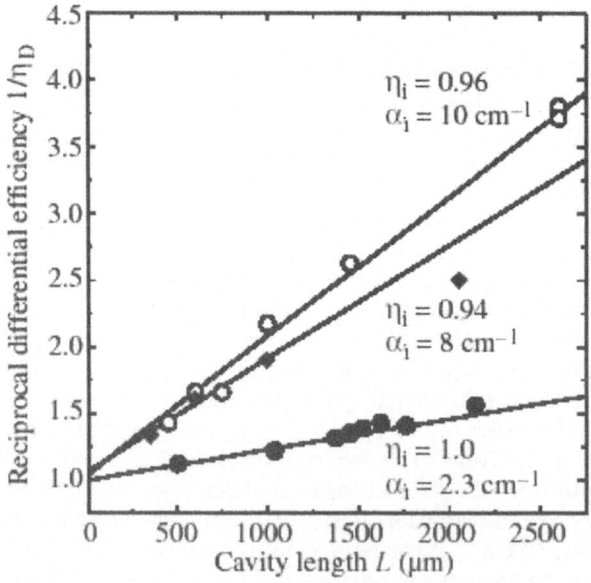

Fig. 9. Reciprocal differential efficiency versus cavity length for the SML QD (•), SK QD (°) and QW (♦) lasers.

The extremely low internal loss of 1.3 – 1.5 cm^{-1} has been reported for 1.24 μm self-organized QD lasers [15]. In these long-wavelength QD lasers, the pile-up of carriers in the waveguide layer is effectively suppressed through deep localization in the quantum dots. However, in the present case of a 0.94 μm QD laser, one could expect a stronger effect of free carrier absorption. The lower optical gain of the SK QD laser forces the quasi-Fermi levels to get higher, thereby increasing the concentration of free carriers in the waveguide layer. Thus, the better uniformity of the QD array, which can be achieved by using the sub-monolayer growth mode, also results in a lower internal loss. In the case of QW, internal losses are as high as 8 cm^{-1}. We believe that this is associated with the self-absorption in the QW states lying higher than the Fermi energy.

The power of light emitted from two uncoated facets (P_{out}) is shown in Fig. 10 as a function of the drive current (I) for 1 mm SML QD and SK QD diodes operating in CW mode at 10 °C heat-sink temperature.

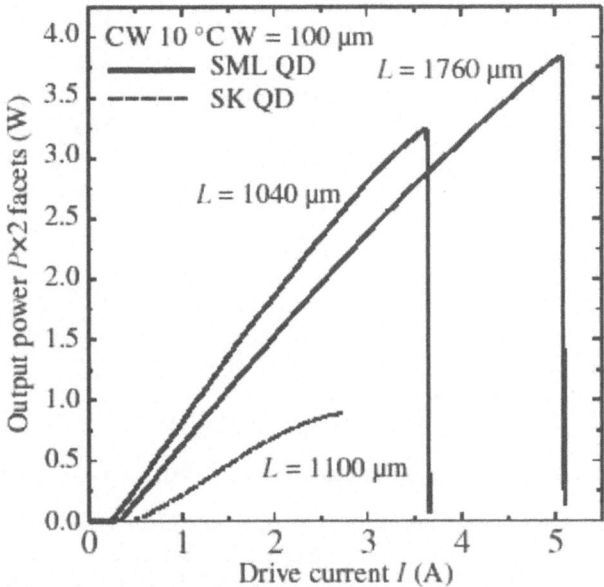

Fig. 10. Power of light emitted from two uncoated facets versus drive current for 100 μm wide SML QD (full curve for different stripe lengths) and SK QD (broken curve) laser structures operating in CW mode at 10 .C.

The maximum output power of the SK QD laser is 0.9–1 W. The rollover is caused by the inefficient power conversion resulting from the low slope efficiency. The SML QD laser demonstrates much higher output power. The maximum output power achieved for a 1.04 mm long stripe is 3.2 W. This value is limited by the catastrophic optical mirror damage. The output characteristics of a longer (1.76 mm) SML QD diode are also presented in Fig. 10. In this case, the maximum output power is even higher, 3.85–3.9W, owing to the better heat dissipation. The total power conversion efficiency (η_C) has been calculated from an experimental $P_{out}(I)$ dependence and a measured diode voltage. Figure 11 shows η_C, as a function of drive current for 1 mm, long SML QD and SK QD laser diodes. It can be seen that the SK QD diode has relatively low power conversion efficiency with a maximum value of 21% at 0.7 W output power. On the contrary, the SML QD laser exhibits much better efficiency. The device has a maximum total power conversion efficiency of 59% at a total power output of 1.5W. To the best of our knowledge, this is the highest efficiency ever reported for any diodes based on self-organized QDs. The figure also shows that the SML QD device has a power conversion efficiency exceeding 50% over a wide range of P_{out}, from 0.55 to 3 W. This result unambiguously demonstrates the potential of QD lasers based on uniform QD arrays for high power applications.

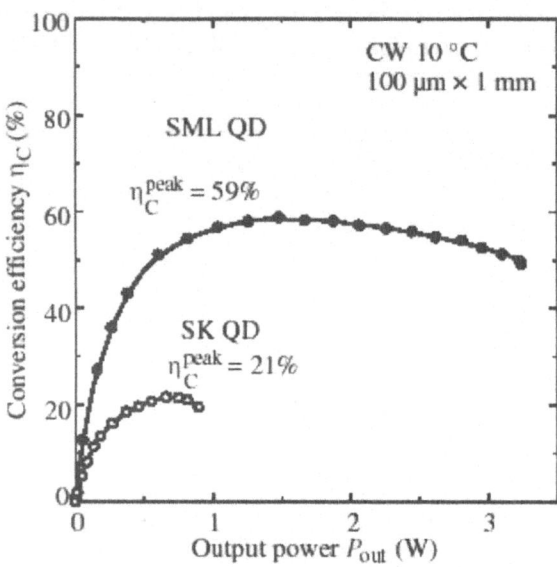

Fig. 11. Total power conversion efficiency versus drive current for SML QD (●) and SK QD (°) lasers.

Thus, owing to the better uniformity of sub-monolayer QDs, the SML QD laser shows two times higher differential gain and much lower internal loss, as compared to the SK QD laser. This results in superior power characteristics which compare well with the best achievements of QW lasers. In particular, a maximum output power of 3.9W and a peak power conversion efficiency of 59% were achieved in the SML QD laser.

3.2. 1.3 MICRON QUANTUM DOT LASERS

A possibility to extend the wavelength range of GaAs-based diode lasers to 1.3 μm has been recently demonstrated by the use of self-organized In(Ga)As QDs, [16]. Moreover, these long-wavelength QD lasers have demonstrated threshold current densities as low as 20 A/cm^2 [17] owing to low transparency current density and deep localization of the active region. However, the low surface density of QDs that provides the low transparency current also results in the small maximum gain. The gain saturation forces the use of low-loss cavity designs, e.g. very long (several mm) cavities, in combination with HR/HR facet coatings or four-cleaved facets. Such designs restrict the light output and, therefore, the reported external differential efficiencies were typically very small.

Promising efficiency characteristics (57%) have been reported when the QD gain was enhanced by the use of three planes of QDs [18]. In this work, we report on further progress in long-wavelength QD lasers. Use of up to ten QD planes each capped with external InGaAs quantum well enables to improve significantly the internal differential efficiency (~95%), decrease the internal loss (to 1.2 cm^{-1}), and increase the maximum optical gain (to 24 cm^{-1}). As a result, the maximum *external* differential efficiency as high as 88% was achieved in combination with acceptable characteristic temperature (T_0 ~ 150 K).

The external differential efficiency (η_D) was estimated from the room temperature light-current characteristics measured on devices of different length (L). The internal loss (α_i) and the internal differential efficiency (η_i) were then estimated from the experimental $1/\eta_D$-L dependence assuming a 32% reflectivity for the as-cleaved facets. An example of the $1/\eta_D$-L dependence, and the best-fit parameters, are shown in Fig. 12 for the laser structure based on ten QD planes. In this particular case, α_i and η_i were estimated to be 93% and 1.2 cm^{-1}, respectively. All over the range of the ground-state lasing the external differential efficiency exceeds 75%. In shorter diodes ($L<0.5$ mm), a transition to the excited-state lasing is observed as a blue shift of the lasing wavelength from ~ 1.29 to ~ 1.17 μm. The maximum η_D of 88% is achieved near the shortest L of the ground state-lasing. To our best knowledge, this is among the best efficiency characteristics ever reported for QD lasers of any spectral range [19]. In structures based on less number of QD planes, the maximum output loss of the ground state-lasing decreases (the minimum L increases). Correspondingly, the maximum η_D slightly decreases to 84 and 80% (N=5 or 2) due to earlier transition to the excited-state lasing. This highlights the importance of a high saturated gain to achieve good external differential efficiency. In particular, a modal gain as high as 24 cm^{-1} was measured in the structure based on 10 QD planes.

The values of α_i and η_i, for the structures with different number of QD planes, are summarized in Fig. 13 as functions of the QD surface density (assuming 5×10^{10} cm^{-2} in each plane). For comparison, the results are also shown which were measured on laser structures with QDs in a GaAs matrix (no QW cap layer was deposited) and emitting near 1-1.1 μm. It is seen that the internal loss regularly decreases as the QD surface density increases. A noticeable decrease of α_i is also observed in QD-in-QW structures as compared to that of their QW-free counterparts. The results obtained can be explained under assumption that the internal loss in QD lasers is mainly contributed by free-carriers in the waveguide region. As the localization of the active region gets deeper (QD-in-QW), the population of GaAs matrix is strongly reduced. Also, higher gain of the multiplane structure provides inverse population of the active region at lower carrier concentrations thereby decreasing free-carrier absorption. The internal differential efficiency does not demonstrate a regular dependence on the QD surface density. On the other hand, η_i increases from about 70% to 95% as a result of capping QD plane by QW. It was shown in [20] that the efficiency of carrier injection into the

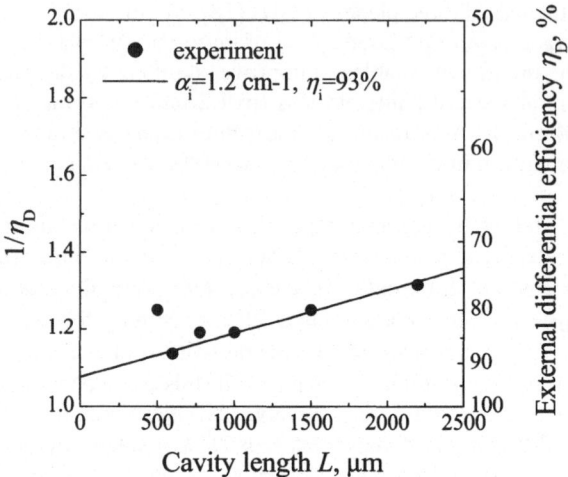

Fig. 12. Dependence of external differential efficiency on cavity length for the structure based on ten QD-in-QW planes (*N*=10). Symbols: experiment; line: best fit (η_i=93%, α_i=1.2 cm^{-1}).

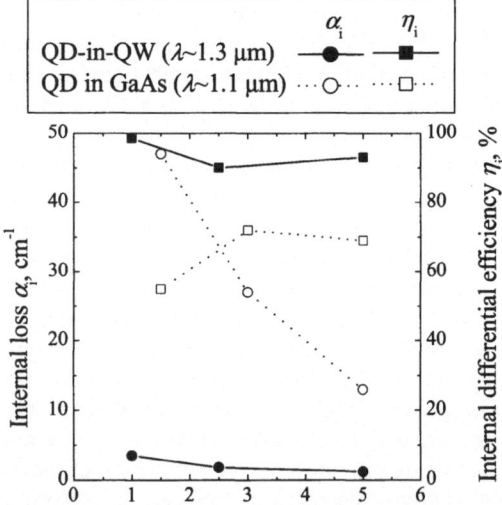

Fig. 13. Dependence of internal loss (α_i, circles) and internal differential efficiency (η_i, squares) on total density of quantum dots in the laser active region (n_{QD}) for laser structures of similar design and based on either QDs in QW (solid symbols) or QDs in bare GaAs matrix (open symbols).

active region is commonly the main limiting factor for the internal differential efficiency of a QW laser. This is especially true in case of a QD laser because of a small fill factor of the QDs in the *p-n* junction plane. On the other hand, an external quantum well surrounding a QD plane provides an efficient path for carrier capture to the QD states.

More, number of QD plane intrinsically corresponds to higher transparency current density. As a result, the minimum threshold current density (J_{th}) slightly increases from 40 A/cm^2 (N=2) to 65 A/cm^2 (N=10). The value of transparency current density J_{tr} was deduced from the experimental relationship between modal gain (G) and J_{th} by extrapolating the gain-current dependence to the point of zero gain. An example of the evaluation procedure is shown in Fig. 14 for the structure based on 10 QD planes. In this structure, the J_{tr} was estimated to be 58 A/cm^2, i.e. about 6 A/cm^2 per QD plane. Nearly the same values were found for the other numbers of QD planes. This is in excellent agreement with a simple estimate of J_{tr} from the QD surface density and the radiative recombination lifetime (1 ns). In its turn, this confirms that the contribution of any non-radiative recombination channel to the threshold current density is negligible. We would like to mention that partial substitution of a GaAs cap layer, with an InGaAs QW, prevents point defect formation in low-temperature material owing to higher surface mobility of In adatoms.

Efficiency and threshold characteristics of QD diode lasers were correlated to active region design. As the ground-state localization energy increases (QD-in-QW), the carrier concentration in the laser waveguide decreases. This results in a significant decrease in the internal loss (to 1.2 cm^{-1}) and improves the characteristic temperature (T_0~150 K). Also, the internal differential efficiency is improved to ~95% owing to better carrier injection efficiency to the QD states localized in the external QW. The use of several QD-in-QW planes increases the maximum modal gain (to 24 cm^{-1}). Owing to mutual effect of high internal differential efficiency, low internal loss and extended range of the ground-state lasing, the laser structure

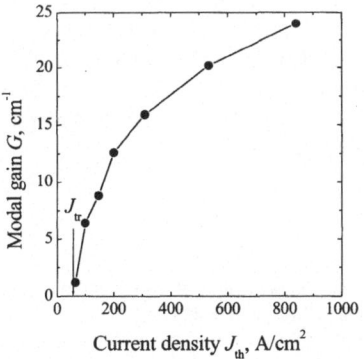

Fig. 14. Relationship between modal gain (G) and threshold current density (J_{th}) for the structure based on ten QD-in-QW planes (N=10). Vertical arrow indicates the transparency current density (J_{tr}).

based on ten QD-in-QW planes demonstrates an external differential efficiency as high as 88%. No evident effect of non-recombination is observed (transparency current density is as low as 6 A/cm^2 per QD plane).

4. Conclusions

The possibility to control size, shape, uniformity, and surface density of self-organized quantum dots leads to significant improvement of laser characteristics. Stacking of InAlAs and InGaAs quantum dots increases the surface density of QDs which results in the considerable increase in the QD laser output power. On the other hand, significant improvement in the output power is achieved by using the SML QDs demonstrating much better uniformity as compared to that of SK quantum dots. Overgrowing the InAs QDs with InGaAs ternary leads to effective increase in the QD volume and increase in the emission wavelength of a laser. Deep carrier localization achieved in these lasers in combination with optimized stacking of several planes of QDs results in very high external differential efficiency which is well comparable to that of record quantum well lasers.

5. Acknowledgement

This work in different parts was supported by Volkswagen Foundation, NanOp CC, INTAS, and the Program of the Ministry of Science of Russia "Physics of Solid State Nanostructures.

The author is grateful to the Directors and Organizing Committee of the NATO Advanced Study Institute "Synthesis, Functional Properties and Applications of Nanostructures" (Heraklion, Crete, Greece, July 26, 2002 – August 4, 2002) for financial support.

The author also acknowledges the "Russian Science Support Foundation".

The author gratefully acknowledges his colleagues whose contribution to this work was extremely important: A.Yu.Egorov, A.E.Zhukov, A.R.Kovsh, N.A.Maleev, S.S.Mikhrin, A.P.Vasil'ev, D.A.Livshits, E.S.Semenova, E.V.Nikitina, M.M.Kulagina, Yu.M.Shernyakov, Yu.G.Musikhin, I.P.Soshnikov, M.V.Maximov, A.F.Tsatsul'nikov, N.N.Ledentsov, and Zh.I.Alferov – Ioffe Institute, St.Petersburg, Russia; D.Bimberg – Technical University of Berlin, Germany.

6. References

1. Ledentsov, N.N., Grundmann, M., Heinrichsdorf, F., et al, (2000) Quantum-Dot Heterostructure Lasers, *IEEE J. Sel. Topics on Quantum Electronics* **6**, 439-451
2. Dingle, R. and Henry, C.H. (1976) Quantum Effects in Heterostructure Lasers, *U.S.Patent 3982207*
3. Bimberg, D., Grundmann, M., Ledentsov, N.N. (1999) *Quantum Dot Heterostructures*, Chichester: Wiley
4. Alferov, Zh.I. (1996) The history and future of semiconductor heterostructures, *in Proc. 99-th Nobel Symp.*
5. Goldstein, L., Glas, F., Marzin, J.Y., et al, (1985) Growth by molecular beam epitaxy and

characterization of InAs/GaAs strained layer superlattices, *Appl. Phys. Lett.* **47**, 1099-1101

6. Ruvimov, S.S., Werner, P., Scheerschmidt, K., et al, (1995) Structural characterization of (In,Ga)As quantum dots in a GaAs matrix, *Phys. Rev. B* **51**, 14766-14769.

7. Ustinov, V.M., Egorov, A.Yu., Zhukov, A.E., et al, (1996) Formation of stacked self-assembled InAs quantum dots in GaAs matrix for laser applications, *Mat. Res. Soc. Symp. Proc.* **417**, 141-146.

8. Kovsh, A.R., Zhukov, A.E., Egorov, A.Yu., et al, MBE growth and characterization of composite InAlAs/In(Ga)As vertically aligned quantum dots, *Mat. Res. Soc. Symp. Proc.* **571**, 109-114.

9. Mikhrin, S.S., Zhukov, A.E., Kovsh, A.R., et al, (2000) 0.94 micron diode lasers based on Stranski-Krastanow and submonolayer quantum dots, *Semicon. Sci. Technol.* **15**, 1061-1064

10. Egorov, A.Yu., Zhukov, A.E., Kop'ev, P.S., et al, (1996) Optical emission range of structures with strained InAs quantum dots in GaAs, *Semicond.* **30**, 707 – 710.

11. Ustinov, V.M., Maleev, N.A., Zhukov, A.E., (1999) InAs/InGaAs quantum dot structures on GaAs substrates emitting at 1.3 mm, *Appl. Phys. Lett.* **74**, 2815-2817.

12. Ustinov, V.M., Zhukov, A.E., Kovsh, A.R., et al, (2000) Long-wavelength emission from self-organized InAs quantum on GaAs substrates, *Microelectronics Journal* **31**, 1-7.

13. Zhukov, A.E., Kovsh, A.R., Ustinov, V.M., (1999) Gain characteristics of quantum dot injection lasers, *Semicond. Sci. Technol.* **14**, 118-123.

14. Kovsh, A.R., Zhukov, A.E., Livshits, D.A., (1999) 3.5 W CW operation of a quantum dot laser, *Electron. Lett.* **35**, 1161-1163.

15. Zhukov, A.E., Kovsh, A.R., Ustinov, V.M., (1999) Continuous-wave operation of long-wavelength quantum dot diode laser on a GaAs substrate, *IEEE Photonics Technology Letters* **11**, 1845-1847

16. Ustinov, V.M. and Zhukov, A.E., (2000), GaAs-based long-wavelength lasers, *Semicond. Sci. Technol.* **15**, R41-R54

17. Liu, G.T., Li, H., Malloy K.J., et al, Extremely low room-temperature threshold current density diode lasers using InAs dots in InGaAs quantum well, *Electron. Lett.* **35**, 1163-1164

18. Zhukov, A.E., Kovsh, A.R., Maleev, N.A., et al, Long-wavelength lasing from multiply stacked InAs/InGaAs quantum dots on GaAs substrates, *Appl. Phys. Lett.* **75**, 1926-1928

19. Zhukov, A.E., Kovsh, A.R., Mikhrin, S.S., (1999) 3.9 W CW power from sub-monolayer quantum dot diode laser, *Electron. Lett.***35**, 1845-1846

20. Smowton, P.M. and Blood, P., (1997) The differential efficiency of quantum well lasers, *IEEE J. Selected Topics in Quantum Electron.* **3** 491-498.

NANOCLUSTERED FILMS AND NANOWIRES

FEDOSYUK V.M.
Institute of Solid State Physics and Semiconductors of the Belorussian Academy of Sciences, P Brovki str 17, 220072 Minsk, Belarus

1. Introduction

Among the magnetic nanostructures [1-17], granular (or nanoclustered) films, which represent nanoinclusions of ferromagnetic material (Co, Fe, Ni) in dia - (f.e.Cu) or paramagnetic (f.e.Re) matrix, are extremely promising. Originally, the interest in inhomogeneous films, formed from alloys of transition metals of the iron group with Cu, arose from the possibility of applying these materials as substitutes for various, currently used, conventional, magnetoresistive sensors, which are based on permalloy transducers. In comparison with the latter, these inhomogeneous, granular alloy films are far more promising owing to the fact that they possess, under certain conditions, an isotropic giant magnetoresistance (GMR), In addition to significantly lower noise; this latter property is due to the absence of domain walls in these materials which is contrary to the situation for ferromagnetic alloys. However, quite recently, granular coatings have also come to be regarded as a material for super high-density magnetic recording [18]. In this context, a comprehensive investigation of these granular films becomes highly appropriate.

A variety of techniques has been applied for the preparation of these materials including MBE, sputtering, vapour deposition, and laser ablation . While these state-of-the-art techniques are very effective in producing good quality films and multilayers, they are extremely sophisticated and therefore costly in comparison with the relatively simple technique of electrodeposition (ED). Although ED has been known for many years, it is only fairly recently that it has been applied with any great success to the production of magnetic nanomaterials [1-5]. One of the advantages and features of the electrodeposition method is the possibility of deposition under both equilibrium (as we have discussed above) and quasi- or non-equilibrium conditions. It creates the possibility to form both homogeneous and heterogeneous (as we want) coatings from completely insoluble element pairs (Co-Cu, Fe-Ag, Ag-Co) and soluble clement pairs (Co-Re, Ni-Cu). Furthermore, it is also possible to regulate the growth direction and the rate of the initially deposited crystallites by either activating or passivating the substrate or initial crystallites. This is usually accompanied by an appropriate organic component in the electrolyte. In this case, we can either enhance or suppress the film growth in the direction perpendicular to the substrate, thus, obtaining effectively two-dimensional growth; this is advantageous for the production of multilayers. Therefore, it is possible, using ED, to grow multilayers, granular

T. Tsakalakos, et al. (eds.) pgs. 561 - 582
Nanostructures: Synthesis, Functional Properties, and Applications;
© *2003 Kluwer Academic Publishers.*

alloyed and homogeneous alloyed films, and to go permanently from one group to another one.

This means that, in principle, granular alloyed films, as a new kind of low-dimensional structures, can be prepared as multilayers by decreasing the cobalt deposition pulse duration. However, electrodeposition, in the stable regime, is more simple.

1.1 NANOCLUSTERED CUCO FILMS

The CuCo system is one which has, in the past, attracted considerable attention. This is partly due to the fact that at room temperature there is almost zero solubility of Co in Cu and the system is therefore well-suited as a candidate for the production of metastable alloys. As early as 1958 Knappwost and Illenger [19] reported magnetic measurements on Co particles precipitated from Cu in bulk samples and Kneller [20] reported results on thin films of CuCo evaporated at room-temperature. More recently, Berkowitz et al [21] and Xiao et al [22] have observed GMR in heterogeneous CuCo alloys and thin films respectively, both produced by dc sputtering. We also [23-30] have investigated a range of Cu and Co-based nanogranular produced by ED.

In previous works on CoFeP and CoW , which are only partially soluble, and also some other systems where ED films had been produced under constant deposition conditions, it was found that all films were homogeneous. Thereby, it seemed likely that in the CuCo system, produced under similar conditions, we should again obtain the same degree of homogeneity. However, as we show in the case of the CuCo system, we produced samples with a range of inhomogeneities.

The films were electrodeposited on to both Cu- and Al-foil substrates and also on to ceramic substrates which had previously been coated with a non-magnetic layer of NiP. The electrolytic composition was $CuSO_4 \cdot 5\ H_2O$ - 30 g/l, $CoCl_2 \cdot 6\ H_2O$ - 3.3 g/l, H_3BO_3 - 6.6 g/l, $MgSO_4 \cdot 7\ H_2O$ - 23.3 g/l, $CoSO_4 \cdot 7\ H_2$ - O 10 to 30 g/l. The composition range of the samples investigated was obtained by varying the $CoSO_4$ concentration in the electrolyte while all other chemical concentrations were held constant. The electrolyte had a pH value of 6.0 and deposition was performed at 20 °C with a current density of 5 mA/cm^2.

The film composition was determined by both X-ray and chemical analysis. For the X-ray analysism we used CoK_α, radiation with a graphite monochromator and a Dron-3M instrument. Transmission electron microscope (TEM) measurements were made using an EMV-1004M and a 300 keV Jeol 3010 machines with an accelerating voltage upto 300keV.

Magnetic measurements, which were mainly made on films deposited onto Cu substrates, were performed in the temperature range 2-300 K and in fields of up to 5 T using a Quantum Design SQUID magnetometer "MPMS5". Fields were applied in the plane of the film and could be set to an accuracy of $\pm\ 10^{-6}$ T, temperatures could be controlled to within $\pm 10^{-2}$ K. Measurements of magnetic moment as a function of field for various temperatures were made, together with the technique of zero-field-cooled (ZFC) and field-cooled (FC) low-field susceptibility measurements [30-34]. In this latter type of measurement, the sample was cooled in zero field from room-temperature to 2 K and then a measuring field of 5 mT was applied. Measurements were then made in this constant field while the temperature was increased. The sample was then cooled down to 2 K in the same field (5 mT) while, again, measurements were made of the moment.

In order to estimate the diamagnetic contribution of the Cu substrate, an uncoated substrate of known mass was measured. Subsequent measurements on ED films were then corrected for the substrate contribution by weighing the whole sample and then subtracting off an appropriately scaled value for the uncoated substrate. It was assumed that the film itself made a negligible contribution to the total mass. Measurements were made on the film immediately after preparation (no anneal) and then after annealing for 30 minutes at 200, 400 and 600 °C. For these anneals, the samples were sealed-off in quartz ampoules in a vacuum of better than 10^{-3} Pa.

Since we wanted to make a precise check on the samples to see whether they satisfied the criteria of SPM, careful measurements were also made on identical Cu sheets on which no films had been deposited and also on the actual substrates used for deposition after mechanically removing the CuCo film. These results were used to correct for the effects of the substrate, which made a small contribution to the SQUID signal, especially at high fields and low temperatures. In order to investigate whether the films exhibited any anisotropy due to the directional nature of their growth, measurements were also made at 300 K whilst rotating the sample from an orientation with the field in-plane to that with it perpendicular to the plane of the sample.

Four different samples were investigated, their compositions were determined by X-ray and chemical analysis and the results are given in Table 1. In all samples investigated, the X-ray diffraction patterns showed the distinct lines corresponding to the f.c.c. Cu structure. In only one sample, 718, was it possible to detect a very weak satellite line corresponding to pure Co. This was the sample with the highest Co concentration and it is understandable since we expected the Co inclusions to be the largest of any of the samples. The presence of only a faint Co line suggests that the sizes of the Co regions are very small, of the order 100 nm or less.

Table 1. Summary of the results obtained for a series of CuCo films produced by electrodeposition.The values quoted for the Co concentrations are the averages of chemical and X-ray analyses.

Sample	T_B, K.	diameter (nm)	at% Co.
725	55+5	7, 6	6
722	80 ± 10	8, 7	8
715	210 + 20	12, 0	11
718	260 ± 25	12, 8	20

Magnetization loops, with similar characteristics, were measured for all samples in the temperature range from 5 to 300 K. Some typical results, for sample 715, are given in Fig. 1. The hysteretic behaviour at low temperatures, combined with the reversible part of the magnetization curve, which extends over a wide field-range (Fig. 1a), suggests that we have a mixture of ferromagnetic and superparamagnetic particles.

564

Whereas, at higher temperatures (Fig. 1b) above the blocking temperature, we observe only superparamagnetism. It is also important to note, in this context, that no sample could be magnetically saturated, even at a temperature of 5 K and the maximum applied field of 5 T.

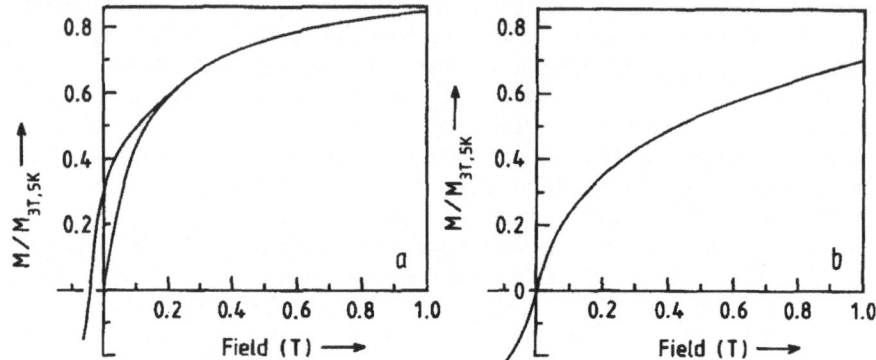

Fig. 1. Typical reduced magnetization curves for sample 715: (a) - 5 K, (b) - 300 K.

The temperature dependence of the remanence (Fig.2) is different from that observed by Thompson et al [35] in CoAg alloys in that, in the present work, we observe curves of continuously varying slope which also differ from sample to sample. This suggests that we have a range of blocking temperatures, depending upon the Co concentration. We have attempted to estimate the maximum blocking temperature for each sample by extrapolating the remanence curves to zero, the maximum blocking temperature increases with increasing Co concentration. The fact that all the remanence curves, except that for sample 718, show strong curvatures, suggests a range of blocking temperatures for each sample and a corresponding range of particle sizes present in the films, which we are able to control by varying the deposition conditions. This also indicates, in our opinion, that we have been successful in producing, immediately after deposition, inhomogeneous alloy systems. The linear variation of the remanence for sample 718 suggests an approximately uniform size of particles in that film.

Earlier work on the system CoNi (Ni 40 to 70 at%) has shown that Co retains its h.c.p. structure, whereas small additions of Fe cause the f.c.c. phase to be stable . We believe that an amount of the order of 3 to 5 at% of Cu is sufficient to stabilize the f.c.c. phase and it thus seems very likely that, in our present measurements, the Co is in the f.c.c. phase.

In order to estimate the maximum particle size in each sample, we have used the well-known expression [36]: $-K_A v = 25 k_B T_B$, where K_A is the anisotropy energy density, v is the critical particle volume and T_B is the blocking temperature. A value of K_A appropriate for f.c.c. Co has been assumed [37]. The results are summarized in Table 1.

The temperature dependence of the magnetization is given in Fig. 3. It is interesting to note that for the sample with the highest Co concentration, 718, it is quasi-linear. Such a linear dependence of magnetization on temperature can be an indication of two-dimensional magnetism [38]. This suggests that the ferromagnetic inclusions of Co in the Cu matrix could be in the form of long needle-like inclusions.

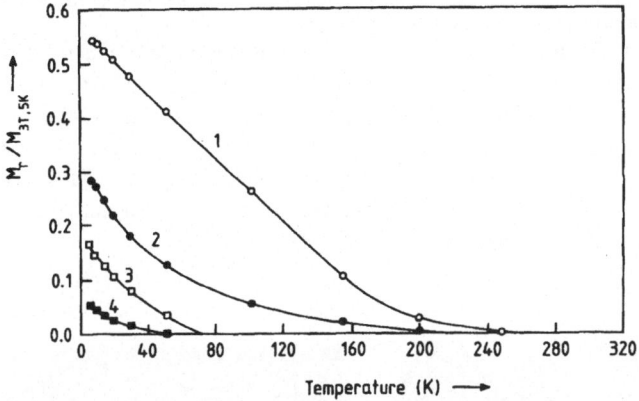

Fig. 2. Temperature dependence of the remanence for the various samples investigated; values have been normalized with respect to the magnetization at 3 T and 5 K. (1) - 718, (2) - 715, (3) - 722, (4) - 725.

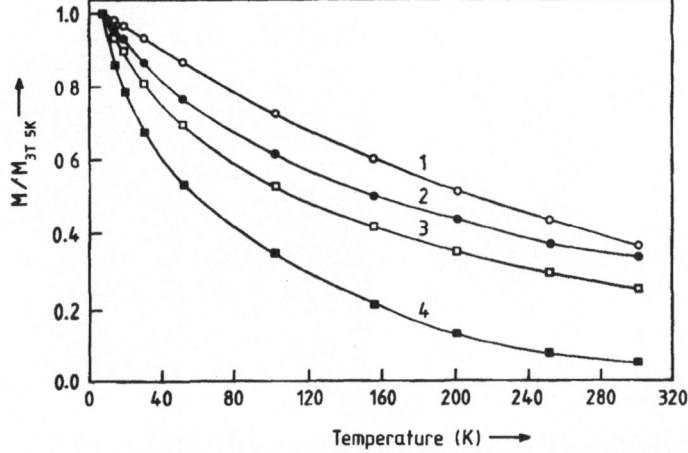

Fig 3. Temperature dependence of the magnetization of the various samples investigated as measured in a field of 3 T and normalized to their values at 5 K. (1) - 718, (2) - 715, (3) - 722, (4) - 725.

1.2. MECHANISM OF STRUCTURE FORMATION OF GRANULAR CU-CO FILMS

Here we present structural investigation results which confirm this supposition. A copper film deposited from an electrolyte without any cobalt ions and organic additives is polycrystalline with ≈700 nm crystals and laminated structure within crystals. If 3.3 g/l $CoCl_2$ $6H_2O$ was added to the solution, then the structure of the film is more complex. There are both large (≈700 nm) crystals, and smaller ones with the size ≈50 nm. The large crystals form twins. The structure of the copper deposit became small-dispersive (≈50 nm) when 10 to 30 g/l CoSO$_4$, 7H$_2$O were added to the solution. Small cobalt particles are observed (Fig. 4a) which are positioned between the copper layers within a copper grain and at the copper grain boundaries. Fig. 5a shows a scheme of such a position of cobalt particles.

The X-ray and electronmicroscopic investigations show f.c.c. Cu reflections for cobalt content in the alloy ≤8 at%. An additional h.c.p. Co(101)+(100) X-ray reflection appeared as the cobalt content in alloy increases to ≈ 8 to 10 at%. If the content of Co reaches *≈20 at%*, the additional reflections disappear in the X-ray spectrum and their intensity increases in the microelectronogram. Moreover, light pieces with size ≈10 nm, which are determined as cobalt deposited particles, are observed in the dark-field microphotography (Fig 4b).

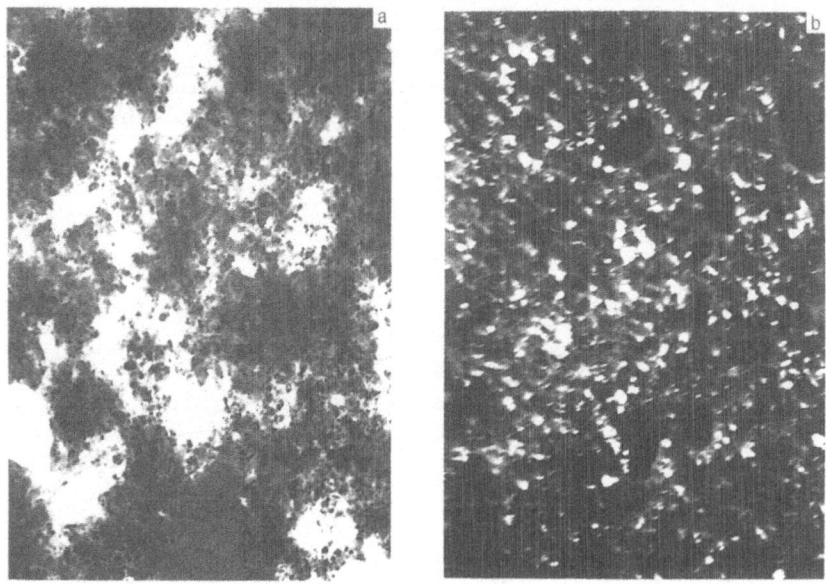

Fig. 4. a) Bright-field and b) dark-field micrographs of Cu$_{89}$Co$_{11}$ sample (x30000)

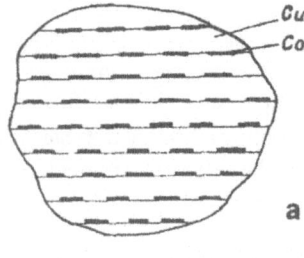

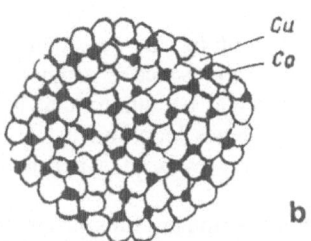

Fig. 5. Schematic representation of cobalt particle inclusion in the Cu matrix at different $CoSO_4 \cdot 7H_2O$ concentrations in the solution a) less than 15 g/1, b) more than 20 g/1

We propose the formation mechanism of distinct cobalt phase granules in the alloyed film deposited from a single solution. The reduction of copper ions on the cathode is accommodated by adsorption of different impurities and saccharine used as surface active substance. The copper laminated structure forms as a "packet" of monatomic (111) f.c.c. Cu monolayers. During the "packet" growth process, impurities are involved into the macrolayer volume and others are forced out to the "packet" boundaries. It leads to blocking of the forming Cu layer and, as a result, polarization is increased. The polarization increase leads to a rising crystallite growth rate, i.e. a new macrolayer is formed. By the way, the polarization increases also at the "packet" boundaries. If the reduction potential achieves ≈0.8 V on these places, the cobalt ion reduction is more probable. Therefore, it is most probable that the cobalt atoms grow at the boundaries between copper laminated grains and between copper layers or twins within the Cu grains. This is also evident from the X-ray measurements. If the concentration of the cobalt salt is large (≈30g/1), then a decreasing copper interplane distance (111) is observed because the radius of cobalt atoms which arc introduced into the Cu crystals is less than the copper atomic radius ($r_{Co} = 0.125$ nm < $r_{Cu} = 0.128$ nm).

It is known that thin (≈20 nm) cobalt layers are formed as f.c.c. unstable phase as the result of epitaxial growth on f.c.c. copper. We have described above that there are non-equilibrium conditions on the copper layers and grain boundaries where cobalt atoms are reduced. So, it is more probable, that a h.c.p. Co phase grows in such conditions and we observed additional (101) + (100) h.c.p. Co reflections at a cobalt content of ≈10 at% in the alloy. Therefore, the intergrain and interlayer ranges have the h.c.p. Co structure. The size of such particles is ≈10 nm (light pieces in the dark-field microphotograph in Fig 4b). If the cobalt content in the alloy reaches ≈20 at% the (101) and (100) h.c.p. Co planes became perpendicular to the film surface, that is not favourable to X-ray reflection, as a consequence, the intensity of the (101) and (100) reflections decreases. But such plane orientation is favourable to transmission electron diffraction and the intensity of (101) and (100) reflections is seen to increase in the electronogram.

Finally, we can draw some conclusions about the cobalt particle formation in the alloyed system Cu-Co electrodeposited from a single solution in stable regime and non-equilibrium conditions. When the copper matrix has laminated structure, which is obtained at small cobalt ion concentration in the solution, the cobalt particles have the shape of pieces situated at the Cu grain boundaries and consist of stable h.c.p. Co phase, their size is $\approx$10 nm (Fig 5a). If the fine-crystalline copper structure is produced (addition in the solution of 1 g/l saccharine and 30 g/l $CoSO_4 \cdot 7H_2O$), the copper crystallites have spherical shape as shown in the scheme Fig 5b. This means that the cobalt and copper crystal lattices coincide in all directions {hkl} and cobalt impurities are also formed as spherical particles. Their size reaches $\approx$30 nm and the structure is a mixture of h.c.p. and f.c.c. cobalt.

1.3 THE INFLUENCE OF ANNEAL AND COMPOSITION ON THE SUPERPARAMAGNETISM OF NANOCLUSTERED CUCO FILMS

Many of the earlier papers have reported the observation of pure superparamagnetism (SPM) in granular systems [1-9] which arises due to the presence of magnetic particles in a non-magnetic matrix. There are two experimental conditions that have to be satisfied for true SPM: First, there must be no hysteresis loop above the maximum blocking temperature. Second, the magnetization curves must superimpose when plotted as a function of reduced field.

We have recently reported measurements on granular films produced by the comparatively simple and economically attractive technique of electrodeposition [14]. In contrast to other workers, however, we have found that, even in electrodeposited films of relatively dilute Co concentration, e.g. $Cu_{0.94}Co_{0.06}$, pure SPM is not exhibited but rather that particle interactions play an important role. In this paragraph, we report on the influence of anneal and of composition on these interactions.

Fig. 6 shows the magnetization for an unannealed $Cu_{0.94}Co_{0.06}$ film as a function of reduced field and as measured at different temperatures well above the maximum

blocking temperature of the sample, 50 ± 5 K [14]. These curves do not superimpose and we conclude that the sample does not exhibit pure SPM. Such behaviour was typical for the whole range of concentrations investigated both before and after anneal.

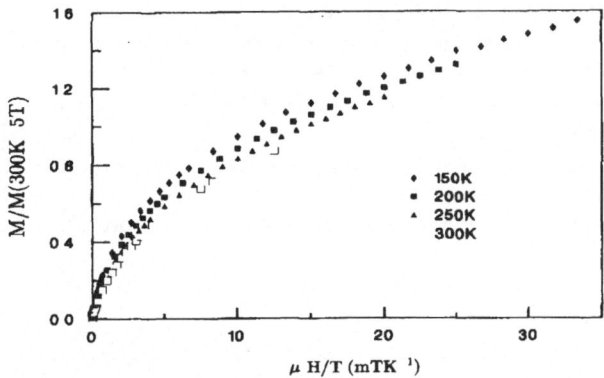

Fig. 6. Dependence to magnetization of an unannealed $Cu_{0.94}Co_{0.06}$ films a function of reduced magnetic field for the temperatures indicated. The magnetization is normalized to its value at 5 T and 300 K.

In Fig 7, the temperature dependence of the reciprocal of the FC susceptibility is plotted as a function of anneal for sample $Cu_{0.94}Co_{0.06}$. On the assumption that at high temperatures the susceptibility can be represented by a Curie-Weiss law $\chi \propto (T-T_{int})$, where T_{int} is an interaction temperature, T_{int} is obtained by extrapolating the linear part of the curve. While some caution must be exercised over the interpretation of the results [39], values obtained for T_{int} are: 20 ± 2 K for the unannealed sample, and 35 ± 4 K and 50 ± 5 K after anneal at 200 and 400°C. On anneal of the sample, there is a gradual increase in curvature of the χ^{-1} plots. Measurements were also made after anneal at 600°C, but in this case, we were unable to cool from above the maximum blocking temperature.

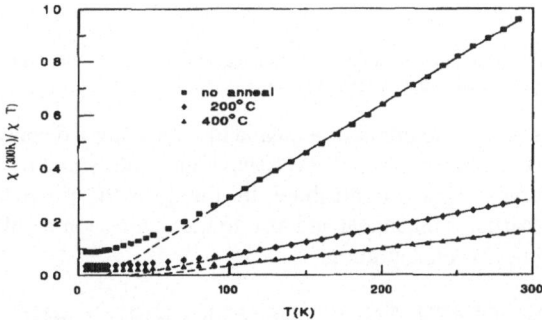

Fig. 7. Low field reciprocal susceptibility as measured in a held of 5 mT for a $Cu_{0.94}Co_{0.06}$ film as a function of anneal at the temperatures indicated.

Fig 8 gives the variation of the reciprocal susceptibility for a range of unannealed $Cu_{1-x}Co_x$ samples. Here again, it can be seen that there is a strong increase in curvature of the plots with increasing Co concentration which reflects the increasing particle interactions. This makes extrapolation of the curves to determine T_{int} rather difficult. However, by extrapolating from the high-temperature region, we are able to obtain values of 40 ± 5 K and 120 ± 10 K for the compositions $x = 0.11$ and 0.20 respectively. The sample with $x = 0.35$ shows a different temperature dependence to the other concentrations. This corresponds, however, to a composition region in which, for a CuCo dc sputtered sample, Childress and Chien [40] observed re-entrant ferromagnetism and determined a Curie temperature of about 350 K. All our other samples fall into the concentration region where Childress and Chien observed spin-glass behaviour.

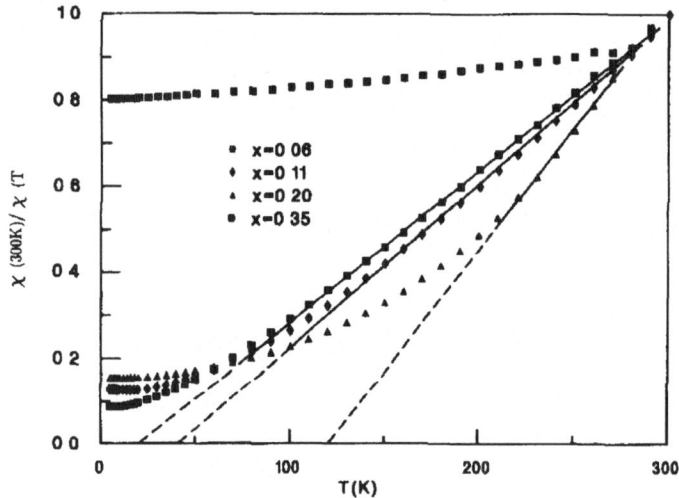

Fig. 8. Low field reciprocal susceptibility for a range of unannealed $Cu_{1-x}Co_x$ samples with concentrations as indicated. The results are normalized to their respective values at 300 K.

In summary, we see that the effect of anneal and increasing Co concentration is to produce additional deviations from pure SPM behaviour as measured by an increase of inter-particle interactions. This we attribute to the growth of particle size and nucleation of Co-rich particles on anneal and also to the increase in number and size of particles with increasing Co concentration.

1.4. TEMPERATURE DEPENDENCES OF REMANENCE AND MAGNETIC SUSCEPTIBILITY IN ELECTRODEPOSITED GRANULAR CU-CO FILMS

The temperature dependences of coercivity H_c (Fig. 9) were used to determine the blocking temperature T_b. We obtained $T_b = 50, 80, 210$, and 260 K for films with 6, 8, 11, and 20 at. % Co [14]. Fig. 10 shows the temperature dependences of remanence M_r

measured after the film was held in a magnetic field. The blocking temperature is difficult to assess from these data, because the temperature-dependent remanence measured in this way is highly sensitive to the interaction of magnetic cobalt particles in the nonmagnetic copper matrix [41, 42]. The unusual behavior of curve 5 in Fig. 2 (35 at. % Co) warrants attention: M_r rises somewhat near 10 K. A similar anomaly was observed earlier for Ag-Co films [43]. One possible reason for this anomaly is that a small amount of cobalt oxide can be formed during electrodeposition at a high electrolyte acidity (pH ~ 6) [44, 45]. The higher the concentration of $CoSO_4 \cdot 7H_2O$, in the electrolyte, the larger the amount of cobalt oxide present in the films. Accordingly, the effect in question becomes more pronounced as the cobalt content of the films rises. Cobalt oxide, a ferromagnetic material, is capable of increasing the magnetic moment of Cu-Co films in some temperature range.

Figure 11 shows the temperature dependences of initial magnetic susceptibility after zero-field and low-field cooling (5 mT) runs for all the compositions studied. Most of the ZFC curves pass through a maximum. For the most dilute alloy, $Cu_{94}Co_6$, the point at which the ZFC and FC curves diverge (T_b^{max}) lies at -250 K, in rough agreement with the blocking temperature found from the $M_r(T)$ curves (Fig. 10). At higher cobalt concentrations, T_b^{max} exceeds room temperature. As distinct from conventional dependences of remanence and coercivity (Fig. 9), which approach zero at these blocking temperatures, M_r, is not zero at room temperature. It is likely that the blocking temperature evaluated from the $M_r(T)$ curves corresponds to point T_b^{max}, which represents the energy of interaction of magnetic cobalt inclusions in the copper matrix. The energies of interaction determined from inverse magnetic susceptibility curves [41,42] are given below in table 2.

Table2. The energies of interections of nanoclustered CuCo filmms with different concentration of cobalt.

$E \times 10^3$, eV	1.7	3.3	4.7	14
at. % Co	6	8	11	20
10^3, eV %Co				

The shift in peak susceptibility observed in the ZFC curves as a function of cobalt concentration suggests that the average blocking temperature rises. The maximum blocking temperature T_b^{max}, which can be determined from ZFC and FC curves, also increases with cobalt content.

Thus, the present results demonstrate that the blocking temperature T_b, determined from the temperature dependence of remanence (Fig. 10) correlates to the maximum blocking temperature T_b^{max} [46,47] found from ZFC and FC curves. This finding probably stems from the fact that both quantities characterize the interaction of magnetic particles in the nonmagnetic matrix; that is, the system as a whole exhibits superparamagnetic behavior [47].

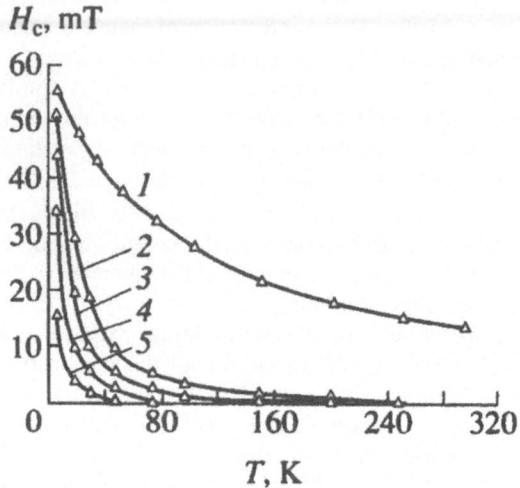

Fig. 9. Temperature dependences of coercivity for Cu-Co films containing (*l*) 35, (2) 20, (J) 11, *(4)* 8, and *(5)* 6 at. % Co.

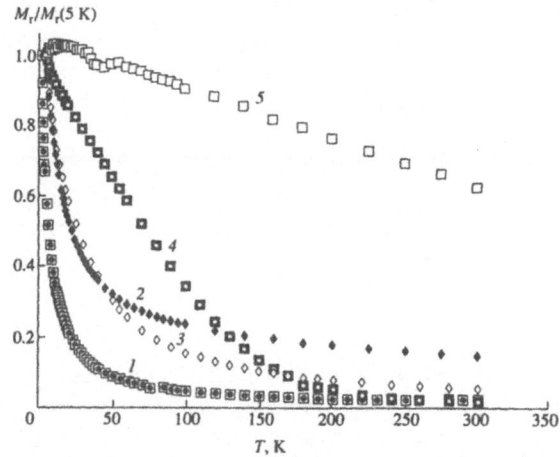

Fig. 10. Temperature dependences of remanence for Cu-Co films measured after the film was held in a magnetic field of 5.5 T ;at 2 K: (1) 6, (2) 8, *(3)11, (4)* 20, (5) 35 at. % Co.

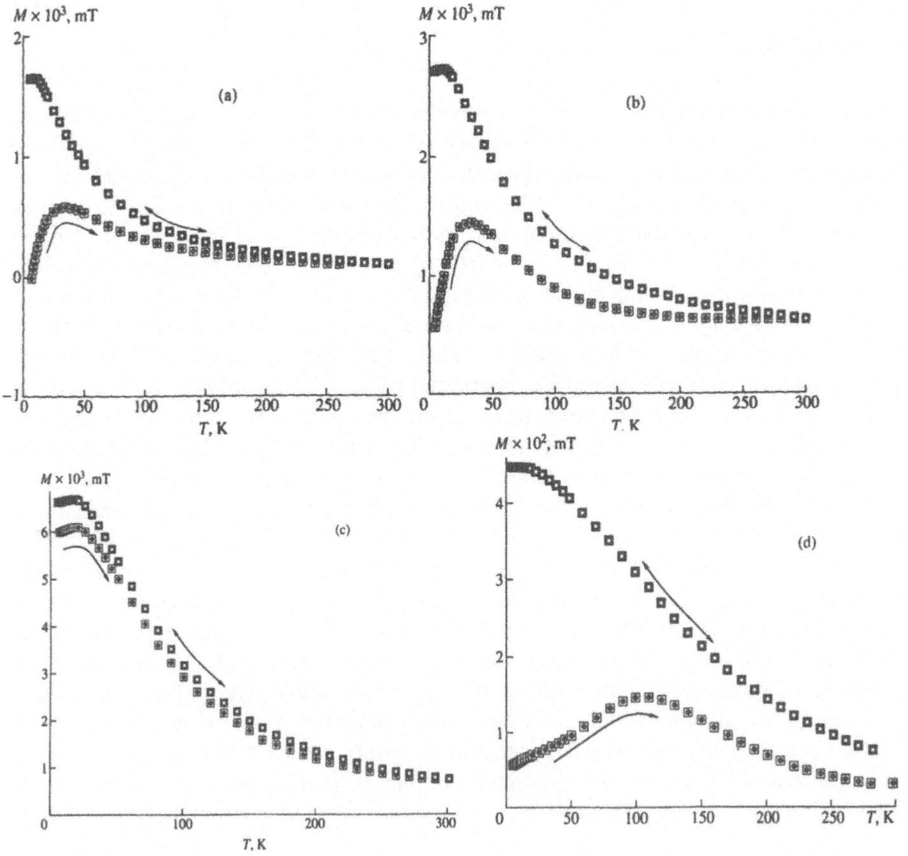

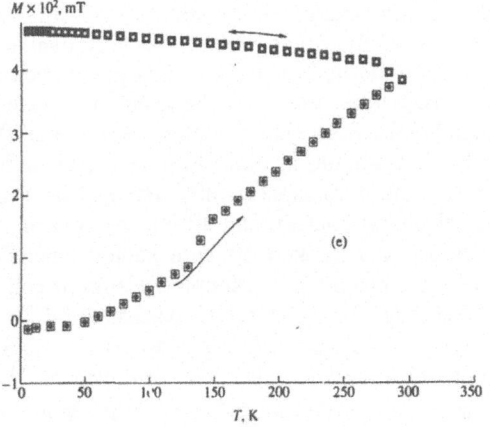

Fig. 11. Temperature dependences of magnetization measured after zero-field and low-field cooling for Cu-Co films containing (a) 6, (b) 8, (c) 11, (d) 20, and (e) 35 at. % Co.

2. Nanoclustered Nanowires

2.1 INTRODUCTION

Template electrodeposition, by which is meant electrodeposition in natural or artificial holes in an insulating layer on a conducting substrate, offers an inexpensive route to the fabrication of patterned metal films, in some cases with structures that cannot be produced by any other method. For example, we show that it is possible to prepare nanowires of Co-Cu granular alloy by template electrodeposition in porous aluminium oxide, and that the GMR of the wires increases, as expected, on annealing Heterogeneous alloys consisting of ferromagnetic grains with dimensions in the nm-range surrounded by a non-magnetic matrix are of interest because they can exhibit giant magnetoresistance (GMR) [48,49]. They are more straightforward to prepare than other GMR materials, as unlike superlattices [50] and spinvalves [51] they do not require the control of layer thicknesses to sub-nm precision. They may therefore prove to be suitable field sensing elements for magnetic sensors where a low unit cost and a large response are more important than sensitivity to small field changes.

In the present chapter we demonstrate a further advantage of electrodeposition, namely that it can easily be used to grow material on patterned substrates. In particular, we show that it is possible to prepare heterogeneous alloys in the form of nanowires having lengths of several μm or tens of μm, and diameters of only a few tens of nm. These nanowires exhibit GMR, the magnitude of which increases on annealing. GMR in electrodeposited nanowires has been reported previously [52,53], but for wires consisting of multilayers, rather than a heterogeneous alloy. Multilayered nanowires are prepared by switching the electrodeposition potential between two values, with precise control of the time at each potential to ensure constant layer thicknesses, but heterogeneous alloy nanowires are much simpler to prepare as they can be grown galvanostatically (at constant current).

2.2 EXPERIMENT

Co-Cu heterogeneous alloy nanowires having lengths of several tens of μm and diameters of 10 or 300 nm were deposited in the pores of commercially available anodic aluminium oxide membranes . Unlike the nuclear track-etched polycarbonate membranes used in previous studies of rianowire, these can be used in annealing studies. Prior to sample growth, all membranes were coated with an evaporated Au layer on the side that was not exposed to electrolyte, in order to have a conducting substrate for electrodeposition. Each membrane was subsequently mounted on a Cu plate, with which the Au coating made electrical contact, and all but the area of the membrane in which deposition was desired was masked off with kapton tape. For deposition, the membrane acted as the working electrode in a simple 2-electrode cell in which a Pt plate acted as counter electrode. The electrolyte consisted of 30 g $CuSO_4 \cdot 5H_2O$ (Cu sulphate), 6.6 g H_3BO_3 (boric acid), 120 g $Na_3C_6H_5O_7 \cdot 5.5H_2O$ (sodium citrate) and 50 g $CoSO_4 \cdot 7H_2O$ (Co sulphate) per liter of purified H_2O (conductivity $< 10^{-18}$ Ω^{-1} m^{-1}). Deposition was carried out at room temperature and electrolyte pH = 5.7, using a galvanostat (EG&G model 363) to control the current density. Sufficient metal was deposited for the wires to emerge from the ends of the membrane pores, forming approximately hemispherical caps [54].

2.3 RESULTS AND DISCUSSION

Following growth, the room temperature magnetoresistance (MR) of the nanowires was measured with one contact made using Ag DAG to the caps at the tops of the nanowires and the other to the Au layer at the base of the wires. This configuration meant that generally the MR of several nanowires would be measured in parallel. Figure 1 shows the measured MR for Co-Cu nanowires electrodeposited in an aluminium oxide membrane with quoted thickness 60 μm, pore diameter 200 nm and pore density 10^9 cm^{-2}. In fig. 12(a) the applied magnetic field was parallel to the long axis of the wires, and in fig. 12(b) it was perpendicular. Since the current was always along the wires, these configurations correspond to measuring the longitudinal and transverse MR, respectively. It is immediately apparent that although the absolute magnitude of the MR is small (less than 1%), in both configurations it is negative, showing that we do indeed have GMR, as expected for a heterogeneous Co-Cu alloy. There appear to be symmetric "shoulders" either side of the central peak in both MR curves, which may be related to details of the magnetization reversal mechanism. This is still being investigated.

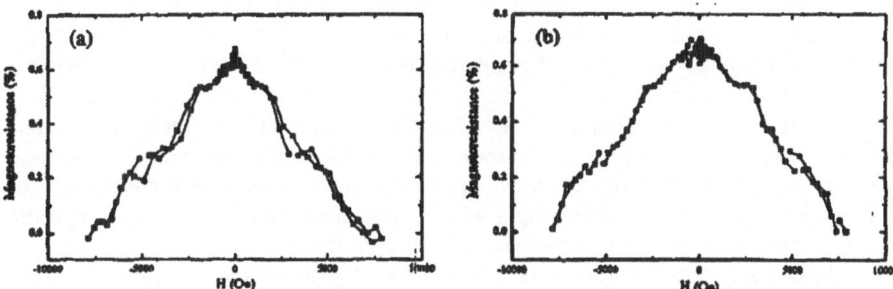

Figure 12. Percentage magnetoresistance for Co-Cu heterogeneous alloy nanowires electrodeposited in an aluminum oxide membrane with quoted pore diameter 200 nm and pore density 10^9 cm^{-2}, measured at room temperature with the magnetic field applied (a) parallel and (b) perpendicular to the long axis of the wires.

For the sample of fig.12, the nominal deposition current density (assuming the quoted pore diameter and density, and that electrodeposition took place in all pores) was 50 mA·cm^{-2}. Samples were also prepared in membranes with quoted pore diameter 20 nm and density 10^{10} cm^{-2}. Figure 13 shows the longitudinal (a) and transverse (b) MR for Co-Cu nanowires grown with a nominal deposition current density of 400 mA·cm^{-2} in one such membrane. Despite the current density for this sample being a factor of 8 larger than that for the sample of fig. 1, the measured compositions of both samples were similar, i.e. Co$_{19}$Cu$_{81}$. The GMR for this sample is also significantly less than 1%.

Interestingly, the GMR of electrpdeposited continuous Co-Cu heterogeneous alloy films of a given composition was also reported to be insensitive to current density, although the current density did have a strong effect on the GMR of electrodeposited continuous Co-Ag films [55].

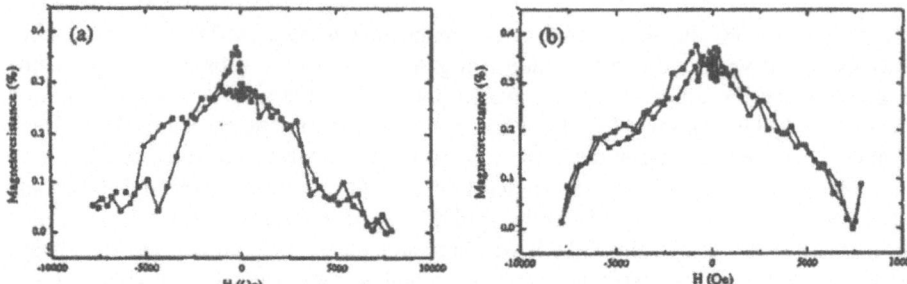

Figure 13. Percentage magnetoresistance for Co-Cu heterogeneous alloy nanowires electrodeposited in an aluminum oxide membrane with quoted pore diameter 20 nm and pore density 10^{10} cm^{-2}, measured at room temperature with the magnetic field applied (a) parallel and (b) perpendicular to the long axis of the wires.

The GMR of continuous Co-Cu heterogeneous alloy films grown by a variety of methods has been shown to increase on annealing, due to phase separation, which leads to an increase in the number and size of the Co-rich particles [48,49]. We therefore investigated the effect of annealing on the GMR of our electrode-posited heterogeneous alloy nanowires. Samples were annealed while still in the aluminium oxide membranes for 30 minutes at 200° C or 400° C, in a 10^{-5} torr vacuum to reduce the risk of oxidation. In all cases, the room temperature GMR was found to increase significantly. For example, fig. 14 shows the longitudinal (a) and transverse (b) MR for the same, sample as fig. 13 following a 400° C annealing, and it can be seen that the maximum MR in each configuration has more than doubled.

A particularly interesting feature of fig. 14 is that the shapes of the MR curves with the applied field (a) parallel and (b) perpendicular to the long axis of the wires are distinctly different. This apparent anisotropy in the magnetization reversal process suggests that the nanowire geometry does influence the magnetic properties of the heterogeneous alloy, possibly because some of the Co-rich particles approach the nanowire diameter (20 nm) in size [14,23-27], or possibly because even if the nanowire diameter is much larger than the particle dimensions, it may be less than the range over which magnetic interactions are important. Alternatively, during electrodeposition and subsequent annealing, the nanowire geometry could influence the shapes and distributions of the Co-rich magnetic particles, or could lead to stress-induced anisotropy in these particles [56].

Although porous aluminium oxide membranes do allow annealing studies, they have the disadvantage that it is relatively difficult to dissolve the aluminium oxide to release the nanowires for structural studies. Some nanowires were therefore electrodeposited in nuclear track-etched polycarbonate membranes which can easily be

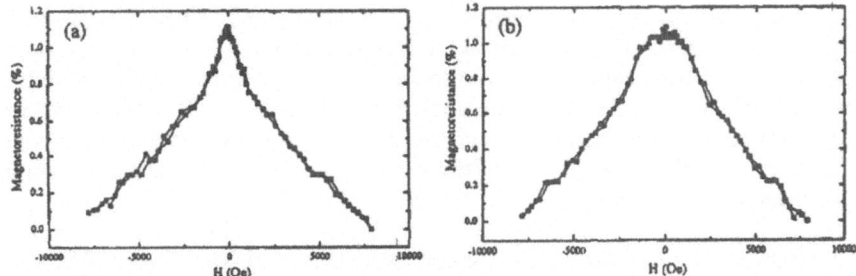

Figure 14. Percentage magnetoresistance for the same sample as Figure 2, but following an annealing at 400°C for 30 minutes, measured with the magnetic field applied (a) parallel and (b) perpendicular to the long axis of the wires

dissolved using chloroform to facilitate characterization by transmission electron microscopy (TEM). Fig.15 shows a TEM image of heterogeneous alloy nanowires

electrodeposited at a nominal current density of 3.3 A cm^{-2}, and having a measured room temperature GMR of under 0.5%. It is apparent that the nanowires are highly polycrystalline. No evidence of oxide, reported to be present in some continuous electrodeposited Co-Cu films, was found in these as-deposited nanowires.

Fig.15. Transmission electron micrograph of Co-Cu heterogeneous alloy nanowires electrodeposited in a nuclear track-etched polycarbonate membrane with quoted pore diameter 10 nm and pore density 6×10^8 cm^{-2}.

In conclusion, we have shown that it is possible to prepare heterogeneous Co-Cu alloy in the form of nanowires using constant current electrodeposition. The room temperature GMR of the as-deposited wires is small, and appears to be rather insensitive to the nominal current density, but further work is needed to establish the dependence of the composition and microstructure on the electrodeposition conditions. Annealing the nanowires leads to a significant increase in the GMR, presumably as a result of increased phase segregation, and the shapes of the MR curves measured with applied field parallel and perpendicular to the long axis of the wires reveals anisotropy in their magnetic properties.

2.4. NANOCLUSTERED GRANULAR AGCO AND AGCUCO NANOWIRES

With very few exceptions [57], previous studies have concentrated on systems with Cu as the non-magnetic component. We therefore decided to investigate the properties of electrodeposited granular alloy films with Ag as a nonmagnetic component since such films grown by sputtering showed interesting magnetotransport behaviour [58]. In contrast to previous work on the Ag-Co system, we have prepared these films in the form of nanowires, which will allow an investigation of how their properties are affected by this extreme form of nanoscale patterning.

Magnetic granular alloy films often exhibit their most interesting properties when the magnetic component is present in the range $\sim$ 10-30 atomic%. To achieve this composition range in electrodeposited CuCo and CuFe films, the magnetic component should also constitute 10-30% of the metal ions in the electrolyte [14,23-25]. In order to obtain AgCo alloy films, however, the relative Co concentration has to be greatly increased owing to the extremely high rate of Ag deposition compared to that of Co. The electrolyte used to electrodeposit AgCo nanowires had the following composition: $CoSO_4 \cdot 7H_2O$ 10g/1; 60% $HClO_4$ 50 ml/1; thiourea 0.4 g/1; $AgNO_3$ 0.8 g/1. Growth was carried out at room temperature, and the electrolyte pH was < 1.

Nanowires were electrodeposited in porous anodic Al oxide membranes with a thickness of 60 μm and quoted pore qiameter and density 20 nm and 10^{10} cm^{-2}, respectively. Prior to deposition the membranes were coated with Au on one side to create a conducting substrate suitable for electrodeposition. The wires were grown at a constant current density of 3.75 mA per cm^2 of membrane. In this respect heterogeneous alloy nanowires differ from the multilayered nanowires grown by others) for CPP-GMR studies [59] - to grow the latter, the deposition potential is switched between two values at intervals that need to be precisely controlled in order to ensure constant layer thicknesses.

Following deposition, the magnetotransport properties of the wires were measured at room temperature by making one contact to the hemispherical caps formed when the nanowires emerge from the porous membrane templates and a second contact to the Au substrate. The as-grown AgCo nanowires exhibited a negative magnetoresistance (MR) for applied field both parallel and perpendicular to the current, indicative of GMR. The magnitude of the effect was, however, small with the maximum measured percentage change in the MR being lately 0.2%.

In addition to these AgCo nanowires, some heterogeneous alloy nanowires containing Ag and Cu as well as Co were grown. The electrolyte for these nanowires had the following composition: $CoSO_4 \cdot 7H_2O$ 50g/l; $CuSO_4 \cdot 5H_2O$ 30 g/l; $Na_3C_6H_5O_7 \cdot 2H_2O$ 120 g/l, H_3BO_3 6.6 g/l; $AgNO_3$ 0.5 g/l. Deposition was again carried out at room temperature and the electrolyte pH was 5.17.

The as-grown AgCuCo nanowires also showed GMR and the maximum measured percentage change in the MR was approximately 0.2%. The nanowire composition was measured by energy dispersive X-ray fluorescence. These nanowires were subsequently annealed for 30 min at 400°C. As was found to be the case for CuCo heterogeneous alloy nanowires [60], there was a significant increase in the MR after annealing, probably due to increased phase segregation of the Co.

2.5 SQUID STUDIES OF CO-CU NANOGRANULAR ALLOYED NANOWIRES

We have reported in the paragraph 3.1.3 the first electrodeposition of heterogeneous alloy nanowires. Arrays of such nanowires may be considered as a highly non-uniform distribution of magnetic fine particles. For example, their extremely high aspect ratio (ratio of length to diameter), will make the demagnetizing field highly anisotropic, unless the spacing of the wires is extremely small, so that the array of wires becomes magnetically equivalent to a conventional heterogeneous alloy film. Here, we present a preliminary study of the magnetic properties of heterogeneous alloy nanowires measured by SQUID magnetometry, with particular emphasis on initial susceptibility measurements after cooling in zero field and on cooling in low magnetic field [31-34].

Fig. 15 is a transmission electron micrograph of Co-Cu heterogeneous alloy nanowires electrodeposited in a nuclear track-etched polycarbonate membrane with thickness 6 μm, quoted pore diameter 0,01 μm and pore density 6×10^8 cm^{-2}. Although his membrane thickness and pore diameter and density are rather different to those of the aluminium oxide membranes, nanowires grown in both kinds of membrane showed GMR at room temperature. The figure shows that these heterogeneous alloy nanowires are highly polycrystalline with very small grain sizes.

X-ray fluorescence compositional analysis of a number of samples showed that despite significant differences in the nominal current densities, the composition remained approximately constant at $\sim Co_{20}Cu_{80}$. Continuous Co-Cu films with a similar composition required an electrolyte less concentrated in Co (30g $CoSO_4 \cdot 7H_2O$ l^{-1}), but were also grown with rather lower current densities (typically 5 mA cm^{-2}) [60].

Fig. 16 shows *M-H* curves for nanowires grown in an aluminium oxide membrane with quoted pore diameter 0.02 μm. Curves are presented for the applied field (a) parallel and (b) perpendicular to the long axis of the wires. They show only a relatively small amount of hysteresis at 5 K and less at 300 K. If the average composition is $Co_{20}Cu_{80}$, one can assume that the average magnetization of a single heterogeneous alloy nanowire will be $\sim 1/5$ that of Co. This would give a demagnetizing field of ~ 1.7kOe when the magnetization is saturated perpendicular to the long axis of the wires and ~ 0 when it is parallel, leading to a shape anisotropy which is clearly seen in fig. 16.

Fig. 17 shows the ZFC/FC initial magnetic susceptibilities of heterogeneous alloy nanowires. This sample was grown in the same kind of membrane as the sample in Fig. 16, but with twice the deposition current. ZFC measurements were made after cooling the sample to 5 K in zero field and measuring up to 300 K in a field of 50 Oe parallel to the long axis of the wires. The FC measurements were then made by cooling the sample in the same field. Since the ZFC initial susceptibility in Fig. 17 shows no clear peak, but rather keeps increasing with increasing T, it is apparent that the Co clusters in these heterogeneous nanowires have a range of blocking temperatures which extends at least to room temperature. This is different to what was observed for continuous electrodeposited $Co_{20}Cu_{80}$ films [23-27], for which there was a peak in the ZFC initial susceptibility at $T \sim 107$ K.

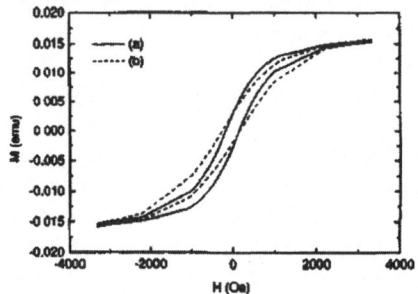

Fig.16 Hysteresis loops measured at 5 K for nanowires grown in an aluminium oxide membrane with quoted pore diameter 0.02 μm. (a) Applied field parallel to the long axis of the wires (i.e. perpendicular to the membrane in which they are contained). (b) Applied field perpendicular to the long axis of the wires (i.e. parallel to the membrane). The lines are guides to the eye, joining measured data points.

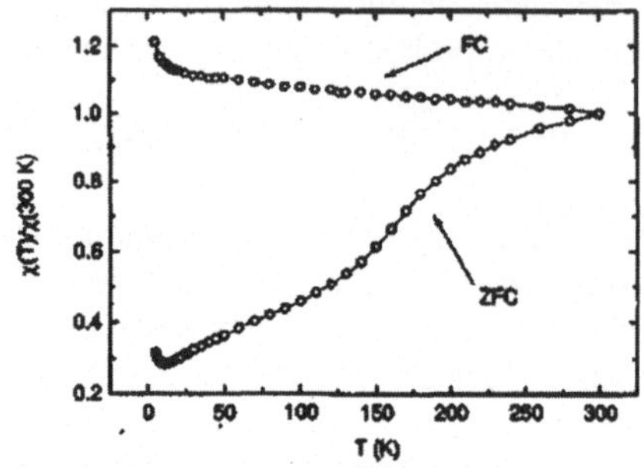

Fig.17 Normalized initial magnetic susceptibility of heterogeneous alloy nanowires after cooling in zero magnetic field (ZFC) and in an applied field of 50 Oe (FC). The nanowires were grown in an aluminium oxide membrane with quoted pore diameter 0.02 μm, and were not removed from the membrane prior to measurement. The lines are guides to the eye.

Another difference between these heterogeneous nanowires and continuous films is the presence of a rise in the ZFC/FC curves in the low-temperature range 5-20 K, which is absent for the continuous films. This feature is also absent from ZFC/FC curves measured for heterogeneous nanowires electrodeposited in pores of larger diameter. A similar feature was previously observed for electrodeposited Co-Re films and attributed to the presence of extremely small clusters [29]. In the case of the heterogeneous nanowires, this explanation is supported by the observation that the sharp rise at low temperatures disappears or is reduced on annealing the sample at 400°C for 30min (fig. 18),   presumably as a result of agglomeration of the extremely small clusters with larger clusters. A comparison of fig. 17 and fig.18 also shows that the shoulder seen in both ZFC curves moves to higher temperatures for the annealed sample, indicative of higher blocking temperatures, consistent with the annealed wires containing larger ferromagnetic particles resulting from agglomeration. The absence of the sharp rise in the FC/ZFC susceptibility for continuous heterogeneous alloy films and larger diameter heterogeneous alloy nanowires suggests that these contain many fewer extremely small clusters, possibly as a consequence of their being grown at lower current densities.

In summary, electrodeposited Co-Cu heterogeneous alloys may be prepared in the form of nanowires, as well as continuous films. From the magnetic measurements, there is evidence that the Co-rich clusters in the smallest-diameter granular alloy nanowires have a broader range of sizes than in the corresponding continuous films.

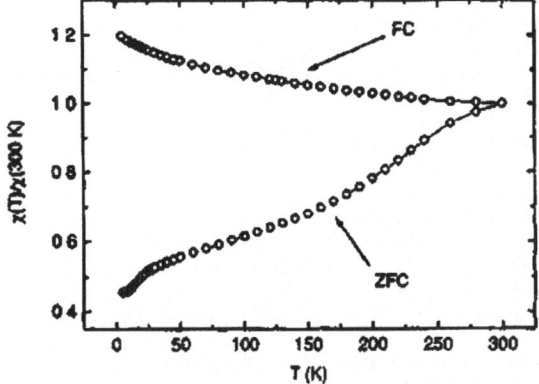

Fig. 18 Initial magnetic susceptibility after cooling in zero magnetic field (ZFC) and in an applied field of 50 Oe (FC) for nanowires from the same sample as those of Fig. 17, but annealed at 400°C for 30 min. The nanovires were not removed from the membrane prior to measurement, and the lines are guides to the eye.

3. Acknowledgments

I wish to express my special gratitude to my collegues and frends W.Schwarzacher from Bristol and H.J.Blythe from Sheffield Universities, O. I. Kasyutich from Nanomagnetic LTD (Bristol) in cooperation with whom this work has been done.

The present work was partioly surpoted by INTAS grant N 00-0761

4. References

1. Fedosyuk, V.M., Tochitskii, T.A. (2002) Electrodeposited nanostructures, BSU, Minsk, ISBN 985-476-005-7, 353p.
2. Fedosyuk, V.M. (2000) Nanoscaled magnetic electrodeposited structures on the basis of ion group metals:preparation, structure, magnetic and magnetoresistive properties. In "Nanostructured Films and Coatings", editted by G.M. Chow, I.A. Ovid'iko and T.Tsakalakos, Kluwer Academic Publishers NATO ASI series **78**, p.85.
3. Fedosyuk, V.M. (2000) Multilayered Magnetic Structures, BSU, Minsk, ISBN 985-445-415-0, 197p.
4. Moffat, T.P. Mat. (1997) Res. Soc. Symp. Proc. **457**, p.413 .
5. Tochitskii,T.A, Blythe, H.J., Jones, G.A., Fedosyuk, V.M. (2001) J. Magn. and Magn. Matter. **224**, p.221.
6. Schwarzacher, W., Fedosyuk ,V.M., Kasyutich, O.I. (1999) J. Magn. and Magn. Matter. **188-189**, p.185.
7. Fedosyuk, V.M., Kasyutich ,O.I., Blythe, H.J., Ravinder, D., (1996) J. Magn. Magn. Mater. **156**, p.345.
8. Schwarzacher, W., Lashmore, D.S. (1996) IEEE Trans. Magn. **32** , p.3133.
9. Blythe, H.J., Fedosyuk, V.M., Jones, G.A. (1998) J. Magn. Magn. Mater. **184**, p.28.
10. Fedosyuk ,V.M., Tochitskii,T.A,, Kasyutich, O.I. (1997) J. Phys. Status. Solidi (a) **16**, 2, p.631.
11. Ross, C.A. (1994) Annu. Rev. Mater. Sci. **24**, p.159.
12. Fedosyuk, V.M., Blythe, H.J., Kasyutich ,O.I. (1997) J. Func. Mater. **4**, 4, p.505.
13. Fedosyuk, V.M., Blythe, H.J., Kasyutich, O.I. (1997) J. Func. Mater. **4**, 1, p.69.
14. Blythe, H. J. and Fedosyuk, V. M. (1994) J. Phys. Status. Solidi (a) **146**, k13 .
15. Fedosyuk, V.M. (2002) Nanoscaled electrodeposited alloys. in "Encyclopedia of Nanoscience and Nanotechnology", editted by W.Nalwa, American Scientific Publishers.
16.Fedosyuk, V.M. (2002) Structure and mechanism formation of nanogranular films, in "Atomistic aspects of Apitaxial Growth." Editted by M. Kotrla, N.I. Papanicolau, D.D.Vvedenski and L.T.Wille Kluwer Academic Publishers, **65**, pp.535-550.

582

17. Fedosyuk, V.M., Sharko, S. in Proceed of "Magnetic Materials and Applications, Minsk, 2002, September 29-October 3, p.37.
18. Mizuseki, H., Kikuchi, K., Tanaka, K. (1996) Jap.J.Appl.Phys. **37**, 48, p.2155.
19. Knappwost, A. and Illenger, A. (1958) Naturwissenschaften **45**, p.238.
20. Kneller, E. (1962) J. Appl. Phys. **33**, p.1355.
21. Berkowitz, A.E., Mitchel, J.R., Carey, M.J., Young, A.P., Zhang, S., Spada, F. E., Parker, F.T., Hutten, A., and Thomas, G. (1992) Phys. Rev. Letters **68**, p.3745.
22. Xiao, J. Q., Jiang, J. S., and Chien, C. L. (1992) Phys. Rev. Letters **68**, p.3749.
23. Blythe, H.J.,.Jones, G.A, Fedosyuk, V.M. (1996) J. Mater.Sci **31**, 24, p.6431.
24. Blythe, H.J. and Fedosyuk ,V.M. (1995) J. Phys. Cond. Mat. **7**, p.3461.
25. Blythe, H.J., Fedosyuk, V.M., Jones, G.A. (1996) J. Mater. Science, **31**, p.6431.
26. Blythe, H.J., Fedosyuk, V.M., and Kasyutich, O. I. (1996) Mater. Sci. Letters **26**, p.69.
27. Fedosyuk, V.M., Blythe, H. J., and Kasyutich ,O. I. (1999) Phys. Low-Dim. Structures **9/10**, p.159 .
28. Fedosyuk, V.M., Kasyutich, O.I., and Blythe, H.J. (1997) J. Functional Materials 4, p.512.
29. Blyte, H.J. and Fedosyuk, V.M. (1996) J. Magn. Magn. Mater. **155**, p.352 .
30. Blythe, H.J., Fedosyuk, V.M., Schwarzacher, W., Kasyutich, O.I. (2000) J.Magn.and Magn.Matter. **208**, p.251.
31. O'Grady, K., El-Hilo, M. and Chantrell, R.W. (1993) IEEE Trans. Magn. **29**, p.2608.
32. El-Hilo, M., O'Grady, K.and Chantrell, R.W. J. (1992) Magn. Magn. Mater. **117**, p.21.
33. El-Hilo, M., O'Grady, K. and Chantrell ,R.W. (1992) J., Magn. Magn. Mater. **114**, p.295.
34. Chantrell, R.W., El-Hilo,M.and O'Grady, K. (1991) IEEE Trans. Magn. MAG. **124**, p.3570 .
35. Thompson, S.M., Gregg, J.F., Staddon ,C.R., Daniel, D., Dawson, S.J., Oundajela, K., Hammann, J., Fermon, C., Saux, G., Coey, J. M. D. and Fagan, A. (1993) Phil. Mag. B **68**, p.923.
36. Bean, C.P. and Livingstone, J.D. (1959) J. Appl. Phys **30**, p.120.
37. Ilyushenko,L.F.(1979)Electrodeposited Magnetic Films, Nauka (Science), Minsk 275p.
38. Rau ,C. (1990) Appl. Phys. A **49**, p.579.
39. O'Grady, K., El-Hilo, M., Chantrell ,R.W. (1993) IEEE Trans. Magn. **29**, p.2608.
40. Childress, J.R., Chien, C.L. (1991) Phys. Rev. B **43**, p.8089.
41. O'Grady, K. and Watson, M.L. (1994) Nippon Oyo Jiki Gakkaishi, **18**, 91, pp. 379-384.
42. El-Hillo, M., O'Grady K., and Chantrell, R.W. (1992) J. Magn. Magn. Mater., **117**, pp. 21-28.
43. Conde, F., Gomes-Polo, C., and Hernando, A., (1994) J. Magn. Magn. Mater., **138**, pp.123-131.
44. Fedosyuk, V.M. and Shadrow, V.G., (1989) Phys. Status Solidi, **115**, pp. 279-284.
45. Fedosyuk, V.M. and Kasyutich, O.I. (1992) J. Inf. Rec. Mater., **20**, pp.255-264.
46. Fedosyuk, V.M., Blythe, H.I., and Kasyutich, O.I. (1996) Mater. Sci. Lett., **26**, 1/2, pp. 69-72.
47. Fedosyuk, V.M., Blait, Kh.I., and Kasyutich, O.I. (1995) Pis'ma Zh. Tekh. Fiz., **21**, 18, pp.52-55.
48. Berkowitz, A.E., Mitchell, R., Carey, M.J., Young, A.P., Zhang, S., Spada, F.E., Parker, F.T., Hutten, A. and Thomas, G. (1992) Phys. Rev. Lett., **68**, p.3745.
49. Xiao, J.Q., Jiang, J.S. and Chien, C.L. (1992) Phys. Rev. Lett., **68** , p.3749.
50. Baibich, M.N., Broto, J.M., Pert, A., Nguyen Van Dau, F., Petroff, F., Etienne, P., Creuzet, G., Friederich, A. and Chazelas J. (1988) Phys. Rev. Lett., **61**, p.2472.
51. Dieny, B., Speriosu, V.S., Parkin, S.S.P., Gurney, B.A., Wilhoit, D.R., and Maun, D. (1991) Phys. Rev. B **43** p.1297.
52. Piiaux, L., George, J.M., Despres, J.F., Leroy, C., Ferain, E., Legras, R., Ounadjela, K., and Fert, A. (1994) Appl. Phys. Lett., **65** , p.2484.
53. Blondel, A., Meier, J.P., Doudin, B. and Ansermet ,J.-Ph. (1994) Appl. Phys. Lett., **65**, p.3419.
54. Whitney, T.M., Jiang,J.S., Searson, P.C.and Chien, C.L. (1993) Science, **261**, p. 1316.
55. Zaman, H., Yamada, A., Fukuda, H.and Ueda, Y. (1998) J. Electrochem. Soc., **145**, p. 565.
56. Meier, L., Doudin, B., and Ansermet, J.-Ph. (1996) J. Appl. Phys., **79** , p.6010.
57. Zaman, H., Yamada, A., Fukuda, H., Ueda, Y. (1998) J. Electrochem. Soc. **145**, p. 565.
58. Barnard, J.A., Waknis, A., Tan, M., Haftek, E., Parke,r M.R., Watson, M.L. (1992) J. Magn. Magn. Mater., **114** , p.L230.
59. Schwarzacher, W., Lashmore, D.S. (1996) IEEE Trans. Magn. **32** , p.3133.
60. Schwarzacher, W., Kasyutich, O.I., Evans P.R., Darbyshire, M.G., Ge Yi, Fedosyuk, V.M., Rousseaux, F., Cambril, E., Decanini, D. (1999) J. Magn,. Magn. Mater., **198-199**, p.185.

583

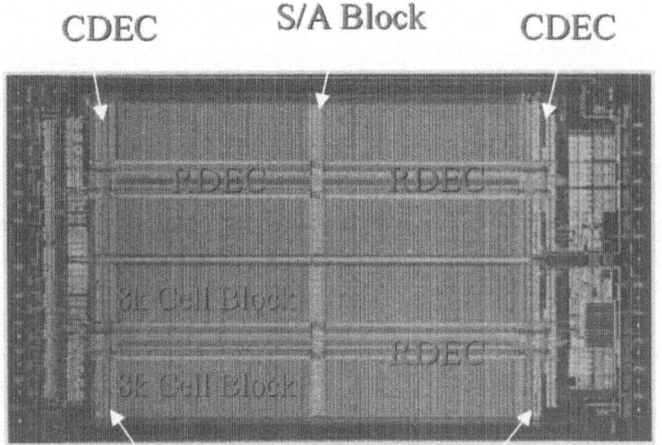

CDEC   S/A Block   CDEC

Reference Cell Blocks

Fig. 1.1 Chip photograph of 64kb FRAM (DJ Jung, Samsung Electronics, VLSI '97), 1T/1C, open bit line architecture, density/org:8k*8, chip size: 7.4*3.4 mm^2, cell size: 16.9*7.6 µm^2

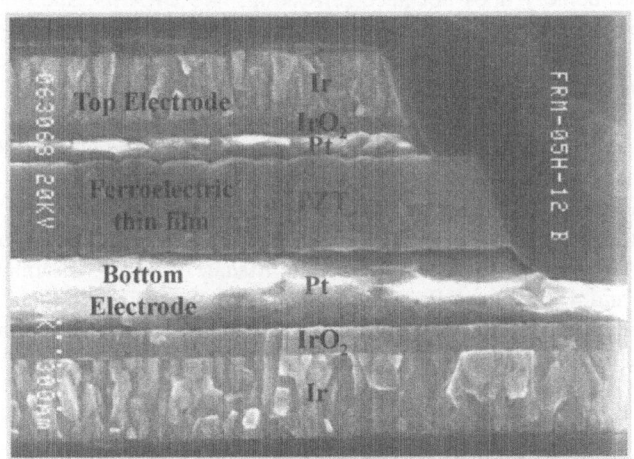

Fig. 1.2 TEM cross-section of ferroelectric cell in Fig.1.1

NANO-FERROELECTRICS

J. F. SCOTT
Symetrix Centre for Ferroics, Earth Sciences Department
Cambridge University

1. Introduction and Deposition Techniques

Beginning in the 1950s every large US microelectronics company (Bell Labs, IBM, Ford, RCA, etc.) was involved in ferroelectrics research. The main driving force was the idea that the +P polarization state and the –P polarization state of a ferroelectric could be used to encode the "1" and "0" of the Boolean algebra in which modern digital computers operate. At that time, however, ferroelectrics were available only as single crystals or rather thick ceramics. Since a typical coercive field for switching a ferroelectric from +P to –P (or vice versa) is ca. 40 kV/cm, a 1 mm thick device would have an operating voltage of 4000 Volts! Moreover, the devices were expensive. Therefore as silicon DRAM (dynamic random access memories) devices developed rapidly, ferroelectric RAMs were left on the back-burner as objects of mere academic novelty. This changed rapidly through the 1980s as oxide films as thin as 20 nm were fabricated in pinhole-free 6" commercial wafer form. At that point the advantages of ferroelectric memories over Si DRAMs was recognized once again: They are non-volatile (the memory does not need refreshing, like DRAMs, and does not forget if power is interrupted); they are radiation hard, no single event upset – SEU; and they are lighter in weight than Core magnetic memories, and 1000x faster to erase and rewrite than are EEPROMs – electrically erasable programmable read-only memories).

The result has been a ferroelectrics renaissance. Ferroelectric RAMs are now used in smart debit cards at the 16 kbit and 64 kbit level (Figs.1.1-1.3); in SONY Playstation 2 (Fig. 1.4; memory actually made by Toshiba) and telecommunications. The highest density ferroelectric (FE) chips available are 4 Mbit from Samsung (using chemical solution deposition lead zirconate titanate – PZT – ceramics ca. 40 nm in grain size) and 4 Mbit from Panasonic (using strontium bismuth tantalate – SBT). A fully commercial 8 Mbit ferroelectric RAM is scheduled for production by Siemens (Japan) and Toshiba on 1 September 2003, using sputtered PZT.

T. Tsakalakos, et al. (eds.); pgs 583 - 600
Nanostructures: Synthesis, Functional Properties and Applications;
© *2003 Kluwer Academic Publishers.*

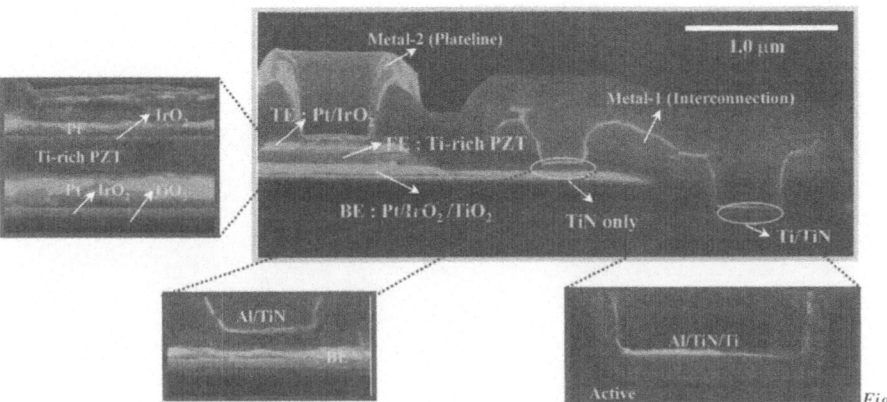

1.3 Fully Integrated Robust Capacitor for a Highly Reliable 1T/1C FRAM (DJ Jung, Samsung Electronics, VLSI '97)

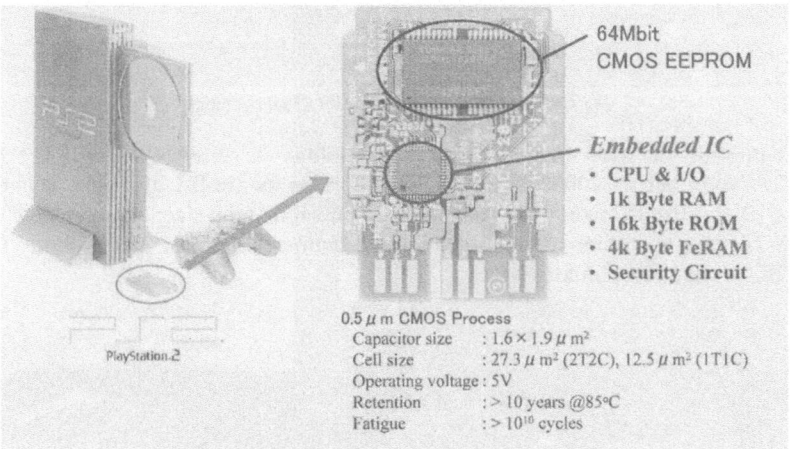

Fig. 1.4 Sony Playstation2 with Toshiba 32 kb FRAM

586

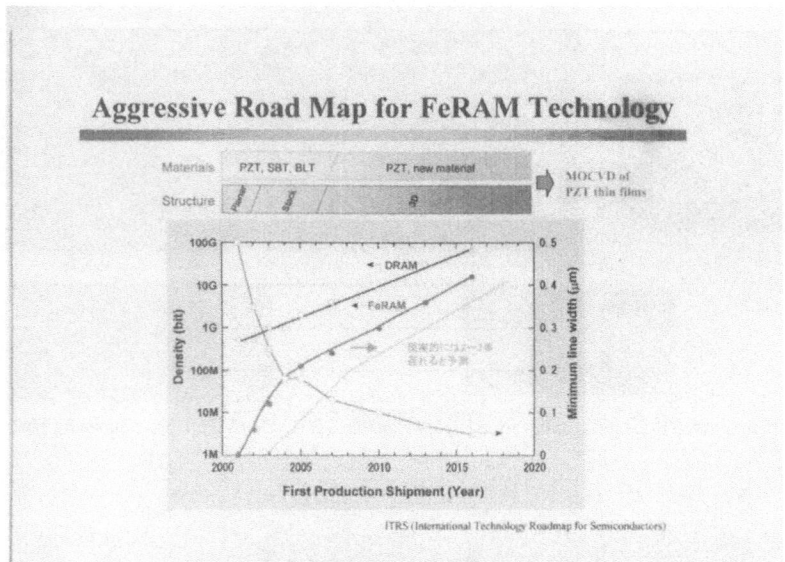

Fig 1.5 Aggressive roadmap for FRAM technology

At present the road map for FRAM technology is summarized in Fig.1.5 and Table I. Note that by 2008 the linewidth requirements are 0.1 microns; technology node is 70 nm; feature size F is 0.13 microns; 256 Mbit is the size; and complete cycle time is 16 ns. Therefore for this technology, nano-scale is not just a trendy buzzword; it is a very strict imperative.

Table I. Road map for FeRAM Technology

Road Map for FeRAM Technology

Year of first product shipment	2000	2001	2002	2003	2004	2005	2008	2011	2014	2017
Technology node	180nm	130nm	130nm	130nm	100nm	100nm	70nm	50nm	35nm	25nm
1) Feature size (μm): F	0.6	0.5	0.35	0.25	0.18	0.18	0.13	0.1	0.07	0.05
2) FeRAM generation (bit) (Standard memory samples)	256 kb	1 Mb	4 Mb	16 Mb	64 Mb	128 Mb	256Mb	1 Gb	4 Gb	16 Gb
3) Access time (ns)	120	100	80	65	40	30	20	10	8	6
4) Cycle time (ns)	180	160	130	100	70	50	32	16	12	10
6) Cell area factor: a	81	60	40	24	16	10	10	8	8	8
7) Cell size (μm^2)	29.0	15.0	4.90	1.50	0.52	0.32	0.17	0.08	0.04	0.02
8) Total cell area (mm^2) for standard memory	7.60	15.73	20.55	25.17	34.79	43.49	45.37	85.90	168.36	343.60
10) Projected capacitor size (μm^2)	3.00	2.00	1.00	0.50	0.25	0.13	0.07	0.03	0.015	.0075
11) Capacitor area (μm^2)	3.00	2.00	1.00	0.50	0.25	0.13	0.088	0.069	0.055	0.044
12) Capacitor structure	planar	planar	stack	stack	stack	stack	3D	3D	3D	3D
12.5) Ferro. depo. method	sputter, CSD					sputter, CSD, MOCVD		MOCVD, new method		
13) 2T2C, 1T1C	2T2C	2T2C	2T2C	1T1C	1T1C	1T1C	1T1C	1T1C	1T1C	1T1C
14) Ferro. material			PZT, SBT			PZT, SBT		PZT, new materials		
18) Vdd: Voltage applied to ferro. capacitor (V)	5.0	3.0	3.0	2.5	1.8	1.5	1.2	1.0	0.7	0.7
19) Switching charge @Vdd (μC/cm^2)	20	20	20	20	20	30	40	40	40	40
19.5) Switching charge @Vdd (fC/cell)	600	400	200	100	50	39.0	35.4	27.8	22.0	17.5
21) Retention @85°C (Years)	10	10	10	10	10	10	10	10	10	10
22) Fatigue with assuring retention	10^{10}	10^{12}	10^{15}	10^{15}	10^{16}	10^{16}	10^{16}	10^{16}	10^{16}	10^{16}

☐ An implementation method already exists.

☐ An implementation method is only proposed.

▨ An implementation method should be developed.

$$(0.36 \ \mu m)^2 = 1.13 \ \mu m^2$$

5), 9), 15), 16), 17), 17.5) deleted.

588

1.1. NANOPHASE DEPOSITION TECHNIQUES

The three main techniques for 20 nm – 100 nm ferroelectric cells are: electron-beam direct writing (EBDW);[29] focussed-ion-beam deposition;[30] and spontaneous self-patterning.[31]

1.1.1 *EBDW*

EBDW consists of the use of a cannibalized SEM electron microscope gun to cut out array patterns of ferroelectric capacitors. The results, illustrated in Figs.1.6 and 1.7, are very impressive, but it must be emphasized that a single pattern, ca. 20 x 20 μm in size, takes about 24 hours to fabricate. Therefore this technique is not at present suitable for commercial production.

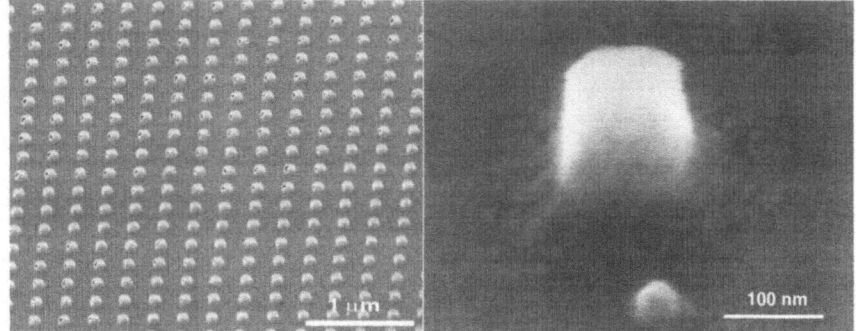

Fig.1.6. PZT nanoscale array produced by e-beam direct writing. Left: Array on 1 micron scale (bar at lower right); right: single cell at 100 nm scale (M. Alexe, private communication).

Fig.1.7. PZT arrays (EBDW) of varying capacitor cell sizes (Alexe).

These images may be contrasted with the results of focussed ion beam patterning (FIB) in Fig.1.7 (PZT), and with self-patterning, which in bismuth-excess SBT and bismuth titanate (both PLD and CVD deposition), Fig. 1.8.

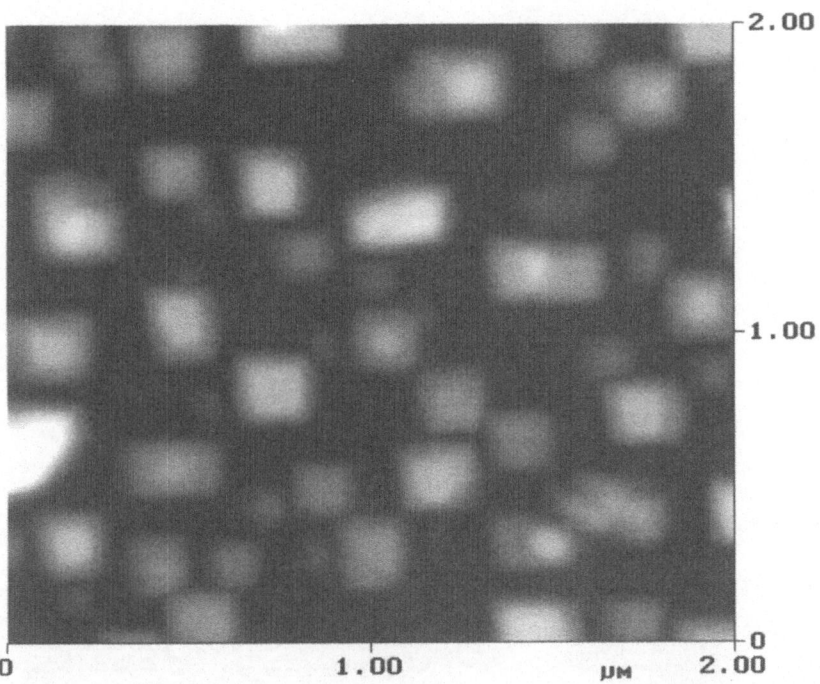

Fig.1.8. Spontaneous self-patterning of δ-Bi₂O₃ nanophase electrodes on the surface of PLD-deposited bismuth titanate. These are square pyramids with edges along the [100], [010] axes of the underlying silicon substrate, atop a metallic Bi wetting layer, and satisfy in detail the theory of Shchukin et al.,[31]

1.2 . FOCUSSED-ION-BEAM PROCESSING:

The FIB processing of nanoscale ferroelectric capacitors is analogous to that of EBDW, except that an ion beam is rastered rather than an electron beam, as illustrated in Fig.1.9. The resulting depth profile is shown schematically in Fig. 1.10. Note that lateral sizes as small as 20 nm on edge can be produced in this way. Figs.1.11a,b,c show actual devices fabricated in different diameters.

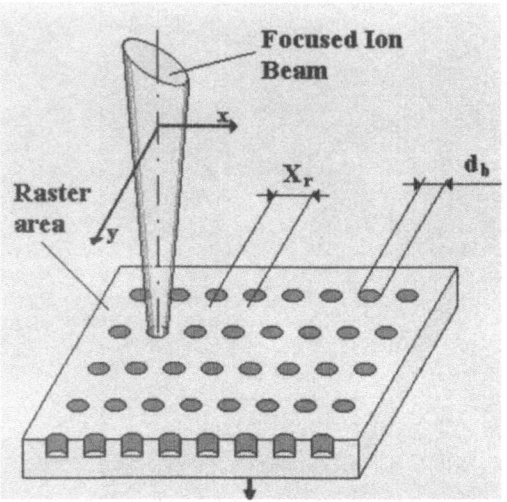

Fig. 1.9. Schematic diagram of the focussed ion beam process (R. Ramesh, private communication).

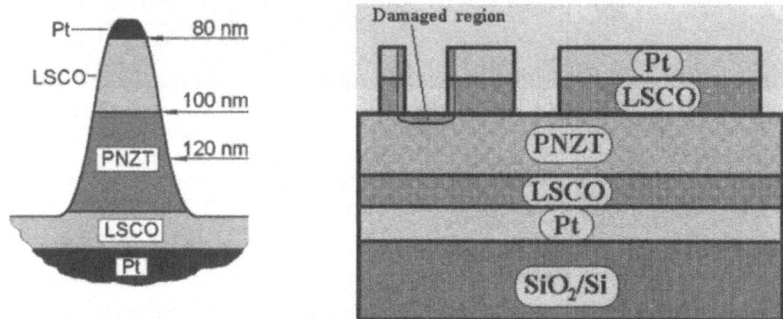

Fig. 1.10 (a) (left) Schematic cross section of typical FIB depth profile; (b) (right) Integration into a nanophase FRAM device.

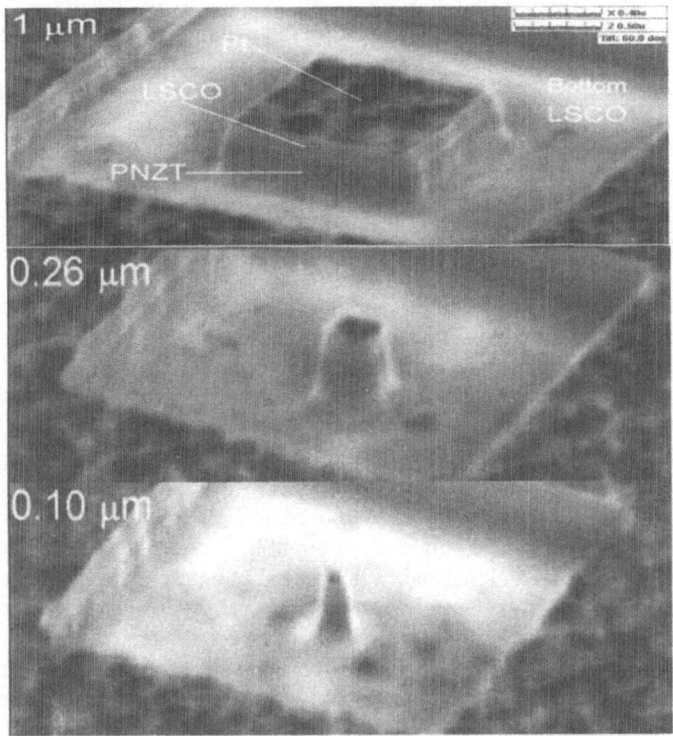

Fig.1.11 a,b,c: Top to bottom, FIB PZT devices of successively smaller lateral dimensions: 1.0 micron; 260 nm; 100 nm.

2. Commercial Devices and Electrical Characterization

The recent development of 4 Mbit ferroelectric random access memories (FRAMs) by both Samsung[1] and Panasonic/Matsushita[2] has shown that extremely high-density ferroelectric memory devices (ULSI) are commercially feasible within the next few years. Concomitant with this industrial progress is the demonstration in research laboratories that patterned array of cells as small as 20 nm on a side[3] is possible (sufficiently small for Gbit device fabrication), and that these nanoscale devices possess most of the performance parameters, such as switched charge density 2P$_r$ and coercive field E$_c$, of large-area capacitors of the same thickness.[4] Thus it seems timely to provide an update to the author's earlier monograph[5] on the topic of ferroelectric memories which summarizes the state of the art for nanophase devices and at the same time makes contact with the related problem of ferroelectric gates as replacements for metal gates in field effect transistors (FETs), which are also nanoscale (< 0.1 microns).   Such "1T" (one-transistor per bit) devices offer great advantages over the usual "1T-1C" (one-transistor/one-capacitor per bit) pass-gate architecture in that they use up much less chip area and eliminate the need for a destructive READ operation (and reset), which exacerbates the problem of fatigue in existing devices.[6] The present chapter is an attempt to review these things; not included, but emphasized in a second book by the author (Ferroelectric Memories II: Nano-phase Devices), is the use of modern optics to characterize ferroelectric arrays, including three-dimensional [3D] "photonic crystals" (porous Si or Al_2O_3 with tubes that can be filled with ferroelectrics).[7] After a period of great success in the 1960s with $LiNbO_3/LiTaO_3$ planar waveguides,[8] optics had passed through a quiet period in the study of ferroelectrics, punctuated by occasional bursts of activity (phase conjugate optics,[9] thermal focussing and optical bistability,[10] photorefractive effects) but now seems poised for a minor renaissance, particularly in the case of photonic crystal applications.[11]

In all of this work it has been the author's view that ferroelectric oxides must be treated as other wide-bandgap semiconductors, such as GaN, ZnSe, or other III-V or II-VI semiconductors. Thus we need to know everything about them that we know about II-VI or III-V compounds: Bandgaps, effective masses, mobilities, donor and acceptor levels, etc. This will take a number of years and a number of Ph.D. theses. However, the alternative – making a thousand-million-dollar technology out of materials that are poorly understood – is folly. In the past ferroelectrics were the laboratory curiosities of universities and not used in applications that actually required switching. Studies emphasized single crystals and bulk ceramics, and hence transport properties, both electrical and thermal, were viewed as uninteresting. The result was a thousand papers on the free energy expressions describing phase transitions in these materials and indeed nearly a thousand on the pseudo-spin model of hydrogen (proton) ordering in KH_2PO_4, but virtually no papers on leakage currents or breakdown fields in ferroelectrics that might help engineers make integrated devices (in Si or GaAs) from them.

We single out Prof. V. Fridkin, among a handful of exceptions, who recognized early the importance of treating ferroelectrics as wide-gap semi-conductors. His book on ferroelectric semiconductors was seminal.

2.1. CONFINEMENT ENERGIES IN FERROELECTRIC NANODEVICES

Confinement energies are a trendy topic in nanoscale semiconductor microelectronics devices.[12] The basic idea is that in a system in which the electron mean free path is long with respect to the lateral dimension(s) of the device, a quantum-mechanical increase in energy (and of the bandgap) in the semiconductor will occur. In general confinement energies exist only in the ballistic regime of conduction electrons, that is, where the electron mean free path exceeds the dimensions of the crystal. This usually requires a high-mobility semiconductor at ultra-low temperatures. Such effects are both interesting and important in conventional semiconductors such as Si or Ge, GaAs and other III-Vs, and perhaps in II-VIs. However, despite the fact that the commonly used oxide ferroelectrics are wide-bandgap p-type semiconductors (3.0 eV $< E_g < 4.5$ eV),[13] neither their electron nor hole mean free paths are sufficiently long for any confinement energies to be measured. Typically the electron mean free path in an ABO_3 ferroelectric perovskite is 0.1 to 1.0 nm,[14] depending on applied electric field E, whereas the device size d is at least 20 nm. Therefore any confinement energy (which scales as d^{-2}) might be a meV or two, virtually unmeasurable, despite a few published claims[15-18] reporting extraordinarily large effects. In the case of Bi_2O_3 and $SrBi_2Ta_2O_9$ (SBT) these claims may arise from two-phase regions at the sample surfaces.[19] This is theoretically interesting and very important from an engineering device point of view; if it were not true the contact potential at the electrode interface in a 1T-1C device, or at the ferroelectric-Si interface in a ferroelectric-gate FET, would depend critically on the cell size, which would add a very undesirable complication to device design.

2.2 COERCIVE FIELDS IN NANODEVICES

One of the most pleasant surprises in the research on small-area ferroelectrics is the observation, shown in Fig.2.1, that the coercive field is independent of lateral area.[20] Coercive fields in nanophase ferroelectric cells have generally been measured via atomic force microscopy (AFM).[21]

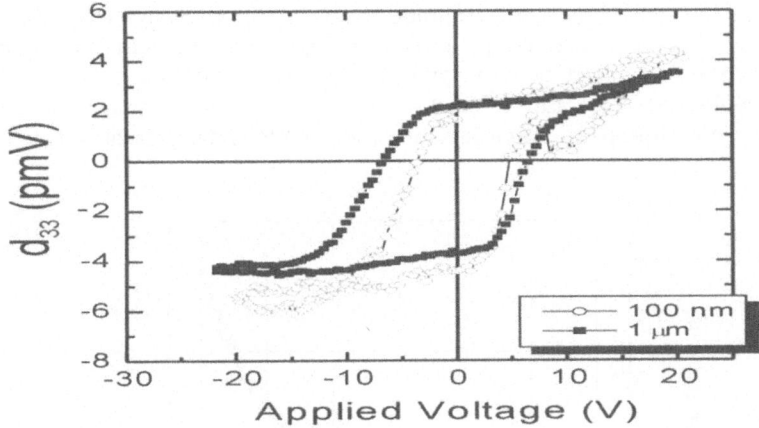

Fig.2.1 Hysteresis in PZT capacitors of 1.0 and 0.1 micron lateral size, showing that P_r and E_c are nearly independent of area.

2.3 ENHANCED LEAKAGE CURRENTS

It is well known that ionic leakage currents are greatly enhanced in nanophase structures,[22-24] due primarily to space charge effects. In many cases this increase in leakage current density is sufficient to take the device from the Schottky limited current regime to the space-charge-limited current regime (SCLC), as shown in Fig.2.2a for PZT. The latter case is most easily illustrated by graphing the square root of current density versus applied voltage (Fig.2.2b). Fig.2.2c shows similar results for bismuth titanate nanostructures ($Bi_4Ti_3O_{12}$). It is also well known that ferroelectric films such as BST exhibit significant space-charge limited ionic conduction,[25] in addition to their electronic conduction. What has not yet been established is whether the electronic conduction in nanophase ferroelectrics is also enhanced. We note in this context that very recently Gruverman et al. have shown[26] that their AFM switching data for PZT, and in particular the cathode-anode polarity asymmetry, are compatible only with the model of the present author in which an n-type inversion layer is postulated at the ferroelectric-electrode interface. In particular, they found that the fact that the ferroelectric-electrode interface region always has an n-type inversion layer in a nominally p-type ferroelectric gives rise to a built-in asymmetry for positive/negative voltage polarity applied to the bottom electrode. This result is not compatible with Tagantsev's "seed" model of preexisting domains and adds credibility to the basic space-charge depletion width model of the present author.[25]

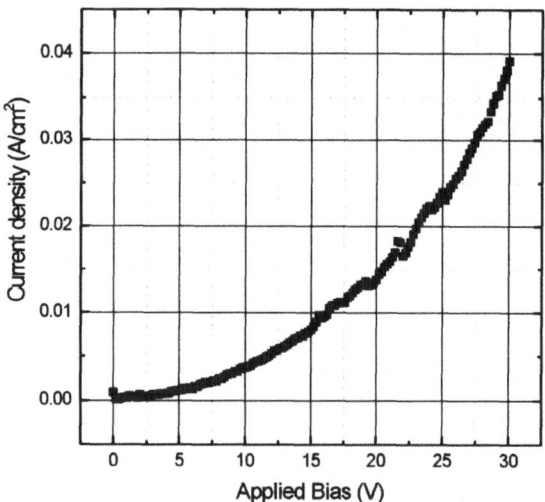

Fig.2.2a Quadratic leakage current (SCLC) in nanophase PZT capacitor.

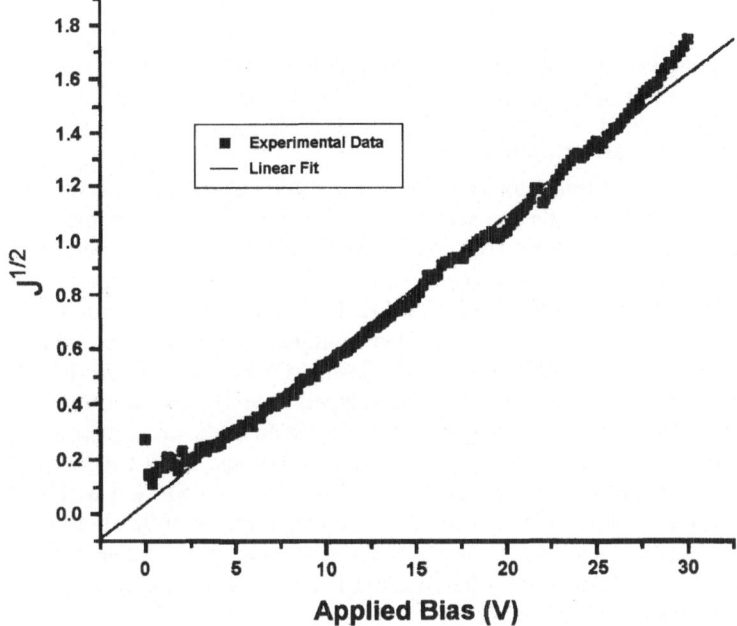

Figure 2.2b Leakage current data of Fig.1.2a for PZT replotted on square-root graph, showing space-charge-limited current (SCLC).

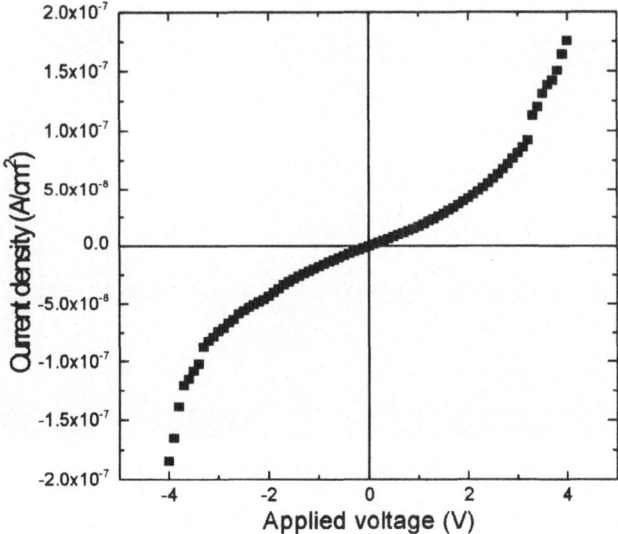

Figure.2.2c Leakage current data for bismuth titanate nanophase capacitor, showing quadratic space-charge-limited current.

2.4 REAL DEVICES

In the final analysis it is the production of real commercial devices on a 0.1-micron scale that is the test of the models based on data such as those in Figs.2.1 and 2.2abc. A comprehensive review of ferroelectric RAM technology was presented this year in Japan by Takashima[27] (Toshiba). His novel "chain-FeRAM" consists of a memory cell comprised of one transistor and one ferroelectric capacitor connected in parallel, and a block of multiple cells and selecting transistor in series. With this geometry he has achieved in a 0.5 micron two-metal CMOS device an impressive 37-ns access time and an 80-ns READ/WRITE cycle time. Takashima's paper appears in a special issue of IEICE Trans. Electron., edited by H. Ishiwara, which is recommended to the readers of this book; it gives an excellent comparison of all types of nonvolatile memories, including MRAM, EEPROM, flash, MNOS, etc. In the FeRAM area, several important products have been developed recently by Samsung, which are briefly described below. In this regard the present state of the art is represented by the 4 Mb FRAM from Samsung, as reported in the 1999 and 2000 IEDM Conferences (International Electron-Device Meeting) (Fig.2.3). This is a 512 k x 8 chip. First we see in Fig.2.3 a 512-kb block, sectioned into sixteen sections of 32 kb each. This is a 1T-1C open bit-line architecture in which the final chip size is 12.96 x 5.93 cm and the cell size is 3 x 3 microns. Metallization is Al/W, with aluminum plate line and signal line but tungsten bit line.

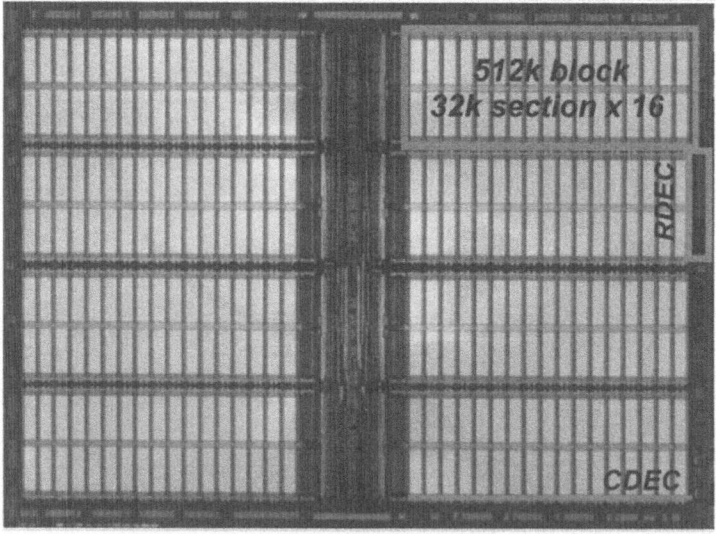

Figure.2.3. Samsung 4 Mb FRAM 512k memory block.

An SEM cross section of this device is shown in Fig.2.4. The ferroelectric capacitor used is PZT and the electroding is iridium/iridium oxide, as illustrated in Figs.2.5; the latter has an enlarged scale in which the electrode-ferroelectric sandwich structure can be seen clearly. The 64k device was a double-level metallization chip (rather than triple-metal) and was perhaps the first fully commercial 64 kb FRAM

[1997 and 1998 VLSI (Very Large-Scale Integrated Circuit Confs.)]. Both top (TE) and bottom electrodes (BE) were a Pt/IrO$_2$ bilayer combination, but the chip metallization (as opposed to the FE capacitor electrodes) was an aluminum/titanium-nitride (AlTiN) combination. Note that Metal 2 is Pt; metal 2 is mostly TiN but not uniformly so everywhere (only TiN to contact the bottom electrode but Ti/TiN to the active line). These should be contrasted with the tungsten and aluminum metallization employed for the 4 Mb FRAM device.

The Samsung 4 Mb FRAM uses a halo-type n-channel MOSFET (Jpn. J. Appl. Phys. (1996)). Its characteristics (Fig.2.5) are slightly different from that of a more conventional n-chalnnel MOSFET. Note that the depletion level extends to within 0.1 microns of the gate in the halo device, and the zero-potential contour extends well up into the gate, unlike the case with the conventional MOSFET device. The gate region is 6 x 10^{12} cm^{-2} boron-implanted at 50 keV.

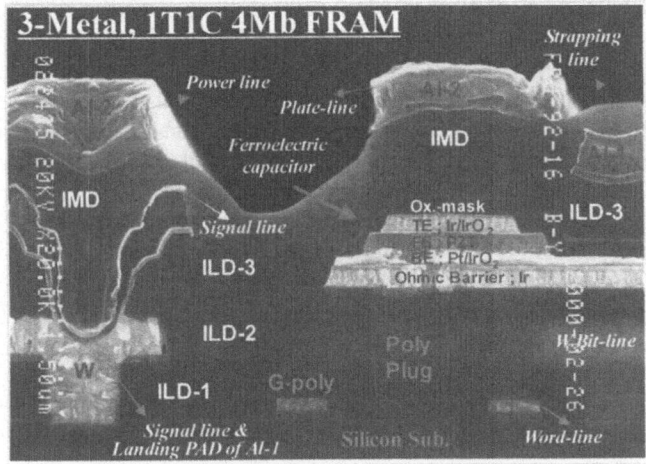

Figure.2.4. SEM micrograph of Samsung 4 Mb FRAM.

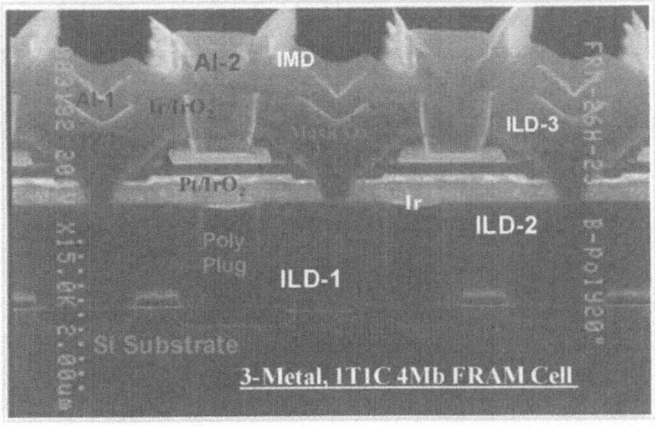

Figure.2.5. Ferroelectric/electrode cross-sectional view of the Samsung 4 Mb FRAM.

2.5 PRESENT DESIGN APPROACHES TO GBIT FRAMS:

Takashima has given a detailed account of the Toshiba strategy for achieving 1 to 16-Gbit FeRAMs. Firstly, in order to achieve the fastest speeds, half the nominal V_{dd} bias is applied to the ferroelectric capacitor without driving the plateline, because the cell plate is fixed at this value and the bitline is pre-charged at ground. This eliminates the cell-plateline delay. This eliminates the need for refresh, because all data are held by short-circuiting the ferroelectric capacitors (thus the "1" data do not decay due to junction leakage currents). The second idea is to connect a number of cells in series. For a bit-line capacitance of 300 fF, Toshiba finds that 1024 cells can be connected to one sense amplifier (compared with 256 cells in a normal geometry), thus reducing the number of sense amplifiers x4. The series cell configuration reduces chip size to 56% of the conventional cell size.

The present DRAM layout is an $8F^2$ cell size. Takashima shows that FeRAM stacked capacitor cells will also realise $8F^2$, but that the chain FeRAM can obtain $6F^2$, and one can obtain a $4F^2$ size using cross-point geometries with self-aligned technology.

The main problem he addresses for Gbit FRAM development is the apparent decrease in cell polarization as capacitor sizes shrink (that is, switched charge – not charge density, which stays approximately constant[27]), and he suggests that this limits [2D] capacitor structures for FeRAMs to 256 Mbit, but that it can be overcome by drastically reducing the bitline capacitance. For example, for a realizable bitline capacitance of 10 fF, a large ±350 mV signal can be read out. By amplifying the cell signal via a gain transistor, one can achieve a stable FRAM memory with 35 ns access time and 70 ns cycle time, even with a rather small cell polarization of 8 fC. I point out for readers that a PZT capacitor with remanent polarization 2Pr of ca. 80 $\mu C/cm^2$ produces in a nanophase 0.1 x 0.1 μm^2 cell 8 x 10^{-15} C [or approximately the 8 fC quoted, i.e. 50,000 electrons].

2.6 OTHER STATE-OF-THE-ART FERROELECTRIC THIN-FILM DEVICES:

Before ending this chapter, it would be remiss not to mention the other principal area in which thin-film ferroelectric devices are making an impact – that of 10-30 GHz microwave devices, especially phased-array radar. An up-date of that area is given in Table II.1, following F. Miranda.[28]

	Ferroelectric	Semiconduct	Ferrite	MEMS
Cost	Low	expensive	Very expensive	Low
Reliability	Good after $>10^5$ bias cycles	Very good (if properly packaged)	excellent	Questionable (further tests required)
Power handling	Good, $>1W$	Low power, few tens of watts	Very high (Kilowatt	Low power $<1W$
Switching speed	Very fast ($<10^{-10}$ sec)	Fast at low power ($<10^{-9}$ s)	Slow (inductance)$< 1\mu s$	Slow (mechanical; ~5 μs)
Radiation tolerance	Excellent	Poor (good if radiation hardened)	Excellent	Excellent (mechanical; no solid state juncts. involved)
DC Power consumption	Low (hundred volts, no current)	Low (few mW)	High current hundred	negligible
RF loss	~ 5 dB/360° @ K-band	~1.5 dB/bit @ K-band	<1dB/360° @ X-band	~ 2.3 dB/337.5° @ Ka-band
Switching energy	unknown	Few Joules	Tens of Joules	10 nano-Joules
Linearity	unknown	IMD intersect +35 to +40	unknown	IMD intersect +80 dBm

Table II. Comparison of ferroelectric thin-film devices for microwave applications with competitive technologies (F. Miranda, Integ. Ferroelec. 2001).

3. Summary

In this chapter I have tried to review the basic physics issues related to the development of nanoscale (<100 nm) ferroelectric capacitors for memory devices. This includes submicron ferroelectric FET gates as well as FRAMs. Emphasis is upon deposition and the dependence (or independence) of electrical characteristics. I stress that there are no measurable confinement energies. More detailed discussions

600

of space charge, voltage drops within the electrodes, oxygen vacancy gradients, and related interface problems are given in my book "Nano-Ferroelectrics" due out in 2003 in the Springer series on Microelectronics.

4. References

1. Samsung 4 Mb FRAM: D. Jung et al., IEDM (1999, 2000); VLSI (1997, 1998); Integ. Ferroelec. (2001); Jpn. J. Appl. Phys. (1996).
2. Panasonic 4 Mbit FRAM: T. Otsuki et al., Integ. Ferroelec. (2001); see also J. F. Scott, Ferroelectrics 2000, ed. N. M. Alford and E. Yeatman (IOM Commun., Oxford, 2000), pp.1-12.
3. S. Aggarwal, C. S. Ganpule, I. G. Jenkins, B. Nagaraj, A. Stanishevsky, J. Melngailis, E. Williams, and R. Ramesh, Integ. Ferroelec. **28**, 213-225 (2000); C. S. Ganpule, A. Stanishevsky, Q. Su, S. Aggarwal, J. Melngailis, E. Williams, and R. Ramesh, Appl. Phys. Lett. **75**, 409-411 (1999).
4. M. Alexe, C. Harnagea, D. Hesse, and U. Goesele, Appl. Phys. Lett. **78**, 1793-1795 (1999).
5. J. F. Scott, Ferroelectric Memories (Springer, Berlin, 2000).
6. H. Ishiwara, Jpn. J. Appl. Phys. **32**, 442 (1993); H. Ishiwara, T. Shimamura, and E. Tokumitsu, Jpn. J. Appl. Phys. **36**, 1655 (1997).
7. J. Schilling, F. Mueller, S. Matthias, R. B. Wehrspohn, U. Goesele, and K. Busch, Appl. Phys. Lett. **78**, 1180-2 (2001); S. Ottow, V. Lehmann, and H. Foell, Appl. Phys. **A63**, 153-159 (1996).
8. P. K. Tien and A. A. Ballman, J. Vac. Tech. **12**, 892 (1975).
9. J. Feinberg, Optical Phase Conjugates (Academic, New York, 1983); J. Feinberg,. D. Heiman, A. R. Tanguay, Jr., and R. W. Hellwarth, J. Appl. Phys. **51**, 1297 (1980); D. L. Staebler and J. J. Amodei, J. Appl. Phys. **43**, 1042 (1972); S. K. Kurtz, S. D. Kozikowski, and L. J. Wolfram, Electrooptic and Photorefractive Materials, ed. P. Gunter (Springer, Berlin, 1987), p.110.
10. R. A. O'Sullivan, K. W. McGregor, and J. F. Scott, J. Phys. Condens. Mat. **13**, R195-234 (2001).
11. A. Birner, R. B. Wehrspohn, U. Goesele, and K. Busch, Adv. Mater. **13**, 3776 (2001).
12. P. M. Petroff, A. Lorke, and A. Imamoglu, Physics Today **54**, 46 (2001).
13. R. Waser and D. M. Smyth, Ferroelectric Thin Films: Synthesis and Basic Properties, eds. C. Paz de Araujo, J. F. Scott, and G. W. Taylor (Gordon & Breach, Amsterdam, 1996), pp.47-92.
14. A. J. Dekker, Phys. Rev. **94**, 1179 (1954).
15. B. Yu, C. Zhu, and F. Gan, J. Appl. Phys. **82**, 4532 (1997).
16. S. Kohiki, S. Takada, A. Shimizu, K. Yamada, and M. Mitone, J. Appl. Phys. **87**, 474 (2000).
17. J. F. Scott, J. Appl. Phys. **88**, 6092 (2000); J. F. Scott, Proc. AMF-3 (Hong Kong, Dec. 2000); Ferroelectrics (in press, 2001).
18. S. Kohiki et al., J. Appl. Phys. **88**, 6092 (2000).
19. W. Zhou, J. Sol. St. Chem. **101**, 1 (1992); J. A. Switzer, M. G. Shumsky, and E. W. Bohannan, Science **284**, 293-296 (1999).
20. M. Alexe, A. Gruverman, C. Harnagea, N. D. Zakharov, A. Pignolet, D. Hesse, and J. F. Scott, Appl. Phys. Lett. **75**, 1158-60 (1999).
21. A. Gruverman, Ferroelectrics (in press 2001); A. Gruverman, O. Auciello, and H. Tokumoto, Appl. Phys. Lett. **69**, 3191-3 (1996).
22. J. Maier, Electrochemistry **68**, 395 (2000).
23. J. Maier, Sol. St. Ionics **131**, 13 (2000).
24. J. Maier, J. Jamnik, and M. Leonhardt, Sol. St. Ionics **129**, 25 (2000).
25. S. Zafar et al., J. Appl. Phys. **82**, 4469 (1997); J. F. Scott, Ferroelec. **232**, 25-34 (1999).
26. A. Gruverman, A. Kholkin, A. Kingon, and H. Tokumoto, Appl. Phys. Lett. **78**, 2751 (2001) and in ISIF -- Colorado Springs, March 2001 (Integ. Ferroelec., in press).
27. D. Takashima, IEICE Trans. Electron. E84-C, 747 (2001); this special issue gives an outstanding coverage and comparison of all types of nonvolatile memories; 0.3-micron Ir/PZT/Ir cells on W plugs: T. S. Moise, S. R. Summerfelt, G. Xing, L. Colombo, and T. Sakoda, IEDM Tech. Dig. 941 (1999); 0.07-micron cells: M. Alexe,C. Harnagea, D. Hesse, and U. Goesele, Integ. Ferroelec. **27**, 104 (2000).
28. F. Miranda, Integ. Ferroelec. (in press, 2001).
29. M. Alexe, C. Harnagea, W. Erfurth, D. Hesse, and U. Goesele, Appl. Phys. **A70**, 247-251 (2000); M. Alexe et al., Ferroelectrics (in press 2001); M. Alexe and J. F. Scott, J. Eur. Ceram. Soc. (in press 2002)..
30. C. S. Ganpule, R. Ramesh, et al. (private communication).
31. Experiment: J. F. Scott, M. Alexe, N. D. Zakharov, A. Pignolet, C. Curran, and D. Hesse, Integ. Ferroelec. **21**, 1 (1998); Theory: Shchukin et al., Phys. Rev. Lett. (1995).

OPTICAL CHARACTERIZATION OF MECHANICAL PROPERTIES OF THIN FILMS AND STRUCTURES

S.TAMULEVIČIUS, L.AUGULIS
Kaunas University of Technology
Institute of Physical Electronics
Savanoriu av.271Kaunas LT-3009
Lithuania

1.Introduction

The developments of miniaturized tests are vital task in materials science and is currently important problem in many domains in science and technology. In addition to the microelectronics or optoelectronics, traditional technologies like microlitography and microfabrication are rapidly finding applications in many areas, from sensors and actuators to biomedical devices. An explosive demand is expected for microelectromechanical systems (MEMS) for use in industries such as automotive, chemical, aircrafts etc. MEMS will also find applications for in-situ process monitoring, and numerous other sensor and actuators system [1]. As a rule microsystems use association of various components such as films, membranes, wires and shells. Evaluating the mechanical properties (Young's modulus, biaxial modulus, shear modulus, flexural modulus) of such components is a new challenge in mechanical engineering [2-6]. On the other hand scaling down the size brings that film thickness is comparable to microstructural dimensions (grain size, dislocation spacing), where film properties can significantly different from the volume ones. The study of the mechanical properties of thin films requires a high degree of accuracy for the measurements of the stress-strain or strain-time data.

Film testing requires that one differentiates between the properties of the film, which are adherent to the substrates and those that are free standing, that is, they have been separated from their substrates [7].

Almost all the films on substrate are stressed (under compression or tension) - the elastic stress in thin film is an inherent part of the deposition process. Typical values of the stress in thin metallic films can change in the range $10^7 - 10^9$ Nm^{-2}, in the insulator films these stresses are bit lower and as a rule in many cases they are compressive. The stress can change mechanical as well as electrical, optical and other properties of thin films. There are many examples illustrating shifts in bandgap in semiconductors,

T. Tsakalakos, et al. (eds.) pgs. 601 - 618
Nanostructures: Synthesis, Functional Properties, and Applications;
© *2003 Kluwer Academic Publishers.*

transition temperature for semiconducting films, or expected magnetic anisotropy induced by mechanical stress in thin films [8]. Resultant stress in thin film σ is equal to the sum of all external stress applied to σ_{ext} the film, thermal stress σ_{th} arising due to difference of thermal coefficients of the thin film and substrate as well as intrinsic stress σ_{in}. The stress may arise for many reasons: the presence of defects during or after depositions (like vacancies, dislocations), incorporation of atoms from residual gases, variation of the interatomic spacing with crystal size, microscopic voids and special arrangements of dislocations, phase transformations, from a mismatch in the coefficient of thermal expansion between the substrate and the film if deposited at high temperature and measured at a lower temperature or from a mismatch in the elastic modulus if deposited on a substrate under stress, occurrence of solid state transformations, etc. [9-12]. It is very important to be able to monitor and control stress in the growing film and substrate to avoid undesirable effects and investigate the stress dependence on technological parameters.

X-ray diffraction is one of the most powerful methods for microstructural characterization [13-17]. Characteristic parameters of the real crystalline structure in the conventional x-ray diffractometry can be obtained from the position, width and intensity of the diffraction peaks. Applying grazing incidence or measuring the strain with simultaneous scanning of measurement direction makes this method flexible and informative when stress distribution versus film thickness or components of the strain are required.

The use of optical probe techniques for the real - time monitoring of different processes is favored because of their nondestructive character and their potential use in real time feedback control. Depending on the properties of the object to be investigated (opaque or transparent) and problems to be solved, the different experimental techniques can be applied: laser reflective interferometry, laser spot scanning interferometry, optical leverage with a laser beam [18-23] etc. A common feature for all the mentioned techniques is a registration of strain during bending, tension or compression of freestanding films (stress- strain experiments) or registration of the strain of the film-substrate structure due to residual stress that is related to technological steps of thin film formation.

The aim of this article is to make a survey of different optical experimental techniques as well as theoretical approaches that are used in the analysis of mechanical properties of thin films.

2. Optical Measurement Techniques of Strain and Internal Stress in Thin Films

The formulas that have been used in virtually all-experimental determinations of film stress are variants of an equation first given by Stoney in 1909 [8]. Stoney's equation for a film whose thickness is small as compared to the substrate thickness, relates average stress in the film and variation of the curvature of film – substrate system:

$$\sigma_f = \frac{1}{6} \frac{E_s h_s^2}{(1-v)h_f} \left(\frac{1}{R_2} - \frac{1}{R_1} \right) \tag{1}$$

Where E_s is Young's modulus of the substrate, h_s is the thickness of the substrate, v is Poisson's ratio of the substrate, h_f is the thickness of film, R_2 and R_1 are the radii of the substrate after and before thin film deposition respectively.

As one of the low-cost methods for determining film stress, an optically levered laser technique is commonly used to measure a radius of substrate curvature induced by a deposited film. One end of a sample is clamped onto a translation stage and a lateral shift of a light spot reflected by its surface is measured on a distant screen, as the sample is translated. One can understand that measuring the angle of reflection as a function of position on substrate, enables to control the slope s of the system (Fig.1):

$$s = \frac{du_z}{dy} = -\frac{\alpha}{2} \tag{2}$$

or curvature:

$$\frac{1}{R} = \frac{d^2 u_z}{dy^2} = \frac{ds}{dy} = -\frac{1}{2} \frac{d\alpha}{dy} \tag{3}$$

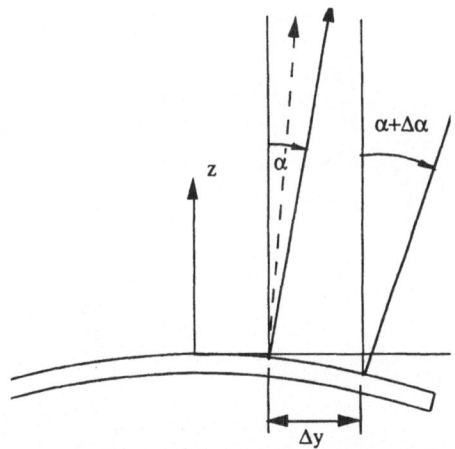

Figure 1. Schematic presentation of the optical level technique

The cantilever technique in combination with different registration methods enables to monitor in real time deposition, ion beam/plasma etching or sputtering, thermal heating etc., to control composition, structure, stress and strain dynamics for a wide range of materials: metals, semiconductors, insulators in the form of thin and thick layers or multilayer structure [24,25]. For in-situ analysis, deflection of the free end of thin plate (or beam) is measured as the other end of the beam is imagined to be clamped rigidly. This technique is used often because the deflection can be monitored

604

continuously during deposition of the film and thus information about the stress distribution can be obtained. The formalism of the technique is based on the simple relation between the cantilever's free end displacement δ and radius of the curvature of the film- substrate system R :

$$R = \frac{L^2}{2\delta}$$

(4)

Where L is the length of cantilever.

In addition to the discussed material, it should be noted that (4) is valid under some conditions - the beam should be gripped in the clamp that no sliding would take place - rigid gluing sometimes is necessary or sophisticated technology including micromachining technique of oxidized layer or epilayers to produce cantilever as one component of micromechanical device should be involved. The second area where attention should be paid when the cantilever technique is applied to monitor thin film deposition kinetics, is the substrate temperature stabilization during deposition. The temperature difference along the length of the beam can give rise to non-uniform bending of the composite giving rise to the smaller stress as given by (4). The next thermal effect arises from the temperature gradient through the thickness of the substrate that acts as the equivalent force acting on the free end of cantilever. Typical example of the scheme used in the experiments in-situ is demonstrated in Fig.2.

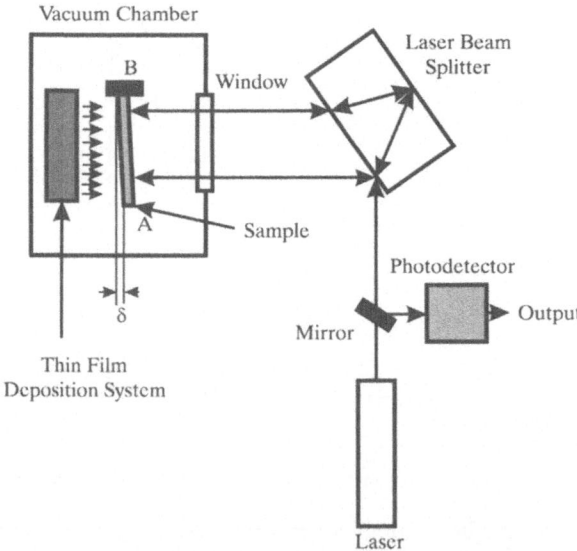

Figure 2. The double-beam differential laser interferometer [24]

The interferometer is mounted on the window of the vacuum chamber. It can be used to measure the displacement δ of the free end of the cantilever (sample) in the process of thin film deposition and etching (in situ). The sample is rigidly damped at point B keeping free the end A. Measurements are provided from the "back" side of the coated sample. In this way, the influence of thin film deposition, ion implantation or etching effects on the reflectivity of the surface is eliminated. The sensitivity of this type of interferometer depends on distance between the interfering beams.

The sensitivity of the cantilever technique in the strain measurements can reach up 10^{-6} and this fact allows applying this method for control of the nucleation and coalescence of thin films that can be provided both in –situ or ex-situ. As example stress dependence in the CdS films deposited on crystalline (100) GaAs by Susequent Ionic Layer Adsorbtion Reaction technique (SILAR) [26,27] are presented in Fig.3. One can see the correlation between the mode of growth (the AFM photos) and stress level in the film.

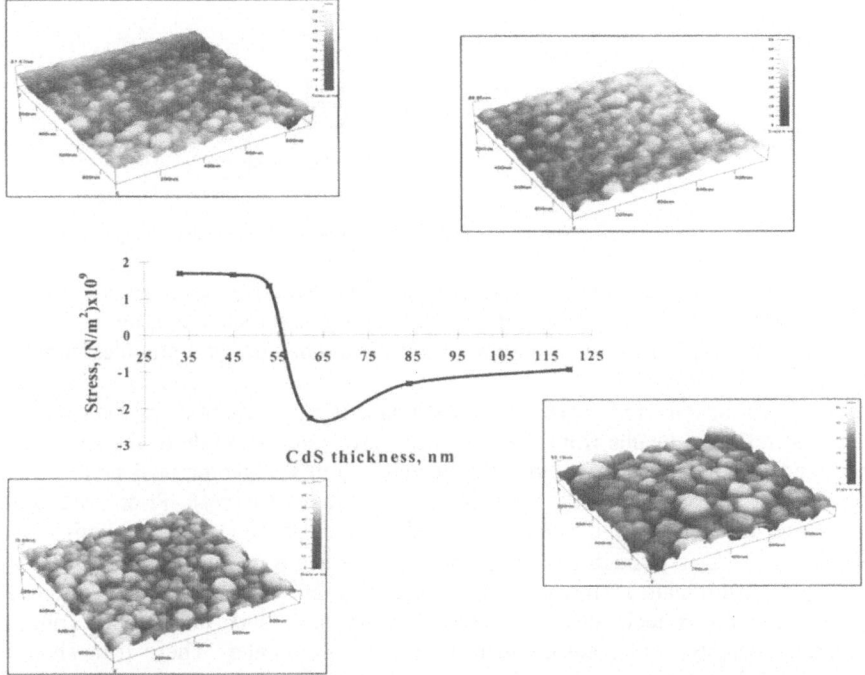

Figure 3. Stress dependence versus thickness in the CdS thin films SILAR deposited crystalline GaAs (the AFM photos are given for the disperse (tensile stress region) and continuos films (compressive stress region).

From this figure one can see that tensile stress dominates in the region of 30-50 nm. Increase in the thickness of films results in the change of the sign of stress from tensile to compressive. In the small thickness region this stress is a result of film – substrate interaction and it appears due to molecular forces acting at the three-dimensional islans and GaAs surafce. The further deposition results in growth of crystallites and decrease of the roughness of the surface. We can conclude that two-dimensional growth takes place from the thickness 60 nm upwards and is responsible for the compressive stress.

Optical technique is most promising experimental method when two-dimensional strain distribution is of interest. In this case wave front of the reflected beam from the

606

sample under analysis is compared with the one reflected from the reference surface. Depending on the light splitting method, interferometers are commonly classified as wave front-dividing or amplitude-dividing. In most cases the amplitude-dividing interferometers for stress determination are used. As an example, the modified Michelson interferometer is presented in Fig. 4.

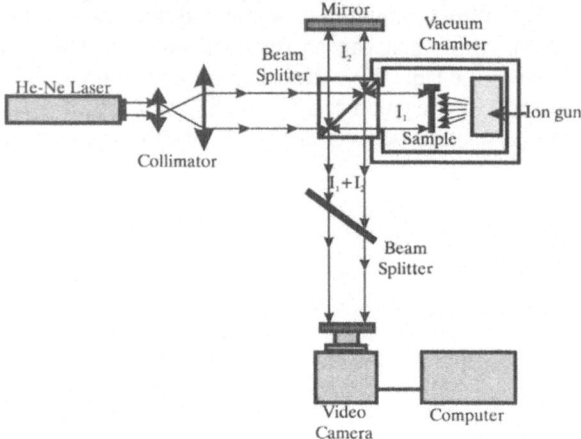

Figure 4. The Michelson type interferometer for deformation measurements in-situ [25]

This kind of equipment allows to provide either strain measurements during thin film deposition, ion bombardment, thermal annealing or for static measurements. Laser interferometer is installed in a window of a vacuum chamber and provides monitoring of 2D surface profile of substrate, which curvature can vary due to additional ion bombardment or thin film deposition. Interference fringes generated by the sum of two expanded reflected beams from the reference mirror and backside of the substrate are registered by CCD and recorded by computer. From the geometrical point of view, interference fringes represent lines of equal altitude (height). Prior dark fringes correspond to the height difference of half wavelength. So simple counting of the interference fringes through the entire substrate surface provides accuracy of 0.15 μm for height (co-ordinate z) measurements (for the He-Ne laser used in the experiment). An image analysis technique is used to identify and mark interference fringes of specified intensity, and measure their x and y co-ordinates. These data should be arranged as the series of xyz values describing 2D surface profile related to the reference surface. A real 2D-strain profile can be obtained as a difference of two such profiles, before and after exposure to thin film deposition. Stress can be defined from the differential Stoney's equation.

3. Electronic Speckle Pattern Interferometry in Thin Film Analysis

3.1 OPTICAL ARRANGEMENT

Electronic Speckle Pattern Interferometry (ESPI) (or TV Holography) combines different aspects and properties typical for the optical interference and holographic methods [28,29]. The principle of speckle interferometry uses the ability of laser light of interference (space coherence of the light waves). The object to be analyzed is

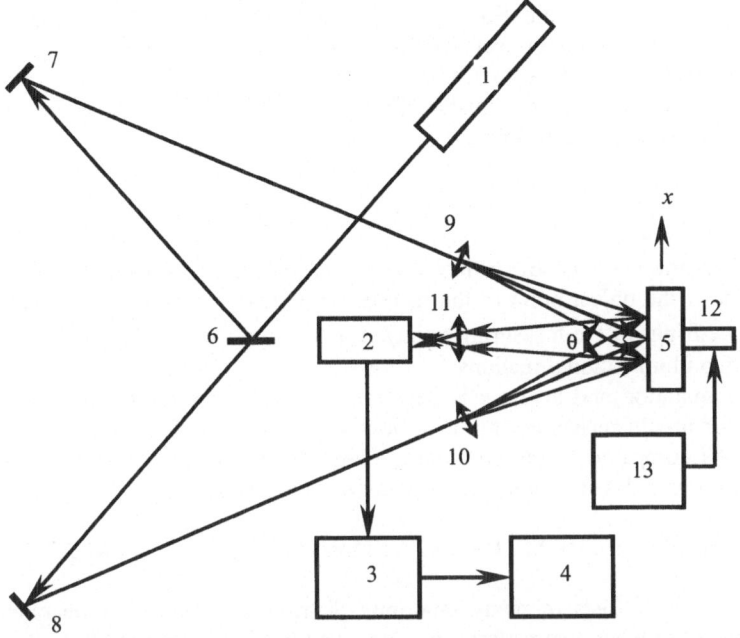

Figure 5. Schematic of the ESPI system [32]. 1 - laser, 2 – video camera, 3 - frame grabber, 4 – PC, 5 – object, 6 – beam splitter, 7,8 – mirrors, 9,10,11 – lenses, 12 – heating element, 13 – source for the heating element, x- measurement direction.

entirely illuminated with laser light. The light waves, which are reflected by the single points on the object's surface, interfere and produce so called speckle pattern. This speckle pattern is superposed by reference light, which is produced according to the used measuring technique by different means. The interference between the reference light and the speckle pattern produces a new speckle pattern in which additionally shape and position of the object's surface is contained. This speckle pattern is stored in the image – processing computer as a reference image. When the object is moving the speckle pattern is changing. By the comparison with the reference pattern, correlation fringes are produced which represent the displacement respectively deformation component of the object's surface. The main progress in the developing of such measurement technique was done in the nineties, when fast computers used in data acquisition and analysis were employed. The software can be used to filter, to count the correlation fringes, and to transform them automatically into a quantitative set of deformation. In this method a complicated interference fringe pattern due to interference between the waves reflected from the reference and measured surfaces is registered in a limited aperture optical system. Areas of correlated motion of the surface are used to define tangential (in plane) or normal (out of plane) displacement due to vibration, thermal or mechanical exposure of the object. Application of the different type of optical schemes allows defining 3D measurements. Diverse applications of ESPI including [30-31] studies of thin films mechanical properties were reported. Fig. 5 presents an example of the ESPI optical arrangement used for the thermal strain measurements. The used optical scheme was sensitive to the displacement in the direction x that corresponds to the rows in the image matrix. Displacement of the individual points was defined using (5) if the displacement was accompanied by changes of some correlation fringes.

$$\Delta x = Nd = \frac{N\lambda}{2\sin(\Theta/2)} \tag{5}$$

where N is the number of fringes, d is the measuring sensitivity (deformation component of the object point in the measuring direction in plane, corresponding one interference fringe at the measuring point), λ is the applied wavelength, Θ is an angle between two illumination directions.

Data acquisition and analysis is performed in real time, so this method becomes versatile for in-situ measurements to follow strain kinetics of the diffusive reflecting surfaces. As compared to the considered classical optical interferometry this feature allows expanding class of the materials under consideration.

3.2 PRINCIPLES OF THE SPECKLE PATTERN CORRELATION ANALYSIS

Speckle pattern consists of many randomly distributed speckles. Appearance of the correlation zones when two patterns are combined can be understood from the analysis of intensity redistribution of two individual points of this pattern.

Let's note waves produced at the plane of CCD on the dual illumination scheme are:

$$E_1 = a_1 \exp[i(\omega t + \varphi_1)], \quad E_2 = a_2 \exp[i(\omega t + \varphi_2)]; \tag{6}$$

where ω is the cyclic frequency of the light, φ_1 and φ_2 are the initial phases, a_1 and a_2 are the amplitudes of the waves.

Resultant intensity due to superposition of these coherent waves (reference intensity I_R) can be expressed by well-known formula:

$$I_R = a_1^2 + a_2^2 + 2a_1 a_2 \cos(\varphi_1 - \varphi_2). \tag{7}$$

Due to deformation or thermal expansion of the object phases of the waves will be changed by $\Delta\varphi_1$ and $\Delta\varphi_2$ and resultant intensity will be expressed as:

$$I_D = a_1^2 + a_2^2 + 2a_1 a_2 \cos(\varphi_2 - \varphi_1 + \Delta\varphi_2 - \Delta\varphi_1) \tag{8}$$

One can understand that difference of the intensities of two interference patterns will depend on the extra change of phases of two point sources:

$$\Delta I = |I_D - I_R| = 2a_1 a_2 |\cos(\varphi_2 - \varphi_1 + \Delta\varphi) - \cos(\varphi_2 - \varphi_1)|, \tag{9}$$

or

$$\Delta I = 4a_1a_2 \left| \sin\left(\frac{\Delta\varphi}{2}\right) \sin\left(\varphi_2 - \varphi_1 + \frac{\Delta\varphi}{2}\right) \right|; \qquad (10)$$

were $\qquad \Delta\varphi = \Delta\varphi_2 - \Delta\varphi_1 .$

I.e. points where $\left| \sin\left(\frac{\Delta\varphi}{2}\right) \right| = 1$ will have maximum intensity and points

where $\left| \sin\left(\frac{\Delta\varphi}{2}\right) \right| = 0$ will have minimum intensity.

From this simple analysis one can understand that correlation fringes appearing in the difference of intensities of two images consisting of huge numbers of speckles can be explained as zones of equal phases (like in classical interferometric fringe analysis).

Unfortunately the real two-dimensional speckle distribution is more complicated as was discussed above – it consists of many speckles of different intensities, distribution of which can be expressed by a Gauss law [28]. In this case extra steps are needed to increase the visibility and ratio signal/noise. In most cases automatic analysis of differences images requires significant processing to enhance the contrast of fringe regions. Correlation fringes obtained by different steps (matrix subtraction + filtering, or correlation method) can be used to define individual point displacement of the analyzed object. Fringe density and fringe distribution are the key parameters in this analysis. Fig. 6 shows typical correlation fringes in the electronic speckle patterns of aluminum surface during thermal heating obtained by simple subtraction of the intensities of two-dimensional image matrices and applying the correlation calculations (distance between two lines of the black grid on the picture is equal to 1 cm)

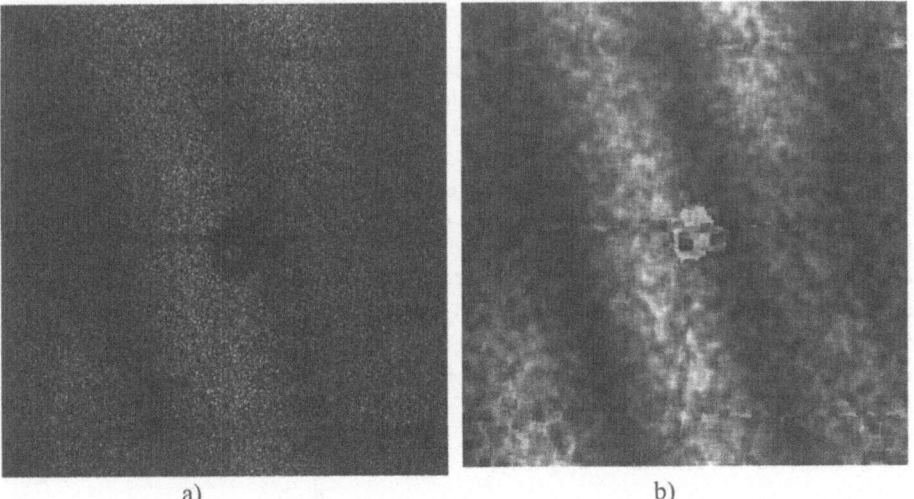

a)   b)

Figure 6. Correlation fringes appearing in the difference of intensities of two images (a) and obtained by the correlation method (b) [33]

3.3 MICROMECHANICAL ESPI MEASUREMENTS

Most of the methods reported above are focused on the measurements of objects with typical linear dimensions much bigger than 1 mm, and, as a thumb rule, surfaces under investigation are diffuse reflective surfaces. These physical limits restrict the application of the ESPI with respect to analyses of complicated shape or small dimension objects. The solution of these problems could be achieved by application of the new optical setup or creation of new methods of data analysis, combining good resolution power and high degree of localization (or small area of measurement). The last method was reported in [34] as "a microscopic ESPI method". In this work, we present a new optical setup that could be used for the analysis of displacement of objects with diffuse as well as mirror-like reflecting (specular) surfaces. The algorithm of data analysis, based on the correlation matrix calculation on a microscopic scale, is proposed as well. Combination of these two new approaches allows to minimize area of analysis as well as to extend the variety of the materials to which the ESPI can be applied.

The experimental set-up used in the measurements includes typical components of the optical setup such as: the He-Ne laser (wavelength 0.6328 μm, power 10 mW), beam expander, mirror system, and registration equipment (CCD camera, frame grabber, PC). Typical sensitivity ranges for the normal and tangential components of the displacements for the optical setup used were 0.15 μm and 0.31 μm respectively. The typical data acquisition rate was 6 images per second, which allowed measuring displacement rates of different points at the surface up to 1 μm/sec.

In addition to the classical components, the new element – an optical splitter with variable diffuse transmission and reflection – was introduced. The principles of the optical scheme including the original element for the normal displacement measurements are presented in Fig. 7. The expanded laser beam (1) due to a beam splitter (2) is directed on the object (3) under analysis. The matte surface of the splitter is covered by the light scattering material (in our experiments, aluminum powder was used). By adding or removing powder one can change intensity of the scattered light. In this way, two scattered waves from the splitter and surface under analysis are produced. The wave, reflected from the splitter surface, is used as a reference wave that interferes with the wave reflected from the object. The resulting interference picture is registered with an optical system including the aperture (4), lens system (5), and CCD camera (6). Two points A and B on the surface will be displayed as the resulting intensities at the points A' and B' in the camera (Fig. 7). One can understand that intensities of the light at the points A' and B' depend on the phase shift between two waves as well as on the reflectivity of the surfaces producing these two waves. If any deformation of the object (3) Δx takes place, the optical path difference between two waves in the direction of measurement changes as well, and the intensity of the light at any point of the CCD screen can be presented as following:

$$I = I_o + I_m \cos\left(\frac{4\pi\Delta x}{\lambda} + \Delta\varphi_0\right), \qquad (11)$$

where λ is the wavelength, $\Delta\varphi_0$ the initial phase shift, I_0 this intensity of the light (that doesn't depend on the phase difference), and I_m the amplitude of intensity in the interference fringe system. If we denote the reference wave intensity as I_r and the object wave intensity as I_{ob}, than, according to the combination of intensities in the interference fringe system $I_0 = I_r + I_{ob}$ and $I_m = 2\sqrt{I_r I_{ob}}$. One can understand that the maximum visibility of the interference fringes is registered in the case when $I_r \approx I_{ob}$. This can be done choosing the right reflectivity of the splitter or changing the angle α (this angle in the experiment was changed within a range close to the 90°).

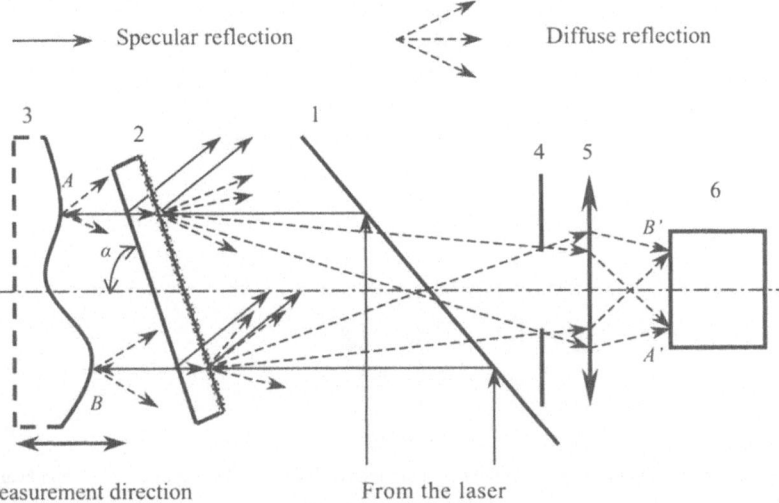

Figure 7. The principles of the ESPI with variable transmittance of the splitter: 1- the expanded laser beam, 2- beam splitter, 3- object, 4- aperture, 5- lens system, 6- CCD camera. [35]

The complicated two-dimensional interference fringe system during motion of the surface under the measurement was registered as series of two-dimensional matrices of intensities (Z=768x576 pixels) with 256 gray levels. The novelty of the data analysis in our case was a minimization of the area of analysis up to dimensions corresponding to the statistical linear dimensions of the single speckle. In this case, motion of an individual point (or small area of the surface under analysis) can be obtained analyzing time dependence just of part (nxn) of two-dimensional matrices. As a result of the measurement and chosen area of analysis, the new (nxn) matrices $\mathbf{A}_m$ are formed. The size of matrices (or minimum dimensions of analysis) is defined by the aperture number of the optical system and surface properties. The next step in the data analysis includes calculation of the correlation coefficients $M_{i,j}$ in between of subsequently registered matrices $\mathbf{A}_m$. The new matrix of correlation coefficients $\mathbf{M}$ is formed that is used to define the displacement of the object at any time.

The discussed method was applied to measure thermal strain kinetics in the Zr/Si system. Thin Zr films (thickness 7 μm) were deposited by magnetron sputtering on the crystalline silicon beam (12x4 mm). The deposition rate was 0.055 μm/s, the residual gas pressure 2-10 Pa. According to x-ray diffraction data the thin Zr films can be described as polycrystalline films.

In the strain measurements, the cantilever technique was applied, and time dependence of co-ordinates of three points A, B, and C (Fig. 8) was measured as a function of temperature. A thermostat including temperature control was employed in the experiment. In addition to the control of individual points, the motion of correlation areas was registered (classical ESPI approach) as well.

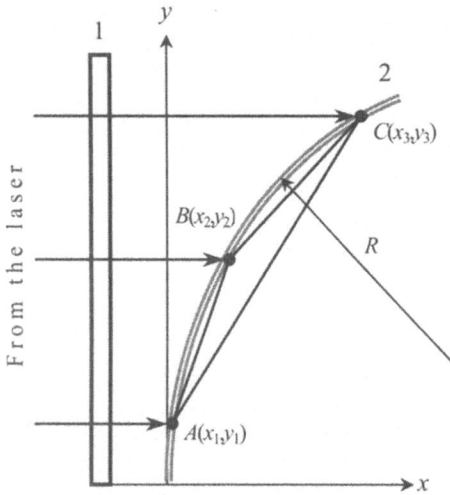

Figure 8. Schematic illustration of the curvature measurement 1– beam splitter, 2– silicon beam with the thin film

One can understand that due to thermal heating of the structure within a small temperature region intrinsic and external stress are constant values and residual stress changes due to thermal component σ_{th} only. Change of the thermal stress can be expressed as:

$$\Delta\sigma_{th} = E_f\left(\alpha_s - \alpha_f\right)\Delta T, \tag{6}$$

where E_f is thin film Young's modulus, α_s, α_f α_s are the thermal expansion coefficients of the substrate and thin film, $\Delta T = T_2 - T_1$, (T_1 is the initial temperature and T_2 is the final temperature of the sample). One can conclude as well that curvature variation of the film–substrate system related to the induced thermal stress according to the Stoney equation can be expressed as following:

$$\Delta \sigma_f = \Delta \sigma_{th} = \frac{E_s h_s^2}{6(1-v)h_f}\left(\frac{1}{R_2} - \frac{1}{R_1}\right) = \frac{E_s h_s^2}{6(1-v)h_f}\Delta\left(\frac{1}{R}\right) \qquad (7)$$

where E_s is Young's modulus of the substrate, h_s is the thickness of the substrate, v is Poisson's ratio of the substrate, R_2 and R_1 are the radii of substrate corresponding to the temperatures T_1 and T_2, and $\Delta\left(\frac{1}{R}\right)$ is the change of curvature corresponding to the temperature range ΔT.

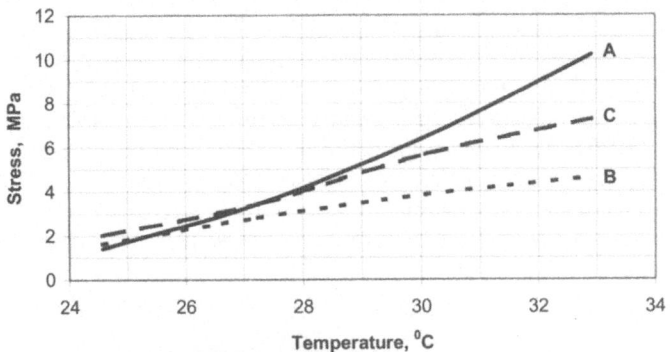

Figure 9. Dependence of the thermal stress versus the ΔT for different substrate temperatures during film deposition: A- 150 °C, B- 300 °C, C- 400 °C

The thermally induced residual stresses in thin Zr films as a function of ΔT are shown in Fig. 9. One can see a linear dependence of the thermal stress versus temperature (as it is implied by equation (6)) as well as its dependence on the substrate temperature during deposition. Measuring the thermally induced strain in the thin film-substrate system, one can determine the elastic properties of the thin film. I.e. knowing thickness and thermal expansion coefficient of the thin film, as well as Young's modulus and Poisson's ratio of the substrate, one can calculate the elastic modulus of thin film as:

$$E_f = \frac{1}{\alpha_s - \alpha_f}\frac{E_s d_s^2}{6(1-v)d_f}\frac{\Delta\left(\frac{1}{R}\right)}{\Delta T}. \qquad (8)$$

For example, choosing $\alpha_f = \sim 10^{-6}$ 1/K, one can calculate Young's modulus $E_f \approx 100$ GPa which is close to the value of the bulk material.

The given example illustrates the suitability of the optical method that allows to expand the variety of the materials under investigation as well as to perform analyses of the motion of individual point of the substrate. Application of the splitter with variable reflectance can be employed to measure strain kinetics of different surfaces (with diffuse and specular reflection). Minimization of the area available for the analysis enables to apply the ESPI for the micromechanical measurements i.e. to control motion of the systems having typical dimension less than 1 mm.

3.4 THE ESPI IN UNIAXIAL TENSILE TESTING

Thin films are particularly well suited for the systematic study of a central problem in materials science, which is the relation between the length scales of the microstructure and the physical properties. Important material properties that have to be characterized: elastic modulus, Poisson's ratio, fracture stress, yield stress can be obtained in the uniaxial tension testing. In principle, this procedure is analogous to conventional tensile testing methods of bulk materials; but, the fragility of the films and their extreme sensitivity to small flaws in the samples render uniaxial tensile testing a difficult task.

Speckle interferometry is a well-established optical technique for measuring in-plane displacements. In contrast to grid and moiré techniques, the speckle technique does not require the preparation of a grating on the specimen surface, as soon as its roughness is sufficient enough. In addition, ESPI method is very suitable for computerized specimen deformation measurement [36].

Fig.10 presents a general view of the system of piezo-actuated microtensile set up and ESPI that are mounted on a 25 cm x15 cm x2 cm stainless steel baseplate.

Figure 10. General view of the microtensile equipment combined with the ESPI

The moving plate (2) is supported by two rods (3) and a piezoelectric stack 'PZTech' (4) fixed on the base (1). A spring (5) ensures the stability of the moving plate on the three contact points. The moving plate assembly includes a miniature load cell SENSOTEC (9) and a specimen holder (8). The specimen (thin film, or freestanding film with a support holder) (6) is glued onto the fixed (7) and moving (8) grips.

The electronic speckle pattern interferometer is used to measure in-plane displacement of different points of sample gauge surface under tensile testing conditions A diode laser (wavelength 0.675 μm, output power 5 mW) is used for lightening (10) and the dual illumination method is used [10]. With this technique, the speckle pattern is produced by simultaneous illuminations of the specimen (6) with two laser beams, which are symmetric with respect to the observation direction. Here, the laser beam is split (11) and the two obtained beams are then oriented to the specimen using mirrors (12, 13). One of them is mounted on a moving support (14), which is used to modulate the phase of light. The resultant speckle pattern is focused by an objective (15) toward a CCD camera (16) having a resolution of 640x480 pixels. The digital speckle pattern images (frames) are sequentially stored for further analysis. An original program developed using the National Instruments LabView software controls the tensile deformation system. Monitoring of piezoelectric stack (via an amplifier) and phase modulation is achieved through a multifunctional interface. Signal from the load cell (LC) and attenuated amplifier signal are collected via the same interface.

Images of selected sample areas acquired at each step of PZ displacement are stored as matrices from which smaller sub-matrices (M) are extracted for further processing. Correlation coefficients are calculated between each sub-matrices, which allow for the determination of a temporal correlation matrix (C). Frequency and phase of the correlation fringes are then calculated by Fast Fourier Transform of each row of the (C) matrix, yielding the displacement of one point of the sample surface.

The tensile tests on foils of 10 □m thick aluminum of commercial purity have been performed. The gauge length of the specimens was 6 mm, *i. e.*, the distance between moving and fixed grips, with a gauge width of 1 mm. A typical stress-strain curve obtained by tensile testing of the above mentioned foils is presented in Fig. 11, for a nominal strain-rate of 8.3×10^{-6} s^{-1}. The early portion of the curve often exhibits a noticeable deviation from linear elastic behavior that can be attributed to some kind of specimen "unwrinkling". This part has been ignored for Young's modulus determination, which values range between 54 and 65 GPa. A direct comparison with Young's modulus values reported in the literature is not very easy since one to one correspondences in specimen size, chemical composition, etc, are never met. However, it must be noticed that a good agreement is obtained with values commonly accepted for aluminum thin foils, *i.e.* about 60 MPa (for a review [37]).

616

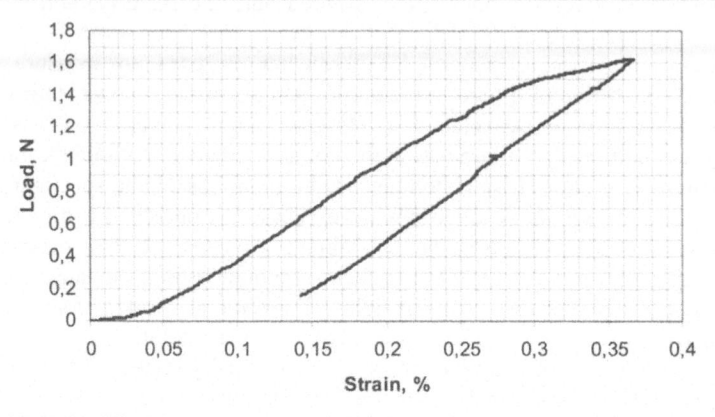

Figure11. The load - strain dependence for the aluminum foil measured with microtensile equipment and ESPI

One may also remark in Fig. 11 the absence of a well-defined yield stress. The gradual transition between elastic and plastic stages suggests that, like in bulk aluminum specimen, micro yielding occurs at low stresses in thin foils. A significant deviation from linear behavior is detectable at a stress level of nearly 130 MPa. It does not necessarily correspond to the actual transition stress between micro and macro plastic stages, but it can be used as a characteristic value of the flow stress. Indeed, determination of a 'proper' yield stress is not an easy task and usually requires additional information such as the strain dependence activation volume values, load-unload cycle which is not yet available.

4.Concluding Remarks

Physical principles of the conventional method for the internal stress measuring - the cantilever technique and electronic speckle pattern interferometry were discussed and related to the strain control in thin film – semiconductor substrate system. Stress kinetics of the thin film structure can be monitored in – situ allowing to control this process at the nucleation stage of the film. The main advantage of the electronic speckle pattern interferometry as compared to the classical interferometry and holographic methods is ability to measure strain of the real diffusive surfaces. Development of the electronic speckle pattern interferometry allows to apply it to the small size (hundreds of micrometers) interrupted physical systems (microelectromechanical devices, microstructures etc.) to monitor and control variations of geometrical dimensions of the different components. Development of the new analysis method is prospective for the new class of materials – freestanding films. Application of the newly developed optical method with the microtensile machine allows defining elastic properties of thin film and influence of different technological conditions during deposition.

5. Acknowledgements

This work was partially supported by the Lithuanian State and Studies Foundation

6. References

1. *Handbook of Microlithography, Micromachining, and Microfabrication*, vol.1: *Microlithography*, ed. P. Rai-Choudhury, SPIE Optical Engineering press, IEE, London UK, 1997.
2. Nix, W.D. (1989) Mechanical properties of thin films, *Metallurgical Transactions A,* , vol.20A, p.2217-2245.
3. Read, D.T., Dally, J.W. (1993) A new method for measuring the strength and ductility of thin *films J. Mater. Res.* , vol.8, p.1542-1549.
4. Poilane, C., Delobelle, P., Bornier, L., Mounaix, P., Melique, X., Lipens, D. (1999) Determination of the mechanical properties of thin polyimide films deposited on a Ga As substrate by bulging and indentation tests *Materials Science and Engineering*, vol.A262, p.101-106.
5. Read, D.T. (1998) Piezo –actuated microtensile test apparatus *Journal of Testing Evaluation, JTEVA,* , vol.26, No.3, p.255-259.
6. LaVan, D.A., Sharpe ,W.N. (1999) Tensile testing of microsamples *Experimental Mechanics*, vol.39, No.3, p.210-216.
7. Brotzen, F.R. (1994) Mechanical testing of thin films *International Materials Reviews,* , vol.39, No.1, p.24-45.
8. Tamulevicius, S (1998) Stress and strain in the vacuum deposited thin films, *Vacuum*, vol. 51No 2 p.127- 139.
9. Smidt, F.A. (1990) Use of ion beam assisted deposition to modify the microstructure and properties of thin films *International materials review*, vol.35, No2, p.61-128;
10. Hoffman, R.W, (1976) Mechanical properties of non-metallic thin films, in: *Physics of nonmetallic thin films*, NATO Summer School on Metallic and Nonmetallic thin films, Corsica 1974, Ed. By C.H.Dupuy, A.Cachard, Plenum press, p.273 – 353.
11. Harper, James M.E, Cuomo, Jerome J., Gambino, Richard J., Kaufman Harold R. (1985) Modification of thin film properties by ion bombardment during deposition *Nuclear Instruments and Methods in Physics Research* B7/8, , p.886-892;
12. Esinger, W. (1992) Ion sources for ion beam assisted thin film deposition, *Rev.Sci., Instrum.*, 63 (11), p.5217 – 5233.
13. Badawi, K.F., Goudeau, PhPacaud, J., Jaounen, C. (1993) X-Ray diffraction study of residual stress modification in Cu/W superlattices irradiated by light and heavy ions, *Nuclear Instruments and methods in Physics Reasearch*, B vol. 80/81, p.404-407.
14. Evans, H.E., Huntz, A-M., (1994) Methods of measuring oxidation growth stresses, *Materials at high temperatures*, vol. 12, No. 2-3, p. 111-117,.
15. Malhotra, S.G., Rek, Z.U., Yalisove, S.M., Bilello, J.C., (1997) Analysis of thin film stress measurement techniques, *Thin Solid Films*, vol. 301 (1-2), p. 45-54.
16. Noyan, I.C., Cohen, J.B. (1987) *Residual stress : Measurement by diffraction and interpretation*, Springer-Verlag, New York,
17. Ballard, B.L., Predecki, P.K. (1994) Stress-depth profiles in magnetron sputtered Mo films using grazing incidence x-ray diffraction, in: *Advances in x-ray analysis*, ed. Gilfrich, et al.,vol. 37, Plenum Press, New York, p.189 - 196,.
18. Wu, C.H., Weber, W.H., Poter, T.J. Tamor, M.A. (1993) Laser reflective interferometry for in-situ monitoring of diamond film growth by chemical vapor deposition , *Journal of Applied Physics*, , v;73, No6, p. 2977 – 2982.
19. Sternheim, M., W.van Gelder, Hartman, A.W. (1983)A laser interferometer system to monitor dry etching of patterned silicon, *J.Electrochem.Soc.: Solid State Science and Technology*, v.130, No 3, p.655-658.
20. Kempf, J., Nonnenmacher, M., Wagner H.H. (1993) Electron and Ion Beam Induced heating Effects in Solids by Laser Interferometry, *Appl.Phys.* A56, No4, p.385 – 390.
21. Kempf, J. (1982) Optical in situ sputter rate measurements during ion sputtering, *Surface and Interface Analysis*, vol. 4, No3, p.116-119.

618

22. Leusink, G.J., Oosterlahen, T. G., Janseun, A.M., Radelaar S. (1992) In situ sensitive measurement of stress in thin films, *Rev.Sci.Instrum.*, 63(5) p.3143- 3145.
23. Takeshi Aoki, Yasuo Nishikawa, Seichi Kato (1989) An improved optical lever technique for measuring film stress, *Japanese Journal of Applied Physics*, vol.28, No2, , p.299 –300.
24. Tamulevičius, S.,.Augulis, L Laukaitis G.(2001) Optical measurements of strain and stress in thin films in: *International Conference on Solid State Crystals 2000: Epilayers and Heterostructures in Optoelectronics and Semiconductor Technology Proceedings of SPIE* , J.Rutkowsky, J.Wenus, L.Kubiak, Editors, ISSN 0277-786X, The Society of Photo-Optical Instrumentation Engineers.,vol.4413 p.242- 247.
25. Tamulevičius, S. (2000) Optical measurements of strain and stress in thin films in: Proceedings of the International Conference, *"Thin Film Deposition of Oxide Multilayers. Industrial Scale Processing"*, Vilnius, Lithuania, 28-29 September 2000, eds.B.Vengalis, A.Abrutis, ISBN 9986-19-394-X, Vilnius University Press, p.108 –112.
26. Laukaitis, G., Tamulevičius, S., Valkonen, M.P. Laskela, M., Lindroos. S. (1999) Stress and surface studies of SILAR grown CdS thin films on (100)GaAs substrates *Thin Solid Films* v.355- 356, p.430- 431.
27. Laukaitis, G., Lindroos, S., Tamulevicius, S., Leskela, M. (2001) Stress and morphological development of CdS and ZnS thin films during the SILAR growth on (100)GaAs, *Applied Surface Science* v.185, p.134 – 139.
28. *Digital Speckle Pattern Interferomertry and Related Techniques*, ed. P. K. Rastogi, John Wiley & Sons Ltd, Chichester, 2001.
29. R. Jones and C. Wykes, (1983) *Holographic and Speckle Interferometry,* Cambrige University Press.
30. Bonnotte, E., Delobelle, P., Bornier, L., Trolard, B., Tribillon G. Two interferometric methods for the mechanical characterization of thin films by bulging tests. *Journal of Materials Research*, 1997, vol.12/6, p.2234-2248.
31. Read, D. T. Young's modulus of thin films by speckle interferometry. *Measurement Science* and Technology, 1998, vol.9, p.676-685.
32. Tamulevicius, S., Augulis, L., Augulis, R., Zabarskas,, V. Levinskas, R., Poskas, B., (2001) Thermal strain measurements in graphite using electronic speckle pattern interferometry *IAEA Technical Committee Meeting on "Nuclear Graphite Waste management"*, held 18-20 October 1999 in Manchester, UK, International Atomic Energy Agency, Vienna Austria TCM-Manchester 99, p. 125-133. (http://www.iaca.or/at/inis/aws/htgr/abstracts/abst_manchester-10.html)
33. Tamulevičius, S., Augulis, L., Augulis. R., Zabarskas, V. (1999) Thermal strain measurements using electronic speckle pattern interferometry *Medžiagotyra (Materials Science)* ISSN 1392-1320, Kaunas:Technologija. nr.4(11), p.20-26
34. Lokberg, O. L., Seeberg, B. E., Vestli, K. (1997) Microscopic video speckle interferometry. *Optics and Lasers in Engineering*, , vol.26, p.313-330.
35. Tamulevičius, S., Augulis, L., Laukaitis, G., Žadvydas, M. (2002) Electronic Speckle Pattern interferometry for micromechanical measurements, *Advanced Engineering Materials*, v4, No 08 p. 546-550
36. Augulis, L., Tamulevičius, S., Puodžiukynas, L., Augulis, R., (2000) Optimization of the correlation method in electronic speckle pattern interferometry, *Lithuanian Journal of Physics* ISSN- 1392-1932, Vilnius : Fisica., vol.40 Nr6 p.425-430.
37. Augulis, L., Tamulevicius, S., Bonneville, J., Templier, C., Goudeau,Ph. (2002) Testing of the Mechanical Properties of Thin Films: a Review, *Materials Science (Medžiagotyra)* ISSN 1392-1320, Kaunas:Technologija., vol.8, No1, p.3-9.

.

OPTICALLY ACTIVE SILICON NANOSTRUCTURES PREPARED FROM IMPLANTED Si BY ANNEALING AT HIGH HYDROSTATIC PRESSURE

A. MISIUK[1] AND I.E. TYSCHENKO[2]
[1]Institute of Electron Technology, Al. Lotnikow 32/46, 02-668 Warsaw, Poland
[2]Institute of Semiconductor Physics, RAS, Pr. Lavrentieva 13, 630090 Novosibirsk, Russia

Abstract. Nano-clusters composed mostly of implanted atoms as well as structural defects are created in single crystalline silicon implanted with hydrogen (Si:H), helium (Si:He) and oxygen (Si:O) as well as in silicon dioxide implanted with silicon (SiO_2:Si) or germanium (SiO_2:Ge) and annealed at up to 1400 K under high hydrostatic pressure, up to 1.5 GPa. Such materials can exhibit photoluminescence, PL, at infrared (Si:H, Si:He, Si:O), visible (Si:H, Si:He, SiO_2:Si,Ge) or ultraviolet (SiO_2:Si,Ge) regions. Recent works reporting PL from high temperature – pressure treated Si:H, Si:He, Si:O and SiO_2:Si,Ge are reviewed and new results presented.

Key words: silicon, implantation, Si:H, Si:He, Si:O, SiO_2:Si,Ge, annealing, high pressure, nanostucture, photoluminescence

1. Introduction

Silicon is leading semiconductor in the microelectronics used for production of integrated circuits, ICs. Current progress in communication technology results in the increased demands for optoelectronic components integrated with ICs, desirably also produced on the base of silicon. However, because of its indirect band gap, Si is a poor light emitter, unsuitable for typical optoelectronic applications.

It is a reason for numerous attempts to design the silicon based optoelectronic materials compatible with current Si based ICs processing. After the discovery of efficient visible photoluminescence (PL) from porous silicon, pSi [1], much interest has been initially devoted to this material.

It is generally accepted that optical activity of pSi in the visible light region arises from the quantum confinement effect in the presence of an enlarged band gap of nanosized Si while the role of other structural defects can not be neglected [2, 3]. However, many problems connected with the instability of pSi have been encountered in its practical application. It is the reason why other ways to produce optically active Si (Ge) based materials with silicon (germanium) clusters / particles of nano dimensions are considered.

PL in the visible light region has been reported for the RTA treated hydrogen implanted silicon on insulator structures subjected to hydrogen implantation [4].

T. Tsakalakos, et al. (eds.) pgs. 619 - 638
Nanostructures: Synthesis, Functional Properties, and Applications;
© *2003 Kluwer Academic Publishers.*

Dislocations and nanosized oxygen clusters can be created in bulk Si and in silicon implanted with oxygen by their mechanical deformation or appropriate annealing [5]. It has been suggested [6] that dislocation – related photoluminescence at infrared light region (IR) can be exploited in Si based light emitting diodes.

Besides mentioned, the following methods to produce the optically active structures were proposed: crystallization of amorphous Si by rapid thermal annealing (RTA), spark processing of silicon, chemical vapor deposition, laser pyrolysis of silane, thermal and electron beam evaporation, laser ablation (laser vaporization), radio - frequency magnetron and ion beam sputtering [7] as well as ion beam synthesis of Si/Ge nanocrystals and nanoclusters (SiO_2:Si,Ge structures [8]). Just the last approach to produce nanosized structures in the form of Si or Ge nanocrystallites embedded in a dielectric medium, for example in SiO_2 layer grown on Si substrate, is now considered as the most promising one for future technological applications. To produce the nanosized SiO_2:Si,Ge structure, SiO_2 needs to be implanted with Si or Ge and subjected to appropriate thermal treatment procedure.

While most attention was paid to the temperature and duration of the thermal treatment of SiO_2:Si,Ge, until recently not much was known on the effect of external (hydrostatic) pressure during the treatment on resulting properties of the treated material. However, just enhanced hydrostatic pressure (HP) of gas ambient at annealing (HT – HP treatment) exerts pronounced effect on the properties of silicon, oxygen and of other components in the Si – O and similar systems. It concerns the diffusivity of components [9], nucleation of (optically active) clusters [10] as well as of the shear stress at the cluster / defect / matrix boundary [11]. Just the shear stress at the defect / matrix boundary is responsible for creation of dislocations and defects in such material systems [12]. Typically more nanosized clusters and defects are created under HP [13] and they are more stable in respect of changing preparation and annealing conditions.

It is the reason why the effect of HT – HP on photoluminescence (PL) of silicon implanted with hydrogen (Si:H), helium (Si:He), oxygen (Si:O) and of silicon dioxide implanted with oxygen (SiO_2:Si) or germanium (SiO_2:Ge) deserves attention. Hydrogen / helium implantation into Si has become recently a topic of remarkable interest, also because of potentials of implantation - induced nanosized platelets and microcavities for creation of gettering - active areas and for Si layer splitting, known as the Smart - Cut process [14, 15]. The HT – HP treatment was proven to affect markedly photoluminescence of such nanostructures [16].

Oxygen implantation in silicon is widely applied for the SIMOX process (creation of a buried Si dioxide layer in the Si substrate by O^+ implantation followed by high temperature annealing). The HT – HP treatment affects strongly creation of SiO_2 nanoclusters in such systems as well as PL from Si:O [17].

In what follows, the recently published papers concerning PL from high temperature – pressure treated Si:H, Si:He, Si:O and SiO_2:Si,Ge are reviewed and some new results presented.

2. Experimental Details

The Si:H samples were prepared by H^+ or H_2^+ implantation (hydrogen doses, $D = 7.5 -$ 300 x 10^{15} cm^{-2}, energy, $E = 20 - 135$ keV, H atom projected range, $R_p = 0.1 - 0.65$ μm) into the Floating zone (Fz) or Czochralski (Cz) grown 001 or 111 oriented silicon wafers of about 0.6 mm thickness with the interstitial oxygen concentration, c_o, up to $1 \times 10^{18} cm^{-3}$ (determined by Fourier Transform Infrared Spectroscopy, FTIR). The Si:He samples were prepared by He^+ implantation ($D = 5 - 50 \times 10^{15} cm^{-2}$, $E = 150 - 300$ keV, He atom projected range, $R_p = 0.9 - 1.35$ μm) into 001 oriented Fz-Si or Cz-Si.

To prepare the Si:O samples, the 001 or 111 oriented Fz-Si or Cz-Si wafers with c_o $\leq 2 \times 10^{16} cm^{-3}$ and $8 \times 10^{17} cm^{-3}$, respectively, were implanted with O^+ ($D = 1 - 20 \times 10^{17} cm^{-2}$, $E = 50 - 200$ keV, $R_p = 0.12 - 0.65$ μm). To prepare the SiO_2:Si,Ge structures, the n-type, 5-10 Ω cm, 001 oriented Si wafers were oxidized at 1270K in a wet oxygen atmosphere to grow a 500 nm thick surface SiO_2 layer. The oxide layer was implanted with Si^+ or Ge^+ ions.

Si^+ ions were implanted into SiO_2, at first at $E = 100$ keV and then at $E = 200$ keV using, respectively, silicon doses $D = 1.2 \times 10^{16}$ cm^{-2} and 2×10^{16} cm^{-2} (LD – low dose), 1.8×10^{16} cm^{-2} and 3×10^{16} cm^{-2} (MD – medium dose), 2.3×10^{16} cm^{-2} and 4.4×10^{16} cm^{-2} (HD – high dose) or 3.9×10^{16} cm^{-2} and 6.3×10^{16} cm^{-2} (VHD – very high dose). In the case of implanted Ge^+ ions, the double implantation at $E = 450$ keV and then at 230 keV to $D = 1.1 \times 10^{15}$ cm^{-2} and 6.6×10^{14} cm^{-2} (0.1 at. % Ge), 6.6×10^{15} cm^{-2} and 4.0×10^{15} cm^{-2} (0.67 at. % Ge), and 3.0×10^{16} cm^{-2} and 1.8×10^{16} cm^{-2} (3 at. % Ge) was carried out. During implantation the ion current density was kept at 0.5 - 1 μA cm^{-2} and the substrate temperature was maintained at 120 - 130 K by means of a LN_2 - cooled stage. After implantation the Si:H, Si:He, Si:O and SiO_2:Si,Ge samples were cleaved into pieces of about 8×12 mm^2 dimension and subjected to annealing at 10^5 Pa (atmospheric pressure) or to the HT – HP treatment [5] in argon atmosphere at up to 1400 K, 1.5 GPa for up to 5 h.

The concentration profiles of implanted atoms were determined by Secondary Ion Mass Spectrometry (SIMS) or Rutherford Back Scattering (RBS, back scattering of 1.5 MeV He^+ ions), while the sample structure was viewed by Transmission Electron Microscopy (TEM).

The 265 nm line of a YAG laser or 337 nm line of a N_2 laser were employed for PL excitation at room temperature (PL determination in the visible / ultraviolet light region) while PL at IR of the HT – HP treated Si:He and Si:O samples was measured at 10 K using excitation with the 488 nm line of an Ar laser.

Raman spectra of the SiO_2:Si,Ge samples were measured at room temperature in the back scattering geometry with a DFS-52 spectrometer (excitation with the 457 nm, 476 nm, 488 nm, 496 nm and 514 nm lines of an Ar laser).

3. Optical Activity of HT – HP Treated Si:H, Si:He, Si:O and SiO₂:Si,Ge

3. 1. EFFECT OF HT – HP TREATMENT ON PHOTOLUMINESCENCE OF Si:H

During annealing of Si:H at atmospheric pressure hydrogen atoms segregate near the projected ion range, R_p. Hydrogen - filled defects and defect clusters are then formed (Fig. 1). The pressure in the H_2 - filled platelets, bubbles and microcavities in annealed Si:H can reach a Giga - Pascal range [18]. Oxygen gettering [19] in the damaged layer

622

was detected even for the as implanted oxygen - containing Cz-Si:H samples for temperature of substrate during implantation $\geq$ 570 K (Fig. 2A).

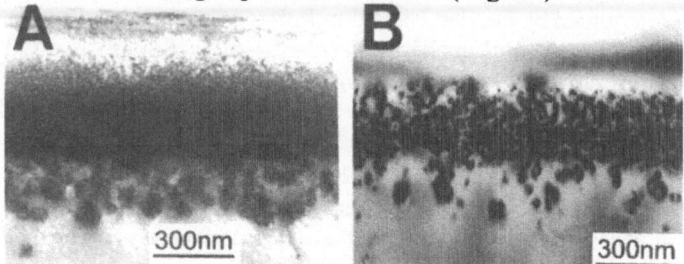

Figure 1. TEM image of Fz-Si:H ($D = 6 \times 10^{16}$ cm^{-2}, 135 keV). A: as - implanted; B: annealed at 920 K – 10^5 Pa for 10 h.

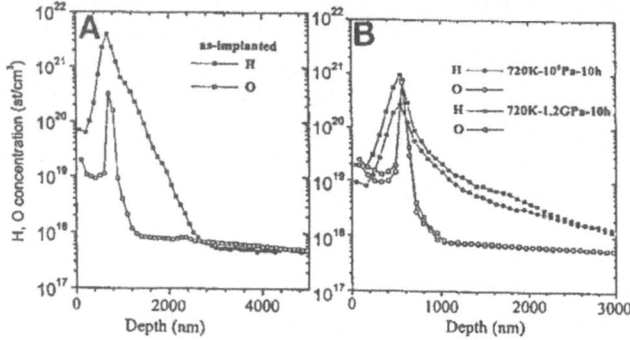

Figure 2. Depth profiles of hydrogen and oxygen in as – implanted and annealed / HT – HP treated Cz:Si:H samples (oxygen concentration in Cz-Si, $c_o = 8 \times 10^{17}$ cm^{-3}, H_2^+ implantation done at 650 K, $D = 6 \times 10^{16}$ cm^{-2}, 135 keV, R_p = 0.5 μm). A: as - implanted; B: annealed at 10^5 Pa or HT – HP treated at 720 K for 10 h.

Annealed or HT – HP treated Si:H may be considered as an example of the solid foam (porous) - like Si structure (Fig. 1).The total volume of hydrogen filled, TEM – detectable platelets in Si:H was reported as constant, of about 8×10^{-8} cm^3/cm^2, at least for the samples implanted with $D = 3 \times 10^{16}$ cm^{-2} and annealed at $\leq$ 1170 K [20]. Assuming that valid for Si:H prepared by implantation with other H$^+$ doses, the relative volume, V_r, of the platelet - like defects in the Si:H samples can be estimated from:

$$V_r = \pi d_p <r_p^2> f / t \qquad (1)$$

where: d_p - density, r_p - radius, f – thickness of platelets and t - thickness of the disturbed layer [20].

For $t \approx 100$ nm and for the uniform distribution of hydrogen – filled defects, the V_r value would be equal to about 1% for Si:H implanted with $D = 3 \times 10^{16}$ cm^{-2} and $\approx$ 10 % for that with $D = 3 \times 10^{17}$ cm^{-2}. However, not all defects are TEM detectable. Hydrogen atoms occupied at least 10 % of the disturbed layer volume near R_p in the as implanted sample with $D = 6 \times 10^{16}$ cm^{-2} as it follows from comparison of the peak hydrogen concentration (of about 5×10^{21} cm^{-3} - Fig. 2A) with the concentration of Si atoms in the host lattice (5×10^{22} cm^{-3}). So the porosity of that sample would correspond at least to that 10 %, if the smallest H – filled microcavities would be also taken into account.

The hydrogen – containing layer in the annealed or HT – HP treated Si:H sample can be considered as the foam – like structure composed of hydrogen – filled platelets and microcavities of different dimensions, as well as of other defects. The as – implanted (if implantation done at $\geq$ 520 K) and HT – HP treated Cz-Si:H samples contain also oxygen atoms, gettered within the implantation – disturbed areas (Fig. 2).

Enhanced hydrostatic pressure during annealing (HT - HP treatment) of Si:H results in retarded hydrogen out - diffusion (Fig. 2B) and creation of more numerous defects in the sub - surface Si layer [21]; the implantation - damaged area in the Si:H samples treated at 720 K - 1.2 GPa for 1 h indicate the presence of microdefects also below the defect area which is more extended.

PL spectra at IR of the HT – HP treated Si:H samples ($D = 4 \times 10^{16}$ cm^{-2}, 45 keV) HT – HP treated at 720 K - HP for 10 h and at 870 K - HP for 1 h indicate the presence of a broad PL band at about 0.79 eV, of unknown origin [23]. The wide peak at about 0.92 - 1.01 eV may also be related to dislocations (superposition of the D3 and D4 lines) or to defects induced by stresses around the hydrogen - related defects [23]. Annealing at 10^5 Pa of Si:H and of the hydrogen implanted silicon – on - insulator structures ($D = 1 - 3 \times 10^{17}$ cm^{-2}, 24 keV) at 470 - 970 K has been reported to result in a formation of nanostructure and in PL peaking at about 420 nm [4].

The as - implanted Si:H samples ($D = 7.5 - 30 \times 10^{15}$ cm^{-2}, 20 – 45 keV) did not exhibit visible PL (Fig. 3); the samples implanted with $D = 1 - 3 \times 10^{17}$ cm^{-2}, 24 keV indicated a very weak PL at 400 – 700 nm [4]. The treatments at 720 K - 1.2 GPa for 2 / 10 h of the low dose implanted Si:H sample resulted in a weak PL band at about 400 nm (corresponding to 3.1 eV – Fig. 3), while no PL was detected for the sample annealed at 720 K - 10^5 Pa for 10 h.

The PL spectra of the annealed and HT – HP treated Si:H samples ($D = 4 \times 10^{16}$ cm^{-2}, $E = 135$ keV) are presented in Fig. 4. The samples show distinct PL at about 440 nm, of the highest intensity after the treatment at 920 K - 1.2 GPa and 1470 K - 1.5 GPa.

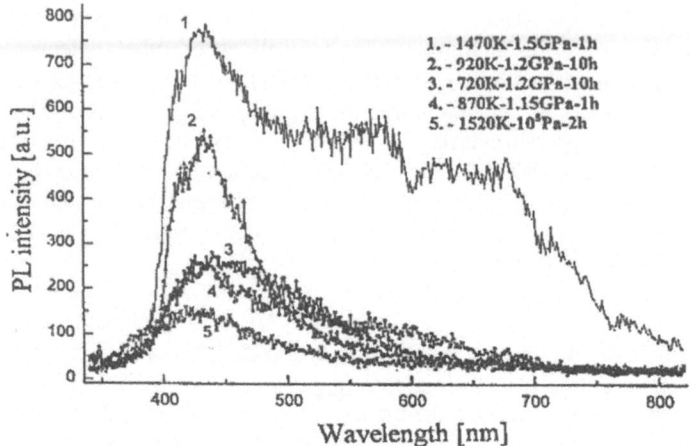

Figure 3. PL spectra of as-implanted, annealed and HT – HP treated Cz-Si:H samples (D = 7.5x10^{15}cm^{-2}, E = 20 keV). Samples were prepared by H$^+$ implantation through 200 nm thick SiO$_2$ layer; R_p = 0.2 µm. PL measurement at 295 K, excitation by YAG laser, λ = 265 nm. Annealing / treatment parameters are indicated.

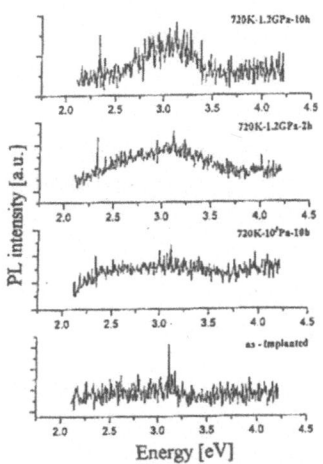

Figure 4. PL spectra of annealed and HT – HP treated Cz-Si:H (D = 4x10^{16}cm^{-2}, E = 135 keV). H$_2^+$ was implanted through 200 nm thick SiO$_2$ layer and sample preannealed at 720 K – 10^5 Pa for 15 min. PL measurement at 295 K, excitation by N$_2$ laser, λ = 337 nm. Annealing / treatment parameters are indicated.

Hydrogen - filled cavities are created near R_p in the Si:H samples during implantation and in effect of subsequent annealing / HT – HP treatment. The TEM – detectable nanometer sized (6 - 12 nm) platelets and bubbles are subjected to evolution with HT, HP and treatment time [4]. Typically more numerous but smaller defects are created under HP.

The distinct visible PL band peaking at 400 - 440 nm was detected for the Cz-Si:H samples, of the larger intensity for those prepared with higher implanted hydrogen doses and HT - HP treated at 720 – 1470 K. No marked shift of PL with changing treatment conditions was observed. Just the optically active Cz-Si:H samples contained hydrogen and oxygen atoms near R_p, in the concentrations, respectively, up to 10 at. % and 1 at. % (Fig. 2). Accounting for the presence of hydrogen and oxygen near R_p, it is reasonable to suppose that visible PL from the Cz-Si:H samples is related to the HT – HP induced creation of the nanometer – sized Si crystallites as well as to some other optically active chemical compounds, containing hydrogen and oxygen. The Si nanocrystallites, H- and O- containing compounds and non – radiative recombination centres (also produced during implantation and treatment) are most probably responsible for the wavelength position and intensity of PL detected for the HT – HP treated Si:H samples.

3. 2. EFFECT OF HT – HP TREATMENT ON PHOTOLUMINESCENCE OF Si:He

During annealing of Si:He at atmospheric pressure, similarly as in the case of the hydrogen - containing Si:H samples, He atoms segregate in gas - vacancy complexes and form bubbles just after implantation, depending on implantation parameters (first of all on ion dose, D, and energy, E). Besides creating voids or bubbles, typically of up to a few tens of nanometer dimension, numerous point and extended defects are formed in Si:He at annealing. More prolonged annealing can result in complete outdiffusion of helium from Si:He. Enhanced hydrostatic pressure of ambient atmosphere at annealing of Si:He results in decreased dimension of post - implantation defects, retarded He outdiffusion [21] and in gettering of oxygen in the case of He implantation into oxygen containing silicon, Cz-Si [19].

Annealing of Si:He under 10^5 Pa results in detectable PL at IR, at about 0.80 eV (dislocation – related PL or PL related to specific structural defects [23]), at 0.935 eV (supposedly originating from vacancies filled with helium), at 1.011 eV and 1.019 eV (attributed, respectively, to divacancies filled with He atoms and to divacancies [24]). TEM images of the HT - HP treated Si:He samples are presented in Fig 5. The Si:He sample treated at 720 K - 1.1 GPa for 10 h indicates the presence of numerous very small voids (of about 5 nm dimension) within the wide He - containing disturbed layer (Fig. 5A) while the treatment at 870 K - 1.1 GPa results in a creation of voids while the sample structure is even more disordered (Fig. 5B). The treatment of Cz-Si:He at $\geq$ 870 K - HP resulted in an accumulation of oxygen in the damaged buried layer [19].

PL spectra at IR of the Fz-Si:He samples annealed or HT – HP treated at 720 K and 870 K are presented in Fig. 6. All spectra indicate the presence of PL peak at about 0.88 eV, probably corresponding to the dislocation - related D2 line. The PL line at about 0.94 nm (attributable to helium filled vacancy [24]) was of the highest intensity for the Si:He sample treated at 870 K - 1.15 GPa for 1 h.

The treatment of Si:He at 870 K - 1.15 GPa for 10 h resulted in a marked decrease of the intensity of PL at 0.88 eV, while the PL line at 0.79 eV reached the high intensity. It is reasonable to assume that the PL line at 0.79 eV is related to the presence of specific point defects [24]. Still, similarly as in the case of earlier discussed effects of the HT - HP treatment on the Si:H samples, the origin of PL in the Si:He samples in the IR region needs to be clarified.

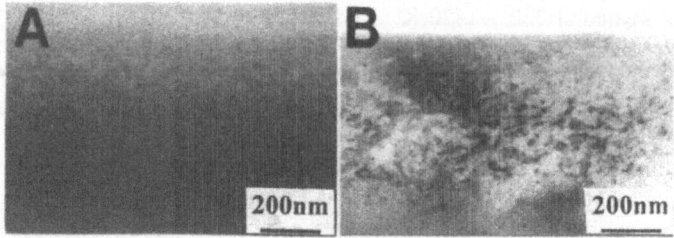

Figure 5. TEM images of Cz-Si:He samples (He$^+$ dose 5×10^{16}cm^{-2}, $E = 150$ keV, $R_p = 0.98$ μm), HT – HP treated under 1.1 GPa for 10 h at 720 K (A) and at 870 K (B).

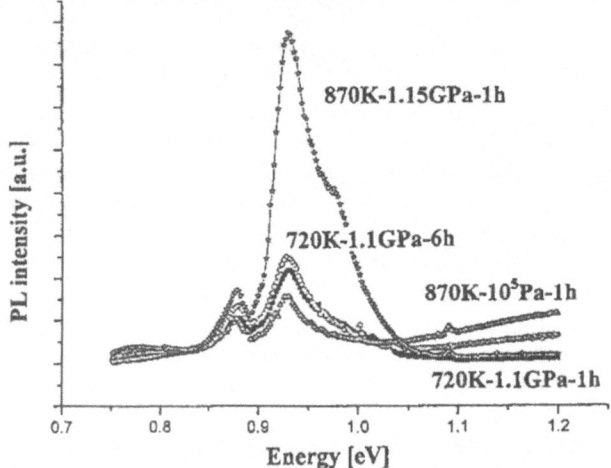

Figure 6. PL spectra at IR of annealed and HT – HP treated Fz-Si:He samples ($D = 5 \times 10^{16}$cm^{-2}, $E = 150$ keV, $R_p = 0.98$ μm). PL measurement at 10 K, excitation by λ = 488 nm. Annealing / treatment parameters are indicated.

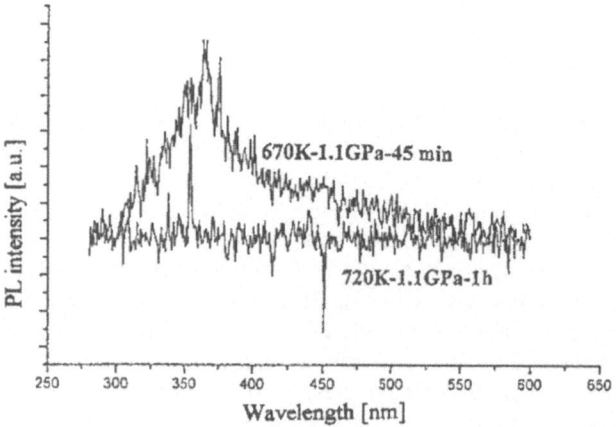

Figure 7. PL spectra at visible light region of HT – HP treated Fz-Si:He samples ($D = 2 \times 10^{16}$cm^{-2}, $E = 150$ keV, $R_p = 0.98$ μm). PL measurement at 295 K, excitation by YAG laser, λ = 265 nm. Treatment parameters are indicated.

Ultraviolet PL peaking at about 360 nm was detected for the Si:He sample (D = $2x10^{16}cm^{-2}$, E = 150 keV) treated at 670 K – 1.1 GPa for 45 min. (Fig. 7) as well as for the Si:He samples implanted with the higher helium dose (D = $5x10^{16}cm^{-2}$, E = 150 keV) and treated at 720 K - 1.1 GPa for 10 h. No PL was observed for the similar samples but treated at higher temperatures (for PL excited with λ = 265 nm and measurement done at 295 K). The PL band at 360 nm seems to be related just to the presence of nanometer sized Si crystallites in the disturbed layer, produced in Si:He by implantation and the subsequent HT - HP treatment. The hydrogen - and oxygen - related species (responsible for visible PL in the HT - HP treated Si:H samples) could not be created in the disturbed buried layer (no hydrogen and oxygen atoms were present in remarkable quantities in floating zone grown Fz-Si). The HT - HP induced effects in Si:He are related, at least partially, to retarded outdiffusion of helium at HP. Visible PL, resembling that of porous silicon, can be considered as an evidence of creation in Si:He of optically active nanometer - sized structural features. Still the origin of that PL in HT – HP treated Si:He demands further clarification.

3. 3. EFFECT OF HT – HP TREATMENT ON PHOTOLUMINESCENCE OF Si:O

Oxygen implantation into single crystalline silicon with subsequent annealing is widely used to create insulating SiO_2 precipitates or continuous layer embedded in Si (SIMOX technology). To produce the SIMOX structure, oxygen implanted silicon (Si:O) is subjected to high temperature (HT) treatment, at up to 1670 K. Depending on implanted oxygen dose and energy as well as on the annealing parameters, dispersed SiO_2 precipitates (in fact of the SiO_{2-x} composition) or continuous SiO_2 layer are created. Individual SiO_{2-x} clusters and precipitates are produced for $D \leq 2x10^{17}$ cm^{-2}. The Si:O structure with continuous buried SiO_2 layer is formed by implanting Si with a high oxygen dose, $D \geq 1x10^{18}$ cm^{-2}.

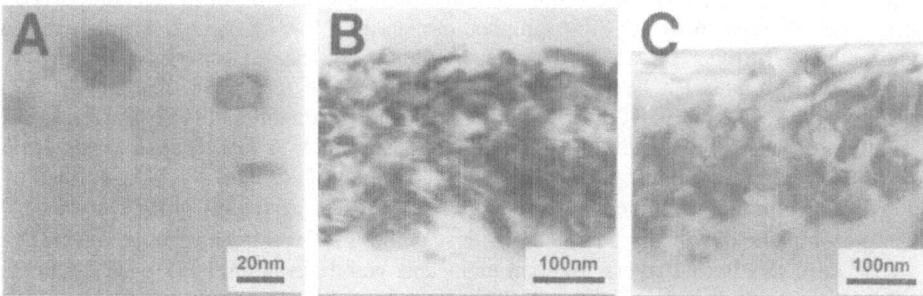

Figure 8. TEM images of annealed / HT – HP treated Si:O samples. A: Cz-Si:O (D = $1x10^{16}cm^{-2}$, 200 keV, R_p = 0.4 μm), treated at 1400 K – 1.2 GPa for 5 h; B: Cz-Si:O (D = $1x10^{17}cm^{-2}$, 200 keV, R_p = 0.4 μm), annealed at 1400 K – 10^5 Pa for 5 h; C: as B but treated at 1550 K – 1.5 GPa for 10 min.

Creation of the SiO_{2-x} clusters and precipitates in Si:O is concomitant with stress, both at the implantation stage and during subsequent treatment, because the SiO_2 volume is about twice that of Si, and their thermal expansion coefficients are much different.

External stress created by enhanced hydrostatic pressure (HP) of gas ambient at annealing (HT - HP treatment) can exerts dramatic effect on Si:O [25, 26]. It is possible to tune the strain and so the creation of dislocations and of other structural defects at the Si / SiO$_{2-x}$ boundary by the HT – HP treatment because dislocations at the SiO$_{2-x}$ / Si boundary are created only if the misfit at the precipitate / matrix boundary would reach the critical value [27].

The misfit, ε, and so the shear stress at the SiO$_{2-x}$ precipitate / Si matrix boundary are dependent on temperature, HT, and hydrostatic pressure, HP, during the treatment, in accordance with:

$$\varepsilon = \varepsilon_0 + \frac{K_{SiO2-x}}{3K_{SiO2-x} + 4G_{Si}} [\Delta HT (\beta_{SiO2-x} - \beta_{Si}) + HP (1/K_{Si} - 1/K_{SiO2-x})] \qquad (2)$$

where: ε_0 - initial misfit at the SiO$_{2-x}$ precipitate / Si matrix boundary at 295 K, 10^5 Pa; β_{Si} and β_{SiO2-x} - coefficients of volume thermal expansion; K_{Si} and K_{SiO2-x} - bulk moduli; G_{Si} - shear modulus (the bottom indexes denote the respective material); $\Delta HT = T_{exp.} -$ 295 K.

The shear stress at the SiO$_{2-x}$ / Si boundary decreases typically with HT and HP and so less or even no dislocations are created at the appropriately chosen HT – HP conditions (at 298 K: $K_{Si} = 98$ GPa; $K_{SiO2} = 40$ GPa; $\beta_{Si} \approx 1.3 \times 10^{-5}$ K^{-1}; $\beta_{SiO2} = 0.16 \times 10^{-5}$ K^{-1}; $G_{Si} = 68$ GPa) [27].

TEM images of the Si:O samples ($D \leq 1 \times 10^{17}$ cm^{-2}), HT - HP treated at ≥ 1400 K are presented in Fig 8. Creation of dislocations was strongly suppressed for the treatment temperatures ≥ 1400 K. The treatment of the Si:O sample with $D = 1 \times 10^{16}$ cm^{-2} at 1400 K – 1.2 GPa for 5 h resulted in creation of individual SiO$_2$ platelets of dimensions not exceeding 20 nm (Fig. 8A); no dislocations were detected. The treatment of the higher dose implanted sample ($D = 1 \times 10^{17}$ cm^{-2}) at 1550 K – 1.5 GPa for 10 min. also did not produce dislocations (Fig. 8C) while annealing of that sample at 1400 K – 10^5 Pa resulted in creation of numerous dislocations and other defects (Fig. 8B). The PL spectra of the Si:O samples, prepared by low oxygen dose implantation and subjected to the HT – HP treatment at 1400 K, indicate that the intensities of the dislocation – related PL lines (e.g. of the D1 dislocation related PL line at 0.81 eV) decrease with HP (Fig. 9). It means that the HT – HP treatment results in suppression of creation of dislocations in Si:O, as compared to the effect of annealing at atmospheric pressure (the highest intensity of the dislocation related D1 PL line).

Quasi - continuous embedded SiO$_2$ layers are formed in the HT – HP treated Si:O samples implanted with $D = 3.5 - 6 \times 10^{17}$ cm^{-2} while, for the samples prepared with $D \geq 1 \times 10^{18}$ cm^{-2}, that buried layers are continuous and well defined. The HT – HP treatment of the samples with $D > 1 \times 10^{17}$ cm^{-2} did not suppress fully creation of dislocations but the dislocation density remains to low and so the intensity of the dislocation – related PL has been reported to be almost negligible [17].

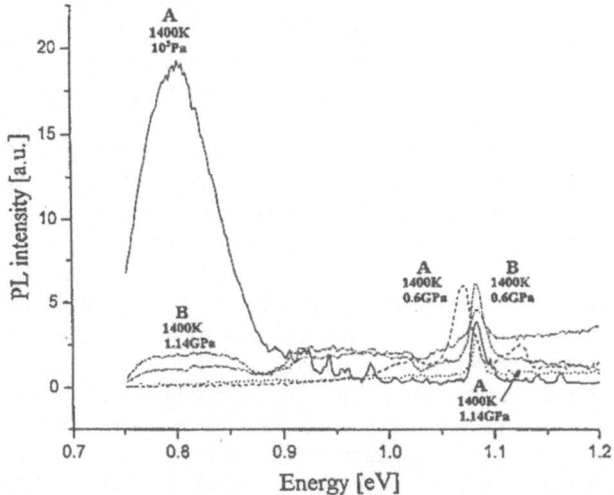

Figure 9. PL spectra of Cz-Si:O samples annealed / HT – HP treated for 5 h, A ($D = 1 \times 10^{16}$cm^{-2}, 200 keV, R_p = 0.4 μm) and B ($D = 1 \times 10^{17}$cm^{-2}, 200 keV, R_p = 0.4 μm). Treatment conditions are indicated.

It is interesting to note that just such Si:O samples, prepared by implantation with high ($D \geq 3.5 \times 10^{17}$ cm^{-2}) oxygen dose but at comparatively low energy ($E \leq 50$ keV) indicate, after the specific HT – HP treatments, PL in the visible light region. For example, the Si:O samples, prepared by oxygen implantation with $D = 2 \times 10^{18}$ cm^{-2}, 50 keV indicated visible PL at 3.5 eV after the treatment at 1400 K – 1.1 GPa for 5 h. Most probably this effect is related to oxygen deficient centres, ODC [28] created near the Si surface in the shallow oxygen implanted silicon samples.

Similar explanation can be suggested to explain visible PL detected in the HT – HP treated bulk silicon [29].

3. 4. EFFECT OF HT – HP TREATMENT ON PHOTOLUMINESCENCE OF Si:O₂:Si,Ge

Photoluminescence of the SiO$_2$:Si and SiO$_2$:Ge structures is related to the presence of Si or Ge nanocrystals dispersed in the amorphous SiO$_2$ matrix. Contrary to the case of Si with forbidden gap equal to about 1.1 eV, silicon dioxide is transparent for visible light because of its much wider (≥ 4 eV) forbidden gap.

The technique of ion beam synthesis of the SiO$_2$:Si and of SiO$_2$:Ge structures involves typically two stages: high dose (about 10^{16} - 10^{17} cm^{-2}) implantation of Si$^+$ or Ge$^+$ into the SiO$_2$ dielectric matrix and subsequent annealing of the implanted samples. Just ion implantation is perfectly compatible with the current Si based IC technology because of possibility to place a controlled number of implanted ions to a definite sample depth.

The parameters (temperature, time, ambient gas) of subsequent annealing play also an important role in the structural and light emitting properties of the SiO$_2$:Si and SiO$_2$:Ge nanostructures. Also enhanced HP of ambient gas during annealing of SiO$_2$:Si and SiO$_2$:Ge was proven to be important for formation of the light emitting nanocrystals and nanoclusters [4, 13]. This influence is related to changed diffusivity of the sample components under external stress, of crucial importance if the formation of nanocrystals is limited by diffusion of intrinsic defects (e. g. vacancies and interstitial atoms) [30].

PL properties of the SiO$_2$ films implanted with Si$^+$ ions were investigated as a function of hydrostatic pressure applied during low temperature (at 670 K and 720 K) and high temperature ($\approx$1400 K) anneals [31, 32].

In the case of VHD implantation, the as implanted SiO$_2$:Si samples exhibit two PL peaks at about 3.44 eV (ultraviolet) and 2.70 eV (blue). PL of high intensity was also seen at about 2.07 eV (red). Annealing at 670 K – 10^5 Pa for 10 h resulted in the decreased intensity of the both ultraviolet and blue PL peaks. Intensity of the red PL peak at 2.07 eV did not change after annealing.

The PL intensity of all observed PL peaks was found to increase for the SiO$_2$:Si samples HT – HP treated at 1.2 GPa. No change in the PL peak position was observed. The main excitation peak was detected at the same wavelength as in the case of SiO$_2$:Si sample annealed at atmospheric pressure [33]. The PL peak 2.70 eV is associated with the neutral oxygen vacancy in the silicon dioxide network (i. e. with $\equiv$Si-Si$\equiv$ centre [34]) while the 2.07 eV PL band is connected with the radiative recombination of an electron-hole pair within the non-crystalline nanoclusters [35].

The PL spectra from non-implanted silicon dioxide and these recorded from the SiO$_2$:Si samples before and after the treatment at 670 K, 10 h under argon pressures of 10^5 Pa and 1.2 GPa are presented in Fig. 10. The intensity of PL, especially of that at 2.7 eV and at 3.44 eV, is markedly increasing with HP.

Subsequent annealing at 670 K - 10^5 Pa of the samples treated at first under HP = 1.2 GPa reduced again the intensity of the PL peaks of high energy. At the same time the intensity of the red PL band at 2.07 eV showed no marked changes.

The dependence of the above-described PL peaks on the HP value was studied also after the HT – HP treatment at 720 K for 10 h [31]. A near-logarithmic dependence of the intensity of all PL peaks on HP was found within the 10^5 Pa - 1.2 GPa range. The intensity of PL bands increased after the treatment at 720 K – HP with increased Si$^+$ ion dose while annealing at 720 K under 10^5 Pa resulted in PL decreasing with the Si concentration [34].

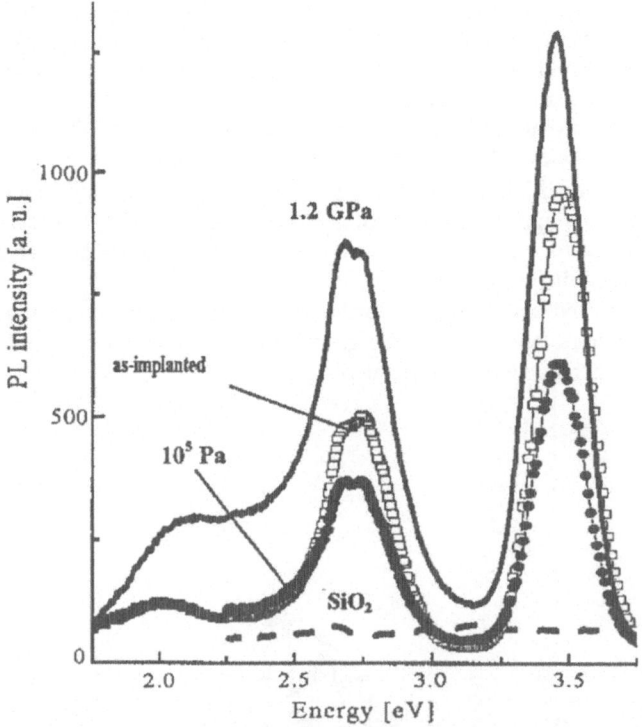

Figure 10. PL spectra from SiO$_2$ film and from SiO$_2$:Si samples (VHD implantation with Si$^+$: D = 3.9×10^{16} cm^{-2} at E = 100 keV and D = 6.3×10^{16} cm^{-2} at 200 keV), before (as-implanted) and after annealing / treatment for 10 h at 670 K under 10^5 Pa and 1.2 GPa.

The effect of HP applied during the HT – HP treatment on the intensity of short wavelength PL can be related to HP - stimulated formation of the light emitting centres or / and to the HP – induced decrease in concentration of the nonradiative recombination centres. The increasing intensity of the blue and violet PL bands with the increased dose of implanted Si atoms after the HT – HP treatment but decreased PL after annealing at atmospheric pressure speak in favour of the HP - stimulated formation of the light emitting centres. Furthermore, the invariability of the energy position of PL peaks implies that the treatment at 670 K – HP does not effect in creation of some other type of light emitting centres, but rather stimulates further formation of the ≡Si-Si≡ centres. The mechanism of this effect is, however, far from being understood. Possibly, it is related to reduced activation energy for creation of the ≡Si-Si≡ centres because of the HP – induced change of short range atom ordering in the implantation - disturbed SiO$_2$ matrix.

The HT – HP treatment of SiO$_2$:Si prepared by low, medium or high dose implantation (LD, MD and HD – see Section 2) resulted in the same effects as in the VHD samples but the PL intensities were much lower, almost negligible for the LD samples [31].

632

Annealing of the VHD implanted SiO_2:Si samples at 1400 K – 10^5 Pa for 5 h resulted in high intensity PL peaking in the near - IR spectral region at 1.46 - 1.77 eV (Fig. 11). This PL is associated with the radiative recombination of the quantum confined electron-hole pairs in silicon nanocrystals [35].

Enhanced pressure used during the treatment leads to a blue shift of PL while the PL intensities decrease with HP. The samples treated at 0.6 GPa exhibited the two broad PL bands at 1.90 eV and 2.25 eV. Increased HP up to 0.9 GPa resulted in a disappearance of red PL and in an appearance of violet PL peaking at 2.95 eV. Most probably the silicon nanocrystals formed under HP are smaller as compared to those created at 10^5 Pa because of retarded diffusion of Si atoms to nucleation sites in SiO_2.

As it seems, increased uniform stress during the treatment results in promoted formation of small Si complexes and prevents the growth of larger nanoclusters and nanocrystals. This supposition demands to be confirmed.

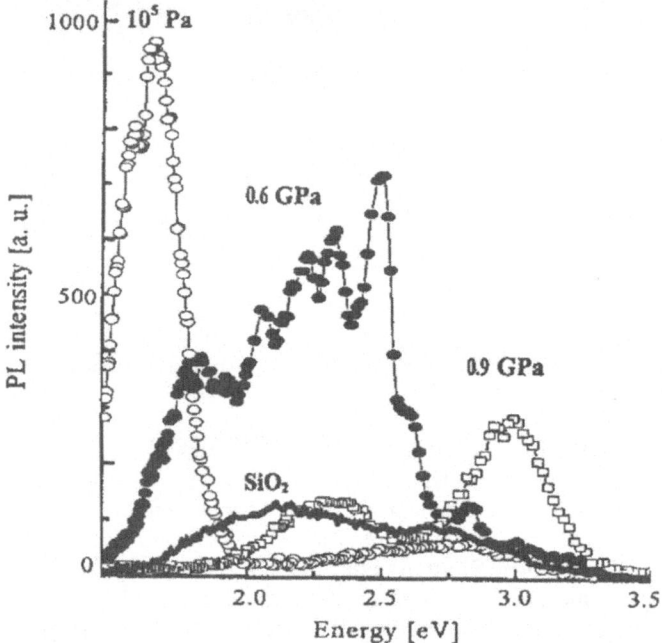

Figure 11. PL spectra from SiO_2:Si samples (VHD implantation with Si^+: $D = 3.9\times10^{16}$ cm^{-2} at $E = 100$ keV and $D = 6.3\times10^{16}$ cm^{-2} at 200 keV), as-implanted and after annealing / treatment for 5 h at 1400 K - 10^5 Pa, 0.6 GPa and 0.9 GPa.

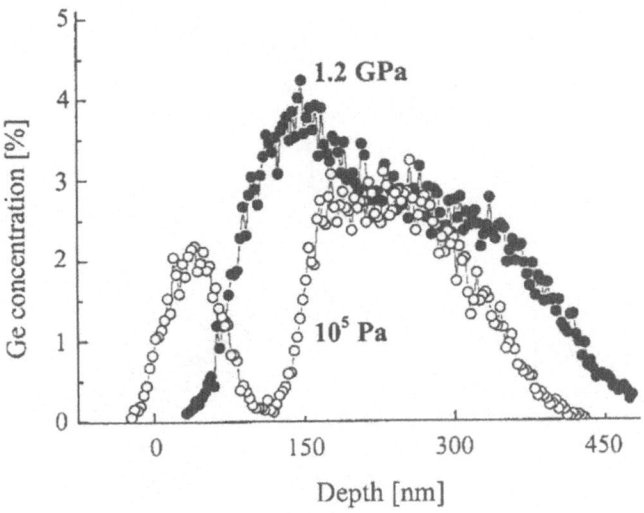

Figure 12. Depth distribution of Ge atoms in SiO$_2$:Ge, determined by RBS. Ge$^+$ implantation: D = 3.0×10^{16} cm^{-2} at E = 450 keV and D = 1.80×10^{16} cm^{-2} at E = 230 keV. Samples were annealed / treated for 5 h at 1400 K under 10^5 Pa and 1.2 GPa.

Enhanced hydrostatic pressure of ambient atmosphere affects also the properties of SiO$_2$ films implanted with Ge$^+$ ions, SiO$_2$:Ge [36]. Ge atoms did not change their positions in the SiO$_2$:Ge sample subjected to the treatment at 1400 K - 1.2 GPa for 5 h (Fig. 12) while annealing of SiO$_2$:Ge under atmospheric pressure leads to outdiffusion of Ge atoms from the ion implanted region toward the near surface layer of SiO$_2$.

Position of the Raman peaks from the HT – HP treated SiO$_2$:Ge samples is dependent on the concentration of embedded Ge atoms. The broad Raman peak centred at 300 cm^{-1} was observed for the SiO$_2$:Ge sample with a 0.1 at. % Ge content. Its halfwidth (FWHM) was equal to about 15 cm^{-1}. The peak position and its width suggest that this peak originates from the Ge-Ge mode in the noncrystalline Ge matrix. As the Ge concentration increased to 0.67 at. %, the Raman spectrum indicated a double peak feature. The peak of about 10 cm^{-1} width was also centred at 300 cm^{-1}. The next very sharp Raman peak of much lower intensity was also detected at about 306.5 cm^{-1}. The peak position and width show that it is also related to the Ge-Ge mode in the Ge crystalline matrix of nanometric dimension. The peak of higher frequency, at about 306.5 cm^{-1} corresponds possibly to stressed Ge-Ge bonds. It would mean that the Ge nanocrystals, grown at HT – HP, are stressed.

Strong enhancement of the Raman peak intensity from the stressed Ge nanocrystals was observed for SiO$_2$:Ge prepared by implantation with the highest Ge$^+$ dose (3 at. % of Ge). No shift in this peak position was detected. The intensity of the 300 cm^{-1} Raman peak was slightly lower. The Raman peak at 300 cm^{-1} was of the highest intensity for the SiO$_2$:Ge samples prepared by annealing under atmospheric pressure. The FWHM value of this peak was equal to about 5 cm^{-1}.

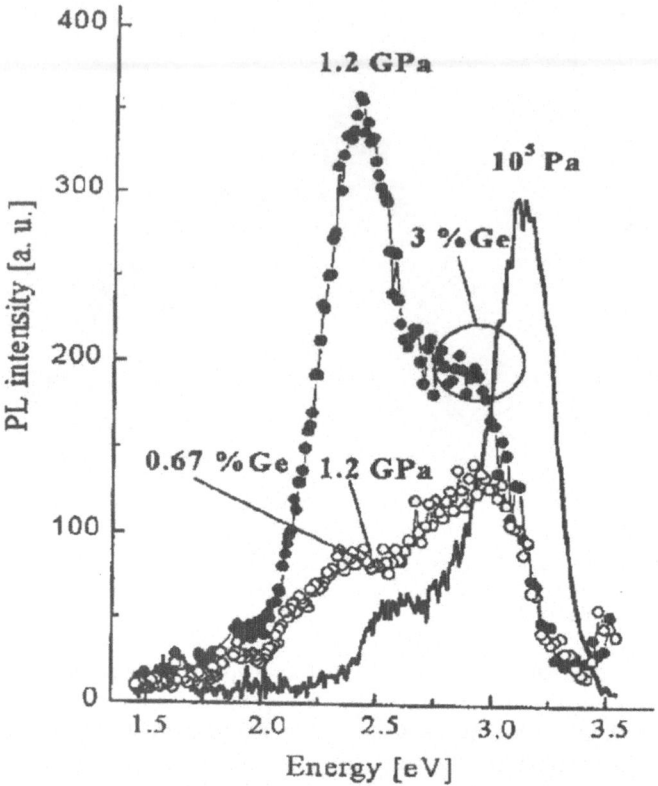

Figure 13. PL spectra from SiO_2:Ge samples with 0.67 % and 3 % Ge contents, annealed / treated at 1400 K for 5 h under 10^5 Pa and 1.2 GPa.

It means that unstressed Ge nanocrystals are created in SiO_2:Ge at 10^5 Pa. The Raman and RBS data may suggest that Ge nanocrystals are formed by diffusion mechanism; Ge diffusion is retarded at HP. A broad PL peak centred at about 3.1 eV was observed in the case of SiO_2:Ge sample with a 3 at. % Ge content, annealed at 1400 K under atmospheric pressure (Fig. 13). The PL peak at about 2.70 eV of lower intensity was also detected. These two PL peaks were reported previously [36]; they are probably related to radiative recombination at the $\equiv$Ge-Si$\equiv$ and $\equiv$Ge-Ge$\equiv$ centers, respectively. The PL peak at 2.95 eV has been also reported for the SiO_2:Ge samples treated at 870 K – 1.2 GPa [36].

The HT – HP treatment of the SiO_2:Ge samples results in dramatic changes of their PL spectra; the detected PL peaks are of considerable intensity. For example, the PL peak at 2.38 eV was detected for the case of SiO_2:Ge treated at 1400 K - HP (Fig. 13). Its intensity increases for the SiO_2:Ge samples with Ge concentration exceeding 0.1 at. %. It means that the presence of this PL band correlate with the concentration of germanium nanocrystals.

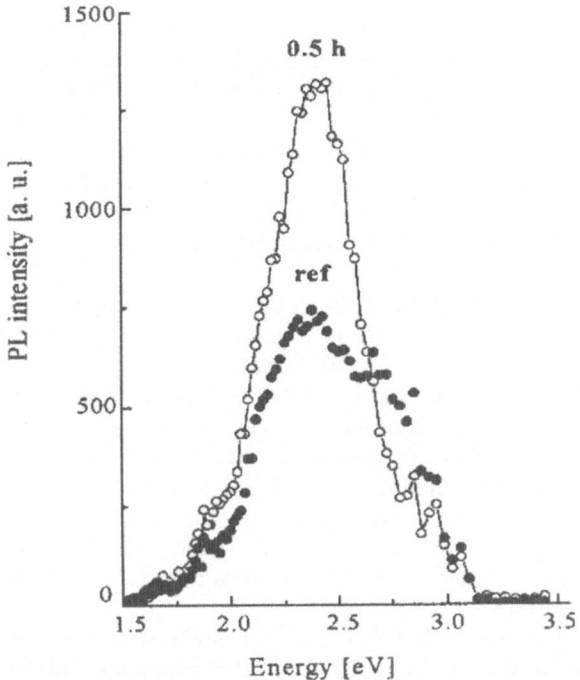

Figure 14. PL spectra from SiO₂:Ge sample with 3 % Ge content treated at 1400 K - 1.2 GPa for 5 h (PL spectrum after that treatment was marked as ref, full circles) and subsequently annealed at 1400 K – 10^5 Pa for 0.5 h (empty circles).

Hitherto performed investigations (in fact, preliminary) prove the potential of HT – HP treatment for preparation of the optically active SiO_2:Si and SiO_2:Ge nanostructures, exhibiting photoluminescence of high intensity at visible / ultraviolet light region. However, almost nothing is still known on possible effect of treatment time and of the combined treatment, under pressure and at 10^5 Pa, on optical activity of SiO_2:Si and SiO_2:Ge. Just such combined treatment can result in improved optical activity (Fig. 14): the SiO_2:Ge samples treated at 1400 K – 1.2 GPa and subsequently annealed at 1400 K under atmospheric pressure indicate increased intensity of PL at 2.38 eV. Investigations of this effect and of the origin of PL at 2.38 eV are in progress.

4. Conclusions

The high temperature - pressure (HT - HP) treatment of Si:H, Si:He and SiO_2:Si,Ge can result in a strongly enhanced photoluminescence at the infrared, visible and even ultraviolet regions, while the HT – HP treatment of Si:O results typically in a decreased PL intensity at IR.

The HT - HP treatment of Si:H and Si:He results in creation of nanostructured buried porous – like layers containing gas - filled bubbles, platelets, cavities as well as numerous extended and point defects. Such structures exhibit both visible and infrared photoluminescence. However, the potential of Si:H and Si:He in respect of possible application is limited, also because their main constituent, silicon, is not transparent for the visible and ultraviolet light.

The HT – HP treatment of the SiO_2:Si structures at about 670 K results in enhanced concentration of the light emitting centres (e.g. of $\equiv$Si-Si$\equiv$) formed in the thermally grown SiO_2 films implanted with Si^+ ions. The HT – HP treatment at about 1400 K leads to creation of modified silicon nanocrystals: the blue shift of the PL spectra was detected.

In the case of Ge^+ ion implantation (SiO_2:Ge structures), the HT – HP treatment results in formation of the stressed Ge nanocrystals and in the enhanced intensity of blue - green PL without change of the PL peak position.

The nature of PL active species created under HT - HP in SiO_2:Si and SiO_2:Ge is still not known definitely and demands future research.

The SiO_2:Si and SiO_2:Ge structures are most promising in respect of possible practical application. Photoluminescence of the SiO_2:Si and SiO_2:Ge structures is related to the presence of Si or Ge nanocrystals embedded in the SiO_2 matrix. Such structures are mechanically reliable and optically stable, suitable for device applications. Contrary to the case of Si with forbidden gap equal to about 1.1 eV, SiO_2 is transparent for visible and ultraviolet light.

The HT - HP treatment of SiO_2:Si,Ge and of similar structures can be considered as an unique tool to produce optically active Si based structures with no use of chemical reagents, and so potentially compatible with IC Si based technology. Still the hitherto performed investigations can be considered as the preliminary ones and so much more work needs to be done to solve emerging problems.

5. Acknowledgement

The authors thank Dr I.V. Antonova, Dr. V.P. Popov, Dr. A.B. Talochkin and Dr K.S. Zhuravlev from the Institute of Semiconductor Physics, RAS, Novosibirsk, Dr L. Rebohle, Dr W. Skorupa, and Dr M. Voelskow from Institute of Ion Beam Physics and Materials Research, Research Centre Rossendorf, Dresden, Dr L. Bryja from the Wroclaw Institute of Technology, Prof. V. Raineri from the CNR IMETEM Catania, Dr A. Romano – Rodriguez from the Barcelona University, m. sc. A. Wnuk from the Warsaw Institute of Technology as well as Prof. A. Barcz and Mr M. Prujszczyk from the Institute of Electron Technology, Warsaw for preparation and measurements of some samples used in this study.

This work was partially supported by the Polish Committee for Scientific Research (grant no. 4T08A 034 23).

6. References

1. Canham, L.T. (1990) Silicon quantum wire array fabrication by electrochemical and chemical dissolution of wafers, *Appl. Phys. Lett.* **57**, 1046-1048.
2. Canham, L.T. (ed.) (1997) *Properties of Porous Silicon,* The Institution of Electrical Engineers, London.
3. Cullis, A.G., Canham, L.T., and Calcott, P.D.J. (1997) The structural and luminescence properties of porous silicon, *J. Appl. Phys.* **82**, 909-965.
4. Tyschenko, I.E., Talochkin, A.B., Zhuravlev, K.S., Obodnikov, V.I., Popov, V.P. (2002) Structural and photoluminescence properties of H$^+$ ion – implanted silicon – on – insulator structures formed by hydrogen slicing, *Solid State Phenomena* **82-84**, 509-514.
5. Misiuk, A. (2000) High pressure – high temperature treatment to create oxygen nano – clusters and defects in single crystalline silicon, *Mater. Phys. Mech.* **1**, 119-126.
6. Steinman, E.A., Kveder, V.V., Vdovin, V.I., Grimmeiss, H.G. (1999) The origin and efficiency of dislocation luminescence in Si and its possible application in optoelectronics, *Solid State Phenomena* **69-70**, 32-32.
7. Yue, L., and He, Y. (1997) Studies on room temperature characteristics and mechanism of visible luminescence of Ge-SiO$_2$ thin films, *J. Appl. Phys.* **81**, 2910-2912.
8. Mutti, P., Ghislotti, G., Bertoni, S., Bonoldi, L., Cerofolini, G.F., Meda, L., Grilli, E., and Guzzi, M. (1995) Room-temperature visible luminescence from silicon nanocrystals in silicon implanted SiO$_2$ layers, *Appl. Phys. Lett.* **66**, 851-853.
9. Antonova, I.V., Popov, V.P., Fedina, L.I., Shaimeev, S.S., Misiuk, A. (1996) A DLTS study of the evolution of oxygen precipitates in Si at high temperature and high pressure, *Semiconductors* **30**, 760-764.
10. Misiuk, A., Jung, W., Surma, B., Jun, J., Rozental, M. (1997) Effect of stress induced defects on electrical properties of Czochralski grown silicon, *Solid State Phenomena* **82-84**, 393-398.
11. Misiuk, A. (1991) Evolution of process – induced defects in silicon under hydrostatic pressure, *Solid State Phenomena* **19&20**, 387-392.
12. Misiuk, A., Bak-Misiuk, J., Kaniewska, M., Zhuravlev, K.S., Raineri, V., and Antonova, I.V. (2002) Nanostructured layers in high temperature – pressure treated silicon implanted with hydrogen / helium, *Mater. Phys. Mech.* **5**, 31-38.
13. Misiuk, A., Rebohle, L., Iller, A., Tyschenko, I.E., Jun, J., Panas, A. (2000) Photoluminescence from pressure – annealed nanostructured silicon dioxide and nitride films, in G.M.Chow et al. (eds) *Nanostructured Films and Coatings,* Kluver Academic Publishers, Dordrecht, pp. 157-170.
14. Raineri, V. (2000) Voids in silicon substrates for novel applications, *Mater. Sci. Engineer.* **B73**, 47-53.
15. Xiang Lu, Cheund, N.W., Strathman, M.D., Chu, P.K., Doyle, B. (1997) Hydrogen induced silicon surface layer cleavage, *Appl. Phys. Lett.* **71**, 1804 – 1806.
16. Misiuk, A., Bak-Misiuk, J., Barcz, A., Romano-Rodriguez, A., Antonova, I.V., Popov, V.P., Londos, C.A., Jun, J. (2001) Effect of annealing at argon pressure up to 1.2 GPa on hydrogen-plasma-etched and hydrogen-implanted single-crystalline silicon, *Intern. J. Hydrogen Energy* **26**, 483-488.
17. Misiuk, A., Bak-Misiuk, Bryja, J., Katcki, J., Ratajczak, J., Jun, J., Surma, B. (2002) Oxygen precipitation in Si:O annealed under high hydrostatic pressure, *Acta Phys. Polon. A* **101**, 719-727.
18. Cerofolini, G.F., Calzolari, G., Corni, F., Nobili, C., Ottaviani, G., Tonini, R. (2000) Ultradense gas bubbles in hydrogen-or helium-implanted (or coimplanted) silicon, *Mater. Sci. Engineer.* **B71**, 196-202.
19. Misiuk A., Barcz, A., Raineri, V., Ratajczak, J., Bak-Misiuk, J., Antonova, I.V., Wierzchowski W., Wieteska, W. (2001). Effect of stress on accumulation of oxygen in silicon implanted with helium and hydrogen, *Physica B* **308-310**, 317-320.
20. Grisolia, J., Christiano, F., Assayag, B., Claverie, A. (2001) Kinetic aspects of the growth of platelets and voids in H implanted Si, *Nucl. Instrum. Methods Phys. Res. B* **178**, 160-164.
21. Misiuk, A., Bak-Misiuk, J., Antonova, I.V., Raineri, V., Romano-Rodriguez, A., Bachrouri, A., Surma, H.B., Ratajczak, J., Katcki, J., Adamczewska, J., Neustroev, E.P. (2001), Effect of uniform stress on silicon implanted with helium, hydrogen and oxygen, *Comput. Mater. Sci.* **21**, 515-525.
22. Mudryi, A.V., Korshunov, F.P., Patuk, A.I., Shakin, I.A., Larionova, T.P., Ulyashin, A.G., Job, R., Fahrner, W.R., Emtsev, V.V., Davydov, V.YU., Oganesyan, G. (2001) Low-temperature photoluminescence characterization of defects formation in hydrogen and helium implanted silicon at post-implantation annealing, *Physica B* **308-310**, 181-184.
23. Ulyashin, A.G., Job, R., Fahrner, W.R., Mudryi, A.V., Patuk, A.I., Shakin, I.A. (2001) Low-temperature photoluminescence characterization of hydrogen- and helium-implanted silicon, *Mater. Sci. Semicond. Processing* **4**, 297-299.

638

24. Raineri, V., Coffa, S., Shilagyi, E., Gyulai, J., Rimini, E. (2000) He vacancy interactions in Si and their influence on bubble formation and evolution, *Phys. Rev. B* **61**, 937-945.

25. Misiuk, A., Barcz, A., Ratajczak, J., Lopez, M., Romano-Rodriguez, A., Bak-Misiuk, J., Surma, H.B., Jun, J., Antonova, I.V., Popov, V.P. (2000) Effect of external stress on creation of buried SiO_2 layer in silicon implanted with oxygen, *Mater. Sci. Engineer.* **B73**, 134 – 138.

26. Misiuk, A., Barcz, A., Ratajczak, J., Katcki, J., Bak-Misiuk, J., Bryja, L., Surma, B., Gawlik, G. (2001) Structure of oxygen – implanted silicon single crysals treated at $\geq$ 1400 K under high argon pressure, *Cryst. Res. Technol.* **36**, 933 – 941.

27. Misiuk, A. (2002) Application of high temperature – pressure treatment for investigation of defect creation in basic materials of modern microelectronics: Czochralski silicon and silicon containing films, in H.D. Hochheimer et al. (eds) *Frontiers of High Pressure Research II: Application of High Pressure to Low-Dimensional Novel Electronic Materials*, Kluver Academic Publishers, Dordrecht, pp. 275-289.

28. Skuja, L.(2000 Optical properties of defects in silica, in G. Paccioni et al. (eds) *Defects in SiO₂ and Related Dielectrics: Science and Technology*, Kluver Academic Publishers, Dordrecht, pp. 73-116.

29. Karwasz, G.P., Misiuk, A., Ceschini, M., and Pavesi, L. (1996) Visible photoluminescence from pressure annealed intrinsic Czochralski-grown silicon, *Appl. Phys. Lett.* **69**, 2900-2902.

30. Aziz, M.J. (2001) Stress effects on defects and dopant diffusion in Si, *Mat. Sci. Semicond. Proc.* **4**, 397-403.

31. Tyschenko, I.E., Rebohle, L., Yankov, R.A., Skorupa, W., Misiuk, A. (1998) Enhancement of the intensity of the short-wavelength visible photoluminescence from silicon-implanted silicon - dioxide films caused by hydrostatic pressure during annealing, *Appl. Phys. Lett.* **73**, 1418-1420.

32. Zhuravlev, K.S., Tyschenko, I.E., Vandyshev, E.N., Bulytova, N.V., Misiuk, A., Rebohle, L., Skorupa, W. (2002) Effect of hydrostatic pressure on photoluminescence spectra from structures with Si nanocrystals fabricated in SiO_2 matrix, *Acta Phys. Polon.* **A 102**, 337-344.

33. Rebohle, L., Tyschenko, I.E., Fröb, H., Leo, K., Yankov, R.A., Von Borany, J., Kachurin, G.A., Skorupa, W. (1997) Blue and violet photoluminescence from high-dose Si^+- and Ge^+-implanted silicon dioxide layers, *Microelectronic Engineer.* **36**, 107-110.

34. Skorupa, W., Yankov, R.A., Tyschenko, I.E., Fröb, H., Böhme, T., Leo, K. (1996) Room -temperature, short-wavelength (400-500 nm) photoluminescence from silicon-implanted silicon dioxide films, *Appl. Phys. Lett.* **68**, 2410-2412.

35. Kachurin, G.A., Rebohle, L., Skorupa, W., Yankov, R.A., Tyschenko, I.E., Fröb, H., Böhme, T., Leo, K. (1998) Short-wavelength photoluminescence from SiO_2 layers implanted with high doses of Si^+, Ge^+, and Ar^+ ions, *Semiconductors* **32**, 392-396.

36. Tyschenko, I.E., Rebohle, L., Talochkin, A.B., Kolesov, B.A., Voelskow, M., Misiuk, A., Skorupa, W. (2002) Blue-green photoluminescence from silicon dioxide films containing Ge^+ nanocrystals formed under conditions of high hydrostatic pressure annealing, *Solid State Phenomena* **82-84**, 607-612.

NANOELECTRODES ON SILICON FOR ELECTROCHEMICAL APPLICATIONS

A. I. KLEPS, A. ANGELESCU, M. MIU, M. SIMION,
A. BRAGARU, M. AVRAM
*National Institute for Research and Development in Microtechnologies
(IMT-Bucharest) P.O.Box 38-160, 72225 Bucharest Romania*

1. Introduction

Different science fields, such as nanofabrication technology, electrochemistry and surface and material sciences, and electronics are involved in nanoelectrode fabrication and applications. Due to the recent progress in nanoscience and nanotechnology, the nanoelectrode domain was approached by many groups [1-4].

In this paper it is described a new technology for fabrication pyramidal nanoelectrodes and nanoelectrode arrays as working electrodes in an electrochemical cell. One nanoelectrode element has one dimension lower than 500 nm. Nanoelectrodes have various domain of applications: analytical chemistry (environment pollution control) [5-7], medicine [8-9], biology and biochemistry [9-10], scanning electrochemical microscopy [12-13],   etc.

2. Nanoelectrode Properties

The nanoelectrode, in a conducting liquid, represents a point with *spherical (radial)* diffusion, of the active species; this leads to the decreasing of the surface to volume ratio and to the increasing of the current density at the electrode surface.

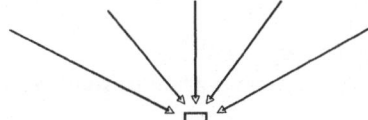

Fig. 1. Radial diffusion of the electroactive molecules towards the nanoelectrode

Nanoelectrodes offer higher sensitivity than macroelectrodes of conventional size, because an electroactive molecule can approach the nanoelectrode from every direction (Fig. 1). Therefore, the flux of electroactive molecules toward the electrode is much greater for a nanoelectrode than for a macroelectrode, for which the diffusion is planar.

Mathematical solutions for each kind of electrode response (macro and nano) predict the superior electrode sensitivity of nanoelectrodes [14].

T. Tsakalakos, et al. (eds.) pgs. 639 - 648
Nanostructures: Synthesis, Functional Properties, and Applications;
© *2003 Kluwer Academic Publishers.*

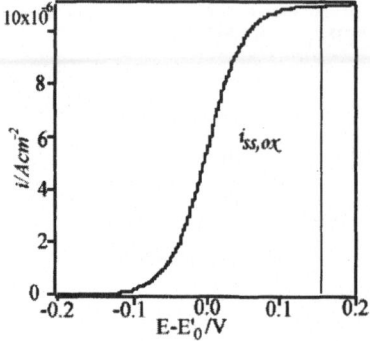

Fig. 2. The voltammogram for the nanoelectrode

In Fig. 2 it is presented the cyclic voltammogram of the nanoelectrode with spherical diffusion. The shape of the wave indicates the current reaches a steady state in which the current is independent of time.

The study of the nanoelectrode behaviour, involves not only the electrode size and shape, but also the kinetic parameters of the electroanalytical technique employed that governs the thickness of the diffusion layer developed (e.g., the potential scan rate for cyclic voltammetry).

Analyzing the diffusional regimes for nanoelectrode arrays, there are observed three situations depending on the kinetic parameters of the electro-analytical technique employed: (i) linear regime (high scanning rate, short time); at a sufficiently short time, any planar electrode behaves as an infinitely large planar electrode (Fig. 3 a); (ii) radial regime (low scanning rate, long time) (Fig. 3 b); (iii) overlapping regime (very low scanning rate, very long time); the value of the overlap factor depends on the electrode geometry and the time of electrolysis (Fig. 3 c) [15].

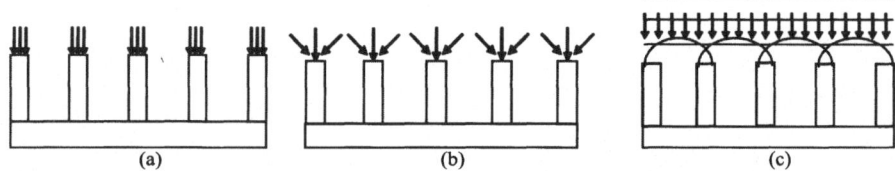

Fig. 3. Diffusional regimes for nanoelectrode arrays: (a) linear regime; (b) radial regime; (c) overlapping

As the distance between two electrodes increases, the current densities decrease. Only at low current density, it is an optimum current response which corresponds well to the theory. The main properties of nanoelectrodes are:

- A steady state for a faradaic process is attained very rapidly.
- Small size of the electrodes permits measurements on very limited solution volumes.
- Electrochemical experiments can be conducted in highly electrically resistive media;
- The faradaic-to-charging current ratio, I_F/I_C, is improved, as the charging current decreases in proportion to decreasing electrode area, while the steady-state faradaic current is proportional to its characteristic dimension.
- The limiting current density increases as the electrode size decreases;
- The mass transport rate to an nanoelectrode is higher than that which can be achieved using larger electrodes;
- The detection limits are lower, due to the higher signal-to-background ratio;

- Chemical or electrochemical processes which are too fast for larger electrodes can be studied using an nanoelectrode; the quick response of nanoelectrodes allows monitoring of the low-frequency fluctuation of signals and rapid recording of steady-state polarization curves;
- The very low level of Faraday current results in the beneficial effect of very small ohmic potential drop;
- The current output is practically insensitive to conventional flow in solution.

3. Pyramidal Nanoelectrodes Made by Silicon Etching

The nanofabrication of sensor elements is considered the most significant development in chemical sensor technology, in the last years. In this paper, it is presented the design and the experimental technology for the working electrode based on metal/dielectric silicon nanostructures. This electrode is the main part of a system utilized for the detection of heavy metals in liquid media by cyclic voltammetry method. Different technologies are developed for nanoelectrode fabrication: (i) lithographic techniques [16]; template-synthesized nanomaterials [17-19]; single-walled carbon nanotubes (SWCNTs) / fullerenes; chemical linking of nanoparticles to carbon nanotubes; synthetic nanowires [20-26]; embedded nanoparticles [27-28]; pyramidal nanoelectrodes made by silicon etching [29-31].

A very simple fabrication technology for a new electrode array architecture with pyramidal micro- or nano-electrodes is described. This technology is based on standard processes used in silicon device fabrication such as: silicon etching, metallic and dielectric film deposition. In order to release the top of the pyramidal electrode, the photolithographic exposure takes place on the whole silicon surface. The nanoelectrode proposed to be realised has the following schematic structure (Fig. 4).

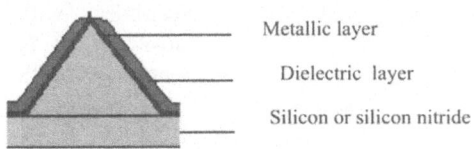

Fig. 4. Nanoelectrode schematic structure

The nanoelectrodes array consists of a series of individual nanoelectrodes, with equal distances between them, separated by an insulating material; these spaces are high enough to avoid the overlapping of diffusion layers.

3.1 NANOELECTRODE ARRAY DESIGN

In an NE array, for each singular nanoelectrode, the Fick's diffusion law is applied with modified mathematical form related to the edge effects, so in addition to the mass transport perpendicular to the electrode, it develops a parallel one; for disk nanoelectrodes the current equation is the sum of liniar and radial diffusion, and the steady-state limiting current is:

$$i = 4nFDC\ r_o$$

where: ro = electrode radius, and the other symbols have their usual meanings.

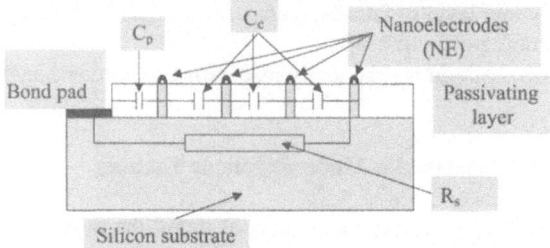

Fig. 5. Nanoelectrode array basic structure, with parasitic elements

Parasitic elements are determined by the radius of the metal electrode (r_o=5-200 nm), the distance between the centers of two electrodes (d = 12–24 μm), and the isolated layer thickness (h=0,5-4 μm): Cp-capacitance between the contacting metallic pad and electrolyte (< 0.1 pf); C_c -capacitance of the dielectric region between the electrodes (< 10 pF); R_s -silicon substrate resistance (< 2 Ωcm). In Fig. 6 it is presented the lay-out of the metal/dielectric test nanostructures.

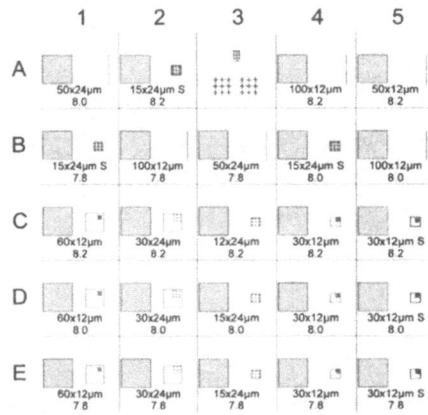

Fig. 6. Lay-out of the metal/dielectric test nanostructures

Different nanostructure experimental types with variable electrodes number and position in the array are realized: (a) array of 60 x 60 electrodes, with 12 μm distance between them, (b) array of 30 x 30 electrodes, with 24 μm distance between them. (c) array of 30 x 30 electrodes, with 12 μm distance between them; (d) array of 15 x 15 electrodes, with 24 μm distance between them; (e) array of 30 x 30 electrodes, with grid and 12 μm distance between them; (f) array of 15 x 15 electrodes, with grid and 24 μm distance between them.

3.2. NANOELECTRODE ARRAY FABRICATION

The substrate material was a p-type (100) silicon wafer, of 6-10 ohmcm resistivity. Pyramidal structures were realized by Si etching through a 0.6 μm silicon dioxide mask. An additional oxidation process was applied in order to sharp the structure. After etching, the substrate was again oxidized to ensure the isolation of the electrode layer from the semiconducting substrate (Fig.7 - 8).

The electrode material was deposited on the whole Si surface. The most suitable materials for electrode fabrication are gold, and platinum. We have used either a gold (Au) film (200 nm) deposited by evaporation method, or a platinum (Pt) film (100-300 nm) deposited by MOCVD using platinum bisacetonate as precursor. The insulating material was inorganic films SiO_2 or SiO_2/Si_3N_4 sandwich. The compositional analyses on the nanoelectrode top, and bottom, and between the pyramidal structures show a continuous deposition on the whole Si/SiO_2 surface (Fig. 9-10). The optimum support oxide thickness resulted from many experiments is 500-700 nm; after the photolithographic process, silicon was isotropic etched in a solution: $25:10:1(HNO_3: CH_3COOH: HF)$. Thermal oxidation conditions for silicon tips sharpening were: O_2 atmosphere, $T=950^0C$, $t=30$ min. After the Au or Pt film vacuum deposition, an isolating material was deposited on the whole surface: SiO_2 or PSG, $d = 1\mu m$, Si_3N_4, $d = 200$ nm. A short UV exposition process was performed without any mask in order to release the top of the electrodes. Finally, contact wires were bonded and the structures were isolated with silicon resin.

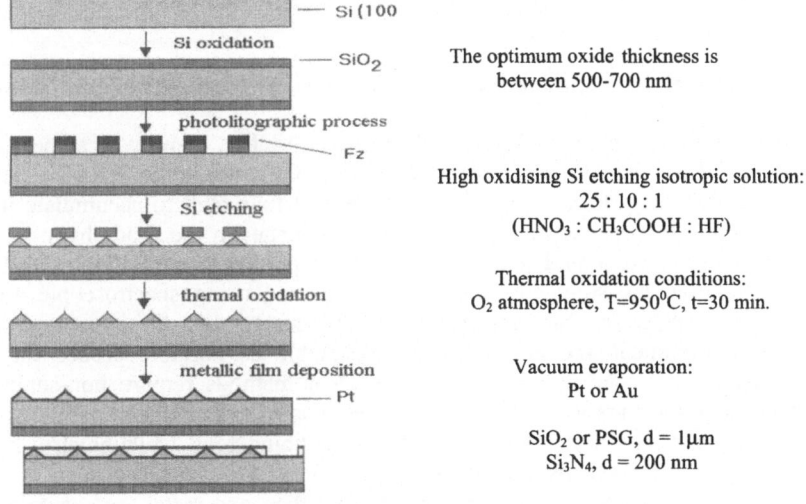

The optimum oxide thickness is between 500-700 nm

High oxidising Si etching isotropic solution:
$25:10:1$
$(HNO_3 : CH_3COOH : HF)$

Thermal oxidation conditions:
O_2 atmosphere, $T=950^0C$, $t=30$ min.

Vacuum evaporation:
Pt or Au

SiO_2 or PSG, $d = 1\mu m$
Si_3N_4, $d = 200$ nm

Fig. 7. Nanoelectrode array fabrication. Final insulated structure

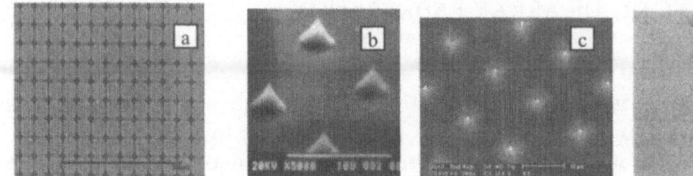

Fig. 8. Experimental gold nanoelectrode array on silicon (a) photolithographic process; (b) pyramidal structures; (c) and (d) insulated structures

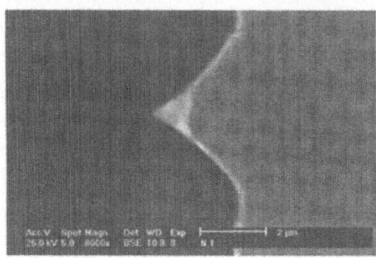

Fig. 9. SEM image of MOCVD -Pt film

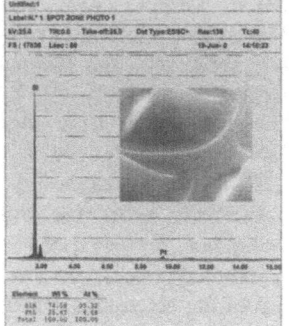

Fig. 10. Compositional analysis between the pyramidal structures

4. Electroanalytical applications. Metallic trace electrochemical analysis

At trace levels, most of elements are essential to life. However, some of them, especially heavy metals are health hazardous; they are toxic even at very low concentrations, they do not degrade with time and thus tend to accumulate in living organisms and to concentrate during transfers of matter in the food chain. Maximum heavy metal contents should be in the range 1 to 50 ppb (parts per billion or μg/L) [32].

Quantitative measurements are usually obtained using spectroscopic analytical techniques such as atomic absorption or fluorimetry, but these methods require expensive equipment and are not easily adaptable as portable probes for *in situ* determinations. On the contrary, electrochemical methods require non-sophisticated instrumentation and are well adapted to the quantitative measurements [33].

For trace metals determining in water or in liquid food, voltammetry, including polarography presents many advantages:

• the simultaneous determination of more than one metal (as example is very common the determination of lead and cadmium in food by anodic stripping voltammetry -AVS). Copper, cadmium, lead and zinc can be determined during the same polarographic scan. The time saving of determining two or many metals simultaneously is evident.

• The trace metal technique (<100 ppb) is performed with the same instrumentation that is used to determine many other ions, such as ascorbic acid or nitrate.

4.1. EXPERIMENTAL MEASUREMENT CONDITIONS

Electrochemical analysis offers a good sensitivity and can be used *in situ* analysis; the working electrodes based on micro and nano-arrays realized by standard processes used in silicon device manufacturing systems can detect trace metals with superior sensitivity compared to macro-electrodes. Heavy-metals like Cd, Pb, Zn, Fe are detected by Square-Wave Voltammetry method with remarkable sensitivity than Cyclic Voltammetry method [34]. The cyclic voltammograms were accomplished with two types of voltammetric system: (i) an electrochemical system Voltamaster 1, for qualitative determinations; (ii) a measurement system that includes Trace Master 5; POL 150 Polarographic Analyser (Radiometer Copenhagen), for quantitative determinations. The square wave voltammograms were measured with the polarographic and voltametric system: Trace Master 5 and POL 150 Polarographic Analyser (Radiometer Copenhagen) (Fig.11-12).

The electrochemical cell contains three electrodes immersed in the electrolyte solution: the nano-electrode arrays as working electrodes, a Pt wire as auxiliary electrode and an Ag/AgCl as reference electrode. This three-electrode configuration offers more accurate measurements of the peak currents related to the working electools, maintains control of the electrochemical cell between experiments, preventing accidental damage to sensitive systems. This software was used for both type of voltametric measurements (CV and SWV). The dissolved oxygen from the analysed solution was eliminated by bubbling a pure argon stream during 5-15 min. The solution was stirred with a PTFE-coated stirring bar.

4.2. PB AND CD DETERMINATION

Lead and cadmium are determined simultaneously. Nanoelectrode arrays were used in combination with differential pulse measurement. During electrolysis, Pb^{2+} and Cd^{2+} are reduced on the working electrode. The measurement is then performed by potential scanning in the anodic direction. The amalgamated metals are stripped back into solution in their ionic form. The resulting stripping current is proportional to the concentration of the metal ions in solution. All measurements were carried out at 25^0C in 0.1M, HCl or 1M, KNO_3 as supporting electrolytes.

Cell parameters are: stirrer 400 rpm, purge time 300s, and electrolysis time 45 s; wait time 10s. Signal parameters: technique used was SWV, E initial – 800 mV, E final + 0mV, pulse amplitude +50 mV. For the experimental voltammogrames and calibration curves, the measurements were made in the same conditions for two types of nanoelectrode arrays in a solution CdPb.

The peak current measured with 225 nanoelectrode array, with 24 µm, the distance between electrodes, was smaller compared to the one measured with 900 nanoelectrode array, with 12 µm, the distance between electrodes (nA compared to µA).

4.3. CALIBRATION CURVE

For the calibration curves, we have chosen Pb and we performed the measurements for different concentrations. The detection limit with Cyclic Voltammetry method was 10^{-3} M. To detect lower concentrations we used the more sensitive technics, the Square - Wave Voltammetry method.

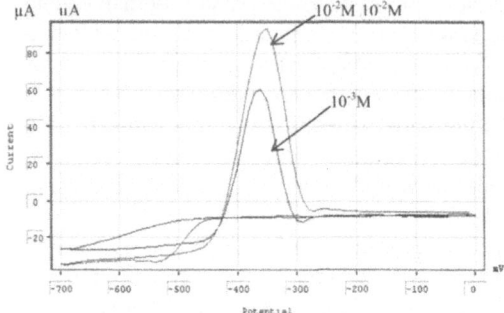

Fig. 11. Cyclic voltammogrames for 900 nano-electrode array in 10^{-3} M and 10^{-2} M CdPb solutions with 0.1M - HCl as supporting electrolyte

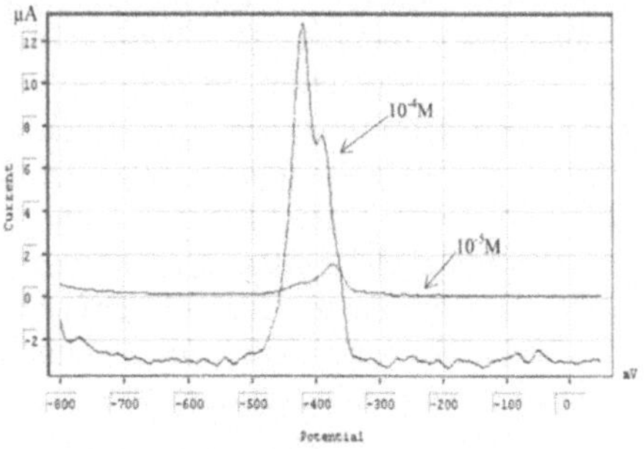

Fig. 12. Square-wave voltammogrames for 900 nanoelectrode array in 10^{-4} M and 10^{-5} M CdPb solutions with 0.1M - HCl as supporting electrolyte

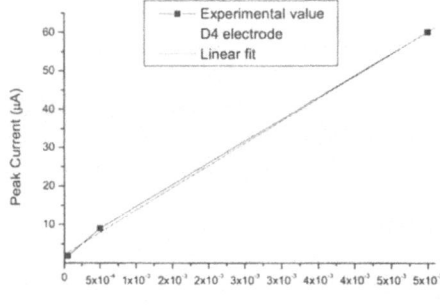

Fig. 13. Lead calibration curve

From the experimental voltammograms, the cyclic voltammograms and the square-wave voltammo-grams, measured with nanoelectrodes, (Figs. 11-12) the cathodic peak at the potential correspond-ding to each metal was measured and a calibration curve was plotted as the peak current versus Pb concentration. The calibration curves allow the interpolation of the results for different concentrations of metallic ions. The calibration curve for Pb resulted from voltammograms with 900 nanoelectrode array presents linearity between peak currents and ion concentrations (Fig. 13).

5. Conclusions

Nanoelectrodes have important applications in the field of electroanalytical analyses. The steady state for the faradaic process with nanoelectrodes is attained very rapidly.

Pyramidal nanoelectrodes present important structural characteristics with applications in measurements of metallic pollution in liquid media. The proposed fabrication technology is based on standard silicon processes and equipments; different metal electrode can be used. The test structure was designed with variable electrodes number in the array at the same potential, and related to different bio-medical applications, the nanoelectrode could be individual addressed. The measured current is three orders magnitude higher for nanoelectrode array with micrometric distances between two adjacent nanoelectrodes, than for nanometric distances; the detection limits are low due to the high signal-to-background ratio. The square-wave voltammetry measurement sensitivity for CdPb solution was 10^{-5} M, using an array of 900 nanoelectrodes as working electrode, in 0.1M – HCl supporting electrolyte. Small size of the electrodes permits measurements on very limited solution volumes.

6. Acknowledgement

Some of the technological processes were done in ICTIMA-Padova in the frame-work of the Romanian-Italian Bilateral Research Program and in IM Mainz under the EMERGE project no. HPRI-CT-1999-00023.

7. References

1. Bard, A. J. and Zoski, C. G. (2000) Voltammetry: A Retrospective Report, Analytical Chemistry, (invited), 72 346A-352A.
2. Green, S. J. Stokes, J. J. Hostetler. M. J. Pietron, J., Murray, R. W. (1997), Three Dimensional Monolayers: Alkanethiolate-stabilized Gold Cluster Molecules Are Nanometer-sized Electrodes, J. Phys. Chem. B, 101, 2663-2668.
3. Chen, S., Ingram R. S., Hostetler, M. J., Pietron, J. J., Murray, R. W., Schaaff, T. G., Khoury, J. T., Alvarez, M. M., Whetten R. L. (1998) Gold Nanoelectrodes of Varied Size: Transition to Molecule-like Charging, Science, , 280, 2098-2101. (NSF, ONR).
4. Jeoung E., Galow T.H., Schotter J., Bal M., Ursache A., Tuominen M.T., Stafford C.M., Russell T.P., Rotello V.M. (2001) Fabrication and characterization of nanoelectrode arrays formed via block copolymer self-assembly. Langmuir 17: 6396-6398.
5. Demaille, C., M., Brust, M. T. and Bard, A. J. (1997). Fabrication and Characterization of Self-Assembled Spherical Gold Ultramicroelectrodes, Anal Chem 69(13): 2323-2328.
6. Amatore, C., (2000) Les Ultramicroélectrodes: De Nouveaux Horizons pour l'Electrochimie Moléculaire? Spectra, 151, 1990, 43-46.
7. Amatore, C., (1995) Electrochemistry at Ultramicroelectrodes, in "Physical Electrochemistry: Principles, Methods and Applications" (Rubinstein, I., .Ed.), M. Dekker, New York.. Chap.4. pp.131-208.
8. Amatore, C., Kelly, R.S., Kristensen, E.W., Kuhr, W.G., Wightman, R.M. (1986) Effect of Restricted Diffusion at Ultramicroelectrodes in Brain Tissue. The Pool Model: Theory and Experiment for Chronoamperometry. J. Electroanal. Chem., 213, , 31-42.
9. Amatore, C. (1996) Ultramicroelectrodes - Their Basic Properties and Their Use in Semiartificial Synapses, C R Acad Sci Ser Ii B 323(11): 757-771.
10. Brunetti, B.; Ugo, P.; Moretto, L.M.; Martin, C.R. (2000) Electrochemistry of Phenothiazine and Methylviologen Biosensor Electron-transfer Mediators at Nanoelectrode Ensembles, J. Electronanal. Chem., 491, 166-174.
11. Kleps, A. I., Angelescu, A. S. and Dascalu, D. (2000) New micro- and nanoelectrode arrays for biomedical applications, BioMEMS & BIOMEDICAL NANOTECHNOLOGY WORLD 2000 Conference, Sept, , Ohio, USA.
12. Macpherson, J. V., Jones, C. E. , Barker, A. L. and Unwin, P. R. (2002) Electrochemical Imaging of Diffusion through Single Nanoscale Pores, Anal. Chem., 74, 1841.

648

13. Macpherson, V. (2001) Electrochemical-AFM in Encyclopedia of Electrochemistry. Volume 3. Instrumentation and Electroanalytical Chemistry: Bard, A. J. and Stratmann, B. (eds), Wiley-VCH Verlag GmbH, Weinheim.

14. Aoki, K. (1998) Electroanalysis 5, 627–639.

15. Bellomi, S. (1995), Caratterizzazione elettroanalitica e voltammetria di scambio ionico su ensemble di nanoelectrodi (Tesi di Laurea), 10.

16. Schoer, J. K., Zamborini, F. P., and Crooks, R. M. (1996) Scanning Probe Lithography; Nanometer-Scale Electrochemical Patterning of Au and Organic Resists in the Absence of Intentionally Added Solvents or Electrolytes, J Phys Chem 100 (26): 11086-11091.

17. Cepak, V. M., Hulteen, J. C. Guangli Che, Jirage, K. B.,. Lakshmi, B. B., Fisher E. R., and Martin, C. R. (1997) Chemical Strategies For Template Synthesis of Composite Micro and Nanostructures, Chem. Mater.9, 1065-1067

18. Martin, C. R.,Mitchell, D. T. (1998) Nanomaterials in Analytical Chemistry, Anal. Chem., , 9, 322A-327A.

19. Masuda, H., M. Yotsuya and M. Ishida (1998). Spatially Selective Metal-Deposition into a Hole-Array Structure of Anodic Porous Alumina Using a Microelectrode, Jpn J Appl Phys Pt 2 37 (9AB): L1090-L1092.

20. Campbell, J. K., Sun, L., and Crooks, R. M. (1999) Electrochemistry Using Single Carbon Nanotubes, J Am Chem Soc 121(15): 3779-3780.

21. Zhang, X. J., Zhang, W. M., Zhou, X. Y., and Ogorevc, B. (1996) Fabrication, Characterization, and Potential Application of Carbon-Fiber Cone Nanometer-Size Electrodes, Anal Chem 68(19): 3338-3343.

22. Cliffel, D. E. (1998) Chemical reactions of electrochemically reduced fullerene and combining scanning electrochemical microscopy with other analytical methods, Univ. of Texas,Austin,TX,USA.: 142 pp.

23. Che, G., Lakshmi, B. B., Martin, C. R., Fisher, E. R., and Ruoff, R. S. (1998) Chemical Vapor Deposition (CVD)-Based Synthesis of Carbon Nanotubes and Nanofibers Using a Template Method, Chem. Mater. 10, 260-267.

24. Che, G., Lakshmi, B. B., Fisher, E. R., and Martin, C. R. (1998) Carbon Nanotubule Membranes and Possible Applications to Electrochemical Energy Storage and Production, Nature 393, 346-347.

25. Li, J., Cassell, A., Delzeit, L., Han, J., and Meyyappan, M. (2002) Novel Three Dimensional Electrodes: Electrochemical Properties of Carbon Nanotube Ensembles, Journal of Physical Chemistry B.

26. DuVall, S.H., McCreery, R.L. (2000) Self-catalysis by Catechols and Quinones during Heterogeneous Electron Transfer at Carbon Electrodes, J. Am. Chem. Soc., 122, 6759.

27. Demaille, C., M. Brust, M. T. and Bard, A. J. (1997) Fabrication and Characterization of Self-Assembled Spherical Gold Ultramicroelectrodes, Anal Chem 69 (13): 2323-2328.

28. Daniele S., Bragato C., Battiston G. A., Gerbasi R. (2001) Voltammetric characterisation of Pt–TiO2 composite nanomaterials prepared by metal organic chemical vapour deposition (MOCVD), Electrochimica Acta 46, 2961–2966.

29. Kleps, A. I., Angelescu, A. S, Miu, M., Avram, M., Simion M. (2001) Technology of silicon nano- and microelectrode arrays for pollution control, Proceedings of the International Conference on Semiconductors CAS, Sinaia, Romania.

30. Avram, M., Angelescu, A. S., Kleps, A. I., Miu, M., Popescu, A. (2001) Fast scan cyclic voltammetry simulation for silicon nanoelectrodes, Proceedings of the International Conference on Semiconductors CAS 2001, Sinaia, Romania;

31. Kleps, A. I., Angelescu, A. S, Miu, M., Avram, M., Simion M. (2001) Nanostructures of pyramidal shape, technology and applications, Micro and Nano Engineering, MNE'2001 Grenoble, France

32. (1998) Journal Officiel des Communautés Européennes.

33. Herdan, J., Feeney, R., Kounaves, S.P., Flannery, A.F., Storment, C.W., Kovacs G.T., and Darling, R.B. (1998) Environ. Sci. Technol. 32, 131.

34. Simion M., Iorgulescu, E. , Kleps, A. I., Angelescu, A. S, Miu, M. (2002) Micro and nanoelectrode calibration for voltammetric measurements, Proceedings of the International Conference on Semiconductors CAS 2002, Sinaia, Romania.

EFFECT OF GAS ENVIRONMENT ON THE RECOMBINATION PROPERTIES OF NANOSTRUCTURED LAYER- SILICON INTERFACE

S.LITVINENKO, A.KOZINETZ, V.SKRYSHEVSKY, O.TRETYAK
Kiev Taras Shevchenko National University, Radiophysical Faculty
01033 Volodimirska 64, Kiev, Ukraine

1. Introduction

Recently, porous silicon has demonstrated the attractive physical properties which offer a potential to design antireflection coatings for solar cells [1,2,3], light emitting diodes [4], for gettering of metal impurities [5], as electro- and thermoinsulating material and other applications. It is also considered a material for gas sensors due to its adsorptional ability, strong dependence of luminescence, and conductivity on the gas ambient. In spite of such advantages of the material, it has not yet been applied widely for the gas sensing perhaps because the nature of the adsorption mechanisms has not been studied enough.

Remarkable properties of porous silicon are mainly conditioned by its nanostructure. Unlike the traditional semiconductors, it consolidates the crystalline structure and huge square of the surface ($200 - 600$ m^2cm^{-3}), that can enlarge the effects of molecular adsorption [6]. Thus, the possibility to use the effect of the adsorption for sensor applications depends on the understanding of nanopores and nanocrystallites formation and on the design of the reliable technology of porous silicon with controlled parameters.

Anodic etching of silicon, in HF solution, is known to produce nanoporous silicon with pores of 2-4 nm diameter. Figure 1 presents a scanning electron microscopy image of the pores and crystallites profile in PS layer for the investigated samples. The pores reveal the dendritic structure and the porosity decreases towards the interface with the silicon substrate. Hence, the experimental results confirm that the material of porous silicon consists of nanosized pores and crystallites and the square of its physical surface is several orders larger than the square of the substrate. Another question at gas sensor design, is to analyze the influence of the adsorption on the device properties.

For example, the influence of the H_2O_2, O_2, C_2H_5OH molecules adsorption on the spectra and intensity of the photoluminescence of porous silicon has been investigated [7]. The shift up to 100 nm in the PL spectra, during the adsorption, was observed. The adsorption of the H_2O_2, O_2, C_2H_5OH, $C_2(CN)_4$ molecules was shown to change the concentration of the dangling bonds in the porous silicon [8]. This conclusion was a result of electron paramagnetic resonance and photoluminescence investigations.

T. Tsakalakos, et al. (eds.) pgs. 649 - 654
Nanostructures: Synthesis, Functional Properties, and Applications;
© *2003 Kluwer Academic Publishers.*

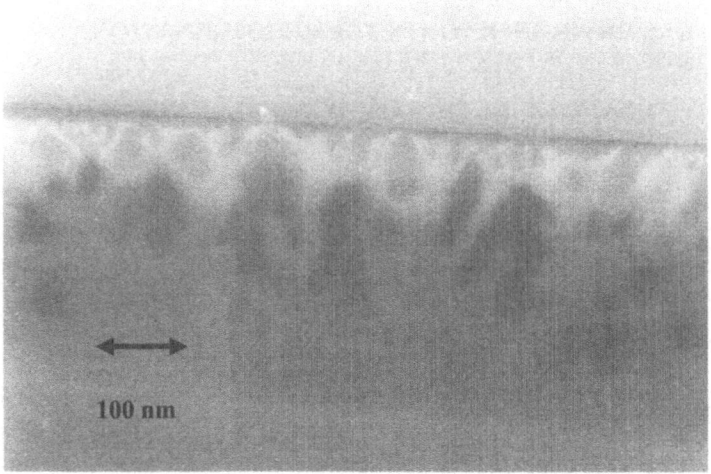

Figure 1. SEM image of pores profile in PS layer

In most such cases, the average "bulk" properties of the porous silicon dependent on the adsorption were investigated. However, the adsorption in the "bulk" of the porous silicon can influence the properties of the interface between the PS layer and its silicon substrate as well. Taking into account this consideration, we have set a task to reveal the parameters of the interface that are sensitive to the adsorption and to study the corresponding behaviours.

2. Modelling and Experiment

Surface recombination velocity of minority carriers at the silicon interface has been proposed to be consider as a principal parameter which is sensitive to adsorption. This parameter is expected to change under the adsorption in the porous silicon since the density, energy level, and charge of the recombinative electronic, states at the interface can change as well as band bending near the semiconductor surface. In turn, the change of surface recombination velocity will influence the characteristics of certain semiconductor device (for example, diode, transistor) and thus could be measured experimentally.

A test photovoltaic structure has been designed for our experiment with photocurrent strongly dependent on surface recombination velocity (Figure 2). Let us consider sharp, and shallow n+ p junction, with homogeneous distribution of doping impurities. In the case of illumination, from thin n+ emitter, it is a common solar cell structure optimized for maximum efficiency of PV energy conversion. When the incident light is from thick p-base region, the expression for the spectral response SR calculated for solar cell [9] is

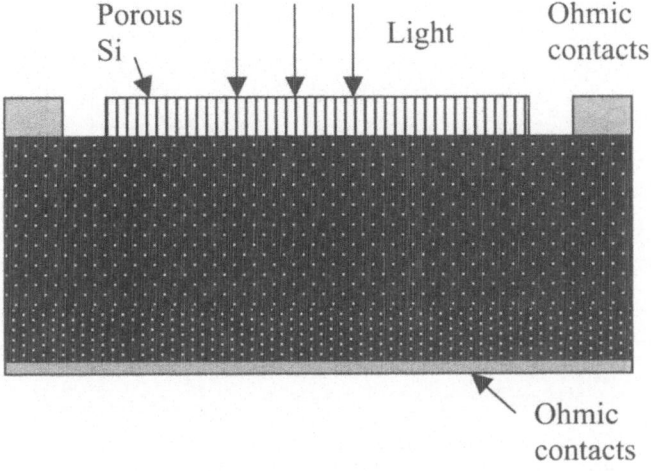

Figure 2. Structure based on p-n junction proposed for sensor application.

still valid with corresponding replacement of the n+ emitter parameters by the parameters of p-base: thickness W, surface recombination velocity S_2, diffusion length of minority carriers in the base L_n, diffusion coefficient D_n. Then, the spectral response under high light absorption ($\alpha(h\nu)W \gg 1$, $\alpha(h\nu)Ln \gg 1$) for the structure of Figure 2 can be approximated as:

$$SR_2 = \frac{1 + S_2 \alpha(h\nu) L_n/D_n}{S_2 L_n/D_n sh(W/L_n) + ch(W/L_n)} \qquad (1)$$

Calculations show that the spectral response in the last case is smaller, but much more sensitive to the surface recombination velocity S_2, at the base region interface than the spectral response in the common solar cell configuration. The test structure has a nanoporous silicon layer as an active gas adsorbing region assigned to influence the S_2 value. Therefore, such device structure may be interesting for the detection of molecules adsorption.

The behaviour of spectral response, when the recombinational properties of the reactive interface with nanoporous layer are changed, have been simulated by the help of PC – 1D software that is available for the calculations of solar cell characteristics. This software tool make it possible to obtain more precision results than the mentioned formula since it can take into account more realistic dopant profiles in the P-N junction. The calculated dependencies of SR_2 on S_2, for high and low absorption (λ=620 nm, 500 nm and λ=1000 nm correspondingly), are shown in Figure 3.

652

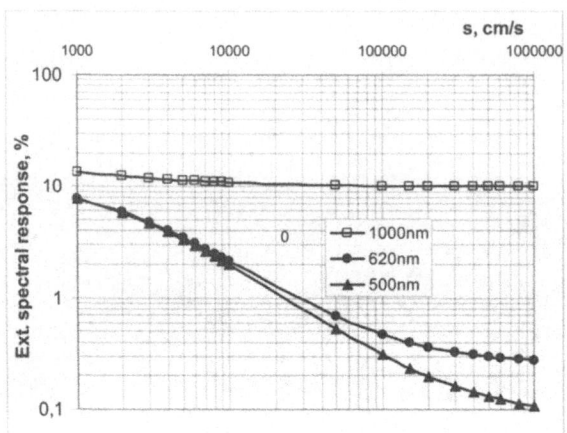

Figure 3. Calculated dependence of PV response for silicon p-n
junction on surface recombination velocity under monochromatic

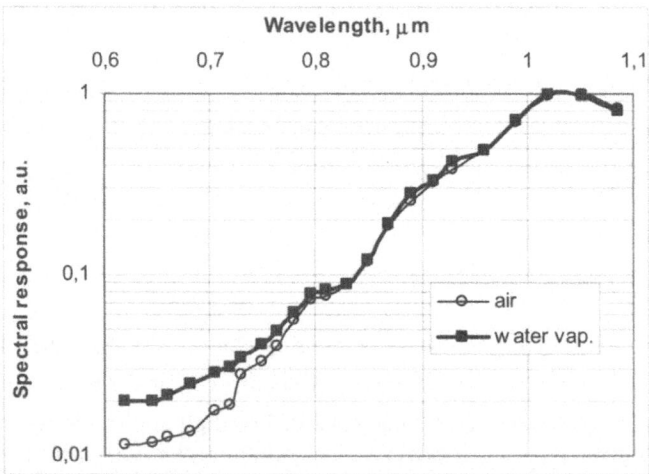

Figure 4. Spectral response at air and water vapour ambient (experimental)

The interpretation of these results is clear: when the device structure is illuminated from the base side, the influence of the surface recombination on the photocurrent increases for light with high absorption coefficient, if the most of the excess charge carriers are generated close to the surface. If the n$^+$ emitter is illuminated, the excess charge carriers are generated close to the junction and separate easy without significant influence of the surface recombination.

Porous silicon layer on the base side of the device has been created by standard anodization in $HF:C_2H_5OH:H_2O$ electrolyte. The current density in the Galvaostatic mode was 15 $mAcm^{-2}$, anodization time 1-5 min. It results in the 1 μm porous layer thickness. Silver paste was used for Ohmic contacts formation.

The spectral response of the test structure in dry and humid air were measured by the setup consisted of tungsten lamp, monochromator, chopper, selective nanovoltmeter. The area with porous silicon layer was illuminated through diaphragm to avoid the influence of monocrystalline silicon surface. Typical results are shown in Figure 4.

In the short wavelength region, the 20-30% increase of PV response is observed., however, some samples demonstrated the decrease of PV response in the presence of water vapor. The increase of PV response in this spectral region corresponds to the reduction of surface recombination velocity at the silicon substrate – porous silicon interface. This change may be caused by the re-charge of the porous silicon close to the interface that leads to the alteration of band bending; and directly by the change of recombination centers density at the interface. Surface recombination velocity is known to depend in non-monotonous manner on the surface band bending [10]. This model can explain that both increase and reduction of surface recombination velocity observed experimentally under water vapour exposure are caused by different initial surface barrier heights. The S_2 parameter has been estimated by fitting the experimental curves of spectral response by the curves calculated by PC-1D. So far $S_2 = 10^6$ cm/s for dry air and $S_2 = 10^5$ cm/s were obtained. Such high values confirm the supposition about the large physical area of the investigated interface.

3. Conclusions

i) The influence of adsorption from gas ambient (dry and humid air) on the surface recombination velocity that characterize the interface between silicon substrate and nanoporous/nanocrystalline silicon layer has been investigated.
ii) P-N junction with nanoporous silicon layer created at the p-base side is shown to be a device suitable for the detection of molecular adsorption. Photocurrent induced by short wavelength light illumination from the base side is an experimental parameter sensitive to the adsorption.

4. References

1. Hamilton B. (1995) Porous silicon. Topical Review // Semicond.Sci.Techn.-.-V.10.-P.1187-1207.
2. Skryshevsky V.A., Laugier A., Litvinenko S.V., Strikha V.I. (1998) Towards a Stable Porous Silicon Layers for Silicon Solar Cells // Proc.2nd World Conf. PV Solar Energy Conversion, Vienna (Austria). P.1611-1614.
3. Skryshevsky V.A., Kilchitskaya S.S., Kilchitskaya T.S., Litvinenko S.V., Nechiporuk A.I., Fave A., Laugier A. (2000) Impact of recombination and optical parameters on silicon solar cell with the selective porous silicon antireflection coating // Proc.16th European PV Solar Energy Conf., Glasgow (United Kingdom). P.1634-1637.

654

4. Cullis A.G., Canham L.T, Calcott P.D.G. (1997) The structural and luminescence properties of porous silicon // J.Appl.Phys.- V.82, N 3.- P.909-965.

5. Tsuo Y.S., Menna P., Pitts J.R., Jantzen K.R., Asher S.E., Al-Jassim M.M., Ciszek T.F. (1996) Porous silicon gettering // Record 25th IEEE Photovoltaic Specialists Conf. –Washington (USA). P.461-464.

6. Bisi O., Ossicini S., Pavesi L. (2000) Porous silicon: a quantum sponge structure for silicon based optoelectronics // Surface Sci.Reports.- V.38.- P.1-126.

7. Th.Dittrich, E.A.Konstantinova, V. Yu.Timoshenko. (1995) Influence of molecule adsorption on porous silicon photoluminescence. Thin Solid Films 255, 238-240.

8. E.A.Konstantinova, Th.Dittrich, V. Yu.Timoshenko, P.K.Kashkarov. (1996) Adsorption-induced modification of spin and recombination centers in porous silicon. Thin Solid Films 276, 265-267.

9. Sze, S.M. (1981) Physics of Semiconductor Devices, A Wiley-Interscience Publication, New York. Chichester. Brisbane. Toronto. Singapore.

10. Stevenson D.T., Keyes R.I., (1954) Physica, 20, 1041.

SYNTHESIS AND NOVEL APPLICATION OF NANOMATERIALS IN TUNGSTATE, TITANIA AND SILICON NITRIDE SYSTEMS

CSABA BALÁZSI

Hungarian Academy of Sciences, Research Institute of Technical Physics and Materials Science, Ceramics Department, 1121 Budapest, Konkoly-Thege M. út. 29-33

Abstract. This paper summarizes some recent advances made in synthesis and application of nanomaterials in tungstate, titania and silicon nitride systems. Firstly, the room temperature acidic precipitation and ageing process of tungsten oxide dihydrate grains are presented. Fourier transform infrared spectroscopy studies have shown that structural stability of Freedman type grains can be altered by presence of Li^+ and NH_4^+ ions. Water molecules and OH^- groups present in structure play an important role in ion insertion process during ageing. As regards titania system a mechanochemical coating process is highlighted. Afterwards, a study about silicon nitride ceramics with well known properties as low density, high strength and toughness is presented. As an alternative way to improve the mechanical and thermal properties of silicon nitride matrix, we performed the preparation and examination of nano-crystalline carbon added silicon nitride ceramic matrix composites.

Keywords tungsten oxide hydrates, titania, nanoparticles, silicon nitride, ceramic matrix composites

1. Case of Tungsten Oxide Hydrates

Tungsten oxide hydrates are gaining increasing attention in electro-, photo- and gasochromic materials research. Due to the wide interest in the field of chromism and large applicability in modern technology, numerous investigations were made on these materials. Several decades earlier the layered structure tungsten oxide dihydrate ($H_2WO_4.H_2O$) was successfully applied as the parent phase for metastable hexagonal tungsten oxide representing an optimal substance in intercalation chemistry [1]. Starting from the same basic material (Zocher and Jacobson type tungsten oxide dihydrate) hexagonal-WO_3 with variable levels of residual sodium was prepared and with deliberately added organic contamination cubic-WO_3 was also observed in our laboratory [2,3,4]. In the latest reports new interesting applications came to light of already realised or prospective tungsten oxide monohydrate based devices, which could easily be synthesised starting from the sol-gel method. Recently, an all-plastic $WO_3.H_2O$/polyaniline flexible electrochromic device was reported, using Freedman's acidic precipitation method for tungsten oxide monohydrate preparation [5]. Other up to

T. Tsakalakos, et al. (eds.) pgs. 655 - 673
Nanostructures: Synthesis, Functional Properties, and Applications;
© *2003 Kluwer Academic Publishers.*

date photo-, electro- and gasochromic applications utilising tungsten oxide as the basic-active material have attracted much interest recent years [6-11]. However, some uncertainty still persists relating the mechanism of chromism phenomenon (favourable or non-favourable role of water molecules), sensitivity against environmental conditions (especially high humidity, water formation on WO_3 pores) and long-term degradation [12,13].

Electro-, photo- and gasochromic devices are realised mainly by thin film processing, assuring porous structures and columnar morphologies for insertion host, enhancing ion mobility. Some of the authors have found direct correlation between the ion insertion (lithiation) mechanism and bonding state of water to the tungsten oxide network [5,8-10]. High diffusion coefficient was observed (considered as protonic diffusion) in the case of samples containing certain amount of humidity [9,10]. Structural water molecules have been found directly involved in lithiation process of tungsten oxide films prepared by sol-gel synthesis [8]. Fourier transform infrared spectroscopy, FTIR measurements have shown a Li/H_2O exchange with the formation of the bronze phase and partial elimination of water molecules from the WO_3 network. An electrochemical degradation study of tungsten oxide hydrate films have been performed by Judeinstein et al. [14] They have found that ion exchange properties of electrochromic $WO_3.nH_2O$ layers decreased when the amount of adsorbed water molecules decreased. Faster response times and shorter lifetimes were obtained when the water content increased. The same conclusion was confirmed and the beneficial role of structural water on lithium insertion kinetics into tungsten oxide was demonstrated by Marcel et al. [5]. For the preparation of $H_2WO_4.H_2O$, the parent phase of the above mentioned materials the acidic preparation of sodium tungstate solution is used. Freedman's method [15] is more general, but for the preparation of metastable WO_3 phases Figlarz and co-workers [Ref.1 and see other references in Figlarz's paper] suggested the earlier method of Zocher and Jacobson [16]. The main difference between these two methods is in the rate of precipitation caused by differing pH in the reaction solution. In paper [2] we showed that "spindle shape" morphology of the grains typical and unique for $H_2WO_4.H_2O$ prepared according to the method of Zocher and Jacobson is forming at the washing steps applied in the method. In papers [17] and [18] we showed that pH in the aqueous ambient directly controls the morphology and structural stability of Zocher and Jacobson type grains. In this paper Freedman type grains are handled in aqueous environments containing various cations (Li^+ and NH_4^+) with different concentrations. The results on long-term ageing observations are presented.

1.1 METHOD FOR TUNGSTEN OXIDE HYDRATE PREPARATION

Freedman (F) type preparation. Tungstic acid hydrate samples were prepared following the preparation route of Freedman [18] (Table 1.). Washing of these precipitates was carried out by 0.1N HCl solution on filter. Stirring and centrifuging were not used for the preparation of these samples. For long-term ageing investigations the tungsten oxide hydrate gels were mixed with water or tungstate solutions and allowed to stay under solutions in polypropylene containers. The following tungstates were deliberately added to solutions: lithium tungstate (Li_2WO_4) and ammonium metatungstate (($NH_4)_6H_2W_{12}O_{40}.2-6H_2O$, with $WO_3\%$:91.3%, $NH_3\%$: 3.35%). The suspensions in containers were regularly re-mixed by hand shaking. Four months and 24 months ageing times at room temperature were chosen for the F suspensions.

Characterisation of samples. The pH values of the washing liquids were determined. The procedures of the residual sodium analysis have been described earlier [2,3]. Infrared absorption spectra were taken by BOMEM MB-102 FTIR spectrophotometer equipped with deutero-triglicine-sulfate detector, at a resolution of 4 cm^{-1}, in the range of 400-4000 cm^{-1}; 2mg/g KBr pellets were used.

Table 1. Summary of the preparation conditions for typical F type grains.

Sample	Precipitation conditions			Washing conditions		Product of preparation	
	Quantity of starting Na_2WO_4. $2H_2O$ [g]	Temperature of precipitation [°C]	Number of washing steps	Volume of washing liquid [ml]	Time of interaction in one step [h]	XRD Guinier	[Na] in solid [ppm]
F	16.5	25	3 by 0.1N HCl on filter	2000	0.12	H_2WO_4. H_2O	< 10

1.2   OBSERVATIONS ON TUNGSTANE OXIDE HYDRATES

Details about F type particles ageing characteristics can be found in Table 2. The morphology of F type particles is shown in Figure 1. Adhered irregularly shaped submicron grains of freshly prepared F preparation are shown in Figure 1a. Well developed rectangular platelets similar in shape to the Zocher type particles, can be observed after two month ageing in alkali ion free aqueous ambient (Figure 1b). At ageing in ammonium tungstate solution the grains transform into new phase(s) as it is shown in Figure 1c. This morphology is characteristic for the sample aged in a solution with ion concentration, $S_1/[NH_4^+]$=0.095 N, ageing time, 24 month.

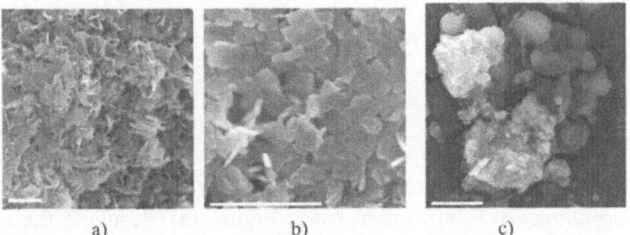

a)   b)   c)

Figure. 1. Morphology development for F type tungsten oxide hydrate particles in aqueous ambient. a) Irregularly shaped crystalline particles of $H_2WO_4.H_2O$, Joint Committee on Powder Diffraction Standards, JCPDS 18-420 precipitated by Freedman method, see Table 1. b) Rectangular platelets of $H_2WO_4.H_2O$ developed in undoped aqueous ambient during two month ageing process. c) New phases (other than $H_2WO_4.H_2O$) evolving in ammonium tungstate solution (sample $S_1/[NH_4^+]$), ageing time 24 months. Bar denotes 1 μm for SEM pictures.

In the same time the δ(O-W-O) bending mode present at 430 cm^{-1} (Figure 2b) has disappeared from the infrared spectra (Figure 2c). Moreover, in the case of sample with higher concentration (Figure 2d) even after four months ageing a strong peak at 1400 cm^{-1} has been developed. In this case the vibration changes below 1200 cm^{-1} - presenting degradation of material- are more clearly showed. According to [20,21,22], we could assign this peak at 1400 cm^{-1} with bending modes of adsorbed OH$^-$, the pair of these vibrations (3220 cm^{-1}) being overlapped with stronger vibrations of structural water in the case of our samples (in the range of 3000-3600 cm^{-1}). In the works of Cheng et al. [23] and Gui et al. [24] the infrared spectra data for ammonium tungstate compounds, $(NH_4)_x.WO_3$, $(NH_4OH)_x.WO_3$ ($0.13<x<0.33$) and $(NH_4)_{0.25}WO_3.1/3H_2O$ have been published. These compounds are characterised by a weak peak (in the case of $(NH_4)_x.WO_3$) or reveal a well developed peak at 1400 cm^{-1}. Our sample (Figure 2c), has the NH_4^+ ion incorporated in structure. It is characterised by a single peak in the range 900-600 cm^{-1}, corresponding to ν(O-W-O) stretching modes. On the basis of infrared spectroscopic measurements and literature data it is reasonable to consider that this peak at 1400 cm^{-1} (accompanied by the ν(O-W-O) stretching mode vibrations merging into one single broader band) is an indicator for some kind of NH_4^+ ions present in the structure.

Table 2. Summary of ageing conditions for F type particles.

Sample	$H_2WO_4 \cdot H_2O$ g	Added alkali tungstate solution		pH immediately after mixing	pH after 2 months
		Volume ml	Concentration of alkali ion, N		
$S_{undoped}$	0.72	76	-	2.34	2.23
$S_1/[NH_4^+]$	0.72	50	0.095	2.49	2.59
$S_2/[NH_4^+]$	0.72	28	0.017	2.12	2.00
$S_1/[Li^+]$	0.72	50	0.095	5.96	6.33
$S_2/[Li^+]$	0.72	30	0.016	3.31	5.06

The same ageing time but lower ion concentration ($S_2/[NH_4^+]=0.017$ N) keeps intact the starting morphology of F type grains (as in Figure 1a). Similarly the sample $S_2/[Li^+]$ with a concentration of Li$^+$ ($S_2/[Li^+]=0.016$ N) has preserved the starting morphology even at ageing time as long as 24 months. Room temperature FTIR absorption band assignations of the ageing products are presented in Table 3. At lower ion concentration and after four months contact at room temperature between tungsten oxide hydrate grains and NH_4^+ solution a peak at 1400 cm^{-1} has appeared in the infrared spectra

(Figure 2b). This peak can be observed after 24 months contact at the same ion concentration as well (Figure 2c). At longer interaction time a change at lower vibration modes (lower than 1200 cm^{-1}), breaking of stretching bonds v(O-W-O) at 709 cm^{-1} and 632 cm^{-1} transforming to a broad single v(O-W-O) band at 673 cm^{-1} can also be noticed.

Table 3. Room temperature FTIR absorption band assignations of the ageing products.

Assignations Sample	vHOH (H$_2$O)	δHOH (H$_2$O)	vW=O	vO-W-O	δO-W-O	(δOH) W-OH or (W-NH$_4$OH)	Reference
S$_{undoped}$ 3536		1605	1005	682	425	-	[20,21]
4 months 3374			940	630	377		
ageing 3177			910				
S$_{undoped}$ 3424		1626	945	665	-	-	[20,21]
24 months							
ageing							
S$_2$/[NH$_4^+$] 3549		1609	1010	709	430	**1400**	[20-24]
4 months 3391			947	632	378		
ageing 3187			922				
S$_2$/[NH$_4^+$] 3549		1630	1010	673	380	**1400**	[20-24]
24 months 3407			954				
ageing 3204			939				
S$_1$/[NH$_4^+$] 3557		1622	1010	775	426	**1400**	[20-24]
4 months 3395			949	717	374		
ageing 3169			908	648			

We found the peak at ≈1400 cm^{-1} evolved in the case of S$_2$/[Li$^+$] sample as well, but with lower intensity than presented in the NH$_4^+$ case (Fig. 2). In Figure 3b and Figure 3c the spectra of sample with lower ion concentration (S$_2$/[Li$^+$]=0.016 N) aged for four and 24 months can be seen. Comparing the infrared spectra with the reference sample (Figure 3a) only one difference can be observed after four months ageing, namely that a peak at 1414 cm^{-1} appears; it can be found after longer ageing time as well. However, after long-time ageing the same merging process of v(O-W-O) stretching modes can be noticed (Figure 3c), as presented earlier in the case of NH$_4^+$. From stretching bonds v(O-W-O) at 721 cm^{-1} and 667 cm^{-1} only the first one can be found at 718 cm^{-1}, meanwhile the second one is transforming to a weak shoulder at 634 cm^{-1}.

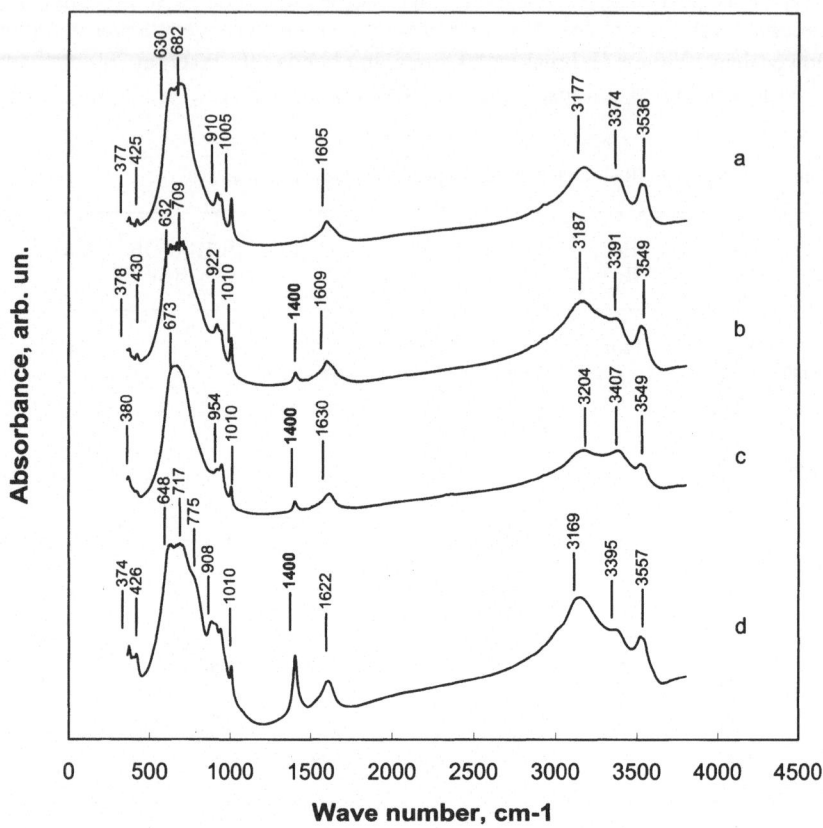

Figure 2. FTIR spectra of $H_2WO_4.H_2O$ aged in NH_4^+ containing aqueous solution. a - $S_{undoped}$ 4 months ageing, $H_2WO_4.H_2O$. b – S2/[NH_4^+]=0.017 N, 4 months ageing. c – S2/[NH_4^+]=0.017 N, 24 months ageing. d – S1/[NH_4^+]=0.095 N, 4 months ageing.

At this stage of examinations it is interesting to recall some of the earlier results reported about electrochemical lithium intercalation into tungsten oxide films. Badilescu et al. [8] have found that for all of tungsten oxide films prepared by different methods (thermally evaporated, sputtered, sol-gel) the IR peak at 1440 cm^{-1} changed accordingly to the lithiation process. The more amount of Li$^+$ was intercalated the higher the peak at 1440 cm^{-1} became. We think that the peak at 1414 cm^{-1} observed in the spectra of our Li aged sample S2/[Li$^+$] is essentially the same as the peak at 1440 cm^{-1} reported by Badilescu.

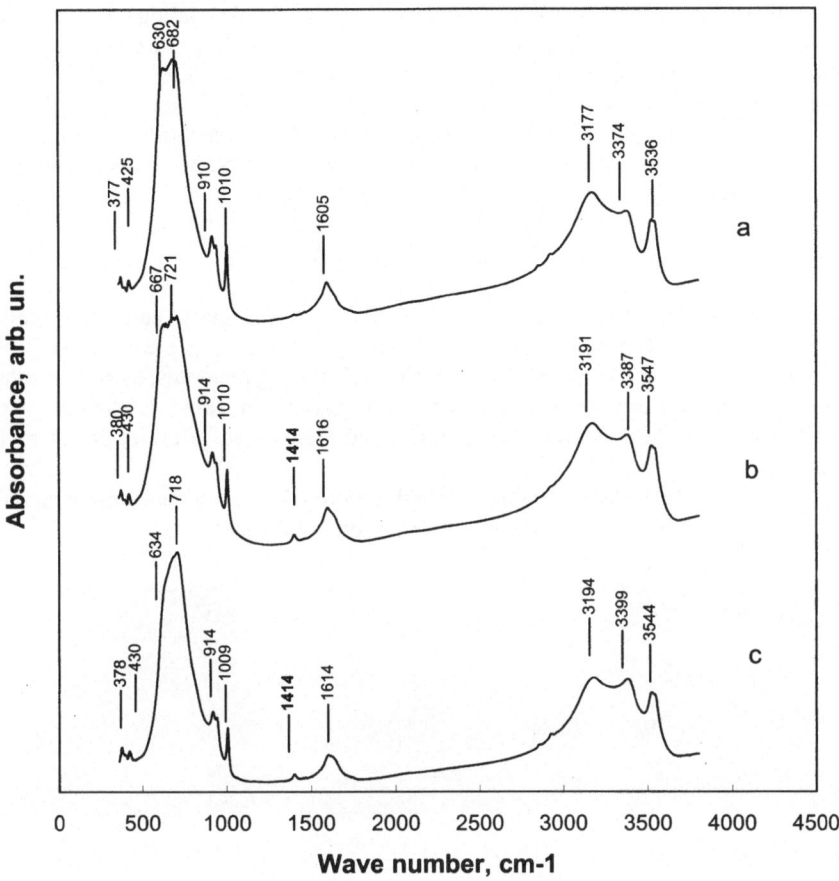

Figure 3. FTIR spectra of $H_2WO_4.H_2O$ aged in Li^+ containing aqueous solution. a - $S_{undoped}$ 4 months ageing, $H_2WO_4.H_2O$. b – S2/[Li^+]=0.016 N, 4 months ageing. c – S2/[Li^+]=0.016 N, 24 months ageing.

For the assignation of this peak however, in our opinion it is more likely an OH vibration influenced by incorporated Li^+ ion similarly to the influence of NH_4^+ ion. The literature data and FTIR results of this work indicate that ageing at room temperature in alkali and NH_4^+ solutions results in the incorporation of the alkali ions and NH_4^+ into the tungstate structure. NH_4^+ ion represents the most relevant case in this series of experiments.

2. Case of Titania Coatings on Silica

Titanium dioxide (TiO_2) coatings have been intensively studied for a wide variety of applications, such as antibacterial, detoxification uses because of their photocatalytic properties or as novel nano-crystalline solar cells due to their photovoltaic properties.

Sol-gel methods have a large application in processing of titania coatings. However, as a new method, attempts have been made to produce titania coatings through mechano-chemically way. In this study an investigation was made for surface coating of silica powders with titania nano-particles involving mechano-chemical activation. Several starting materials with different molar ratio were reacted in a planetary ball-mill. Dry and wet conditions were applied. Structural measurements (SEM, EDX) reveal a successful coating process.

2.1. METHOD FOR TITANIA COATING ON SILICA

The starting materials (Figure 4) were activated in a planetary ball-mill (Model Pulverisette-5, Fritsch, Germany). The experiments were planned as presented in Table 4. Dry and wet conditions were applied. Quantities of starting materials, as well as other characteristics of each experiments have been summarised in Table 4. Samples were taken after each stop of 5, 30, 60, 120 minutes and were dried in exsiccator at room temperature.

In order to study stability of coatings, samples were taken from detaching materials from balls during ultrasonic agitation in ethanol or distilled water.

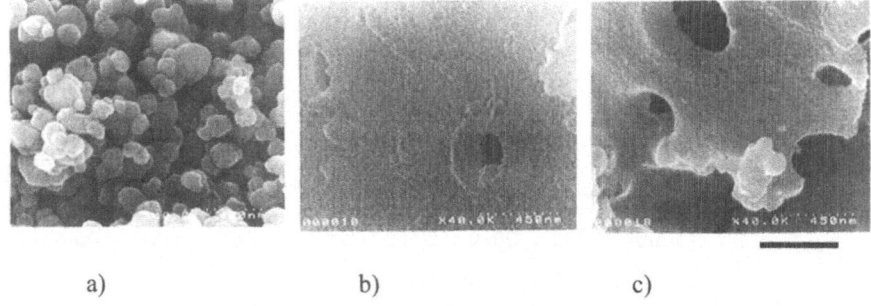

a)   b)   c)

Figure 4. Some starting materials: a) titania anatase form. b) silica as received. c) silica coated. Bar: 450 nm.

Table 4. Preparation conditions of titania coatings on silica

Sample	Ti subst.	Quantity of starting materials [g/g]	Molar ratio	Temperature of preparation [°C]	Milling balls, balls/mixture weight ratio	Volume of distilled water	Time of treatment [minutes], sample [g]	Rotation [rpm]
VI.	TiO₂ anatase.	SiO₂/TiO₂ 0.75/1	1	20°C	teflon, φ = 8 mm, 40pcs., balls/mixture = 13/1.75 one jar	10 ml		
VI/5							5 min/2g	330
XI.	Ti(OH)₄ cryst.	SiO₂/ Ti(OH)₄ 0.75/1.45	1	20°C	teflon, φ = 8 mm, 40pcs., balls/mixture = 13/2.25 one jar	20 ml		
XI/120							120 min/all	330

2.2   OBSERVATIONS OF TITANIA COATING ON SILICA

Energy dispersive X-ray spectra of starting materials can be followed on Figure 5. Beside of main chemical components of starting materials (Ti, Si, O) only Pt and Pd energy values are detected which were used as sputtering agents.

Figure 5. presents the EDX spectra of starting materials (TiO₂ anatase and SiO₂ porous) and end product VI/5 (as in Figure 4c), prepared in wet conditions (Table 4). From this figure an observation can be made, that the end material is a mixture of starting materials (TiO₂ anatase and SiO₂ porous). We obtained similar data (SiKb1 and TiKa1 are present in all spectras of end material, Figure 5c, 5d, 5e) from three different parts of sample.

Scanning electron micrographs presented on Figure 6. are containing the same information about morphology as discussed in previous case (Figure 4c). Using other titania starting materials (as in Table 4, sample XI.) the effects are the same as presented earlier (case of sample VI/5, EDX observation results).

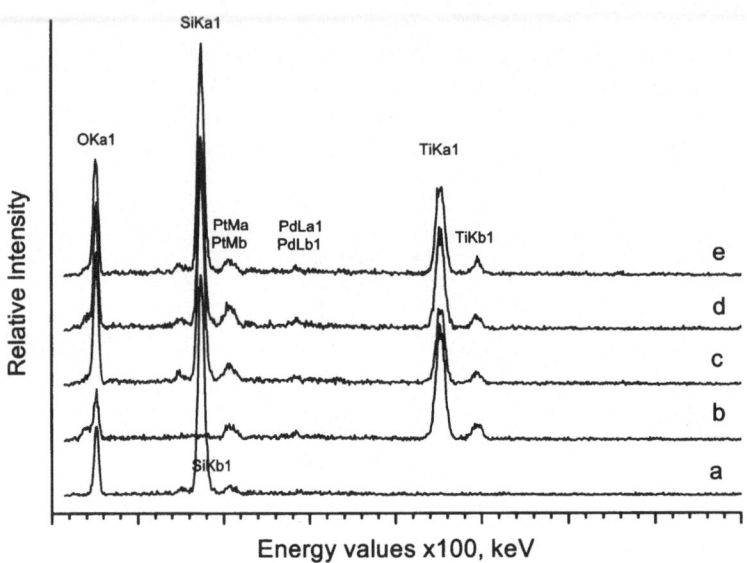

Figure 5. Energy dispersive X-ray spectra (EDX) of starting materials and end product VI/5 (as in Figure 4c, preparation as in Table 4). Observation: End material VI/5 is a mixture of starting materials. a – SiO₂ porous. b – TiO₂ anatase. c, d, e – coated material showing both of the SiKb1 and TiKa1 peaks.

In the same manner as in anatase case, we can find adhering titania grains to silica surface, but in the same time we can follow the developing new morphology, a new fibrous structure. In the same time in very explicit way comes to light all details of deposition mechanism of titania to silica surface, as follows. Because of inhomogeneous parts of silica starting material, first only a small amount of silica shows morphological transition to smashed-fibrous state.

The remaining non activated silica parts, especially the edges of cavities are presenting an ideal place for settling of titania particles from aqueous slurry. First the settling of titania particles occurs on the edges. By increasing the activation time all silica bodies together with titania particles on surface transform to smashed-fibrous end material (as in Figure 6b, 6c). As result we have the titania particles embedded in silica surface with a totally new morphology.

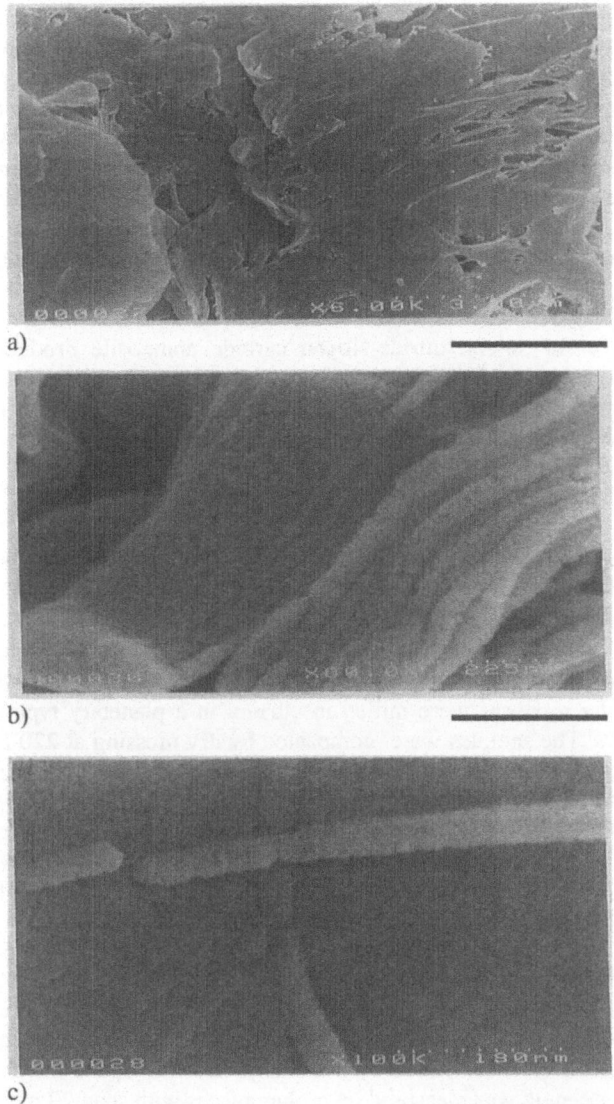

Figure 6. Scanning electron micrographs of silica coated with titania (sample XI/120). a) appearing new fibrous morphology, bar 3 μm, b) –c) coated parts at higher magnifications, bar 225 nm in b) and bar 180 nm in c).

3. Case of Silicon Nitride

The processing of new materials with extreme mechanical, thermal and electrical properties is of great technological interest. In the research field of ceramics many recently developed preparation routes were determined to produce ceramic matrix composites with superior technological characteristics [35]. Different strategies for the synthesis of silicon carbide-silicon nitride composites from pre-ceramic polymers are presented by Xujin Bao et al [36]. The use of a polymeric precursor route to synthesise ceramic composites offers processing at low temperature and the possibility to optimise the new composite materials production by tailoring the composition and molecular structure of the polymers. A low cost silicon carbide-silicon nitride composite processing route has been reported by Hnatko et al [37]. In this case, the formation of bulk silicon nitride based composite is realised by carbothermal reduction of SiO_2 by carbon in the Y_2O_3-SiO_2 system at the sintering temperature. Although the mechanical properties of as-prepared samples should be further optimised, this process seem to be a perspective choice for silicon nitride-silicon carbide composite production. At our department silicon nitride based composites were prepared through carbon phase addition by mechanochemical synthesis and HIP sintering. In this work results about FTIR measurements and mechanical properties are presented.

3.1   METHOD OF COMPOSITE PREPARATION

Details about sample preparation can be followed in Table 5. The compositions of the starting powder mixtures of the four materials were the same: 90 wt.% Si3N4 (Ube SN-ESP), 4 wt.% Al2O3, and 6 wt.% Y2O3. In addition to batches carbon black (Taurus Carbon black, N330, average particle size between ~50-100 nm), graphite (Aldrich, synthetic, average particle size 1-2 μm) and carbon fiber (Zoltek, PX30FBSWO8) were added. The powder mixtures were milled in ethanol in a planetary type alumina ball mill for 150 hours. The samples were compacted by dry pressing at 220 MPa. Carbon fibers were added to mixtures only before dry pressing (samples 629 from Table 5.) Samples from 644 and 645 batches were collected during milling process after several stops. These samples were passed to FTIR examinations. Infrared absorption spectra were taken by BOMEM MB-102 FTIR spectrophotometer equipped with deutero-triglicine-sulfate detector, at a resolution of 4 cm^{-1}, in the range of 400-4000 cm^{-1}; 2 mg/g KBr pellets were used. The materials were sintered at 1700°C in high purity nitrogen by a two-step sinter-HIP method using BN embedding powder. The dimensions of the as-sintered specimens were approximately 3.5 x 5 x 50 mm. After sintering the weight-gain values were determined. The density of the as-sintered materials was measured by the Archimedes method. The elastic modulus and the four-point bend strength were determined by a bending test with spans of 40 mm and 20 mm. Three-point strength was measured on broken pieces with span 20 mm.

Table 5. The composition of starting powder mixtures

Material	Composition, wt%			Added carbon			
				carbon black	graphite	carbon fiber	
	Si_3N_4	Al_2O_3	Y_2O_3	C/Si_3N_4 molar ratio	C/Si_3N_4 molar ratio	wt% mixture	to
642	90	4	6	-	-	-	
644	90	4	6	3	-	-	
645	90	4	6	-	3	-	
629	90	4	6	-	-	1	

3.2 OBSERVATIONS MADE ON COMPOSITES WITH SILICON NITRIDE MATRIX

Infrared spectroscopy measurements are presented on Figure 7 and Figure 8. Samples extracted from 644 batch during long duration milling experiment can be followed on Figure 7c, Figure 7d and Figure 7e. Infrared spectra of yttria and alpha silicon nitride starting powders are presented on Figure 7a and Figure 7b. On yttria powder spectra we can distinguish two characteristic peaks at 606 and 631 cm^{-1}. The mixture of powders mechanically activated for 1 hour (Figure 7c) is characterised mainly by alpha silicon nitride vibration modes (as on Figure 7b). O-H stretching modes at 3432 cm^{-1} and O-H bending at 1634 cm^{-1} can be found on infrared spectra. These vibrations are assigned with N-H vibrations in case of inert atmosphere working conditions [38] . At the end of mechanical activation (Figure 7d) as result we have a powder mixture with dominant alpha silicon nitride vibration modes. The vibration modes considered to be Si-N at 600 cm^{-1} and yttria at 640 cm^{-1} have been appeared. At the end of activation O-H bonds disappeared from spectra. After oxidation at 800°C for 2 hours (Figure 7e) peaks characteristic to alpha silicon nitride and yttria can be observed on infrared spectra. Alumina which has a broad band at 798 cm^{-1} (not presented on figure) and presents 4 wt% of the mixture can not be seen and has no substantial effect to the spectra of mixtures.

668

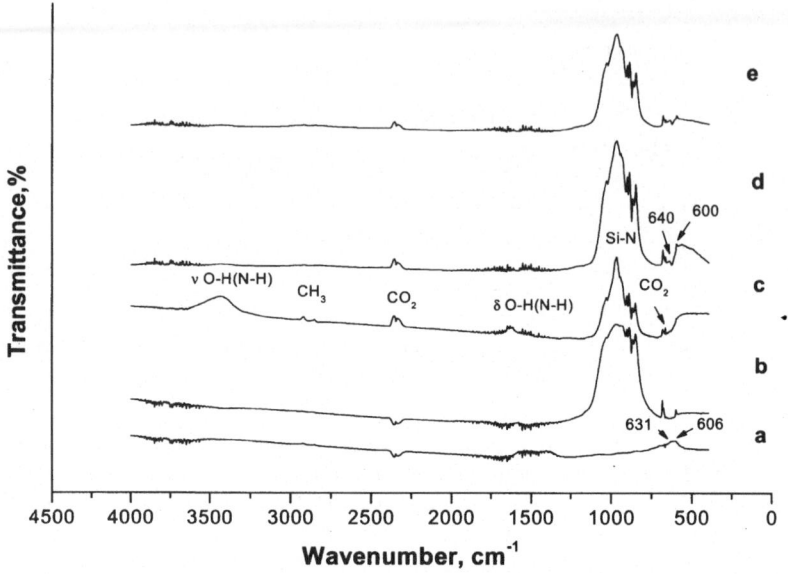

Figure 7. FTIR spectra of starting materials and milling products (batch 644). a – Y_2O_3 starting sintering additive powder. b – αSi_3N_4 starting powder. c – Resulting mixture after 1h activation. d – Resulting mixture after 150h activation. e – Resulting mixture after 150h activation and oxidation at 800°C for 2 h.

Samples gained from 645 batch during long duration milling experiment can be followed on Figure 8a, Figure 8b and Figure 8c. The mixture of powders mechanically activated for 1 hour (Figure 8a) is characterised mainly by alpha silicon nitride vibration modes (as on Figure 7b). In this case O-H stretching modes at 3432 cm^{-1} can be found on infrared spectra even after oxidation at 800°C (Figure 8c). After 150 hours of mechanical activation (Figure 8b) as result we have a mixture with dominant alpha silicon nitride vibration modes. At 2094 cm^{-1} a weak shoulder has appeared. According to Ref. 38 we assigned this peak with Si-H vibration. The vibration modes Si-N at 600 cm^{-1} and yttria at 640 cm^{-1} are more developed in this case. In agreement with earlier observations (batch 644), after oxidation at 800°C for 2 hours (Figure 8c) peaks characteristic to alpha silicon nitride and yttria can be found on infrared spectra.

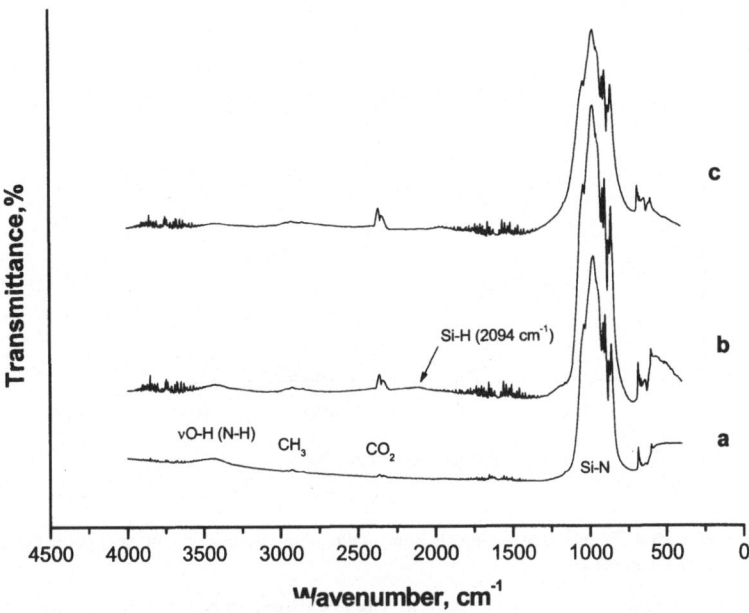

Figure 8. FTIR spectra of milling products (batch 645). a – Resulting mixture after 1h activation. b – Resulting mixture after 150h activation. c – Resulting mixture after 150h activation and oxidation at 800°C for 2 h.

On the basis of infrared spectroscopic measurements a confirmation has been made that no considerable differences occurred during milling process of batches 644 and 645. At the end of this process we obtained a milling product that could be characterised by the same structural characteristics, same vibration modes. In these series of experiments Si-C-N bond formation was not observed. After performing the mechanical activation of powder mixtures, rectangular samples were obtained by dry pressing at 220 MPa. The as-obtained samples were oxidised at different temperatures. Oxidising at different temperatures resulted in samples with different carbon content as presented in Figure 9.

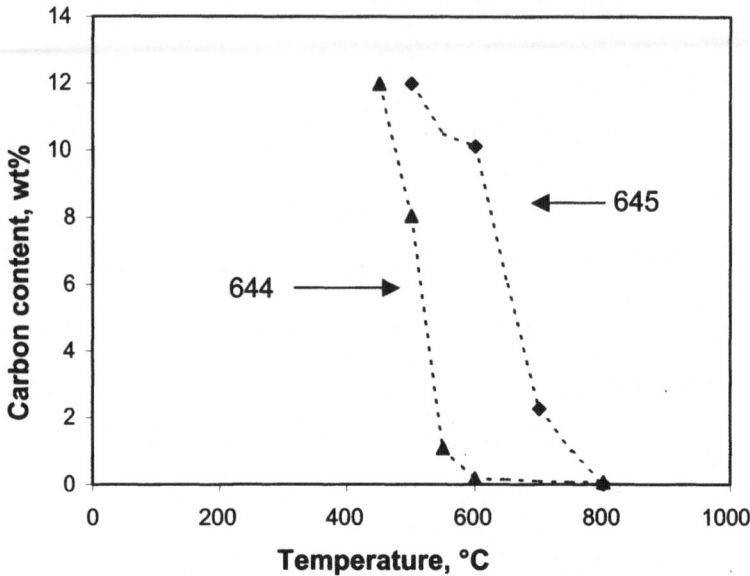

Figure 9. Samples with different carbon contents after oxidation in atmosphere. In each point the average value of four samples are presented.

An interesting remark can be added to observations presented on Figure 9. The behaviour of nanocrystalline carbon black and graphite added to silicon nitride matrix was found to be sharply different regarding the oxidation process. From 450°C up to 600°C the carbon black content has a decreasing tendency, at 600°C having a maximum of 0.4 wt% contribution to structure. At this stage the graphite content is around 10 wt%. From this point the graphite content has also a decreasing tendency with increasing temperature. Continuing the oxidation process above 800°C, according to weight losses we obtained a carbon free structure. Relation between apparent density of samples containing carbon (644 and 645) and reference (642 and 176) carbon free samples and modulus of elasticity after sintering are presented on Figure 10. The 642 reference samples are characterised by a higher apparent density and higher modulus than samples with carbon content. Samples 644 (nanocrystalline carbon black) have higher modulus values than samples with graphite (645). Linearly fitted straight line 176 were added to Figure 10. from Ref. 39. These line have been produced to study partial and final sintering processes. From Figure 10. seems that carbon content variation has the same role as partial sintering. The same curves (regression lines) as in the case of line 176 , but different gradients can be observed at samples 644 and 645. The linear regression line 176 has the intersection with x axis is at 1,88 value.

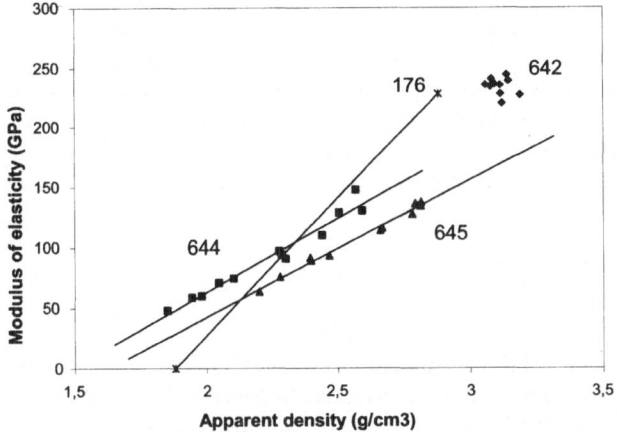

Figure 10. Relation between apparent density and modulus of elasticity after sintering. 644 and 645 samples containing carbon, 642 and 176 carbon free samples.

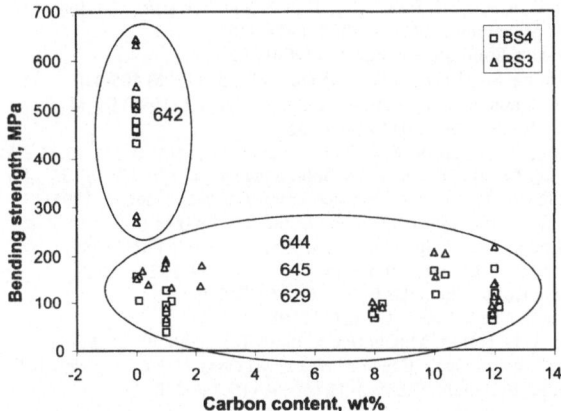

Figure 11. Relation between carbon content and four point bending strength (BS4) and three point bending strength (BS3). Carbon contents for samples 644 and 645 resulted from oxidation process as in Figure 3. Samples 629 are characterised by 1 wt% carbon content as in Table 5.

As conclusion can be drawn that samples 644 and even the prognosis part of regression line for 645 at this point has a higher value. This means that at the initial stage of sintering a certain amount of carbon addition has a beneficial role to the modulus of elasticity. This tendency is maintained till 2,338 value obtained from intersection of regression lines 176 and 644. Above this value, although modulus of batches 644 and 645 have an increasing tendency, compared with reference 176 carbon

672

content has a detrimental role to modulus. A comprehensive view about the carbon content effect to mechanical properties can be seen on Figure 11. Samples with added carbon (629, 644, 645) present lower values for mechanical properties as compared with 642 reference samples . To clarify the role of added carbon in future work we should consider the complex sintering process [41], the formation of rod like β-Si₃N₄ (which assure the good mechanical properties) and new sintering glassy phases and the possible carbothermal reaction with SiC grains as result [37]. Further structural observations are needed to clarify the exact mechanism of sintering and of carbothermal reduction (if any) in the presence of carbon.

4. Acknowledgement

Csaba Balázsi thanks for OTKA Postdoctoral Research Grant (D38478), OTKA, grant no. T 32730 and János Bolyai Research Grant. Support of Dr. J. Pfeifer, Prof. M. Yoshimura, Dr. P. Arató, Mr. F. Wéber are highly acknowledged.

5. References

1. M. Figlarz, Prog. Solid State Chem., 19 (1989) 1.
2. Cs. Balázsi, J. Pfeifer, Solid State Ionics, 124 (1999) 73-81.
3. J. Pfeifer, Cs. Balázsi, B. A. Kiss, B. Pécz, A. L. Tóth, J. Mat. Sci. Lett., 18 (1999) 1103.
4. Cs. Balázsi et al. Solid State Ionics, 141-142 (2001) 411-416.
5. C. Marcel, J.-M. Tarascon, Solid State Ionics, 143 (2001) 89-101.
6. C. Bechinger, B. A. Gregg, Sol. Energy Mat. and Sol. Cells, 54 (1998) 405-410.
7. O. A. Harizanov, K. A. Gesheva, P. L. Stefchev, Ceramics Int., 22 (1996) 91-94.
8. S. Badilescu et al. Thin Solid films, 250 (1994) 47-52.
9. J-G. Zhang, C. E. Tracy, D. K. Benson, S. K. Deb, J. Mater. Res., Vol. 8, No. 10. (1993)
10. S. Papaefthimiou, G. Leftheriotis, P. Yianoulis, Solid State Ionics, 139 (2001) 135-144.
11. C. G. Granquist, Handbook of Inorganic Electrochromic Materials, Elsevier, 1995.
12. A. W. Czanderna et al., Sol. Energy Mat. and Sol. Cells, 56 (1999) 419.
13. C. M. Lampert, Sol. Energy Mater. Sol. Cells. 52 (1998) 207. and 6 (1999) 449.
14. P. Judenstein, R. Morineau, J. Livage, Solid State Ionics, 51 (1992) 239.
15. H. Zocher, K. Jacobson, Kolloidchem. Beih., 28, (6), 167 (1929)
16. M. L. Freedman, J. Amer. Chem. Soc., 81, 3834 (1959)
17. Cs. Balázsi, J. Pfeifer, A. L. Tóth, J. Mihály, Proceedings Transcom'99, S6, 59-62.
18. Cs. Balázsi, Materials Structure, Bull. Czech-Slovak Cryst., Ass. Vol. 6, 2 (1999) 135.
19. K. Furusawa, S. Hachisu, Sci. Light, (Tokyo), 15 (1966) 115. I and II.
20. M.F. Daniel et al. J. Solid State Chem., 67, 235-247 (1987)
21. A. Chemseddine, F. Babonneau, J. Livage, J. Non-Crystalline Solids, 91 (1987) 271-278.
22. C. Guery et al. J. Solid State Electrochem. 1 (1997) 199-207.
23. K. H. Cheng, A. J. Jacobson, M. S. Whittingham, Solid State Ionics, 5 (1981) 355-358.
24. C. Gui, Y. Shang-Ci, Z. Juan, J. Cent. South Univ. Technol., Vol. 31, 6 (2000) 502-505.
25. K. Schronert, R. Weichert, Chem. Ing Tech. 41 (1969) 295.
26. S. Suda, S. Ichikawa, N. Wada, T. Umegaki, J. of Materials Sci. 35 (2000) 3023.
27. S. Rajesh Kumar, C. Suresh, Asha K. Vasudevan, N. R. Suja, P. Mukundan, K. G. K. Warrier, Materials Letters 38 (1999) 161-166.
28. W. P. Hsu, R. Yu, E. Matjevic, J. Colloid and Interface Sci. 156 (1993) 56-65.
29. M. Ocana, J. V. Garcia-Ramos, C. J. Serna, J. Am. Ceram. Soc. 75 [7] 2010-12 (1992).
30. H. Shin, R. J. Collins, M. R. De Guire, A. H. Heuer, C. N. Sukenik, J. Mater. Res. Vol. 10, No. 3. Mar 1995.
31. H. Imai, H. Hirashima, J. Am. Ceram. Soc., 82 [9] 2301-304 (1999)

32. N.V. Kosova, A. Kh. Khabibullin, V. V. Boldyrev, *Solid State Ionics*, 101-103 (1997) 53-58.
33. K. Schrijnemakers, N. R. E. N. Impens, E. F. Vansant, *Langmuir* 1999, 15, 5807-5813.
34. Y. Zhu, L. Zhang, C. Gao., L. Cao, *J. of Materials Science* 35 (2000) 4049-4054.
35. C. G. Papakonstantinou, P. Balaguru, R. E. Lyon, Composites: Part B 32 (2001) 637.
36. X. Bao, M. J. Edirisinghe, Composites: Part A 30 (1999) 601.
37. M. Hnatko, P. Sajgalik, Z. Lences, F. Monteverde, J. Dusza, P. Warbichler, F. Hofer, Key Eng. Mat. Vols 206-213 (2002) 1061.
38. G. Yu, M. J. Edirisinghe, D. S. Finch, B. Ralph, J. Parrick, J. Eur. Ceram. Soc. 15 (1995) 581.
39. P. Arató, E. Besenyei, A. Kele, F. Wéber, J. Mat. Sci. 30 (1995) 1863.
40. X. Junmin, J. Wang, Materials Letters 49 (2001) 318.
41. M. D. Alcala, J. C. S. Lopez, C. Real, A. Fernandez, P. Matteazzi, Diamond and Rel. Mat. 10 (2001) 1995.

APPLICATIONS OF FUNCTIONAL NANOCOMPOSITES

T. TSAKALAKOS[1], R. L. LEHMAN[1], T. N. NOSKER[1], J. D. IDOL[1], R. RENFREE[1], J. LYNCH, K. E. VAN NESS[2], M. DASILVA, S. WOLBACH, E. LEE
[1]*Rutgers University, Piscataway, NJ 08854, Department of Ceramic and Materials Engineering Rutgers University, 607 Taylor Road, Piscataway NJ 08854-8065* [2] *Washington and Lee University Lexington Virginia*

Abstract. The length scales defining structure and organization determine the fundamental characteristics of a material. Traditional polymeric materials exhibit organization on two length scales: the molecular scale, e.g., the unit cell of the crystal through folding chain or the local arrangement of amorphous polymer phase, and the scale particles/phases within the composite typically much longer length and on the order of micrometers or greater. In immiscible polymer blend, materials (IMPB), however, that melt domains in the nanoscale range, have been observed with ratios that are capable of being generated by manipulating shear rate, temperature, and viscosity during melt processing and by selecting the polymer pair to conform to the requisite viscosity/volume fraction relationship. Properties of IMPB's have recently shown remarkable enhancements. Nanotechnology of dispersion of nanoparticles in IMPB's has also been found to be critical component in fabricating nanocomposites of extraordinary structural and functional performance. Among the various methods that are currently used, are functionalization techniques such as coating of nanoparticles with proper material in order to maximize homogeneity of dispersed nanoparticles in the polymer matrix. This approach provides an opportunity for the processing of polymer/ceramic composites at the nanoscale level. Specific examples of ceramic nanoparticle nanocomposite will be discussed with emphasis on mechanical and magnetic properties.

1. Introduction

There is a general concensus, among industry and academia, that most of the structural and functional new applications for nanomaterials will come from composites. These nanocomposites will have an immense impact in the future, job growth. These applications vary from use in polymers, biological applications such as drug delivery batteries, electronics, cosmetics, sensors, fuel cells, and catalysis to coatings on metals and computer screens and other displays.

Polymers with infiltration of nanoparticles are on the verge of commercialization. Compared with traditional fillers, nanocomposites generally offer enhanced physical

T. Tsakalakos, et al. (eds.) pgs. 675 - 689
Nanostructures: Synthesis, Functional Properties, and Applications;
© *2003 Kluwer Academic Publishers.*

features, such as; increased stiffness, strength, barrier properties, and heat resistance, without loss of impact strength and with improved aesthetics in a very broad range of common thermoplastics and thermosets. And, because particle sizes are on the order of the wavelengths of visible light, they do not change optical properties such as transparency that is often required in these applications.

The major problems with nanocomposite development, apart of cost effectiveness, are agglomoration and dispersion of nanosized reinforcements. Most of these problems have been solved successfully through the functionalizaton of nanoparticles and formation of second and third generation of nanoparticles core-shell hybrid structures. One of the most promising nanocomposite is Immiscible Polymer Blends (IMPB).

The novel properties and performance of these IMPBs are derived from high mechanical performance, at very low cost, resulting from non-isotropic behaviour of the different phases with directional properties (determined during melt-to-solid processing). Tougher, stronger engineering plastics, derived from enhanced interfacial forces with volume fractions adjusted to generate two solid state interpenetrating phases, are also an important factor.

The aim of the advanced functional nanomaterials development is to engineer "unit cell" building blocks with the dimensions of small-scale physics to: assemble these building blocks into 3-D large-scale bulk materials, to utilize inorganic nanoparticle technology to provide functional particle compositing capabilities that enable exploitation of the immense interfacial surface area of IMPBs to produce new classes of novel functional and structural composites, and to develop and evaluate alternative structural IMPB designs for recycled materials (to make more and bigger all-plastic structures)

2. Immiscible Polymer Blends (IMPB).

Blends of two or more polymers exist in miscible or immiscible forms. When the cohesive energy density of the component pure polymers differs by less than a few percent, a miscible blend results. Miscible blends, or alloys, of pure polymers are convenient economic path to plastics compositions with cost effective property combinations not offered by single pure polymers. They are widely used in automobiles, appliances, medical devices, and in many other engineering plastic and packaging applications.

If the cohesive energy density of the pure polymers, which depends on the chemical structure and composition, differs by more than a few percent, an immiscible polymer blend (IMPB), consisting of two distinct phases, is formed. In these blends, the disperse domains have been chemically modified to enhance interfacial attraction with resulting increased impact strength to acceptable levels, as shown in Figure 1.

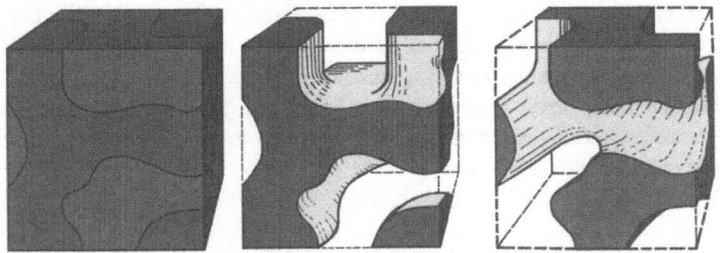

Figure 1. Proposed Structure of Immiscible Polymer Blends

However, when the polymer pair volume fractions (v) and melt viscosities (η) are interactively balanced, that is $v_a/v_b \approx \eta_a/\eta_b$, a unique immiscible polymer blend can form having unexpectedly high properties and a co-continuous morphology. This structure provides for a more efficient stress transfer between the immiscible phases without the need for chemical modification. Demonstrated examples are the patented high performance HDPE/PS, HDPE/PP/ glass and HDPE/PS/glass IMPB composites developed at Rutgers University. Figures 2,3 demonstrate clearly the enhanced modulus of IMPB and its corresponding microstructure.

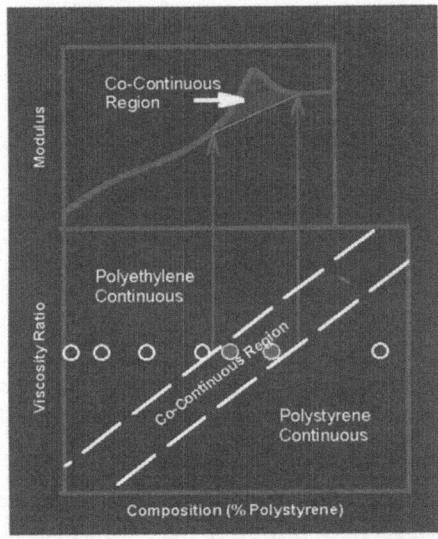

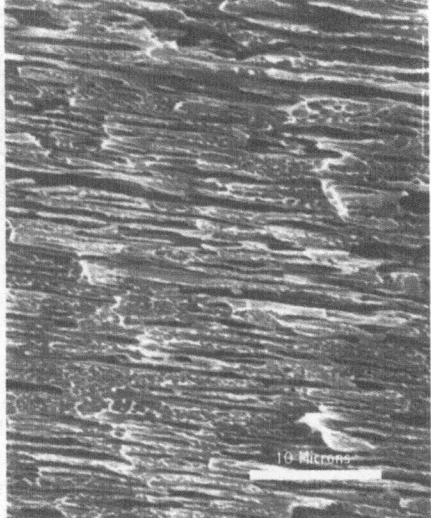

Figure 2 Robust Mechanical behaviour of IMPB's.   *Figure 3* IMPB microstructure

Another patented novel mixing and dispersion method of nanoparticles is Nanomilling™. Nanomilling shears a fluid within the gap of a cylinder spinning inside an outer cylinder. Thus, with a narrow residence time distribution and the shearing of thin layers, the polymer-nanoparticle mixture experiences a uniform shear environment, enabling all nanoparticles to be uniformly deagglomerated. Another advantage of this approach is that it is continuous processing technology that can be integrated with any polymer processing operation. Current research is focused on solvent-based polymer systems for the processing of optical, packaging, piezoelectric and biological.

Other developments include the Virtual Mixedness Software (VMS) that can create images and image analysis data sets of ideal mixtures of nanoparticles in the polymer phase. The overall research target is making light-weight high strength polymeric structures for the many applications such as high strength structures where conductive metal nanoparticles are dispersed in one of the interpretrating immisible phases to create an electrically conductive network.

3. Mechanically Enhanced Polymethylmethacrylate by Infiltration of Alumina/Titania Nanoparticulates.

One of the main areas for improvement in polymer-nanoparticle composite materials is to develop ways to effectively incorporate the nanomaterials in bulk molten polymer systems, as shown schematically in Figure 4, that depicts uniform distribution of Al_2O_3/TiO_2 in PMMA polymer.

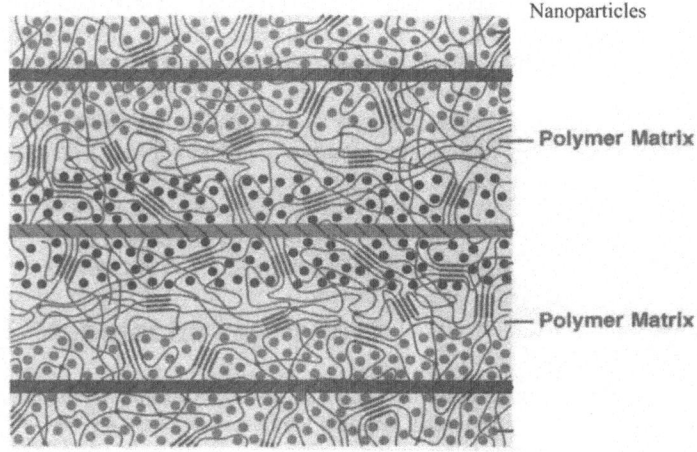

Figure 4. Schematic microstructure of a PMMA–Al_2O_3/TiO_2 nanocomposite. The dramatic enhancement of strain to failure with simultaneous increase of specific modulus and strength makes this Nanocomposite an excellent candidate for light structural par industrial applications

The mechanical properties of a nanocomposite consisting of 5wt% ceramic (87:13wt% Al2O3:TiO2) in polymethylmethacrylate (PMMA) were investigated. In this paper, the adhesion and dispersion of functionalized nanoparticles were examined. The dispersion of the ceramic in the PMMA matrix proved to be more effective than that of adhesion. By proper dispersion, translucent non-porous nanocomposites with increased toughness, strength, hardness, and scratch resistance were fabricated with high reproducibility.

3.1 EXPERIMENTAL MATERIALS AND PROCEDURES

Alumina and titania nanoparticles (39 and 40 nm average diameter respectively) obtained from Nanophase Technologies Corporation, were initially mixed and heated at

250°C to remove any hydroxyl groups present in the as-received powder. Micron sized alumina titania spray dried granules (obtained from Inframat) of ~50μm diameter were also used for comparison with that of the Nanophase Technological particles. These spray dried particles were composed of sub-micron particulates of size ~0.4μm. Under the vigorous processing conditions, the soft spray dried granules were able to break up into their respective nanoparticles. The specific surface area of the nano alumina and titania were 42 and 38 m2/g respectively. SEM and XRD were used for the characterization of the as-received powders and the nanocomposite. The glass transition temperatures (T_g) of the fabricated neat PMMA and nanocomposite were determined through DSC.

In the first part of this study adhesion was considered where three surfactants, dimethylmethoxysiloxane, 3-aminopropyltriethoxysilane, and 3-(2-aminoethylaminopropyltrimethoxysilane), used for the formation of a functionalized monolayer on the surface of the particulates. The amount of surfactant needed to form a monolayer on the ceramic particulates was determined based on the specific surface area of the powder (S_A) and the specific wetting surface of the surfactant (W_S), as shown below. The specific wetting surface of the surfactants (Gelest Inc.) is ~358 m^2/g. $S_A = (6)/(d\rho)$ Amount of Surfactant Needed = $((m)(S_A))/(W_S)$, where d and ρ are the diameter and density of the ceramic respectively. The effectiveness of the surfactants was measured via FTIR.

In the second part of this study, four dispersants were considered: phosphate polyester (BYK), polyimine ester (KD-2), oleyl alcohol, and O-(2-aminopropyl)-O'-(2-methoxyethyl) polypropylene glycol. The processing procedures were distinctly different for the adhesion and dispersing mechanisms. In the first process where adhesion was desired, an in-situ free radical polymerization method was used. In the later method where dispersion was of concern, the integral blend method was used. Respectively, Azobisisobutyronitrile (AIBN) and 1-decanethiol were added as initiator and chain transfer agent. Azobisisobutyronitrile acted as an initiator to the free radical polymerization process. These generated free radicals which are unstable in nature become paired with the σ bond of C=O in the monomer methylmethacrylate. The chain transfer agent (CTA) was used to finely tune the polydispersity index (PDI) of the resulting polymer. The PDI of neat PMMA was about ~1.6 and with the addition of two drops of the CTA, it was reduced to ~1.3. This reduction in the polydispersity index effectively contributes to an increase in the observed mechanical properties.

3.1.1. Method 1: (In-situ Free-Radical Method)

A solution was prepared consisting of ~0.12g (as calculated above) of surfactant and high purity methanol solvent and mixed for 30 minutes. Alumina titania nanoparticles (1.0g) were added and suspended in the solution and mixed for an additional 2 hours. Upon mixing, the solution was placed in an oven set at 60oC for 3 days for drying and to remove all absorbed alcohol. It was extremely important for all absorbed hydroxyl groups from the alcohol be removed from the coated powder. The coated nanoparticles (0.2g) were then placed in 3.8g of monomer methylmethacrylate (Across Organics) which formulated a ~5wt% nanocomposite. Two drops of 1-decanethiol and 0.02g of the initiator azobisisobutyronitrile was added to the solution. At 50 °C, in a silicone oil bath, the solution was mixed with a magnetic stir bar in a sealed scintillation vial for 3 hours or until the viscosity increased to acceptable levels. The viscous liquid was then transferred to another vial (w/o stir bar) of appropriate dimensions and placed in an oven set at 55 oC for an additional 20 hours for full polymerization to occur. Excess monomer was removed by heating the sample at 80oC under vacuum for an additional 6 hours. A similar procedure as stated above was carried out for the spray dried alumina titania micron sized granules.

3.1.2. Method 2: (Integral Blend Method)

A solution consisting of 7.6g of monomer methylmethacrylate, 0.4g of nano alumina titania, 2 drops of 1-decanthiol, and 0.02g of dispersant (phosphate polyester (BYK), polyimine ester (KD-2), oleyl alcohol, or O-(2-aminopropyl)-O'-(2-methoxyethyl) polypropylene glycol) was added into one scintillation vial. This solution, in its entirety, would later designate a 5wt% nanocomposite. The components were mixed with a magnetic stir bar for 15-20 hours, at room temperature, to create well-dispersed slurry. The time needed for sufficient mixing was based on the type and amount of dispersant used. Initiator (0.04g of AIBN) and the chain transfer agent (2 drops of 1-decanethiol) were then added to the solution. This solution was then mixed in a silicone oil bath set at 50°C for 3 hours or until the viscosity reached appropriate levels. The stir bar was then removed and the semi-viscous solution was placed in an oven set at 55°C for an additional 20 hours. Finally, excess monomer was removed by increasing the temperature to 80°C in a vacuum oven for 6 hours. This procedure was then carried out for the spray dried micron sized alumina titania granules.

Upon the formation of the neat PMMA and nanocomposites, mechanical characterization was done via tensile testing. By the determination of the glass transition temperature of the polymer and nanocomposite, the specimens were warm pressed into a flat sheet using a biaxial Carver Press. The sheets were then milled into the required dimensions of the dog-bone specimen. The strength and toughness of the neat PMMA and nanocomposite were effectively determined. Vickers hardness tests were performed on both the neat PMMA and 5wt% filler nanocomposites.

3.2 RESULTS AND DISCUSSION

The infiltration of the nano-sized alumina titania was more effective than that of the micron sized. Less chemically bonded water was discovered in the nano-sized thermal sprayed powder, however, was significant in the spray dried micron sized granules. SEM

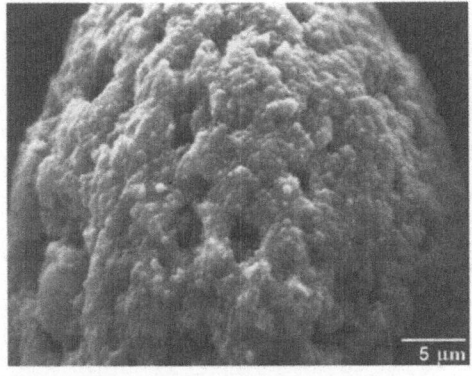
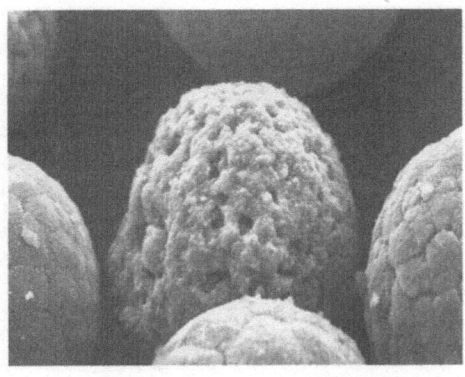

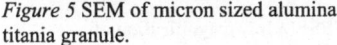

Figure 5 SEM of micron sized alumina
titania granule.

Figure 6 SEM of micron sized alumina
titania granule.

It can be said that these granules are less effective as filler in the formation of a nanocomposite due to high porosity. Within each particle, large pores are observed where chemically bonded water exists due to moisture in the air. In addition, methanol was necessary to be used as a solvent which has a relatively low evaporation temperature compared to other alcohols such as the isopropyl which was initially used for processing. Although extensive heat treatment was used, an insignificant amount of water and/or hydroxyl from the alcohol was able to be removed in its entirety.

Powders obtained from Nanophase Technologies Inc. showed average particle size of 40 nm and severe degree of agglomeration as shown in Figures 7-11. Figures 12-14 depict a smooth and a fractured surface of the polymer nanocomposite.

Figure 7 FESEM of nanosized alumina granule
due to agglomeration

Figure 8 FESEM high resolution of nanosized
alumina nanoparticles 40 nm.

682

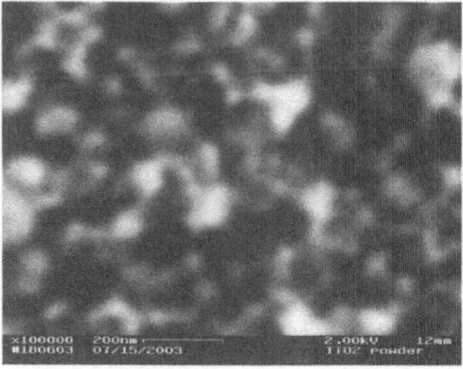

Figure 9 FESEM high resolution of nanosized
Titania nanoparticles 40 nm.

Figure 10 FESEM low magnification of
nanosized Alumina-Titania shown
agglomoration.

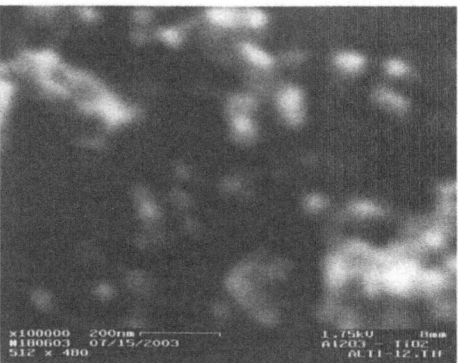

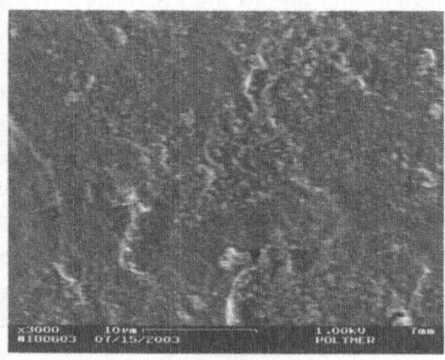

Figure 11 FESEM high resolution of nanosized
Alumina-Titania nanoparticles 40 nm with
minor agglomeration.

Figure 12 FESEM image of fractured surface of
PMMA/ Alumina-Titania polymer nanocomposite.

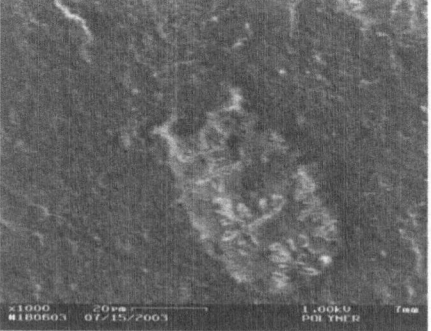

Figure 13 FESEM image of a smooth surface of
PMMA/ Alumina-Titania polymer nanocomposite.

Figure 14 FESEM image of fractured surface of
PMMA/ Alumina-Titania polymer nanocomposite
showing dibree from agglomerate granule.

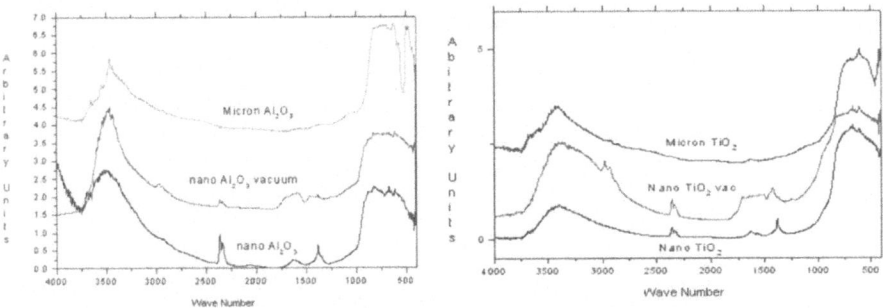

Figure 15 FTIR of alumina powders   *Figure 16* FTIR of titania powders.

These findings were complemented through the FTIR results (Figures 15-16) where a very large peak was present at the wave number range ~3000-3750 cm^{-1} for the micron sized particle. By heating these granules up to 350 °C in a high vacuum, a significant amount of chemically bonded water was able to be removed. It is important to note, however, that within minutes of removal from the vacuum oven significant moisture from the air was absorbed within the surface of the particles. Any water or hydroxyl groups on the surface of these particles are detrimental during the polymerization process. Due to the careful preparation and storage of the nano sized particles, the aforementioned was less of an issue. Upon receiving the nanopowder from the distributor, it was properly secured in a desiccator and never exposed to air during the polymerization process.

From the FTIR results, it is now known that, for micron sized alumina, an alpha phase is present. This has been established from the rather broad peak located at the ~525-925cm^{-1} wave number range and the strong peak located at ~450 cm^{-1}. However, for the nano sized alumina powder, this strong peak is not present. Instead, a peak from the wave number range ~400-950 cm^{-1} exists which designates a definite gamma phase of alumina. Also, for the as-received (w/o heat treatment) gamma alumina nano powder, a rather strong peak is present at wave number 1384 cm^{-1}. This peak represents an organic compound (HO-CH$_2$CH$_3$) attached to the alumina atoms. The as-received powder was exposed to 250°C under vacuum and, as shown from the FTIR results, the organic associated peak was reduced substantially signifying the burn off of organic. It is believed, however, that this organic compound effectively assists in the formation of our polymer ceramic nanocomposite.

For the micron sized titania compound, it has been discovered that a rutile phase exists due to the presence of the strong peak at wave number 421 cm^{-1}. For the nano sized titania powder, however, this peak is not seen which suggests the presence of the anatase phase. This was also confirmed by the much broader peak at the wave number range of ~400-1000 cm^{-1}. As in the case of the as-received nano alumina, the nano titania atoms also has a similar organic compound coated around the particles. This is confirmed by the strong peak at wave number 1382 cm^{-1}. This peak was also reduced when the as-received powder underwent heat treatment at 250°C under vacuum. Due to the peculiarity of the FTIR results, XRD (Figures 17,18) was found necessary to positively confirm these results.

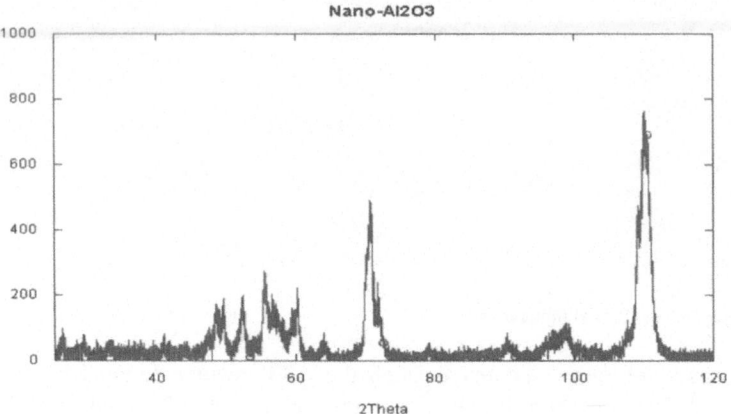

Figure 17. XRD of alumina nanoparticles.

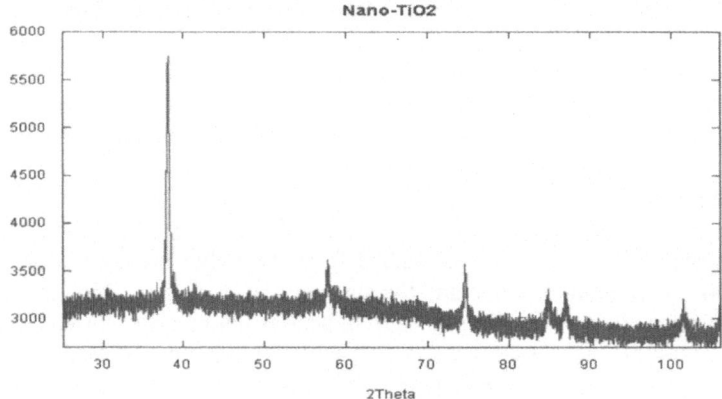

Figure 18. XRD of titania nanoparticles.

As expected, the nano powder consisted of a gamma phase of alumina and an anatase phase of titania. The micron sized powder of alumina and titania consisted of rutile and alpha phases respectively. XRD was also done for the neat polymer and nanocomposite. This is shown in Figures 7 and 8, respectively. As in both cases, the results were nothing more than the expected radial distribution function (RDF).

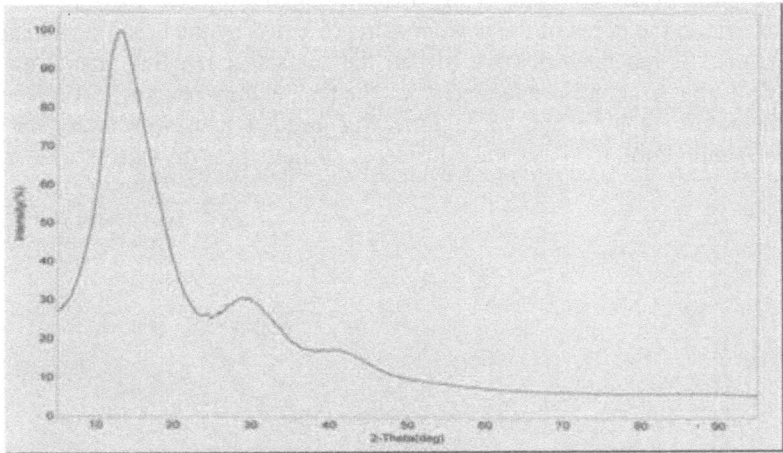

Figure 19. XRD of neat polymethylmethacrylate.

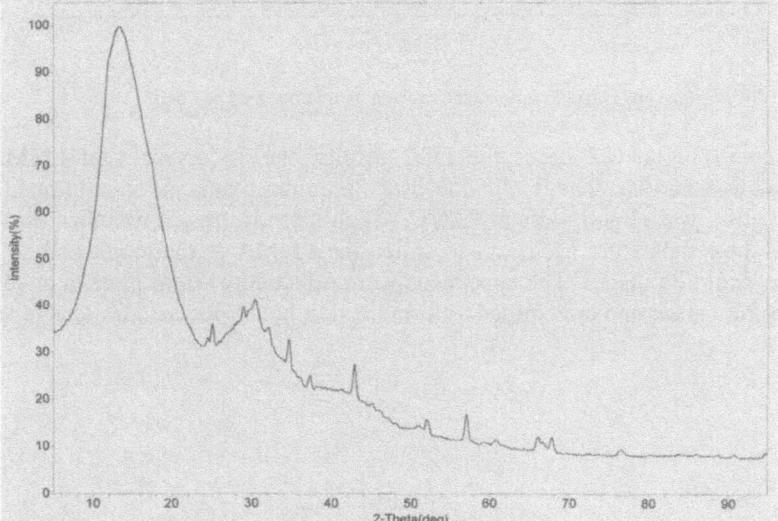

Figure 20. XRD of nanocomposite.

For the nanocomposite spectrum, several peaks superimpose the RDF due to the 5wt% alumina titania filler. For fillers less than 3wt%, no apparent diffraction was observed for the nanocomposite. These spectrums are expected due to the nature of polymethylmethacrylate which is 100% amorphous and therefore unable to cause any diffraction. It is because of this completely amorphous structure, PMMA is one the most transparent materials known. In addition, PMMA can have a thickness of several feet while remaining transparent.

Both siloxane and silane treatments, which was used for adhesion purposes, were proven to effectively coat the ceramic particulates. Upon the coating of the surfactant and necessary drying, the particles were analyzed under FTIR. This was to see not only if the surfactant was able to effectively coat the ceramic particles but, more importantly, to see if this applied surfactant evaporated during the 65°C drying process. This is

686

shown in Figure 21 where again a significant amount of water was present on the surface and within the pores of the powder as represented by the broad peaks and large wave numbers. It has been discovered that the surfactant remains intact only when heated to 65°C, but becomes evaporated when the temperature reaches 75°C. Due to the lower temperature requirement, more time was needed than what was first found necessary. A minimum of 3 days was considered sufficient for drying.

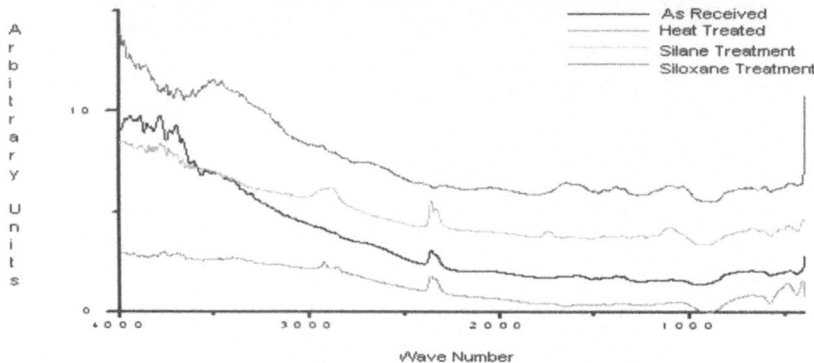

Figure 21 FTIR of silane and siloxane surfactant treatment on micron sized powders.

As seen in Figure 22, from the DSC results, the T_g of the neat PMMA was established at 96.25°C. The T_g for the 5wt% alumina titania filler, infiltrated in the PMMA matrix, was slightly less at 92.5°C. With these results, it was then possible to determine the temperature necessary to soften the PMMA (a thermoplastic) so tensile specimens could be made. The specimens were effectively warm pressed at 110°C to form a uniform sheet and then milled to form the dog-bone specimen for tensile testing.

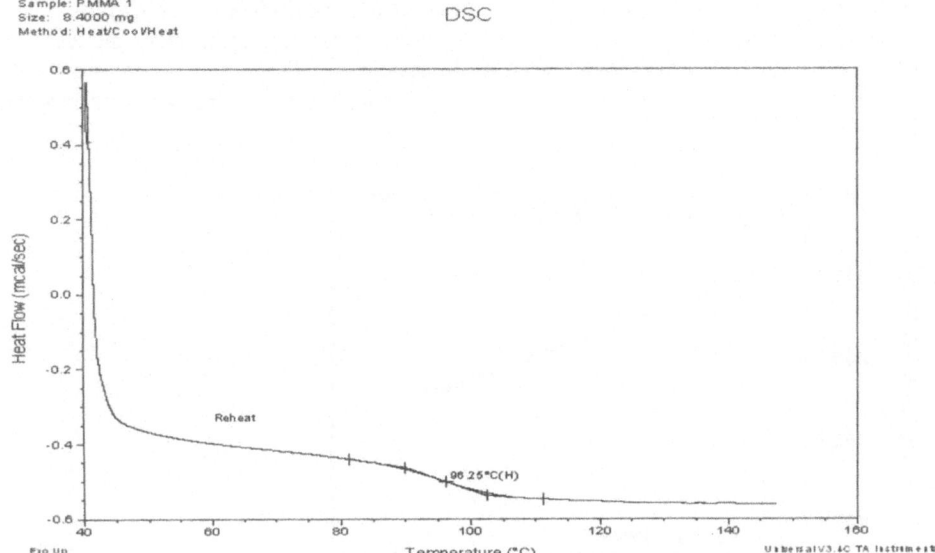

Figure 22. DSC of neat PMMA.

From tensile testing curves, as shown in Figure 23, can be obtained for the neat PMMA and nanocomposite. As shown, the fracture toughness, tensile strength, and strain to failure, can be dramatically increased by the careful dispersing of 5wt% filler alumina titania nanoparticles.

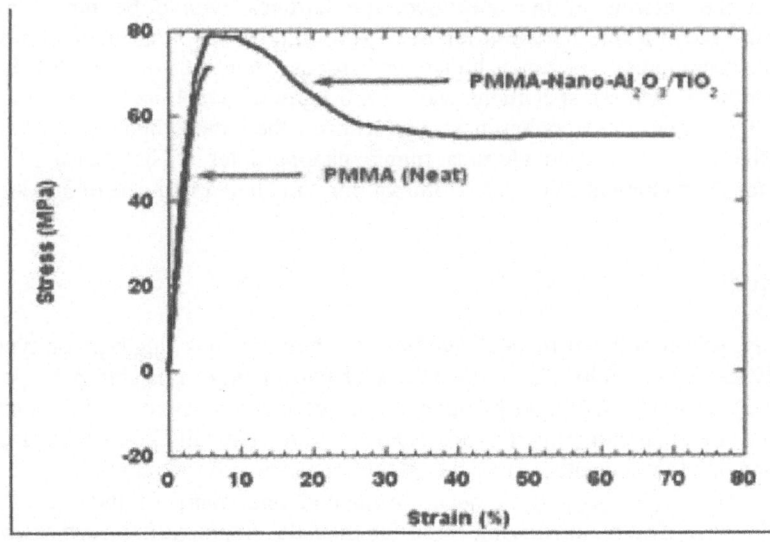

Figure 23. Stress strain curve for neat PMMA and nanocomposite.

The Vickers hardness of neat PMMA (Figure 24) was significantly less than the nanocomposites. The average Vickers hardness for neat PMMA was ~72 kg/mm^2. This value for the hardness had a variance of 12 kg/mm^2, depending on whether the test was set forth on an interior or exterior region of the specimen. This result is consistent with the literature where several scientists have reported Vickers hardness values ranging from

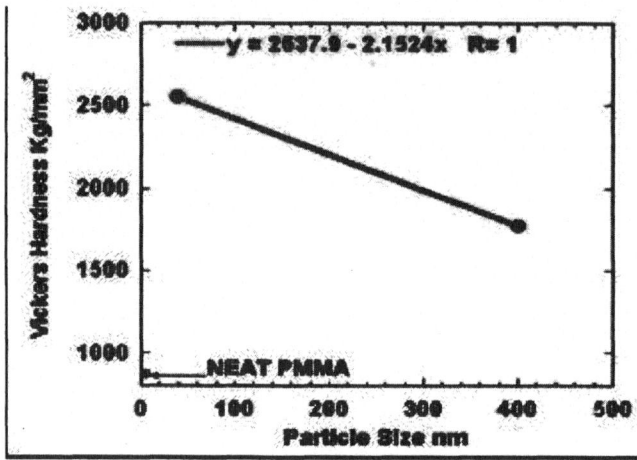

Figure 24. Microhardness test for neat PMMA and the as-received thermal sprayed and spray dried powders infiltrated in the PMMA matrix to form the nanocomposite.

40-80 kg/mm^2, depending on the processing conditions. A lower polydispersive index is known to increase the hardness of a material. In addition, if any excess monomer is left within the interior of the specimen, the hardness would be much less than anticipated. The Vickers hardness for the 5wt% filler of alumina titania spray, dried micron sized granules, was much higher in hardness than that of neat PMMA. The Vickers hardness of these specimens was ~1750 kg/mm^2 which had a variance of 200 kg/mm^2. The highest Vickers hardness was seen for the nanocomposite that had 5wt% filler of the thermal sprayed alumina titania nanoparticles. The measured Vickers hardness for the nanocomposite was ~2580 kg/mm^2 and had a variance of 350 kg/mm^2.

4. Summary

Historically, useful polymer blends have been the homogenous single-phase type, more commonly known as "alloys". These alloys of two or more miscible pure polymers, with closely matching solubility parameters, are of marked scientific and commercial interest because the blend properties are usually a mathematically predictable average of those of the pure components. Over the last thirty years, a large number and variety of such polymer blends/alloys have been formulated and many of these are in wide commercial use. The major driving force for their development has been to avoid the much higher costs and long lead times associated with the discovery, manufacturing process design and scale up, plant construction, and commercial introduction expense for a new polymer. Innovative IMPB technology, for strengthening the continuous phase or enhancing the interfacial forces between the immiscible domains, will likely elevate the performance of an IMPB to the vicinity of an alloy or a more expensive pure polymer. This concept offers the advantages of plastics lightweight, low toxicity, potentially easy recyclability, and others, at a lower cost or higher level of versatility than offered by metals, wood, concrete, glass, or many composites. Some notable, already developed, IMPB successes are now in-place in important industrial applications. The incorporation of nanoparticles into PMMA greatly enhances the mechanical performance of PMMA, with respect to strain to failure. The high surface area of functionalized nanoparticles provides critical modification of the polymer structure. Au-nanoparticle coated PLLA-mPEG-nanospheres copolymer, loaded with protein and superparamagnetic magnetite nanoparticles, were successfully produced as biocompatible drug vehicles for effective drug delivery. A newly PMMA ferrite fabricated nanocomposite showed excellent radar wave absorption. Building future nanocomposites will require a full understanding of the associated processes

5. Acknowledgements

The support of AMIP-The Center of Advanced Materials via Immiscible Polymer Processing and the New Jersey Commission on Science and technology is greatly appreciated.

6. References

[1.] Konstantin, Hadjiivanov. FTIR Study of CO and NH_3 co-adsorption of TiO_2 (rutile). Applied Surface Science, 135 (1998) 331-338.

[2.] Yingchun, Zhu. Structural Characterization of TiO_2 Ultrafine Particles. Journal of Materials Research, vol. 14, No 2, Feb. 1999.

[3.] Nyquist, R.A. Infrared Spectra of Inorganic Compounds, Academic Press (1971).

[4.] Croft, M., Zakharchenko I., Gulak Y., Zhong Z., Hastings J., Hu J., Holtz R., DaSilva M., and Tsakalakos T., Scattered Intensity Profiling with Energy Dispersive x-ray Scattering. Journal of Applied Physics, vol. 92, No 1, July 1, 2002.

[5.] Siegel, R.W. Mechanical Behavior of Polymer and Ceramic Matrix Nanocomposites. Scripta Materialia, vol. 44 (2001) 2061-2064.

[6.] Ash, B.J. Glass Transition Behavior of alumina/polymethylmethacrylate Nanocomposites. Materials Letters vol. 55 (2002) 83-87.

Subject Index

Author Index